Ecological and Health Effects of Building Materials

Junaid Ahmad Malik · Shriram Marathe
Editors

Ecological and Health Effects of Building Materials

Springer

Editors
Junaid Ahmad Malik
Department of Zoology
Government Degree College
Bijbehara, Kashmir (J&K), India

Shriram Marathe
Department of Civil Engineering
NMAM Institute of Technology (VTU, Belagavi)
Nitte, Karnataka, India

ISBN 978-3-030-76072-4   ISBN 978-3-030-76073-1 (eBook)
https://doi.org/10.1007/978-3-030-76073-1

This Springer imprint is published by the registered company Springer Nature Switzerland AG
The registered company address is: Gewerbestrasse 11, 6330 Cham, Switzerland

Foreword

Although our forefathers lived in buildings made of natural materials, today's homes contain a high concentration of chemicals and heavy metals, which contaminate indoor air or pollute tap water, resulting in a variety of health issues such as asthma, soreness, itchy eyes, itchy skin or skin irritation, respiratory tract inflammation, anxiety, depression, drowsiness, exhaustion, and reproductive impotence. Aside from the toxicity of construction materials indoors, the risk for toxicity during chemical processing must also be considered.

Concrete is an integral part of modern structures and construction, providing optimum prosperity. Because of its cost-effectiveness, resilience, and flexibility, concrete is used for the majority of construction elements in roads, walls, trenches, reservoirs, roof tops, and electricity systems. Research and development in the field of sustainable construction and building materials is accelerating, particularly in nations such as India and China, where large quantities of concrete are needed and waste or by-products are plentiful. More analysis leads to a deeper and more comprehensive view of the effect of materials on concrete output and environmental impact.

This book provides in-depth information on a few hot topics in the field of knowledge that fall under the umbrella of "Sustainable Construction and Green Buildings". The volume provides a detailed outlook on the issues associated with the toxicity and ecological impacts from the conventional building materials, and their resulting human health impacts. This also provides the details on the novel solutions for the construction of the engineering structures, which can be used as the potential alternatives for preserving the ecology, by sustained human health.

This book majorly deals with the present adverse effects of using precarious building materials on the ecology and human health. In the subsequent chapters, a detailed discussion on the novel and greener construction materials, with their utilization as an alternative to the existing harmful conventional methods and materials were presented. This book helps to fill the research gaps in the existing prior-art knowledge in the field of *Sustainable Construction and Green Building* materials and methods giving a due importance to ecology and health, specifically to the fields of sustainable structural engineering, sustainable geotechnical engineering, etc. This book also covers few recent and interesting research studies such as impact of construction materials on environment, impact on health of construction workers, wildlife conservation, embodied carbon, etc., which provides the modern touch to the scope of this book. This book helps out in achieving a sustainable environment through possible adoption of innovative and ecological construction practices.

The editors of this book and the contributing authors of various chapters needed to be appreciated for their sincere efforts in the production of this book, which would definitely help the concerned stakeholders.

Megh R. Goyal, Ph.D., P.E.
Retired Professor in Agricultural and Biological Engineering
University of Puerto Rico
Mayaguez- Puerto Rico, USA

Preface

It is a well-established fact that the constructions of the engineering structures consume more and more earth resources than any other human activities in the world. In addition, the construction-related activities produce several million tons of greenhouse gases, toxic emissions, water pollutants, and solid wastes. This creates a huge impact on the environment and causes severe health issues for humans and animals. The call for the day is to create an eco-friendly construction environment that can satisfy ecological and health requirements. Hence, while choosing building materials for any construction sector, one must consider their potential toxicity and environmental impacts.

The WHO has revealed that nearly one-third of the buildings completed over the past 30 years in the industrialized nations manifest problems capable of harming their occupants, and 40% of the materials used in the construction sector are potentially aggressive in nature. One of the finest customs to track sustainability in construction technology is by the usage of eco-friendly building materials and methods. A new invention of more eco-friendly building construction materials which can assist in solving more health-related burning difficulties in the construction industry will enhance the sustainability in the recent construction practices. The use of sustainable materials and methods has the supplementary benefit of defending the hygienic environment by tumbling the "carbon footprint" of the buildings. These techniques will encourage a cleaner globe and a prospect of eco-friendliness through being aesthetically pleasing and much-added efficiency in conserving the human health impacts.

Consequently, many researchers are trying to find out the solutions which tend to reduce/replace the negative impacts of construction materials considering the preservation of surrounding ecology and human health. As a result, various ecological remedial and innovative products are under development, which shall essentially be used as potential alternatives for the ecological impacts of the conventional building materials.

The impact of construction materials and practices on the health and wellbeing of occupants is often underestimated. This book acts as an essential guide in comprehending and evading harmful materials in construction. The book also covers a range of topics starting with a description of sustainability in construction which can be influenced by securing the health of users, occupants and atmosphere. This book is an innovative spot under the scope of Sustainability aspects in construction practices, which focuses on creating a healthy environment based on ecological rules. The sustainable construction method majorly spotlights to reuse, conserve, renew/recycle, and conserve the sustainable atmosphere, creating eminent and non-toxic construction materials and methods. All the major aspects pertaining to the ecological and health aspects of the construction sector were covered in the subsequent chapters. The book wraps the aspects such as the discussions on major toxic construction materials which come under Volatile Organic Compounds (VOCs), heavy metals, nanomaterials, asbestos, polymers, etc., and their health impacts; few additional attributes such as soil contaminations, water-pollutions from the construction sector are also covered in the book. Further, the book also covers the major aspects such as atmospheric emissions from the construction sector, reuse of industrial wastes in producing building materials, and sustainable civil engineering practices which could help to sustain the ecosystem along with human health. Few recent advances in sustainable and healthy construction practices such as thermal insulations in buildings, sustainable method of curing, bio-concrete, eco-friendly geopolymer masonry, eco-efficient plasters, sustainable building waste disposal, etc., definitely add to the value of this book. Some real-time problems which adversely affect the health and ecology due to construction practices such as health problems of construction workers, ecological impacts of land conversion on wildlife conservation, and embodied carbon in construction and its ecological implications are also covered.

Moreover, this book is presented in a lucid and reachable approach, which clearly provides obligatory opinion and information to anyone aspiring for a better understanding of healthier building construction practices, through giving due consideration to ecological conservation. Hence, this book will act as a mandatory reading volume for the practicing civil engineers, architects, researchers, surveyors, public health professionals, facility managers, and environmentalists, who are concerned and willing to work towards a healthier construction industry.

Bijbehara, India Dr. Junaid Ahmad Malik
Nitte, India Dr. Shriram Marathe

Contents

About the Editors

Dr. Junaid Ahmad Malik received B.Sc. (2008) Science from the University of Kashmir, Srinagar, J&K; M.Sc. (2010) in Zoology from Barkatullah University, Bhopal, Madhya Pradesh; and Ph.D. (2015) in Zoology from the same university. He completed his B.Ed. program in 2017 from the University of Kashmir, Srinagar, J&K. He started his career as Lecturer in School Education Department, Government of J&K for 2 years. Dr. Malik is now working as a Lecturer in Department of Zoology, Govt. Degree College, Bijbehara, Kashmir (J&K) and actively involved in teaching and research activities. He has more than 8 years of research experience. His areas of interest are ecology, soil macrofauna, wildlife biology, conservation biology, etc. Dr. Malik has published more than 20 research papers in various national and international peer-reviewed journals. He has authored 4 books, 16 book chapters, edited 7 books, and more than 10 popular editorial articles. He is also serving as editor and reviewer of several journals with a reasonable repute. He has participated in several State, National, and International conferences, seminars, workshops, and symposia and more than 20 conference papers are to his credit. He is the life member of SBBS (Society for Bioinformatics and Biological Sciences) with membership id LMJ-243.

Readers may contact him at: malik.junaidahmad@gmail.com

Dr. Shriram Marathe is serving as an Assistant Professor in the Department of Civil Engineering at NMAM Institute of Technology, NITTE. He completed his Ph.D. degree at VTU Belagavi in the area of Alkali Activated Concrete Pavements. He accomplished his M.Tech. (Transportation Engineering) from National Institute of Technology, Surathkal, Karnataka, and Master Facilitator Degree (MFLHRD) in Human Resources Development from CLHRD, Mangalore. His areas of research include Alkali activated Concrete applications to pavements, Stabilization of Sub-grade soil, Road Safety, Pavement Material characterization, and Pavement Design. He has guided several projects for B.E. and M.Tech. dissertations. He also successfully executed one funded research project on "Study and to Develop Cost-Effective and Green Masonry Block Using Industrial Waste Materials", funded by NITTE Education Trust. Till date, he has published a total of 50 research papers out of which 10 articles are published in Scopus indexed journals (Three Q1 articles). Out of his research contribution, 24 research articles are published in international journals, 06 technical papers in national journals, and also conference proceedings papers at 13 National/International conferences and 5 papers are under review. Further, he also filed a patent on his invention entitled, "Sustainable Pervious Alkali Activated Concrete Paver Block Pavement for Ground water Recharge" on 29-07-2020. He also served as the conference secretary and successfully completed the CTCS-2020, International Conference held at NMAMIT, Nitte during Decemeber 2020. Further, he is also serving as a potential reviewer for "International Journal of Pavement Research and Technology" published by Springer Nature Singapore, "Construction and Building Materials" published by Elsevier B V, "Case Studies in Construction Materials" published by Elsevier B V, and "Cogent Environmental Science" published by Taylor & Francis Online. Being a resource person, he had delivered several technical talks at various graduate schools. He is a life member of Indian Roads Congress (IRC) New Delhi, Indian Geotechnical Society (IGS) New Delhi, Indian Society for Technical Education

(ISTE), Indian Society of Systems for Science and Engineering (ISSE), Technical Institute for Engineers (India), and Kannada Sahithya Parishadh, Karnataka. He worked as Departmental Co-ordinator (Civil Engineering) for NAAC, NBA, IQAC, Industry Institute Interaction, ISTE, IE(I), and IOV related works.

Chapter 1
Potentially Toxic Construction Materials: An Introduction

Aadil Gulzar, Tabasum Hassan, and Ruquia Gulzar

Abstract For the development of a nation, building materials are important. After the Industrial Revolution in the early nineteenth century, more building activities were generated with the production of construction materials to enhance economic development and jobs. Thousands of synthetic construction materials have been developed and produced worldwide. They come in various shapes, sizes and amounts that serve various forms and functions. Toxic chemicals may be found in common construction goods, such as solvents, paints and varnishes, or dust from building materials. But in the modern era, large numbers of toxic construction materials are being used for construction, a purpose which later gets released in the surroundings, polluting air, water and soil. These toxins are responsible for causing a lot of health issues such as asthma, burning eyes itchiness, inflammation of the nose and throat, headache, skin irritations or rashes, dizziness, fatigue, reproductive dysfunction, nausea, endocrine system disruption, impairment of infant growth and birth defects, suppression of the immune system and cancer. It is not possible to see or smell any of these products, but they are able to cause damage. They can emit low-level toxic exposure or produce cancer-causing carcinogenic substances. The effects may cause short-term effects that can cause death, such as skin allergies or long-lasting health effects.

Keywords Asbestos · Building materials · Health issues · Phthalates · Toxic

1.1 Introduction

Today the world faces a number of major challenges that may lead to the end of our society if it is not solved, or if it is just postponed. One of the major challenges is the consumption of many forms of energy, including non-renewable ones used by humans, releasing a certain amount of waste in the consumption process, and a large

A. Gulzar (✉)
Department of Environmental Science, University of Kashmir, Srinagar 190006, J & K, India

T. Hassan · R. Gulzar
Department of Botany, University of Kashmir, Srinagar 190006, J & K, India

J. A. Malik and S. Marathe (eds.), *Ecological and Health Effects of Building Materials*,
https://doi.org/10.1007/978-3-030-76073-1_1

amount of waste products in the surrounding environment making our planet worse (Meadows et al. 1972). The immediate effects of consuming practices of societies, with impacts for environments and future centuries, have not yet been thoroughly articulated in intergenerational and inter-geographical dimensions (Stern et al. 2006).

In order to achieve sustainable and ecofriendly construction, the European Union has provided guidelines regarding the use of raw materials and production of construction wastes. According to these guidelines, the raw material consumption should not exceed 30% and the waste production by 40%. Not only the use of sustainable raw materials is essential for construction purposes but also the production of toxic wastes is equally significant. The people in the early centuries used to construct buildings made of raw materials that do not release any toxic product to the environment. However, in today's world, a vast amount of hazardous building materials are used for construction, which are then emitted into the environment, polluting the air, water, and soil. Asthma, burning eyes, itchiness, itching of the nose and mouth, nausea, itchy skin or rashes, lightheadedness, weakness, reproductive dysfunction, fatigue, endocrine system dysfunction, deficiency of child development and congenital abnormalities, neuroinflammation, and cancer are all caused by these toxins. In addition to the toxicity of indoor building materials, the potential for toxicity during the processing of such chemicals must not be ignored. For example, Bhopal disaster in India in 1984, in which nearly 15,000 deaths and health problems were caused by a cloud of methyl isocyanide in almost 200,000 human beings (Varma and Mulay 2006; Satyanand 2008). Hazardous waste is produced during the manufacture of chemical materials and that impact must be connected to building materials containing these chemicals. There are a lot of hazardous chemicals used in the construction industry, and some of them are explored in this chapter.

1.2 Polyvinyl Chloride (PVC)

Any material or substance consisting of PVC composition, i.e. mixture of a vinyl chloride polymer or copolymer with different additives, is PVC or polyvinyl chloride (Titow 2012). The only important plastic used in buildings that contain chlorine is PVC. Most of the chemicals, especially polyvinyl chloride utilized during manufacturing and installation of construction materials possess significant toxicity, thus come under severe examination. Vinyl 3 which is another name for Polyvinyl chloride (PVC) is one of the extensively used chlorine-containing plastic polymers in the United States with about 14 billion pounds manufactured in the U.S each year (Sass et al. 2005). More than 75% of PVC use is accounted by the construction industry (Jebens et al. 2013). PVC is used in the building sector for window frames, doors, roller shutters, drinking pressure tubing, wall covering, reservoir lining, fencing, etc. (Patrick 2005). The plastic industry uses a group of chemicals to make PVCs flexible and functional, the majority of which pose concerns about human health and the environment. Throughout the lifecycle of PVC and other chlorinated plastics, the chlorine content has the ability to generate dioxins through processing and disposal (Thornton 2002). PVCs act as a source of Dioxins which are known for their bioaccumulative potential, thus making them a global concern.

Dioxins are inevitable chemicals produced during the production, burning and removal of chlorine-containing materials. Dioxins are one of the most well-known carcinogens in humans (Karstensen 2008). These are the chemical wastes produced during the industrial process of chlorine emerging industries as are produced during the manufacture of PVCs. Since dioxins and other chemicals like furans are very toxic and bio-cumulative; contaminate all food chains, resulting in hazardous effects on biodiversity (Tillitt et al. 1993). In the last four centuries, according to Thornton (2000), the concentration of both dioxin and furans rose from zero to almost 100% in two German lakes and the Baltic region. In addition, in the northern Pacific Ocean, chemical analysis of dolphins showed dioxin and furan levels between 13 and 37 million times more than in surrounding water (Thorton 2002). Several groups of scientists have already proposed that industrial processing of chlorine should be banned (Flores et al. 2018). In fact, dioxins belong to one of only 12 chemicals or families of chemicals targeted for removal by "The Stockholm Convention on Persistent Organic Pollutants (POPs), which is an international treaty. Some of the dioxins are not only strong carcinogens, but also behave as reproductive and growth toxicants. Many of them affect the endocrine and immune systems badly.

1.3 Phthalates

In PVC plastics, phthalates are used as plasticizers. Since phthalate plasticizers are not chemically bound to PVC, they are able to leach, migrate or evaporate into the atmosphere and indoor air, food, other products, etc. (Schindler and Hauser 2004). In health care environments, phthalates are also present in PVC plastic which became another source of exposure to this substance. Inherently rigid, PVC includes additives due to which it became flexible, therefore can be used in bags, flooring wall coverings, upholstery and shower curtains. Established reproductive and developmental toxicants are certain phthalates used for softening PVC. Phthalates cannot bind with PVC, therefore they are released into the environment e.g., into air, water and soil, causing many respiratory problems in adults and children, such as pneumoconiosis, rhinitis, asthma and both insulin resistance and obesity in adults. In the United States, PVC processing uses a large percentage of phthalates and risk evaluations have been performed on phthalates through numerous expert panels in both America and Europe (Heudorf et al. 2007).

1.4 Organic Compounds

1.4.1 Volatile Organic Compounds (VOCs)

The carbon-containing compounds that have the ability to vaporize at room temperature are known as Volatile Organic Compounds (VOCs). Therefore, these compounds can evaporate eventually from a construction product into the air which becomes a source of inhalation to human beings. Chemicals of VOC-type are utilized during the processing of certain plastics and for items such as structural wood or insulators for preparation of binders and other resins; also increase stain repellent and water resistance when used in paints, adhesives, coatings and other treatments (Brown et al. 1994). Formaldehyde, toluene, isocyanates, acetaldehyde, benzene and xylene are some common troublesome VOC compounds released from building materials. When a product is first mounted, VOCs are released in large amounts and then over time in lower amounts, connected to the amount of moisture present in items that are wet at the beginning and then eventually dry. Solid materials such as fabrics, furniture, flooring and furnishings release VOCs more steadily and for a longer period of time, therefore, retaining a low level of emissions. When exposed, construction materials coated in plastic at the point of production and uncovered at the development site or construction site will release concentrated VOCs. Most of these VOCs have significant health implications as well. Some of the VOCs are responsible for causing signs of temporary acute sick construction syndrome and other long-lasting serious health implications, like liver, nervous system and kidney damage, and also increase the risk of cancer (Salasar 2007). Formaldehyde, a recognized human carcinogen, is one of the VOCs of major concern. The possible ecological and health impacts of formaldehyde have generated such a high alarm that many foreign and other national bodies have placed bans on the use of products where formaldehyde can usually be found and discharged (Hileman 1984). Several countries, including Japan, Germany, Netherlands, Norway and Finland have already taken measures to limit the emissions of formaldehyde in textiles. Other VOCs, like xylene, benzene, toluene, and acetaldehyde also pose health and ecological issues, in addition to those caused by formaldehyde. In certain processes e.g., composite wood formation, fabric manufacturing and batt insulation, formaldehyde is used as a binder to prevent shrinkage of the fabric, for improving crease-resistance, for providing stability to dimensions and improving color quality. It is also utilized to increase stain resistance as a part of certain finishing treatments (Schindler and Hauser 2004).

1.4.2 Semi-volatile Organic Compounds (SVOCs)

Organic compounds with higher vapor pressures and released as gas more slowly than VOCs from materials containing them are known as semi-volatile compounds. These compounds have more possibility to be transferred to humans through touch or by binding to dust or through ingestion (Lucattini et al. 2018). In building materials, semi-volatile organic compounds are used for many benefits like enhanced flexibility, protection from water and stains. Halogenated flame retardants prevent combustion or flame propagation. In comparison to VOCs, which appear to be released quickly in the first few hours or days after a product is installed and then emit slowly with time, products possessing SVOCs release them more gradually and over longer periods of time. A variety of chemicals used in construction materials are showing up in higher amounts in human milk, tissue samples, and blood, raising fears about their increasing ability to cause cancer or other significant health effects (Rumchev et al. 2007). Certain VOCs have also been discovered in household dust emitted from construction materials into the atmosphere (Xu and Little 2006). The possible risk of exposure to such dust compounds can be equal to or greater than that of food intake exposure in infants and adults (Hwang et al. 2008). While several SVOCs occur in construction materials, phthalates (softeners used in plastic PVC), halogenated flame retardants (chemicals applied to inhibition products) and perfluorochemicals (added to stain resistance or water repellency products) are of particular concern.

1.5 Heavy Metals

1.5.1 Asbestos

Asbestos can be found in building materials such as walls, floors, and ceilings. Asbestos contains different fibers with a length of 5 mm and a diameter of 3 mm. The various fibers present in asbestos include crocidolite, chrysolite, anthophyllite, amosite, actinolite and tremolite. Until 1960, the effects of asbestos on health were not recognized in the scientific world and these problems were not taken seriously up to 1980. Therefore, most of the building structures formed between 1920s to late 1980s contain mostly asbestos. Since 1980 the scientific world recognizes the ill effects of asbestos and its problems, due to which it was taken seriously.

Asbestos is of different types like white, brown and blue asbestos. White asbestos mostly consists of chrysotile fibers while brown and blue consist of amosite and crocidolite fibers respectively. Among them, blue and brown are highly toxic whereas white asbestos is nontoxic. Blue and brown asbestos is responsible for causing pleural mesothelioma in which patients die after 12 months of exposure (Bianchi et al. 1997; Jarvholm et al. 1999; Azuma et al. 2009). Other health hazards associated with asbestos include serious health issues like asbestosis (a disease in which there is the accumulation of acid which leads to lung damage. This acid is mainly produced in the body to dissolve the fibers of asbestos), lung cancer and other types of cancers. Asbestos is highly toxic even a little exposure to its small quantity can lead to various types of diseases.

1.5.2 Lead

Lead has a very important role in construction. Lead is considered to be an important component of roofs, tank lining and electrical gadgets used in construction. Since ancient times lead has been used in water pipes due to its less corrosive properties (Hodge 1981; Dutrizac et al. 1982; Nriagu 1983). But some quantity of corrosiveness fatal to health still is presented in lead. Various authors are of the view that the use of lead in tank linings and water pipes can lead to corrosion which can be washed away by water with due course of time, hence can lead to water contamination (Zietz et al. 2009). So lead contaminated water is very fatal to human beings. It can cause damage to the central nervous system, kidneys, cardiovascular system and reproductive system. Lead is very poisonous as it directly enters into the bloodstream causes calcium simulation which enables it to cross the brain and blood barrier in children and infants. It causes behavioral problems (Pocock et al. 1994; Canfield et al. 2003; Wilhelm and Dieter 2003). Troesken (2006) is of the view that during the past two centuries the use of lead used in plumbing is as huge as the Bhopal disaster issue to which thousands of children in the USA have lost their lives.

1.5.3 Cadmium

Cadmium is an insoluble metal that is resistant to corrosion. Hence this metal is used to coat steel and iron to protect them from corrosion. The compounds of cadmium are also used for plastic stabilization and glass coloring as these are available in various colors including red, orange, and yellow pigments. This chemical is also used in alloys, solar cells and electroplating. Cadmium and its compounds are highly toxic and can cause cancer and other body problems including cardiovascular, neurological, respiratory and renal by causing damage to the heart, brain, lungs and kidney (Hayat et al. 2019).

1.5.4 Mercury

Mercury is considered to be one of the most influential neurotoxins causing damage to the brain particularly in fetuses and children (Trasande et al. 2005). Therefore, the use of mercury in building materials has been discouraged for the past many years. But in most construction materials mercury is still being used.

1.5.5 Silica

Silica is commonly found in stones, clay, sand, tiles, concrete and bricks. Therefore, is considered to be an important component in building materials. e.g. granite contains almost 15–30% of silica whereas the quantity of silica in sandstone is greater than 70%. Silica is considered to cause a high risk of lung diseases in construction workers after asbestos, as it is inhaled directly after construction or after grinding, cutting, or blasting stones. Long exposure to silica can lead to cancer and many respiratory tract diseases. In causing carcinomas it ranks as one of the high and influential elements (Hoy and Chambers 2020).

1.6 Other Sources of Toxic Materials

1.6.1 Wood Treatment Chemicals

Wood is an important building material for construction purposes but at the same time, it is most vulnerable to fungus and insects (Morrell 2002). Therefore various treatments are used to preserve the wooden materials. For this number of chemicals are used that are water-soluble e.g., creosotes (includes wood tar creosote, coal tar creosote and oil tar creosote), arsenic, copper and chrome. These chemicals are very toxic and are responsible for contaminating the environment. Creosote contains cancer-causing agents (ATSDR 2002; Smith 2008) therefore its use for wood preservation is banned. Creosote is also used for the construction of cross ties in railways. As it is the most toxic element, its use must be minimized and can no longer be reused (Pruszinski 1999). In addition to creosote, arsenic can also be used to prevent the wood from insect attacks. These chemicals cause a lot of health issues including abdominal pain, vomiting, diarrhea, heart diseases, thickening of the skin and even cancer. In addition to arsenic pentachlorophenol, another wood preservative, can cause kidney and liver damage.

1.6.2 Bisphenol A

Bisphenol A (BPA) is an organic synthetic compound with two hydroxyphenyl groups that belong to the diphenylmethane variants and bisphenols groups. This chemical is used in building materials to produce plastic polymers and surface coatings, mostly polycarbonates and epoxy resins. BPA also serves as the basic material to be used in making epoxy coatings of paint, adhesive and many other products. The composition of epoxy products cannot be easily understood, nonetheless, epoxy resins have mostly 2 chemicals used in their manufacture first is BPA and the second one is epichlorohydrin.

There are many health hazards related to bisphenol A, the most common health hazards are related to male and female fertility, hormonal imbalance including polycystic ovarian syndrome (PCOS), carcinomas of breast and prostate glands. Therefore users are at high risk of BPA. Hence their use in construction materials should be minimized. A study was carried out in Japan in which it was found that exposure to epoxy resins can lead to hormonal imbalance in male workers. Epoxy resins on entering into the human body can lead to many health-related issues (Hanaoka et al. 2002).

The use of BPA in the manufacture of polycarbonate plastics should be minimized as it can leach from the bottles and can liners formed from polycarbonate plastics which can become one of the main reasons for endocrine imbalance (Rubin 2011).

1.6.3 Materials Releasing Toxic Fumes on Fire

There are some materials used in building materials that release toxic fumes when these materials catch fire. Large numbers of deaths during fires are caused due to inhalation of these materials. There is an increase in such incidents since 1980 due to the large usage of these materials in buildings from the past couple of years (Gann et al. 1994; Hall and Harwood 1995; Wu 2001; Levin and Kuligowski 2005). Some of the elements have a very high toxic index. Elements like polyethylene and polyurethane foam have a toxicity index greater than 10 (Liang and Ho 2007), so there is a greater recommendation not to use these elements because of their highly toxic nature. Therefore these materials should be covered by non-combustible substances (Liang and Ho 2007). Another material polystyrene is highly combustible and releases very large toxic fumes, therefore their use should be avoided.

There is an initiative for the usage of a large number of flame retardants, which are the chemicals that are used in construction materials to control the spread of flames. But these retardants also release toxic chemicals upon degradation in gaseous forms, which can easily be inhaled. Upon inhalation of these elements, a number of hormonal imbalances can occur in humans especially in children.

1.6.4 Formaldehyde

Formaldehyde is a colorless gas. It is used in many construction materials as binders or adhesives in various woods and carpet products. It is responsible for causing a number of respiratory problems and dermatitis (Kim et al. 2011).

1.6.5 Fiberglass

Fiberglass is used for roofing and as an insulator of heat. Exposure to this material causes skin irritation, bronchitis and asthma to workers on inhalation after cutting, chopping and trimming of these fibers (Neghab and Alipour 2009).

1.7 Conclusion

Most of the substances used in construction materials possess certain degree of toxicity. These toxic substances are released at the stage of their production, during fires and after an installation of the project. These substances have a negative impact on both humans and the surrounding environment. Most of the health issues like cancer, kidney damage, cardiac arrests are associated with these substances. Therefore, there is a dire need to utilize sustainable materials instead of such toxic materials. In addition to this, legal regulations and education to the common masses can help a lot.

References

ATSDR (2002) Toxicological profile for creosote. Agency for toxic substances and disease registry. US Department of Health and Human Services, Public Health Sector, Atlanta, GA, p 11

Azuma K, Uchiyama I, Chiba Y, Okumura J (2009) Mesothelioma risk and environmental exposure to asbestos: past and future trends in Japan. Int J Occup Environ Health 15:166–172

Bianchi C, Giarelli L, Grandi G, Brollo A, Ramani L, Zuch C (1997) Latency periods in asbestos-related mesothelioma of the pleura. Eur J Cancer Prev 6:162–166

Brown SK, Sim MR, Abramson MJ, Gray CN (1994) Concentrations of volatile organic compounds in indoor air—a review. Indoor Air 4(2):123–134

Canfield RL, Henderson JCR, Cory-Slechta DA, Cox C, Jusko TA, Lanphear BP (2003) Intellectual impairment in children with blood lead concentrations below 10 mg per deciliter. N Engl J Med 348:1517–1526

Dutrizac J, O'Reilly J, Macdonald R (1982) Roman lead plumbing: did it really contribute to the decline and fall of the empire? CIM Bull 75:111–115

Flores N, Sharif F, Yasri N, Brillas E, Sirés I, Roberts EP (2018) Removal of tyrosol from water by adsorption on carbonaceous materials and electrochemical advanced oxidation processes. Chemosphere 201:807–815

Gann RG, Babrauskas V, Peacock RD, Hall JR (1994) Fire conditions for smoke toxicity measurements. Fire Mater 18:193–199

Hall J, Harwood B (1995) Smoke or burns–which is deadlier? Nat Fire Protect Assoc J 38:38–43

Hanaoka T, Kawamura N, Hara K, Tsugane S (2002) Urinary bisphenol A and plasma hormone solvents in male workers exposed to bisphenol A diglycidyl ether and mixed organic solvents. Occup Environ Med 2002(59):626

Hayat MT, Nauman M, Nazir N, Ali S, Bangash N (2019) Environmental hazards of cadmium: past, present, and future. In: Cadmium toxicity and tolerance in plants. Academic Press, pp 163–183

Heudorf U, Mersch-Sundermann V, Angerer J (2007) Phthalates: toxicology and exposure. Int J Hyg Environ Health 210(5):623–634

Hileman AR (1984) Weather and its effect on air insulation specifications. IEEE Trans Power Appar Syst 10:3104–3116
Hodge A (1981) Vitruvius, lead pipes and lead poisoning. Am J Archaeol 85:486–491
Hoy RF, Chambers DC (2020) Silica-related diseases in the modern world. Allergy 75(11):2805–2817
Hwang HM, Park EK, Young TM, Hammock BD (2008) Occurrence of endocrine-disrupting chemicals in indoor dust. Sci Total Environ 404(1):26–35
Jarvholm B, Englund A, Albin M (1999) Pleural mesothelioma in Sweden: an analysis of the incidence according to the use of asbestos. Occup Environ Med 56:110–113
Jebens AM, Kälin T, Kishi A, Liu S (2013) Chemical economics handbook. Denver, CO, HIS, USA, pp 583–0100
Karstensen KH (2008) Formation, release and control of dioxins in cement kilns. Chemosphere 70(4):543–560
Kim KH, Jahan SA, Lee JT (2011) Exposure to formaldehyde and its potential human health hazards. J Environ Sci Health Part C 29(4):277–299
Levin B, Kuligowski E (2005) Toxicology of fire and smoke. In: Salem H, Katz S (eds) Inhalation toxicology. Boca Raton, CRC Press, FL, pp 205–228
Liang H, Ho M (2007) Toxicity characteristics of commercially manufactured insulation materials for building applications in Taiwan. Constr Build Mater 21:1254–1261
Lucattini L, Poma G, Covaci A, de Boer J, Lamoree MH, Leonards PE (2018) A review of semi-volatile organic compounds (SVOCs) in the indoor environment: occurrence in consumer products, indoor air and dust. Chemosphere 201:466–482
Meadows DH, Meadows DL, Randers J, Behrens WW (1972) The limits to growth. New York 102(1972):27
Morrell J (2002) Wood-based building components: what have we learned? Int Biodeterior 49:253–258
Neghab M, Alipour A (2009) Evaluation of respiratory toxicity of fiberglass dust. Iran Occup Health 6(2):19–25
Nriagu J (1983) Saturnine gout among roman aristocrats. Did lead poisoning contribute to the fall of the empire? N Engl J Med 308:660–663
Patrick S (2005) Practical guide to polyvinyl chloride. iSmithers Rapra Publishing
Pocock S, Smith M, Baghurst P (1994) Environmental lead and children's intelligence: a systematic review of the epidemiological evidence. BMJ 309:1189–1197
Pruszinski A (1999) Review of the landfill disposal risks and the potential for recovery and recycling of preservative treated timber. Environmental Protection Agency Report
Rubin BS (2011) Bisphenol A: an endocrine disruptor with widespread exposure and multiple effects. J Steroid Biochem mol Biol 127(1–2):27–34
Rumchev K, Brown H, Spickett J (2007) Volatile organic compounds: do they present a risk to our health? Rev Environ Health 22(1):39
Salasar C (2007) Study about VOCs emission from solvent and water based paints. Master thesis, University of Londrina, Brazil (Only in Portuguese)
Sass JB, Castleman B, Wallinga D (2005) Vinyl chloride: a case study of data suppression and misrepresentation. Environ Health Perspect 113(7):809–812
Satyanand T (2008) Aftermath of the Bhopal accident. The Lancet 371:1900
Schindler WD, Hauser PJ (2004) Chemical finishing of textiles. Elsevier, Woodhead Publishing, p 224
Smith P (2008) Risks to human health and estuarine ecology posed by pulling out creosote-treated timber on oyster farms. Aquat Toxicol 86:287–298
Stern NH, Peters S, Bakhshi V, Bowen A, Cameron C, Catovsky S, Crane D, Cruickshank S, Dietz S, Edmonson N, Garbett SL (2006) Stern review: the economics of climate change, vol 30, Cambridge University Press, Cambridge, p 2006
Thornton J (2000) Pandora's poison: chlorine, health, and a new environmental strategy. MIT Press, Cambridge, MA

Thornton J (2002) Environmental impacts of polyvinyl chloride (PVC) building materials. Healthy Building Network, Washington, DC

Tillitt DE, Kubiak T, Ankley GT, Giesy JP (1993) Dioxin-like toxic potency in Forster's tern eggs forms Green Bay, Lake Michigan, North America. Chemosphere 26:2079–2084

Titow MV (2012) PVC technology. Springer, Netherlands, p 1284

Trasande L, Landrigan PJ, Schechter C (2005) Public health and economic consequences of methyl mercury toxicity to the developing brain. Environ Health Perspect 113(5):590–596

Troesken W (2006) The great lead water pipe disaster. MIT Press, Cambridge

Varma R, Mulay S (2006) The Bhopal accident and methyl isocyanates toxicity. Toxicol Organophosphate Carbonate Compd 7:79–88

Wilhelm M, Dieter H (2003) Lead exposure via drinking water-unnecessary and avoidable. Bleiexposition ueber das Trinkwasser-unnoetig und vermeidbar. Umweltmedizin in Forschung und Praxis 8:239–241

Wu C (2001) Discussion on fire safety factors from case studies of building fires. Master thesis, University of Tainan, Taiwan

Xu Y, Little JC (2006) Predicting emissions of SVOCs from polymeric materials and their interaction with airborne particles. Environ Sci Technol 40(2):456–461

Zietz BP, Lass J, Dunkelberg H, Suchenwirth R (2009) Lead pollution of drinking water in lower Saxony from corrosion of pipe materials. Gesundheitswesen (Germany) 71(5):265–274

Chapter 2
Atmospheric Emissions from Construction Sector

Idrees Yousuf Dar, Zaiema Rouf, Maheen Javaid, and Mohmad Younis Dar

Abstract Atmospheric pollution created by the construction industry has both direct and indirect effects on the general environment. The proper assessment and mitigation of the burdens of the environment from construction activities is the need of the hour. There must be a comprehensive evaluation of the impacts that need to be taken care of at all the construction activities. During construction, direct atmospheric gaseous emission, emission of particulate matter and other trace gases are released by the machineries and equipments, which have serious environmental impacts that affect the local air quality to a greater extent. As a result of these atmospheric emissions, the flora and fauna of the area do not grow well, causing significant loss to biodiversity and disruption of the food chain. Further, the various types of equipment used in the construction sector are particularly very noisy, which can cause people living near construction sites to experience varied levels of health disturbances. The construction sector should also share the responsibility of monitoring and limiting the quantity and quality of pollution it collectively generates. Thus, the development projects must be prepared in such a way that it has reduced minimum negative impacts on the environment.

Keywords Atmosphere · Construction · Developmental projects · Green · Monitoring

2.1 Introduction

The contribution of construction sector to the total direct and indirect global greenhouse gas (GHG) emissions in 2010 was 18% (IPCC 2014), and it was also the biggest user of materials in 2005 having direct impacts on the utilization of use and greenhouse gas emissions (Krausmann et al. 2009). Among different important atmospheric gases emitting sectors, there are greater abatement opportunities to a greater extent for reduction of emission in the construction sector and the sector offers large opportunities in the short-term due to its cost importance and relationship with GHG

I. Y. Dar (✉) · Z. Rouf · M. Javaid · M. Y. Dar
Department of Environmental Science, University of Kashmir, Hazratbal, Srinagar 190006, India

J. A. Malik and S. Marathe (eds.), *Ecological and Health Effects of Building Materials*,
https://doi.org/10.1007/978-3-030-76073-1_2

emissions by different chains in construction supply (Giesekam et al. 2015). As a matter of fact, construction sectors are bigger contributors to atmospheric pollution; sectors within the industry have a common responsibility for reducing the amount of waste they produce. In fact, the majority of the policies, acts and regulations mainly focus on decreasing direct atmospheric emissions from the construction sector, in recent years the basic research has paid greater attention to the life-cycle of GHG emissions into the atmosphere of the construction sector (European Commission 2014). The latest review of life-cycle energy use in construction sector reported that embodied energy is between 5 and 100% of total life-cycle energy utilization (equal to 10–97% of total life-cycle emission of carbon emissions) dependent on functions of building, place, use of materials and tentative assumptions regarding the life service and energy usage. These proportions increase as construction changes from conventional to modern, less energy and nearly zero energy constructed buildings (Chastas et al. 2016). Acquaye and Duffy (2010) researched that about 11.7% of national emissions of Ireland in 2005 were from the construction industry, and 71% of emissions were from indirect sources. Meanwhile, the construction industry in Norway produced GHG emissions of 4.2 metric tonnes and 5.3 metric tonnes of CO_2 in 2003 and 2007, respectively, with embodied atmospheric emissions accounting for the majority of total atmospheric emissions (Huang and Bohne 2012). The use of energy in the construction industry was nearly about 50% of total energy use in China in 2007 and the largest contributors to embodied energy use in construction were products for construction, warming, fossil fuels and electricity (Chang et al. 2010). Chen et al. (2017) also found that the construction sector is contributing to about 66.5% of total carbon emissions of China and was the biggest carbon producer of all other industries in 2009 in China, out of which indirect (embodied) carbon emissions were 96.6% with the highest contribution from electricity, water and gas supply sector. Further research studies of Ireland's and Norway's construction sector emissions, as well as others, identified related future areas for emission control across various phases, including increasing the percentage of reusable energy, increasing machinery and equipment maintenance, minimising operations, reducing the amount of carbon-using substances needed, and reducing the distance for transportation (Acquaye and Duffy 2010; Chang et al. 2010).

Atmospheric emissions at the construction sector are produced from activities related to the construction phase of a project. The various construction activities are typically very short-term or temporary in duration. The construction activities includes various types of operation of large on-road and off-road instruments for soil disturbance or hauling of soil and delivery of materials, moving, piling of the construction materials, piles with open storage and both inactive and active disturbed land areas. The emissions of atmospheric pollutants and GHG into the atmosphere may be because of the result of the mentioned onsite activities. Moreover the emission from construction of a project may have a significant effect with regard to atmospheric air quality and global climate change. Fugitive dust along with engine combustion emissions is generated with the usage of large equipment and soil moving operations at construction of buildings that can have substantial temporary effects on regional air quality. PM with a diameter size that is less than ten microns in size may also have their origin from fugitive dust including open fields, roadways, piles of storage, soil work, etc. The various sources of fugitive dust emissions includes building demolition, soil excavation, land clearing, fill and cut operations and use of traffic equipment on roads that are temporary at construction sites. As a matter of fact most of the construction machines use diesel fuelled engines. Exhaust from diesel engines is the source of emission that is having a significant impact on human well-being and health. CARB (California Air Resources Board) in July 1999 listed diesel particulate matter (DPM) at construction sites as a very toxic air contaminant, having both chronic and carcinogenic human health risks. DPM in addition to diesel exhaust also includes atmospheric emissions of certain other pollutants e.g., NOx and ROG (reactive organic gases like benzene and carbon monoxide) as well as GHG. Atmospheric emissions from the construction sector are mostly produced with the use of huge, diesel-fuelled scrapers, excavators, heavy loaders, bulldozers, haul trucks, large compressors, diesel fuelled generators and other large equipment. Atmospheric emissions at construction sector from both fugitive dust and from combustion sources may vary to a greater degree daily depending on the level and nature of activity, the type of operation, use of dust reducers, moisture content of soil, and onsite prevailing weather conditions.

As the construction sector is showing rise in growth, it will have a deleterious effect on the environment. As per the U.K. Green Building Council, about 400 million tons of materials a year are utilised by the construction industry and most of them can have varied adverse implications on the environment. More ever the materials used during various construction activities can also have a negative impact on the surrounding environment because of the extraction of raw materials. According to the Environmental Protection Agency (EPA 2008), in the United States, a number of equipment and materials daily used by construction workers and building firms, such as various chemicals onsite can significantly be harmful to public health and to the environment. Furthermore, the United States construction industry is accounting for about 160 million tons (25%) wastes other than industrial wastes generated a year. In another research by United States Green Building Council (USGBC), the construction industry uses 40% of energy worldwide, with estimates that the atmospheric emissions from building of commercial value will raise by 1.8% by 2030. According to the Environmental Protection Agency, construction activity can change the land surface mainly because of vegetation clearing and excavating. According to the EPA, this means that surrounding environments of the construction site can be heavily polluted, which may experience a rise in atmospheric pollution. As per Kleiwerks International the construction material from construction sector like cement, sand, concrete, aluminium and steel, are responsible for huge quantities of CO_2 emission because of higher concentration of "embodied energy content", having 9.8 mt (million tons) of gaseous CO_2 created from the creation of 76 million tons of finished concrete in the US. This research further adds that the current practices of the construction industry at reducing atmospheric emissions are greatly non-effective and may even produce higher levels of greenhouse gas pollution. It further says that the construction sector activities utilize various types of materials from nature and the construction sector accounts for one-sixth of world freshwater consumption, one-fourth of wood use, and generates one-fourth of global waste. However, Environmental Protection Agency's regulations are clear and the rules say that at the outset of any construction project the protection of the environment should be first priority.

Global warming is the result of increasing average atmospheric temperature, and drives a lot of changes to the globe's weather systems and climate. Heat-trapping greenhouse gases (GHG) are emitted in the atmosphere as a result of swift changes by humans in the atmosphere (Alhorr et al. 2014). Among atmospheric emissions, emission of carbon dioxide (CO_2) into the atmosphere is the important man-made greenhouse gas because of its increased concentration in the atmosphere and its property to remain in the air for a longer period of time (Riffat and Mardiana 2015). CO_2 emission in the atmosphere is both from natural and anthropogenic sources. Urbanization process is one of the main sources of CO_2 production. Urbanization in real terms is a continuous process that converts rural places into urban places with a large number of persons in urban areas and the increase of the built environment both vertically and horizontally. In urban cities, the built environment refers to the developed surroundings that create infrastructure and different services for human society, and the built environment is one of the most important components of a country's socioeconomic growth. Therefore, the increased urbanization has played a pivotal role in production CO_2 emissions in the construction sector.

The building sector in general is from construction to operation that may be again divided into two components; residential buildings and non-residential buildings. The construction sector includes the processes of making structures in areas of a place and the operation, maintenance and service of the constructed object. With the construction sector showing growth in development, a major direct and indirect effect of the construction sector on the environment has been seen. It is also taken as one of the important utilizing and waste producing portions of the economy (Bilal et al. 2020). Various environmental impacts of the construction sector can be differentiated into ecosystem impacts, people impacts and natural resource impacts (Zolfagharian et al. 2012). The building sector also consumes significant energy and production of atmospheric emission, for example GHG emissions, PM, oxides of sulfur, carbon monoxide, and oxides of nitrogen (Sandanayake et al. 2019). Due to energy consumption by the building sector, the ambient CO_2 level has increased (Adams and Nsiah 2019; Chang et al. 2019). The major sources of CO_2 emissions into the atmosphere by the construction sector is from the energy consumption that is needed for the manufacturing and transportation of different construction materials to the processing of different resources, building waste disposal, and the need for construction equipment (Yan et al. 2010). The building sector also utilizes a major percentage of energy that is non-renewable and that results in generation of a huge concentration of CO_2 in the atmosphere (Huang et al. 2018). Building sector contributes to about 39% of the global annual CO_2 (IEA 2019) (Fig. 2.1). Furthermore it has been found that in developed and developing nations more than one-third of the use of total energy and CO_2 production is from the building sector (Klufallah et al. 2014). Therefore, atmospheric CO_2 emission controlling measures are crucial (Langevin et al. 2019). In order to create mitigation of CO_2 emission, planning on

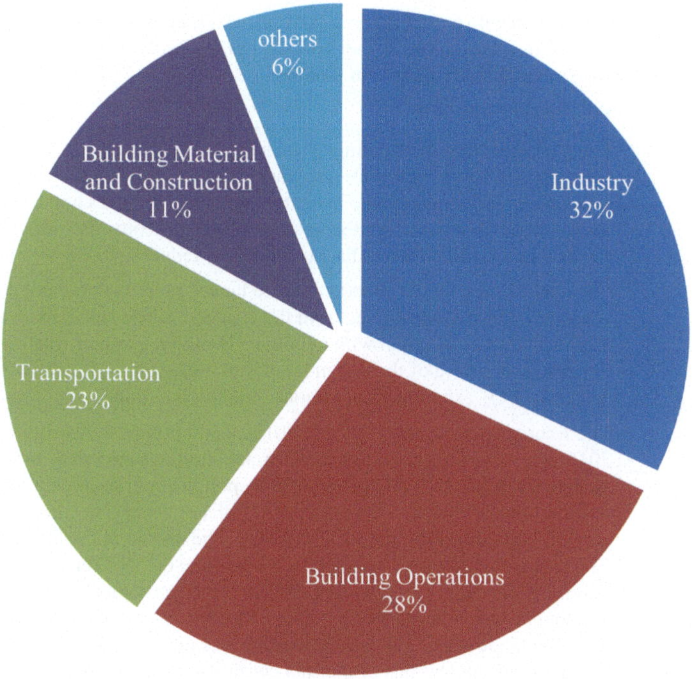

Fig. 2.1 Contribution of CO_2 emission from various sectors (Adapted after: IEA 2019)

energy conservation and implementation of effective strategies to decrease potential emission mitigation must be at first priority (Ma et al. 2019).

No doubt urbanization is going at an increased rate in these times than in the past era. The construction sector has a crucial role in the production of various pollutants particularly carbon dioxide (CO_2) into the environment. In fact building construction, onsite construction operation, and use of the built environment has been found to increase atmospheric emissions into the ambient air, huge amounts of CO_2 and other harmful gases. Different types of challenges and issues are rising from the construction sector in decreasing atmospheric emissions. Overuse of energy from non-renewable resources, weak construction design, and absence of sustainability design in the construction sector is the main reason that atmospheric emission mitigation measures are not working up to the international standards. Now therefore the atmospheric emission control schemes and different plans are important along with standard guidelines and standard frameworks. The various strategies to mitigate atmospheric emissions from the construction sector are the set policies and standards, doing impact assessment, applying low carbon emitting technology, and reducing utilization of energy. All the stakeholders in the construction sector need to play their respective roles effectively to decrease atmospheric emissions and help to fight global warming and climate change (Table 2.1).

Table 2.1 Standards associated with reducing concentration of CO_2 emissions in different buildings on Nationally Determined Contribution (NDC) set up in 2015 (The Paris Agreement Commitment and the United Nations Sustainable Development Goals)

Nation	Policies and standards
China	Energy Utilization of Buildings standard was enacted by the Ministry of Housing and Urban-Rural Development in the year 2016. The mentioned standard includes energy use indicators for different types of buildings. It has the main purpose to reduce the quantity of energy consumption of building sector energy of the nation that subsequently reduces the emission of CO_2
Australia	National Carbon Offset Standards were launched by Australian Federal Government for Building sector in the year 2017. The standards were established in association with the Green Building Council Australia. The important objective of these standards is to measure, mitigate, offset, report, and audit CO_2 exhausts from various building operations
India	The Energy Conservation Act of 2001 is a part policy and was introduced in 2016, whose main aim was at commercial buildings under the Perform, Achieve, and Trade (PAT) program. The policy has conserved almost 9 million tons of oil that is equivalent (MTOE) of energy, resulting in reducing annual CO_2 emissions almost 23 $MtCO_2$. The update in the Energy Conservation Building Code (ECBC) was done in 2017 for commercial buildings that admit improvement efforts for decarbonization. The first national model building energy code called the Energy Conservation Building Code for Residential Buildings was introduced in 2018 with much simpler implementations of thermal comfort and passive system improvement
European Union	The European Commission as part of Cleaner Energy for all the European policy packages set in 2016, targets to mitigate climate change done by GHGs, including emission of CO_2, by proposals for an efficient energy market, and strategies for renewable energy. Control of the Energy Performance of Buildings Directive (EPBD) was done in 2018 to get high-energy efficiency and decarbonisation by 2050
Sweden	The Centre for Sustainable Construction in 2016 was formed under a policy of Swedish Government to enhance the use of materials that are sustainable and energy-efficient renovations that would also reduce CO_2 emissions. A certification scheme was introduced in 2019 addressing the environmental effects of recent buildings
Japan	In 2017, The Act for the Improvement of Consumption of Energy Performance of Buildings (Building Energy efficiency Act) was included in the year 2017, which has regulatory measures for mandatory compliance with energy efficiency standards for non-residential buildings. To be achieved by 2030, the act is part of the Japanese government policy on the zero-energy-building [ZEB]/zero-energy-house [ZEH] system
Canada	Tighter energy performance standards were introduced in 2016 for energy-using product categories in buildings. In 2022, new building energy codes have been planned to be introduced as part of the Pan-Canadian Framework on Clean Growth and Climate Change to increase efficient energy in existing buildings. The Canadian Government in 2019 was working to produce a net-zero-energy-ready building code

(continued)

Table 2.1 (continued)

Nation	Policies and standards
Germany	A package of emission control measures in 2019 was formed by the German government in the building sector to meet the requirement of Agenda in the year 2030
USA	The California 2019 Building Energy Efficiency Standards was formed in 2018 as the first code in the United States of America. In 2018, the New York State Energy Research and Development Authority were formed to enhance the sustainability of buildings and efficiency of the buildings
Nigeria	The first building energy code was established in the year 2017 with a contribution between the German Development Agency (GIZ) and the Nigerian Energy Support Program having the aim of establishing minimum standards for efficient energy building construction in Nigeria
Singapore	The Code on Environmental Sustainability Measures for Buildings was launched in 2016 for existing non-residential buildings within Singapore's Building Control Regulations
Switzerland	Switzerland's new Energy Act came into force in 2018, for increased energy efficiency in buildings towards decarbonisation. It also includes the usage of a CO_2 tax on standing fuels (heating and industry). Under this Act, CO_2 tax and subsidizing of geothermal energy have been included. A central Act on Reducing the Emission of CO_2 was revised in 2019 to implement NDC in the building sector

2.2 Various Construction Activities Contributing to Atmospheric Pollution

2.2.1 Use of Onsite Vehicles and Plants

Use of plants and vehicles depends upon the onsite construction activities and also includes various types of machinery for example excavators, bulldozers, and other heavy vehicles. Various Machinery and plants that are used on construction sites are not properly governed by the authorities. Due to the higher degree and type of construction projects, equipment is running continuously and polluting the atmosphere over a longer period of time. Due to very heavy equipment machinery, and related vehicles onsite, operating more on diesel based engines, they release various types of atmospheric pollutants. This may include various types of gases like oxides of carbon (carbon monoxide and carbon dioxide), oxides of nitrogen and sulphur and other hydrocarbons.

2.2.2 *Building Demolition and Land Clearing*

The land for construction activities is to be cleared and made stable for construction; the process of clearing should be completed so that it makes sure that it has the minimal effects on the atmosphere. Higher levels of dust are generated with the construction of buildings to a greater extent because of disruption and demolition of existing structures.

2.2.3 *Chemicals*

Various types of hazardous chemicals are used at construction sites. These chemicals include various types of paints, glues, oils, thinners and plastics, which produce various types of noxious vapours and other volatile gases.

2.2.4 *PM_{10}*

The huge amount of construction dust produced from cement, concrete used, silica and wood from construction sites are together classified as PM_{10}. PM_{10} is particulate matter having size of less than or equal to 10 micrometres in diameter that is not visible to the naked human eye. The gaseous exhaust from diesel engines of the plant at construction sites and other machines and vehicles is also a huge contributor to PM_{10}. More precisely, this PM is also known as diesel particulate matter (DPM) as it contains sulphates and silicates that add pollutants to the atmosphere.

2.3 Issues and Challenges

The biggest confrontation in the sustainable advancement of the construction sector is the continuous increase in CO_2 emissions because of usage of un-sustainable sources of energy in processes like organization, construction, and working of buildings (Huang et al. 2018). Further CO_2 emissions also result from the wide usage of land in the process of urbanization (Klufallah et al. 2014).

Fossil fuel based energy is unsustainable, but still it contributes to a huge proportion of used energy during the activities of construction and working. Those sources of energy which are sustainable or renewable are responsible for only 6% of the overall energy utilized in this sector, whereas the utilization of fossil fuels in the construction processes is responsible for 40% of global greenhouse gas emissions. Even if various new methods are being devised for reducing the CO_2 footprint of the construction sector, especially in urban—communities with too high density, yet a lot needs to be done (Yim et al. 2018). The use of energy that is a non-sustainable source impacts the atmosphere directly, and it is in proportion to the quantity used directly. Building construction results in the emission of CO_2 either in direct or indirect ways. CO_2 is emitted directly from the combustion of diesel, fuel oil, natural gas, and other oil—based equipment, on the other hand CO_2 is emitted indirectly from utilization of electricity. Worldwide, the indirect emissions of CO_2 contribute about 85% of the overall CO_2 generated whereas indirect emissions accounts for only 14%. As per the statement of the 2020 Climate and Energy Framework, 27% of the energy ought to be procured from sustainable sources of energy, along with that there should be 27% rise in energy efficiency or productivity (Pal et al. 2017). Moreover, numerous challenges are there in achieving solutions that are sustainable to very low achievement and high efficiency. The one possible solution can be enumeration of the processes of operation and construction in order to obtain a detailed evaluation. Construction involves the gathering of the construction material, establishment of foundation and structure, and the working and transport of equipment. The strategy includes the preservation side of a constructed building and its foundation. The prerequisite for evaluating the life cycle is the comprehensive listing of these activities during all the stages pertaining with the life cycle of a building.

2.4 The Importance of Building Green

The various processes by building green utilise those materials in construction activities which can preserve 250 metric tons of CO2 emissions on an annual basis, as per the statement of environmental group LEED. Moreover as per the latest report given by the Dodge Data and Analytics, there is a regular doubling of green building every three years, along with that it is expected that 60% of the construction a will be activities by 2018 will be green and about 70% of the survey respondents are of the view that the highest benefit of green building is the less operating cost. The research concludes that the construction firms that are increasingly being told to construct projects which are sustainable and as well as efficient in energy. The increasing trend towards the construction of building green projects has directed the Environmental Protection Agency towards the instigation of an adequate research in this field, involving the collaboration with the National Institute of Building Sciences in the formulation of Building Green Construction Code, that throws a detailed light on the approaching way of construction firms towards the green building by the incorporation of different federal rules and regulations. There are various programmes given by EPA which are given below:

Energy Star Program—Such as the Environmental Protection Agency and the Department of Energy have come together for the creation of the Energy Star program, that is responsible for promoting the usage of materials having high energy efficiency in buildings throughout the United States, according to the website of EPA.

Industrial Recycling Program—Further EPA holds specific initiatives such as the Industrial Recycling Program of the EPA, which gives awareness about how the recycling of demolition debris and construction can be done in order to lessen impacts of the construction sector on the environment. This programme also includes the recycling of industrial materials in construction applications utilised in situ by the contract workers besides improving the product quality.

EPA's Environmentally Preferable Purchasing Program—The programme helps in improving the environment at the hands of construction companies by suggesting them to buy those products which will help in improving the quality of environment.

EPA's GreenScapes Program—Offering solutions which have cost efficiency and are eco-friendly is what this programme is meant for. The principal objective of this programme is to smother wastes along with pollution and work for the protection of natural resources throughout the construction process.

2.5 Impacts in General

The atmospheric emissions from the construction sector influencing the natural surroundings are not only contributed from the operational stage, but also incorporate those that are embodied in the whole life cycle, both from construction as well as from demolition of cities and constructed buildings. Worldwide chain supplies, including brick-making, excavation, demolition, and transportation can be hazardous for the environment, and 'build in' embodied emissions from a building. From construction, atmospheric particles of dust, such as silica dust or hardwood are also known to cause adverse health impacts including asthma, heart disease, and silicosis (Safety and Health 2015). Dust of silica that is generated during the preparation of concrete and exposure to this substance which is potentially toxic can cause threats to health across the built environment globally. This fact is evident that emissions of CO_2 result in climate change and global warming that has a tendency to pose serious impacts on human health and environment. The emissions of CO_2 in atmosphere function like blankets that absorb heat, and consequently warming up the planet (Klufallah et al. 2014). This is the layer which is responsible for preventing the earth from cooling effects, and hence elevating worldwide temperatures. Global warming has serious consequences on environmental conditions, the supply of food and water, the pattern of weather conditions, along with sea levels. The NOAA Global Climate Summary states that the temperature of ocean and land taken together from 1980 has shown an increase the average rate of which is 0.07 °C per decade. The release of CO_2 in the atmosphere results in acid rain that in turn damages trees physically (Paoletti and Manes 2003) and the built environment (Cellura et al. 2018; Bravo et al. 2006). These consequences of Atmospheric gases from the construction sector can be clearly noticed. These emissions extend enormously beyond increasing the global temperatures that are influencing ecosystems and communities all around the globe.

2.6 Consequences of Atmospheric Pollution from Construction Sector

2.6.1 Construction Workers

It has been found by research that PM_{10} can pass into lungs so deep into the persons those who take PM polluted air. Workers working at construction sites on a daily basis are at very higher risk getting complications of health. Substandard air quality because of atmospheric pollution can create the below mentioned health risks:

1. Coughing, wheeze and breathing shortness
2. Heart and other respiratory complications
3. Cancer of lungs and other organs
4. Heart and other Strokes
5. Aggravation in asthma.

In Fact in the construction sector 56% of the cancers are occurring in occupational men. For example mesothelioma, that is a cancer type that is caused due to exposure to asbestos developing on the inner lining of the two lung lobes and chest. Continuous exposure to the dusts produced and fibres generated, for example silica and asbestos, and to the fumes and gases produced by various vehicles and machines is the common reason among construction workers that lung cancer is very common. The construction workers doing their job at the construction sites are generally exposed to different carcinogenic compounds as a result of various construction activities.

2.6.2 Residents of Locality

The effects of atmospheric pollution to a greater extent are felt by people living near construction sites. People living in not in close proximity of construction sites as construction workers to the atmospheric pollutants but they may experience different effects of poor air quality. PM_{10} and known atmospheric pollutants are also dispersed by air to the near atmosphere and get settled later on. The residents near construction sites not knowingly often breathe PM and can then experience different health complications such as cough, breath shortness as a short-term health consequence.

2.6.3 Environmental Effects

Apart from adverse effects on human well being, there is an urgent need of awareness about the adverse implications of atmospheric pollutants upon the environment. Construction sites cause 14% PM 2.5 (particulate matter having 2.5 micrometers of diameter) and about 8% of Particulate Matter 10 exhausts. Most emissions originate from the machines used in the construction sector along with generators which operate on fuel such as diesel, only 1% is represented by demolition and other activities occurring at the sites. This poses a severe threat for the survival of plants as well as animals and ultimately results in the disruption of food chain biodiversity loss.

2.7 Prevention of Atmospheric Pollution from Construction Sector

The construction process utilizes many chemicals, the majority of which if not managed or handled properly can prove detrimental to both the workers and the environment. Hence EPA laid down the recommendations for designing, installing, implementing and maintaining effective pollution prevention strategies, throughout the project course to ensure the safe and proper discharge of pollutants with less negative effects on the atmosphere. The regulations state that it is to make sure that the minimisation of production of pollutants emitted from various instruments utilized or observed at construction sites, including the vehicles at sites, the wastewater from wheel wash, and other related chemicals. Further the regulations further maintain that it needs to reduce the exposure of construction materials, end products, building wastes materials, associated products in precipitation as well as snow water. Environmental Protection Agency further stated that it is not mandatory for those construction sites where the water sources and the atmosphere around the sites of construction are not at risk due to pollutants generated

2.7.1 Pollution Prevention Strategies

The management of the amount of pollutants you generate as a firm or individually are very imperative. Strategies for pollution control are having a much positive effect on the business of construction besides curbing the adverse effects on workers, nearby residents, as well as the environment. As per the statement of the Environmental Damage (Prevention and Remediation) (England) Regulations 2015, firms are being made to pay if they cause any damage to land, water, air, biodiversity in England. The regulations have enforced an enforcement of a principle known commonly as the principle of 'polluter pays'. This policy makes businesses of construction responsible for the emissions they are creating by encouraging businesses to reduce their atmospheric impact with incentives of monetary nature. This principle is meant for holding the firms liable for the pollutants they generate by preparing them to lessen their impact on the environment via financial incentives. These are usually referred to as enforcement undertakings since these are alternatives to prosecutions, and the money is given to the projects that help wildlife. No doubt that the construction phases produce a variety of atmospheric pollution, that it is manageable and avoidable. The atmospheric emissions produced by construction sector can be reduced by following the given suggestions:

- **Don't burn construction waste materials**. This reduces smoke and releasing of poisonous vapours for example carbon monoxide into the air.
- **Hybrid technology adaptation**. Diesel engine based excavators and diggers should be replaced hybrid prototype machines that work on electric power should be used.
- **Use low sulphur diesel**. Fuel especially diesel low in sulphur should be used to run various equipment and vehicles.
- **Improvement of existing equipment**. PM filters and catalyst converters for control of atmospheric pollutants need to be used.
- **Use water sprays or sprinklers**. These should be used to minimize different types of dust by stopping its further spreading.
- **Source local materials**. Materials from locality should be used to avoid the transportation of materials from large distance.
- Use of natural and artificial renewable and sustainable construction materials.
- **Wearing proper Personal Protection Equipment**. Such as the correct type of respiratory protective equipment (RPE) depending on the task.

2.7.2 Mitigating Atmospheric Pollution by Cause

- **Emissions from in-use buildings**: **S**ourcing of energy from renewable. Decrease emissions of operational carbon by focusing on total zero carbon building performance, which requires optimal energy efficiency for building systems and fabric.
- **Emissions from building life-cycle**: Sources from local, recycling or reuse of materials all reduce pollution produced by, transportation, demolition processes and construction.
- **Priority on production of brick**: production of higher proficient technologies, especially during brickfiring, can decrease emission of atmospheric pollutant.
- **Short Lived Climate Pollutants**: The production of Short-Lived Climate Pollutants from lighting, heating should be reduced. In addition, developments in construction quality can enhance heat well-being and eliminate demand for warming. Constructions planned according to weather, onsite usage of power and light from renewable sources is an effective result for reducing large-scale and site specific atmospheric pollutants from the construction sector.
- **Hydrofluorocarbons**: With majority of people all over the globe having health threats because of limited access to cooling in buildings for main needs of health, it is mandatory to promote accessible and sustainable cooling means.
- **Passive strategies for design**: This includes buildings that have energy efficient fabric material, ventilation and vegetation that can decline requirements for cooling in buildings and thus maintain comfortable conditions for living.

- **Dust from Construction**: Generation of dust from construction sites should be properly managed with strict organisational and national regulation, appropriate policy and practice employed on site, and other strategies for dust reduction. Off-site modular construction practices can be preferred due to more controlled dust production and lower volume.
- **Reducing Waste**: The overall process of construction can cause the generation and withdrawal of a huge amount of atmospheric pollutants. But, enhancing the work efficiency, prioritise the technologies which can effectively reduce production of waste and enhancing supplies and materials for construction helps to a great extent.
- **Exploration of Recycling Options**: Finding options for industrial recycling needs to be prioritized completely.
- **Usage of Eco-Friendly Materials/Products/Tools**: Construction industry must consume products, tools, and materials that are environmentally friendly and are designed for reduction of air pollution and consumption of energy on the construction sites.
- **Protection of Earth Resources**: Any sort of work related to construction must be aimed for protection of not only protection of environment but also include protection of plants and the animals in a given area.

Indoor Environment

- **Infiltration of Pollutants**: On an average we spend of about 90% of our overall time within the building, it is comprehensible that the major part of our exposure to outdoor pollution appears inside. In the present situation, where 91% of populace live in polluted outdoor air environments, thus it is advised to have a conscientious approach of ventilation strategy (WHO 2018).
- **Focus on fabric of buildings**: A superior building fabric can be successful way to minimize the exposure of atmospheric pollutant to infiltration and to create a more pleasant indoor air environment with expenditure of minimum energy. Fully insulated walls can efficiently work for all climates; trapping of heat can eventually keep an indoor air cool or warm as well as declining other wellbeing threats like, noise discomfort.
- Moreover, activities of people for reduction of their individual share atmospheric outdoor pollution are an appropriate way to minimizing the outdoor quality of air than that in the buildings.
- **Air Ventilation**: Enhanced levels of ventilation, with adequate screening are compulsory and are an important strategy for cleaning IAPs by exchanging clean air with fresh air, which can promote to prevent or reduce the negative health impacts. Minimum ventilation systems and plans vary according to quality of outdoor air and climate, moreover in some areas with greater concentrations of air-based particulate matter; more filtration of air is often required to keep the indoor environment healthy.
- **Mould**: Walls with mould are often present in moist, temperate climates, or cold regions as consequences of the infiltration of cold air in outdoor environment

through the cracks present in the fabric of building, often shown by a poor envelope of buildings, which after condensation forms moist layer when exposed to considerably warmer inside materials.

- In the cold regions, enhancing air tightness of building and material quality of insulation can decrease the chance of mould build up and consequent threats to health, also enhancing thermal comfort and efficient energy of the environment in the construction sector.
- In hot climates, concentration on proper ventilation system to reduce and eliminate clear condensation and stale air is important that can possibly be increased with proper air conditioning or utilization of dehumidifying equipments. If such technologies are proven to be highly efficient appliances, energised by renewable forms of energy sources, then we can potentially reduce the threat of increasing atmospheric pollution from energy generation upwards.
- **VOCs**: These are produced from a variety of regular products that includes aerosols, varnishes solvent-based cleaning products, paints, and various preservatives. The knowledge about the VOC exposure related health impacts is rising, low-VOC products or the products that can capture VOCs are getting easily available for local public, the workers of construction industry, and design professionals.
- **Toxic materials**: Varied exposure to materials that are highly toxic for example asbestos is already outlawed by local and national building codes in various regions across the globe. Countries where it has not been the case, training for architects, awareness campaigns, designers, and policy updates for the general public are beneficial strategies for reducing health threats.

The building and construction sectors must identify the liability it has to supervise and reduce the abundance of atmospheric pollutants it collaboratively creates on the environment. Of the one easiest measure to take is to be aware of the quantity of the pollutants and waste that construction activities generate and the consequences these pollutants have as far as the environment is concerned. No matter if you are an employer or construction worker in the construction sector, there are policies and regulations that decline the quantity of atmospheric pollution generated that needs to be enforced and encourage others to do the same. But ironically, the atmospheric pollution generated by construction sectors is directly affecting the environment and its potential to do its job in a sustainable manner. Anywhere in the construction sector outside the environment is polluted, passive or natural ventilation plans are not suitable. Energy-utilizing air filtration is sometimes used but this can elevate more the utilization of energy from the construction sector (unless the energy used is produced by renewable energy sources or utilizing systems which are highly efficient), that can result in a synergistic effect. Globally the demand of energy from air conditioning is estimated to triple by the year 2050, as consequences the negative effect on global air quality is bound to enhance (IEA 2018). Moreover, during construction activities within the buildings having less vulnerability to toxic chemicals or materials, the majority of the risk is from outside atmospheric pollution. When we are inside buildings, a major portion of exposure to the outdoor atmospheric pollutants happens due to addition through cracks in the building fabric aperture or windows, apertures (Allen et al. 2017).

2.8 Stakeholder Roles in Reduction of Atmospheric Pollution

2.8.1 General Public

- Eco-clean energy should be favoured for transportation and power and to promote conservation of energy as far as possible.
- Enhance the quality of building construction and restrain unhealthy toxic compounds in furnishings; choose products with low-VOC where possible for materials example carpets or paints.
- Affective ventilation must be guaranteed for access to clean and fresh air.
- Analyse investing in an IAQ monitoring.
- Utilize a team for service management and/or landlord to deliver a better air environment for residents and occupants.

2.8.2 Business

- Cleaner energy must be chosen for transportation and power, and should enhance energy conservation.
- Good IAQ indoor air quality should be maintained with proper ventilation strategy, healthy materials and utilisation of real time monitoring of indoor air environments.
- Priority should be given to liable provision for buildings—to prioritise ethical, recycled and local, materials with potentially no or moderately low VOC concentrations that lead to emissions.
- Promote the initiatives of sustainable finance worldwide for green buildings specifically micro-financing schemes in the developing countries.

2.8.3 Government

- The authorities should prioritize investment in green energy, carbon depletion and promote decentralised renewable networks of energy in rural areas.
- The government should support efficient forms of energy by enhancing standards related to building and should be encouraging the retrofit programmes.
- More secure and sustainable construction methods should be incentivised.

- There should be implementation of national guidelines for IAQ and building ventilation.
- Utilisation of recognized toxic materials be discouraged, and administration should legislate a minimum standard for contaminants with potential high risk.
- The government authorities should supervise the outdoor air environment and to disclose data publicly, and encourage IAQ monitoring in high occupant areas such as hospitals, offices and schools.

2.9 Conclusion

Everywhere in the world there is some sort of construction activity going on that is very helpful in developing the nations and increasing the standards of living. The construction industry plays a crucial role in the emissions of various pollutants in the atmosphere. The humongous production and release of pollutants from the construction sector have been found to have severe consequences and impacts contributing to global warming and climate change. The various deleterious impacts of the non-sustainable construction activities have not only put a stress on the environment but have also impacted humanity. Energy generated from fossil fuels is no doubt non-sustainable, but it contributes for a higher percentage of the energy utilized in the construction and operation processes. The various basic strategies to reduce atmospheric emissions from the building industry are the policies and enforcing standards, conducting impact assessment, adoption of low carbon technology, and minimizing use of energy. If we humans continue with the current policies and approach in the reduction of atmospheric emission from the construction sector, it will be very late to rectify and undo the mistakes that have been done in the past. We will fail to achieve the goals of global sustainable development and the near future of sustainable communities and sustainable cities will remain uncertain. The construction sector must be provided with enough attention and care so that it can reduce and curtail the atmospheric emissions effectively. A very comprehensive analysis is needed to study the nature of atmospheric emission, rate of emissions, quantity, quality and controlling measures in the construction sector, and local and world organizations must frame a sustainable inclusive framework to handle the issue of harmful emissions from the construction sector. For a sustainable future of the world, it is very necessary to impose necessary actions and measures to curtail emissions from the construction sector and that will lead to contribute in the fight of combating climate change.

References

Abergel T, Dean B, Dulac J (2017) Towards a zero-emission, efficient, and resilient buildings and construction sector. Global status report 2017. UN Environment and International Energy Agency, Paris, France

Acquaye AA, Duffy AP (2010) Input–output analysis of Irish construction sector greenhouse gas emissions. Build Environ 45:784–791

Adams S, Nsiah C (2019) Reducing carbon dioxide emissions: Does renewable energy matter? Sci Total Environ 693:

Ahmed AK, Ahmad MI, Yusup Y (2020) Issues, impacts, and mitigations of carbon dioxide emissions in the building sector. Sustainability 12(18):7427

Alhorr Y, El Iskandarani E, Elsarrag E (2014) Approaches to reducing carbon dioxide emissions in the built environment: low carbon cities. Int J Sustain Built Environ 3:167–178

Allen JG, Bernstein A, Cao X, Eitland E, Flanigan S, Gokhale M, Yin J (2017) The 9 foundations of a healthy building. School of Public Health, Harvard

Berg F, Fuglseth M (2018) Life cycle assessment and historic buildings: energy-efficiency refurbishment versus new construction in Norway. J Archit Conserv 24:152–167

Bilal M, Khan KIA, Thaheem MJ, Nasir AR (2020) Current state and barriers to the circular economy in the building sector: towards a mitigation framework. J Clean Prod 276. Article no 123250

Bravo AH, Soto AR, Sosa ER, Sánchez AP, Alarcon JAL, Kahl J, Ruiz BJ (2006) Effect of acid rain on building material of the El Tajin archaeological zone in Veracruz, Mexico. Environ Pollut 144:655–660

Cellura M, Guarino F, Longo S, Tumminia G (2018) Climate change and the building sector: modelling and energy implications to an office building in southern Europe. Energy Sustain Dev 45:46–65

Chang CT, Yang CH, Lin TP (2019) Carbon dioxide emissions evaluations and mitigations in the building and traffic sectors in Taichung metropolitan area, Taiwan. J Clean Prod 230:1241–1255

Chang Y, Ries RJ, Wang Y (2010) The embodied energy and environmental emissions of construction projects in China: an economic input-output LCA model. Energy Policy 38:6597–6603

Chastas P, Theodosiou T, Bikas D (2016) Embodied energy in residential buildings-towards the nearly zero energy building: a literature review. Build Environ 105:267–282

Chen J, Wu Y, Yan H, Shen L (2017) An analysis on the carbon emission contributors in the Chinese construction industry. In: Wu Y, Zheng S, Luo J, Wang W, Mo Z, Shan L (eds) Proceedings of the 20th international symposium on advancement of construction management and real estate. Springer Singapore, Singapore, pp 1197–1206

Climate and Clean Air Coalition (2019) Bricks [online]. http://www.ccacoalition.org/en/initiatives/bricks

Climate and Clean Air Coalition (2019) Clean cooling technology in Jordan is a first for the Middle East. http://www.ccacoalition.org/en/news/clean-cooling-technology-jordan-first-middleeast

European Commission (2014) Communication from the commission to the European Parliament, the Council, the European Economic and Social Committee of the regions on resource efficiency opportunities in the building sector

Giesekam J, Barrett JR, Taylor P (2015) Construction sector views on low carbon building materials. Build Res Inf 44:423–444

Giunta M, Lo Bosco D, Leonardi G, Scopelliti F (2019) Estimation of gas and dust emissions in construction sites of a motorway project. Sustainability 11(24):7218

Huang L, Bohne RA (2012) Embodied air emissions in Norway's construction sector: input-output analysis. Build Res Inf 40:581–591

Huang L, Krigsvoll G, Johansen F, Liu Y, Zhang X (2018) Carbon emission of global construction sector. Renew Sustain Energy Rev 81:1906–1916

IEA (International Energy Agency) (2019) World energy statistics and balances (Database). IEA. www.iea.org/statistics

IPCC (Intergovernmental Panel on Climate Change) (2014) Climate change: synthesis report. Contribution of working groups I, II and III to the fifth assessment report of the intergovernmental panel on climate change. IPCC, Geneva, Switzerland

IEA (International Energy Agency) (2018) The future of cooling. Opportunities for energy-efficient air conditioning. http://www.oecd.org/about/publishing/TheFutureofCooling2018CorrigendumPa

Klufallah MM, Nuruddin MF, Khamidi MF, Jamaludin N (2014) Assessment of carbon emission reduction for buildings projects in Malaysia—a comparative analysis. In: E3S web of conferences, vol 3. EDP Sciences, Bangi, Malaysia

Krausmann F, Gingrich S, Eisenmenger N, Erb KH, Haberl H, Fischer-Kowalski M (2009) Growth in global materials use, GDP and population during the 20th century. Ecol Econ 68:2696–2705

Langevin J, Harris CB, Reyna JL (2019) Assessing the potential to reduce US building CO_2 emissions 80% by 2050. Joule 3:2403–2424

Ma M, Ma X, Cai W, Cai W (2019) Carbon-dioxide mitigation in the residential building sector: a household scale-based assessment. Energy Convers Manag 198. Article no 111915

Pal SK, Takano A, Alanne K, Siren KA (2017) Life cycle approach to optimizing carbon footprint and costs of a residential building. Build Environ 123:146–162

Paoletti E, Manes F (2003) Effects of elevated carbon dioxide and acidic rain on the growth of holm oak. Developments in environmental science. Elsevier, Amsterdam, Netherlands, pp 375–389

Riffat SB, Mardiana A (2015) Building energy consumption and carbon dioxide emissions: threat to climate change. J Earth Sci Clim Chang 2015:S3

Safety and Health (2015) Silicosis: what it is and how to avoid it. https://www.safetyandhealthmagazine.com/articles/12507-silicosis-what-it-is-and-how-to-avoid-it

Sandanayake M, Zhang G, Setunge S (2019) Estimation of environmental emissions and impacts of building construction—a decision making tool for contractors. J Build Eng 21:173–185

Sesana E, Bertolin C, Gagnon AS, Hughes J (2019) Mitigating climate change in the cultural built heritage sector. Climate 7:90

UN Environment and International Energy Agency (2019) Towards a zero-emissions, efficient and resilient buildings and construction sector. In: 2019 global status report for buildings and construction. UN Environment and International Energy Agency, Paris, France

US Environmental Protection Agency (EPA) (2008) Determination of PEMS measurement allowances for gaseous emissions regulated under the heavy-duty diesel engine in-use testing program. Revised final report EPA420-R-08–005, US EPA, Arlington, VA, USA

WHO (2018) Ambient (outdoor) air quality and health. https://www.who.int/news-room/fact-sheets/detail/ambient-(outdoor)-air-quality-and-health

Yan H, Shen Q, Fan LC, Wang Y, Zhang L (2010) Greenhouse gas emissions in building construction: a case study of one Peking in Hong Kong. Build Environ 45:949–955

Yim S, Ng ST, Hossain U, Wong JMW (2018) Comprehensive evaluation of carbon emissions for the development of high-rise residential building. Buildings 8:147

Zolfagharian S, Nourbakhsh M, Irizarry J, Ressang A, Gheisari M (2012) Environmental impacts assessment on construction sites. Construction research congress. American Society of Civil Engineers, West Lafayette, IN, USA, pp 1750–1759

Chapter 3
Polyvinyl Chloride (PVC), Chlorinated Polyethylene (CPE), Chlorinated Polyvinyl Chloride (CPVC), Chlorosulfonated Polyethylene (CSPE), Polychloroprene Rubber (CR)—Chemistry, Applications and Ecological Impacts—I

Shelley Oberoi and Monika Malik

Abstract As every coin has two sides, similarly polymers are playing an important role in our daily life as well as are creating pollution in the environment. Polymers have been part and parcel of emerging fields of science and technology. The word polymer is derived from the Greek words "poly" and "meres" which means 'many' and 'parts'. Polymers are macromolecules that are produced by the repetition of small molecules called monomers. Cellulose, protein, starch, and natural rubber are the examples of natural polymers. Polyvinyl chloride is a product of free radical polymerization of vinyl chloride. Substitution of hydrogen atoms in high-density polyethylene by chlorine atoms will produce chlorinated polyethylene. Chlorinated polyvinyl chloride is a thermoplastic polymer that is produced by chlorination of polyvinyl chloride by free radical chlorination reaction. Commercially available polymer Teflon is chemically chlorosulfonated polyethylene. Polychloroprene rubber is synthesized by emulsion polymerization of chloroprene. Different polymers have different aspects as polyvinyl chloride is applied for the making of rigid pipes, flooring, etc. Chlorinated polyvinyl chloride is compounded with other ingredients to get desired properties according to various applications. The properties of chlorinated polyethylene depend on the chlorination of polyethylene, which makes it suitable in the production of wires, cables, coal mine cables, etc. Chlorosulfonated polyethylene has been used widely for protective coatings, electrical cable jacketing, roof shielding, etc. Polychloroprene is being used in car fan belts, gas kits,

S. Oberoi (✉)
Department of Humanities and Applied Sciences, K.C. College of Engineering & Management Studies & Research, Thane, Maharashtra, India
e-mail: shelly.oberoi@kccemsr.edu.in

M. Malik
Department of Applied Science, Galgotias College of Engineering and Technology, Greater Noida, Uttar Pradesh, India

J. A. Malik and S. Marathe (eds.), *Ecological and Health Effects of Building Materials*,
https://doi.org/10.1007/978-3-030-76073-1_3

mouse pads, corrosion resistance coating, etc. Some of these synthetic polymers have adverse effects on the environment, flora, and fauna.

Keywords Chlorination · Monomers · Macromolecules · Polymerization · Synthetic

3.1 Introduction

Polymers have become an integral part of our lives. The word polymer evolved from a Greek word that means many parts. Proteins, carbohydrate, starch, cellulose, and rubbers are those natural polymers which have an existence from the origin of life. These polymers are categorized as natural and synthetic based on their origin. Polyethylene, polyvinyl chloride, polystyrene, bakelite, and synthetic rubbers are very well-known polymers in the world of plastics. Polymers are those macromolecules that are formed by the repetition of macromolecules known as monomers. The process of repetition of monomers is known as polymerization. These macromolecules are exhibiting exclusive properties of high molecular weights, viscoelasticity, and glass transition temperature. The credit to discover the term polymers goes to a Swedish chemist, Jöns Jacob Berzelius in 1833 (Mustafa et al. 2013).

Many years have been dedicated by scientists to create the engrossing world of polymers. In 1820, rubber was considered a polymer but its properties like fluidity and blending were discovered by Thomas Hancock for molding purposes. After some years in 1839, Charles Goodyear worked on the advancement of properties of the rubber by heating with sulphur. This innovation was patented by Goodyear in 1844.

In 1846, an invention of Christian Friedrich Schönbein came into the picture in the form of cellulose nitrate or Gun Cotton which was an explosive polymer with good molding properties at high temperatures. Later on, methyl rubber was manufactured from 2,3-dimethyl butadiene as the first synthetic rubber in Germany. A new concept was proposed by Hermann Staudinger in 1920 that polymers are macromolecules with covalent bonds. The Nobel Prize for chemistry was awarded to Paul Flory for his incredible contribution in the field of polymeric science (Young and Lovell 2011).

3.2 Types of Polymers

There are different types of polymers which are playing different roles according to need. Polysilanes, polysilazanes, polysulfides, polyphosphazenes, polyborazylenes, polythiazylsn and polysiloxanes are inorganic polymers in which the polymeric backbone is made up of inorganic atoms. Organic polymers are those polymers in which the backbone is composed of carbon and other atoms. These organic polymers are further classified into synthetic and natural polymers:

Natural Polymers—These are the organic polymers that are found in nature. The basic composition of the human body is based on natural polymers like protein and nucleic acids. Organic polymer cellulose is the fundamental structure of plants similarly the main component of food is starch.

Synthetic Polymers—Synthetic polymers are those polymers that can be synthesized. Thermoplastics and thermosetting are further classifications of synthetic polymers. Thermoplastics can change their shape at a particular temperature. Polyvinyl chloride, nylon, polyethylene, polypropylene, and polystyrene are some examples of thermoplastics. Thermosetting are those polymers that cannot be reheated, reshaped, and reused. Bakelite is a common example of thermosetting (Gad 2014).

3.3 Polyvinyl Chloride (PVC)

Polyvinyl chloride (PVC) is the most commonly used synthetic polymer with CAS number 9002-86-2. It is produced by the addition of the vinyl chloride monomer. On large scale, polyvinyl chloride is manufactured from vinyl chloride monomer in presence of free-radical initiators by using the suspension, bulk, solution, and emulsion polymerization. In comparison to most plastics, PVC is much dense at 20 °C; the density is 1.37–1.43 g/cm^3; 1.53 g/cm^3 (crystalline); 1.373 g/cm^3 (amorphous), melting point is 103–230 °C and glass transition temperature is 87 °C. Rigid PVC is very hard with extremely good tensile strength.

In 1835 and 1872, Henri Victor Regnault and Eugen Baumann were acknowledged for the discovery of polyvinyl chloride. Later on, in 1913 the first patent for the polymerization of vinyl chloride in the presence of sunlight to form PVC was granted to the German chemist Friedrich Heinrich August Klatte. Waldo Semon, in 1926 gave new dimensions to PVC by discovering the process of plasticization (Wypych 2016).

Polyvinyl chloride is a thermoplastic material and has been part of our lives. There are many methods of its synthesis given in the literature. Pure PVC is a mechanically tough, rigid, electrically insulating material that shows good water, weather, and chemical resistance, but it is unstable towards light and heat. In the presence of UV light and heat, hydrogen chloride (HCl) formation occurs due to loss of chlorine, which can be prevented by using stabilizers. Generally, these stabilizers are composed of salts of metals like calcium, barium, cadmium, or lead [COM (2000) 469-C5-0633/2000-2000/2297(COS)].

PVC is synthesized by the chain polymerization of the monomer, vinyl chloride. Vinyl chloride is prepared from the reaction of chlorine (57 wt%, manufactured by the chlor-alkali electrolysis) and ethylene (43 wt%) (Pascault et al. 2012). PVC shows good thermal, mechanical stability and poor shielding ability towards ultraviolet (UV) light because of this reason the applications of polyvinyl chloride (PVC) are limited. To improve the characteristics of PVC, a hybrid nanostructure containing layered double hydroxides (LDHs) and α-Manganese oxide (α-MnO_2) nanorods were used. In the first step with the help of bio-safe molecules, LDH and α-MnO_2 nanorods were modified to make them adaptable to the PVC matrix. In the second step, different weight percentages of the resulting nanohybrid like 5, 10, including 15 wt% were introduced to the polyvinyl chloride. In the third step, different analyses of resulting nanocomposite films were identified. It was observed that UV-visible absorption spectra of nanocomposites show an enhanced absorption peak as compared to the pure polyvinyl chloride (Mallakpour and Naghdi 2020).

3.3.1 Applications

PVC is a widely used valuable polymer. Polyvinyl chloride (PVC) products are physically and chemically recyclable. PVC is a thermoplastic used to make various household products such as kid's houses, shoe stands, laptop stands, kid's scooters, bookshelves, etc. The improved properties like high impact resistance, good temperature capability, lightweight, etc. make PVC capable of replacing building materials such as clay, metal, concrete, and wood used traditionally in many applications. Nowadays, PVC acts as a 'wonder material' which can be used in different ways to make useful products, and one of the most popular products among them is PVC furniture (Khan and Malvi 2016).

PVC can be treated easily, and has excellent basic properties. It is flexible and inexpensive hence suitable for the piping industry. Thermally PVC is a sensitive thermoplastic therefore a large number of compounds like heat stabilizers, processing aids, pigments, lubricants, impact modifiers, and fillers must be added to stabilize it. U-PVC (Unplasticized polyvinyl chloride) has great resistance towards chemicals so it is suitable for merging with the polished surfaces of the internal pipe wall to minimize scaling for a better service life of pipe with excellent flow characteristics (Walsh 2011). PVC is biologically and chemically resistant which makes it desirable for most household corrosion-resistant sewerage pipe applications (Ameer et al. 2013).

Nearly 25% of plastic materials used as medical products are made of polyvinyl chloride. Plasticized PVC has transparency retention and good clarity which allows monitoring of fluid flows in tubes and its resistance towards kinking. It prevents the risk of interruption in fluid flow through tubes (McKeen 2014).

A systematic study has been carried out in Las Vegas to use electroactive and transparent plasticized polyvinyl chloride (PVC) gel as a soft actuator for tiny mechanical devices, synthetic muscle applications, Opto-electro-mechanical devices, and optics (Hwanget al. 2019). Phthalates are plasticizers of PVC which are used widely in food packaging applications. This plasticized PVC can also be used to make certain food processing equipment like tubing and conveyor belts. However, due to the hazardous effects of phthalates, other compounds can be used to replace these phthalates (Carlos et al. 2018).

3.3.2 Ecological Impacts

Polyvinyl chloride (PVC) is mainly used as a flooring material, building material, food packaging material, piping material, and various household materials. At different intervals in the vinyl lifecycle, many harmful by-products like organochlorine get formed and escape into the environment which can harm human health and the environment.

Numerous health hazards like birth defects, dysfunction of the endocrine system, cancer, neurotoxicity, and suppression of the immune system have been caused due to feeding stocks and by-products produced at different intervals in the vinyl lifecycle. Many bioaccumulative substances are also found in the different stages of the PVC lifecycle, which are oil-soluble in nature, and can intensify negative effects as they move to higher levels in the food chain. These bioaccumulative substances easily cross the placental barrier and are able to deposit in the mother's milk of mammals (Thornton 2007).

PVC is a major cause of dioxin formation, which is a true global pollutant, now the presence of dioxin has been investigated in animals, humans, in the tissues of whales and polar bears. Dioxins hold the potential to cross the placental barrier easily and to be deposited in a mother's milk. As a result, infants consume high doses of dioxin in comparison to adults. During formulation, phthalates are mixed with polyvinyl chloride which causes various health effects like reduction in sperm count, testicular damage, infertility, and damaged reproductive system (Thornton 2007).

The locally manufactured pipes were investigated to observe the movement of its vinyl chloride monomer (VCM) from unplasticized polyvinyl chloride. For this purpose, a study was carried out at different time intervals using variables such as total dissolved solids, pH, and temperature in water. The strength of VCM in the running water of pipes was evaluated using the head-space technique or gas chromatography. After exposure, at 45 °C for 30 days the VCM concentration was detected at more than 2.5 ppb in water. It was observed that the migration of VCM remains unaffected to water temperature unless it was raised to 45 °C (higher value) whereas total dissolved solids (TDS) and pH of water were identified to affect the movement of PVC monomer (vinyl chloride) from unplasticized polyvinyl chloride pipes (Muhammad et al. 2000).

An experiment was carried out in Romania to observe the impacts of PVC on environmental factors by the production process of PVC. The study showed that the process hurts the environment with photochemical ozone creation potential (POCP), high contribution to AP (acidification potential), GWP (global warming potential), and HTP (human toxicity potential). The main contributors to HTP are dioxins emitted during the production of vinyl chloride whereas emissions of the volatile organic compound during the processing of crude oil lead to photochemical ozone creation. During various stages in the process, the consumption of energy contributes to climate change while the emitted substances cause acidification (Comaniţă et al. 2020).

The health and environmental impacts of vinyl chloride monomer (VCM) in the PVC industry is classified as Group A by the EPA (US Environmental Protection Agency) as a carcinogen to humans (US EPA 2000).

3.4 Chlorinated Polyethylene (CPE)

This thermoplastic is produced by chlorination of polyethylene in solution or powdered form at elevated temperature. Chlorinated Polyethylene (CPE) was discovered by Fawcett, EW; Gibson, RO; Perrin MW in the year 1939 (Wypych 2016).

There are various ways to chlorinate polyethylene, chlorination can be performed in a solution where solvent should be non-reactive to chlorine. Polyethylene can also be produced by aqueous suspension, solvent suspension, melting, solid-phase, and block chlorination. Various forms of polyethylene like powdered, granulated, fibrous, and thin-film are the preferred choices. Carbon tetrachloride, chlorobenzene, and ethane tetrachloride are commonly used solvents for chlorination. Chlorination through a diluted solution of polyethylene is the best way to get chlorinated polyethylene of desired properties with uniformly distributed chlorine. The JCJ company of Great Britain invented the process of chlorination in the gaseous form of chlorine by utilizing low-density polyethylene suspension in presence of carbon tetrachloride or acetic acid solvents and metal chloride (Donskoi et al. 2003).

Researchers put much effort into this polymer to make it more usable. To obtain various CPI of different properties like hard, brittle to elastic at different conditions, a mixed solution suspension technique was used to chlorinate Low-density polyethylene (LDPE) at three different temperatures. Through various characterizations (static mechanical, chemical and thermal), two different crystal structures were predicted which disappeared during chlorination (Akovali and Vatansever 1986).

Graft copolymerization is a way to form a polymer of distinctive properties. Hence to intensify, thermal and mechanical properties of CPE, chlorinated polyethylene-chlorinating/grafting-poly (acrylic acid) and its sodium salt ionomer were prepared by in-situ chlorination graft co-polymerization. This Grafted copolymer was fabricated with a CPE backbone and polyacrylic acid branches. This Grafted copolymer and its carboxylated ionomer were subjected to Fourier Transform-Infrared (FTIR), to get gel permeation chromatography and degree of grafting report. The effect of various factors like reaction temperature, chlorine content, monomer concentration of the product was studied to describe the reaction (Wang et al. 2012).

In the same way, a dedicated team of Wang et al. (2014) worked on gas solid-phase chlorination to get grafted polymer of maleic anhydride for low-density polyethylene. This synthesis was patented (in situ chlorinating graft copolymerization) in which two problems were sorted out with a novel approach. The first problem was chlorination and grafting feasibility at the same time on low-density polyethylene molecular chains and the second was to regulate control of highly chlorinated polyethylene synthesis. Fourier transform and 1-hydrogen nuclear magnetic resonance (1H-NMR) techniques were applied to check the feasibility of chlorination. Gel permeation chromatography, stereoscopic microscopy, and chlorine content versus time curve were used to analyze the sample's molecular weight, distribution thermal properties, and agents that affect the chlorination process. The degree of grafting was observed at 2%. Adhesion and impact strength tests were also applied to check the feasibility of the polymer (Wang et al. 2014).

In the presence of ultraviolet light, high-density polyethylene (HDPE) was chlorinated (radical chlorination) by an aqueous slurry process under moderate pressure. In the experiment, two temperature conditions were applied in two steps, below and above the crystalline melting point. Three products were obtained with a chlorine concentration of 6.5, 12 and 34% by weight of HDPE. The FTIR spectrum of all three samples was taken and compared with polyethylene and polyvinyl chloride to confirm reaction progress. Flask combustion method (DIN EN ISO 1158)—an elemental analyzing technique was applied to check chlorine content (Razmirad and Moradi 2012).

In this study, the composition and microstructure of chlorinated polyethylene were determined by the pyrolysis gas chromatographic method. Pyrolysis of CPE polymer generates aromatic compounds through dehydrochlorination of the trimer at high temperature. The different levels of formed ethylene and vinyl chloride trimmers decided composition and microstructure. 13-C NMR analysis is helpful for the structure elucidation of CPE polymer having 25–48% chlorine (Wang and Smith 1997).

Blending is a technique to get a better product in polymer science. To this, an effort has been made to synthesize blends of CPE, PVC, and poly alpha-methylstyrene-acrylonitrile in a ratio of (70/30).The outcome of the investigation was that the addition of CPE improved the toughness and heat stability without impacting the heat resistance (Zhang et al. 2010).For the advancement of physico mechanical characteristics of polyvinyl chloride, a blend of PVC and CPE was prepared. The result is that the addition of CPE improved the impact strength, electrical properties, and fire resistance but the brittle point was not improved. These changes were dependent upon the ratio of PVC and CPE in blends and chlorine content of CPV. In this study, factors like thermo rheological, residual crystallinity, compatibility, applicability, and principle of temperature–time suspension were studied. Some modifiers were added to upgrade elasticity, impact strength and to reduce elastic modulus and strength of quasi-static tension. The impact strength of PVC and CPE could also be enhanced by adding polystyrene. Morphological study of blended samples confirmed that nano-sized polystyrene will create a spatial network (Maksimov et al. 2003).

For manufacturing of plastic runway or rubber hose, composite of chlorinated polyethylene/industrial waste red mud/carbon black were synthesized where ratios of red mud/carbon black were 40/20, 20/40, 10/50, 5/55 and 0/60. All synthesized samples were subjected to testing the mechanical and thermal properties. The addition of red mud to chlorinated polyethylene/carbon black raised the tensile strength of a composite by 10–14 MPa and thermal degradation was increased by 10 °C. After three days, hardness and tensile strength of different composite samples were further increased by 10 shore A and 2 MPa. The study concluded that composite sample was more stable thermally when red mud (20 phr) and carbon black ratio was 20/40. The Payne effect was also reported by the rubber process analyzer. It was found that composite prepared by 20 phr red mud has good properties (Qiu et al. 2020).

3.4.1 Applications

CPE, a good general-purpose oil, has the resistance against heat and ozone. The CPE is synthesized by chlorination of HDPE by the free radical aqueous process. The product has chlorine content of 36–48%. A research was conducted on the blending of chlorinated polyethylene with polyvinyl chloride, polypropylene, and styrene-acrylonitrile to upgrade mechanical properties like impact strength, breaking elongation, and tensile strength (Iranmanesh and Shafiei 2012).

A thin film of 8–10 μm is fabricated by randomly prepared CPE (71% chlorine by weight of PE). Here, high chlorine content is accountable to increase glass transition temperature (110 °C) and Young's modulus (2.6 GPa), these increased parameters can also elevate breakdown strength. For restricted loss at elevated temperature and well-maintained medium permittivity (3–4), random dispersion of chlorine atoms is playing a pivotal role. The changed parameters resulted in high energy density (12 J/CM3) and high discharging efficiency (83%) under an electric field of 700 MV/m. So these parameters can make CPE a polymeric dielectric material for high pulse metalized film capacitors (Zhao et al. 2018).

It was reported for the first time that UV radiation can accelerate the rate of chlorination. A vibrated-bed reactor was used to study the kinetics of thermal and UV enhanced gas–solid HDPE chlorination at that temperature which was below the melting point. The kinetics of chlorination was recorded with the help of a UV-Visible spectrophotometer. Even chlorinated products had no chlorinated crystal structures with small content of chlorine. In this mechanism, to minimize residual crystallinity and to increase the homogeneity of chlorination, multi-stage chlorination was suggested. The presence of –CH=CH– bonds in CPE assured that chlorination occurred at 150 °C. One more observation was remarked by differential scanning calorimetry that when the temperature was beyond the melting point, melting enthalpy was decreased effectively (Zhang et al. 2018).

The most common application of CPE is as an impact modifier for PVC which can increase the toughness of PVC and another chemical Dioctyl phthalate (DOP) is known to increase the plasticity of PVC. The samples of varied CPE and DOP were subjected to tensile strength and breaking elongation testing at changed hot-pressing temperatures. The result of the study was in reduction of tensile strength and up-gradation of breaking elongation. Mechanical properties and blending effects of CPE and PVC samples were raised by raising the hot-pressing temperature. DOP has the potential to precipitate at hot pressing temperatures (Xie et al. 2011).

3.4.2 Ecological Impacts

Many studies have been conducted to inspect the adverse effect of polymers on the marine ecosystem. The water pollution generated by plastics is a major concern for the environmentalists. A study has been conducted on the presence of microplastics (MPs) in the Gulf of Guinea (Ogun and Osun Rivers) in Nigeria. The 29 samples of three insect species have been analyzed by micro-Fourier-transform infrared (μFTIR) spectrophotometer and digital microscope. Results showed that CPE was found in *Chironomus* sp. of Ogun River (Akindele et al. 2020).

Hydrochloric acid and carbon monoxides are combustion and degradation products of CPE which are known to produce carcinogenic activities. Small particles of CPE are also injurious for eyes and skin so precautionary measures are required to work with CPE. To enhance the properties of CPE, additives are added which can release volatile organic compounds at high temperatures (Akovali 2012a, b).

3.5 Chlorinated Polyvinyl Chloride (CPVC)

The synthesis of chlorinated polyvinyl chloride (CPVC) is an example of hydrogen atom substitution from a polyvinyl chloride (PVC) molecule by a chlorine atom. This substitution is followed by a free radical mechanism. Heat and ultraviolet radiation are required to initiate the reaction. CPVC was invented by Schoenburg of IG Farben Industries in 1934, who also demonstrated that chlorinated PVC contained 64–68% of chlorine. The trade name of this thermoplastic is Geon (Wypych 2016). This thermoplastic has also been an interesting topic to researchers for its excellent properties. Many methods of its synthesis have been suggested from time to time.

Lu et al. (2011) worked on the plasma-assisted synthesis of CPVC (containing 67% of chlorine) by the gas–solid contracting process. The produced CPVC was found to be comparable to the commercial CPVC in mechanical properties, thermal stability, and microstructure. In this mechanism, the cold plasma generated free radical chlorine played an active role in surface activation of PVC. This reaction was decoupled into two steps. In step one, on the particle's surface chlorination occurred which was enhanced by plasma and in the second step, migration of chlorine occurred from the surface to the core in the dielectric barrier discharge plasma fixed bed reactor. After the complete cycle of 3 h, CPVC was produced with 67% of chlorine content. The characterization of the sample was carried out with the help of scanning electron microscopy, thermogravimetric analysis, and Raman spectra (Lu et al. 2011). In another new approach, PVC was subjected to preliminary foaming treatment for the chlorination process. The reaction was accomplished in the presence of supercritical carbon dioxide and acetone. The experiment resulted successfully in the form of increased content of chlorine as 0.66 g/g for PVC which was treated before chlorination. On the other hand, non-pretreated PVC had only 0.60 g/g chlorine content. The result was compiled based on gel permeation chromatogram, differential scanning calorimetric analysis and 13-C carbon nuclear magnetic resonance (13-C NMR) spectrum. CPVC which was pretreated had given the proof of fine thermal properties, narrow molecular weight distribution, and uniform distribution of chlorine (Qian et al. 2017).

Yang et al. (2015) used plasma circulating fluidized bed reactors for plasma-assisted synthesis of CPVC. CPVC has been advised to be superior thermoplastic than PVC for heat stability, mechanical and flame retardation properties because of the enhanced polarization effect of chloro group on PVC. These two stepped decoupled chlorination processes performed as a slow diffusion mechanism. In both steps, polymer surface and reactive gas activation occur by plasma, and migration of absorbed chlorine proceeds to the core of the polymer. This was a novel approach to synthesize CPVC in plasma circulating fluidized bed reactor rather than the traditional way to synthesize by an aqueous suspension. The final product was characterized by scanning electron microscopy and Raman spectrum which indicated the presence of fine microstructure and uniform chlorine distribution.

A methodological work has been done to identify the ability of CPVC in making heterojunction with Tin (Sn) and Titanium (Ti) compounds. This CPVC/SnS_2/TiO_2 (Chlorinated Polyvinyl Chloride/Stannous Sulphide/Titanium dioxide) heterojunction has been recognized as an innovative, highly effective, visible-light-driven ternary photocatalyst. The photocatalytic test signified that CPVC/ SnS_2/TiO_2 heterojunction is an effectual photocatalyst for photocatalytic reduction of aqueous chromium salt and photocatalytic degradation of methylene blue in electromagnetic radiation of visible region (Liu and Zhang 2021).

Merah et al. (2013) tested the tolerance power of CPVC pipes against the worst environmental conditions. Natural and accelerated artificial weathering is a major cause to affect the mechanical, physical, and chemical properties of CPVC pipes. Various observations have been noticed by applying different testing-sample of CPVC pipes kept in natural outdoor environments for duration of two weeks to eighteen months under UV exposure of 100–3000 h. THE same CPVC sample was examined for the tensile strength test. The outcome of the study was, the strength and stiffness of CPVC samples have been affected to a small extent by natural and accelerated weathering. Artificial UV exposure of 100 h and natural weathering (duration 15 days) caused a reduction in fracture strain. The physical examination of the CPVC sample approved that both types of weathering generated gradual discoloration in samples. Attenuated total reflection infrared (ATR-FTIR) and Ultraviolet-Visible (UV-Visible) spectroscopy confirmed that dehydrochlorination was responsible for the degradation of CPVC (Merah et al. 2013).

3.5.1 Applications

CPVC is a polymer of substantial importance in industries because of its excellent chemical resistance, anti-corrosion, good mechanical, physical, low flame, and low smoke spreading properties. CPVC is a popular thermoplastic in making chemical processing pipes, pulp processing pipes, paper processing pipes, water treatment pipes, sewage treatment pipes, food processing pipes, residential hot–cold water plumbing pipes, industrial-residential sprinkler pipes, fittings, and valves. The uniqueness of CPVC is to provide a higher temperature service range. Many CPVC products can be operated at 200 °F. For hot corrosive liquids, CPVC is an ideal material at 210 °F. CPVC acts as glue for solvent cement. At 73 °F, CPVC and PVC reveal the same physical properties (Walsh 2011). CPVC resin is considered a multipurpose material for various industrial applications because of its anti-corrosion, high mechanical, smoke suppression, and heat retardant properties (Xin et al. 2017).

CPVC pipes are designed to resist the worst conditions of temperature and pressure. Research has been performed in Saudi Arabia to observe the adverse effects of environmental conditions on CPVC pipes. The samples were prepared from locally available CPVC pipes, which were subjected to standard tensile and single edge notched tension (SEN) fracture toughness tests for duration of 1–9 months. The result revealed that adverse environmental conditions affected tensile strength and elasticity minutely, but the surface of the sample was affected severely. The study concluded the environmental tolerance of CPVC at high temperature and pressure conditions. Therefore, the use of CPVC is gaining popularity in the plumbing industry (Merah 2007). CPVC has been a well-recognized polymer in US markets since 1982. The proposal was submitted to the California Building Standards Code in February 2007 to get approval for the installation of chlorinated polyvinyl chloride pipes in the domestic water supplying system. The limited use of CPVC pipes has been approved based on an environmental impact report issued by the department of housing and community development (Martins et al. 2009).

On controversial notes, the use of steel pipes has faded away in fire protection systems because of these CPVC pipes which are low priced and can be installed very easily. On the other side of the coin, the use of CPVC pipes is unsuccessful in comparison to steel pipes. The impact, environmental stress cracking, high pressure, and manufacturing defects are some challenges in the path of CPVC pipes applications. As a consequence, more research is required to overcome these issues for application purposes (Hayes et al. 2010).

3.5.2 Ecological Impacts

CPVC pipes are a desired product of the market for drinking water distribution systems but the monomer vinyl chloride used in the synthesis of CPVC is carcinogenic in nature. This vinyl chloride can leach from CPVC pipes into drinking water. In a study by Walter et al. (2011), different levels of vinyl chloride accumulation were analyzed from samples of different sources. Pieces of evidence showed that vinyl chloride can also be accumulated as a disinfection by-product. The presence of organotins in water is another threatening issue which is caused by plastic pipes.

Keeping this threat in view, a study was carried out to check the possibility of organotins leaching of heat stabilizers and organic compounds into the water. These organic compounds are the main component of organic solvents, and these organic solvents are used to seal CPVC pipes. Various analytical techniques were applied to check the availability of these toxins in the water. In experimental work, samples of PVC and CPVC pipes were subjected to hydride generation atomic absorption spectroscopy, gas, and mass chromatography. There was an observation that leaching of organotins and organic compounds occurred in a biphasic manner. The outcome of the study was that the concentration of dimethyl tin dichloride, butyl tin dichloride, methyl ethyl ketone, tetrahydrofuran, and cyclohexanone obtained 10 ppm to 10 ppb in exposed water samples which is hazardous for health and creating new problem statements for researchers (Boettner et al. 2002).

3.6 Chlorosulfonated Polyethylene (CSPE)

Chlorosulfonated polyethylene (CSPE) was discovered in 1940 by McQueen DM at DuPont (DuPont 2007). This chlorosulfonated polyethylene contains 24–43% chlorine and 1.0–1.4% sulfur. Acsium and Hypalon are the trade names for CSPE and the CAS number is 9002-88-4. At 20 °C, the density of CSPE is 1.0–1.27 g/cm^3, melting point is 87–140 °C and glass transition temperature ranges from 7 to −27 °C. It is white to slightly yellow in color with an ether-like smell and flame resistance. In comparison to neoprene and butyl rubber, it possesses better weather and ozone resistance. CSPE elastomers exhibit much better abrasion resistance and mechanical properties than CPE (Wypych 2016). This is a valuable and marketable polymer. Many investigations have been conducted on this polymer.

Preparation of chlorosulfonated polyethylene using a gas, solid, and liquid three-phase reaction: In this invention, the preparation of CSPE comprises the addition of chlorinated polyethylene powder into a multilayer stirred fixed bed reactor along with an anti-adhesive agent, followed by the exposure to atomized sulfonyl chloride under ultraviolet irradiation. During the process according to the weight of the chlorinated polyethylene, a total of 4 to 15% sulfonyl chloride was added by maintaining the temperature below 80 °C. A finished product of CSPE was obtained after rinsing with water, centrifugal dehydration followed by drying in hot air (CN102153683A).

In this methodology, the synthesis of chlorosulfonated polyethylene was performed by the suspension method. Chlorinated polyethylene with chlorosulfonation reagent was suspended in a solvent. The resulting suspension was induced either by ultraviolet light or an initiator at a certain temperature to form chlorosulfonated polyethylene as a product which on filtration followed by washing and drying results in white powder (CN104725534A).

CSPE was synthesized in the presence of an initiator by the reaction of a combination of chlorine and sulfur dioxide, chlorine and sulfuryl chloride, or a weak base and sulfuryl chloride with polymers. The resulting polymer may contain 1–5% sulfur and 20–60% chlorine as sulfonyl chloride groups. The parent polymer crystallinity has been destroyed by the chlorine atoms which resulted in better chemical properties like resistance towards heat, ozone, oxidizing chemicals, and oil. Due to the presence of the sulfonyl chloride group, chlorosulfonated polyethylene can combine with bivalent metal sulfides, radical traps, or oxides to form stable compounds with cross-linkage (Ennis 2000).

Chlorosulfonated polyethylene was synthesized using a mixture of dioxide sulfur and chlorine along with chlorinated polyethylene in a tank reactor. The results of the investigation showed that elevated reaction temperature from 30 to 50 °C directly leads to accelerate the rate of chlorosulfonation with higher sulfur content (Zhao et al. 2001).

3.6.1 Applications

Chlorosulfonated polyethylene has many advantages over other polymers like polyolefins and polyvinyl chloride. It is a synthetic rubber produced from polyethylene and is popular for its excellent resistance towards ultraviolet light, chemicals, and temperature (30–130 °C) (Akovali 2012a, b). The resistance to a wide range of temperature makes it suitable for various industrial applications.

Chlorosulfonated polyethylene is widely used in construction and industrial applications where high performance is needed. In the automobile sector for the year 2008, its consumption was nearly one-fourth of total world consumption. In the construction sector, it is used as liners and roofing membranes for reservoirs and ponds (Akovali 2012a, b).

A systematic work has been carried out on fast-changing technology for electric vehicles using polymeric insulated cable made of chlorosulfonated polyethylene sheath and ethylene-propylene-diene insulation having an excellent thermal performance with high current carrying capacity. By this application, the thermal conductivity was modified for both the materials using hybrid boron nitride. The outcome of the study appears as a significant improvement in the current-carrying capacity with excellent mechanical and electrical–mechanical properties (Du et al. 2019).

Bulgakov et al. carried out a study to observe the improvement in the adhesive strength of rubbers with the use of modified chlorosulfonated polyethylene. This modification was made with the use of amino-containing compounds. It was analyzed that the adhesive strength of the compositions to resins increases by two to five times with these modifications. Chlorosulfonated polyethylene has important properties like resistance towards the fire, abrasion, atmospheric action, and chemicals which made it a better choice for the composition of coatings, enamels, sealants, prime coatings, and adhesives (Bulgakov et al. 2012).

To enhance the strength of adhesive joints between vulcanized rubbers, the adhesives of chlorosulfonated polyethylene was modified with the help of the products of aniline and glycidyl ester of methacrylic acid (Kablov et al. 2012).

Forty years ago, in the United States, a geomembrane was first developed as a roofing membrane and pond liner material made of chlorosulfonated polyethylene. In different regions of the world, geomembranes have been used as floating covers for the storage of municipal potable water and containment in industries. These chlorosulfonated polyethylene have proven to be a good choice of material as geomembranes. These floating covers are designed for the prevention of contamination and debris from infiltration to the reserved source of water (Fraser et al. 2019).

As Chlorosulfonated polyethylene possesses excellent resistance towards irradiation, heat, weathering, sunlight, and ozone, therefore, it is widely used as sheeting material for cable construction in various fields like in nuclear energy plants (Simendić 2017).

To find out the solution to problems related to the capping of steep slopes during the closure of hazardous waste piles and old municipal landfills, CSPE is considered an excellent material for geotextile and geocomposites. This product can overcome the problems related to surface friction and also has resistance towards puncture and localized subsidence. A sheet of chlorosulfonated polyethylene on lamination with a nonwoven geotextile either on one or both sides produces the desired product (Frobel and Taylor 1991).

3.6.2 Ecological Impacts

In general, chlorosulfonated polyethylene is considered a safe material. Due to the presence of residual carbon tetrachloride (CCl_4) and chloroform some acute or chronic potential health effects of chlorosulfonated polyethylene have been reported (DuPont 2007). This chemical can enter into the body either by breathing or through the skin, which may irritate skin, eye, nose, throat, lungs. In Some cases, it shows adverse effects on the liver, central nervous system, and kidneys also. If additives are used in the formulations of chlorosulfonated polyethylene for the purpose to enhance additional properties, then their selection should be done very carefully as these additives can cause harmful emissions at higher temperatures. Thermal decomposition and combustion of chlorosulfonated polyethylene can produce hydrochloric acid (HCl), sulfur dioxide, and carbon monoxide, which are highly toxic and irritant (Akovali 2012a, b).

3.7 Polychloroprene Rubber (CR)

Polychloroprene is a synthetic material composed of polymerized chloroprene and is commonly known as chloroprene rubber (CR) or neoprene with CAS number 9010-98-4. Among all the vulcanized elastomers, the polychloroprene exhibits good performance. It has better resistance than natural rubber towards oils, water, solvents, and heat. At 20 °C the density of CR is 1.22–1.25 g/cm^3, melting point is 40 °C, and glass transition temperature −20 °C. It is odorless, white to grey in color, its own remarkable mechanical and physical properties show resistance towards chemicals, acids, alkalis, sunlight, ozone, oil, heat, and fire. Polychloroprene never becomes soft on exposure to heat. Polychloroprene was discovered by Wallace Carothers and Julius Arthur Nieuwland in 1930. CR is considered to be a high demand product in the world market of elastomers (Wypych 2016).

Polychloroprene is an important diene-based elastomer, synthesized by free-radical emulsion polymerization from its monomer 2-chloro-1,3-butadiene which is known as chloroprene, (Campbell 2000). The monomer of polychloroprene is chloroprene, prepared by the reaction of dimerized acetylene with hydrochloric acid within a sealed system. In the presence of free radicals, formed monomer chloroprene is converted into polychloroprene polymer (PCP) by emulsion polymerization (Lynch 2001).

Polychloroprene can be vulcanized using metal oxides like zinc oxide and magnesium oxide. However, it can be vulcanized without metal oxides in the presence of zinc oxide. But to provide resistance, it is essential to add magnesium oxide (Coran 2013).

3.7.1 Applications

Polychloroprene has wide applications in different industries such as the automotive industry, construction industry, as adhesives and for missile launchers in the form of liner pads (Meier et al. 1971). The adhesive produced from polychloroprene is a solution type adhesive, made by the mixing of neoprene, anti-aging agent, magnesium oxide, filler and antioxidant. This adhesive can work in a wide range of temperatures from −50 to +80 °C, it possesses good resistance towards alcohol, oil, weak-acid, weak-alkali, aliphatic hydrocarbon, and water. It is commonly used for bonding structures or different materials. Polychloroprene on mixing with oil-soluble phenolic resins gives a better performance which can be applied to bind metals like steel, copper, aluminum, and nonmetals including ceramic, cement fiber boards, and plastics (Li 2011).

Polychloroprene has excellent mechanical properties, having good resistance towards tear, oil, weather, heat, chemical, ozone, and low flammability. These properties make it a choice next to nitrile rubber. Polychloroprene is widely used in various engineering applications like in the production of transmission/conveyor belts, gloves, as the material for gaskets, tubing, wraps and sheets, weather stripping, and hose (for gasoline), in cable coatings, as modifiers of bitumen, etc. Polychloroprene possesses low permeability to water which makes it suitable for sealer-type finishes used in masonry and concrete. Preformed chloroprene rubber joint seals are used for pavements of concrete (Akovali 2012a, b).

Polychloroprene is used as a raw material for contact adhesives, which are used for binding furniture, high-pressure laminates, kitchen cabinets, automotive trim, custom display cabinets, roofing-membrane attachment, wall partitions, and interior and exterior panels (Akovali 2012a, b).

In many nuclear plants located in Foramark, Ringhals, and OKG chloroprene rubber is used in membrane valves to form membrane diaphragms. To control the flow of gas and water in power plants, membrane valves are used. (Rosato et al. 2004).

Neoprene fabrics are stretchable and thick enough to avoid wear and tear. Neoprene is a suitable material to be used for the manufacturing of wetsuits. The fabric made of neoprene has soft synthetic rubber foam which can be prepared thicker in comparison to other fabrics; therefore these wetsuits can provide heat insulation and protection from natural injury (Oh et al. 2019).

3.7.2 Ecological Impacts

Pure polychloroprene is not considered carcinogenic. All major polychloroprene are approved by the FDA. Whereas the processed form of polychloroprene may contain a variety of ingredients that are harmful and cause problems. Mainly chloroprene, toluene, and butadiene are volatile ingredients of polychloroprene along with thiourea and lead (Report on Carcinogens 2011).

The monomer chloroprene is a highly reactive and volatile material that is identified as carcinogenic to humans having 4.8 h as estimated residence time in the atmosphere. It is also supposed to be a strong toxicant to the neuro, endocrine, blood, and cardiovascular systems. Acute exposure affects the liver and kidneys also. Chloroprene is also identified as a toxic air contaminant. At low concentrations, the vapors of chloroprene cause irritation in the eyes and respiratory tract whereas at high levels these vapors affect central nervous system depressants.

EPA (An official website of the US Government) included Butadiene in the list of toxic pollutants. Butadiene is used in the manufacturing of polychloroprene. It was concluded from the study which was conducted at the EPA's Laboratory that during the processing of polychloroprene the emission levels of butadiene were 2–40% (US EPA 1985). Acute high-level exposure to butadiene may cause nausea, damage to the central nervous system, and lowering of the pulse whereas irritation of the throat, eye, and respiratory tract was caused by its low-level exposure. Other volatile ingredients like lead and thiourea compounds are already considered hazardous to human health. Lead and its compounds especially Lead oxide whether it is water-soluble or not, are poisonous in nature. Lead acts as a cumulative poison and is able to accumulate in soft tissues, the liver, and kidneys (Akovali 2007).

The small amount of rosin or colophony in polychloroprene adhesives is a skin contact sensitizer. It was stated by the EU that if the colophony level in polychloroprene is 0.1% or greater then it must be labeled as a potential skin contact sensitizer (EU Dangerous Preparations Directive 1999/45/EC 2002). On another side, polychloroprene shows very less oral toxicity, even if a person comes in direct contact or from gloves, clothing, boots, etc. This polymer is also responsible for producing adverse effects in this way. Irritation and skin allergic reactions were reported due to unused thiourea left after vulcanization. The burning of polychloroprene liberates hydrogen chloride gas to the environment which is an irritant to the eye and respiratory tract. The chlorine content present in polychloroprene is also accountable for the release of dioxin throughout its lifecycle. This dioxin is a highly toxic chemical and carcinogenic to humans (IARC Group 1 carcinogen) (Akovali 2007).

References

Akindele EO, Ehlers SM, Koop JHE (2020) Freshwater insects of different feeding guilds ingest microplastics in two Gulf of Guinea tributaries in Nigeria. Environ Sci Pollut Res 27:33373–33379

Akovali G (2007) Plastics, rubber, and health. iSmithers Rapra Publishing, Shrewsbury, UK

Akovali G (2012a) Plastic materials: chlorinated polyethylene (CPE), chlorinated polyvinyl chloride (CPVC) chlorosulfonated polyethylene (CSPE), and polychloroprene rubber (CR). In: Pacheco-Torgal F, Jalali S, Fucic A (eds) Toxicity of building materials. Woodhead Publishing, pp 54–75. https://doi.org/10.1533/9780857096357.54

Akovali G (2012b) Plastic materials: chlorinated polyethylene (CPE), Chlorinated polyvinyl chloride (CPVC) chlorosulphonated polyethylene (CSPE), and polychloroprene rubber (CR). In: Pacheco-Torgal F, Jalani S, Fucic A (eds) Toxicity of building materials. Woodhead Publishing, pp 59–60. https://doi.org/10.1533/9780857096357.54

Akovali G, Vatansever A (1986) Notes on chlorinated polyethylene: its preparation, some mechanical and thermal properties. Polym Eng Sci 26(17):1195–1199

Ameer AA, Abdallh MS, Ahmed AA, Yousif EA (2013) Synthesis and characterization of polyvinyl chloride chemically modified by amines. Open J Polym Chem 03(01):11–15. https://doi.org/10.4236/ojpchem.2013.31003

Boettner EA, Ball GL, Hollingsworth Z, Aquino R (2002) Organic and organotin compounds leached from PVC and CPVC pipe. Michigan University Ann Arbor Department of Environmental and Industrial Health

Bulgakov AV, Kablov VF, Keibal NA et al (2012) Improvement of adhesion strength of compositions based on chlorosulfonated polyethylene to rubbers. Polym Sci Ser D 5:49–52. https://doi.org/10.1134/S1995421212010030

Campbell IM (2000) Introduction to synthetic polymers, 2nd edn. Oxford University Press, Oxford, UK

Carlos KS, de Jager LS, Begley TH (2018) Investigation of the primary plasticisers present in polyvinyl chloride (PVC) products currently authorised as food contact materials. Food Addit Contam Part A 35(6):1214–1222

Comaniţă ED et al (2020) Environmental impacts of polyvinyl chloride (PVC) production process. In: Paper presented in E-health and bioengineering conference (EHB), Iasi, Romania, 19–21 November 2015

Coran AY (2013) Vulcanization. In: Mark JE, Erman B, Roland CM (eds) The science and technology of rubber, 4th edn. Academic Press, pp 337–381. https://doi.org/10.1016/B978-0-12-394584-6.00007-8

Donskoi AA, Donskoi SMA, Zaikov G (2003) Fire resistance and thermally stable materials derived from chlorinated polyethylene. Taylor & Francis, CRC Press, FL, USA

Du BX, Kong XX, Li J, Xiao M (2019) High thermal conductivity insulation and sheathing materials for electric vehicle cable application. IEEE Trans Dielectr Electr Insul 26(4):1363–1370

DuPont MSDS (Material Safety Data Sheets) (2007) https://www1.mscdirect.com/MSDS/MSDS00004/63415905-20150718.PDF

Ennis R (2000) Chlorosulfonated polyethylene. John Wiley & Sons. https://doi.org/10.1002/0471238961.0308121505141409.a01

Ethylene Thiourea (2011) Report on carcinogens, 12th edn. US department of health and human services, Public health program, National toxicology program. pp 191–194. https://www.ashlandmass.com/DocumentCenter/View/442/National-Toxicology-Program-Report-on-Carcinogens-PDF

Fraser B, Rhodes S, Neal M, Gersch A (2019) CSPE performance & history in long term potable water storage applications. In: Paper presented in Geosynthesis 2019 conference, Houston, Texas, USA

Froebel KR, Taylor RT (1991) CSPE/geotextile geocomposites. In: Koerner MR (ed) Landfill Closures, Elsevier, pp 61–75. https://doi.org/10.1016/B978-1-85166-644-7.50012-4

Gad SE (2014) Polymers. Encyclopedia of toxicology, 3rd edn. Elsevier, pp 1045–1050. https://doi.org/10.1016/B978-0-12-386454-3.00912-X

Hayes MD, Hanks M, Hagan FE, Duvall DE (2010) Challenges in Investigating chlorinated polyvinyl chloride pipe failures. J ASTM Int 8(1):1–12

Hwang T, Frank Z, Neubauer J, Kim KJ (2019) High-performance polyvinyl chloride gel artificial muscle actuator with graphene oxide and plasticizer. Sci Rep 9(1):1–9. https://doi.org/10.1038/s41598-019-46147-2

Iranmanesh F, Shafiei BM (2012) A review on mechanical properties and applications of chlorinated polyethylene. Adv Environ Biol 6(2):530–533

Jie B (2011) Method for preparing chlorosulfonated polyethylene through gas, solid and liquid three-phase reaction. China Patent CN102153683 A

Kablov VF, Bulgakov AV, Keibal NA et al (2012) Chlorosulfonated polyethylene modified with amino derivatives of glycidyl methacrylate. Polym Sci Ser D 5(2):102–105. https://doi.org/10.1134/S1995421212020050

Khan A, Malvi CS (2016) PVC pipe designer furniture. J Polym Compos 4(2):29–33

Li Y (2011) Adhesives. In: Ren S (ed) Building decorative materials, 1st edn. Woodhead Publishing, pp 325–341. https://doi.org/10.1533/9780857092588.325

Liu S, Zhang Y (2021 in the process) Synthesis of CPVC-modified SnS_2/TiO_2 composite with improved visible-light-driven photocatalysis. Mater Res Bull 135 111125.https://doi.org/10.1016/j.materresbull.2020.111125

Lu W, Cao T, Wang Q, Cheng Y (2011) Plasma assisted synthesis of chlorinated polyvinyl chloride using a gas-solid contacting process. Plasma Process Polym 2(8):94–99

Lynch M (2001) Manufacture and use of chloroprene monomer. Chem Biol Interact 135–136:155–167. https://doi.org/10.1016/s0009-2797(01)00188-0

Maksimov RD, Ivanova T, Zicans J, Negreeva SN, Plume E (2003) Thermal and elastic properties of poly (vinyl chloride) (PVC) + chlorinated polyethylene (CPE) blends. Mater Res Innov 7(5):326–330. https://doi.org/10.1007/s10019-003-0272-0

Mallakpour S, Nadia M (2020) A green strategy toward the preparation of poly (vinyl chloride) nanocomposites reinforced with MnO_2 @ layered double hydroxide nanohybrids as efficient UV shielding materials. New J Chem 44(27):11566–11576. https://doi.org/10.1039/D0NJ02135K

Martins JN, Freire E, Hemadipour H (2009) Applications and market of PVC for the piping industry. Polímeros 19(1):58–62. https://doi.org/10.1590/S0104-14282009000100014

McKeen LAW (2014) Plastics used in medical devices. In: Modjarrad K, Ebnesajjad S (eds) Plastics design library. Handbook of polymer applications in medicine and medical devices, 1st edn. William Andrew Publishing, pp 21–53. https://doi.org/10.1016/B978-0-323-22805-3.00003-7

Meier JF, Minter HF, Connors HJ (1971) Molded neoprene liner pads for the Poseidon missile launcher. J Appl Polym Sci 15(3):619–640. https://doi.org/10.1002/app.1971.070150310

Merah N (2007) Natural weathering effects on some properties of CPVC pipe material. J Mater Process Technol 191(1–3):198–201

Merah N, Bazoune A, Fazal A, Khan Z (2013) Weathering degradation mechanisms of chlorinated PVC. Int J Plast Technol 17(2):111–122

Muhammad HA, Sheikheldin SY, Fayad NM, Naseemuddin K (2000) Effect of water quality parameters on the migration of vinyl chloride monomer from unplasticized PVC pipes. Water Air Soil Pollut 120(1):195–208. https://doi.org/10.1023/A:1005249808374

Mustafa NS, Omer MAA, Garlnabi MEM, Ismail HA (2013) Reviewing general polymer types, properties and application in the medical field. Int J Sci Res 5(8):212–221

Oh H, Oh KW, Park S (2019) A study of the improvement of foam material sealing technology for wetsuits. Fash Textiles 6(1):25. https://doi.org/10.1186/s40691-019-0181-5

Pascault JP, Hofer R, Fuertes P (2012) Mono-, di-, and oligosaccharides as precursors for polymer synthesis. In: Matyjaszewski K, Moller M, McGrath JE, Hickner MA, Hofer R (eds) Polymer science: a comprehensive reference. Elsevier, pp 59–82. https://doi.org/10.1016/B978-0-444-53349-4.00254-5

Qian YH, Cao GP, Li XK, Wang N (2017) Synthesis of chlorinated poly(vinyl chloride) with uniform distribution of chlorine assisted by supercritical carbon dioxide and a co-solvent. Ind Eng Chem Res 23(56):6562–6571

Qiu L, Phule AD, Han Y, Wen S, Zhang ZX (2020) Thermal aging, physico-mechanical, dynamic mechanical properties of chlorinated polyethylene/red mud composites. Polym Compos 41(11):4740–4749

Razmirad MR, Moradi A (2012) Characterization and structural study of chlorinated polyethylene production in suspension phase. Chem Technol 7(1):1–8

Rosato DV, Rosato DV, Rosato MV (2004) Plastic property. In: Rosato DV, Rosato DV, Rosato MV (eds) Plastic product material and process selection handbook. Elsevier, pp 40–129. https://doi.org/10.1016/B978-185617431-2/50005-0

Simendić JB (2017) The properties of gamma-irradiated elastomeric nanocomposites based on chlorosulfonated polyethylene. Contemp Mater 8:73–79

Thornton J (2007) Environmental impacts of polyvinyl chloride (PVC) building materials a briefing paper for the healthy building network. In: Proceedings of environmental IO 2007

US EPA (1985) A summary overview of health effects associated with chloroprene. US Environmental Protection Agency EPA, 600/8–85/011F

US EPA (2000) Toxicological review of vinyl chloride. EPA: US Environmental Protection Agency, EPA/635R-00/004

Walsh T (2011) The plastic piping industry in North America. In: Kutz M (ed) Applied plastics engineering handbook, 1st edn. Elsevier, pp 585–562

Walter RK, Lin PH, Edwards M, Richardson RE (2011) Investigation of factors affecting the accumulation of vinyl chloride piping used in drinking water distribution systems. Water Res 45(8):2607–2615

Wang FCY, Smith PB (1997) Composition and microstructure analysis of chlorinated polyethylene by pyrolysis gas chromatography and pyrolysis gas chromatography/mass spectrometry. Anal Chem 69(4):618–622. https://doi.org/10.1021/ac960947c

Wang S, LI M, Wang NA, Zhao J (2012) Preparation and properties of chlorinated polyethylene-chlorinating/grafting-poly (acrylic acid) and its sodium-salt ionomer. J Macromol Sci Part B 51(7):1401–1414

Wang Y, Liu L, Jing Z, Zhao J, Feng Y (2014) Synthesis of chlorinated and anhydride-modified low-density polyethylene by solid-phase chlorination and grafting-improving the adhesion of a film-forming polymer. RSC Adv 4(24):1–23

Wypych G (2016) Handbook of polymers, 2nd edn. Elsevier, Chem Tech Publishing, p 67

Xie XL, Li WH, Wei YH (2011) Effect of blending modification on tensile performance of CPE/PVC. Adv Mater Res 284–286:1732–1735

Xin ML, Li MD, Yang B, Li SP, Wu YC (2017) The method to process chlorinated polyvinyl chloride adopts water phase suspension. In: The 2nd annual 2016 international conference on mechanical engineering and control system, Wuhan, China, 15–17 April 2016, pp 139–145. https://doi.org/10.1142/9789813208414_0018

Yang Q, Lu W, Mao D, Yang R, Cheng Y (2015) Plasma-assisted synthesis of chlorinated polyvinyl chloride (CPVC) using a plasma circulating fluidized bed reactor. Plasma Process Polym 13(3):387–396

Young RJ, Lovell PA (2011) Introduction to polymers. CRC Press, FL, USA, Taylor & Francis group, pp 3–4

Zhang Z, Chen S, Zhang J (2010) Influence of chlorinated polyethylene on poly (vinyl chloride)/poly (α-methylstyrene-acrylonitrile) blends: mechanical properties, morphology and thermal properties. Polym Test 29(8):995–1001

Zhang X, Yang Q, Mao D, Cheng Y (2018) Experimental study on thermal and UV-enhanced gas-solid chlorination of high-density polyethylene. Int J Chem Reactor Eng 16(6):1–11. https://doi.org/10.1515/ijcre-2017-0254

Zhao R, Cheng S et al (2001) Preparation of chlorosulfonated polyethylene by the gas-solid method. Inc J Appl Polym Sci 81(14):3582–3588. https://doi.org/10.1002/app.1815

Zhao Y, Li Q, Zhang X, Li H, Lu J, Zhang Z (2018) High energy density and discharging efficiency achieved in chlorinated polyethylene films for high energy-storage applications. Macromol Chem Phys 219(8):1–8

Chapter 4
Polyvinyl Chloride (PVC), Chlorinated Polyethylene (CPE), Chlorinated Polyvinyl Chloride (CPVC), Chlorosulfonated Polyethylene (CSPE), Polychloroprene Rubber (CR)—Chemistry, Applications and Ecological Impacts—II

Najla Bentrad

Abstract A lot of plastics are used in the construction industry and are known to release volatile substances into indoor air or leach organotin. Both of these phenomena have an impact on health and environmental ecosystems. It can be considered as materials that can have significant negative impacts on some levels, and the construction industry poses a major risk to consumers and highlights many problems. Some of these elements can be eliminated while others seem to be inherent in the material itself and are therefore inevitable. These emissions depend on the composition of the plastics concerned and various related parameters. Adverse health impacts associated with plastics show a number of issues; some of them could be eliminated through design, but some are inherent to the material itself and therefore unavoidable. These plastics include PVC, chlorinated polyethylene, chlorinated polyvinyl chloride, chlorosulphonated polyethylene, and polychlorinated propylene rubber, all of which are discussed in this chapter. We briefly describe their chemical properties, uses, and effects on human health, as well as some other alternative materials to be used for environmental sustainability issues.

Keywords Environmental development · Health toxicity · Chlorinated polyethylene · Chlorinated polyvinyl chloride · Chlorosulfonated polyethylene · Polyvinyl chloride · Polychloroprene rubber · Sustainable construction

N. Bentrad (✉)
Department of Biology and Physiology of Organisms, Faculty of Biological Sciences, University of Sciences and Technology Houari Boumediene (USTHB), Algiers, Algeria

J. A. Malik and S. Marathe (eds.), *Ecological and Health Effects of Building Materials*,
https://doi.org/10.1007/978-3-030-76073-1_4

4.1 Introduction

Due to their efficiency and economy, different plastics are used in buildings for several purposes and used in different mechanical, electrical, and plumbing systems. However, people are deeply worried about the possible health impacts of fibers, composite products, and building chemicals. Any of which emit volatile organic compounds (VOC) into indoor air, altering indoor air quality, impacting human comfort and productivity. These VOCs depend on the form of plastic used and the different relevant parameters (i.e. the purity and type of additives used in the preparation process, if not the purity) and other parameters such as temperature and relative humidity during the application process and also on the surface. The consequences of VOC range from a relaxation of mucous membranes to the introduction of multiple respiratory illnesses and other adverse health effects, as well as degradation of endocrine hormones in the human body and can cause cancer in many cases. Besides, various studies have shown that many harmful effects of organic pollutants in house dust, such as various polycyclic aromatic hydrocarbons (PAH), polychlorinated biphenyls (PCB), polybrominated diphenyl ethers (PBDE), and some other categories of which were not widely used in electrical insulation until recent times.

Many heavy metals and toxic compounds (such as lead, cadmium, chromium, mercury, bromine, tin, antimony, etc.) are used as additives (Pigments, fillers, UV stabilizers, and flame retardants for plastics). Although these compounds are encapsulated as a suspension in a polymer matrix, they are not chemically combined with polymer molecules and may gradually be formed into the contact environment during the lifetime of the plastic object. Most of these elements are considered to be toxic to the human body and may therefore cause serious health and environmental problems. There are several examples of these additives which can be leached from the plastic matrix to the contact environment. Most of these migrations may be direct, but they still rely on weather patterns, temperature, and environmental characteristics. Also, the most direct example is plastic water pipes that utilize organotin stabilizers as additives, by poisonous organotin compounds which are leached through the water even if they are drinking (Forsyth and Jay 1997). The presence of chlorine dioxide in the water even accelerates these migrations, resulting in a very fast deterioration of the pipeline content (Yu et al. 2011).

Similarly, as plastic waste is disposed by incineration or landfill, hazardous metal contaminants may be emitted from plastic and may enter the atmosphere or fall into the environment. Among the various additives mentioned above, there is also a plasticizer, which is usually a low-volatility organic ester, which can migrate directly as a result of leaching. Most of the plasticizers are carcinogenic, however the DEHP (diethylhexyl phthalate), a phthalate plasticizer recommended and widely used in PVC are non-toxic compound.

In this context, it is important to note that 25 years ago the Bhopal tragedy in India when more than 40 tons of extremely poisonous methyl isocyanate gas (used in the manufacture of pesticides and rubber) (polyurethane) emerged from the Parr factory, killing more than 5,000 people in a matter of days; so according to local officials, complications from cyanide exposure raised the death toll by nearly 10,000 (Bangla 2010).

Concerning the health effect on the climate, the impact of chlorofluorocarbons (CFC) on the ozone layer is an another well-known example. Until recently, plastic processors have also used CFC as a blowing agent in the manufacture of polystyrene and polyurethane foam. While not directly toxic, they have indirect health effects due to their degradation of the ozone layer and their impact on global warming. The CFC and other halogenated ozone-depleting compounds, such as carbon tetrachloride and trichloroethane (halogenated alkanes), are mostly caused by a rise in ultraviolet radiation caused by the ingestion of the following substances. The ozone layer has a variety of biochemical consequences, such as skin cancer, cataracts, etc. (McFarland and Kaye 1992).

The United Nations placed an end to the development and usage of ozone-depleting compounds, including CFC, and facilitated the search for ozone-friendly alternatives (UNEP 2006). Among the plastics most widely used in construction are polyethylene (PE) and polyvinyl chloride (PVC) and some of their derivatives, such as chlorinated polyethylene (CPE), chlorosulfonated polyethylene (CSPE), chlorinated ethylene (CPVC), and polychloroprene (or neoprene, CR), and their potential human health effects would be the key themes. These plastic materials are commonly viewed to be one of the most challenging materials to accept in the area of green building certification (often they are even referred to as "red list materials"). The aim of this book chapter is to determine Chemical characteristics, applications, and ecology impact of PVC (the poison plastic), Chlorinated Polyethylene (CPE), Chlorinated Polyvinyl Chloride (CPVC), Chlorosulfonated Polyethylene (CSPE), Polychloroprene Rubber (CR).

4.2 Chemistry, Applications and Ecological Impacts of Plastic Materials

Today, plastics are very important for all social interactions and are commonly used in different locations. They are one of the most commonly available products for product creation, providing many advantages to humanity and allowing scientists to suggest future medical and technical advancements (Osama et al. 2020).

It is estimated that the amount of plastic used today is 20 times that of 60 years ago. Since plastics are inexpensive, flexible, versatile, lightweight, and durable materials, they can be molded into a variety of items and can be used in a wide range of applications (Osama et al. 2020). Today, the properties of plastics support some health innovations that once seemed futuristic and have become an indispensable part of daily treatment to help save countless lives, prevent disease and prevent injury. Many of these materials are prepared by an appropriate method of alteration on the respective base plastics.

4.2.1 Polyvinyl Chloride (PVC)

Poly (vinyl) chloride, known by the acronym PVC (an acronym from the English name polyvinyl chloride) (ISO 1043–1) is a mass-consumption thermoplastic polymer, amorphous or low-crystalline, the major representation of the fluoropolymer band. Polyvinyl chloride (PVC) is an organic complex of plastics. The chemical formula is-(CH2-CHCl) in-which is formed by radical polymerization of vinyl chloride monomer (abbreviated to VCM, formula CH2 = CHCl). The main polyvinyl chloride (PVC) particles are exposed to artificial seawater or air heat and ultraviolet B (UVB) solar radiation. Using electron scanning micrographs, Brunauer–Emmett–Teller (BET) complex surface area analyzer and spectroscopic image analysis allow to determine the surface and chemical properties of fresh and damaged infrared particles (Tang et al. 2018). Thermal and UVB degradation produces a special morphology of PVC.

Also, dehydrochlorination and oxidation during the degradation process confirmed an increase in functional groups which change the chemical properties of PVC (Yu et al. 2016). On the other side, under sunlight with or without seawater, surface irregularities have been reported which seem to have been caused the deterioration of the PVC and, no new functional groups have been identified. This suggests that the chemical properties of PVC in the aquatic environment have been stable for a long time (Tang et al. 2018). The Polyvinyl chloride (PVC) formulation is an important plastic resin that can be used for construction, pipes, coatings, and other purposes. Exposure to these vinyl chloride monomer at the early stages of development has led to a significant consequences: an excessive amount of rare liver cancer, hepatic angiosarcoma, has been detected in facilities around the world. According a PVC research, a large-scale epidemiological trials to discover biomarkers in molecular pathways, has provided useful information on occupational cancer pathogenesis (Lewis 1999).

The most common and most useful application of plastic products is in hospitals and healthcare systems. Therefore, studies on plastic products and their others categories, as well as plastic products and equipment used in the manufacturing of hospitals and medical systems. Also, medical treatment and some legislation influences its design and production. So, plastics are pushing advances in once futuristic healthcare industry and have become an integral part of everyday treatment to save countless lives, avoid illness, and prevent injuries (Osama et al. 2020). Besides, analyzes the effect of plastic waste on hospitals and treatment facilities, as well as the recent results on minimizing the use of plastic products, and outlines the benefits and concerns focus on the use of plastic products in hospitals and healthcare and on future goals, problems, and opportunities (Osama et al. 2020).

The numerous components of the PVC system, including residual vinyl chloride monomer and some additives, can pose a danger to human health. Since 1974, the chloride monomer content of PVC resin has been significantly reduced, decreasing the incidence of cancer for users of PVC products. Further study is expected on the effects of PVC dust on the lungs and other potential health effects. The PVC plastic system was characterized by the interaction with its components and whole PVC system and the biological system (Karstadt et al. 1976).

The marine ecosystem is contaminated with resin and synthetic particles could threaten the health of aquatic species. A research analyzed the influence of polyvinyl chloride (PVC) particles on blood parameters, white blood cells, lipid peroxidation, and antioxidant processes (brain and gills) of *Clarias gariepinus* in freshwater as indicator species widely used as an ecotoxicological test model. Fish specimens were subjected to a diet high in PVC particles (95.41 ± 4.23 μm) at the following concentrations: 0.50%, 1.50%, and 3.0%, and a 45-day control diet and then accompanied by a 30-day purification procedure. In all concentration treatment groups, mean cell volume and cell hemoglobin were significantly reduced and time-dependent. The number of neutrophils decreased as the exposure time to PVC increased, while the values of lymphocytes and monocytes between control and exposed fish groups did not differ significantly. The activity of glutathione peroxidase in the brain and of the exposed group significantly changed compared with the control group, during different exposure periods, the activity of superoxide dismutase in the brain and the exposed group was inhibited. The activity of catalase dropped considerably in the brain of the population exposed to 0.5% PVC and decreased over time, although the activity of catalase did not change significantly (Iheanacho et al. 2020).

The amount of lipid peroxidation in the brain of the PVC exposure community increased significantly with an increase in dosage. Changes in hematology, antioxidant enzymes, lipid peroxidation, and acetylcholinesterase function suggest oxidative stress and neurotoxicity in fish *C. gariepinus* is an important biological measure for assessing the environmental effect of PVC particles (Iheanacho et al. 2020). However, PVC does not induce drastic changes in the levels of lipid peroxidation in nude fish but raises the levels of lipid peroxidation as exposure time increases. As exposure time increased, the activity of acetylcholinesterase in the brain of the exposed fish.

Over the past decade, plastics deposition in flora, fauna, and humans has been the subject of global interest. To determine the possible toxic effects of PVC (polyvinyl chloride) microplastics on freshwater fish embryos, carp larvae were used for long-term dietary exposures of 30 to 60 days (Xia et al. 2020), four separate dietary forms of microplastic and non-plastic management therapies (10%, 20%, and 30% respectively). The study indicates that the control sample, microplastics greatly inhibited weight growth and weight gain in all PVC treatments. The superoxide dismutase (SOD) and catalase CAT actions were analyzed and an antagonist interaction was found between them. After 30 days of exposure as the concentration of PVC rises, the activity of glutathione peroxidase (GPx) first increases and then decreases, then after 60 days of exposure, a dose-dependent downward trend was observed (Xia et al. 2020).

Indeed, the malondialdehyde (MDA) levels are greatly decreased when exposed to varying concentrations of microplastics in different tissues and changes in the expression of antioxidant-related genes have been shown in the liver of larvae exposed to PVC microplastics.

Particullary, CYP1A and GST genes transcription increased and then decreased at higher plastic concentrations after 30 days of exposure. In addition, histological tests have demonstrated that the cytoplasm of the liver is vacuolated when exposed to 20 and 30% of microplastics. This research offers basic toxicological evidence to explain and measure the impacts of PVC microplastics on freshwater ecosystems (Xia et al. 2020). This study examined the effects of polyvinyl chloride (PVC) particles on hepatic antioxidant enzyme activity, serum biochemistry, and liver histology of juvenile *Clarias gariepinus*. A total of 180 (mean weight 25.15 g) of *C. gariepinus* was fed a diet supplemented with PVC (95.41 $\pm$ 4.23 μm) with inclusion levels of 0.5, 1.5, and 1.5. A fish sample from each treatment was collected every 15 days for biochemical characterization of hepatic antioxidant enzyme, and histopathological analysis. Compared to the control group, glucose and triglyceride levels in the PVC-treated group increased significantly. During 15 days and 30 days of exposure, protein levels in the PVC-treated group decreased significantly from 3% to 0.5%, respectively, while serum enzyme production in all PVC-treated groups increased significantly over time (Iheanacho and Odo 2020).

Over time, the function of antioxidant enzymes (superoxide dismutase, glutathione peroxidase, and catalase) in the liver of the patient community also decreased gradually and the amount of lipid peroxidation increased in the community treated with PVC. Histopathological examination of the fish liver revealed that in the PVC control community, the intake of glycogen, fat vacuoles, and liver cell degeneration and necrosis are moderate to severe. As result, this study has shows that microplastic PVC can cause oxidative damage and histopathological changes in the liver of exposed fish (Iheanacho and Odo 2020).

4.2.2 Chlorinated Polyethylene (CPE)

Chlorinated polyethylene (CPE) is a convenient polyethylene with a chlorine content of 34–44%. It is mixed with polyvinyl chloride (PVC) since the soft, rubber chlorinated polyethylene is integrated into the PVC matrix to improve impact resistance and increases temperature tolerance. The PVC sheets are often used for softening without the possibility of plasticizer migration. The CPE properties depend on the type and quality of the polyethylene material, the content and the uniformity of chlorine substitution. In addition, the chlorination process can be adjusted to produce amorphous products or products that contain increased levels of residual crystallinity in the polyethylene. The CPE can be interlinked with peroxide to form an elastomer and can be used in the cable and rubber industries. When the chlorination level reaches 55%, the durability and durability of chlorinated polyethylene tends to improve consistance. There are two primary types of chlorination methods, the solution phase, and the suspension phase. Radical chlorination of polyethylene film in the heterogeneous solid–gas process is another reaction tool that has recently gained interest.

If the content of the chlorine increases, the polymer becomes brittle. Thermoplastic CPE is combined with other polymers for white and brown PVC profiles. If chlorination is raised to a degree where the polymer is only semi-compatible with

PVC, a high impact strength mixture can be produced and the content can be separated into impact alteners. Chlorinated polyethylene elastomers from a high-density polyethylene backbone that has been chlorinated by an aqueous free radical suspension method. Some polymers have a different chlorine content (36–48%), molecular weight, and intrinsic crystallinity.

China is currently the world leader in the production of CPEs (83%), while China and the United States remain the two largest users in 2008 (Ormonde and Klin 2009). Approximately 75–80% of CPE is used in impact resistance applications, particularly in China, for modification of doors/windows and pipes/drainage profiles and some styrene copolymers. Also, it is a type of plastic used in flexible cloth, wires, and cables with multiple applications (such as pipes and gaskets). Chlorinated polyethylene elastomers are relatively new polymers derived from the high-density polyethylene backbone that has been chlorinated by an aqueous free radical suspension method.

Unfortunately, the properties of these simple polymers are not adequate to explain some properties of chlorinated polyethylene elastomers. Current research characterizes chlorinated polyethylene elastomers as general purpose for environmental resistant (oil, heat, ozone...) to these elastomers. Since the polymer chain is saturated, a radical traitement method is required to offers an effective balance of efficiency to the mixer, including real resistant temperatures from −60 °F to 300 °F (Sollberger and Carpenter 1975). The chlorination process can be adjusted to produce amorphous products or products that contain increased levels of residual crystallinity in the polyethylene.

Chlorinated polyethylene, as it is mixed with many other polymers (especially polyvinyl chloride), has many uses in blends, wire, and cable coatings, adhesives, floor tiles, film, and can even be used as a thermoplastic elastomer (Salamone 1996). The efficiency and mechanical properties of CPE are strongly dependent on the degree of chlorination, the microstructure of the polymer chain, the processing, this type of polyethylene, and the used solvent. Other important applications for CPE include wire and cable sheathing, roofing membranes, geomembranes, pipes and tubing, coated fabrics, molded formwork, extrusion profile slabs, and wet underfloor casting. Concrete floorboards may be used for the construction of composite building envelopes and joints.

CPE is the ideal cost/performance for vinyl coating substrates with low-temperature endurance, ductility, and high wire acceptance. It has excellent impact tolerance in vinyl enclosure substrates, good wire acceptability, and ductility that can also be used for storage applications. All of these applications currently occur in the construction sector. The health effects of CPE may be listed as follows.

CPE should be produced in hard or soft plastics without plasticizers, otherwise, special additives (such as UV and thermal stabilizers, antioxidants, etc.) can be used.

This can occur in processing equipment where careful maintenance and dust management is necessary to ensure safe handling. These dust particles may cause eye irritation (i.e. chemical goggles must be worn), skin contact must be avoided. Thermal degradation of CPE (thermal decomposition and combustion, although CPE is difficult to burn) can produce HCl (hydrochloric acid) and CO (carbon monoxide) in addition to dioxins that are highly irritating and toxic. Indeed, if we consider only the base polymer (PE), its thermal degradation primarily creates carbon monoxide (CO) which is known to be a systemic toxin and asphxiant gas (Ammala et al. 2011).

4.2.3 Chlorinated Polyvinyl Chloride (CPVC)

CPVC plastics consist of about 85% of resin and 15% additives, such as antioxidants, lubricants, stabilizers, dyes, pigments, enhancing additives, and manufacturing supplements. Depending on its composition, concentration, and conditions, each of these additives can pose a health risk when thermal stabilizers, suitable thermal stabilizers include predominantly phosphate stabilizers (such as disodium phosphate), maleimides, sulfur compounds, and alkyl tin compounds is evaluated. In the manufacture of PVC and CPVC plastics, various organotin compounds (mainly mono-and displaced alkyl tin) are the most commonly used as heat stabilizers (organotin compounds account for 2–4% of CPVC valves), including pipes. These structure come into contact with drinking water, they quickly get quickly into the water (Boettner et al. 1982). According an Analysis performed by Health Canada has shown that monobutyltin and dibutyltin in PVC and CPVC drinking water are recorderd at ng/L (WHO 2004).

Related experiments have shown that the quality of most samples is below the detection limit (in particular 0.5 ng/L). Its mechanical properties can be modified more extensively using traditional composite manufacturing, and can have several different chemical compositions, so other additives shall be included (Noveon 2003).

CPVC is a polymer used in the construction sector and described to be a secure material. The use of CPVC as an internal corrosion coating for tanks and vessels is very common. It is also often used in adhesive applications in PVC, as a sealant for plasticizers and PVC for wire and cable coatings. The probable health effect (acute or chronic) of the transformed CSPE content is due to the evolution of CCl_4 (according to Dupont's Material Safety Data Sheet 2008). This chemical elements reaches the human body through the skin or through breathing, which may irritate the skin, eyes, nose, throat, and/or lungs. The last target organs are the liver, central nervous system, and kidneys.

Organotins are classified as an antigen and, following oral administration to mice, appear to be spread mainly to the liver and kidneys (Ehman et al. 2007) and the thyroid (Pennings et al. 1987). The German Federal Institute for Public Health Safety and Veterinary Medicine has stated that the appropriate daily consumption of butyltin compounds is 0.25 μg/kg body weight (Rudel 2003).

The PVC is a CPVC matrix extending to the environment, but drinking water is included if we consider the use of plumbing. The Vinyl Chloride Monomer (VCM) is listed as a carcinogen by OSHA (Occupational Safety and Health Administration) and NIEH (National Environmental Health Institute) (NIEH Research 1997), which induces tumors in the liver, brain, and lung, lymphatic and hematopoietic processes. The study found that, after 30 days of exposure, the concentration of VCM leached from PVC pipes is typically higher than 2.5 μg/L (Al-Malack et al. 2000), which is above the known (safe) value (0.5–2 μg/L). The CPVC made from PVC is supposed to have the same monomers and related risks. Third, thermal decomposition, oxidation, or pyrolysis of CPVC fire materials (Fardell 1993) involves some hazardous gasses such as sulfur monoxide, carbon dioxide, hydrogen chloride, and carbon dioxide. In addition to organotin compounds and hydrocarbons, it also includes a small number of benzene, aromatic (and aliphatic) hydrocarbons, and a small volume of carbon chloride and carbon tetrachloride. When CPVC is heated in thermal phase, an annoying peroxide vapor is formed. Also, the PVC and CPVC are well known to have limited thermal stability due to their dehydrochlorination process. In particular, the decomposition/thermal degradation products of CPVC are identical to PVC, so, in addition to these chemicals, some other compound must be added.

VCM is considered carcinogenic and systemic toxins, dioxins, furans, PBT (persistent and bioaccumulative hazardous substances), and PCB (polychlorinated biphenyls) are irritating to mucous membranes and respiratory tract, all of these elements are potentially carcinogenic (Ammala et al. 2011). The inhalation of decomposition (and/or combustion) products can irritate the respiratory tract, eyes, and skin. Depending on the severity of exposure, other physiological response will be coughing, pain and inflammation. A long-term exposure to fumes and vapors generated by heating or thermal decomposition of CPVC may cause an asthma syndrome due to inhalation of process vapors or fumes. Since CPVC can have a health impact, and its use in PVC cable is declining due to flame retardants. Most solvents used in pipe cement, primers, and cleaners are considered as irritating substances.

4.2.4 Chlorosulfonated Polyethylene (CSPE)

Chlorosulfonated polyethylene (CSPE or CSM rubber) is a synthetic PE-based rubber noted for its high chemical resistance. It is a new form of water-soluble polymer containing sulfonic acid groups that will be produced; Chloridric acid (HCl) will be eliminated from the polymer backbone by substituting some groups of H atoms, chlorine, and sulphonyl chloride (SO_2Cl). The sulpho chlorinated may be done on solid surfaces or in a solution.

CSPE/CSM rubber is used in a wide range of manufacturing and building applications and required high performance. The automobile market is the important field behind CSPE's worldwide production. In 2008, almost a quarter of the total global consumption of CSPE was in the automotive sector. The second main customer for CSPE is the architectural application in roofing membranes and coatings for ponds and tanks. So, CSPE has the largest market share, followed by PVC and ethylene-propylene-diene monomer terpolymer (EPDM) (Griffin 1982). Besides, if additives are used in CSPE formulations with other required properties, it may be of great concern to VOCs emission which should be released at high temperatures and must be carefully selected and monitored. In addition to construit membrane (monolayer) applications, the CSPE/CSM rubber is also used for geosynthetic applications such as coatings, caps, and float covers and insulation for vessels, gaskets, sealing strips, wire, and cable.

This book chapter addresses chlorosulfonated polyethylene (CSM) formed by a liquid–solid bulk reaction with sulfonyl chloride (SO_2Cl_2). The basic structure of the CSM liquid–solid method was contrasted with that of the FT-IR and the 1H NMR. The thermal stability of the CSM liquid–solid system is greater than that the CPE method owing to the inclusion of the chlorosulfonyl community in the CSM molecule, whereas the distribution of chlorine is deteriorating. Compared to the industrial product, CSM provided by the solution process liquid–solid method has higher tensile strength and elongation at deformation (Han et al. 2018). The fact that the solvent solubility of High density polyethylene (HDPE) is not taken into account in the preparation of CSM by a liquid–solid process and that HDPE with a higher molecular weight and a narrower molecular weight range may be chosen. The Differential Scanning Calorimetry (DSC) and Dynamic Mechanical Analysis (DMA) measurements revealed that the liquid–solid CSM method was lower and the modulus higher and that the thermal stability was very comparable. The CSM prepared using the liquid–solid process satisfies some specifications and this introduces a new definition for the industrial development of chlorosulfonated polyethylene (Han et al. 2018). Thermal oxidation (thermal decomposition and combustion, while CSPE begins with difficulty) will generate primarily HCl, SO_2, and CO in addition to dioxins or related products which are extremely irritating and poisonous. The carcinogenicity of the CSPE is not evident. It is a material listed by IARC stands for the International Agency for Cancer Research and American Conference of Governmental Industrial Hygienists, but there is no proof to suggest that it is carcinogenic to humans and there is no proof (or insufficient evidence) to prove its effectiveness in laboratory animals (Akovali 2007; DeBono et al. 2020). A chemical reaches the human body through the skin or through breathing, which may irritate the skin, eyes, nose, throat, and/or lungs. The last target organs are the liver and central nervous systems as well as the kidneys.

4.2.5 Polychloroprene Rubber (CR)

CR (chloroprene, polychloroprene, poly [2-chloro-1,3-butadiene] rubber, also known as the first "duprene" and later "neoprene" consists predominantly of the following formula Trans 1,4-polychloroprene composition) [CH_2–CCl = CH–CH_2]n. This is a major diene-based elastomer formed by radical emulsion polymerization of its chloroprene monomer, 2-chloro-1,3-butadien. Polymerization tends to take place almost entirely in the trans 1,4 shape with some cis composition, and CR is a crystallized elastomer. CR can be vulcanized on its own by heating with zinc and magnesium oxides to reorganize and extract chlorine. This is the same approach used by several other chlorinated polymers (i.e. CSPE). CR vulcanization is distinct from traditional methods with the inactivation of the double bonds and the α-methylene groups.

CR has perfect mechanical properties, chemical and heat resistance, and low flammability. It has a high resistance to ozone and temperature. It has a very high damping capacity and can withstand various environmental factors, and is generally resistant to ketones, chlorinated solvents, and gamma rays.

In general, civil engineering and architectural research require elastomers with excellent weather resistance, long-term flexibility, and good mechanical properties. CR is widely used in general engineering applications, such as the production of cured products, shells, and sheets, as an asphalt modifier, as a base material for adhesives, and in various cable coatings and conveyors in various molded belts and profile products. The CR Latex is used for adhesives, adhesives, coatings, tempering additives, modified asphalt and concrete, and foam. Highly crystalline CR polymers are typically favored for adhesives, whereas softer and more compact polymers are preferred for applications with mechanical properties, which can be mixed and formed into rubber products.

If pure, the CR is unclassified for human carcinogenicity effect and is not considered to be toxic. However, various refined CR products, including solid rubber products and adhesives, can contain several additives that may be stored or released (gasses form) which may be toxic or cause skin irritation as follows:

The volatile ingredients which may exist in CR include primarily chloroprene monomer, toluene, and butadiene, in most cases as an additive to certain lead-containing products (i.e. lead oxide) used as a mixing agent, and lead to from the plastic system (Ethylene Thiourea, Carcinogens Report 2002), and all of these ingredients may remain in the organism even in small amounts (trace).

Low acute exposure causes inflammation of the eyes, mouth, and respiratory tract, and high acute exposure can cause nausea, decreased heart rate, and central nervous system damage. Besides, CR adhesives can contain rosemary or rosin, which is used as skin sensitizer. The EU stipulates that a rosin amount of 0.1% or more must be labeled for all of these substances are toxic and carcinogenic (Akovali 2007). Rubber adhesives and gaskets typically contain solvents such as hexane, naphtha, acetone, and zinc oxide. Updated and novel CR adhesives that contain water as a solvent do not face any such problems. Although CR has very low oral toxicity, it can mostly cause skin inflammation and allergic reactions or , i.e. dermatitis by close contact or wearing garments, hats, boots, or PC fabric carriers. (Akovali 2007).

4.3 Control Agents and Other Non-Toxic Alternative Compounds

The accumulation of vast volumes of petroleum-derived plastics in the atmosphere has posed environmental and health issues. Giacomucci et al. (2019) studied the biodegradation capacities of five bacterial strains, namely *Pseudomonas chlororaphis,Pseudomonas citronellolis,Bacillus subtilis,Bacillus flexus,and Chelatococcus daeguensis*, to polyethylene, polypropylene, polystyrene, and polyvinyl chloride films under aerobic conditions. After 45 days of incubation, a fragmentation of the film was observed showing the degrading structure PVC. Chemical analysis of incubated films confirmed the biodegradation of PVC plastic waste as seen by a gravimetric measurement and the weight loss is recorded up to 19% after 30 days of incubation.

Since it substitutes different rubbers and plastics as building materials in terms of durability and efficiency, it is recommanded to use safer plastics, carefully chosen and prepared to achieve the lowest possible VOC emissions:

The TPO (thermoplastic polyolefin) and EPDM (ethylene-propylene-diene monomer, saturated polyethylene chain) meet the safety criteria. Materials such as PP (polypropylene), HDPE, and Ethylene propylene diene monomer (EPDM) are common replacements for CSPE and CR in geomembrane applications. The use of solvent-based PC (polyurethane) adhesives can be avoided as far as possible and all water-based adhesives can be substituted. Alternatively to the CR adhesive, the copolymer and a styrene block can be considered.

4.4 Conclusion

A variety of plastic materials used in the building industry are considered either to emit volatiles to indoor air or to leach organotin, all of which can impact comfort, health, and productivity. Their characteristics, applications, and health effects are presented briefly and several alternate materials for their use are presented.

These polymers plastics are used as major or minor components in a wide range of industrial applications. Other important end applications for CPE include roofing membranes, wire and cable ducts, geomembranes, automotive and industrial pipes and tubes, coatings, molded shapes, extruded profiles and use as base polymers. For example the use of CPE as an impact modifier, mainly in PVC formulation, accounted for nearly 74% of global consumption in 2017. Actually, China was the largest user of CPE, accounting for approximately 82% of global demand and four largest producers operate 34% of the world's CPE production, led by other Asian nations, the United States and Europe. In overall, emissions or leaching of toxic particules can depend on the structure of the plastic material used, along with the different parameters associated with it and should be controlled and removed from the environment or other alternative materials for sustainability purposes.

References

Akovali G (2007) Plastics, rubber and health. iSmithers-Rapra, Shrewsbury, UK

Al-Malack MH, Sheikheldin SY, Fayad NM et al (2000) Effect of water quality parameters on the migration of vinyl chloride monomer from unplasticized PVC pipes. Water Air Soil Pollut 120(1–2):195–208

Ammala A, Bateman S, Dean K, Petinakis E, Sangwan P, Wong S, Yuan Q, Yu L, Patrick C, Leong KH (2011) An overview of degradable and biodegrad- able polyolefin. Prog Polym Sci 36:1015
Boettner EA, Ball GL, Hollingsworth Z, Aquino R (1982) Organic and organotin compounds leached from PVC and CPVC pipe. EPA Summary Project, February 1982
DeBono NL, Logar-Henderson C, Warden H, Shakik S, Dakouo M, MacLeod J, Demers PA (2020) Cancer surveillance among workers in plastics and rubber manufacturing in Ontario, Canada. Occup Environ Med 77(12):847–856. https://doi.org/10.1136/oemed-2020-106581
Ehman KD, Phillips PM, McDaniel KL, Barone S Jr, Moser VC (2007) Evaluation of developmental neurotoxicity of organotins via drinking water in rats: dimethyl tin. Neurotoxicol Teratol 29:622–633
Ethylene Thiourea (2002) Report on carcinogens, 10th edn. US Department of Health and Human Services, Toxicology Program, December 2002
Fardell PJ (1993) Toxicity of plastics and rubber in fire. Rapra report 69, Vol. 6, no. 9, iSmithers-Rapra, Shrewsbury, UK
Forsyth DS, Jay B (1997) Organotin leachates in drinking water from chlorinated poly(vinyl chloride)/(CPVC) pipe. Appl Organomet Chem 11:551–558
Giacomucci L, Raddadi N, Soccio M, Lotti N, Fava F (2019) Polyvinyl chloride biodegradation by *Pseudomonas citronellolis* and *Bacillus flexus*. N Biotechnol 52:35–41. https://doi.org/10.1016/j.nbt.2019.04.005
Griffin W (1982) Manual of built-up roof systems. McGraw Hill, New York
Han L, Luo L, Wang P et al (2018) Preparation of chlorosulfonated polyethylene by liquid-solid method and comparison with industrial products. J Polym Res 25:237. https://doi.org/10.1007/s10965-018-1624-1
Iheanacho SC, Odo GE (2020) Dietary exposure to polyvinyl chloride microparticles induced oxidative stress and hepatic damage in *Clarias gariepinus* (Burchell, 1822). Environ Sci Pollut Res Int 17:21159–21173. https://doi.org/10.1007/s11356-020-08611-9
Iheanacho SC, Igberi C, Amadi-Eke A, Chinonyerem D, Iheanacho A, Avwe Moya F (2020) Biomarkers of neurotoxicity, oxidative stress, hepatotoxicity and lipid peroxidation in *Clarias gariepinus* exposed to melamine and polyvinyl chloride. Biomarkers 25(7):603–610. https://doi.org/10.1080/1354750X.2020.1821777
Karstadt M (1976) PVC: health implications and production trends. Environ Health Perspect 17:107–115. https://doi.org/10.1289/ehp.7617107
Lewis R (1999) Vinyl chloride and polyvinyl chloride. Occup Med 14(4):719–742
McFarland M, Kaye J (1992) Chlorofluorocarbons and ozone. Pergamon Press, Oxford, UK
NIEH (National Institute of Environmental Health) (1997) Eighth report on carcinogens, perspectives, 105(9), September 1997. https://europepmc.org/backend/ptpmcrender.fcgi?accid=PMC1470364&blobtype=pdf
Nikolaou AD, Gatidou GM, Golfi Nopoulos SK, Thomaidis N, Lekkas TD (2007) A one-year survey of organotin compounds in the reservoirs supplying the drinking water treatment plants of Athens, Greece. Desalination 210:24–30
Noveon (2003) US patent to Noveon, USA, Plastic pipes and fi fittings for home and industrial use, 20030157321 https://patents.google.com/patent/US20030157321A1/en
Ormonde E, Klin T (2009) Chlorinated polyethylene resins and elastomers. Chemical Industries Newsletter, SRI Consulting, March 2009
Osama M, Basmage MSJ, Hashmi (2020) Plastic products in hospitals and healthcare systems. Encyclopedia Renew Sustain Mater 1:648–657. https://doi.org/10.1016/B978-0-1oo2-803581-8.11303-7
Sadiki A, Williams D (1999) A study on organotin levels in Canadian drinking water distributed through PVC pipes. Chemosphere 38:1541–1548
Salamone JC (1996) Polymeric material encyclopedia. CRC Press, Boca Raton FL, p 1235
Sollberger LE, Charpentier CB (1975) Polyéthylène chloré - une caractérisation complète. Journal Des Élastomères Et Des Plastiques 7(3):233–257. https://doi.org/10.1177/009524437500700302

Tang CC, Chen HI, Brimblecombe P, Lee CL (2018) Textural, surface and chemical properties of polyvinyl chloride particles degraded in a simulated environment. Mar Pollut Bull 133:392–401. https://doi.org/10.1016/j.marpolbul.2018.05.062

UNEP (United Nations Environment Programme) Ozone Secretariat (2006) Handbook for the montreal protocol on substances that deplete the ozone layer, 7th edn. In: US environmental protection agency (1985) 'A summary overview of health effects associated with chloroprene', EPA 600/8–85/011F.

WHO report (2004) Dialkyl Tins in drinking-water', background document for development of WHO (Guidelines for drinking-water quality). WHO/SDE/ WSH/03.04/109

Xia X, Sun M, Zhou M, Chang Z, Li L (2020) Polyvinyl chloride microplastics induce growth inhibition and oxidative stress in *Cyprinus carpio* var. larvae. Sci Total Environ 716:136479. https://doi.org/10.1016/j.scitotenv.2019.136479

Yu W, Azhdar B, Andersson D, Reitberger T, Hassinen J, Hjertberg T, Gedde UW (2011) Deterioration of polyethylene pipes exposed to water containing chlorine dioxide. Polym Degrad Stab 96(5):790–797

Yu J, Sun L, Ma C, Qiao Y, Yao H (2016) Thermal degradation of PVC: a review. Waste Manag 48:300–314. https://doi.org/10.1016/j.wasman.2015.11.041

Chapter 5
Volatile Organic Compounds Emission from Building Sector and Its Adverse Effects on Human Health

Zaiema Rouf, Idrees Yousuf Dar, Maheen Javaid, Mohmad Younis Dar, and Arshid Jehangir

Abstract The volatile organic compounds (VOCs) from building sector include aliphatic hydrocarbons, halo-hydrocarbons, and aromatic hydrocarbons. The recent research findings conducted in the non-residential indoor environments have demonstrated that the concentration of indoor VOC often exceeds the outdoor concentrations. The recent evidence insinuates that the significant amount of VOCs can clearly have both acute and chronic adverse effects on human well-being including cardiovascular and nervous system, increased mortality, etc., and may even cause cancer. Furthermore, VOCs have a significant effect in "sick building syndrome" reported by the WHO. Volatile organic compounds generally occur as liquids or as vapours at normal room temperature but may also exist in solid form such as bathroom deodorants, dichlorobenzene, naphthalene, and para-mothballs. Due to the presence of plenty of VOCs it is not practicable to handle all the compounds and their toxicity. It is evident that while hundreds of VOCs may exist in any type of environment, the substantial health effects of the exposure of VOCs are yet to be identified. Thus to improve the air quality of the indoor building environment it is mandatory to expertise the better understanding of the mechanism and basic characteristics of VOC generation from the building sectors and construction materials.

Keywords Benzene · Building sector · Carcinogenic · Formaldehyde · Indoor environment · VOCs

5.1 Introduction

Despite the fact that humans are becoming more opulent and developing towards modern life, we spent an estimate of 80% of our time in the indoor air environment (Poljansek et al. 2017). Whereas, in environmental equity dialogue the indoor environment has been scarcely mentioned (Adamkiewicz et al. 2011). There has been a constant increase in the tendency of constructing the structures that are more

Z. Rouf · I. Y. Dar (✉) · M. Javaid · M. Y. Dar · A. Jehangir
Department of Environmental Science, University of Kashmir, Hazratbal, Srinagar 190006, J&K, India

J. A. Malik and S. Marathe (eds.), *Ecological and Health Effects of Building Materials*,
https://doi.org/10.1007/978-3-030-76073-1_5

focussed on the extravagance and consolation of the inhabitant as well as towards the energy conservation leading to the construction of buildings which are more airtight and have a lesser number of operable windows and meagre airflow (Chang and Gershwin 2004). As per the documentation of EPA, the concentrations of the VOCs are comparatively 2–5 times more than the concentrations present outdoors and occasionally much higher (EPA 2016). Selectively, the newly constructed buildings occupy the topmost level so far as VOC off-gassing in the indoor environment is concerned, which tends to the increased number of new substances producing VOC particles collectively within a short space of time (Wang et al. 2007). In the course of certain activities, the concentrations of VOCs in the indoor environments can achieve 1,000 folds than the outside environment. Meanwhile, various studies have depicted that when VOCs are considered separately in the indoor environment the emissions are not very high. Moreover, the total VOC (TVOC) concentrations in an indoor air may be comparatively as much as five times higher than the concentration of VOCs present outside. Apart from new buildings, various consumer products are responsible for the emission of VOCs thus elevating the inside concentrations (Wang et al. 2007; Jones 1999).

The regularly found VOCs within the indoor environment were aliphatic aldehydes, alkanes, benzenes, substantial chlorinated aliphatic hydrocarbons, and terpenes. Though there are significant sources of VOCs present indoor, thus the concentrations of VOCs are considerably greater than that of the outdoor environment (Meciarova et al. 2017). This insinuates that the inhabitant maybe intermittently exposed to miscellaneous VOCs at the higher level concentration in a short span period, moreover depending upon the duration of exposure, occupant activity location, and type of the VOC present (Tsai 2019). In connection with the winter season, the extent of VOCs in indoor air is about three to four times greater as compared to concentrations of the summer season (Barro et al. 2009). The quality of air within the building sector is not so clean and safe particularly in buildings that involve the utilisation of hazardous chemicals or wherein processes of combustion take place. The indoor air pollution generally depicts the production and transportation of pollutants in the inner side of different indoor environments where people reside and carry out their work, like workplaces of industries, hospitals, schools, apartments, and homes of the private sector (Heinsohn and Cimbala 2003). The indoor air quality (IAQ), or indoor environmental quality (IEQ), in broader terms, has received significant awareness from the common public along with practitioners including the researchers (Lai et al. 2009). The organic compounds having boiling points less than 250 °C at (101.325 kPa) atmospheric pressure are generally referred as "volatile organic compounds" (Huang et al. 2020). VOCs are referred as the wide range of compounds with an organic origin to which people are exposed on regular intervals. Materials from the buildings such as furniture, cleaning products, paint, and cosmetics are the predominant VOC sources. These regular products cause emissions of VOCs in the form of gases that are ultimately inhaled by the people (Lim et al. 2014).

There are various causes responsible for the potential sources of indoor pollution in school buildings as in case utilization of materials having high emission for the construction and furnishing of buildings, restricted landscaping with an inadequate drainage system, the pattern of air conditioning, heating and ventilation, the absence of preventative measures, cramped conditions and products of cleaning which are responsible for releasing chemicals into the air (Godwin et al. 2007; Moriske et al. 2008). The pollutants present in indoor environments expose the occupants of the building to the adverse health threats (Chan et al. 2015). The VOCs have constantly been placed among five leading environmental public health risks (EPA 2015). Though the children have comparably higher sensitivity to pollutants than that of the adults, yet they spend more time in the school indoor environment where they are easily exposed to an unspecified amount of building pollutants (De Gennaro et al. 2013). The enormous amount of released VOCs comprises of alcohols, aldehydes, alkanes, alkynes, aromatics, alkenes, esters, halocarbons, sulphur/nitrogen containing VOCs, and ketones, while as the environmental impression of these compounds is characteristically reliant on the effectiveness of the particular VOC. Assume that hundreds of VOCs may exist in all forms of environment; the precise effects of VOCs exposure on humans are still unidentified. The prominent levels of VOCs are contributed by the minimal pace of air exchange system in between the indoor air environment and the outdoor as consequences of tightly-shut windows and continuous increase in the utilisation of humidification devices (Schlink et al. 2004). Commonly utilized products that carry the VOCs are glues (ethyl benzene, toluene, n-hexane, xylene, vinyl chloride) inks paints, and varnishes, products used to remove stains (for example carbon tetrachloride, trichloroethylene, trichloroethane, tetrachloroethylene), propellants for aerosol (dichloromethane, dichlorodifluoromethane), cosmetics (acetone, ethyl acetate, propylene glycol, phenolic compounds), household products commonly used for cleaning (acetone, chloroform, phenol, dimethyl ammonium chloride, ethanol), wood preservation products (p-dichlorobenzene, pentachlorophenol) (Maji and Ashok 2003; Wille and Lambert 2004). As documented by WHO, VOC plays a significant role in SBS (Ten Brinke et al. 1998). While considering the objective to estimate potential health effects correlated with the exposure of VOCs, it is obligatory to understand and estimate both the core sources of pollution as well as pollutant concentration with an appropriate indoor monitoring.

5.2 Sources of VOCs in Building Sectors

VOCs, a significant group of compounds could be important as these compounds are considerably available in construction materials and that way can be ceaselessly released in a slow manner over an extended period of time that can pose a considerable threat to human wellbeing and health when there is a notable exposure towards them (Rumchev et al. 2007). Considering VOCs that are present in the building environment, the significant sources may incorporate building occupants who spend an extended amount of time in the buildings, microbial sources, emissions from automobiles and new construction as well as renovations (Brown 1999) (Table 5.1).

Table 5.1 Commonly known volatile organic compounds and their potential sources

Chemicals	Major sources of exposure
Acetone	Cosmetics
Alcohols (isopropanol, ethanol)	Cleansers, spirits
Aromatic hydrocarbons (xylenes, ethyl benzene, trimethyl benzenes, toluene)	Adhesives, combustion sources, gasoline, paint
Aliphatic hydrocarbons (decane, undecane, octane)	Paints, combustion sources, gasoline, adhesives
Benzene	Smoking, pumping gas, driving, passive smoking, auto exhaust
Carbon tetrachloride	Global background, fungicides
Chloroform	Dishes, washing clothes, showering (10 min average)
p-Dichlorobenzene	Moth cakes, room deodorizers
Formaldehyde	Pressed wood products
Methylene chloride	Solvent use, paint stripping
Styrene	Smoking
Tetrachloroethylene	Visiting dry cleaners, storing or wearing dry-cleaned clothes
I, I, I-Trichloroethane	Aerosol sprays, storing or wearing dry-cleaned clothes
Trichloroethylene	Unknown (correction fluid, cosmetics, electronic parts)
Terpenes (limonene, ex-pinene)	Scented deodorizers, fabrics, food, fabric softeners, cigarettes, polishes, beverages

Adapted after: Burn et al. (1993)

Wood being a biological and natural material, when burned, emits various profiles of organic chemical substances majorly being formaldehyde, very volatile organic compounds (VVOCs), and volatile organic compounds (VOCs). The concentrations of these compounds vary in comparison to the emissions that occur within indoor spaces of buildings, thus making huge concentrations of VOCs indoors comparatively to that of outdoors (Skulberg et al. 2019). The indoor variety of VOCs is extensive and includes personal products, products utilised in the household, and building materials (Rumchev et al. 2016). Any sort of material within a building can emit a notable amount of organic chemicals into the air. There are a wide variety of sources through which VOCs are emitted among commercial and residential buildings that include a wide range of construction material, and indoor samples of air may thus carry hundreds of VOCs (Rumchev et al. 2007). Mainly, the sources VOCs in building air quality include anthropogenic activities, products for personal care, smoking, products utilised for house cleaning, building products, and pollution from outside (Sarigiannis et al. 2011). Bartzis et al. (2015) also considered the emissions of NMVOC occurring due to plug-in air fresheners, personal care products and kitchen cleaning agents. Outdoor sources mainly include petrol stations, traffic sources and various chemical industries that include those dealing with oil, coal, chemicals, and paints (Godish et al. 2015). Other uses of VOCs in the building sectors as demonstrated by WHO (2010) include:

- Do-It-Yourself (DIY) items such as wallpapers, adhesives, paints, glues, lacquers and varnishes,
- Household agents for cleaning for example detergents, disinfectants, shoe products, carpet cleaners and softeners,
- Cosmetics including shampoos, liquid soap, nail hardeners and varnishes,
- Electronic equipment, such as photocopiers and computers,
- Miscellaneous consumer items such as paper products as well as insecticides.

These selected VOCs that are widely available in the air of building sectors can be classified into several subgroups that include aromatic hydrocarbons, aliphatic, oxygenated and chlorinated hydrocarbons. Many hundred diverse VOCs have been determined in the air of building sectors by government organizations and academic researchers (Wolkoff and Nielsen 2001). Ketones and alcohols are usually used in manufacturing of cosmetics and products used for personal care such as colognes, hair spray, nail paints, rubbing alcohol, nail paint removers, and perfumes. Ketones are also consumed in paint thinners, aerosols, adhesives, and varnishes, window cleaners. Alcohol based VOCs comprises benzyl alcohol, ethyl alcohol, isopropyl alcohol while ketone based VOCs include acetone, ethyl acetate, methacrylates (methyl or ethyl) and methyl ethyl ketone.

5.3 General Classification of VOCs

The categories and the overall concentration of VOCs in the air environment have been reviewed (Brown 1999). Generally, VOCs include acetone, benzene, ethanol, ethylbenzene, dichloroethylene, n-decane, xylenes, ethyl acetate, n-nonane, nonanal, methyl ethyl ketone, tetrachloroethylene, 1, 2, 4-trimethylbenzene, p-dichlorobenzene, 1, 2- *n*-hexane, 1, 1, 1-trichloroethane, dichloromethane, limonene, and toluene. These VOCs have been further divided into four categories: aldehydes (i.e. HCHO), chlorinated aromatic compounds (such as 1, 4-dichlorobenzene and 1, 2-dichlorobenzene), non-chlorinated aromatic compounds (that is, benzene, ethylbenzene, styrene, toluene and xylenes), chlorinated aliphatic compounds (for example carbon tetrachloride, tetrachloroethylene, dichloromethane, chloroform, and trichloroethylene) (Tsai 2019). Furthermore discussions on the four mentioned categories of VOC are summarized as under.

5.3.1 Formaldehyde (H-CHO)

H-CHO is a colourless, gaseous compound having a different, foul odour. It is a highly volatile and very reactive compound because it rapidly reacts with hydroxyl radicals to form formic acid at a normal room temperature (Wolkoff 2013). Perhaps, HCHO might be most significant VOC pollutant in the air environment of building sectors because its key sources are the materials that are used in building and furnishings for an instance carpet, medium-density fibreboard, particleboard, and plywood (Godish et al. 2015). Among all, the aldehydes are one of the significant pollutants in building sectors, which are commonly emitted from materials which are utilised for decoration and are caged in airtight and compact buildings (Klett et al. 2014; Mitsui et al. 2008). Aldehydes such as formaldehyde are emitted from various industrial products, such as adhesives, cleaning agents, cigarette smoke, cosmetics, carpeting, plywood, construction materials, disinfectants, treated wood resins, medium-density fibreboard, plastic particleboard, and fabrics (Li et al. 2014b). The exposure of humans to low concentration of aldehyde may cause irritation in eyes, throat irritation, breathlessness, and chest tightness (Zhu and Wu 2015). Whereas, the constant exposure to considerably high concentration of aldehyde intensifies risk of acute poisoning, whereas prolonged exposure can mark negative impacts on human comfort and health that eventually can result in chronic toxicity (Main and Hogan 1983; Andersen et al. 2008; Liotta 2010; Zhu and Wu 2015). Long-term exposure of formaldehyde can also give rise to the nasal tumours and irritation of few sensory organs such as mucous membranes of eyes, skin, and also in respiratory system (Collins et al. 2001; Yu and Crump 1998). As reported, formaldehyde may cause sensory irritation under certain occupational and environmental conditions (Wolkoff 2013).

5.3.2 Chlorinated Aromatic Compounds

The group of chlorinated volatile compounds include 1, 4-dichlorobenzene (p-dichlorobenzene), and 1, 2-dichlorobenzene (o-dichlorobenzene). These chlorinated VOCs are broadly utilised in industries and in commonly used products like chemical dye stuffs, odour-masking agents, and pesticides (Godish et al. 2015). 1, 4-Dichlorobenzene, which is the most significant volatile compound out of three dichlorobenzenes, generally is colorless to white solid compound at the normal conditions having a strong pungent odour. Moreover, it is poorly soluble in water and comparatively has high volatility. As consequence, it is chiefly applied as a deodorant in indoor spaces of residential and office areas such as urinal deodorizers, room/restroom deodorizers, and also as an insecticide fumigant for the moth control and toilet bowl blocks (WHO 2010).

5.3.3 Non-chlorinated Aromatic Compounds

The non-chlorinated VOCs arise usually from outdoor air environment and petroleum-related products that are utilised in the indoor, including adhesives, coatings, paints, enamels, household cleaners, fuels, varnishes, glues, lacquers, gasoline, lubricants, and building materials like wallpaper and carpets (Rosch et al. 2014). Aromatic non-chlorinated compounds are not only toxic but also carcinogenic (Kim and Shim 2010). Compounds such as ethyl benzene, benzene, and toluene are used in several formulations and products such as paint, petrochemicals, detergents, and medicine (Özçelik et al. 2009). Materials that are consumed in construction, decorating and remodelling contributes enormously to the concentrations of benzene in the indoor environment. Benzene is also available in caulking, paints, flooring adhesives, particleboard furniture, fiberglass, wood panelling, plywood, and paint remover (WHO 2010).

5.3.4 Chlorinated Aliphatic Compounds

Generally, chlorinated VOCs include chloroform, trichloroethylene, carbon tetrachloride, tetrachloroethylene (also referred as perchloroethylene), and dichloromethane, are primarily applied as solvents, that are commonly used ingredients of oil and fat degreasers, water repellents, shoe polishes, paint remover, epoxy paint sprays, and dry cleaners in laundries (WHO 2010). Other halogenated VOCs consist of dichloroethane, chlorobenzene, tetrachloroethane, trichloroethane, tetrachloroethylene, and trichloroethylene (Giraudon et al. 2008). In general, such chemical compounds are applied in processing of paints, adhesives, and chemical extracting agents, polymer syntheses, manufacturing of drugs, as cleaning agents, and as solvents in chemical reactions (Aranzabal et al. 2014). Through the drinking water humans can be susceptible to chlorinated aliphatic VOCs by, inhalation, drinking water and adsorption during swimming (Huang et al. 2014).

5.4 Nature and Types of VOCs

By nature, VOCs are a general category of those organic compounds having over 10.3 Pa of Reid vapor pressure at normal temperature (293.15 K) and pressure (101.325 kPa). Thus VOCs consist of a wide array of chemicals that contain carbon and can easily evaporate at room temperature (Olsen and Nielsen 2001; Li et al. 2009; Ojala et al. 2011).

The VOCs are the most common pollutants that are present in the air of building sectors and are generally defined as the organic compounds having a boiling point that possibly ranges from 50 to 260 °C (Sarigiannis et al. 2011). The moderately low

Table 5.2 WHO classification system for organic indoor pollutants

Group	Boiling point (°C)	Example compounds
VVOC	<0 to 50–100	Propane, Butane, methyl chloride,
VOC	50–100 to 240–260	Formaldehyde, ethanol (ethyl alcohol), acetone, hexanal, toluene, d-Limonene, 2-propanol (isopropyl alcohol)
SVOC	240–260 to 380–400	Plasticizers (phthalates), pesticides (DDT, chlordane, fire retardants (PCBs, PBB))

Adapted after: WHO (1989)

boiling point of VOCs allows evaporation or sublimation from the solid phase into the gaseous form. This peculiarity of VOCs is at times described as 'off-gassing'. The variety of VOCs that are identified in the air environment of building sectors is usually greater than that of outdoor air (Rumchev et al. 2016). VOCs in certain instances are classified by the ease of their emission (EPA 2017). For example, WHO categorizes indoor organic pollutants as:

- Very volatile organic compounds (VVOCs)
- Volatile organic compounds (VOCs)
- Semi-volatile organic compounds (SVOCs).

The documented organic pollutants are categorised into subgroups in several ways. Whereas, WHO classified the organic pollutants of indoor environments (Table 5.2) on the basis of their respective boiling points.

Typically the nature and type of VOCs in the indoor of building sectors is based on source of emission. Some examples of VOCs are aromatics, alkanes, alcohols, aldehydes, olefins, ketones, esters, ethers, paraffins, halogenated hydrocarbons, nitrogen and sulfur containing compounds (Khan and Ghoshal 2000; Carpentier et al. 2002; Miranda-Trevino and Coles 2003; Ozturk and Yilmaz 2006; Soylu et al. 2010; Doggali et al. 2012; Yosefi et al. 2015).

5.5 Health Effects

Undoubtedly, the sprays containing aerosols that are utilized frequently in the building sectors contain an appreciable amount of VOCs that are chemicals of organic origin that can vaporize readily at room temperature and have significant contribution to air pollution in indoor air. Furthermore, they can have an immediate health effect on humans as they spend enormous time indoors and thereby shorten the expectation of life as consequences of cardiovascular and respiratory ailments and from weak functions of lung (Dales et al. 2008; Pharmacy Magazine 2019) (Table 5.3). The Indoor air contamination resulting from VOC might affect well-being of human at the domestic level, in open areas and also in the commercial buildings (Rumchev et al. 2007).

Table 5.3 Threshold limit values (TLV) of indoor regulated VOCs and their TLV basis (ACGIH 2017)

Designated indoor VOCs	TLV (ppm)	TLV basis
Carbon tetrachloride	5	Liver damage
Dichloromethane	50	COHb; CNS impairment
Ethyl benzene	20	URT irritation; cochlear damage; kidney damage (nephropathy)
Styrene	20	CNS impairment; peripheral neuropathy; URT irritation
Chloroform	10	CNS impairment; liver and embryo/foetal damage;
1,2-Dichlorobenzene	25	Eye irritation; liver and URT damage
1,4-Dichlorobenzene	10	Kidney damage; eye irritation
Benzene	0.5	Leukaemia
Toluene	20	Visual impairment; pregnancy loss; female reproductive
H-CHO	0.1	Eye and URT irritation; URT cancer
Xylenes	100	CNS impairment; URT and eye irritation
Tetrachloroethylene	25	CNS impairment
Trichloroethylene	10	Cognitive decrements; CNS impairment; renal toxicity

CNS Central Nervous System; *H-CHO* Formaldehyde; *URT* Upper Respiratory Tract; *COHb* carboxyhemoglobin

According to the documentation of the US EPA, other health effects comprise irritation in the eyes, throat, and nose; serious harm to kidneys, CNS, and liver; loss of coordination, headache, and nausea. Whereas, some identified Volatile organic compounds can cause cancers; also some of the VOCs are alleged or assumed to cause cancer in humans. Generally, the principal symptoms or initial indications correlated with the vulnerability of VOCs may include irritation in the conjunctiva, skin allergies, throat and nose discomfort, skin allergic reaction, shortness of breath or dyspnea, dizziness, nausea, emesis or vomiting, epistaxis or nose bleeding, and decline in levels of cholinesterase serum (Kim et al. 2010; EPA 2019). The capability of organic chemicals to trigger adverse impacts on health differs substantially from the ones that are considerably toxic, to those with unknown health impacts (EPA 2019). Environment that is substantially built in the building air plays an essential part in general safety since of both the sum of time we spend inside and capacity of buildings to emphatically or contrarily affect our wellbeing (Allen et al. 2016).

VOCs categorically represent one of the most important categories of harmful pollutants, which influence quality of air and also human well-being and health in building environments. The VOCs depicted as one of the foremost critical groups of pollutants, impacting the standard conditions of air and health of humans and well-being in the indoor situations. Nevertheless, recent shreds of evidence propose that a considerable amount of given VOCs can lead to unfavourable impacts on human health and even can cause cancer (Rumchev et al. 2007). Wolkoff et al. (1997) and Weschler (2004) conducted various research studies and documented that the VOCs respond to oxidants for example ozone to generate compounds that are reactive and can likely cause unfavourable sensory effects.

VOCs can be absorbed through lungs and are then transferred through bloodstream to every organ of the body and hence have the potential to influence various other organs and tissues as either parent compound or bioactivated metabolite (Boeglin et al. 2006). A few VOCs, for instance limonene and styrene, can readily react with ozone or with oxides of nitrogen to generate secondary aerosols and new oxidation products, which can cause complaints regarding sensory irritation (Wolkoff et al. 2006).

Formaldehyde is labelled as one of the foremost common VOCs known and is additionally demonstrated as a very familiar respiratory irritant. It gets absorbed and deposited in the upper respiratory tract while reacting with the biological macromolecules (Maroni et al. 1995). In addition to the sensory and respiratory irritation caused by the chemicals present in the building sectors, exposure to volatile compounds can also influence the nervous system, which results in a lack of focus and concentration. Numerous mechanisms of the pathways could be involved in human health impacts, a few of which are the immune system's response, sensitization, impacts on the heart rate, irritation in the sensory organs, and also psychological and social stress reactions. Even though there are no visible threat for leukemia and/or lymphoma at standard concentrations of VOCs such as benzene, and HCHO in the indoor air environment (WHO 2010), these are widely presumed or known to be carcinogenic and/or of toxic quality for kidney, upper respiratory tract liver, and CNS (Tsai et al. 2018).

In general terms, the adverse health impacts incited by groups of VOCs can be categorised as carcinogenic or non-carcinogenic.

5.5.1 Carcinogenic Effects

Primarily, the carcinogenic impacts of VOCs are on the lung, liver, blood (leukemia and non-Hodgkin lymphoma), kidney, and biliary tract cancer. As an illustration, the International Agency for Research on Cancer (IARC) has specifically classified benzene as a carcinogen for humans under group 1, whereas other VOCs such as tetrachloroethylene and ethylbenzene are reviewed likely to be carcinogens for humans and are labelled under Group 2A or 2B (De Gennaro et al. 2013). Several identified VOCs are also presumed to act as carcinogens for humans such as vinyl chloride. Other groups of VOCs are considered as carcinogens for animals and may also function as human carcinogens such as methylene chloride, chloroform, trichloroethylene, dichlorobenzene, tetrachloroethylene. Another group of VOCs are mutagens for example a-pinene and are also called carcinogens for weak animals such as limonene (Burn et al. 1993).

5.5.2 *Non-carcinogenic Effects*

In the buildings, health-related issues usually result from the complaints of odour. Most VOCs with unpleasant and smelly properties have odour limits just below the levels that are considered toxic (Loftness et al. 2007). Nonetheless, it is mandatory to prioritize protection from odours for a better quality of life, and efficiency reasons, whereas, several VOCs particularly with respect to styrene and formaldehyde, have authenticated guidelines which are based on the odour (WHO 2000). The recent examinations for VOCs levels that were measured in residential areas and workplaces have demonstrated that average levels of VOCs are unlikely to surpass the odour thresholds (Hodgson and Levin 2003; Rumchev et al. 2007).

5.6 Sick Building Syndrome (SBS)

The ailments that have been ascribed to the habitation of building sectors have been demonstrated since the 1970s (Lyles et al. 1991). In the mid-1980s, the SBS was first observed and documented by the researchers (Finnegan et al. 1984; Hicks 1984) and came to attention after it was sensationalized by the media (Carey et al. 1985). Several VOCs may be related to the diversity of chronic health effects and symptoms such as reactions caused by allergies and asthma. These symptoms are frequently observed in newly constructed buildings or in buildings that are under construction or renovations (Berglund et al. 1984). Moreover, numerous VOC based research on office employees have clearly reported a sound relationship between illnesses like irritation in the mucous layer and CNS symptoms, and overall exposure of humans to VOCs. These noticeable symptoms and signs are very familiar to symptoms that are associated with the SBS (Mølhave et al. 1986; Hodgson et al. 1991).

5.7 Exposure of VOCs to Humans

Most often, the constant exposure of humans to VOCs in the air environment of building sectors is in turn because of the inhalation of VOCs floating in the air. The standard of the air that we breathe can exhibit considerable effect on human sense of comfort and health (Loftness et al. 2007). Subsequently, the discharge levels of VOCs from the building materials have relevance in the connection to their adverse health impacts, whereas various European nations have established nationwide prohibitions for VOC discharge levels of VOCs from several materials of buildings (ECA 2013). Besides, these pollutants can be emitted and absorbed by the walls of the building environment. The commonly used compounds to which occupants regularly get in touch also include VOCs, a few of which are regarded as perilous for human well-being. The time duration of the exposure represents an essential factor despite the fact that the pollutant concentration is moderately low (Edwards et al. 2001).

Standards have been set for the estimation of emissions or discharge from building materials within indoor air (CEN 2017). Other estimations reported that the occupants with smokers have considerably greater degree of VOCs that includes toluene, ethylbenzene, styrene, and benzene that are related to environmental smoke of tobacco (Wallace and Pellizzari 1987; Hajimiragha et al. 1989).

According to the EPA, "among the immediate symptoms that some people have experienced soon after exposure to some organics include":

- Respiratory tract, eye irritation
- Headaches
- Dizziness
- Memory impairment and visual disorders.

The aromatic group of VOCs can result in confusion, weakness, nausea, in-appetence and loss of memory, tiredness, and loss of vision even at low-level exposure. At high level exposure inhalation of volatile compounds can result in dizziness, unconsciousness, and even death (Kim and Hong 2002; Suib 2013).

5.7.1 Steps to Reduce Exposure by EPA (2017)

- To increase proper ventilation when using products that emits VOCs.
- Meet or exceed any label precautions.
- Do not store opened containers of unused paints and similar materials within the school.
- Formaldehyde, one of the best known VOC, is one of the few IAPs that can be readily measured.
- Identify, and remove the source, if possible.
- If not possible to remove, reduce exposure by using a sealant on all exposed surfaces of panelling and other furnishings.
- Use integrated pest management techniques to reduce the need for pesticides.
- Use household products according to manufacturer's directions.
- Provide plenty of fresh air when using these products.
- Throw away little-used or unused containers safely; buy in quantities that you will use soon.
- Keep out of reach of pets and children.
- Never mix products used for household care unless directed on the label.

5.8 Indoor Exposure Guidelines

The United Nations' Sustainable Development Goals (UN SDGs) have brought up again specific focus on environmental issues such as pollutant emissions and climate change (EPA 2015). For an instance, SDG number 12 that focus on responsible production and consumption, and targets significant decline in emission of various chemicals to air, water, and soil in order to reduce or decline their detrimental effects on environment and human health (EPA 2015; Le Blanc 2015). A number of countries have established the labelling and guideline scheme for the construction and building products that targets the emission control from such sources (Loftness et al. 2007). In 2002, Canada established "Guidelines for Volatile Organic Compounds in Consumer Products" that targets VOC content limits specifically for 23 consumer product categories. Other indoor environment related scheme, for example Indoor Climate Labelling Scheme in Denmark, were originally introduced to reduce the emissions from the building material and are built not on the legislation but on market forces. Various countries have now building rules and regulations at minimum requirements for ventilation that can eventually minimize the exposure to a variety of IAs, including VOCs (Witterseh 2002). It is easily understood, inhalation is the primary passage of hazardous VOC exposure in the building air environment, and on the contrary absorption through the skin and intake from consuming water, beverages, and food are noticeable. As in IAQ guidelines and standards, about 13 VOCs were designated as IAPs (Tsai 2019).

According to the, 7th Environmental Action Programme (EAP) for EU, "Inventorying chemicals emissions and characterising their impacts can help measure progress towards the objectives for environmental quality and public health as well as towards a non-toxic environment" (EU 2013; Persson et al. 2019). U.S. Green Building Council (USBGA) have credited numerous designs that not only aims to have better environmental performance and energy efficiency but also includes the rules and guidelines for improving filtration and, ventilation, controlling chemical and pollutant sources, to utilise materials with low emissions, improving lighting and thermal conditions, and to provide daylight sights to inhabitants of the building (USGBC 2014).

As a consequences of the health impact of VOCs, the international organizations like World Health Organisation (WHO), administrative agencies such as the Word related Security and Wellbeing Organization within the Joined together States; and cosmopolitan association of scientists and professionals in particular the American Industrial Hygiene Association have prescribed instructions that aims to regulate the limiting values of human exposure to specific airborne pollutants that are present within the building. For an instance, the Ministry of Health, Labor and Welfare (MHLW) in Japan has chosen eight targets VOCs which includes HCHO, acetaldehyde, ethylbenzene, p-dichlorobenzene, styrene, toluene, tetradecane, and xylenes as the components of IAQ guidelines (Noguchi et al. 2016). According to the IAQ guidelines and standards systematic compilation, there has been no documented or recorded census given by several official organizations based on the IAQ guidelines, the components of IAPs, and their suggested and the restricting values (Abdul-Wahab et al. 2015; Luengas et al. 2015).

It can be assumed from the fact that various guidelines and standards solely focus on the reduction and prevention of irritations that are caused by several IAPs whereas other guidelines are solely based on health related concerns (Luengas et al. 2015). Generally, the guidelines provided by WHO is considered as valid and authorized (WHO 2010). The non-governmental organization (NGO), and scientific organization such as ACGIH that is committed to promoting health and safety within offices and other workplaces. According to ACGIH, the TLV limits are based on health and thus are not designed to be utilized as guidelines, because the standards have to be set by reviewing the authorized literature from numerous disciplines of science, including epidemiology, including occupational medicine, industrial hygiene, and toxicology (ACGIH 2017).

5.9 Conclusion and Solution

Taking IAQ into consideration, all those organic compounds that have the property to evaporate under normal temperature and pressure are referred to as VOCs and it is mandatory to assess the potential effect of IAQ on human well-being. Though IAQ has been persuading the environment related issues from past two decades, a wide gap of knowledge still exists with respect to unfavourable and antagonistic impacts on human health for various known indoor air components. Though it is commonly known that the main route of exposure to VOC is inhalation, in addition to this, the intake of food, water and the beverage consumption is readily noticeable. When utilizing the products that emit VOCs, the ventilation in the residential areas or within the office should be enhanced and the opened containers of unutilised paints should be discarded and other materials that emit VOCs. Formaldehyde is one of widely known VOC pollutants, present in the indoor environments of building sector that can be readily measured. If it is the potential pollutant indoors, the identification and removal of the source is essential. Formaldehyde which is also consumed in the permanent mattress ticking and press fabric has adverse effects on human health; the sensitive individuals must avoid using such products.

Any sort of activity in declining the emission of aerosols in the indoor sector can have a rapid impact on IAQ and also on human well-being since the majority of the aerosol sprays are used in the indoor building sectors. Therefore, to prevent human health and well-being, it is mandatory to estimate the average concentrations of VOCs from the building environment to which people are exposed. Subsequently, the appropriate and reliable standards have been difficult to set up, whereas the indoor air regulation and building materials have been agitated with inaccurate information and confusions. It is an essential goal to reduce the concentrations of VOCs inside as well as outside environmental and health. Howsoever, it is compulsory to take note that some VOCs of concern are present indoors as well as outdoors that don't have an impact on photochemical oxidation and thus are not governed by EPA. It is mandatory to create an appropriate distinction and understand the strategies while using it in improvement of IAQ.

References

Abdul-Wahab SA, En SCF, Elkamel A, Ahmadi L, Yetilmezsoy K (2015) A review of standards and guidelines set by international bodies for the parameters of indoor air quality. Atmos Pollut Res 6(5):751–767

ACGIH (American Conference of Governmental Industrial Hygienists) (2017) TLVs and BEIs: based on the documentation of the threshold limit values for chemical substances and physical agents. Cincinnati, USA

Adamkiewicz G, Zota AR, Fabian MP, Chahine T, Julien R, Spengler JD, Levy JI (2011) Moving environmental justice indoors: understanding structural influences on residential exposure patterns in low-income communities. Am J Public Health 101(S1):S238–S245

Allen JG, MacNaughton P, Satish U, Santhanam S, Vallarino J, Spengler JD (2016) Associations of cognitive function scores with carbon dioxide, ventilation, and volatile organic compound exposures in office workers: a controlled exposure study of green and conventional office environments. Environ Health Perspect 124:805. https://doi.org/10.1289/ehp.1510037

Andersen ME, Clewell HJ, Bermudez E, Willson GA, Thomas RS (2008) Genomic signatures and dose-dependent transitions in nasal epithelial responses to inhaled formaldehyde in the rat. Toxicol Sci 105(2):368–383

Aranzabal A, Pereda-Ayo B, Gonzalez-Marcos MP, Gonzalez-Marcos JA, Lopez-Fonseca R, Gonzalez-Velasco JR (2014) State of the art in catalytic oxidation of chlorinated volatile organic compounds. Chem Pap 68(9):1169–1186

Barro R, Regueiro J, Llompart M, Garcia-Jares C (2009) Analysis of industrial contaminants in indoor air: part 1. Volatile organic compounds, carbonyl compounds, polycyclic aromatic hydrocarbons and polychlorinated biphenyls. J Chromatogr 1216(3):540–566

Bartzis J, Wolkoff P, Stranger M, Efthimiou G, Tolis EI, Maes F, Fernandes O (2015) On organic emissions testing from indoor consumer products use. J Hazard Mater 285:37-45

Baur X, Budnik LT, Zhao Z, Bratveit M, Djurhuus R, Verschoor L, Jepsen JR (2015) Health risks in international container and bulk cargo transport due to volatile toxic compounds. J Occup Med Toxicol 10(1):1–18

Berglund B, Lindvall T,Sundell J (1984) Indoor air 1. Recent advances in the health science and technology

Boeglin ML, Wessels D, Henschel D (2006) An investigation of the relationship between air emissions of volatile organic compounds and the incidence of cancer in Indiana counties. Environ Res 100(2):242–254

Brandys RC, Brandys GM (2010) International indoor air quality standards and guidelines for over 2000 chemicals and biological substances. OEHCS Publications, Las Vegas, NV

Brinke JT, Selvin S, Hodgson AT, Fisk WJ, Mendell MJ, Koshland CP, Daisey JM (1998) Development of new volatile organic compound (VOC) exposure metrics and their relationship to "sick building syndrome" symptoms. Indoor Air 8(3):140–152

Brown SK (1999) Occurrence of volatile organic compounds in indoor air. In: Salthammer T (ed) Organic indoor air pollutants: occurrence measurement evaluation. Wiley-VCH, Weinheim, Germany, pp 170–184

Bruno P, Caselli M, De Gennaro G, Iacobellis S, Tutino M (2008) Abstract. Indoor Air 18(3):250–256

Burn J, Henk J, Bloemen T (eds) (1993) Chemistry and analysis of volatile organic compounds in the environment. Blackie Academic & Professional, London
Carey J, Hager M, King P (1985) Beware 'Sick-Building Syndrome'. The deadliest pollutants of all may be the ones you breathe at home or at work. Newsweek 105(1):58–60
Carpentier J, Lamonier JF, Siffert S, Zhilinskaya EA, Aboukaıs A (2002) Characterisation of Mg/Al hydrotalcite with interlayer palladium complex for catalytic oxidation of toluene. Appl Catal A General 234(1–2):91–101
CEN (2017) EN 16516 construction products: assessment of release of dangerous substances—determination of emissions into indoor air. European Committee for Standardization, Brussels
Chan SH, Van Hee VC, Bergen S, Szpiro AA, DeRoo LA, London SJ, Sandler DP (2015) Long-term air pollution exposure and blood pressure in the sister study. Environ Health Perspect 123(10):951–958
Chan WR, Parthasarathy S, Fisk WJ, McKone TE (2016) Estimated effect of ventilation and filtration on chronic health risks in US offices, schools, and retail stores. Indoor Air 26(2):331–343. https://doi.org/10.1111/ina.1218
Chang C, Gershwin ME (2004) Indoor air quality and human health. Clin Rev Allergy Immunol 27(3):219–239
Chithra VS, Nagendra SS (2012) Indoor air quality investigations in a naturally ventilated school building located close to an urban roadway in Chennai, India. Build Environ 54:159–167. https://doi.org/10.1016/j.buildenv.2012.01.016
Collins JJ, Ness R, Tyl RW, Krivanek N, Esmen NA, Hall TA (2001) A review of adverse pregnancy outcomes and formaldehyde exposure in human and animal studies. Regul Toxicol Pharmacol 34(1):17–34
Dales R, Liu L, Wheeler AJ, Gilbert NL (2008) Quality of indoor residential air and health. CMAJ 179(2):147–152
Dales RE, Cakmak S, Leech J, Liu L (2013) The association between personal care products and lung function. Ann Epidemiol 23(2):49–53
De Gennaro G, Farella G, Marzocca A, Mazzone A, Tutino M (2013) Indoor and outdoor monitoring of volatile organic compounds in school buildings: indicators based on health risk assessment to single out critical issues. Int J Environ Res Public Health 10(12):6273–6291
Doggali P, Teraoka Y, Mungse P, Shah IK, Rayalu S, Labhsetwar N (2012) Combustion of volatile organic compounds over Cu–Mn based mixed oxide type catalysts supported on mesoporous Al_2O_3, TiO_2 and ZrO_2. J Mol Catal A Chem 358:23–30
ECA (2013) Report No 29. EUR 26168 EN. Harmonisation framework for health based evaluation of indoor emissions from construction products in the European Union using the EU-LCI concept. ECA, Luxembourg
Edwards RD, Jurvelin J, Saarela K, Jantunen M (2001) VOC concentrations measured in personal samples and residential indoor, outdoor and workplace microenvironments in EXPOLIS-Helsinki, Finland. Atmos Environ 35(27):4531–4543
EPA (2016) Volatile organic compounds' impact on indoor air quality. EPA
EPA (2017) Research impacts report 2017. https://www.epa.gov/research/epa-research-impacts-report-2017
EPA US (2015) why indoor air quality is important to schools. Overviews and factsheets. US EPA. https://www.epa.gov/iaq-schools/why-indoor-air-quality-important-schools
EPA US (2019) Volatile organic compounds' impact on indoor air quality. Recuperado de. https://www.epa.gov/indoor-air-quality-iaq/volatile-organiccompounds-impact-indoor-airquality#intro
EU (2004) Occurrence of 13 volatile organic compounds in foods from the Canadian total diet study. Off J Eur Union 143:87–96
EU (2013) Decision No 1386/2013/EU of the European Parliament and of the Council of 20 November 2013 on a General Union Environment Action Programme to 2020 "Living Well, within the Limits of our Planet"

Federal Environment Agency UBA (2008) Guidelines for indoor air hygiene in school buildings. In: Moriske HJ, Szewzyk R (eds) KOMAG mbH, Berlin, Germany. http://www.gesundheitsamtbw.de/MLS/Documents/Guidelines_Air_Quality_Schools_2008.pdf

Finnegan MJ, Pickering CA, Burge PS (1984) The sick building syndrome: prevalence studies. Br Med J (Clin Res Ed) 289(6458):1573–1575

Giraudon JM, Nguyen TB, Leclercq G, Siffert S, Lamonier JF, Aboukaïs A, Su BL (2008) Chlorobenzene total oxidation over palladium supported on ZrO2, TiO2 nanostructured supports. Catal Today 137(2–4):379–384

Godish T, Davis WT, Fu JS (2015) Air quality, 5th edn. CRC Press, Boca Raton, FL

Godwin C, Batterman S (2007) Indoor air quality in Michigan schools. Indoor Air 17:109–121

Hajimiragha H, Ewers U, Brockhaus A, Boettger A (1989) Levels of benzene and other volatile aromatic compounds in the blood of non-smokers and smokers. Int Arch Occup Environ Health 61(8):513–518

Heinsohn RJ, Cimbala JM (2003) Indoor air quality engineering: environmental health and control of indoor pollutants. CRC Press, Boca Raton, FL

Hicks JB (1984) Tight building syndrome: when work makes you sick. Occupational health & safety (Waco, Tex) 51–56

Hodgson MJ, Frohliger J, Permar E, Tidwell C, Traven ND, Olenchock SA, Karpf M (1991) Symptoms and microenvironmental measures in nonproblem buildings. J Occup Med 33(4):527–533

Hodgson AT, Levin H (2003) Volatile organic compounds in indoor air: a review of concentrations measured in North America since 1990

Huang B, Lei C, Wei C, Zeng G (2014) Chlorinated volatile organic compounds (Cl-VOCs) in environment—sources, potential human health impacts, and current remediation technologies. Environ Int 71:118–138

Huang S, Yang D, Tang Q, Deng W, Zhang L, Jia Z, Guo L (2020) Pt-loaded ellipsoidalnanozeolite as an active catalyst for toluene catalytic combustion. Microporous Mesoporous Mater 305:

Huang Y, Ho SSH, Lu Y, Niu R, Xu L, Cao J, Lee S (2016) Removal of indoor volatile organic compounds via photocatalytic oxidation: a short review, and prospect. Molecules 21(1):56

Hyttinen M, Masalin-Weijo M, Kalliokoski P, Pasanen P (2010) Comparison of VOC emissions between air-dried and heat-treated Norway spruce (*Picea abies*), Scots pine (*Pinus sylvestris*) and European aspen (*Populus tremula*) wood. Atmos Environ 44(38):5028–5033. https://doi.org/10.1016/j.atmosenv.2010.07.018

IARC (International Agency for Research on Cancer) (2018) IARC monographs on the evaluation of carcinogenic risks to humans. IARC Web. https://monographs.iarc.fr/ENG

Jones AP (1999) Indoor air quality and health. Atmos Environ 33(28):4535–4564. https://doi.org/10.1016/S1352-2310(99)00272-1

Kamal MS, Razzak SA, Hossain MM (2016) Catalytic oxidation of volatile organic compounds (VOCs)–a review. Atmos Environ 140:117–134

Khan FI, Ghoshal AK (2000) Removal of volatile organic compounds from polluted air. J Loss Prev Process Ind 13(6):527–545

Kim SB, Hong SC (2002) Kinetic study for photocatalytic degradation of volatile organic compounds in air using thin film TiO_2 photocatalyst. Appl Catal B Environ 35(4):305–315

Kim JJ, Jung SK, Kim JT (2010) Wireless monitoring of indoor air quality by a sensor network. Indoor Built Environ 19(1):145–150. https://doi.org/10.1177/1420326X09358034

Kim SC, Shim WG (2010) Catalytic combustion of VOCs over a series of manganese oxide catalysts. Appl Catal B Environ 98(3–4):180–185

Klett C, Duten X, Tieng S, Touchard S, Jestin P, Hassouni K, Vega-González A (2014) Acetaldehyde removal using an atmospheric non-thermal plasma combined with a packed bed. Role of the adsorption process. J Hazard Mater 279:356–364

Lai ACK, Mui KW, Wong LT, Law LY (2009) An evaluation model for indoor environmental quality (IEQ) acceptance in residential buildings. Energy Build 41(9):930–936

Le Blanc D (2015) Towards integration at last? The sustainable development goals as a network of targets. Sustain Dev 23(3):176–187

Li WB, Wang JX, Gong H (2009) Catalytic combustion of VOCs on non-noble metal catalysts. Catal Today 148(1–2):81–87

Li JW, Pan KL, Yu SJ, Yan SY, Chang MB (2014) Removal of formaldehyde over Mnx1-xO2 catalysts: thermal catalytic oxidation versus ozone catalytic oxidation. J Environ Sci 26(12):2546–2553

Lim SK, Shin HS, Yoon KS, Kwack SJ, Um YM, Hyeon JH, Lee BM (2014) Risk assessment of volatile organic compounds benzene, toluene, ethyl benzene, and xylene (BTEX) in consumer products. J Toxicol Environ Health Part A 77(22–24):1502–1521

Lin CC, Yu KP, Zhao P, Lee GWM (2009) Evaluation of impact factors on VOC emissions and concentrations from wooden flooring based on chamber tests. Build Environ 44(3):525–533. https://doi.org/10.1016/j.buildenv.2008.04.015

Liotta LF (2010) Catalytic oxidation of volatile organic compounds on supported noble metals. Appl Catal B Environ 100(3–4):403–412

Loftness V, Hakkinen B, Adan O, Nevalainen A (2007) Elements that contribute to healthy building design. Environ Health Perspect 115(6):965–970

Luengas A, Barona A, Hort C, Gallastegui G, Platel V, Elias A (2015) A review of indoor air treatment technologies. Rev Environ Sci Biotechnol 14(3):499–522

Lyles WB, Greve KW, Bauer RM, Ware MR, Schramke CJ, Crouch J, Hicks A (1991) Sick building syndrome. South Medi J 84(1):65–71

Main DM, Hogan TJ (1983) Health effects of low-level exposure to formaldehyde. J Occup Environ Med 25:896–900

Maji CS, Ashok N (2003) Pollution preventing anionic lithographic inks. J Hazard Mater 105:103–119

Maroni M, Seifert B, Lindvall T (1995) Indoor air quality. A comprehensive reference book. Elsevier

Meciarova L, Vilcekova S, Burdova EK, Kiselak J (2017) Factors affecting the total volatile organic compound (TVOC) concentrations in Slovak households. Int J Environ Res Public Health 14(12):1443

Miranda-Trevino JC, Coles CA (2003) Kaolinite properties, structure and influence of metal retention on pH. Appl Clay Sci 23(1–4):133–139

Mitsui T, Tsutsui K, Matsui T, Kikuchi R, Eguchi K (2008) Catalytic abatement of acetaldehyde over oxide-supported precious metal catalysts. Appl Catal B Environ 78(1–2):158–165

Mølhave L, Bach B, Pedersen OF (1986) Human reactions to low concentrations of volatile organic compounds. Environ Int 12(1–4):167–175

Molina MJ, Molina LT (2004) Megacities and atmospheric pollution. J Air Waste Manag Assoc 54(6):644–680

Nazaroff WW, Weschler CJ (2004) Cleaning products and air fresheners: exposure to primary and secondary air pollutants. Atmos Environ 38:2841–2865

Noguchi M, Mizukoshi A, Yanagisawa Y, Yamasaki A (2016) Measurements of volatile organic compounds in a newly built day care center. Int J Environ Res Public Health 13(7):736–749

Nourian A, Abba MK, Nasr GG (2020) Measurements and analysis of Non-Methane VOC (NMVOC) emissions from major domestic aerosol sprays at source. Environ Int 146:

Ojala S, Pitkäaho S, Laitinen T, Koivikko NN, Brahmi R,Gaálová J, Keiski RL (2011) Catalysis in VOC abatement. Top Catal 54 (16–18):1224

Olsen E, Nielsen F (2001) Predicting vapour pressures of organic compounds from their chemical structure for classification according to the VOC. Directive and risk assessment in general. Molecules 6(4):370–389

Özçelik Z, Soylu GSP, Boz İ (2009) Catalytic combustion of toluene over Mn, Fe and Co-exchanged clinoptilolite support. ChemEng J 155(1–2):94–100

Ozturk B, Yilmaz D (2006) Absorptive removal of volatile organic compounds from flue gas streams. Process Saf Environ Prot 84(5):391–398

Persson L, Arvidsson R, Berglund M, Cederberg C, Finnveden G, Palm V, Wood R (2019) Indicators for national consumption-based accounting of chemicals. J Clean Prod 215:1–12
Pharmacy Magazine (2019) Theair that we breathe. A public health crisis. https://www.pharmacymagazine.co.uk/the-air-that-we-breathe-a-public-health-crisis
Poljansek K, Marín Ferrer M, De Groeve T, Clark I (2017) Science for disaster risk management knowing better and losing less. ETH Zurich
Raffaele E, Behn C, Dix B (2012) On the formaldehyde release of wood particles. Eur J Wood Wood Prod 70(6):911–912. https://doi.org/10.1007/s00107-012-0625-8
Rosch C, Kohajda T, Roder S, von Bergen M, Schlink U (2014) Relationship between sources and patterns of VOCs in indoor air. Atmos Pollut Res 5(1):129–137
Rumchev K, Brown H, Spickett J (2007) Volatile organic compounds: do they present a risk to our health? Rev Environ Health 22(1):39
Rumchev K, Win T, Bertolatti D, Dhaliwal S (2016) Prevalence of respiratory symptoms among children in rural Myanmar-disease burden assessment attributable to household biomass smoke. Indoor Built Environ 25(5):728–736
Santos VP, Pereira MFR, Órfão JJM, Figueiredo JL (2011) Mixture effects during the oxidation of toluene, ethyl acetate and ethanol over a cryptomelane catalyst. J Hazard Mater 185(2–3):1236–1240
Sarigiannis DA, Karakitsios SP, Gotti A, Liakos IL, Katsoyiannis A (2011) Exposure to major volatile organic compounds and carbonyls in European indoor environments and associated health risk. Environ Int 37(4):743–765
Schlink U, Rehwagen M, Damm M, Richter M, Borte M, Herbarth O (2004) Seasonal cycle of indoor-VOCs: comparison of apartments and cities. Atmos Environ 38(8):1181–1190
Schraufnagel DE, Balmes JR, Cowl CT, de Matteis S, Jung SH, Mortimer K, Perez-Padilla R, Rice MB, Riojas-Rodriguez H, Sood A, Thurston GD, To T, Vanker A, Wuebbles DJ (2019) Air pollution and noncommunicable diseases: a review by the forum of international respiratory societies' environmental committee, part 1: the damaging effects of air pollution. Chest 155:409–416
Sclink U, Thiem A, Kohajda T, Richter M, Strebel K (2010) Quantile regression of indoor air concentrations of volatile organic compounds (VOC). Sci Total Environ 408(18):3840–3851
Skulberg KR, Nyrud AQ, Goffeng LO (2019) Health and exposure to VOCs from pinewood in indoor environments. Front Built Environ 5:107
Soylu GSP, Özçelik Z, Boz I (2010) Total oxidation of toluene over metal oxide supported on a natural clinoptilolite-type zeolite. Chem Eng J 162(1):380–387
Srivastava A, Joseph AE, Wachasundar SD (2004) Qualitative detection of volatile organic compounds in outdoor and indoor air. Environ Monit Assess 96(1):263–271
Suib SL (2013) New and future developments in catalysis. Catalysis for remediation and environmental concerns. Newnes
Tsai WT (2019) An overview of health hazards of volatile organic compounds regulated as indoor air pollutants. Rev Environ Health 34(1):81–89
Tsai JH, Yao YC, Huang PH, Chiang HL (2018) Fuel economy and volatile organic compound exhaust emission for motorcycles with various running mileages. Aerosol Air Qual Res 18(12):3056–3067
US Consumer Product Safety Commission (1995) Indoor air pollution: an introduction for health professionals. American Lung Association
USGBC (US Green Building Council) (2014) LEED v4 user guide. USGBC, Washington DC. https://www.usgbc.org/resources/leed-v4-user-guide
Wallace LA, Pellizzari ED (1987) Personal air exposures and breath concentrations of benzene and other volatile hydrocarbons for smokers and nonsmokers. Toxicol Lett 35(1):113–116
Wang S, Ang HM, Tade MO (2007) Volatile organic compounds in indoor environment and photo catalytic oxidation: state of the art. Environ Int 33(5):694–705. https://doi.org/10.1016/j.envint.2007.02.011

Weschler CJ (2000) Ozone in indoor environments: concentration and chemistry. Indoor Air 10(4):269–288

Weschler CJ (2004) Chemical reactions among indoor pollutants: what we've learned in the new millennium. Indoor Air 14(7):184–194

WHO (1989) Indoor air quality: organic pollutants. EURO reports and studies, vol 111. WHO, Copenhagen

WHO (2010) WHO guidelines for indoor air quality: selected pollutants. Geneva, Switzerland

Wille SM, Lambert WE (2004) Volatile substance abuse—post-mortem diagnosis. Forensic Sci Int 142(2–3):135–156

Witterseh T (2002) Status of the indoor climate labelling scheme in Denmark. Proc Indoor Air 612–614

Wolkoff P (2013) Indoor air pollutants in office environments: assessment of comfort, health, and performance. Int J Hyg Environ Health 216(4):371–394

Wolkoff P, Clausen PA, Jensen B, Nielsen GD, Wilkins CK (1997) Are we measuring the relevant indoor pollutants? Indoor Air 7(2):92–106

Wolkoff P, Nielsen GD (2001) Organic compounds in indoor air—their relevance for perceived indoor air quality? Atmos Environ 35(26):4407–4417

Wolkoff P, Wilkins CK, Clausen PA, Nielsen GD (2006) Organic compounds in office environments—sensory irritation, odor, measurements and the role of reactive chemistry. Indoor Air 16(1):7–19

World Health Organization (2000) The world health report 2000: health systems. Improving performance. World Health Organization

Yosefi L, Haghighi M, Allahyari S, Shokrani R, Ashkriz S (2015) Abatement of toluene from polluted air over Mn/Clinoptilolitee-CeO2 nanopowder: impregnation vs. ultrasound assisted synthesis with various Mn-loading. Adv Powder Technol 26(2):602–611

Yu C, Crump D (1998) A review of the emission of VOCs from polymeric materials used in buildings. Build Environ 33(6):357–374

Zhu J, Wong SL, Cakmak S (2013) Nationally representative levels of selected volatile organic compounds in Canadian residential indoor air: population-based survey. Environ Sci Technol 47(23):13276–13283

Zhu Z, Wu RJ (2015) The degradation of formaldehyde using a Pt-TiO_2 nanoparticles in presence of visible light irradiation at room temperature. J Taiwan Inst Chem Eng 50:276–281

Chapter 6
Comprehensive Analysis of Research Trends in Volatile Organic Compounds Emitted from Building Materials: A Bibliometric Analysis

Fatma Nur Eraslan, Mansoor Ahmad Bhat, Eftade O. Gaga, and Kadir Gedik

Abstract Volatile organic compounds (VOCs) are diverse compounds present in elevated indoor air concentrations, generally exceeding the outdoor levels. Indoor VOCs are emitted from various sources and activities, such as cleaning, cooking, paints, adhesives, furniture, carpets, and building materials, including wood and wood-based composite materials, reducing the indoor air quality. VOC emissions can significantly impact human health because people spend more than 80% of their time in the indoor environments and in recent years, the health effects of VOCs released from building materials have drawn tremendous interest. Moreover, there is no published literature showing the scientometric approach to explore the tendencies. This study aims to provide a bibliometric analysis of VOCs' research used in different building materials using the Web of Science Core Collection database from 1986 to 2020. According to bibliometric analysis results, 1,242 documents were printed in 475 different sources. The main language was English by far; China exhibited high collaborations, productivity, and influence on VOC and building material studies; the most productive journal was Building and Environment. Furthermore, "VOC" and "indoor air" were used most frequently in the author keywords in the early years of the research, the keywords "formaldehyde" and "building materials" increased to use more recently.

Keywords Building materials · Human health · Indoor air · Microplastics · Network analysis · Research trend · Scientometrics · VOCs · Web of Science (WoS)

F. N. Eraslan · M. A. Bhat · E. O. Gaga · K. Gedik (✉)
Department of Environmental Engineering, Faculty of Engineering, Eskisehir Technical University, Eskisehir, Turkey
e-mail: kgedik@eskisehir.edu.tr

M. A. Bhat
e-mail: mansoorahmadbhat@eskisehir.edu.tr

J. A. Malik and S. Marathe (eds.), *Ecological and Health Effects of Building Materials*,
https://doi.org/10.1007/978-3-030-76073-1_6

6.1 Introduction

Considering that most people spend their time indoors (households, office buildings, schools, etc.), the importance of indoor air quality is increasing. Various factors can influence indoor air quality, including microbiological, physical, indoor, and outdoor ventilation systems (Jones 1999). Building materials and different associated materials used in the construction industry, one of these factors, cause the release of various pollutants "COx, NOx, particles, volatile organic compounds (VOCs), micro-nanoplastics, etc." into the indoor air (Spengler and Chen 2000). Research has shown that pollutants from floors, ceilings, and walls have a negative impact on Indoor Air Quality (IAQ) in buildings (Harčárová et al. 2020). In most cases, indoor pollutants' level is substantially higher than outdoors (Batterman et al. 2007; Liu and Little 2012). One of the most important air pollutants are VOCs (Mujan et al. 2019; Rösch et al. 2014). VOCs enter the indoor environment as a result of release from the activities of the people (smoking, cooking, cleaning, washing, heating, etc.) (Buonanno et al. 2009; Mishra 2017; Morawska et al. 2017; Morawska and He 2006), home furnishings (Yrieix et al. 2010), household and building materials (sealants, plastics, paints, varnishes, solvents, adhesives, insulation materials, cleaning agents, disinfectants and cosmetics) (Harčárová et al. 2020; Hormigos-Jiménez et al. 2017; Kwok et al. 2003; Seltzer 1997; Spengler and Chen 2000). Also, photocopiers, printers, faxes, toners, and inks are essential sources in office buildings (Arı 2020; Lee et al. 2001). As is the case for construction materials and furniture, the newer the appliances, the higher the VOC emissions' potential rate is. VOC pollution concentrations from machines not requiring chemical suppliers, "e.g. computers" decline dramatically over time (Azimi et al. 2016; Chan et al. 2020; Kowalska et al. 2015). The indoor levels of these VOCs frequently exceed outdoor concentrations due to the widespread use of synthetic materials in the new building and interior finishing (Brooks et al. 1991). The occurrences of formaldehyde and other VOCs in building materials and emissions during the use phase are closely related to the manufacturing procedures. For example, manufacturing of widely used medium-density fiberboard consists of several steps: firstly, wood chips are milled into wood fibers; the wood fibers are then blended with adhesive resins, and the resulting mixture is dried by hot air; finally, the wood fibers after resin application are placed on a conveyer belt and hot-pressed into medium-density fiberboard (He et al. 2012). With low cost and good performance, urea-formaldehyde and phenol-formaldehyde resins are the most commonly used adhesives in wood-based panels (Liu and Little 2012). Acetaldehyde is present in laminates, cork, foam mattresses and linoleum in the form of polyester resins and basic dyes. Phenol is presented in materials, such as wall coverings and vinyl flooring, and dichloromethane is often used in adhesives (Harčárová et al. 2020). Therefore, lead to high emissions of VOCs such as formaldehyde when these products are used indoors. VOCs emanating from building materials includes: Adhesives "alcohols, amines, benzene, dimethylbenzene, formaldehyde, terpenes, toluene, xylenes," caulking compounds "alcohols, alkanes, amines, benzene diethylbenzene, formaldehyde, methyl ethyl, xylenes," ceiling tiles "formaldehyde," clipboard/particle board "alcohols, alkanes, amines, benzene, formaldehyde, terpenes, toluene," floor and wall coverings paints, "acetates, alcohols, alkanes, amines, benzene, formaldehyde, methylstyrene, xylenes," paints, stains & varnishes "acetates, acrylates, alcohols, alkanes, amines, benzene, formaldehyde, limonene, polyurethane, toluene" (Brooks et al. 1991; Kwok et al. 2003; Wallace et al. 1987). Among all the VOCs in the indoor environment, formaldehyde is one of the most common and best-known compound and a priority indoor air pollutant due to its wide distribution in indoor air and its highly toxic nature (Salthammer et al. 2010). The amount and composition of undesirable volatile organic compounds in the indoor environment depend on several factors. The most important ones include the number of people indoors, interior and technical equipment, relative humidity, the age of the building and its location within the urban unit, degree of air pollution, intensity and quality of ventilation, season and type of heat source (Harčárová et al. 2020).

The actual effect of VOCs indoor pollution on human health remains a topic of study, given the significant rise in types of cancer and associated diseases. The integration of new appliances and technology in homes and workspaces has led to higher levels and diversification of VOCs (Enesca and Cazan 2020). Ambient VOCs are usually complex mixtures of species from different sources, which may jointly contribute to the toxic effects (Hua et al. 2018). VOCs' health hazards associated with the use of building materials result from inhalation of fumes or vapours and skin absorption (Kwok et al. 2003). Short exposure to high values of VOCs can result in irritation of the eyes, nose, or mouth, headaches, nausea and vomiting, dizziness, allergy, and asthma (Singh et al. 2016). Long-term high VOC exposure values will lead to increased cancer risk, damage to kidneys, and adverse effects on the central nervous system (De Gennaro et al. 2014). The levels of exposure to VOC and the time spent indoors are causing increasing health hazards of volatile organics (El-Hashemy and Ali 2018).

The use of plastics in construction materials has recently increased tremendously. Plastics are used for the manufacturing of goods and materials, like "insulation products, damp proofing, floorings, roofing, windows, and laminated surfaces of kitchen and other fitments." In addition, buildings have many structures and equipment containing plastics or made of plastics, like pipes. In comparison to plastics used in building materials, plastics are also used in the packaging. Plastics are often used as various types of coverings and in tarpaulins during building. Furthermore, "paints, varnishes, waxes and glues" also contain plastics. Additionally, plastics are usually used in furniture, e.g. cushions, textiles, glues, surface treatments and domestic appliances "refrigerators, washing machines, kitchen appliances, media devices etc." All of these contribute to the overall number of plastics used in the housing and manufacturing industries. Waste plastic bags, which are non-biodegradable, have been recycled to produce floor and wall tiles with lesser flammability and enhanced tensile strength (Dhawan et al. 2019). Plastics are wastes that cannot be decomposed by biological processes (Barnes 2019). Instead of disposing of these non-biodegradable wastes (NBW), reusing them as building materials has been an approach that has attracted attention, especially in recent years (Karthikeyan et al. 2020). Many researchers made concrete by adding NBW as a fiber or powder or in a scrap manner to obtain the best results (Al-hadithi et al. 2019; Foti 2019; Pešic et al. 2016; Silva et al. 2013). The addition of fibers in concrete would act as crack inhibitors and substantially improve the tensile strength, cracking resistance, impact strength, wear and tear, fatigue resistance, and concrete ductility (Malagavelli 2020). Concerning the plastic waste, "polypropylene (PP), polyethylene (PE) represents the most considerable fraction, followed by polyethylene terephthalate, most known as PET." This is due to their chemical resistance, relatively high melting point, and low price (Silva et al. 2013). Alternatively to PP, fibres from low-density PE (Alhozaimy and Shannag 2009) have been used to decrease the cracks of plastic shrinkage from concrete while slightly decreasing compressive power. Recycled polyethylene terephthalate (PET) fibres were also examined but were observed to decay after reaction to the alkalinity of concrete (Jean et al. 2012; Silva et al. 2005). The high-density polyethylene (HDPE), whose physical and chemical properties are most similar to those from PP, is another recyclable polymer candidate for mass fibre processing (Alhassan et al. 2017). It is manufactured in large amounts from plastic bottles which are used as "beverage and mineral water containers" (Karthikeyan et al. 2020). The manufacture of PET bottles has grown tremendously as a consequence of the dramatic rise in drink use, also due to the desirable properties of this plastic, including "low density, high resistance, weight ratio, high toughness, ease of design/manufacturing, and low cost" (Foti 2019). Today, plastics are used in fields such as furniture and carpet manufacturing, as well as building coatings, wall paints, insulation materials, building materials (Lin et al. 2009; Pešic et al. 2016; Silva et al. 2005; Vianello et al. 2019).

The microplastics can get released from the products used in different building materials and are the pollutants of environmental concern. Discussion of the potential negative effects of airborne microplastics on human health has only recently emerged (Prata 2018). It depends on the scale of the probabilities of airborne fibrous microplastics penetrating our air system. First of all, the words respirable and inhalable should be well defined (Gasperi et al. 2018). Plastics and microplastics capable of penetrating nose and mouth and depositing in the upper airways are inhaled, and the deep lung may penetrate and deposit. Deposition in the airway depends on aerodynamic diameter, and deposition falls over 5 μm diameter within the respiratory region (Donaldson and Tran, 2002). Therefore, microplastics can reach the alveoli in the nasal system due to their small diameter (Amato-Lourenço et al. 2020). However, the persistence of inhaled fibrous microplastics is related to the durability and clearance from the lungs (Greim et al. 2008). In the first study on the presence of inhaled microplastics in the lungs, it was emphasized that different types of microplastics are found in the lungs and show little signs of deterioration (Pauly et al. 1998). Fibre sizes play a part in toxicity in addition to persistence. Moreover, it is believed that microplastics that are inhaled or swallowed can be transferred to the circulatory system and other organs (Wright and Kelly 2017).

Bibliometric analysis is a literature analysis method that uses statistical methods (e.g. citation analysis) to analyze books, articles, reports and other publications. This technique has been progressively used to monitor various scientific disciplines' research performance and support appropriate policy action. A bibliometric analysis enables researchers to identify the publications' key research topics and interrelationships based on the collected publications' citations, bibliography, and text information. Bibliometric research is crucial for mapping research-related literature. As a valuable tool for literature analysis, bibliometric can effectively depict the rules of discipline development and has a wide application in different knowledge domains. Nowadays, many bibliometric studies are carried out in an attempt to analyze the process of collaboration between researchers and institutions. Digitalization of literature and the formation of online literature databases allowed researchers to explore the body of literature and research performance smoothly and efficiently. The Web of Science database contains the world-famous "Science Citation Index, Social Sciences Citation Index and Arts, and Humanities Citation Index." Also, the fundamental compendium of "Web of Science 1986–present" has numerous benefits: (1) The sources cited for all articles are completely indexed and accessible; (2) It is possible to get all the information about both authors and their respective affiliations; (3) Visually, the citation monitoring characteristic recognizes citation events and patterns, and (4) Developments of analysis and structures of publishing can be calculated by using the effects of empirical retrieval. According to the author's knowledge, few bibliometric analyses were carried out on the subject of VOCs (Cheng et al. 2019; Zhang et al. 2010), however, there have not been any scientometrics research undertaken in the release of VOCs from the building materials. The main aim of this work is to provide a bibliometric overview of findings of the VOCs used in different building materials by using the "Web of Science Core Collection database" from 1986 to 2020.

6.2 Methods

6.2.1 Search Strategy

All data used in the study have been obtained from the Web of Science (WoS) online database (WoS 2021). In scientometrics, selecting preliminary research data is of utmost importance, as these data directly impact the findings and results. For Bibliometric analysis, all publications from 1986 to 2020 were extracted using the keywords "Volatile Organic Compounds," VOCs*, VOC*, "Building Material," and "Building Materials" (In fact, building materials and VOCs the first relevant article was published in 1986). Before 1986 none of the articles was published regarding the above-mentioned keywords. The publication of articles started in the year 1986. Due to this reason, 1986 was considered as the starting year. The search was conducted in January 2021. Accordingly, the data consisted of 1,242 publications, and for each document, we selected "full records and cited references" and exported the results in a format for further analysis. The records were then downloaded in different files as Web of Science provides just 500 documents per download. Care was taken to analyze the gathered data to ensure their identification. Details on the paper included "authors, title, source, abstract, language, document type, keywords, addresses, cited reference count, times cited, publisher information, ISSN, page count, subject category, and citation report."

The performance review of research in this bibliometric analysis was centered on previously scientometric research (Cañas-Guerrero et al. 2014; Cheng et al. 2019; Geng et al. 2017; Li et al. 2017; Zhang et al. 2010). Web of Science database was employed as the resource to get all the scientometric statistical information concerning the scientific production in the subject of health effects of volatile organic compounds emitted from building materials. The Web of Science website has been chosen since it is the largest comprehensive citation and abstract platform of peer-reviewed articles. While various datasets can provide different findings, the most accurate literature sources for publications and citations are believed to be the WoS repository. Apart from this, the Web of Science is easy to use and has simple and advanced search tools. Moreover, WoS is the most frequently used database for empirical metrology for instance: in many bibliometric investigations, this Web of Science Databases have been commonly employed in different fields including VOCs and building materials (Can-Güven 2020; Cañas-Guerrero et al. 2014; Cheng et al. 2019; Factor et al. 2010; Harzing and Alakangas 2016; Kawuki et al. 2020; Li et al. 2017; Zhang et al. 2010).

6.2.2 Data Analysis

The bibliometric analysis of publications obtained from the database was analyzed using "Biblioshiny application" (using the R-studio) (Kawuki et al. 2020). Preliminary info regarding information includes: "Timespan, sources (journals, books, conferences), documents, average years from publication, average citations per documents, average citations per year per document, references." Meanwhile, the documents categories consist of: "article, article; book chapter, article; data paper, article; early access, article; proceedings paper, editorial material, editorial material; book chapter, meeting abstract, note, proceedings paper, review, review; early access." While as the article information consists of author's keywords. Furthermore, the authors' data include "author appearances, authors of single-authored documents, and authors of multi-authored documents." "Authors collaboration includes single-authored documents, documents per author, authors per document, co-authors per document, collaboration index". Graphics were created by considering the first 10 ranks (country, publication, author, word, journal). Besides, the VOS viewer (version 1.6.15) was used to visualize network analysis (Wang et al. 2020).

In order to include all documents published in all languages, the basic search method was used. Based on this, the 1,242 documents were published in ten languages, such as "English, German, French, Japanese, Chinese, Portuguese, Polish, Czech, Turkish, and Spanish." English was the leading written language. Papers written in the English language were 1,207 publications made up 97% of total publications, followed by German, French, Japanese, Chinese, Portuguese, Polish, Czech, Turkish and Spanish comprising 12, 6, 6, 3, 3, 2, 1, 1, and 1 publication respectively. This was expected because English is the international language of choice in building materials, VOCs, indoor environments and health impacts research. According to the WoS database, owing to English's global superiority, ethnic and language influences are less important in shaping publications, since it is the universal language for performing research investigations and the strict writing style for academic articles. However, topographical influences performed significant character in worldwide publication on building materials, VOCs, indoor environments and health impacts research.

The parameters included in the bibliometric analysis are:

Annual scientific production: It presents the number of publications published each year between the dates determined on the subject of interest. For this study, the yearly scientific output was calculated from (1986–2020); therefore, the relationship between the number of articles and scientific production can be assessed.

Collaborations: The analysis includes countries, institutes, and authors included in each article. It determines the most productive countries, institutes, and authors for research on the subject. The form of collaboration was decided by the addresses of the authors, where it was allocated independently if no partnership was identified. International collaboration was allocated if the article was co-signed by authors from more than one nation.

Journal: In this category, parameters such as the number of annual citations of journals, total citations, and yearly production dynamics were analyzed.

Authors: The authors' productions for the relevant topic over time and their cooperation with countries were examined.

Citations Index: The h-index is a citation index used to express the author's publication's quantity and significance, as first suggested by Hirsch (2005). "When a researcher's h-index is h, which means that there are at least h papers of this researcher that have been cited for h times" (Hirsch 2005). It is also used to explain the importance of h-index journals. In a study, the h-index calculated for journals showed a high correlation with the Thomson Institute for Scientific Information (ISI) impact factors, which are used as a standard to measure journal quality (Hodge et al. 2013). Also, the g-index and m-index of these articles and journals were calculated. These two indexes are used as a variant of the h-index. The g-index, first pioneered by Egghe (2006), is used to resolve the inconsistency in evaluating journals with the same h-index values. The g-index gives more weight to highly cited articles, and the higher it means, the more cited articles of that journal or author. Hence, g-index values are always equivalent to or greater than the h-index (Hodge et al. 2013). The calculated m-index is obtained by dividing the h-index by the number of years since the scientist's first published article. However, Hirsch thinks that the researcher's first published article may not always be a correct starting point (Hirsch 2005). Even though the h-index was considered in many studies (Dettori et al. 2019; Kawuki et al. 2020; Mingers et al. 2012) due to the ease of calculation, h, g, and m-indexes were calculated in this study.

Keyword analysis: The progression of the mainly significant research subjects over time is examined using keywords. In the study, the keywords described by the authors were analyzed separately. These keywords are called authors keywords.

6.3 Results and Discussion

The following topics have been studied by this bibliometric analysis: the history of research interest in the "Building material" and "Volatile Organic Compound" category; trends in publication and citation; the most productive countries and institutes, and their co-occurrence; authors' production over the time, total citations, and their impact values; the evolution of the journals in this category; the most global cited relevant documents and their author; the most important author's keywords. There were a total 1,242 publications that met the selection criteria mentioned, containing 9 document types "article, article-book chapter, article-data paper, article-proceedings paper, editorial material, editorial material-book chapter, meeting abstract, note, proceedings paper, review." The article was the most frequently used document type comprising 73.2% (910) of the total production, followed by proceedings paper 12.8% (159), review 7.4% (92), article; proceedings paper 4.1% (51), article; book chapter 1% (13) and others less than 1%. According to bibliometric analysis, the average years from publication was 8.84 while the average citations per document and the average citations per year per document were 22.6 and 2.57, respectively. Besides, the documents included in the bibliometric analysis contain a total of 34,437 references.

6.3.1 Trends in Publication

When calculating the annual number of publications, the analysis interval was set (1989–2020). Before 1989, there were only 2 articles (one published in 1986 and the other in 1987). The 1989 year was shown as the starting point as the topic's production increased in 1989 (Fig. 6.1). According to Fig. 6.1, while the number of publications had increased until 2001 and touched 27 publications, there was a decrease in 2002. There were almost twice as many publications in 2003 as the previous year. According to the Fig. 6.1, from the year 1989 to 1996, the number of published articles was below ten however after 1997 to 2020 the number of published papers per year was greater than 10. The other noteworthy point in the graph was in 2005–year data. As a result of the bibliometric analysis, it was determined that publications of relevant topics increased rapidly in 2005 due to the 10th International Conference on Indoor Air Quality and Climate (Fig. 6.1) (Indoor Air, 2005). This conference was held 4–9 September 2005 in Beijing, China and was attended by over 1,200 scientists and specialists in the subject of the indoor environment. Multiple keynote lectures were also offered by globally respected academics welcomed as plenary guests. Of all the

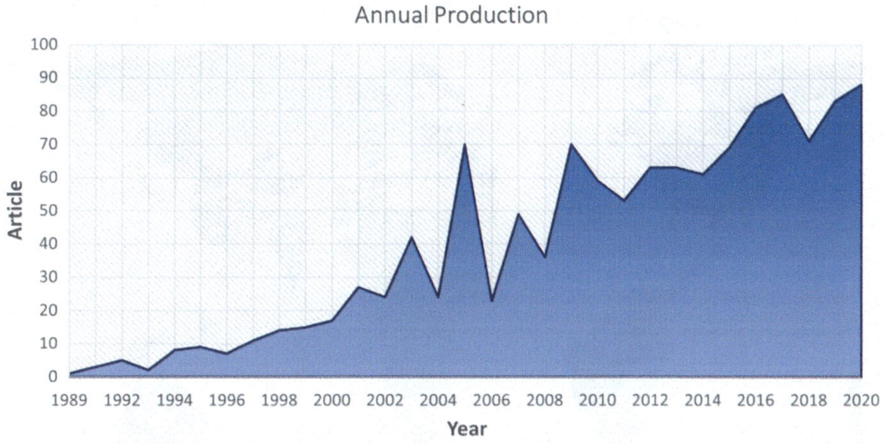

Fig. 6.1 The trend of literature publications annually

800 lectures, papers and exhibitions, the main presentation was the paper entitled "Carbon and cooling in UK office environments." Through this gathering, organizers plan to advance and expand their expertise on indoor environmental science and innovation with the goal of improving quality of life via better conditions in the buildings where people live and work. The total number of publishes articles in this year was 70, which was the highest number of published articles from (1989–2005). The year with the most increased scientific production (88) to date was in 2020. This is shown schematically in Fig. 6.1. Besides the analysis results, the annual growth rate of published articles was 16.1%.

6.3.2 *Most Contributing Countries and Their Collaborations*

Collaboration networks among various countries and institutions were developed on the basis of published studies from countries and institutions. As presented in Fig. 6.2, the outcomes of the study showed links between different countries or various institutions. The red lines are showing us the collaboration between the countries or institutions in terms of the published articles. The width of lines represents the strength in collaborations. The Most cooperation was between China and USA (35), followed by Canada and the USA (17), China and Korea (13), Canada and China (12), United Kingdom and China (12), USA and Denmark (12), Denmark and Germany (8),USA and Korea (7). While the collaborations among China and Japan, France and Italy, France and USA, United Kingdom and Korea, United Kingdom and the USA, USA, and Germany were the same as all these countries published 6 articles (Fig. 6.2). Also, shades of blue were used to show the total number of publications in the countries. In this case, the countries with the highest total number of publications were as follows: China (320), USA (233), South Korea (94), Germany (92), France (77),

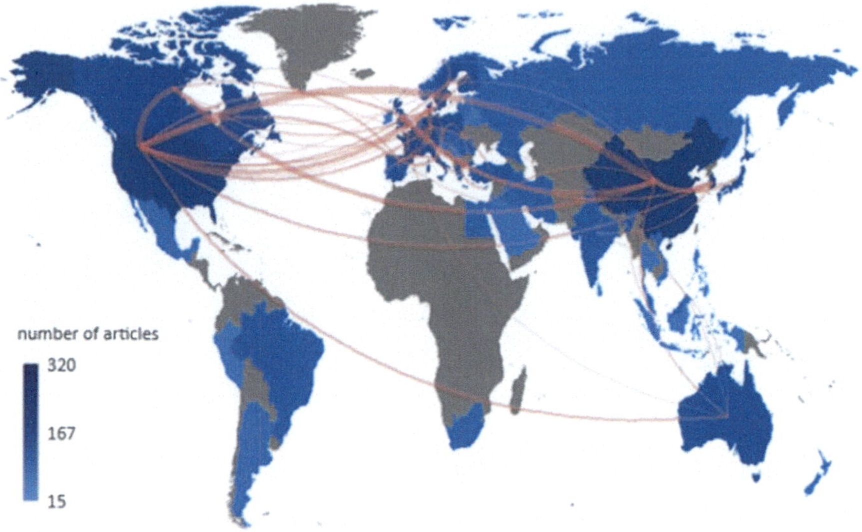

Fig. 6.2 Geographical distribution of scientific publications based on countries and country collaboration

Canada (70), Japan (68), Denmark (62), Sweden (49), and England (39) respectively (Fig. 6.2).

Detailed statistics were given in the top ten countries commonly quoted (Table 6.1). The results revealed that China harbours the highest total citation (5,110), followed by the USA (5,106) and Denmark (2,211). However, Japan had the least number of citations (1,145) among all the countries (Table 6.1). Besides, when the relationship between the total number of publications in countries and citations was examined, it was seen that the country with the highest average article citation was

Table 6.1 Information of the top 10 frequently cited countries

Rank	Country	Documents	Total citations	Average article citations
1st	China	962	5,110	16.59
2nd	USA	626	5,106	28.53
3rd	Denmark	111	2,211	59.76
4th	Germany	206	2,137	26.06
5th	UK	113	1,536	45.18
6th	South Korea	285	1,411	15.85
7th	Canada	140	1,385	28.85
8th	Sweden	115	1,357	33.92
9th	France	235	1,227	19.48
10th	Japan	226	1,145	19.74

Denmark (59.76), followed by UK (45.18) and Sweden (33.92). In contrast, the least average article citations were seen in South Korea (15.85) (Table 6.1). Moreover, the highest numbers of articles were published in China (962) and USA (626) respectively, while at least a number of articles were published in Denmark (115). From Table 6.1 it gets clear Denmark was the most productive country in terms of average article citations plus total citations because its total number of published articles was just 111 which is almost nine times lesser than the articles published by the first rank country China which published 962 articles and six times lesser than the USA which published 626 articles.

From Fig. 6.2 and Table 6.1, it gets clear that China had the highest collaborations with other countries, the highest total number of publications and the highest total citations among all nations. In the future, more contributions are expected to increase from China in terms of collaborations, number of publications and citations. In China, the construction industry is expected to continue its dramatic growth due to ongoing rapid urbanization (Mao et al. 2015), and there is a massive demand for construction projects (Huang et al. 2018). Apart from this, the construction industry in China accounts for around one-half of the world's newly built floor area, which results in large volumes of pollution emissions and resource consumption (Huang et al. 2018). To mitigate these adverse impacts on the environment, Chinese government departments have taken steps to reform the construction industry towards sustainable development (Qi et al. 2010; Yu et al. 2019).

6.3.3 Most Relevant Institutions

This study seeks to assess the classification of the top universities in using keywords for analysis. Figure 6.3 shows the 10 most influential institutions from WoS in terms of published articles. 1,186 published the total of 1,242 articles from (1986–2020). Among the 1,186 institutions that participated in the analysis, the "Tsinghua University" leads the institutions' production with 151 publications and consists of 12% of the total global publications. Simultaneously, as the "Hong Kong Polytech University" had published the 44 articles having 3.54% of total publications. However, Concordia University had published 42 articles contributing 3.38% of total world publications. Meanwhile, the Seoul Natl University, Tongji University and University Calif Berkeley have published the least number of published articles individually (33) among the first ten affiliations. The total contribution of these universities individually is 2.66%. From Fig. 6.3, it gets clear only Tsinghua University contributed more than 10% in terms of total contribution while the contribution by rest nine affiliations was less than 4%. Moreover, only Tsinghua University published more than 100 articles; however, the rest of the universities published articles were less than 50 (Fig. 6.3). The top 10 institutes oscillate between 2.6 and 12% (number of published papers 33 to 151) (Fig. 6.3).

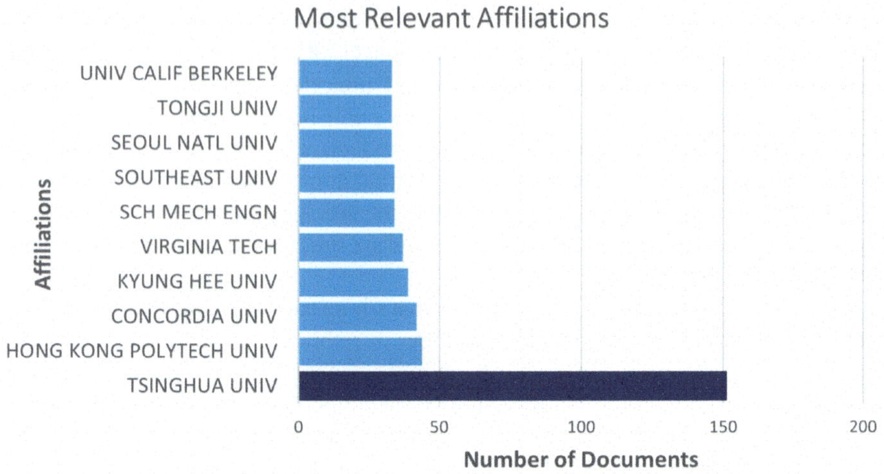

Fig. 6.3 Top 10 relevant affiliations in terms of their published number of documents

6.3.4 Most Contributing Authors

Several criteria are commonly used to assess scientists' influence, such as the total number of citations, and impact values (Patience et al. 2017). The analyzed publications from WoS were produced by 3,308 authors, while the total number of author appearances was 5,070. However, the authors of single-authored documents were 58. Meanwhile, authors of multi-authored documents were 3,250. Besides, single-authored documents were 77, while the documents per author and authors per document were 0.375, 2.66, respectively. On the other hand, the number of-authors per document was 4.08, and the collaboration index was 2.79. The total number of citations and impact values of the authors in the study was calculated. The results of the 10 most influential authors are given in Table 6.2. The three highest cited authors of the research included Wolkoff P (Total Citations, TC = 1,420), Salthammer T (TC = 1,305), and Zhang Y (TC = 1,052), while as the least cited Author was Little JC (TC = 660) (Table 6.2). Further, the results revealed that Zhang Y has the highest h-index and g-index (h = 20, g = 31) followed by Wolkoff P (h = 16, g = 23) and Xiong J (h = 16, g = 25) (Xiong J is not included in the top 10 authors in total citation (TC = 654) rankings). Conceivably for Wolkoff P and Xiong J with similar h-indexes, Xiong J has a higher g-index due to more citations or more paper (Table 6.2). In this case, Xiong J has more effective in terms of overall scientific impact than Wolkoff P. However, Marutzky R was having the least h and g-index (1). The Zhang Y published the highest number of articles (55), while as the least number of articles (1) was published by Marutzky R (Table 6.2).

Moreover, the top 10 authors' production over time was obtained with bibliometric analysis, as shown in Fig. 6.4. The figure shows the number of papers and total citations of the authors by years. The bubbles' size indicates the number of documents

Table 6.2 Most Cited Authors and their impact

Rank	Author	h-index	g-index	m-index	Total citation	Number of paper	Year
1st	Wolkoff P	16	23	–	1,420	23	1995
2nd	Salthammer T	14	21	0.583	1,305	21	1997
3rd	Zhang Y	20	31	–	1,052	55	1999
4th	Jones AP	2	2	0.087	839	2	1998
5th	Mentese S	3	4	0.231	820	4	2008
6th	Marutzky R	1	1	0.091	754	1	2010
7th	Norback D	9	9	0.346	697	9	1995
8th	Nielsen GD	6	6	0.25	697	6	1997
9th	Haghighat F	13	21	0.52	677	21	1996
10th	Little JC	13	22	0.65	660	22	2001

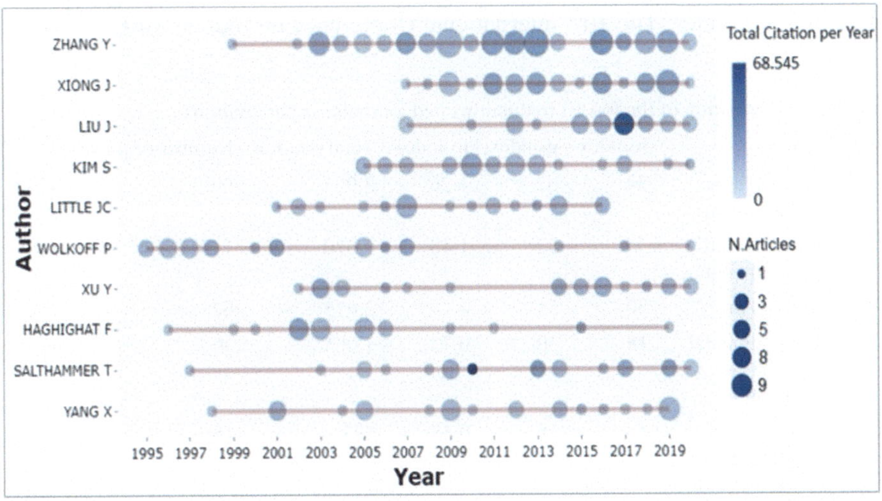

Fig. 6.4 Top-Authors' production over time

per year while with the increasing citations, the color fluctuates from light to dark blue. Salthammer T., published a paper in 2010 (only 1 paper), has the highest total number of citations per year to date (TC per year = 68.545) and followed by, Liu J., with his four papers published in 2017 is second on the total number of citations list (TC per year = 54) (Fig. 6.4). The most paper production per year belonged to Zhang Y in 2009 (number of papers per year = 9, total citations per year = 6.75) and in 2013 (number of papers per year = 8, total citations per year = 19.75). The least number of total articles (21) were published by Haghighat F, Salthammer T and Yang X, while the maximum number of total articles were published by Zhang Y (Fig. 6.4).

6.3.5 Most Productive Journals

From WoS, the analyzed documents were produced by 475 sources (journals, conferences and books). Information of the top 10 frequently cited journals are listed below (Table 6.3). The Atmospheric Environment journal topped the list having the total of 3,903 citations from the year of (1994–2020), which was followed by the Building and Environment journal having the total of 3,670 citations from the year of (1998–2020) however the Journal of Hazardous Materials was having the least number of citations (TC = 434) from the year of (2009–2020). The annual production of the 5 most active resources regarding the research topic is shown in Fig. 6.5. The Building and Environment journal, the second leading journal in terms of the total number of citations, however, it topped in the most relevant resource analysis (Fig. 6.5) with 140 papers. The Atmospheric Environment journal was the second leading journal in terms of published papers (69), followed by Indoor Air journal, consisting of a total 67 published articles, Indoor and Built Environment journal having a total of 44 published articles. The 10th International Conference on Indoor Air Quality and

Table 6.3 Information of the top 10 frequently cited journals: as per citations

Rank	Source	h-index	g-index	m-index	Total citation	Number of paper	Year
1st	Atmospheric Environment	34	62	1.26	3,903	69	1994
2nd	Building and Environment	34	52	1.48	3,670	140	1998
3rd	Indoor Air	30	51	–	2,675	67	1995
4th	Environmental Science & Technology	18	30	0.72	1,065	30	1996
5th	Indoor Air-International Journal of Indoor Air Quality and Climate	13	15	0.48	832	15	1994
6th	Chemical Reviews	1	1	0.09	754	1	2010
7th	Indoor and Built Environment	14	22	0.64	610	44	1999
8th	Science of the Total Environment	13	20	0.57	495	20	1998
9th	Environment International	11	14	0.34	440	14	1989
10th	Journal of Hazardous Materials	11	16	0.92	434	16	2009

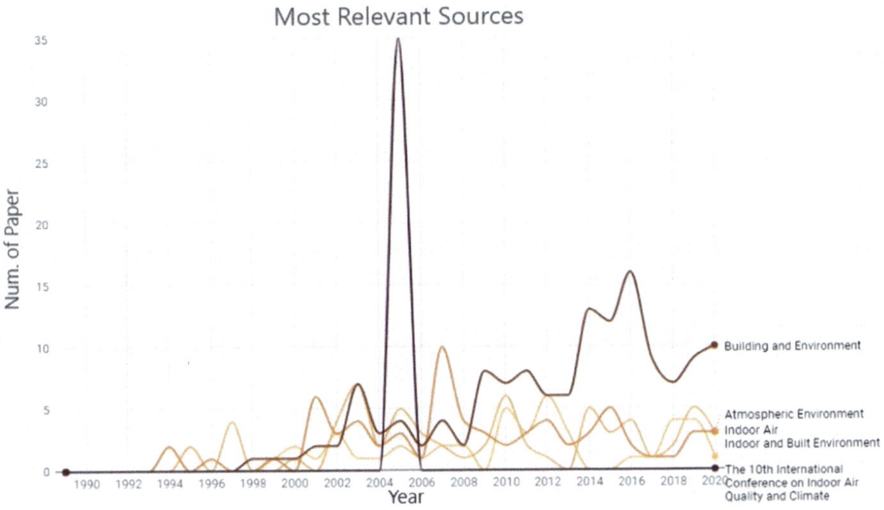

Fig. 6.5 Most relevant sources as per the number of papers published per year

Climate had the least number of published articles (35) (Fig. 6.5). All these articles were published during the year of 2005. The source that has made the most production every year since 2009 is the Building and Environment journal, while as the maximum number of articles (16) were published during the year 2016. However, from 1989–1997 none of the articles was published in Building and Environment journal. While as the least number of articles (1) were published during 1998–2000. From 1998–2020, the articles were regularly published in the Building and Environment journal (Fig. 6.5). In the case of the Atmospheric Environment journal, the maximum number of articles (10) were published during 2007. Meanwhile, during 1989–1993, 1995, 1997–1998 and 2000, none of the articles was published. However, the least number of articles (1) was published during 1996, 1999, 2006 and 2017–2018. From 2001–2020 articles were continuously published in the Atmospheric Environment journal (Fig. 6.5). In the Indoor Air journal, the maximum number of articles (7) were published during 2003. However, during 1989–1994, 1996, 1998, 1999, 2001 and 2013, none of the articles was published. While as during 2000, 2008, 2012 and 2017, the minimum number of articles (1) was published (Fig. 6.5). Meanwhile, during 2002–2012 and later from 2014 until 2020, articles were continuously published. In case of Indoor and Built Environment journal the maximum number of articles (6) were published during 2012 however from 1989–1998, 2009 and 2014–2015 none of the articles was published. While as during 1999, 2001, 2003–2004, 2006, 2016–2017 and 2020 least number of articles (1) were published (Fig. 6.5). It was seen from the last five years that articles have been continuously published in Indoor and Built Environment journal. The 10th International Conference on Indoor Air Quality and Climate was an exception compared to the other four journals. Under this publication house, only 35 articles were published during 2005 while as from 1989–2004 and 2006–2020 none of the articles was published as this conference happened only once (Fig. 6.5).

Further, the results revealed that the "Atmospheric Environment" and "Building and Environment" journals have the highest h-index (34) (Table 6.3). These two sources with similar h-indexes have different g-indexes due to the number of citations and papers. The Atmospheric Environment journal has 62 g-index, while as the Building and Environment journal has 52 g-index (Table 6.3). According to the results, it is determined that the Atmospheric Environment journal has more citations or papers. While as the second leading journal in terms of h-index (30) was Indoor Air. However, in terms of g-index (51), it occupied third place, followed by Environmental Science & Technology having h-index 18 and g-index 30. Meanwhile, Chemical Reviews journal had the least value (1) for both the h and g index (Table 6.3).

6.3.6 *Most Cited Documents*

According to the analysis results, documents had an average of 22.76 citations per document, while with 2.57 average citations per year per document. Relevant top 10 papers consist of 7 reviews and 3 articles (Table 6.4). The most cited article in WoS belonged to Jones and was published in Atmospheric Environment in 1999, under the title "Indoor Air Quality and Health" and was a review article with 790 TC in WoS. This was followed by a paper of Salthammer et al., published in "Chemical Reviews" during 2010 entitled formaldehyde in the indoor environment and was having the 754 total citations. This was also a review article (Table 6.4). These were the two leading articles in terms of total citations. Their total number of citations was almost 2–3 times more than the rest articles, while as in the rest of the articles, the minimum total citations were below 400 (Table 6.4). However, Norback et al., published an article entitled "Asthmatic symptoms and volatile organic compounds, formaldehyde, and carbon dioxide in dwellings" in "Occupational and Environmental Medicine" during 1995 was having the least number 212 of total citations. This was a typical article (Table 6.4).

On the other hand, the most citations per year for two articles in WoS belongs to Salthammer et al. (2010) entitled formaldehyde in the indoor environment, published in "Chemical Reviews" (TC per Year = 68.55), and Yan, 2017 entitled lanthanide-functionalized metal-organic framework hybrid systems to create multiple luminescent centers for chemical sensing, published in "Accounts of Chemical Research" (TC per Year = 67) respectively (Table 6.4). However, concerning other authors, the total citations per year was almost (2–8) times lesser than the above mentioned two authors (Table 6.4). Norback et al. (1995) entitled "asthmatic symptoms and volatile organic compounds, formaldehyde, and carbon dioxide in dwellings" in "Occupational and Environmental Medicine" was having the least number of total citations per year 8.15 (Table 6.4).

Table 6.4 The top 10 most global cited relevant documents from WoS

Author, year	Document title and journal name	Document type	Total citations	Total citations per year
Jones 1999	Indoor air quality and health, Atmospheric Environment	Review	790	35.91
Salthammer et al. 2010	Formaldehyde in the indoor environment, Chemical Reviews	Review	754	68.55
Brown et al. 1994	Concentrations of volatile organic-compounds in indoor air—a review, Indoor Air-International Journal of Indoor Air Quality and Climate	Review	370	13.70
Weschler 2009	Changes in indoor pollutants since the 1950s, Atmospheric Environment	Review	316	26.33
Qıu et al. 2012	Hybrid CuxO/TiO_2 nanocomposites as risk-reduction materials in indoor environments, ACS Nano	Article	269	29.89
Yan 2017	Lanthanide-functionalized metal-organic framework hybrid systems to create multiple luminescent centers for chemical sensing, Accounts of Chemical Research	Review	268	67
Wolkoff and Nielsen 2001	Organic compounds in indoor air—their relevance for perceived indoor air quality? Atmospheric Environment	Article	242	12.1
Korpi et al. 2009	Microbial volatile organic compounds, Critical Reviews in Toxicology	Review	239	19.92
Yu and Crump 1998	A review of the emission of VOCs from polymeric materials used in buildings, Building and Environment	Review	214	9.30

(continued)

6.3.7 Keyword Analysis

Based on the analysis, WoS documents had 2,891 author keywords. The authors' keywords' co-occurrence network was visualized and determined eight clusters through VOS Viewer software, as shown in Fig. 6.6. "Each circle in the network

Table 6.4 (continued)

Author, year	Document title and journal name	Document type	Total citations	Total citations per year
Norback et al. 1995	Asthmatic symptoms and volatile organic-compounds, formaldehyde, and carbon-dioxide in dwellings, Occupational and Environmental Medicine	Article	212	8.15

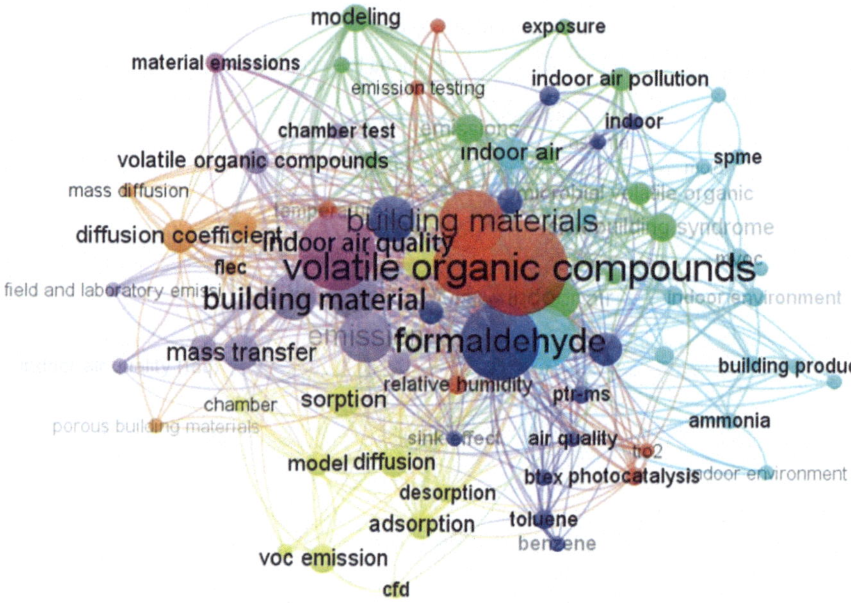

Fig. 6.6 Network visualization map of author's keywords co-occurrence

analysis represents a keyword, and the size of the circle indicates use frequency." The number and color of items in the clusters contain: 16 items in cluster 1 (cyan), 15 items in cluster 2 (dark blue), 11 items in cluster 3 (green), 10 items in cluster 4 (yellow), 10 items in cluster 5 (purple), 9 items in cluster 6 (red), 5 items in cluster 7 (orange), and finally 2 items in cluster 8 (pink) (Fig. 6.6). The co-occurrence keywords analysis revealed that the most frequent author keywords were "volatile organic compound," "building material," "formaldehyde," "indoor air quality," and "indoor air." These keywords formed a strong network and were having strong links with other keywords. While as the keywords which were away from the centre mean they were less used like "building products," "indoor environment," "exposure", "indoor air pollution," "modelling," "material emissions", "mass diffusion," and "VOC emission" etc. The

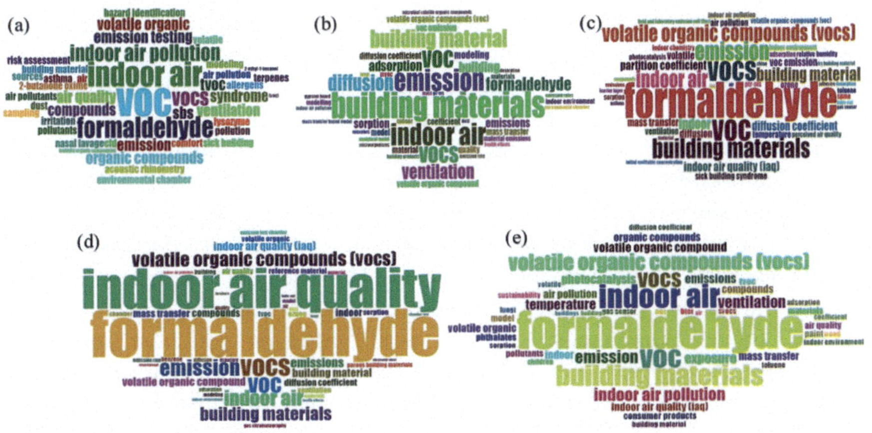

Fig. 6.7 Visualised word-clouds of authors keywords (**a** 1986–2000, **b** 2001–2005, **c** 2006–2010, **d** 2011–2015, **e** 2016–2020)

network formed by these keywords is weak and has weak links with other keywords than the keywords present in the centre (Fig. 6.6).

Besides, a word cloud of author keywords was created. The usage of a keyword depends on its size and centrality, frequency, and magnitude of the word (Kawuki et al. 2020). Figure 6.7 shows the author's keyword clouds for five different periods, and the authors' keywords are shown in Fig. 6.7a belonging to 92 documents published from 1986–2000. However, for another period authors, keywords are shown in Fig. 6.7b present in 187 documents published (2001–2005). While as for third-period authors keywords are shown in Fig. 6.7c existing in 237 documents published from 2006–2010, fourth-period authors keywords are shown in Fig. 6.7d belonging to 309 documents published from 2011–2015. As for final period authors, keywords are shown in Fig. 6.7e present in 409 documents published from 2016–2020. The most frequent authors' keywords used during 1986–2000 were "VOC" (8 documents), "indoor air" (6 documents), "formaldehyde" (5 documents), and "indoor air pollution" (4 documents) (Fig. 6.7a). However, the less regularly used authors keywords were "environmental chamber," "hazard identification," "comfort," "sampling," "volatile organic compound," "risk assessment," "sources," "air pollution," and "syndrome" etc. (Fig. 6.7a). While the most common authors' keywords used during 2001–2005 included "building materials" (19 documents), "emissions" (16 documents), "indoor" (16 documents), "VOC" (15 documents), and "diffusion" (12 documents) as visualized in Fig. 6.7b. However, during this period, the less often used authors' keywords are "dust," "microorganisms," "emission rate," "health effect," and "building product" etc. (Fig. 6.7b).

Whereas the most frequent authors' keywords used during 2006–2010, were "formaldehyde" (25 documents), "VOC" (18 documents), "building materials" (14 documents), and "emission" (13 documents) (Fig. 6.7c). Nevertheless, the least frequently used authors' keywords were "desorption," "asthma," "air," "adhesive," "material," and "gas sensor" etc. (Fig. 6.7c). The maximum current authors' keywords used during 2011–2015 included "formaldehyde" (49 documents), "indoor air quality" (41 documents), "VOCs" (21 documents), "emissions" (17 documents), and "building materials" (14 documents) as visualized in (Fig. 6.7d). However, the less often used authors' keywords during this period were "environment," "furniture," "fungi," "health effect," "chamber," and "gas chromatography" etc. (Fig. 6.7d). While the maximum common authors' keywords used during 2016–2020 were "formaldehyde" (33 documents), "indoor air" (20 documents), "VOC" (19 documents), and "building materials" (18 documents) (Fig. 6.7e). However, the least frequently used keywords during this period were "buildings," "air," "children," "sustainability," "volatile," "fungi," "compounds," "wood," "paint," "diffusion coefficient," and "sorption" etc. (Fig. 6.7e). From Fig. 6.7 it gets clear that literature changed into more specific research methods, from VOC to formaldehyde. The importance of chemicals containing (e.g. formaldehyde) is more influential for research purposes than other types of chemicals existing in building materials. The reason can be that the significant occupational exposure sources for formaldehyde include the "processing of resins for wood products, furniture and fixtures, the wearing of garments and textiles, chemicals and plastics" (Catalani et al. 2019). Furthermore, formaldehyde is revealed to be the cause of nasal cancer in animal studies at comparatively high concentration levels, "i.e., long-term exposures to concentrations greater than about 4 ppm" (Kamata et al. 1997; Kerns et al. 1983), and connections between occupational formaldehyde exposure and multi-type cancer risk are reported (Bachand et al. 2010; Bosetti et al. 2008; Erika et al. 2010). Subsequently, the "National Toxicology Program of the National Institute of Environmental Health Sciences Changed the classification of formaldehyde from anticipated to be carcinogenic in humans to known to be a human carcinogen" (National Toxicology Program 2016).

6.4 Conclusion

Evaluating scientific research production is a challenging task. In this study, a bibliometric method was employed to identify the historical dynamics of studies investigating VOCs' release from building materials. Bibliometric is widely recognized as a well-established research method in information science, particularly for evaluating academics and universities' research performance. It adopts quantitative analysis and statistical methods to analyze the quantitative relation and content information in a given field and further examine the featured research field's detailed characteristics and patterns. The bibliometric analysis assesses and validates the trends of publications covering a given topic and analyses features of articles such as authorships, quotations, and impact factor. In addition, a statistical viewpoint is offered by the scientometric framework to consider better features involved in global construction materials, VOCs, indoor habitats, and studies on human health. One of the most important fields of chemical and environmental science is research into VOCs. For the last few decades, empirical papers on VOC studies have seen an accelerated rise in quantity. In leading scientific journals, numerous articles reporting the latest science accomplishments have been written. Study into VOCs has attracted growing scientific interest, primarily because of their adverse effects on human health and the degradation of the environment. Firstly, in the environment, the office, and consumer goods, VOCs are pervasive and various. These VOCs are issued from various construction materials as well. In recent decades, the building industry has undergone rapid growth, possibly due to social and economic change. The building sector has a significant effect on the economy, on the environment and culture. For example, one of the main consumers of resources, especially oil, is the construction industry. Energy is the main input not only at the operating stage but also during the manufacture of construction materials during the entire life cycle of buildings. However, these building materials release different chemicals, particularly VOCs, affecting indoor air quality and human health.

This study's main contribution is to present an international image of research activity in the field of building materials, volatile organic compounds, human health, and indoor environment in the last few years. A total of 1,242 documents from 1986 to 2020 were collected and analyzed. This bibliometric analysis was presented based on "yearly publications, document types and languages, subject categories, journals, countries, institutions, collaborations, research keywords, authors and their impacts, and trends." The main language was English by far; although another ten languages were still used; consequently, this designated that with "VOCs" released from building materials research became more globally concerned. China exhibited high productivity and influence on VOC and building materials studies, standing "first in both the number of publications and citation times." China was also competitive concerning collaborations. In this study, the 1,242 technical documents were printed in 475 different sources "journals, books, conferences." The most remarkable journals included Building and Environment, Atmospheric Environment, Indoor Air, and Environmental Science & Technology. "Tsinghua University" is the utmost productive university followed by "Hong Kong Polytech University." Additionally, the results showed that Wolkoff P. was the leading author in the total number of citations (TC = 1,420, article no. = 23). Moreover, according to the top 10 authors' production over time Salthammer T., published a paper in 2010 (only 1 paper), has the highest total number of citations per year to date (TC per Year = 68.545), while as the most document production per year belonged to Zhang Y in 2009 (number of papers per year = 9, TC per Year = 6.75). The boom of scientific activity in the last few years has also increased researchers and institutions' interest in publishing their studies. We implemented a step-by-step study of the study emphases and patterns, from the single-dimensional keyword intensity to the keyword matrix and time series analysis, to define trends. The number of studies regarding volatile organic compounds, especially in indoor environments, is expected to increase as their consumption in indoor building materials and products are continuously rising. Apart from this, they create serious human health and indoor air quality issues and make them topics of concern for researchers.

The synthetic analysis of author keywords provided insight into VOC, and building materials research focuses. The time-series distributions of these keywords were divided into five phases: 1986–2000, 2001–2005, 2006–2010, 2011–2015 and 2016–2020. Also, while “VOC” and “indoor air” were used most frequently in the author keywords in the early years of the research, the keywords “formaldehyde” and “building materials” increased more since 2006. The network of the identified keywords showed that “formaldehyde,” “VOC,” “indoor air”, “indoor air quality,” and “building materials” had the strongest centralities and were considered to represent the leading edges of VOC and building materials research. The fields of these leading research keywords showed substantial growth in the number of publications over time and were expected to grow in the future. Furthermore, this study’s results provide insights into patterns and trends in global VOCs, building materials, indoor environments and human health research and help expand our understanding of the situation of global VOCs, building materials, indoor environments and human health research. There has been a lot of research interest in this area, as evidenced by the continuing growth of scientific outputs. Moreover, this study’s bibliometric approach can be applied to trend analyses of other research fields. Additionally, the study results can be of great utility for organization and research planning in terms of the chemicals present in different building materials and their impacts on human health.

References

Al-hadithi AI, Tareq A, Khairi W (2019) Mechanical properties and impact behavior of PET fiber reinforced self-compacting concrete (SCC). Compos Struct 224(4):1–12. https://doi.org/10.1016/j.compstruct.2019.111021

Alhassan M, Al-rousan R, Ababneh A (2017) Flexural behavior of lightweight concrete beams encompassing various dosages of macro synthetic fibers and steel ratios. Case Stud Constr Mater 7(9):280–293. https://doi.org/10.1016/j.cscm.2017.09.004

Alhozaimy AM, Shannag MJ (2009) Performance of concrete reinforced with recycled plastic fibres. Mag Concr Res 4(4):293–298. https://doi.org/10.1680/macr.2008.00053

Amato-Lourenço LF, dos Santos GL, de Weger LA, Hiemstra PS, Vijver MG, Mauad T (2020) An emerging class of air pollutants: potential effects of microplastics to respiratory human health? Sci Total Environ 749:141676. https://doi.org/10.1016/j.scitotenv.2020.141676

Arı A (2020) A comprehensive study on gas and particle emissions from laser printers: chemical composition and health risk assessment. Atmos Pollut Res 11(2):269–282. https://doi.org/10.1016/j.apr.2019.10.013

Azimi P, Zhao D, Pouzet C, Crain NE, Stephens B (2016) Emissions of ultrafine particles and volatile organic. Environ Sci Technol 50(3):1260–1268

Bachand AM, Mundt KA, Mundt DJ, Montgomery RR (2010) Epidemiological studies of formaldehyde exposure and risk of leukemia and nasopharyngeal cancer: a meta-analysis. Crit Rev Toxicol 40(2):85–100. https://doi.org/10.3109/10408440903341696

Barnes SJ (2019) Out of sight, out of mind: plastic waste exports, psychological distance and consumer plastic purchasing. Glob Environ Chang 58:10. https://doi.org/10.1016/j.gloenvcha.2019.101943

Batterman S, Jia C, Hatzivasilis G (2007) Migration of volatile organic compounds from attached garages to residences: a major exposure source. Environ Res 104(4):224–240. https://doi.org/10.1016/j.envres.2007.01.008

Bosetti C, Mclaughlin JK, Tarone RE, Pira E, Vecchia CL (2008) Formaldehyde and cancer risk: a quantitative review of cohort studies through 2006. Ann Oncol 19(4):29–43. https://doi.org/10.1093/annonc/mdm202

Brooks BO, Utter GM, DeBroy JA, Schimke RD (1991) Indoor air pollution: an edifice complex. Clin Toxicol 29(3):315–374

Buonanno G, Morawska L, Stabile L (2009) Particle emission factors during cooking activities. Atmos Environ 43(20):3235–3242. https://doi.org/10.1016/j.atmosenv.2009.03.044

Can-Güven E (2020) Microplastics as emerging atmospheric pollutants: a review and bibliometric analysis Air Qual Atmos Health 14:203–215. https://doi.org/10.1007/s11869-020-00926-3

Cañas-Guerrero I, Mazarrón FR, Calleja-Perucho C, Pou-Merina A (2014) Bibliometric analysis in the international context of the "construction & Building Technology" category from the Web of Science database. Constr Build Mater 53(10):13–25. https://doi.org/10.1016/j.conbuildmat.2013.10.098

Catalani S, Donato F, Madeo E, Apostoli P, Palma GD, Pira E, Mundt KA, Boffetta P (2019) Occupational exposure to formaldehyde and risk of non hodgkin lymphoma: a meta-analysis. BMC Cancer 19:1245–1254. https://doi.org/10.1186/s12885-019-6445-z

Chan FL, Hon C, Tarlo SM, Rajaram N, House R, Chan FL, Hon C, Tarlo SM, Rajaram N, House R (2020) Emissions and health risks from the use of 3D printers in an occupational setting. J Toxicol Environ Health A 4:1–9. https://doi.org/10.1080/15287394.2020.1751758

Cheng S, Zhang J, Wang Y, Zhang D, Teng G, Huang Q, Zhang Y, Yan P (2019) Global research trends in health effects of volatile organic compounds during the last 16 years: a bibliometric analysis. Aerosol Air Qual Res 19:1834–1843. https://doi.org/10.4209/aaqr.2019.06.0327

De Gennaro G, de Gennaro L, Mazzone A, Porcelli F, Tutino M (2014) Indoor air quality in hair salons: screening of volatile organic compounds and indicators based on health risk assessment. Atmos Environ 83:119–126. https://doi.org/10.1016/j.atmosenv.2013.10.056

Dettori JR, Norvell DC, Chapman JR (2019) Measuring academic success: the art and science of publication metrics. Global Spine J 9(2):243–246. https://doi.org/10.1177/2192568219831003

Dhawan R, Bisht BMS, Kumar R, Kumari S, Dhawan SK (2019) Recycling of plastic waste into tiles with reduced flammability and improved tensile strength. Process Saf Environ Prot 124:299–307. https://doi.org/10.1016/j.psep.2019.02.018

Donaldson K, Tran CL (2002) Inflammation caused by particles and fibres. Inhal Toxicol 14(1):5–27. https://doi.org/10.1080/089583701753338613

Egghe L (2006) Theory and practise of the g-index. Scientometrics 69(1):131–152

El-Hashemy MA, Ali HM (2018) Characterization of BTEX group of VOCs and inhalation risks in indoor microenvironments at small enterprises. Sci Total Environ 645:974–983. https://doi.org/10.1016/j.scitotenv.2018.07.157

Enesca A, Cazan C (2020) Volatile organic compounds (VOCs) removal from indoor air by heterostructures/composites/doped photocatalysts: a mini-review. Nanomaterials 10(10):1965–1984. https://doi.org/10.3390/nano10101965

Erika S, Luoping ZT, Smith M, Allan HS, Craig S (2010) Formaldehyde and leukemia: an updated meta-analysis and evaluation of bias. J Occup Environ Med 59(9):1–4. https://doi.org/10.1097/JOM.0b013e3181ef7e31

Factor WI, Factor RI, Rate C, Report C, Factor I, Rate C (2010) Global psychology: a bibliometric analysis of Web of Science publications. Univ Psychol 9(2):553–568

Foti D (2019) Recycled waste PET for sustainable fiber-reinforced concrete. In Pacheco-Torgal F, Khatib J, Colangelo F, Tuladhar R (eds) Use of recycled plastics in eco-efficient concrete. Woodhead Publishing, Elsevier Ltd. https://doi.org/10.1016/B978-0-08-102676-2.00018-9

Gasperi J, Wright SL, Dris R, Collard F, Mandin C, Guerrouache M, Langlois V, Kelly FJ, Tassin B (2018) Microplastics in air: Are we breathing it in? Environ Sci Health 1:1–5. https://doi.org/10.1016/j.coesh.2017.10.002

Geng S, Wang Y, Zuo J, Zhou Z, Du H, Mao G (2017) Building life cycle assessment research: a review by bibliometric analysis. Renew Sustain Energy Rev 76(3):176–184. https://doi.org/10.1016/j.rser.2017.03.068

Greim H, Borm P, Schins R, Donaldson K, Driscoll K, Hartwig A, Oberdörster G, Speit G, Borm P, Schins R, Donaldson K, Driscoll K, Hartwig A (2008) Toxicity of fibres and particles? Report of the workshop held in Munich Germany. Inhal Toxicol 13(9):737–754. https://doi.org/10.1080/08958370118273

Harčárová K, Vilčeková S, Bálintová M (2020) Building materials as potential emission sources of VOC in the indoor environment of buildings. Key Eng Mater 838(4):74–80. https://doi.org/10.4028/www.scientific.net/KEM.838.74

Harzing A, Alakangas S (2016) Google Scholar, Scopus and the Web of Science: a longitudinal and cross-disciplinary comparison. Scientometrics 2:1–18. https://doi.org/10.1007/s11192-015-1798-9

He Z, Zhang Y, Wei W (2012) Formaldehyde and VOC emissions at different manufacturing stages of wood-based panels. Build Environ 47(7):197–204. https://doi.org/10.1016/j.buildenv.2011.07.023

Hirsch JE (2005) An index to quantify an individual's scientific research output. PNAS 102(11):16569–16572. https://doi.org/10.1073/pnas.0507655102

Hodge DR, Lacasse JR, Hodge DR, Lacasse JR, Disciplinary R (2013) Ranking disciplinary journals with the Google Scholar h-index: a new tool for constructıng C-cases for tenure, promotion, and other professional decisions. J Soc Work Educ 47(3):579–596. https://doi.org/10.5175/JSWE.2011.201000024

Hormigos-Jiménez S, Padilla-Marcos MA, Meiss A, Gonzalez-Lezcano RA, Feijo-Munoz J (2017) Ventilation rate determination method for residential buildings according to VOC emissions from building materials. Build Environ 123(7):555–563. https://doi.org/10.1016/j.buildenv.2017.07.032

Hua X, Wu Y, Zhang X, Cheng S, Wang X, Chu J, Huang Q (2018) Analysis on ambient volatile organic compounds and their human gene targets. Aerosol Air Qual Res 18:2654–2665. https://doi.org/10.4209/aaqr.2018.08.0320

Huang L, Krigsvoll G, Johansen F, Liu Y, Zhang X (2018) Carbon emission of the global construction sector. Renew Sustain Energy Rev 81(6):1906–1916. https://doi.org/10.1016/j.rser.2017.06.001

Jean P, Gleize P, Ramos H (2012) Mechanical properties of recycled PET fibers in concrete. Mater Res 15(4):679–686. https://doi.org/10.1590/S1516-14392012005000088

Jones AP (1999) Indoor air quality and health. Atmos Environ 33(5):4535–4564. https://doi.org/10.1016/S1352-2310(99)00272-1

Kamata E, Nakadate M, Uchida O, Ogawa Y, Suzuki S, Kaneko T, Saito M, Kurokawa Y (1997) Results of a 28-month chronic inhalation toxicity study of formaldehyde in male fischer-344 rats. J Toxicol Sci 22(3):239–254. https://doi.org/10.2131/jts.22.3_239

Karthikeyan S, Arun P, Thiyaneswaran MP (2020) Summary of non-biodegradable wastes in concrete. AIP Conf 2235(4):1–7. https://doi.org/10.1063/5.0007586

Kawuki J, Yu X, Musa TH (2020) Bibliometric analysis of ebola research indexed in Web of Science and Scopus (2010–2020). BioMed Res Int 9:1–12. https://doi.org/10.1155/2020/5476567

Kerns WD, Pavkov KL, Donofrio DJ, Gralla EJ, Swenberg JA (1983) Carcinogenicity of formaldehyde in rats and mice after long-term inhalation exposure. Can Res 43(9):4382–4392

Kowalska J, Szewczyńska M, Pośniak M (2015) Measurements of chlorinated volatile organic compounds emitted from office printers and photocopiers. Environ Sci Pollut Res 22(7):5241–5252. https://doi.org/10.1007/s11356-014-3672-3

Kwok N, Lee S, Guo H, Hung W (2003) Substrate elects on VOC emissions from an interior finishing varnish. Build Environ 38(3):1019–1026. https://doi.org/10.1016/S0360-1323(03)00066-0

Lee SC, Lam S, Kin-Fai H (2001) Characterization of VOCs, ozone, and PM10 emissions from office equipment in an environmental chamber. Build Environ 36(7):837–842. https://doi.org/10.1016/S0360-1323(01)00009-9

Li Y, Li J, Xie S (2017) Bibliometric analysis: global research trends in biogenic volatile organic compounds during 1991–2014. Environ Earth Sci 76(1):1–13. https://doi.org/10.1007/s12665-016-6328-4

Lin CC, Yu KP, Zhao P, Lee GWM (2009) Evaluation of impact factors on VOC emissions and concentrations from wooden flooring based on chamber tests. Build Environ 44(3):525–533. https://doi.org/10.1016/j.buildenv.2008.04.015

Liu Z, Little JC (2012) Materials responsible for formaldehyde and volatile organic compound (VOC) emissions. In: Pacheco-Torgal F, Jalali S, Fucic A (eds) Toxicity of building materials. Elsevier, pp 76–121. https://doi.org/10.1533/9780857096357.76

Malagavelli V (2020) Effect of non biodegradable waste in concrete slabs. Int J Comput Civ Struct Eng 1(1):449–457

Mao C, Shen Q, Pan W, Ye K (2015) Major barriers to off-site construction: the developer's perspective in China. J Manag Eng 31(3):1–8. https://doi.org/10.1061/(asce)me.1943-5479.0000246

Mingers J, Macri F, Petrovici D (2012) Using the h-index to measure the quality of journals in the field of business and management. Inf Process Manag 48(3):234–241. https://doi.org/10.1016/j.ipm.2011.03.009

Mishra S (2017) Is smog innocuous? Air pollution and cardiovascular disease. Indian Heart J 69(4):425–429. https://doi.org/10.1016/j.ihj.2017.07.016

Morawska L, Ayoko GA, Bae GN, Buonanno G, Chao CYH, Clifford S, Fu SC, Hänninen O, He C, Isaxon C, Mazaheri M, Salthammer T, Waring MS, Wierzbicka A (2017) Airborne particles in the indoor environment of homes, schools, offices and aged care facilities: the main routes of exposure. Environ Int 108(4):75–83. https://doi.org/10.1016/j.envint.2017.07.025

Morawska L, He C (2006) Indoor particles, combustion products and fibres. In: Handbook of environmental chemistry, vol 5, pp 1–12. https://doi.org/10.1007/698

Mujan I, Andelkovic AS, Muncan V, Kljajic M, Ruzic D (2019) Influence of indoor environmental quality on human health and productivity—a review. J Clean Prod J 217(1):646–657. https://doi.org/10.1016/j.jclepro.2019.01.307

National Toxicology Program (2016) Report on carcinogens, 14th edn. http://ntp.niehs.nih.gov/go/roc

Patience GS, Patience CA, Blais B, Bertrand F (2017) Citation analysis of scientific categories. Heliyon 3(4):1–23. https://doi.org/10.1016/j.heliyon.2017.e00300

Pauly JL, Stegmeier SJ, Allaart HA, Cheney RT, Zhang PJ, Mayer AG, Streck RJ (1998) Inhaled cellulosic and plastic fibers found in human lung tissue. Cancer Epidemiol Biomark Prev 7(5):0419–0428

Pešic N, Zivanovic S, Garcia R, Papastergiou P (2016) Mechanical properties of concrete reinforced with recycled HDPE plastic fibres. Constr Build Mater 115(3):362–370. https://doi.org/10.1016/j.conbuildmat.2016.04.050

Prata JC (2018) Airborne microplastics: consequences to human health? Environ Pollut 234:115–126. https://doi.org/10.1016/j.envpol.2017.11.043

Qi GY, Shen LY, Zeng SX, Jorge OJ (2010) The drivers for contractors' green innovation: an industry perspective. J Clean Prod 18(3):1358–1365. https://doi.org/10.1016/j.jclepro.2010.04.017

Rösch C, Kohajda T, Röder S, Bergen MV, Schlink U (2014) Relationship between sources and patterns of VOCs in indoor air. Atmos Pollut Res 5(1):129–137. https://doi.org/10.5094/APR.2014.016

Salthammer T, Mentese S, Marutzky R (2010) Formaldehyde in the indoor environment. Chem Rev 110(6):2536–2572. https://doi.org/10.1021/cr800399g

Seltzer JM (1997) Sources, concentrations, and assessment of indoor pollution. In: Indoor Air Pollution and Health, pp 11–52

Silva DA, Betioli AM, Gleize PJP, Roman HR, Go LA, Ribeiro JLD (2005) Degradation of recycled PET fibers in Portland cement-based materials. Cem Concr Res 35:1741–1746. https://doi.org/10.1016/j.cemconres.2004.10.040

Silva ER, Coelho JFJ, Bordado JC (2013) Strength improvement of mortar composites reinforced with newly hybrid-blended fibres: influence of fibres geometry and morphology. Constr Build Mater 40:473–480. https://doi.org/10.1016/j.conbuildmat.2012.11.017

Singh D, Kumar A, Kumar K, Singh B, Mina U, Singh BB, Jain VK (2016) Statistical modeling of O_3, NOx, CO, PM2.5, VOCs and noise levels in commercial complexes and associated health risk assessment in an academic institution. Sci Total Environ 572(8):586–594. https://doi.org/10.1016/j.scitotenv.2016.08.086

Spengler JD, Chen QY (2000) Indoor air quality factors in designing a healthy building. Annu Rev Energy Env 25:567–601

Vianello A, Jensen RL, Liu L, Vollertsen J (2019) Simulating human exposure to indoor airborne microplastics using a Breathing Thermal Manikin. Sci Rep 9(1):1–11. https://doi.org/10.1038/s41598-019-45054-w

Wallace LA, Pellizzari E, Leaderer B, Zelon H, Sheldon L (1987) Emissions of volatile organic compounds from building materials and consumer products. Atmos Environ 21(2):385–393

Wang Y, Xue X, Yu T, Wang Y (2020) Mapping the dynamics of China's prefabricated building policies from 1956 to 2019: a bibliometric analysis. Buildi Res Inf 7:1–18. https://doi.org/10.1080/09613218.2020.1789444

WoS (2021) Web of Science. www.webofknowledge.com/WOS

Wright SL, Kelly FJ (2017) Plastic and human health: a micro issue? Environ Sci Technol 51(5):6634–6647. https://doi.org/10.1021/acs.est.7b00423

Yrieix C, Dulaurent A, Laffargue C, Maupetit F, Pacary T, Uhde E (2010) Chemosphere characterization of VOC and formaldehyde emissions from a wood based panel: results from an inter-laboratory comparison. Chemosphere 79(4):414–419. https://doi.org/10.1016/j.chemosphere.2010.01.062

Yu T, Liang X, Shen GQ, Shi Q, Wang G (2019) An optimization model for managing stakeholder conflicts in urban redevelopment projects in China. J Clean Prod 212:537–547. https://doi.org/10.1016/j.jclepro.2018.12.071

Zhang G, Xie S, Ho Y (2010) A bibliometric analysis of world volatile organic compounds research trends. Scientometrics 83(6):477–492. https://doi.org/10.1007/s11192-009-0065-3

Chapter 7
Heavy Metal Contamination from Construction Materials

Ayodeji Ojo Oteyola and Folasade Adesola Ola-Oladimeji

Abstract Construction or building materials can pose a major public health challenge because they can act as pollutant emission sources. Building materials lead to the contamination of heavy metal. Despite several discoveries as regards the adverse health effects of heavy metals, there is an increase in heavy metal exposure especially in underdeveloped and developing countries. Lead pipes are used within buildings. Lead that dissolves into the drinking water from lead pipes is a threat to human health. The peeling and cracking of lead-based paint is also a major health concern as it has been reported that it affects young children's brain and sense organs. In adults, lead accumulation can lead to high blood pressure, reproduction dysfunction and sometimes death. This chapter focuses on the use of lead and other heavy metals in buildings and their toxicity to humans. The assessment of the risk associated with lead in drinking water is also discussed. In addition, lead pipe replacement is also deliberated. The chapter also reflects on other heavy metals which include cadmium, chromium and mercury with respect to their uses, toxicity and health effects on indoor pollution. Biological monitoring of cadmium, chromium and mercury are also explained.

Keywords Building materials · Cadmium · Chromium · Lead · Mercury · Toxicity

7.1 Introduction

Building materials can act as a source of pollutant emission and thereby affect the quality of indoor air (US EPA 1994). The number of compounds emitted from building materials varies. Exposure to these compounds can be hazardous to human health. There have been several struggles and efforts to limit the exposure in the building design industry. The idea is that if in a building design, a building material

A. O. Oteyola (✉)
Departamento de Morphology, Institute of Biological Sciences, Federal University of Minas Gerais, Av. Antônio Carlos, Belo Horizonte, Minas Gerais, Brazil

F. A. Ola-Oladimeji
Department of Zoology and Environmental Biology, Ekiti State University, Ado-Ekiti, Nigeria

J. A. Malik and S. Marathe (eds.), *Ecological and Health Effects of Building Materials*,
https://doi.org/10.1007/978-3-030-76073-1_7

is known to be a source of pollution and needs more ventilation to neutralize its effects, it will be more advisable to use an alternative material that has less emission rate (CIBSE 1996).

Heavy metals pose a high risk of adverse effects on human health, some metals are carcinogenic, neurotoxic and immunotoxic. Metals are usually found in the environment and as a result of the extensive use in the environment, human exposure is unavoidable. Understanding the mechanisms of heavy metal toxicity has been challenging as a result of the association between metals and living systems. Research has shown that metals are compromised in cardio-respiratory related illnesses that are linked to the pollution of the air (Burnett et al. 2000; Vincent et al. 2001; Claiborne et al. 2002; Sorensen et al. 2005). Osornio-Vargas et al. (2003) reported that transition metals are linked to oxidative stress and respiratory dysfunction.

Some metals are indispensable and vital for life as a result of their role in the provision of required enzymes but excessive concentration of these enzymes can be harmful either by choking important functional groups or altering the conformation of protein molecules (Collins and Stotzky 1989). Metals have a profound effect on physiological and biochemical processes which include photosynthesis and respiration inhibition (Vangronsveld and Clijsters 1994). Transition elements are used in making paints and this poses a serious threat to human health as a result of deviation from manufacturer's standard, unsuitable storage, carelessness in handling and transportation. Lead based paint is still being used in a lot of homes today. The peeling and damage of lead-based paint is hazardous and it can amount to threat to human health especially in children. Cadmium, mercury and some heavy metals are naturally occurring but human exposure occurs as a result of anthropogenic activities.

This chapter focuses on the heavy metals which include lead, cadmium, chromium and mercury as it relates to building materials with emphasis on their structure, uses and its toxicological effects. Awareness about environmental pollutants and their adverse effects on human health cannot be over-emphasized as the aim is to protect people from avoidable exposures.

7.2 Lead

Lead metals play an important role in transporting water and its usage can be traced back as far as the Roman times. Roofing materials can also be produced from lead metals. Lead metals decay slowly which justify its high acceptability in usage. However, lead has been known to be toxic; the first reported effect of lead on human health was the fall of Roman Empire which was influenced by lead poisoning (Dutrizac et al. 1982; Hodge 1981; Nriagu 1983). In the home today, health problems from lead can be from lead pipes which are used in connecting homes to the water source; it can also be from lead roofing and guttering; it can also be from lead-based house paint (Wani et al. 2015). Lead absorption in the body accumulates in the blood, bones, liver, kidney, brain and skin. Krzywy et al. (2010) revealed that lead

Table 7.1 Lead pipes occurrence in Europe (Adapted after: Van den Hoven et al. 1999)

Country	Percentage of Pb communication pipe
Belgium	19
Denmark	0
France	39
Germany	3
Greece	<1
Ireland	50
Italy	2
Luxembourg	7
Netherlands	6
UK	40

causes reproductive dysfunction, hepatic dysfunction and also affects the immune and gastrointestinal systems (Krzywy et al. 2010).

7.2.1 Lead Piping and Water Contamination

Lead pipes are used to convey water from the water main source to the apartment buildings. Lead pipes are buried inside the soil thereby making it difficult to establish their presence or absence. In the 1800s, the beginning of urbanization and mass housing brought about the use of lead connection pipes. Despite the caution that was expressed about the use of lead pipes in the nineteen century (Troesken 2006), the use of lead pipes continued till the 1980s in both Europe and North America. The occurrence of lead pipes in different countries in Europe is shown in Table 7.1. The length of the pipe varies from a few meters to hundreds of meters depending on the distance between the dwelling and the water source. The risk is associated with the length of the pipe, the longer the lead pipe, the greater the risk to human well-being.

Hayes and Skubala (2009) reported that about 25% of dwellings in Europe are likely to be supplied with drinking water through lead pipes. About 2–3% of dwellings in the US and Canada use lead pipes (IWA 2010). The corrosion of the pipe and the contact of the water with the lead pipe determine the amount of lead that will be present in the drinking water. Lead concentration in drinking water can vary from <1 to hundred μg/l.

7.2.2 Lead Toxicity

Lead is stored in the liver and also can be found in the kidneys. Its distribution can also be in the brain, adrenals, prostate, ovary, spleen, fat tissues, heart and spinal cord (Seervarisai et al. 2015). Lead is harmful to human health; reports from clinical researches have confirmed this claim (Lanphear et al. 2000; Menke et al. 2006; Jusko et al. 2008; Khalil et al. 2009). Lead affects many biochemical processes such as calcium inhibition and reaction with proteins. Upon absorption, lead substitutes with calcium and this leads to interaction with biological processes and thereby causing abnormalities. As a result of its reaction with protein, Pb causes changes in enzyme structure and causes dysfunction in the binding site. The primary mechanism that leads to lead toxicity is the oxidative stress that is caused by lead.

Research from the World Health Organization has shown that lead affects children's central nervous system and development (WHO 2016). Children less than six (6) years old are very susceptible to the harmful effect of lead because of the brain growth and development (Wani et al. 2015; WHO 2016). Lead can impair memory, lower intelligent quotient (IQ) in children. Reports have shown that lead has profound effects on speech, hearing, vomiting and intestinal pain. At high concentration as shown in Table 7.2, lead poisoning can cause anemia, paralysis, convulsion or death (Giel-Pietraszuk et al. 2012; Rehman et al. 2018). During pregnancy, maternal lead exposure can be transferred to the developing embryo via placenta (Rehman et al. 2018; Omeljaniuk et al. 2018). In addition, during pregnancy, when the calcium intake of the mother is low, the lead that is stored in the mother's bone can be transferred via the blood to the fetus. This confirms that maternal exposure to lead before or during pregnancy can have a devastating effect on the developing embryo as a consequence of the stored lead in the bone (CDC 2010; Bellinger 2013). As a result

Table 7.2 Level of blood lead and degree of lead poisoning and the symptoms in children and adult (Adapted after: Giel-Pietraszuk et al. 2012)

Blood lead level (μg/L)	Degrees of lead poisoning	Synonyms	
		Children	Adults
<10	Low	Growth disorders, lower IQ, memory disabilities, hearing disorder, speech disorder	Movement from the maternal bloodstream to the fetal through the placenta
10–40	Mild	Muscle pain, fatigue, irritability, impaired hemoglobin synthesis, apathy	Increased in blood pressure
40–70	Moderate	Weight loss, constipation, vomiting, headache, fatigue, muscular weakness	Chronic hypertension, fertility abnormalities, reduced mental abilities
70–100	Serious	Anemia, nephropathy, paralysis	Reduction in sex drive, constipation, headaches, nephropathy, insomnia
>100	Acute poisoning	Convulsion, death	Anemia, death (>150 μg/L)

of the toxicity of lead in children, WHO has recommended blood lead concentration of 5 μg/dL or less for children (CDC 2010; Wani et al. 2015; WHO 2016).

Lead also has harmful effects in adults. It can lead to dysfunction in the nervous system (fatigue, drowsiness), circulatory system (increased blood pressure), hormones (fertility abnormalities, decreased libido), cardiovascular system and it can also lead to death (Wani et al. 2015). Individuals with high concentration of lead in their blood (500–870 μg/L) often encounter sinus node dysfunction and also atrioventricular block (Zawadzki et al. 2006). In adults, lead concentration in the blood has been linked to blood pressure. Nevertheless, its effect on blood pressure depends on dosage and exposure time (Wani et al. 2015). Navas-Acien and colleagues confirmed that exposure to lead is linked to hypertension (Navas-Acien et al. 2007). Research has also revealed the relationship between blood lead concentration and cancer especially brain, lung tumor (Steenland et al. 2019).

7.2.3 Replacement of Lead Pipes

There are some speculations that lead pipes have been withdrawn but of a truth, very few have been removed and many are still been used especially in places like Africa. Some water corporations also claim that lead pipes has been removed but this has been less effective because some homes still use lead pipes within their dwellings (DWI 2010). Efforts have been made to remove all lead pipes but internal lead piping has been the challenge as a result of the stress and the cost involved. Considering the various harmful effects of lead on children and adults, the idea is to have a lead pipe-free home. To achieve this, laws should be made that the home-owners should remove all the lead pipes before sale or letting it out. In the meantime, Hayes and Croft suggested that the application of point-of-use filters can be used to remove lead from water (Hayes and Croft 2012). Leaded solder was allowed in the 1980s. Many homes still have a leaded solder and the removal is not realistic, unless under special conditions. This suggests that water supply systems require a strategy to control corrosion so as to minimize these lethal effects. In addition, lead-free brass with about 8% lead has been allowed. However, in the 1980s, brass fittings containing >8% lead were used and many homes still have this internal pipework system. To have effective control, all countries need to take after the US by enforcing the use of brass with <0.25% of lead and ban the use of leaded solder.

7.3 Cadmium

7.3.1 Cadmium History and Application

Cadmium (Cd) was first discovered in Germany by Friedrich Stromeyer in 1817. Cd is readily found in little concentration mostly alongside sulphide of zinc, lead and copper. Associated use of Cd as building materials include pigments, stabilizers of PVC and alloys, protective agents of steel from corrosion, Ni–Cd batteries (WHO 1992; IARC 1993; Hayat et al. 2019). Cd compounds are used as coloring agents in borosilicate glass. Cd is a rare element; its estimation is around 0.1–0.2 ppm which signifies that it is the 67th element in order of abundance (Brewers et al. 1987).

7.3.2 Chemical Forms and Properties of Cd

Cadmium (Cd) is a member of the group XII in the periodic table of chemical elements. Cd atomic mass is as a result of eight stable isotopes. The colour of Cd is silvery-white. Cd properties are synthetically related to zinc and mercury. In the s orbital it has double electrons and a complete d orbital. Cd is resistant to corrosion hence, the main reason it serves as a protective agent. It is insoluble in water; it forms cadmium oxide when burned in air.

7.3.3 Cadmium Toxicity

The US environmental protection agency identifies cadmium as one of the 126 priority pollutants. Cadmium is absorbed in high quantities from air, food and water contaminations which confirms the role that building materials play in its absorption. Cadmium has the potential to accumulate throughout a lifetime which makes it more harmful. Cadmium salts accumulate and lead to the toxicity of many organs which includes the kidney, lungs, brain, testes, and the heart. Cadmium can cause anemia, kidney failure, anosmia, non-hypertrophic emphysema. People that make cadmium products like batteries, plastic, soldering or welding building materials can be exposed to cadmium.

Cadmium inhalation linked with respiratory diseases which include lung cancer. Mona and colleagues revealed an increase in cadmium level of smoker's urine (Mona et al. 2018). In addition, it was reported that the percentage of bone pain of smokers increased (95%) non-smokers (37.7%). This suggests that exposure to cadmium results in an ototoxic effect which can in turn lead to bone tissue loss. The epidemic of Itai-Itai disease was as a result of cadmium exposure (Hagino and Yoshioka 1960).

Cadmium exposure even at low doses can result in abnormalities in male and female reproduction and thereby affect pregnancy and offspring outcomes (Kumar and Sharma 2019). Exposure to cadmium also causes irregularities in the menstrual cycle, alterations in reproductive hormones, pregnancy loss, premature birth and reduction in birth weight (Lokhande et al. 2004). Cadmium is also known to be an endocrine disruptor because it interferes with many hormonal signaling pathways, one of such is that it can bind estrogen receptor α; it can also interfere with signaling transduction through estrogen and mitogen-activated protein kinase (MAPK) signaling pathways (Söderholm et al. 2020).

Cadmium exposure has an epigenetic effect and can also alter gene expression. Exposure to cadmium can result in DNA methylation and histone modifications. Heavy metals can utilize their toxicity through the miRNA. Endogenous miRNA alterations have been associated with impaired signaling pathways which subsequently result in different kinds of cancer and disorders (Wallace et al. 2020). Takiguchi and colleagues (2003) in their study reported that short time exposure (24 h–1 week) results in hypomethylation, whereas long exposure (8–10 weeks) leads to hypermethylation (Takiguchi et al. 2003).

Cadmium has been classified as a cancer causing agent (WHO 1992; IARC 1993). Cadmium toxicity results in oxidative stress that further promotes the spread of tumor through its effect on cell cycle (Rani et al. 2014). Cadmium inhibits DNA repair mechanisms and this can lead to accumulation of damaged DNA hence the initiation of cancer (Giaginis et al. 2006). A study conducted from 2006 to 2014 as reported by Nawrot and colleagues showed that lung cancer is linked to the exposure to cadmium. In this study, it was revealed that cadmium is a risk factor for lung cancer (Nawrot et al. 2015). Furthermore, scientific data also suggests that cadmium exposure can be associated with prostate cancer (Julin et al. 2012), urinary bladder cancer (Feki-Tounsi and Hamza-Chaffai 2014), cancer of the kidney (Song et al. 2015) and pancreas cancer (Chen et al. 2015).

Studies in animal models have shown that antioxidants (vitamin C and E) can protect from toxicity induced by cadmium (Ognjanovic et al. 2003; Beytut et al. 2003). Membrane separation and reverse osmosis are sometimes used to get rid of heavy metals in liquid but it is very expensive and requires large quantities of chemicals.

7.3.4 Biomonitoring of Cadmium

Biomonitoring studies involving cadmium level helps physicians and public health officials to ascertain exposure level of cadmium in individuals and groups of population. Cadmium level in urine and blood help in biological monitoring of cadmium. Blood cadmium usually reveals current exposure and it ranges from 0.2 to 0.8 pg/l. Although, smokers are known to have higher concentrations ranging from 1.4 to 4.5 pg/l as reported by Elinder et al. (1994). Urinary excretion of cadmium for a day is a biological marker of lifespan exposure, whereas cadmium concentration in blood reveals current exposure (Staessen et al. 1996). Data from biomonitoring studies can help researchers to conduct future research on exposure and health effects of cadmium.

7.4 Chromium

7.4.1 Chromium History and Application

Chromium is commonly used together with other metals. It was discovered by Louis Vauquelin in 1797. Louis Vauquelin discovered chromium while he was making an experiment with a material called 'Siberian red lead', which is also known as 'mineral crocoite ($PbCrO_4$)'. Vauquelin produced chromium oxide (CrO_3) by putting together crocoite and hydrochloric acid (HCl) Application of chromium as building materials include dyes and paints, stainless steel, chrome plating and metallurgy.

Chromium is being added to metals to make stainless steel. Its corrosion resistant nature extends the lifespan of products making it an important ingredient in building materials. About 80% of mined chromium is used for metallurgical processes, in which many of the0se are used in making stainless steel. About 15% is used in the manufacturing of chromium chemicals and about 5% is used in refractory applications. The corrosion resistant nature and the attractive nature form the major factor for its selection. Potassium dichromate ($K_2Cr_2O_7$) is used in chemicals for leather tanning and also used for wood preservation (Johnson et al. 2006). Copper chrome arsenic (CCA) is used in the preservation of woods and it contains chromium trioxide, arsenic pentoxide and copper (II) oxide. It is made available in form of paste or water-based concentration and CCA has been used for preservation of wood so as to prevent decay caused by fungi and wood-boring insects (Cocker et al. 2006). Lead Chromate ($PbCrO_4$) has also been known to be used as yellow pigments in paints.

7.4.2 Chemical Form and Properties of Chromium

Chromium is used in manufacturing stainless steel and pressure-treated wood. Data from exposure to chromium from occupation shows that about half a million workers are exposed to chromium in the US and unspecified millions globally (Zhitkovich 2002; OSHA 2006). Exposure to chromium occurs through air emissions (cigarette smoke, industrial, automobile smoke) and water contamination (industrial release, wastewater leaching) (Welling et al. 2015). Cr(0) usually occurs in alloys with metals such as iron in its metallic form. Through welding, Cr(0)is converted to chromium (III) and chromium (VI). Chromium (III) is an indispensable micronutrient that is required in metabolism of glucose (Schwarz and Mertz 1959), but in the present-day, the report is now being debated (Ali et al. 2011; Xia et al. 2016; Vincent 2017, 2019).

Employees are exposed to chromium in more than 80 different industries. In construction companies, tanning, electroplating, metal finishing, welding, wood preservatives and pigments production, workers are exposed to chromium (IARC 1990).

7.4.3 Chromium Toxicity

Regardless of the mode of chromium absorption, the adverse health effects caused by chromium usually depend on the oxidation state, chemical species bioavailability and solubility. After absorption, chromium compounds can undergo alkylation and oxidation state alteration.

Chromium (III) has low membrane permeability, and this can lead to change in the membrane morphology, disruption in cell functions and also can induce DNA damage (Fang et al. 2014). Transferrin, a β-globulin, helps to protect against chromium toxicity that is obtained from food consumption by inhibiting the aggregation of chromium (III) and mediating chromium (III) transmembrane transportation. The transferrin (*Tf*) gene is distinct having about 38 recognized variants, studies linking the recognized variants of transferrin to Cr (III) transport is required to elucidate the effects of the variant (Edward et al. 2019).

Chromium (VI) exposure may affect many systems in the body which include skin, irritation in respiration, kidney failure, hepatic (liver damage and increase in liver enzymes), genetic (genotoxicity, mutagenicity and carcinogenicity) and reproductive effects (birth weight, anovulation, irregular menstrual cycle) (IARC 1990; Teklay 2016; Wang et al. 2017). However, the gravity of these health effects depends on the extent of the exposure to chromium (VI).

Chromium (VI) compounds have been identified as the primary source of lung cancer (IARC 1990). Chromium (IV) has been considered to be carcinogenic (IARC 1990). Low concentration of Chromium (VI) causes DNA breakage. The established limit by the Occupational Safety and Health Administration (OSHA 2006) was 5 $\mu g\ m^{-3}$, recent risk assessment has revealed that of 1000 employees cancer deaths, about 10–50 were exposed to this limit (Beyersmann and Hartwig 2008). Deposits of chromium (VI) are shown in different studies to cause tissue damage which eventually leads to lung cancer. Accumulation of Cr (VI) in tissues leads to DNA impairment, tumor initiation which eventually causes carcinogenesis (Ishikawa et al. 1994; Proctor et al. 2014; Browning and Wise 2017).

In vitro studies showed that chromium (VI) can form a reaction with histone and then be transported to the nucleus (Levina et al. 2006). The reaction and transportation suggests a mechanism of DNA damage. Cr (VI) exposure induces histone modifications which include reduced acetylation and increased biotinylation and this mechanism has been suggested as a mechanism of tumor initiation and progression (Xia et al. 2014).

Cr (VI) compounds are soluble penetrating the cell through a negatively charged ion carrier (Buttner and Beyersmann 1985). On getting to the cell, reduction reaction takes place by various chemical and enzymatic reductants and it eventually forms Cr (III). There is formation of reactive chromium species Cr (II), Cr (IV), Cr (V) during this reduction reaction. This can induce reactive oxidative stress (ROS) (Borthiry et al. 2007).

7.4.4 Biomonitoring of Chromium

The literature has provided enough reports as regards the biomonitoring of chromium level in urine in exposed populations. For instance in population of people working with materials linked to chromium (for instance, welder) (Mutti et al. 1984; Minoia and Cavalieri 1988; Brand et al. 2010). Little has been done regarding investigating residential populations that are potentially exposed to environmental chromium, the reason being the presence of high levels of chromium in urine of both exposed and controlled populations.

In a population of people handling timber treated with copper chrome arsenic (CCA), exposure was confirmed (Garrod et al. 1999). Dermal exposure was the primary route of exposure and exposure through inhalation was observed as relatively low. Chromium concentration in urine of workers that were exposed to wood preservatives is significantly higher than the non-exposed people.

Human biomonitoring results and concentration of metal air of 36 welders welding solid steel were revealed by Mutti et al. (1984). It was revealed in this study that water soluble Cr (IV) concentrations in the air ranged from 0.010 to 0.152 mg m^{-3} with a median of 0.065 mg m^{-3}, while total Cr concentrations in the air ranged from 0.012–0.224 mg m^{-3} with a median of 0.094 mg m^{-3}. The mean urinary chromium was 33.3 μg/g. Gube and colleagues revealed that after exposure, the mean chromium concentration in the urine is 1.61 μg/l (Gube et al. 2013).

7.5 Mercury

7.5.1 Mercury History and Application

Mercury is represented by the symbol "Hg". "Hg" was coined from Latin word "*hydrargyrum*". Mercury is liquid at ambient temperature. The use of mercury as building materials is related to its use in fluorescent lamps. It can also be utilized in technological applications such as thermometers, barometers, and gas meters. Culturally, it can be used as preservatives in home paints and also as antiseptics. The technological and cultural uses of mercury make it an indoor air pollutant in residential places.

Properties such as low melting and boiling points, biological cycle's participation makes mercury (Hg) to be considered as global pollutants. Mercury is released through natural and anthropogenic activities but studies have revealed that it is the anthropogenic emission that contaminates the environment. Estimates revealed that total release of mercury from both sources from past activities is 6000–11000 tons per year (Swartzendruber 2012). It is believed that emission through anthropogenic activities constitute about 80% of the yearly discharge to the environment. Following the industrial revolution, the mercury level of the surface water has tripled. Mercury concentrations are now about 10–12 times higher in marine animals today as compared to the preindustrial times (prior to 1800) (Braune et al. 2005; Lamborg et al. 2014). Mercury compounds are mostly used in paint production. The content of mercury in a commercial paint is usually about 0.05% and this may cause air pollution.

7.5.2 *Chemical Form and Properties of Mercury*

Mercury is an odorless and silvery metallic liquid and it is insoluble in water. It is the 80th member on the periodic table. Mercury is ubiquitous and it is usually in three forms; (1) elemental (2) inorganic and (3) organic mercury. Reports have shown that all the forms of mercury have toxic effects on mammals (Clarkson et al. 2003; WHO 2007). Elemental (metallic) Hgwhich is liquid at room temperature has a high vapour pressure and it is released to the environment in the form of mercury vapour. The methylation of the inorganic form of mercury leads to the formation of the methylmercury. In the environment today, mammalian species are exposed to all the three forms of mercury; elemental (Hg), inorganic [Hg(I)], [Hg(II)] and organic mercuric compounds (WHO 2007).

Human exposure to elemental mercury is from artisanal gold mining. Inhalation of vapourized elemental mercury is usually absorbed into the blood and crosses the blood-brain barrier. Estimation shows that 10–15 million miners (mostly in developing countries) are affected by chronic mercury intoxication (Ha et al. 2017).

7.5.3 *Mercurial Toxicity*

Generally, mercury(II) chloride or mercuric chloride ($HgCl_2$) and methylmercury (MeHg) play a crucial role in inducing many biological processes ranging from increased lipid peroxidation to formation of ROS, depletion of glutathione (GSH) (Berntsen et al. 2003; Crespo-Lopez et al. 2009). Additionally, exposure to $HgCl_2$ and MeHg alters cell signaling (Chen et al. 2006), reduced integrity of cell membrane (Polunas et al. 2011), impaired DNA repair mechanism (Christie et al. 1985; Gadhia et al. 2012; Pieper et al. 2014; Ryu et al. 2014), immunomodulatory impacts (Gallagher et al. 2011) and also altered the DNA methylation (Farina et al. 2011) eventually resulting in adverse health effects. Many literature reviews discuss various mechanisms by which mercury exposure results in neurotoxicity and cell damage (Clarkson 2002; Mergler et al. 2007; Crespo-Lopez et al. 2009; Farina et al. 2011; Antunes et al. 2016).

In vitro studies have revealed the relationship between MeHg intoxication and neuronal deaths. Exposure to MeHg results in neuritic degeneration (Fujimura et al. 2009), oxidative damage (Yin et al. 2007) and this disruption of biological processes leads to neuronal deaths. It has also been well-documented that organic mercury has the ability to bind with thiol residues which eventually lead to dysfunction in protein containing thiol groups. Mercury exposure can also inhibit the activities of enzymes that correct oxidative stress in cells. MeHg and $HgCl_2$ inhibit the action of thioredoxin (Trx), glutathione reductase (GR) thioredoxin reductase (TrxR), superoxide dismutase (SOD), while MeHg inhibit the action of glutaredoxin (Grx), glutathione peroxidase (Gpx), neuronal nitric oxide synthase (nNOS) (Carvalho et al. 2008; Farina et al. 2009; Carvalho et al. 2011; Robitaille et al. 2016).

In vivo studies revealed that methylmercury is neurotoxic and teratogenic (Stern et al. 2004; Karagas et al. 2012). Toxicogenomics and proteomics studies of tissues from whole organisms revealed that exposure to organic and inorganic mercury can change gene expression and proteins that are involved in oxidants and antioxidants defenses. Altered antioxidants enzymes include glutaredoxin (Grx), glutathione peroxidase (GPx), glutathione reductase (GR), superoxide dismutase (SOD), thioredoxin (Trx), thioredoxin reductase (TrxR), nitric oxide synthase (NOS), peroxiredoxin. Metallothioneins (MTs), genes and proteins involved in metabolism were altered in unicellular alga *Chlamydomonas Reinhardtii* (Beauvais-Flickr et al. 2017), yeast (Jin et al. 2008), macrophytes (Dranguet et al. 2017), *Caenorhabditis Elegans* (Ruda Galvyte et al. 2013), Zebra (Ho et al. 2013), pregnant mice (Robinson et al. 2010; Godefroy et al. 2012) and human (Pinheiro et al. 2008; Grotto et al. 2010).

7.5.4 Biomonitoring of Mercury

Excretion of mercury occurs within a week after exposure. It can be found in urine and faeces at low levels after some months (Goldfrank et al. 1994). Mercury levels in blood that are higher than 35 μg/dl and mercury levels in urine that are higher than 150 μg/dl are considered toxic to humans (NPIS 1996). Mercury vapour in air should not exceed 0.1 mg/m^3. A report showed that 200 mg/l of mercury in blood and 50 mg/g in hair are the established limits and a higher concentration is considered toxicity (Harada et al. 1997).

For biological monitoring of mercury exposure, the common biological samples to examine mercury exposure due to occupation are urine and blood. Hair is best known to examine environmental exposure to methylmercury (Satoh 2000). The mean urinary concentration of mercury in the US population is 0.72 μg/l (95% CI, 0.6–0.8), the mean blood concentration is 0.34 μg/l (95% CI, 0.3–0.4) (CDC 2003). Brune and colleagues (1993) reported that in Europe, the blood concentration is higher. Also, exhaled air could be a possible biomarker of exposure to elemental mercury vapour as a result of its excretion through the lungs. The assessment of methylmercury is done through the blood and scalp hair.

7.6 Remedial Actions

Reduction in metal emissions from industries and waste incinerators will help to reduce environmental and human exposure. Older fluorescent bulbs should be changed to the new energy-efficient bulbs, because the older ones contain mercury. In schools, especially in chemistry laboratories, mercury may be present or in health clinics generally where thermometers and thermostats are often used, it should be handled by experts and also with care. Planting of trees and other vegetation should be encouraged in urban areas so as to improve the quality of life. Trees help in particle uptake and hence can be helpful in lessening the toxicity of heavy metals.

7.7 Recommendations

7.7.1 New Buildings

1. Avoid the use of lead piping
2. Avoid connection of new pipeline to an old lead connection pipe
3. Avoid the use of soldiers containing lead
4. Should in case brass fittings are used, ensure that the lead content is less than 0.25%
5. Avoid inhaling dust and smokes (cigarette or automobile)
6. Parents and guardians should monitor infants and toddler and prevent them from ingesting dust from soil through hand-mouth activities.

7.7.2 Old Building

1. In the presence of lead pipes or brass containing high contents of lead, in the morning before taking a drinking water from the pipe, flush the pipework for at least 2 min.
2. Lead pipes or fittings that contain high lead contents should be removed and treated as hazardous waste.

References

Ali A, Ma Y, Reynolds J et al (2011) Chromium effects on glucose tolerance and insulin sensitivity in persons at risk for diabetes mellitus. Endocr Pract: Official J Am Coll Endocrinol Am Assoc Clin Endocrinol 17(1):16–25. https://doi.org/10.4158/EP10131.OR

Antunes SA, Appel Hort M, Culbreth M et al (2016) Methylmercury and brain development: a review of recent literature. J Trace Elem Med Biol 38:99–107

Beauvais-Flickr, Slaveykova VI, Cosio C (2017) Cellular toxicity pathways of inorganic and methyl mercury in the green microalga *Chlamydomonas Reinhardtii*. Sci Rep 7(1):8034

Bellinger DC (2013) Prenatal exposures to environmental chemicals and children's neurodevelopment: an update. Saf Health Work 4:1–11

Berntsen MH, Aatland A, Handy RD (2003) Chronic dietary mercury exposure causes oxidative stress, brain lesions, and altered behaviour in Atlantic salmon (*Salmo Salar*) parr. Aquat Toxicol 65(1):55–72

Brewers JAM, Barry PJ, MacGregor DJ (1987) Distribution and cycling of cadmium in the environment. In: Nriagu JO, Sprague JB (eds) Cadmium in the aquatic environment. Wiley, New York

Beyersmann D, Hartwig A (2008) Carcinogenic metal compounds: recent insight into molecular and cellular mechanisms. Arch Toxicol 82(8):493–512. https://doi.org/10.1007/s00204-008-0313-y

Beytoote YA, Kamiloglu NN, Aksakal M (2003) Role of dietary vitamin E in cadmium-induced oxidative damage in rabbit's blood, liver and kidneys. Int J Vitam Nutr Res 73:351–355

Borthiry GR, Antholine WE, Kalyanaraman B et al (2007) Reduction of hexavalent chromium by human cytochrome b5: generation of hydroxyl radical and superoxide. Free Radic Biol Med 42:738–755

Brand P, Gube M, Gerards K et al (2010) Internal exposure, effect monitoring, and lung function in welders after acute short-term exposure to welding fumes from different welding processes. J Occup Environ Med 52:887–892

Braune BM, Outridge PM, Fisk AT et al (2005) Persistent organic pollutants and mercury in marine biota of the Canadian Arctic: an overview of spatial and temporal trends. Sci Total Environ 351–352:4–56

Browning CL, Wise JP (2017) Prolonged exposure to particulate chromate inhibits RAD51 nuclear import mediator proteins. Toxicol Appl Pharmacol 331:101–107. https://doi.org/10.1016/j.taap.2017.05.030

Brune D, Aitio A, Nordberg G et al (1993) Normal concentration of chromium in serum and urine—a TRACY project. Scand J Work Environ Health 1:39–44

Burnett RT, Brook J, Dann T et al (2000) Association between particulate and gas-phase components of urban air pollution and daily mortality in eight Canadian cities. Inhal Toxicol 12(Suppl 4):15–39

Buttner B, Beyersmann D (1985) Modification of the erythrocyte anion carrier by chromate. Xenobiotica 15:735–741

Carvalho CM, Chew EH, Hashemy SI et al (2008) Inhibition of the human thioredoxin system. A molecular mechanism of mercury toxicity. J Biol Chem 283(18):11913–11923

Carvalho CM, Lu J, Zhang X et al (2011) Effects of selenite and chelating agents on mammalian thioredoxin reductase inhibited by mercury: implications for treatment of mercury poisoning. FASEB J 25(1):370–381

CDC (2003) Centers for Disease Control and Prevention. Second national report on human exposure to environmental chemicals, 4 March 2003. Available online at: http://www.cdc.gov/exposurereport. Accessed 30 Dec 2020

CDC (2010) Centers for Disease Control and Prevention. Guidelines for the identification and management of lead exposure in pregnant and lactating women. Atlanta. https://www.cdc.gov/nceh/lead/publications/leadandpregnancy2010.pdf. Accessed 28 Dec 2020

Chen C, Xun P, Nishio M et al (2015) Cadmium exposure and risk of pancreatic cancer: a meta-analysis of prospective cohort studies and case-control studies among individuals without occupational exposure history. Environ Sci Pollut Res 22:17465–17474

Chen YW, Huang CF, Tsai KS et al (2006) The role of phosphoinositide 3-kinase/Akt signaling in low-dose mercury-induced mouse pancreatic beta-cell dysfunction *in vitro* and *in vivo*. Diabetes 55(6):1614–1624

Christie NC, Sugiyama O, Cattabeni M et al (1985) Differences in the effects of Hg(II) on DNA repair induced in Chinese Hamster ovary by ultraviolet or X-rays. Mol Pharmacol 24:173–178

CIBSE (Chartered Institute of Building Services Engineers) (1996) Editorial. Building Serv J 18(11):7

Claiborne CS, Larson T, Sheppard L (2002) Testing the metals hypothesis in Spokane, Washington. Environ Health Perspect 110(Suppl 4):547–552

Clarkson TW (2002) The three modern faces of mercury. Environ Health Perspect 110(Suppl. 1):11e23

Clarkson TW, Magos L, Myers GJ (2003) The toxicology of mercury–current exposures and clinical manifestations. N Engl J Med 349(18):1731–1737

Cocker J, Morton J, Warren N et al (2006) Biomonitoring for chromium and arsenic in timber treatment plant workers exposed to CCA wood preservatives. Ann Occup Hyg 50(5):517–525

Collins YE, Stotzky G (1989) Factors affecting the toxicity of heavy metals to microbes. In: Beveridge TJ, Doyle RJ (eds) Metal ions and bacteria. Wiley, Toronto, pp 31–90

Crespo-Lopez ME, Macedo GL, Pereira SI et al (2009). Mercury and human genotoxicity: critical considerations and possible molecular mechanisms. Pharmacol Res 60(4):212–220

Dranguet P, Cosio C, Le Faucheur S et al (2017) Transcriptomic approach for assessment of the impact on microalga and macrophyte of *in-situ* exposure in river sites contaminated by chlor-alkali plant effluents. Water Res 121:86–94

Drinking Water Inspectorate (DWI) (2010) Guidance on the water supply (water quality) Regulations 2000 (as amended) in England, September 2010

Dutrizac J, O'Reilly J, Macdonald R (1982) Roman lead plumbing: did it really contribute to the decline and fall of the empire? CIM Bull M75:111–115

Edwards KC, Kim H, Vincent JB (2019) Release of trivalent chromium from serum transferrin is sufficiently rapid to be physiologically relevant. J Inorg Biochem 202:110901. https://doi.org/10.1016/j.jinorgbio.2019.110901

Elinder C-G, Friberg L, Kjellstrom T et al (1994) Biological monitoring of metals. Chemical Safety Monographs, IPCS. World Health Organization, Geneva

Fang Z, Zhao M, Zhen H et al (2014) Genotoxicity of tri- and hexavalent chromium compounds *in vivo* and their modes of action on DNA damage in vitro. PLoS ONE 9(8):e103194. https://doi.org/10.1371/journal.pone.0103194

Farina M, Campos F, Vendrell I et al (2009) Probucol increases glutathione peroxidase-1 activity and displays long-lasting protection against methylmercury toxicity in cerebellar granule cells. Toxicol Sci 112(2):416–426

Farina M, Aschner M, Rocha JB (2011) Oxidative stress in MeHg-induced neurotoxicity. Toxicol Appl Pharmacol 256(3):405–417

Feki-Tounsi M, Hamza-Chaffai A (2014) Cadmium as a possible cause of bladder cancer: a review of accumulated evidence. Environ Sci Pollut Res 18:10561–10573

Fujimura M, Usuki F, Sawada M (2009) Methylmercury exposure downregulates the expression of *Racl* and leads to neuritic degeneration and ultimately apoptosis in cerebrocortical neurons. Neurotoxicology (Little Rock) 30(1):16–22

Gadhia SR, Calabro AR, Barile FA (2012) Trace metals alter DNA repair and histone modification pathways concurrently in mouse embryonic stem cells. Toxicol Lett 212(2):169–179

Gallagher CM, Smith DM, Meliker JR (2011) Total blood mercury and serum measles antibodies in US children, NHANES 2003–2004. Sci Total Environ 410–411:65–71

Garrod ANI, Martinez M, Pearson J et al (1999) Exposure to preservatives used in the industrial pretreatment of timber. Ann Occup Hyg 43:543–555

Giaginis C, Gatzidou E, Theocharis S (2006) DNA repair systems as targets of cadmium toxicity. Toxicol Appl Pharmacol 213:282–290

Giel-Pietraszuk M, Hybza K, Chełchowska M et al (2012) Mechanisms of lead toxicity. Adv Cell Biol 39:17–248

Godefroy D, Gosselin RD, Yasutake A et al (2012) The chemokine CCL2 protects against methylmercury neurotoxicity. Toxicol Sci 125(1):209–218

Goldfrank FL, Flomenbaum NE, Lewin NA et al (1994) Toxicologic emergencies. Appleton and Lange, Norwalk, CT

Grotto D, Valentini J, Fillion M et al (2010) Mercury exposure and oxidative stress in communities of the Brazilian Amazon. Sci Total Environ 408(4):806–811

Gube M, Brand P, Schettgen T et al (2013) Experimental exposure of healthy subjects with emissions from a gas metal arc welding process–part II: biomonitoring of chromium and nickel. Int Arch Occup Environ Health 86:31–37

Ha E, Basu N, Bose-O'Reilly S et al (2017) Current progress on understanding the impact of mercury on human health. Environ Res 152:419–433

Hagino N, Yoshioka Y (1960) A study of the etiology of Itai-Itai disease. J Jpn Orthop Assoc 35:812–815

Harada M, Nakachi S, Cheu T et al (1997) Mercury pollution in Tanzania: relation between scalp hair, mercury level and health. Institute of Molecular Embryology and Genetics, Kumamoto, Japan, p 14

Hayat MT, Nauman M, Nazir N et al (2019) Environmental hazards of cadmium: past, present, and future. Cadmium toxicity and tolerance in plants. Academic Press, Cambridge, MA, USA, pp 163–183

Hayes CR, Skubala ND (2009) Is there still a problem with lead in drinking water in the European Union? J Water Health 4:569–580

Hayes CR, Croft TN (2012) An investigation into the representativeness of random daytime sampling for lead in drinking water, using computational modelling. J Water Supply Res Technol-Aqua 61:142–152. https://doi.org/10.2166/aqua.2012.092

Ho NY, Yang L, Legradi J et al (2013) Gene responses in the central nervous system of zebrafish embryos exposed to the neurotoxicant methyl mercury. Environ Sci Technol 47(7):3316–3325

Hodge A (1981) Vitruvius, lead pipes and lead poisoning. Am J Archaeol 85:486–491

International Agency for Research on Cancer (IARC) (1990) Chromium, nickel and welding. Monographs Evaluat Carcinogenic Risks Humans 49:1–648. https://publications.iarc.fr/67

International Agency for Research on Cancer (IARC) (1993) Monographs on the evaluation of the carcinogenic risks to humans. Beryllium, cadmium, mercury and exposures in the glass manufacturing industry. IARC Scientific Publications, Lyon, France, pp 119–238

International Water Association (2010) In: Hayes CR (ed) Best practice guide on the control of lead in drinking water. IWA Publishing, London

Ishikawa Y, Nakagawa K, Satoh Y et al (1994) "Hot spots" of chromium accumulation at bifurcations of chromate workers bronchi. Cancer Res 54(9):2342–2346. PubMed: 8162579. https://cancerres.aacrjournals.org/content/canres/54/9/2342.full.pdf. Accessed 30 Dec 2020

Jin YH, Dunlap PE, McBride SJ et al (2008) Global transcriptome and deletome profiles of yeast exposed to transition metals. PLoS Genet 4(4):e1000053

Johnson J, Schewel L, Graedel TE (2006) The contemporary anthropogenic chromium cycle. Environ Sci Technol 40:7060–7069

Julin B, Wolk A, Johansson E et al (2012) Dietary cadmium exposure and prostate cancer incidence: a population-based prospective cohort study. Br J Cancer 107:895–900

Jusko TA, Henderson CR Jr, Lanphear BP et al (2008) Blood lead concentrations <10 microg/dL and child intelligence at 6 years of age. Environ Health Perspect 116(2):243–248

Karagas MR, Choi AL, Oken E et al (2012) Evidence on the human health effects of low level methylmercury exposure. Environ Health Perspect 120(6):799–806

Khalil N, Wilson J, Talbott E et al (2009) Association of blood lead concentrations with mortality in older women: a prospective cohort study. Environ Health 8:15

Krzywy I, Krzywy E, Pastuszak-Gabinowska MA et al (2010) Lead—is there something to be afraid of? Ann Acad Med Stetin 56:118–128

Kumar S, Sharma A (2019) Cadmium toxicity: effects on human reproduction and fertility. Rev Environ Health 34:327–338

Lamborg CH, Hammerschmidt CR, Bowman KL et al (2014) A global ocean inventory of anthropogenic mercury based on water column measurements. Nature 512(7512):65–68

Lanphear BP, Dietrich K, Auinger P et al (2000) Cognitive deficits associated with blood lead concentrations <10 microg/dL in US children and adolescents. Public Health Rep 115(6):521–529

Levina A, Harris HH, Lay PA (2006) Binding of chromium (VI) to histones: implications for chromium (VI)-induced genotoxicity. J Biol Inorg Chem 11(2):225–234. https://doi.org/10.1007/s00775-005-0068-3

Lokhande B, Patil PS, Uplane MD (2004) Studies on cadmium oxide sprayed thin films deposited through non-aqueous medium. Mater Chem Phys 84:238–242

Menke A, Muntner P, Batuman V et al (2006) Blood lead below 0.48 μmol/l (10 μg/dl) and mortality among US adults. Circulation 114:1388–1394

Mergler D, Anderson HA, Chan LH et al (2007) Methylmercury exposure and health effects in humans: a worldwide concern. Ambio 36(1):3–11

Minoia C, Cavalieri A (1988) Chromium in urine, serum and red blood cells in the biological monitoring of workers exposed to different chromium valence states. Sci Total Environ 71:323–327

Mona T, Heba MA, Eman S et al (2018) Impact of occupational cadmium exposure on bone in sewage workers. Int J Occup Environ Health 24:101–108

Mutti A, Pedroni C, Arfini G et al (1984) Biological monitoring of occupational exposure to different chromium compounds at various valencystates. Int J Environ Anal Chem 17:35–41

Navas-Acien A, Guallar E, Silbergeld EK et al (2007) Lead exposure and cardiovascular disease—a systematic review. Environ Health Perspect 115:472–482

Nawrot TS, Martens DS, Hara A et al (2015) Association of total cancer and lung cancer with environmental exposure to cadmium: the meta-analytical evidence. Cancer Causes Control 26:1281–1288

NPIS (L) (1996) Monograph: mercury—organic and inorganic. National poison information Service (London) Medical Toxicology Unit, Guy's and St Thomas' Hospital, London

Nriagu J (1983) Saturnine gout among Roman aristocrats. Did lead poisoning contribute to the fall of the empire? N Engl J Med 308:660–663

Occupational Safety and Health Administration (OSHA) (2006) Department of labor. Occupational exposure to hexavalent chromium. Final rule. Fed Reg 71:10099–10385

Ognjanovic BI, Pavlovic SZ, Maletic SD et al (2003) Protective influence of vitamin E on antioxidant defense system in the blood of rats treated with cadmium. Physiol Res 52:563–570

Omeljaniuk W, Socha K, Soroczynska J et al (2018) Cadmium and lead in women who miscarried. Clin Lab 64:59–67

Osornio-Vargas A, Bonner JC, Alfaro-Moreno E et al (2003) Proinflammatory and cytotoxic effects of Mexico City air pollution particulate matter *in vitro* are dependent on particle size and composition. Environ Health Perspect 111(10):1289–1293

Pieper I, Waha CA, Bornhorst J et al (2014) Mechanisms of Hg species induced toxicity in cultured human astrocytes: genotoxicity and DNA-damage response. Metall 6(3):662–671

Pinheiro MC, Macchi BM, Vieira JL et al (2008) Mercury exposure and antioxidant defenses in women: a comparative study in the Amazon. Environ Res 107(1):53–59

Polunas M, Halladay A, Tjalkens RB et al (2011) Role of oxidative stress and the mitochondrial permeability transition in methylmercury cytotoxicity. Neurotoxicology (Little Rock) 32(5):526–534

Proctor DM, Suh M, Campleman SL et al (2014) Assessment of the mode of action for hexavalent chromium-induced lung cancer following inhalation exposures. Toxicology 325:160–179. https://doi.org/10.1016/j.tox.2014.08.009

Rani A, Kumar A, Lal A et al (2014) Cellular mechanisms of cadmium-induced toxicity: a review. Int Environ Health Res 24:378–399

Rehman K, Fatima F, Waheed I (2018) Prevalence of exposure of heavy metals and their impact on health consequences. J Cell Biochem 119:157–184

Robinson JF, Griffith WC, Yu X et al (2010) Methylmercury induced toxicogenomic response in C57 and SWV mouse embryos undergoing neural tube closure. Reprod Toxicol 30(2):284–291

Robitaille S, Mailloux RJ, Chan HM (2016) Methylmercury alters glutathione homeostasis by inhibiting glutaredoxin 1 and enhancing glutathione biosynthesis in cultured human astrocytoma cells. Toxicol Lett 256:1–10

Ruda Galvyte M, VanDuyn N, Aarnio V et al (2013) Methylmercury exposure increases lipocalin related (lpr) and decreases activated in blocked unfolded protein response (abu) genes and specific miRNAs in Caenorhabditis Elegans. Toxicol Lett 222(2):189–196
Ryu TH, An KG, Kim JK (2014) Genotoxicity in earthworm after combined treatment of ionising radiation and mercury. Radiat Prot Dosim 159(1–4):111–117
Satoh H (2000) Occupational and environmental toxicology of mercury and its compounds. Ind Health 38:153–164
Schwarz K, Mertz W (1959) Chromium (III) and the glucose tolerance factor. Arch Biochem Biophys 85:292–295. https://doi.org/10.1016/0003-9861(59)90479-5
Seervarisai J, Kao Jarern S, Chanprasertyothin S et al (2015) Environmental lead exposure, catalase gene, and markers of antioxidant and oxidative stress relation to hypertension: an analysis based on the EGAT study. Biomed Res Int 2015: https://doi.org/10.1155/2015/856319
Söderholm M, Borné Y, Hedblad B et al (2020) Blood cadmium concentration and risk of subarachnoid haemorrhage. Environ Res 180:
Song JK, Luo H, Yin XH et al (2015) Association between cadmium exposure and renal cancer risk: a meta-analysis of observational studies. Sci Rep 5:17976
Sorensen M, Schins RPF, Hertel et al (2005) Transition metals in personal samples of PM2.5 and oxidative stress in human volunteers. Cancer Epidemiol Biomark Prevent 14(5):1340–1343
Staessen JA, Buchet JP, Lauwerys RR et al (1996) on behalf of the working Groups (1996) Public health implications of environmental exposure to cadmium and lead: an overview of epidemiological studies in Belgium. J Cardiovasc Risk 3:26–41
Steenland K, Barry V, Anttila A et al (2019) Cancer incidence among workers with blood lead measurements in two countries. Occup Environ Med 76:603–610
Stern AH, Jacobson JL, Ryan L et al (2004) Do recent data from the Seychelles Islands alter the conclusions of the NRC Report on the toxicological effects of methylmercury? Environ Health 3(1):2
Swartzendruber PJD (2012) Mercury in the environment pattern and process. University of California Press, London, England
Takiguchi M, Achanzar WE, Qu W et al (2003) Effects of cadmium on DNA-(Cytosine-5) methyltransferase activity and DNA methylation status during cadmium-induced cellular transformation. Exp Cell Res 286:355–365
Teklay A (2016) Physiological effect of chromium exposure: a review. Int J Food Sci Nutr Dietet S7(001):1–11. https://doi.org/10.19070/2326-3350-SI07001
Troesken W (2006) The great lead water pipe disaster. MIT Press, Cambridge, MA
US EPA (US, Environmental Protection Agency) (1994) Indoor air pollution: an introduction for health professionals. Washington, DC: US Environmental Protection Agency
Van den Hoven TJL, Buijs PJ, Jackson PJ et al (1999) Developing a new protocol for the monitoring of lead in drinking water. EUR 19087. https://op.europa.eu/en/publication-detail/-/publication/318e84bf-1029-46ed-a49a-a5b9156ad1aa. Accessed 31 Dec 2020
Vangronsveld J, Clijsters H (1994) Toxic effect of metals. In: Farago MG (ed) Plant and the chemical elements. VCH, Weinheim, New York, Basel, Cambridge, Tokyo, pp 149–177
Vincent JB (2017) New evidence against chromium as an essential trace element. J Nutr 147:2212–2219. https://doi.org/10.3945/jn.117.255901
Vincent JB (2019) Effects of chromium supplementation on body composition, human and animal health, and insulin and glucose metabolism. Curr Opin Clini Nutr Metabolic Care 22(6):483–489. https://doi.org/10.1097/MCO.0000000000000604
Vincent R, Kumarathasan P, Goegan P et al (2001) Acute cardiovascular effects in rats from exposure to urban ambient particles. Health Effects Inst Res Rep 104:1–64
Wallace DR, Taalab YM, Heinze S et al (2020) Toxic-metal-induced alteration in miRNA expression profile as a proposed mechanism for disease development. Cells 9:901
Wang Y, Su H, Gu Y et al (2017) Carcinogenicity of chromium and chemoprevention: a brief update. Onco Targets Thera 10:4065–4079. https://doi.org/10.2147/OTT.S139262
Wani AL, Ara A, Usmani JA (2015) Lead toxicity: a review. Int Toxicol 8:55–64

Welling R, Beaumont JJ, Petersen SJ et al (2015) Chromium VI and stomach cancer: a meta-analysis of the current epidemiological evidence. Occup Environ Med 72(2):151–159. https://doi.org/10.1136/oemed-2014-102178

WHO (1992) World Health Organization. Environmental health criteria 134: Cadmium; World Health Organization: Geneva, Switzerland

WHO (2007) Preventing disease through a healthy environment: exposure to mercury: a major public health concern. World Health Organization Mercury Flyer

WHO (2016) Expert consultation. Available evidence for the future update of the WHO global Air Quality Guidelines (AQGs) (2016).https://www.euro.who.int/__data/assets/pdf_file/0013/301720/Evidence-future-update-AQGs-mtg-report-Bonn-sept-oct-15.pdf. Accessed 28 Dec 2020

Xia B, Ren X, Zhuang Z et al (2014) Effect of hexavalent chromium on histone biotinylation in human bronchial epithelial cells. Toxicol Lett 228(3):241–247. https://doi.org/10.1016/j.toxlet.2014.05.010

Xia W, Hu J, Zhang B et al (2016) A case-control study of maternal exposure to chromium and infant low birth weight in China. Chemosphere 144:1484–1489. https://doi.org/10.1016/j.chemosphere.2015.10.006

Yin Z, Milatovic D, Aschner JL (2007) Methylmercury induces oxidative injury, alterations in permeability and glutamine transport in cultured astrocytes. Brain Res 1131(1):1–10

Zawadzki M, Poręba R, Gac P (2006) Mechanisms and toxic effects of lead on the cardiovascular system. Med Pr 57:543–549

Zhitkovich A (2002) Chromium: Exposure, toxicity and biomonitoring approaches. In: Wilson SH, Suk WA (eds) Biomarkers of environmentally associated disease: technologies, concepts, and perspectives. CRC Press, New York, pp 269–287

Chapter 8
Nanoparticles in Construction Industry and Their Toxicity

G. Santhosh and G. P. Nayaka

Abstract Nanoparticles (NPs) are ultrafine particles having size between 1 to 100 nm. For a few decades, the researchers have explored wide scientific applications of NPs in medicine, electronics, construction, manufacturing and in insulating materials. The benefits of using NPs in construction materials are huge; the NPs can modify the physical, thermal, antimicrobial, self-cleaning and self-sensing, and self-healing and chemical properties of construction materials. NPs such as titanium dioxide, carbon nanotubes, clay and aluminium dioxide are widely used in construction materials. The NPs used in the construction materials are released to the environment or the workers handling the materials likely to be exposed to the NPs, the exposure can be detrimental to the environment and human health. NPs can enter the human body through skin, via inhalation or ingestion. However, the rate of exposure to NPs can cause serious respiratory, cardiovascular, skin and nerve related diseases. This chapter briefs about the importance of NPs in construction materials/industry together with their adverse effects on environment and human life.

Keywords Antimicrobial · Construction industry · Environment · Mechanical · Nanoparticles · Toxicity

8.1 Introduction

The recent reports on urbanization indicate that over 54% of the world's population live in cities, this number may increase by 12% by 2050 (United nation 2014). The urbanization needs economic and social reforms as a part of development, these reforms may bring huge stress on the environment. This development in the cities

G. Santhosh (✉)
Department of Mechanical Engineering, NMAM Institute of Technology, Nitte, Mangalore 574110, India

G. P. Nayaka
Physical and Materials Chemistry Division, CSIR-National Chemistry Laboratory, Pune, Maharashtra 411008, India

J. A. Malik and S. Marathe (eds.), *Ecological and Health Effects of Building Materials*,
https://doi.org/10.1007/978-3-030-76073-1_8

Fig. 8.1 Huge quantity of building waste dumped in the open environment

may increase the pollutants produced by the large quantity of constructional wastes (Tixier et al. 2011; Agudelo-Castaneda et al. 2017).

The constructional waste generated also includes unused materials due to excess supply or mishandling (Akhtar and Sarmah 2018). However, the construction waste is inert, the major problem is its volume and weight due its large production throughout the globe (Galvez-Martos et al. 2018). The handling of the constructional waste is the major issue worldwide due to the environmental impacts by their constituent products (Akhtar and Sarmah 2018). The large quantity of construction waste dumped in the open environment is shown in Fig. 8.1. The use of nanoparticles and nanofibers in constructional materials is increasing day by day, the toxicity of nanomaterials on environment through construction materials during their manufacturing and recycling is focus of many researchers these days.

The term 'nano' was coined by famous physicist 'Richard Feynman'. The nanomaterials revolution has increased a variety of industrial products including the construction materials (Sanchez et al. 2010). Because nanomaterials have excellent surface activity and can be used in various applications, especially in the construction sectors. The nanomaterials are engineered either through top-down or bottom-up approach (Xing et al. 2016). The top-down approach involves the breakdown of large structures into smaller materials of nano dimensions whilst the bottom-up approach involves the piecing of nano sized molecules into larger structures or until the required size is achieved (Xing et al. 2016). Further, these nanomaterials are prepared by techniques such as chemical and physical method. Few nanoparticles prepared by these methods are listed in the Table 8.1.

The nanoparticles can be used as additives or reinforcing materials to enhance the overall characteristic of concrete, mortar, paints, and in insulating materials (Ribeiro et al. 2013; Farzadnia et al. 2013; Aziz 2016; Yan et al. 2017; Ribeiro et al. 2015). Nanoparticles such as SiO_2, Al_2O_3, TiO_2, Fe_2O_3, $CaCO_3$, CuO, nano-metakaolin and nano-magnesium calcite are used in cement-based materials to improve mechanical, thermal and electrical properties (Rashad 2013; Silvestre et al. 2016; Abo-El-Enein et al. 2014; Abdel-Gawwad et al. 2019; Noorvand et al. 2013; Mohseni et al. 2015). The use of above-mentioned nanoparticles can change the kinetics and hydration

Table 8.1 Nanoparticles prepared by chemical and physical techniques

Method	Technique	Nanoparticles
Chemical	Sol-Gel	Metals, oxides, carbides, nitrides etc.,
	Vapour phase reaction	
	Evaporation or condensation	
	Reaction in liquid medium	
	Reaction in solid medium	
Physical	Laser	Metal oxides, fullerenes, semiconductors etc.,
	Plasma synthesis	
	Electronic irradiation	
	Physical vapour deposition	
	Densification and consolidation	

of the cement. Silica fume, fly ash and nanosilica particles are also widely used in cement and concrete industries. On the other hand, nano-metakaolin, nanosilica and nano-alumina enhance the physical properties of the mortar due to their high pozzolanic and nucleating activities nanoparticles. The lightweight aggregates such as vermiculite and perlite are used as fire protecting mortars; these aggregates are also suitable to protect other constructional materials as well.

The successful use of nanoparticles in the construction materials makes it more likely for the workers to be exposed to the toxic particles released by the construction materials. The toxicity of the nanoparticles depends on a few conditions, such as nanomaterials concentration, frequency and duration of exposure. However, the use of advanced technologies does not always have a positive impact on the environment. It is very much essential for mankind to address these issues to fully understand the limitations of the nanoparticles on the environment and relevant safety concerns.

8.2 Importance of Nanomaterials in Construction

The nanoparticles are wonderful because of their nanoscale size (1–100 nm); the crucial factor of the nanoparticles is the size, the materials properties significantly improve in the presence of nanoparticles. The use of nanoparticles can bring many important parameters and behaviour in the bulk materials, electrostatic forces and quantum effects start to dominate. These properties of nanoparticles bring extraordinary changes in the physico-mechanical and physical–chemical behaviour of constructional materials (Table 8.2). Typical nanostructures like cylindrical, spherical and sheets with high quantum effects and high surface to volume ratio have changed the entire chemistry of constructional materials. The properties such as increased strength, self-sensing, self-cleaning, antimicrobial, or pollution remediation capabilities can be improved with the addition of nanoparticles in the construction materials.

Table 8.2 Nanomaterials used in construction materials

Construction materials	Nanoparticles	Properties enhanced
Concrete	Silica, Titania, CNTs, Iron oxide	Mechanical strength, Hydration, crack prevention, compressive strength, Abrasion-resistant
Asphalt concrete Timber	Aluminium oxide, Zycosoil	Life, Higher compaction
Bricks mortar	Clay nanoparticles	compressive strength, Increased surface roughness
Steel	Copper	Corrosion resistance, fatigue life

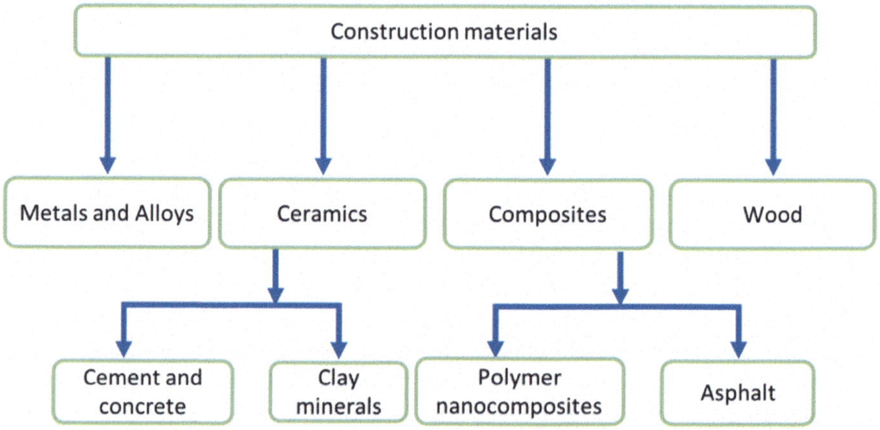

Fig. 8.2 Various materials used in the constructional sectors

The typical construction and building materials used are presented in Fig. 8.2. The construction sector is rapidly growing and there has been a notable change in the industry due to the advancement of nanoscience and technology.

8.3 Nanomaterials in Construction

Nanoparticles can help materials with unique behaviours; the constructional materials can be lighter, stronger with better insulation and acoustic behaviours. The nanoparticle with low density and high strength provides new roots to the market growth for constructional sectors. Nanoparticles have replaced the natural materials by improving the performance of building materials, further, nanoparticles in construction offer better, cheaper, safer and faster approach for the production of building materials. The use of nanoparticles in the raw materials can significantly enhance the product durability and efficiency to newer levels. Nanoscience and technology have greater potential in changing the construction industry into targeted environmental protection.

8.3.1 Concrete and Cement

Concrete and cement are the major construction materials used for decades. These materials are very famous for their performance in buildings. These materials are famous not only because of their high performance; they are famous for their low cost and availability. Further their properties are enhanced by the addition of nanoparticles. The components of cement these days include binders, water and aggregate materials (Nazari et al. 2011). The properties of the concrete can be modified by for particular applications nanoparticles are added (Mohajerani et al. 2019a). It is very much essential to enhance the durability and serviceability of the concrete as it is exposed to different weather conditions. The durability of the concrete can be improved by creating an interface between the concrete components (Salemi et al. 2013). Nanoparticles such as nano-silica, TiO_2 and CNT are used to improve the strength and durability of concrete.

The addition of nano-silica in cement and concrete can significantly reduce the water penetration (Tiwary et al. 2013). Nano-silica, suitable binding materials for cement, can delay the degradation of calcium silicate hydrate (Ca-Si-H), thereby improving the cohesivity between cement and aggregate molecules (Salemi et al. 2013).

CNTs with one-dimensional hollow structure are an emerging filler material in construction sectors. CNTs in concrete can enhance the mechanical property and make concrete impenetrable to salts and water (Konsta-Gdoutos et al. 2010).

Nano-titania (TiO_2) is another nanoparticle which has been widely used in construction materials due to its high surface area. TiO_2 used in concrete can enhance the anticorrosive and photo-catalytic activity, the concrete in the presence of TiO_2 can become self-cleaning and self-disinfecting against the pollutants. TiO_2 in the presence of light breaks the pollutants into water and Co_2 (Khitab 2016). Polycarboxylate ether is a third-generation super plasticizer which is a well-known strength improver that helps in particle dispersion and reduces water content in cement.

8.3.2 Asphalt

Asphalt is the most famous composite material primarily used in road construction. Asphalt can withstand high load for longer duration. Asphalt has very poor response to high and low temperatures, but it becomes vulnerable at high temperatures resulting in melting and cracks can be seen at low temperatures (Rahman et al. 2017). However, this behaviour can be altered by the addition of suitable filler materials/nanomaterials.

Styrene–butadiene–styrene (SBS), polyethylene (PE) and styrene–butadiene–rubber (SBR) are most appreciated modifiers; these can change the elastic and ductile nature of Asphalt to encounter extreme temperature (Zhang et al. 2016). Nano-aluminium oxide (Al_2O_3) is one of the main nanomaterials, with just 5% addition of Al_2O_3 in asphalt cement increases its resistance to high temperatures (Mubaraki et al. 2016). Further, other problems such as stiffness, durability and strength can be altered by adding various nano-modifiers, Zycosoil is an anti-strip agent used to enhance the fatigue life of the concrete or cement (Sarkar et al. 2014).

8.3.3 Bricks

Bricks are the major construction materials for many decades. Bricks consist of clay, sand and few granular materials. These ingredients bind together at a very high temperature to form a brick with good compressive strength (Mohajerani et al. 2016; Mohajerani et al. 2019b). However, these days the addition of nano-clays in the process increases the compressive strength of the brick by 4.8 times (Niroumand et al. 2013). Further to increase the durability of the bricks alko siloxane and silica nanoparticles were added to protect the bricks. The modified bricks showed high resistance to the water uptake with improved durability (Stefanidou et al. 2016).

8.3.4 Mortar

The mortar binder used in the constructional materials has good permeable properties which helps in quick evaporation of moisture from the construction martials. However, water penetration is an important issue that needs to be addressed. The water penetration can cause materials loss hindering the strength of the binder. If the binding material degrades, then the whole material would soon follow. Hence, to use the mortar in water rich areas organic oils or wax are coated.

8.3.5 Steel

Steel is the major and commonly used construction material in all construction works due their durability, high strength to weight ratio and their fire-resistant behaviour. The steel used in huge buildings and bridges face many issues related to strength and corrosion. These issues can be addressed by the use of nanofillers. The copper nanoparticle used as a modifier can enhance the surface roughness of the steel. The anti-corrosion property of the steel can be with just 0.5% of copper nanoparticles solution (Ge et al. 2008).

In addition to the construction materials, various nanoparticles are used in different construction applications and products. Coating nano TiO_2 on walls, and roofs behaves as an anti-fouling agent (Irie et al. 2004; Zhu et al. 2004). Silver nanoparticles in paints can increase antimicrobial property of the (Kumar et al. 2008). CNTs, fullerenes and quantum dots used in cements can enhance the durability and mechanical strength of the cement (Mahendra et al. 2008).

8.4 Environmental Release and Exposure Scenarios

In the name of development, the use and production of nanoparticles has increased, and their release into the environment increases the potential adverse effects on mankind and environment. Exposure analysis is crucial in assessing and preventing the unwanted influence of the nanoparticles on the environment. The risk assessment is very important regardless of nano-toxicity; the poor exposure can prevent the possible health risk. This can be accomplished by proper handling and disposal of nanoparticles in the environment. Further, it is very much essential to analyse the concentration to which the humans are exposed to, it is essential to determine whether nanoparticles retain their structure and reactivity is the big challenge to assess their bioactivity and impact on humans and environment. The manufactured nanoparticles can enter the ecosystem either during manufacturing or when used or when disposed intentionally/unintentionally (Klaine et al. 2008; Wiesner et al. 2006). Despite the awareness of release of nanoparticles, attempts to identify and characterise overriding exposure routes have been quite preliminary. Furthermore, many researchers have studied and evaluated the hazards and impacts of nanoparticles (Griffitt et al. 2007; Lin et al. 2008; Oberdörster et al. 2004; Zhu et al. 2008).

Few studies have shown the environmental implications by nanoparticles fate, transport, transformation, bioaccumulation and bioavailability (Chen et al. 2006a, b; Fortner et al. 2005; Kandlikar et al. 2007; Lee et al. 2008). The studies suggest that the nanoparticles have the ability to affect the environmental and human health (Kirchner et al. 2005; O'Brien et al. 2008), they fail to provide the regulatory guidelines for the safe production, safe disposal of the building nanomaterials. Therefore, understanding the release and reactive behaviours of the nanomaterials represents critical knowledge gaps for the risk assessment (Kartam et al. 2004; Kourmpanis et al. 2008; Poon 2007).

8.5 Toxicity of Nanomaterials

Nanoparticles reinforced construction materials can cause cellular toxicity via multiple mechanisms. The mechanism may include disruption of cell walls, nucleic acid damage, and release of toxic metal ions. Table 8.3 presents the toxic effects of nanoparticles used in building materials.

Table 8.3 Different nanoparticles used in building materials and their toxic effects

Nanoparticle	Effect	Organism	Reference
CNTs	Membrane damage	Bacteria	Ding et al. (2005), Kang et al. (2007), Lam et al. (2006)
Quantum dots	Toxic metal ions can damage DNA	Human cells	Chang et al. (2006), Hoshino et al. (2004), Ryman-Rasmussen et al. (2007), Shiohara et al. (2004)
	Prevents the bacterial growth (Bactericidal)	Bacteria	Kloepfer et al. (2005), Mahendra et al. (2008)
SiO_2	Reactive oxygen species (ROS) are produced and can cause toxicity	Bacteria	Adams et al. (2006)
TiO_2	Growth inhibition, Photosynthesis can be stopped, oxidative damage due to ROS	Bacteria, algae, fish	Blaise et al. (2008), Long et al. (2006), Lyon et al. (2006), Rincon et al. (2004), Wolfrum et al. (2002)

The hazardous effects may range from no damage to sub-lethal effects to mortality. CNTs and TiO_2 nanoparticles are the most widely studied due to their potential hazardous effects. TiO_2 is a prominent photoactive nanoparticle that causes inflammation, cytotoxicity and has the ability to damage DNA of mammalian cells as it can produce ROS (Handy et al. 2008; Karlsson et al. 2008; Oberdorster et al. 1995; Park et al. 2007; Reeves et al. 2008; Sayes et al. 2006; Zhang et al. 1998; Zhu et al. 2008). The major reason for the toxic effect is its morphology, the TiO_2 nanoparticle loose its mobility inside the cells allowing pathogens to generate ROS (Long et al. 2006). CNTs used in the constructional materials can exert pulmonary toxicity (Ding et al., 2005; Jia et al. 2005; Wei et al. 2007). The direct interaction of CNTs with cells can damage the cell walls by creating oxidative stress (Narayan et al. 2005; Kang et al. 2007).Copper based nanoparticles also exhibit strong oxidative stress and have the ability to damage human DNA (Chen et al. 2006a, b; Blaise et al. 2008; Karlsson et al. 2008; Lee et al. 2008). The SiO_2 nanoparticles have categorised as human carcinogens (IARC 1997), the SiO_2 nanoparticles have the ability to induce tumour necrosis (Attik et al. 2008). Ultra-fine nano sized SiO_2 at high concentration can damage water bacteria (Adams et al. 2006). Quantum dots (QDs) are another class of nanoparticles used in various industrial applications. Fluorescent QDs having heavy metals are decorated with organic functionalities to enhance their stability (Yu et al. 2007), the release of heavy metals such as cadmium, lead and zinc can be toxic to bacteria and mammalian cells (Kloepfer et al. 2005; Mahendra et al. 2008; Cha et al. 2007; Derfus et al. 2004; Hardman et al. 2006; Kirchner et al. 2005; Lu et al. 2008; Shiohara et al. 2004). The release of heavy metals can be controlled by surface coating, however, sometimes the surface coatings itself have become toxic to mammalian cells (Hoshino et al. 2004; Lee et al. 2007; Ryman-Rasmussen et al.

2007). On the other hand, the toxicity of QDs can cause oxidative stress, cytotoxicity and nucleic acid damage (Chang et al. 2006; Lin et al. 2008; Lovric et al. 2005).

8.6 Risk Assessment and Analysis

In general, Nanomaterials as we know existed from nature and are used in various applications from ages, however humans are creating a variety of nanoparticles using advanced techniques to meet the current engineering requirements and hence there will be issues regarding their adverse effects on the environment and humans. The nanoparticles are highly active and unstable materials due to their size and high surface to volume ratio makes them suitable materials in various industrial applications. However, the market volumes for the production of nanomaterials have grown tremendously over the last few years.

The use of nanomaterials makes us understand the quantification of exposure and hazard are important in assessing the environmental risk. The critical assessment includes problem identification and problem prioritization based on the impact on the environment (Maynard et al. 2006). These assessments help us address the impacts of one or more nanoparticles. Further, they help us in analysing its worth spending time and money in quantification of risk assessment. This risk assessment process depends on parameters like scientific evidence, general public opinion, advantages and perception of the risk. In other words, industries must have a multidisciplinary approach to process the constructional waste, the waste generated must be reused as input for the new process or to generate new materials and components.

8.7 Critical Knowledge Gaps and Research Needs

Nanoparticles are the most promising and common features in many constructional materials due to their unique and remarkable properties. However, their production and use creates a huge impact on human health and the environment. The adverse effects of nanomaterials used motivating huge research fraternity to concentrate not only on risk assessment, but also on their safe handling and disposal.

Nanoparticles are ultra-fine molecules that have the ability to penetrate into the human cells and to cause many respiratory and skin related issues, but the mode of action of nanoparticles is still a mystery. In particular, the nanoparticles size distribution, chemical stability, bioactivity, and toxicity are not fully understood. Thus, a complete mechanism to understand their structure-reactivity related to immunology and toxicity is necessary. The researchers must expand their research to address the sublethal chronic exposure and their impact on organisms.

The analytical techniques used to quantify the effect on nanoparticles are insufficient in accessing the environmental impacts. Thus, few advanced techniques are necessary to address the metrology of nanoparticles to understand their fate and transformation in different environmental conditions. The improv techniques should be able to monitor the short-term and long-term workers exposure during manufacturing, construction and demolition processes.

Ultimately, the education system must motivate the young minds to research and identify an efficient way to reuse and recycle the constructional materials or constructional wates. The advance research may force the authorities to enforce appropriate guidelines and regulations to mitigate the environmental risks.

8.8 Conclusions

This book chapter presents a brief idea about a few nanomaterials used in construction and building materials. Nanotechnology is so unique in such a way that the use of nanomaterials has created a huge impression in various industrial applications, with the use of nanomaterials new challenges have emerged in terms of biocompatibility, toxicity and their impact on human life. The most widely used nanoparticles in different construction materials such as cement and concrete, asphalt, bricks, mortar and steel are presented. However, nanoscience is still an emerging field with many opportunities and challenges. Each day the use of nanomaterial creates new problems to assess and analyse in the construction industry. Eventually, the effects of nanoparticles on the environment and human health create a huge void in understanding their nature and behaviour. However, the research reveals the negative impact of nanomaterials on environment and human life, stringent action plans and guidelines must be adopted to avoid the environmental risks.

References

Abdel-Gawwad HA, Heikal M, Mohammed MS, Abd El-Aleem S, Hassan HS, Vásquez García SR, Rashad AM (2019) Evaluating the impact of nano-magnesium calcite waste on the performance of cement mortar in normal and sulfate-rich media. Constr Build Mater 203:392–400

Abo-El-Enein SA, Amin MS, El-Hosiny FI, Hanafi S, ElSokkary TM, Hazem MM (2014) Pozzolanic and hydraulic activity of nano-metakaolin. HBRC J 10:64–72

Adams LK, Lyon DY, Alvarez PJJ (2006) Comparative eco-toxicity of nanoscale TiO_2, SiO_2, and ZnO water suspensions. Water Res 40:3527–3532

Agudelo-Castañeda DM, Teixeira EC, Schneider IL, Lara SR, Silva LFO (2017) Exposure to polycyclic aromatic hydrocarbons in atmospheric PM1.0 of urban environments: carcinogenic and mutagenic respiratory health risk by age groups. Environ Pollut 224:158–170

Akhtar A, Sarmah AK (2018) Construction and demolition waste generation and properties of recycled aggregate concrete: a global perspective. J Clean Prod 186:262–281

Attik G, Brown R, Jackson P, Creutzenberg O, Aboukhamis I, Rihn BH (2008) Internalization, cytotoxicity, apoptosis, and tumor necrosis factor-alpha expression in rat alveolar macrophages exposed to various dusts occurring in the ceramics industry. InhalToxicol 20:1101–1112

Aziz H, Ahmad F (2016) Effects from nano-titanium oxide on the thermal resistance of an intumescent fire retardant coating for structural applications. Prog Org Coat 101:431–439
Blaise C, Gagné F, Férard JF, Eullaffroy P (2008) Ecotoxicity of selected nano-materials to aquatic organisms. Environ Toxicol 23:591–598
Cha KE, Myung H (2007) Cytotoxic effects of nanoparticles assessed in vitro and in vivo. J Microbiol Biotechnol 17:1573–1578
Chang E, Thekkek N, Yu WW, Colvin VL, Drezek R (2006) Evaluation of quantum dot cytotoxicity based on intracellular uptake. Small 2:1412–1417
Chen KL, Elimelech M (2006a) Aggregation and deposition kinetics of fullerene (C60) nanoparticles. Langmuir 22:10994–11001
Chen Z, Meng H, Xing G, Chen C, Zhao Y, Jia G, Wang T, Yuan H, Ye C, Zhao F, Chai Z, Zhu C, Fang X, Ma B, Wan L (2006b) Acute toxicological effects of copper nanoparticles in vivo. Toxicol Lett 163:109–120
Derfus AM, Chan WCW, Bhatia SN (2004) Probing the cytotoxicity of semiconductor quantum dots. Nano Lett 4:11–18
Ding L, Stilwell J, Zhang T, Elboudwarej O, Jiang H, Selegue JP, Cooke PA, Gray JW, Chen FF (2005) Molecular characterization of the cytotoxic mechanism of multiwall carbon nanotubes and nano-onions on human skin fibroblast. Nano Lett 5:2448–2464
Farzadnia N, Ali AAA, Demirboga R, Anwar MP (2013) Effect of halloysite nanoclay on mechanical properties, thermal behavior and microstructure of cement mortars. Cem Concr Res 48:97–104
Fortner JD, Lyon DY, Sayes CM, Boyd AM, Falkner JC, Hotze EM, Alemany LB, Tao YJ, Guo W, Ausman KD, Colvin VL, Hughes JB (2005) C60 in water: nanocrystal formation and microbial response. Environ Sci Technol 39:4307–4316
Gálvez-Martos J-L, Styles D, Schoenberger H, Zeschmar-Lahl B (2018) Construction and demolition waste best management practice in Europe. Resour Conserv Recycl 136:166–178
Ge Z, Gao Z (2008) Applications of nanotechnology and nanomaterials in construction. In: First international conference construct developed countries, 235–240
Griffitt RJ, Weil R, Hyndman KA, Denslow ND, Powers K, Taylor D, Barber DS (2007) Exposure to copper nanoparticles causes gill injury and acute lethality in zebrafish (*Danio rerio*). Environ Sci Technol 41:8178–8186
Handy RD, Henry TB, Scown TM, Johnston BD, Tyler CR (2008) Manufactured nanoparticles: their uptake and effects on fish—a mechanistic analysis. Ecotoxicology 17:396–409
Hardman R (2006) A toxicologic review of quantum dots: toxicity depends on physicochemical and environmental factors. Environ Health Perspect 114:165–172
Hoshino A, Fujioka K, Oku T, Suga M, Sasaki YF, Ohta T, Yasuhara M, Suzuki K, Yamamoto K (2004) Physicochemical properties and cellular toxicity of nanocrystal quantum dots depend on their surface modification. Nano Lett 4:2163–2169
IARC Working Group on the Evaluation of Carcinogenic Risks to Humans, International Agency for Research on Cancer, World Health Organization (1997) Silica, some silicates. World Health Organization, Coal Dust and Para-aramid Fibrils
Irie H, Sunada K, Hashimoto K (2004) Recent developments in TiO_2 photocatalysis: novel applications to interior ecology materials and energy saving systems. Electrochemistry 72:807–812
Jia G, Wang H, Yan L, Wang X, Pei R, Yan T, Zhao Y, Guo X (2005) Cytotoxicity of carbon nanomaterials: single-wall nanotube, multi-wall nanotube, and fullerene. Environ Sci Technol 39:1378–1383
Kandlikar M, Ramachandran G, Maynard A, Murdock B, Toscano WA (2007) Health risk assessment for nanoparticles: a case for using expert judgment. Nanotechnol Occup Health 137–156
Kang S, Pinault M, Pfefferle LD, Elimelech M (2007) Single-walled carbon nanotubes exhibit strong antimicrobial activity. Langmuir 23:8670–8673
Karlsson HL, Cronholm P, Gustafsson J, Möller L (2008) Copper oxide nanoparticles are highly toxic: a comparison between metal oxide nanoparticles and carbon nanotubes. Chem Res Toxicol 21:1726–1732

Kartam N, Al-Mutairi N, Al-Ghusain I, Al-Humoud J (2004) Environmental management of construction and demolition waste in Kuwait. Waste Manag 24:1049–1059
Khitab A (2016) Advanced research on nanotechnology for civil engineering applications. IGI Global
Kirchner C, Liedl T, Kudera S, Pellegrino T, Muñoz Javier A, Gaub HE, Stölzle S, Fertig N, Parak WJ (2005) Cytotoxicity of colloidal CdSe and CdSe/ZnS nanoparticles. Nano Lett 5:331–338
Klaine SJ, Alvarez PJJ, Batley GE, Fernandes TF, Handy RD, Lyon DY, Mahendra S, McLaughlin MJ, Lead JR (2008) Nanomaterials in the environment: behavior, fate, bioavailability, and effects. Environ Toxicol Chem 27:1825–1851
Kloepfer JA, Mielke RE, Nadeau JL (2005) Uptake of CdSe and CdSe/ZnS quantum dots into bacteria via purine-dependent mechanisms. Appl Environ Microbiol 71:2548–2557
Konsta-Gdoutos MS, Metaxa ZS, Shah SP (2010) Multi-scale mechanical and fracture characteristics and early-age strain capacity of high performance carbon nanotube/cement nanocomposites. Cement Concr Compos 32:110–115
Kourmpanis B, Papadopoulos A, Moustakas K, Stylianou M, Haralambous KJ, Loizidou M (2008) Preliminary study for the management of construction and demolition waste. Waste Manag Res 26:267–275
Kumar A, Vemula PK, Ajayan PM, John G (2008) Silver-nanoparticle-embedded antimicrobial paints based on vegetable oil. Nat Mater 7:236–241
Lam C-W, James JT, McCluskey R, Arepalli S, Hunter RL (2006) A review of carbon nanotube toxicity and assessment of potential occupational and environmental health risks. Crit Rev Toxicol 36:189–217
Lee HA, Imran M, Monteiro-Riviere NA, Colvin VL, Yu WW, Riviere JE (2007) Biodistribution of quantum dot nanoparticles in perfused skin: evidence of coating dependency and periodicity in arterial extraction. Nano Lett 7:2865–2870
Lee W-M, An Y-J, Yoon H, Kweon H-S (2008) Toxicity and bioavailability of copper nanoparticles to the terrestrial plants mung bean (Phaseolus radiatus) and wheat (Triticum aestivum): plant agar test for water-insoluble nanoparticles. Environ Toxicol Chem 27:1915–1921
Lin P, Chen J-W, Chang LW, Wu J-P, Redding L, Chang H, Yeh T-K, Yang CS, Tsai M-H, Wang H-J, Kuo Y-C, Yang RSH (2008) Computational and ultrastructural toxicology of a nanoparticle, quantum dot 705, in mice. Environ Sci Technol 42:6264–6270
Long TC, Saleh N, Tilton RD, Lowry GV, Veronesi B (2006) Titanium dioxide (P25) produces reactive oxygen species in immortalized brain microglia (BV2): implications for nanoparticle neurotoxicity. Environ Sci Technol 40:4346–4352
Lovrić J, Cho SJ, Winnik FM, Maysinger D (2005) Unmodified cadmium telluride quantum dots induce reactive oxygen species formation leading to multiple organelle damage and cell death. Chem Biol 12:1227–1234
Lu Z, Li CM, Bao H, Qiao Y, Toh Y, Yang X (2008) Mechanism of antimicrobial activity of CdTe quantum dots. Langmuir 24:5445–5452
Lyon DY, Adams LK, Falkner JC, Alvarez BJJ (2006) Antibacterial activity of fullerene water suspensions: effects of preparation method and particle size. Environ Sci Technol 40:4360–4366
Mahendra S, Zhu H, Colvin VL, Alvarez PJ (2008) Quantum dot weathering results in microbial toxicity. Environ Sci Technol 42:9424–9430
Maynard AD, Aitken RJ, Butz T, Colvin V, Donaldson K, Oberdörster G, Philbert MA, Ryan J, Seaton A, Stone V, Tinkle SS, Tran L, Walker NJ, Warheit DB (2006) Safe handling of nanotechnology. Nature 444:267–269
Mohajerani A, Burnett L, Smith JV, Kurmus H, Milas J, Arulrajah A, Horpibulsuk S, Abdul Kadir A (2019a) Nanoparticles in Construction materials and other applications, and implications of nanoparticle use. Materials 12. https://doi.org/10.3390/ma12193052
Mohajerani A, Suter D, Jeffrey-Bailey T, Song T, Arulrajah A, Horpibulsuk S, Law D (2019b) Recycling waste materials in geopolymer concrete. Clean Technol Environ Policy 21:493–515
Mohajerani A, Kadir AA, Larobina L (2016) A practical proposal for solving the world's cigarette butt problem: recycling in fired clay bricks. Waste Manage 52:228–244

Mohseni E, Miyandehi BM, Yang J, Yazdi MA (2015) Single and combined effects of nano-SiO_2, nano-Al_2O_3 and nano-TiO_2 on the mechanical, rheological and durability properties of self-compacting mortar containing fly ash. Constr Build Mater 84:331–340

Mubaraki M, Ali SIA, Ismail A, Yusoff NIM (2016) Rheological evaluation of asphalt cements modified with ASA polymer and Al_2O_3 nanoparticles. Procedia Eng 143:1276–1284

Narayan RJ, Berry CJ, Brigmon RL (2005) Structural and biological properties of carbon nanotube composite films. Mater Sci Eng, B 123:123–129

Nations U, United Nations (2014) World urbanization prospects. Statistical Papers—United Nations (Series A), Population and vital statistics report

Nazari A, Riahi S (2011) Retracted: Al_2O_3 nanoparticles in concrete and different curing media. Energy Build 43:1480–1488

Niroumand H, Zain MFM, Alhosseini SN (2013) The Influence of nano-clays on compressive strength of earth bricks as sustainable materials. Procedia Soc Behav Sci 89:862–865

Noorvand H, Ali AAA, Demirboga R, Farzadnia N, Noorvand H (2013) Incorporation of nano TiO_2 in black rice husk ash mortars. Constr Build Mater 47:1350–1361

O'Brien N, Cummins E (2008) Recent developments in nanotechnology and risk assessment strategies for addressing public and environmental health concerns. Hum Ecol Risk Assess Int J 14:568–592

Oberdörster E (2004) Manufactured nanomaterials (fullerenes, C60) induce oxidative stress in the brain of juvenile largemouth bass. Environ Health Perspect 112:1058–1062

Oberdörster G, Celine RM, Ferin J, Weiss B (1995) Association of particulate air pollution and acute mortality: involvement of ultrafine particles? Inhalation Toxicol 7:111–124

Park S, Lee YK, Jung M, Kim KH, Chung N, Ahn E-K, Lim Y, Lee K-H (2007) Cellular toxicity of various inhalable metal nanoparticles on human alveolar epithelial cells. InhalToxicol 19(Suppl 1):59–65

Poon CS (2007) Management of construction and demolition waste. Waste Manage 27:159–160

Rahman MT, Hainin MR, Wan WA (2017) Use of waste cooking oil, tire rubber powder and palm oil fuel ash in partial replacement of bitumen. Constr Build Mater 150:95–104

Rashad AM (2013) A synopsis about the effect of nano-Al_2O_3, nano-Fe_2O_3, nano-Fe_3O_4 and nano-clay on some properties of cementitious materials—a short guide for civil engineer. Mater Des 1980–2015(52):143–157

Reeves JF, Davies SJ, Dodd NJF, Jha AN (2008) Hydroxyl radicals (OH) are associated with titanium dioxide (TiO_2) nanoparticle-induced cytotoxicity and oxidative DNA damage in fish cells. Mutat Res/fundam Mol Mech Mutagen 640:113–122

Ribeiro MCS, Pereira CMC, Sousa SPB, Nóvoa PRO, Ferreira AJM (2013) Fire reaction and mechanical performance analyses of polymer concrete materials modified with micro and nano alumina particles. Restor Build Monum 19:195–202

Ribeiro MCS, Sousa SPB, Nóvoa PRO (2015) An investigation on fire and flexural mechanical behaviors of nano and micro polyester composites filled with SiO_2 and Al_2O_3 particles. Mater Today: Proc 2:8–19

Rincón A-G, Pulgarin C (2004) Bactericidal action of illuminated TiO_2 on pure Escherichia coli and natural bacterial consortia: post-irradiation events in the dark and assessment of the effective disinfection time. Appl Catal B 49:99–112

Ryman-Rasmussen JP, Riviere JE, Monteiro-Riviere NA (2007) Surface coatings determine cytotoxicity and irritation potential of quantum dot nanoparticles in epidermal keratinocytes. J Invest Dermatol 127:143–153

Salemi N, Behfarnia K (2013) Effect of nano-particles on durability of fiber-reinforced concrete pavement. Constr Build Mater 48:934–941

Sanchez F, Sobolev K (2010) Nanotechnology in concrete—a review. Constr Build Mater 24:2060–2071

Sarkar D, Pal M, Sarkar A (2014) Laboratory evaluation of asphalt concrete prepared with over burnt brick aggregate treated by zycosoil. Int J Civ Environ Struct Constr Archit Eng 8:1302–1306

Sayes CM, Wahi R, Kurian PA, Liu Y, West JL, Ausman KD, Warheit DB, Colvin VL (2006) Correlating nanoscale titania structure with toxicity: a cytotoxicity and inflammatory response study with human dermal fibroblasts and human lung epithelial cells. Toxicol Sci 92:174–185

Shiohara A, Hoshino A, Hanaki K-I, Suzuki K, Yamamoto K (2004) On the cyto-toxicity caused by quantum dots. Microbiol Immunol 48:669–675

Silvestre J, Silvestre N, de Brito J (2016) Review on concrete nanotechnology. Eur J Environ Civ Eng 20:455–485

Stefanidou M, Karouzo A (2016) Testing the effectiveness of protective coatings on traditional bricks. Constr Build Mater 111:482–487

Tiwari AK, Chowdhury S (2013) An overview of application of nanotechnology in construction materials. In: Proceedings of the international symposium on engineering under uncertainty: safety assessment and management (ISEUSAM—2012) pp 483–492

Tixier G, Lafont M, Grapentine L, Rochfort Q, Marsalek J (2011) Ecological risk assessment of urban stormwater ponds: Literature review and proposal of a new conceptual approach providing ecological quality goals and the associated bioassessment tools. Ecol Ind 11:1497–1506

Wei W, Sethuraman A, Jin C, Monteiro-Riviere NA, Narayan RJ (2007) Biological properties of carbon nanotubes. J NanosciNanotechnol 7:1284–1297

Wiesner MR, Lowry GV, Alvarez P, Dionysiou D, Biswas P (2006) Assessing the risks of manufactured nanomaterials. Environ Sci Technol 40:4336–4345

Wolfrum EJ, Huang J, Blake DM, Maness P-C, Huang Z, Feist J, Jacoby WA (2002) Photocatalytic oxidation of bacteria, bacterial and fungal spores, and model biofilm components to carbon dioxide on titanium dioxide-coated surfaces. Environ Sci Technol 36:3412–3419

Xing B, Vecitis CD, Senesi N (2016) Engineered nanoparticles and the environment: biophysico-chemical processes and toxicity

Yan L, Xu Z, Wang X (2017) Influence of nano-silica on the flame retardancy and smoke suppression properties of transparent intumescent fire-retardant coatings. Prog Org Coat 112:319–329

Yu WW, Chang E, Falkner JC, Zhang J, Al-Somali AM, Sayes CM, Johns J, Drezek R, Colvin VL (2007) Forming biocompatible and nonaggregated nanocrystals in water using amphiphilic polymers. J Am Chem Soc 129:2871–2879

Zhang HL, Su MM, Zhao SF, Zhang YP, Zhang ZP (2016) High and low temperature properties of nano-particles/polymer modified asphalt. Constr Build Mater 114:323–332

Zhang Q, Kusaka Y, Sato K, Mo Y, Fukuda M, Donaldson K (1998) Toxicity of ultrafine nickel particles in lungs after intratracheal instillation. J Occup Health 40:171–176

Zhu W, Bartos PJM, Porro A (2004) Application of nanotechnology in construction. Mater Struct 37:649–658

Zhu X, Zhu L, Duan Z, Qi R, Li Y, Lang Y (2008) Comparative toxicity of several metal oxide nanoparticle aqueous suspensions to Zebrafish (Danio rerio) early developmental stage. J Environ Sci Health A Tox Hazard Subst Environ Eng 43:278–284

Chapter 9
Application of Nanoparticles in Construction Industries and Their Toxicity

Vinayaka B. Shet, Lokeshwari Navalgund, Keshava Joshi, and Silvia Yumnam

Abstract Global development is reflected in the growth of the construction field. To improve the construction related activity and impart the beneficial essence of current technology, nanoparticles are used by the industries at different stages. The size dependent properties of nanoparticles in the construction industry is considered for enhancing material strength, crack recovery, self-cleaning applications, antimicrobial coating, energy conservation and restoration of cultural heritage. Nanoparticles used in the construction industries reach the ecosystem through multiple channels and cause environmental implications such as adverse effects on environmentally relevant microbial species, algae, plants, and entry into the food chain. The exposure to nanoparticles by human beings also causes various health implications such as DNA damage, inflammation and cell death. Therefore, it has become crucial to determine toxicity and assessment of risk during the use of nanoparticles. Their toxicity depends on chemical and physical attributes. Uniform global regulatory policy needs to be framed to assess the toxicity, risk and approval of nanoparticles in the construction industries.

Keywords Additive · Antimicrobial · Health · Nanocoating · Nanoparticle · Thermochromic · Toxicity

V. B. Shet (✉)
Department of Biotechnology Engineering, NMAM Institute of Technology (V.T.U., Belagavi), Nitte 574110, Karnataka, India

L. Navalgund · K. Joshi
Department of Chemical Engineering, SDM College of Engineering and Technology (V.T.U., Belagavi), Dharwad 580002, Karnataka, India

S. Yumnam
College of Pharmacy, Gachon University, 191, Hambakmoero, Yeonsu-gu, Incheon 21936, Republic of Korea

J. A. Malik and S. Marathe (eds.), *Ecological and Health Effects of Building Materials*,
https://doi.org/10.1007/978-3-030-76073-1_9

9.1 Introduction

Infrastructure development has gained momentum globally in recent years. At the same time attractive features are incorporated by the construction industries at different stages to enhance the benefits and fulfill the requirements of the end users with the essence of nanotechnology. Richard Feynman introduced the concept of nanotechnology in 1959 while delivering the talk "There's Plenty of Room at the Bottom" at American Physical Society meeting held at California Institute of Technology. Norio Taniguchi coined the word nanotechnology in 1974. The word "nano" is derived from Greek word dwarf (small). The study of objects and phenomena in the range of 1–100 nm is termed as nanoscience. The creation of materials, devices and systems through the arrangement of matter on nano scale is defined as nanotechnology. The nanomaterials can be synthesized using top down and bottom up approaches. Physical, chemical, biological and mechanical methods are widely used for the synthesis of nanoparticles. Most of the synthesized nanomaterials exhibit different properties and effects in comparison to the similar material in a macroscale. The size dependent properties exhibited by nanoparticles in the construction industry are explored for enhancing material strength, crack recovery, acid resistance, self-cleaning applications, antimicrobial coating, fire resistant paints, energy conservation and restoration of cultural heritage. Nanoparticles used in the construction industries reach the ecosystem through multiple channels and cause environmental implications such as adverse effects on environmentally relevant microbial species, algae, plants, and entry into the food chain. The exposure to nanoparticles by human beings also causes various health implications such as DNA damage, inflammation and cell death. Therefore, it has become crucial to determine toxicity and assessment of risk during the use of nanoparticles. Their toxicity depends on chemical and physical attributes such as morphology, composition, uniformity, size, crystallinity, magnetism and surface functionalization. Uniform global regulatory policy needs to be framed to assess the toxicity, risk and approval of nanoparticles.

9.2 Size Dependent Properties

There is an enormous scale difference in the universe. Different forces dominate at different scales and a suitable model explains the phenomena. Nanoscale materials differ from macroscale materials due to the following reasons:

(a) Gravitation force becomes negligible and electromagnetic forces dominate.
(b) Quantum mechanics model is used to describe motion and energy instead of the classical mechanics model.
(c) Greater surface area to volume ratios.
(d) Random molecular motion becomes more important.

The gravity becomes negligible due to the mass of the object at nanoscale. Since the gravitational force is a function of mass and distance, it is weak between nanoscale particles. However, electromagnetic force being a function of charge and distance is not affected by mass, so it can be very strong even when we have nanoscale particles.

The response of material under certain conditions is known as "property". Optical (transparency, color), Chemical (reaction rates, reactivity), Physical (melting point, hardness), Electrical (conductivity), properties are exhibited by nanoscale material compared to its macroscale.

9.3 Nanoparticles in Construction Industry and Application

Nanoparticles exhibit different properties and its usage depends on the specific application. The benefits and applications of various nanoparticles in the construction field are listed in Table 9.1.

9.3.1 Nanoparticles

Widely used nanoparticles in the field of construction are discussed in this section.

9.3.1.1 Silicon Dioxide (SiO_2)

Nano SiO_2 powder is amorphous in nature and appears in white color. Due to the reduced particle size, nano-SiO_2 has many advantages such as greater surface area, enhanced surface adsorption, surface energy, purity and good dispersion. Presence of silicon dioxide in the mortar or cement improves the bonding between aggregates and matrix, reduces porosity, increases the density, particle packing, enhances compressive and flexural strength, facilitates the hydration reaction of concrete and reduces the setting time.

9.3.1.2 Carbon Nanotube

Carbon nanotubes (CNT) are made up of carbon atoms arranged hexagonally. Carbon nanotubes are categorized as single-walled and multi-walled carbon nanotubes. In the field of civil engineering, incorporating CNTs along with cement matrices is exhibiting potential for enhanced mechanical properties and durability. Blending CNT in the cement or mortar improves the tensile strength, flexural strength and compressive strength. CNTs can also have advantages in preventing the microcrack

Table 9.1 Benefits of nanoparticles used in the construction industry

Nanoparticle	Benefit	Application	References
Silicon dioxide	Stability	Additive	Sanchez and Sobolev (2010)
	Resistance to water penetration		
	Protection from fire		
	Self-cleaning		
	Enhanced mechanical properties		
	Frost resistance		
	Refined porosity and microstructure		
	Improved corrosion resistance		
Carbon nanotube	Mechanical strength	Additives	Singh et al. (2017)
	Antibacterial activity		
	Self-cleaning		
	Improved corrosion resistance		
Titanium dioxide	Degradation of air pollutant	Photocatalyst	Colangiuli et al. (2015)
	High compressive and flexural strength	Antibacterial coating	
	Suppress the growth of microorganism	Hydrophobic coatings	
	Prevention of water penetration		
Zirconium dioxide	Tensile strength	Building matrixes	Boulos et al. (2019)
	Suppress the growth of microorganism	Antibacterial coating	
Fe_2O_3	Compressive and flexural strength	Additives	Sanchez and Sobolev (2010)
Vanadium dioxide	Regulate the indoor temperature	Thermochromic smart window	Panagopoulou et al. (2016)
Copper nanoparticle	Corrosion resistance	Additives	Mohajerani et al. (2019)
	Weld ability		
CdSe Quantum Dots	Utilization of solar energy	Photovoltaics	Liu et al. (2020)
Aluminium oxide	Enhanced mechanical properties	Additive	Nazari and Riahi (2011); Behfarnia and Salemi 2013
	Reduction in water absorption		

(continued)

Table 9.1 (continued)

Nanoparticle	Benefit	Application	References
	Refined porosity and microstructure		
	Escalated hydration		
	Frost resistance		
Silver nanoparticles	Antibacterial activity	Additive	Le et al. (2011)
Magnesium oxide	Antibacterial activity	Paint, coating	Huang et al. (2005)
Zinc oxide	Antibacterial activity	Paint, coating	Hochmannova and Vytrasova (2010)

propagation due to the improvement of cementitious materials. It has the chemical resistance property, hence can be used for the constructions prone to chemical exposure. CNT exhibits fire retardant property (Wang et al. 2017).

9.3.1.3 Graphene

Graphene is made up of carbon atoms arranged in the form of sheets. Graphene offers greater tensile strength, better electrical conductivity, anticorrosion and optical properties, hence used for different applications in the construction industry. Flexural strength of the cement can be enhanced by blending graphene. Currently graphene is used as an anti-corrosion coating to protect the steel and photovoltaic panel from the surrounding environment, light emitting diode and touch screen applications inside the building.

9.3.2 Applications

Current section is focused on the potential application of nanoparticles in the construction industry.

9.3.2.1 Additive

Additives are the synthetic or natural material mixed with binding agent, mortars and concrete to achieve desired property in the finished product. Silicon dioxide enhances the stability and prevents the percolation of water. Carbon nanotubes have reported to be enhancing mechanical strength.

9.3.2.2 Self-Healing

The formation of cracks in the concrete structures is a phenomenon that occurs during the service life. Various nanoparticles are adopted to fill the cracks whenever it appears without the external intervention is termed as a self-healing mechanism. Hollow fiber, nanoencapsulation, shape memory materials are explored for self-healing applications. Hollow fibers are filled with healing agents and placed in the concrete structures. During the occurrence of damage to the concrete structure, hollow fibers will break open and release the healing agents to instantly heal the crack (Huseien et al. 2019).

9.3.2.3 Photocatalyst

The concrete mixed with photocatalyst (Titanium dioxide, zinc oxide) will degrade the air pollutants, bacteria, algae and mold using electromagnetic radiation. The photocatalytic reaction takes place in the presence of a light source, when the energy is greater than the band gap of the photocatalyst. The concept is also developed in photocatalytic paint (Dikkar et al. 2020).

9.3.2.4 Antimicrobial

Microorganisms harbouring on the concrete structures such as bacteria, fungi and algae proliferate, propagate and cause various problems. They are responsible for reactions like corrosion, affecting aesthetic appearance, degrading mechanical properties, destroying the internal structure of concrete and reducing durability of concrete. In order to overcome these problems, concrete is blended with nanoparticles exhibiting antimicrobial properties. The nanoparticles of Fe_3O_4, Al_2O_3, CuO, ZnO, TiO_2, $CaCO_3$, and Cu_2O were reported to be effective against microorganisms. Antimicrobial paints and coatings are also available commercially.

9.3.2.5 Nano Paint

As the construction industry is progressing, synergistically the paint industry is growing in terms of the fascinating properties of nanoparticles. To prevent the growth of bacteria, fungus and algae on the building, nanotechnology based paint is gaining importance. The interior and exterior paints are prone to get deteriorated by the microorganism present in the surrounding. Especially during the rainy season due to the moisture condition, different microorganisms grow on the surface. Microorganisms cause damage and distort the appearance of the aesthetics of buildings as well. During the microbial growth, decoloration takes place due to the spore production thus increasing the accumulation of dust. Due to these reasons silver, titanium dioxide, copper in the nanoscale is gaining importance in the paint industry because of its antimicrobial property.

9.3.2.6 Nanocoating

Thin layer of film having nanoscale dimension is used as a protective coating on the construction material. Steel, limestone, concrete, marble and glass are exposed to the environment and prone to undergo corrosion, expansion and growth of microorganisms. The nano coatings are applied to these materials to achieve heat resistance, friction reduction, anti-corrosion and antimicrobial properties. Various coatings are also applied on the window, doors, walls and external surfaces or components to obtain flame-retardant, hydrophilic, hydrophobic, wear-resistant, anti-graffiti, corrosion resistance, photovoltaic properties and provide protection to the base material. Nanoparticles of aluminium oxide, silicon dioxide, titanium oxide and zirconium oxide are incorporated in the coating to enhance the hardness and mechanical properties of the components used in the building materials.

9.3.2.7 Thermochromic

Thermochromic material changes its color reversibly to the corresponding temperature. Indoor solar radiation can be regulated by changing the optical properties of windows with respect to the surrounding temperature. Vanadium oxide exhibits the thermochromic phenomena. Nano crystals of vanadium dioxide are developed by engineering the morphology at nanoscale to improve the performance. In the similar aspectsThermochromic paints are also developed.

9.3.2.8 Photovoltaics

To adopt a sustainability approach in terms of energy generation, building envelopes such as facades, skylights and roofs are integrated with photovoltaics. Parking canopy, wall mountings, cladding are also integrated with photovoltaic.

9.4 Route of Nanoparticle Exposure

As the amalgamation of nanotechnology is escalated, nanoparticles are exposed to the environment and living system causing adverse effects. Nanoparticles are released into the environment at different stages of activity during the manufacturing process, construction site, demolition, recycling process and enter the living system during the consumer use. The nanoparticles released into the environment undergo physical, biological and chemical transformation because of change in their properties.

9.4.1 Manufacturing Process

Synthesis or the manufacturing process of nanoparticles is the primary route for the exposure. As a safety measure, the manufacturing process of nanoparticles is carried out in the enclosed chambers. Hence the chances of nanoparticles released into the air is less and inhalation of nanoparticles during the manufacturing process is not evident. However, during the post manufacturing process such as cleaning or disposal of waste, probably in an unintentional way some nanoparticles get released into the environment. The possibility of dry powder form of nanoparticle releasing inside the laboratory hoods during the handling also is not ruled out.

9.4.2 Construction Site

Nanoparticles released at the construction site will have prolonged exposure in the surrounding atmosphere. People working in the construction site are prone to inhale the nanoparticles causing health implications. There is a necessity that nanoparticle based products need to be handled with utmost care to prevent its dispersion into the air and water. Preparing the silica nanoparticle with cement can be performed in dry or wet form. While preparing the wet form, silica nanoparticles are mixed in water and then dispersed into cement mixture. The process is effective during large scale production and thus avoids the risk of exposure to nanoparticles. Super plasticizers are recommended for preparing the dry form. The coating of nanoparticles in the form of spray must be avoided, since it elevates the risk of exposure. The spray coating will have a higher concentration of nanoparticles when compared to composites.

9.4.3 Demolition

Buildings constructed with nanoparticles incorporated at different stages pose a threat of dispersing nanoparticles to air, water and the surrounding environment during its demolition in future. Currently there is no technology available to recover the nanoparticles embedded with concrete and composite structures. Though it is unavoidable to prevent dispersion of the nanoparticles to the environment during demolition, the coating can be removed to a certain extent.

9.4.4 Natural Phenomena

Due to the continuous exposures of buildings to the environment, damage of buildings, abrasion, and deterioration slowly leads to the release of nanoparticles into the surrounding environment. Natural calamities such as storms, heavy rainfall, floods also cause damage to structures and lead to the leaching of nanoparticles into the water bodies and soil. The accidental fire also releases the nanoparticle into the environment. However, practically it is difficult to analyse the concentration of nanoparticles released into the environment.

9.5 Toxicity

On-going research and development in the field of nanotechnology is exploring different applications in the field of construction industry. At the same time the risk associated with the nano based products on human health and environment is increasing. The study of biological, physical, chemical agents adversely affecting the living system and environment is termed as toxicology. The study of possible adverse effects by the nanomaterials is called nanotoxicology. Toxicity is the degree to which a substance can damage the organism. Nanotechnology has been explored in the various domains due to the impressive benefits. The movement of nanoparticles is much faster than its bulk structure. Due to the dimension of nanoparticles it offers higher surface area, causes reactivity and can diffuse into the cells. Hence it can be more toxic. During the year 1990, the first toxic effect of nanoparticles was reported in which the significant pulmonary inflammation using TiO_2 and Al_2O_3 nanoparticles was reported in comparison with microparticles instilled in rats. Thus focus is shifted to assess the toxicity of the nanomaterials, since more nanotechnology based products are used by mankind.

9.6 Environmental Implications

The nanoparticles are fatal to the growth of Gram negative bacteria due to the presence of thin peptidoglycan layers. In case of Gram positive bacteria, the presence of a thick peptidoglycan layer provides resistance to the nanoparticle effect. Even though the antibacterial property is beneficial during the application point of few, with respect to the environmental perspective it is having a negative impact. In the environment, bacteria has the major role to play in decomposition of organic matter, terrestrial and aquatic food chains, hence negative impact on environmentally relevant bacteria affect the trophic levels. The plants, phytoplanktons exhibit reduced growth rate in presence of higher concentration of nanoparticles by reducing the mitotic index. Mitotic index is the measure of rate of cell division. Copper oxide nanoparticles are reported to induce DNA damage of the plant and reduce the chlorophyll content. Nanoparticles have also been reported to inhibit the seed germination process and root elongation.

9.7 Health Implications

The nanoparticles diffuse into the air much faster than its macro scale particles. The reason is due to the smaller size, shape and greater surface area. Nanoparticles thus present in the air enter into the human body through the inhalation process. Nanoparticles do not have any barrier in the nose to prevent its entry into the human body. They have the ability to cross the cellular barriers and reach the vital organs such as the lung, kidney and liver. Health implications arise due to the damage caused to the mitochondria, DNA by nanoparticles eventually leading to cell death. Nanoparticles can also pass through the blood brain barrier.

The mechanism involved with toxicity of nanomaterials is reported to be through oxidative stress which damages proteins, carbohydrates, DNA, lipids and potential to change the properties of cell membrane disrupting vital cellular functions. The oxidative stress triggered by nanoparticles depends on particle uptake, presence of transition metals and mutagens, solubility, composition, shape, size and aggregation.

The silicon dioxide nanoparticles are reported to accumulate in the brain, kidney, spleen, liver and heart after the entry into the human body through pulmonary exposure and skin. The nanoparticle causes the generation of reactive oxygen species and brings oxidative stress, further leading to cell death. Significant DNA damage in endothelial cells was reported due to the silicon dioxide nanoparticle.

Titanium dioxide is extensively used in nanoparticles in the construction industry. Titanium dioxide exhibits genotoxic, cytotoxic and oxidative effects causing apoptosis and inflammation (Proquin et al. 2017). Titanium dioxide has been listed as a potential cancer causing agent by the International Agency for Research on Cancer. The exposure limit is recommended to be 0.3 mg/m^3 by the Institute of Occupational Safety and Health. The policy focusing on nanotoxicity has to be made uniform across the world to take necessary preventive steps. Risk assessment and management for the building material incorporated with nanoparticles has to be carried out.

9.8 Conclusion

Nanotechnology based products are promising in the field of construction industry. Various nanoparticles have gained the space in paint, coating, additives due to its fascinating properties such as increased mechanical strength, and anti-corrosion, antibacterial properties. Prevention of release of nanoparticles to the ecosystem has to be monitored and regulated. However the standard operating protocol needs to be developed for the safety of stakeholders. Nanotoxicology studies need to be taken up for all the nano based products to assess the risk and instil confidence among the users.

References

Behfarnia K, Salemi N (2013) The effects of nano-silica and nano-alumina on frost resistance of normal concrete. Constr Build Mater 48:580–584. https://doi.org/10.1016/j.conbuildmat.2013.07.088

Boulos L, Foruzanmehr MR, Tagnit-Hamou A, Robert M (2019) The effect of a zirconium dioxide sol-gel treatment on the durability of flax reinforcements in cementitious composites. Cem Concr Res 115:105–115. https://doi.org/10.1016/j.cemconres.2018.10.004

Colangiuli D, Calia A, Bianco N (2015) Novel multifunctional coatings with photocatalytic and hydrophobic properties for the preservation of the stone building heritage. Constr Build Mater 93:189–196

Dikkar H, Kapre V, Diwan A, Sekar SK (2020) Titanium dioxide as a photocatalyst to create self-cleaning concrete. Mater Today Proc. https://doi.org/10.1016/j.matpr.2020.10.948

Hochmannova L, Vytrasova J (2010) Photocatalytic and antimicrobial effects of interior paints. Prog Org Coatings 67:1–5

Huang L, Li DQ, Lin YJ et al (2005) Controllable preparation of nano-MgO and investigation of its bactericidal properties. J Inorg Biochem 99:986–993. https://doi.org/10.1016/j.jinorgbio.2004.12.022

Huseien GF, Shah KW, Sam ARM (2019) Sustainability of nanomaterials based self-healing concrete: an all-inclusive insight. J Build Eng 23:155–171

Le AT, Huy PT, Tam LT et al (2011) Novel silver nanoparticles: synthesis, properties and applications. Int J Nanotechnol 8:278–290. https://doi.org/10.1504/IJNT.2011.038205

Liu D, Liu J, Liu J et al (2020) The photovoltaic performance of CdS/CdSe quantum dots co-sensitized solar cells based on zinc titanium mixed metal oxides. Phys E Low-Dimens Syst Nanostructures 115:113669. https://doi.org/10.1016/j.physe.2019.113669

Mohajerani A, Burnett L, Smith JV, Kurmus H, Milas J, Arulrajah A, Horpibulsuk S, Abdul Kadir A (2019) Nanoparticles in construction materials and other applications, and implications of nanoparticle use. Materials (Basel) 12:3052. https://doi.org/10.3390/ma12193052

Nazari A, Riahi S (2011) Al2O3 nanoparticles in concrete and different curing media. Energy Build 43:1480–1488

Panagopoulou M, Gagaoudakis E, Boukos N et al (2016) Thermochromic performance of Mg-doped VO2 thin films on functional substrates for glazing applications. Sol Energy Mater Sol Cells 157:1004–1010. https://doi.org/10.1016/j.solmat.2016.08.021

Proquin H, Rodríguez-Ibarra C, Moonen CGJ, et al (2017) Titanium dioxide food additive (E171) induces ROS formation and genotoxicity: contribution of micro and nano-sized fractions. Mutagenesis 32:139–149. https://doi.org/10.1093/mutage/gew051

Sanchez F, Sobolev K (2010) Nanotechnology in concrete—a review. Constr Build Mater 24:2060–2071

Singh NB, Kalra M, Saxena SK (2017) Nanoscience of cement and concrete. In: Materials today: proceedings. Elsevier Ltd, pp 5478–5487

Wang Y, Hu H, Rong C (2017) The effect of the diameter of carbon nanotube on the mechanical and electrical properties of cement mortar. In: Key engineering materials. Trans Tech Publications Ltd, pp 479–485

Chapter 10
Potential Environmental Impacts of Nanoparticles Used in Construction Industry

Annika Durve Gupta and Sonali Zankar Patil

Abstract Nanoparticles are ultrafine particles having a size ranging between 1 and 100 nm. Recently, nanoparticles have had various applications in fields like electronics; construction, manufacturing, cosmetics, agriculture, purification, and medicine have increased. Nanomaterials and nanocomposites have distinctive properties (physical and chemical), which have resulted in their use in the construction industry to facilitate innovative applications. The heavily used nanoparticles in the building industry are titanium dioxide, carbon nanotubes, silica, copper, clay, and aluminum oxide. The rising use of these nanoparticles has led to numerous adverse environmental effects. Nanoparticles of carbon, including fullerenes, nanotubes, metal oxides like iron and titanium, and natural inorganic compounds, along with asbestos and quartz, have shown to have toxic effects on the human health and environment. As the size of the particle is very small, they can pass through various mammalian and plant cell membranes and can also be absorbed in them. They can cause various inflammatory reactions and fibrosis in higher organisms and can exert an antioxidant and cytotoxicity effect on unicellular. Numerous respiratory and cardiovascular diseases are associated with continuous exposure to nanoparticles. In this chapter, we are dealing with the various nanoparticles produced and used in the construction industry and their deleterious effects seen in flora and fauna.

Keywords Construction industry · Cytotoxicity · Nanoparticles synthesis · Nanocomposites · Nanotoxicity

A. D. Gupta (✉)
Department of Biotechnology, B. K. Birla College (Autonomous), Kalyan, Maharashtra, India

S. Z. Patil
Department of Bioanalytical Sciences, B. K. Birla College (Autonomous), Kalyan, Maharashtra, India

J. A. Malik and S. Marathe (eds.), *Ecological and Health Effects of Building Materials*,
https://doi.org/10.1007/978-3-030-76073-1_10

10.1 Introduction

Richard P. Feynman, a physicist, gave the notion of nanotechnology in 1959, in his famed address, "There's Plenty of Room at the Bottom", at the American Physical Society at the California Institute of Technology (Feynman 1960). The opinions put ahead by Feynman went unobserved until Norio Taniguchi (1974), who coined the term "Nanotechnology" at the International Conference on Production Engineering (Taniguchi et al. 1974). Drexler defined nanotechnology as the production of materials having sizes between 1 and 100 nm (Drexler 1981). Hence, nanotechnology is quite old but has been explored now-a-days, has had a high demand in the last two decades. In fact, nanomaterials have been utilized unwittingly for thousands of years-nanoparticles of gold were used to stain glasses and helped cure several diseases. Researchers have increasingly been able to research and elucidate the physicochemical properties, which is shape- and size-dependent, of nanoparticles by using sophisticated and creative techniques (Rao et al. 2015; Rai et al. 2015; Abbasi et al. 2016; Giljohann et al. 2010; Pereira et al. 2015). Nanotechnology typically examines the matter that is manipulated at a nano-scale stage of between 1 and 100 nm. In various fields, the development of the nano sector has led to enormous growth viz. food and agriculture, pharmaceuticals, biotechnology, medicine, energy and the environment, material science (Kandasamy and Prema 2015). Nanoparticles can be characterized as particles with at least one dimension varying in diameter from 1 to 100 nm that can alter their physicochemical properties compared to their parent bulk matter. Due to their peculiar characteristics and innovative features, nanoparticles are widely used in various aspects of daily life and energy production. Nanoparticles have physical characteristics in materials science and biology, viz. Homogeneity, conductance and special optical characteristics that make them appealing. Nanoparticles can be synthesized from a variety of materials. Their properties preferably depend on their chemical composition and the size and/or shape of the particles (Kataria et al. 2019).

10.2 Synthesis of Nanoparticles

Two techniques are used to synthesize nanoparticles: the top-down approach and the bottom-up approach.

Several methods have been developed based on these two approaches (Fig. 10.1).

10.2.1 Top-Down Approach

This method includes the breakdown of large bulk material into particles of nano size. This can be achieved by different methods, i.e., Milling, method of attrition and wire technique for electro explosion. This is a fast method of production, but it uses

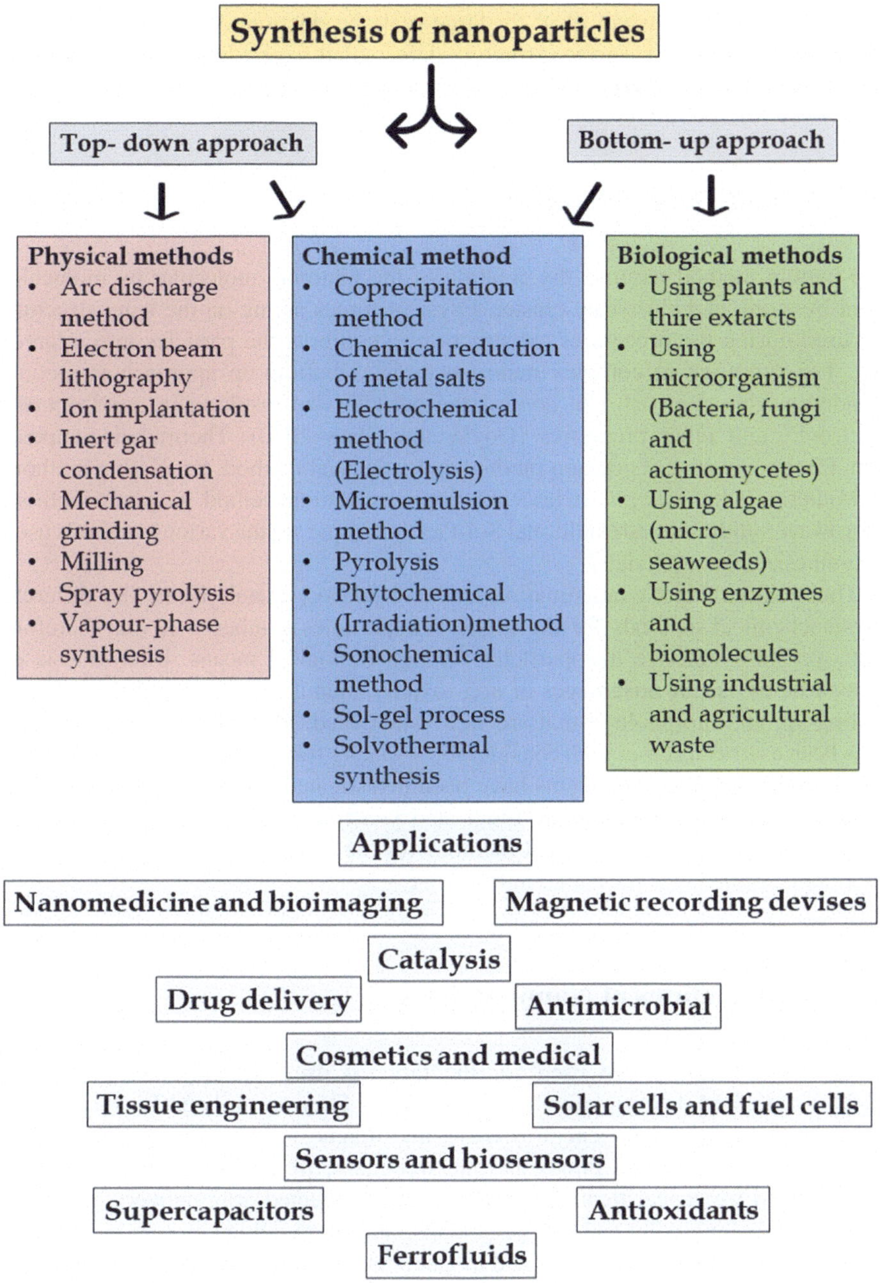

Fig. 10.1 Various methods (Physical, Chemical, and Biological) for nanoparticles synthesis and their applications in a variety of fields

more resources, so it is not ideal for large-scale production. Another downside of the top-down method is surface structure defects that have a huge influence on the physical and other properties of the produced nanoparticles (Thakkar et al. 2010).

10.2.2 Bottom-up Approach

The bottom method requires the creation of the material, molecular by molecule, atom by atom, and cluster by cluster. Physical forces acting on the nano structure are used during the process of construction to combine the particles into a larger one. For synthesis of complex nanostructures, a bottom-up approach is recommended as this approach can control the particle size resulting in good optical, electronic, and other properties (Doyle and Glass 2010). Thermal decomposition, Chemical vapor deposition method, solvothermal method, templating method, hydrothermal synthesis, pulsed laser ablation, combustion method, gas phase method, microwave synthesis, and traditional Sol-Gel technique are the various methods used to synthesize nanomaterials.

The environmental contamination caused by heavy metals limits the different physicochemical methods for the metal nanoparticles synthesis. It was therefore employed to synthesize nanoparticles through biological means. Biosynthesis of nanoparticles has the advantages of non-toxicity, manufacturing duplicability, easy scaling-up, and morphology that are well established. Microorganisms and plants have been used as reserves with considerable capacity for synthesizing nanoparticles. Till date, several microorganisms have been investigated for the synthesis of metal nanoparticles, including bacteria, fungi, and yeast, as well as plants. The various physical, chemical, and biological methods for the synthesis of nanoparticles with their applications were presented in Fig. 10.1.

10.3 Applications of Nanoparticles

Nanoparticles of gold are used for the targeted drug delivery like paclitaxel, methotrexate, and doxorubicin (Rai et al. 2015). They are utilized for the finding of tumour, angiogenesis, genetic disease, and diagnosis of various genetic disorders, photo imaging, and photo thermal treatment. Iron oxide particles of have been utilized for cancer treatment, hyperthermia, targeted drug delivery, repairing of tissue, labelling cells and cell organelles, and immunological assays, purification of biological fluids, imaging, and magnetically reactive drug delivery treatment (Khlebtsov et al. 2011; Huang et al. 2007; Iv et al. 2015). Nanoparticles of silver are used for various antimicrobial, anticancer, anti-inflammatory, and wound medication applications for different purposes (Ahamed et al. 2010). Various properties such as biocompatible, non-hazardous, self-purification, skin-friendly, antimicrobial, and dermatological activities are demonstrated by nanoparticles of titanium and zinc.

They have been used in various fields like cosmetic, biomedical, and as ultraviolet (UV)-preventing agents (Zahir et al. 2015; Ambika et al. 2015). Nanoparticles of palladium and copper are applied in certain polymers, batteries, plastics plasmonic waveguides, and optical restricting gadgets (Momeni et al. 2015; Nasrollahzadeh et al. 2015). Copper and palladium nanoparticles show antimicrobial activity against many pathogenic microorganisms. Metal nanoparticles have been used in the spatial study of different glycosphingolipids, nucleic acids, peptides, lipids, metabolites, and drug particles. The specific properties of nanoparticles make them more acceptable for the design of biosensors and electrochemical sensors (Peng et al. 2011). Nano sensors are designed for the recognition of algal and mycobacteria toxins, along with heavy metals in various sources (Selid et al. 2009). Nanosensors have been created by using nanoparticles for management of various hormones and for the sensing of viruses, crop pests, stress factors, and nutrient levels in the soil (Koren et al. 2015). In many biomedical applications, such as anticancer and antimicrobial applications, biological nanoparticles have been implemented. This is because biologically synthesized nanoparticles, particularly in biomedical applications, show greater efficacy compared to physicochemical nanoparticles (Singh et al. 2016).

10.3.1 Use of Various Nanoparticles in Construction Area

In civil engineering, the use of good building materials plays a very crucial role. Steel, concrete, bricks, stones, wood, glass, and many metallic elements are the key building materials (Khitab 2012). The oldest resources used by primitive humans to build refuge against natural catastrophes and wild animals were stone and timber. Like the evolution of all other species on the planet, the raw material of construction also passed through an evolutionary course. As the modern-day giants in the construction industry, we now have concrete and steel. As the experience grew, engineers and researchers created brand new and smart materials. A recent and new scheme that surfaced at the end of the last centenary has been the utilization of nano technology and nano materials in civil engineering. The use of this new and modern technology has solved several of the construction industry's challenges. This is because of the nanomaterials' higher strength and lower density.

The use of these nanomaterials is expanding in numerous fields, such as automotive, biomedicine, electronics, construction, and robotics due to their unique mechanical, chemical, electronic, and optical properties. Currently, however, the use of nanotechnology and nanomaterials in civil engineering is much less significant, especially in the construction sector (Ge and Gao 2008; Li 2004; Irie et al. 2004). These nanomaterials are mostly used to give sturdier physical composites, lighter structure, improved cement like material properties, low maintenance coating, better-quality pipe joining materials, upgraded heat and sound insulation, better glass reflectivity, water repellents, self-cleaning and antifogging surfaces, ultraviolet light protector, nanosized building safety and structural health monitoring sensors and solar cells (Daniyal et al. 2018) (Table 10.1).

Table 10.1 Several nanomaterials used in the construction Industry

Nanomaterials	Base raw material	Potential advantages	References
Zirconium oxide (ZrO_2)	Concrete and cement	Tensile strength is boosted	Nazari and Shadi (2010), Nazari et al. (2010c)
Silver (Ag)	Paints	Shows antimicrobial properties	Kumar et al. (2008)
Carbon nanotube (CNT)	Concrete and cement	Tensile strength and stability are improved	Małgorzata (2014), Becher (1991), Saafi and Romine (2005), Song et al. (2008), Zhang et al. (2006), Nochaiya and Chaipanich (2011), Chaipanich et al. (2010)
	Ceramics	Thermal and mechanistic properties are enhancement	Becher (1991)
	Nano electrical mechanism systems	Health assessment of structures are carried out in real time	Saafi and Romine (2005), Song et al. (2008)
	Solar cell	Electron mediation efficient	Girishkumar et al. (2005)
Silicon dioxide (SiO_2)	Concrete and cement	Tensile strength and stability are enhancement	Zhang and Li (2011), Hussain and Sastry (2014), Ye et al. (2007)
	Glass	Antireflective and heat isolation	Rana et al. (2009)
Titanium dioxide (TiO_2)	Solar cell	Production of non-utility power	Serpone and Pellizzetti (1989)
	Glass	Antifogging, hydrophilicity, fouling resistance	Ruot et al. (2009), Guerrini (2012)
	Concrete and cement	Self-cleaning, rapid hydration and mechanical strengths are improvement	Mohseni et al. (2015), Ruot et al. (2009), Guerrini (2012)
Calcium carbonate ($CaCO_3$)	Concrete and cement	Tensile strength is boosted	Sato and Diallo (2010), Kawashima et al. (2013), Shaikh and Supit (2014)

(continued)

Table 10.1 (continued)

Nanomaterials	Base raw material	Potential advantages	References
Copper oxide (CuO)	Concrete and cement/steel	Tensile strength and stability are improved, improved weldability and resistance to corrosion by steel	Nazari and Shadi (2011), Nazari et al. (2011)
Aluminium oxide (Al_2O_3)	Concrete and cement	Tensile strengths are boosted	Arefi et al. (2011), Zhenhua et al. (2006)
	Asphalt concrete	Increased serviceability	
Zycosoil	Asphalt concrete	Increased weakened life, higher compaction	Sarkar et al. (2014)
Chromium oxide (Cr_2O_3)	Concrete and cement	Tensile strengths are boosted	Nazari and Riahi (2011a, b)
Ferric oxide (Fe_2O_3)	Concrete and cement	Tensile strengths and stability are improvement	Nazari et al. (2010a, b), Yazdi et al. (2011)
Clay nanoparticles	Bricks mortar	Heightened compressive strength	Daniyal et al. (2018)

The development of laboratories and hospitals free of microorganism, water-resistant homes, metropolitan environmental safety has resulted because of the use of nanoparticles in construction (Lee et al. 2010). Cement nanomaterials have been applied, resulting in a substantial decrease in cementitious composite preliminary setting time as well as finishing setting time. The broad specific surface areas, the larger number of highly unstable and active atoms on the surface, speed up the reaction for hydration of cement and eventually minimize the setting time needed for the cement. The incorporation of nanomaterials also substantially decreases the workability of composites of cement. The substitution of cement with nanoparticles has caused the total surface area to increase. Therefore, more water for lubrication is needed to make wet the particles. The support of high-quality plasticizers and additional cement materials such as silica fume, rice husk ash, fly ash, and nanoparticles have overcome this problem (Daniyal et al. 2018). With the use of sufficient nanomaterial concentrations, the tensile strengths and stability properties of cement composites are dramatically increased. The addition of nanoparticles to it, has produced ultra-high-strength concrete, photocatalytic concrete, self-heating concrete, bendable concrete, and concrete containing CNTs. Instead of micro silica particles, Nano Silica Concrete has nano silica, which has led to higher initial and final compressive forces, greater workability, and lower absorbency. In addition, in the concrete, higher tensile strength and segregation resistance were also seen. The new concrete, Ultra High Strength Concrete, has many advantages viz. the column sections in buildings and amount of steel reinforcement in concrete could be reduced (Li 2004). The calcium nitrite-impeded as well as graphene and nano-TiO_2-mixed cement slurry coating showed significant level of corrosion inhibition of when associated with uninhibited systems (Daniyal et al. 2018).

In asphalt cement, a 5% inclusion of nano-aluminium oxide (Al_2O_3) was tested to verify the effect of the modifier on asphalt cement. This addition demonstrated good high temperature resistance and improved the material's serviceability (Mubaraki et al. 2016; Fang et al. 2016; Mohajerani et al. 2017). Moisture damage is a major obstacle around the world due to the malfunction of the layers made up of hot mix asphalt due to the penetration of water into the sidewalk system, resulting in a loss of hardness, endurance, and strength (Nejad et al. 2012; Jin et al. 2019). Zycosoil has been used to test the asphalt concrete properties when mixed with zycosoil as an anti-strip agent (Sarkar et al. 2014). Owing to the rise of the filler and the reduction of air spaces in the asphalt blend, the fatigue life has been improved. In addition, due to the alteration of the aggregate surface, superior compaction because of the asphalt blend have been obtained (Nejad et al. 2012).

An Investigation to verify the effect of nano-clays on earth bricks was performed out by Niroumand et al. (2013). Nano-clays are nanoparticles coated with mineral silicates, and the utilization of bricks embedded with various nanoparticles is dictated by the chemical composition of the brick. The results showed that a nano-clay addition of 5% can yield a compressive strength of 4.8 times that of regular clay bricks (Niroumand et al. 2013). As compared to standard clay bricks, the nano-clay modifier is the most ecological balanced material. The efficacy of various shielding coatings on bricks was tested by Stefanidou and Kazou (2016) the physical properties of each solution were checked. 1–1.5% silica nanoparticles were modified by addition of silane/siloxane, linseed oil, and alka siloxane. Silica nanoparticles and alkosiloxane have been shown to be the highly efficient method to secure bricks that have shown high endurance to water absorption and a major increase in toughness.

In the building area, timber is also used. A timber coating was made, using silica nanoparticles (SiO_2) and alumina which were integrated with hydrophobic polymers (Mohamed 2015). This coating created an invisible layer, confining water, soil, oil, algae and dust-resistant properties on the surface of the timber. It also provides UV protection and retained surface quality (Mohajerani et al. 2018). Medium density fibreboard (MDF) is a wood processing commodity composed of a synthetic resin merged with fine lignocellulosic fibers that are subjected to stress and high temperature to form panels (Kumar et al. 2013). The method is considered expensive, so reducing the press time would diminish the cost of production and boost the ability of production. The effect of nano-aluminium oxide (Al_2O_3) on the heat transfer step of MDF during the hot-pressing process was studied by Kumar et al. (2013). The results showed an increase in the mechanical and physical traits of MDF panels due to the improvement in heat transfer.

Uneven surface may lead to increasing stress and, hence, cracking due to fatigue in traditional steel; however, surface unevenness can be minimized with the addition of nanoparticles as a modifier and, therefore, cracking can be lowered (Tiwari and Chowdhury 2012). In addition, studies have shown that by strengthening the steel microstructure, the effects of embrittlement of hydrogen and the intergranular cement process are minimized (Mohamed 2015). The effect of colloidal nanoparticles of copper as an enhancer for steel corrosion resistant coatings was studied by Hegazy et al. (2013).

Additional protection of window glass, pavement, walls, and roofs are made by integration or layering with SiO_2 and TiO_2 nanoparticles. The windows could be made fireproof by using silica nano layers which are placed between two glass panels (Mann 2006). Silica nanoparticles added to the windows help monitor external light by acting as an anti-reflection coating and contribute to energy efficiency by having an effect on air conditioning (Rana et al. 2009). TiO_2 is photoactivated to generate reactive oxygen species (ROS) by exposing them to UV light indoors with sunlight, allowing successful eradication of bacterial biofilms and dirt attached to the windows (Irie et al. 2004; Paz et al. 1995). TiO_2 nanoparticles layered on footpaths, roofs and walls serve as an anti-fogging mediator to keep surfaces dirt free when exposed to solar radiation and avoid the accumulation of hydrophobic dust on windows (Zhu et al. 2004; Irie et al. 2004).

In paint, silver nanoparticles can be combined with the paint, to destroy pathogenic microorganisms and provide surfaces (e.g., hospital walls) with antimicrobial properties (Kumar et al. 2008). Due to its photoactivity under UV lighting, titanium dioxide (TiO2) acts as a biological killing agent and can confer antibacterial and antifungal properties on surfaces of the wood (Goffredo et al. 2017). Stronger types of steel can be created by adding nanoparticles to steel coating paints when they are used as concrete building reinforcing bars. These bars are referred to as multi-structural micro-composite formable (MMFX) steel, which is favoured over conventional steel because of its corrosion resistance and robust properties (Mohamed 2015).

In construction buildings, micro and nano-scale sensors and actuators are inserted for precise actual-time observation of structural/material destruction and conditions (e.g., stress, wear, corrosion, and cracking) and environmental circumstances (e.g., temperature, moisture, and smoke) (Saafi and Romine 2005; Song et al. 2008). As the system senses strain inputs, CNT/polycarbonate nanocomposite sensors produce brief alterations in electrical resistance, providing a prompt indication of potential structural damage (Zhang et al. 2006). The extraordinary electron shuttle properties of CNTs and C60 fullerenes are used to improve the efficiency of solar and fuel cells generating renewable energy (Girishkumar et al. 2005; Brown and Kamat 2008).

10.4 Health Effects of Nanoparticles

Nanomaterials are used to improve several different product types (Royal Society 2004), and day by day, the marketing of goods using these specific properties of nanoparticles is growing. These same new features, however, have demonstrated new difficulties in recognizing, predicting, and treating possible harmful health impacts after coming in contact with them (Hood 2004). The extensive use of nanoparticles provides tremendous possibility for human contact and release in the environment. The massive advancement of technology and diverse applications has had an impact on health research and may cause environmental risk. The prospect of nanotechnology, like the development and use of genetically modified organisms, would be highly dependent on public acceptance of the risks and benefits of these nanoparticles. The study shows that in vitro conditions, along with high concentrations of nanoparticles, are highly poisonous to marine organisms, bacteria, and human cells. Also, non-toxic compounds may generally become dangerous at the nanoscale. According to particle physics, nanoparticles size allows them to stay suspended into the air, for days to weeks if released (Elliott 2011; Lidén 2011; Potera 2010). It is possible to inhale nanoparticles, which would result in the accumulation of these particles in the animal or plant respiratory system. This can also result in the bioaccumulation of nanoparticles (Rizzello and Pompa 2014; Yang and Ma 2010; Diao and Yao 2009).

At various stages, nanoparticles may be released into the atmosphere accidentally or incidentally. Some nanoparticles may be counted as possible developing contaminants, growing fears about the related risks to public and environmental health, as their environmental release is presently not regulated. Nanoparticles may go through various alterations once in the environment, which alter their characteristics, effects, and destiny. Therefore, reporting on lifecycle exposure to nanoparticles is important for evaluating possible impacts on ecosystem and human health, as well as minimizing avoidable risks.

Since recent years, there are many innovations in nanotechnology that have created angst regarding the existence, spread, outcome, and transport of nanoparticles into the environment. Nanoparticles are used in an array of manufacturing and civil processes, biomedical functions, consumer goods, food, and drug delivery systems, etc. due to which there is a wide spread of nanoparticles in the atmosphere. The volume of these nanoparticles has risen in the ecosystem, as they are released in the environment as a waste product or as a by-product of some manufacturing procedure, which has resulted in several worries regarding human health and the environment.

Two areas of nanoparticles are important here;

1. In a free shape, where nanoparticles either be discharged into the water or air during production or manufacturing incidents, or as a by-product of manufacturing waste and subsequently deposited in the water, soil, or plants.
2. If they are components of a processed material or commodity in a fixed form, they will eventually have to be reused or removed as waste.

For an extended period of time, nanoparticles emitted at the construction site remain in the air during which employees may be exposed to them. There are no suitable chemical management policies enforced in such areas. Staff inhale dust, which can lead to serious health problems, with the example of dust on worksites (Kumar and Morawska 2014).

10.4.1 Effect of Nanoparticles on Microorganisms

Nanoparticles used in the construction sites have unfavourable effects on the microbial population in the vicinity. These nanoparticles tend to utilize their antibacterial properties via a variety of methods (Aderibigbe 2017; AlMatar et al. 2017; Hemeg 2017; Bassegoda et al. 2018) (Fig. 10.2) such as;

1. Directly interacting with the microbial cell wall,
2. Formation of biofilm suppression,
3. Eliciting of innate and adaptive host immune responses,
4. Production of ROS, and
5. Generation of intracellular effects (e.g., binding with proteins and/or DNA).

The effects of various nanoparticles on microorganisms are shown in Fig. 10.2. Damage to the bacterial cell membrane happens when nanoparticles attach electrostatically to the cell wall and membranes of the bacterial, resulting in the modification of cell membrane potential, depolarization of the membrane, and loss of stability. This results in transport disparity, decreased respiration, cell lysis damage and/or, energy transduction and eventually death of the cell (Pelgrift and Friedman 2013). The most important element for the in vivo *and* in vitro cell-toxicity of nanoparticles is known to be Reactive Oxygen Species (ROS). They are produced indirectly by respiratory chain disruption or directly by the nanoparticles (Nathan and Cunningham-Bussel 2013). A ROS surge occurs due to extreme oxidative stress, which causes injury to all the macromolecules of the cell, resulting in peroxidation of lipid, protein modification, inhibition of enzyme, and damage to DNA and RNA. At high concentrations, ROS can lead to cell death and cause serious DNA damage and mutations at low doses

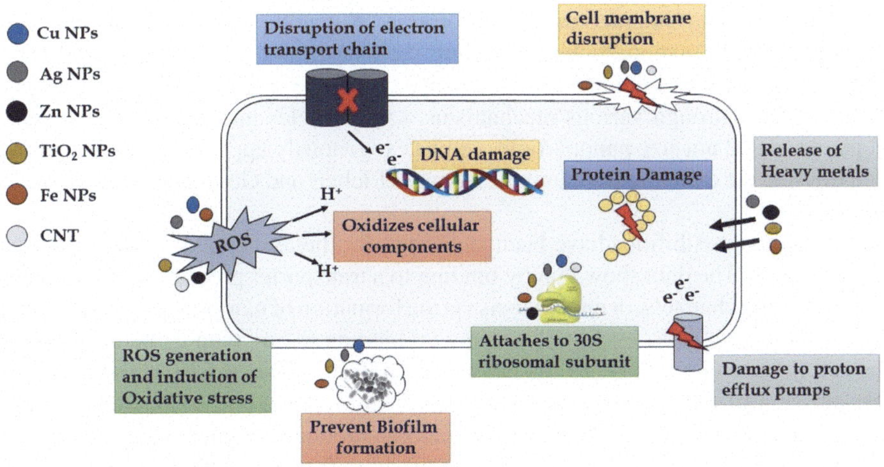

Fig. 10.2 Effects of various nanoparticles on microorganisms

(Pan et al. 2010; Wang et al. 2011). The harmfulness of nanoparticles is a photocatalytic reaction when the production of ROS is caused by visible or UV light. Many other impacts of nanoparticles consist of direct inhibition of vitalenzymes, initiation of nitrogen reactive species (NRS), and programmed death of the cell (Pelgrift and Friedman 2013; Huh and Kwon 2011; Hajipour et al. 2012; Belcher et al. 2011).

Such nanoparticles do not exert the same antibiotic mechanism of action, thus causing more damage. Both lipids and membrane proteins are destroyed by ZnO NPs, which can cause cell death. They result in the formation of hydrogen peroxide (H_2O_2), including Zn^{2+} ions and ROS, which damage the bacterial cell. Copper associates with groups of amines and carbonyls found on microbe's cell surfaces. ROS can be formed by greater concentrations of Cu^{2+} ions. TiO_2 NPs produce ROS, including hydroxyl radicals (·OH) and hydrogen peroxide (H_2O_2), in a photocatalysis phase on contact with UVA and near-UV irradiation (Ruparelia et al. 2008). MgX_2 NPs cause peroxidation of lipids in the microbial cell envelope by generating ROS. MgF_2 NPs can cause peroxidation of lipid and a decline in cytoplasmic pH, which may increase membrane potential (Lellouche et al. 2009, 2012; Belcher et al. 2011). Silver's antimicrobial activity is due to its Ag^+ ions. By binding to them, Ag^+ inhibits the microbe's electron transport chain, destroying RNA and DNA. It also prevents inhibition of cell division by the replication of DNA (Sondi and Salopek-Sondi 2004; Choi et al. 2008). Direct damage to cell walls appears to be the microbial toxicity mechanism of SWNTs, while MWNTs cause toxicity by oxidative stress (Kang et al. 2007, 2008a, b).

10.4.2 Effect of Nanoparticles on Plants

The plant system's interaction and absorption of nanoparticles is determined on the design of the nanoparticles, including the shape, size and amount used. The nanoparticles, which lie in the 40–50 nm size range, are able to easily penetrate the plant tissues. Through various mechanisms, nanoparticles interact with the system of plant roots. 2 nm size nanoparticles can be exogenously applied via leaves, which will traverse the cuticular pore into the stomata (Eichert and Goldbach 2008; Schwab et al. 2016).

Numerous possibilities have been suggested for the absorption of nanoparticles by plant cells. The data show that by binding to a transporter protein, all through the aquaporin, ion channels, or endocytosis via the formation of new pores, nanomaterials could reach plant cells, eventually ending up binding to an organic chemical inside the cell (Maine et al. 2001; Kurepa et al. 2010). In the case of CNTs, this mode of transmission of nanoparticles is mainly seen (Smirnova et al. 2012). Before being transported into the plants; the nanomaterials can produce complexes along with cell membrane carriers or root exudates. Reports show that plants take up most metal-based nanomaterials, including elements for which ion carriers have been recognized (Tani and Barrington 2005). As nanomaterials join plant cells, they can be moved via the plasmodesmata of the cell from one cell to another (Hauck et al. 2008).

At higher concentrations, the plants carry out the accumulation of nanoparticles which can result in an escalation in the activity of superoxide dismutase and catalase enzymes and a reduction in photosynthesis. Silver nanoparticles increase the shoot and root length when used at low intensity but reduce the length at high concentration. In root growth, biomass and root length, a reduction is seen. Broken epidermis and root caps have been found in some cases. Seed germination decreased along with shoot length was affected by different concentrations of silver nanoparticles, and a reduction in plant growth was observed. The reduced mitotic index resulted due to the higher concentration of silver nanoparticles. Cell division processes are disrupted by the silver nanoparticles, resulting in a mixed-up metaphase, chromatin bridge, several chromosomal splits, and eventually cell disintegration. In agricultural and grassland plants, DNA was damaged owing to the existence of copper oxide nanoparticles (Vannini et al. 2013; Krishnaraj et al. 2012; Fageria et al. 1990; Jasim et al. 2017).

Monodisperse particles of nano-zinc oxide inhibit root growth and inhibited chlorophyll synthesis, resulting in decreased photosynthetic effectiveness. When zinc nanoparticles were present, a decrease in germination and simultaneous suppression of root growth was seen (Ramesh et al. 2014; De la Rosa et al. 2013; Zafar et al. 2016). Nanoparticles of copper oxide lowered the chlorophyll content in plants. After the plants' exposure to copper ions and copper nanoparticles, lipid peroxidation was enhanced. Copper/Copper oxide (Cu/Cu_2O) nanoparticles via adsorption could obstruct water channels. The potential for radical penetration into onion roots could also be increased (Fiskesjo 1993; Geremias et al. 2010), which may result in disruption of the entire cell division stages and metabolism. The toxicity study of cobalt and zinc oxide nanoparticles showed that escalating concentrations of the nanoparticles results in the roots elongation inhibition as compared with control plants. The toxicity of nanoparticles of cobalt oxide could be because these nanoparticles can block the water channels through adsorption, while the zinc oxide nanoparticles penetrate drastically into the roots of the plant and destroy the whole metabolism and cell division stages. ZnO nanomaterials are one of the most toxic nanomaterials that could result in termination of root growth (Stella et al. 2010; Ma et al. 2009; Huang et al. 2002). Aluminium oxide results in decreased plant growth and development (Foy and Fleming 1982). Figure 10.3 shows the interaction of nanomaterials in the environment and plants.

Inhibitory effects on the development are seen by the presence of ferrous nanoparticles (Fageria et al. 1990). The surplus iron oxide (Fe_3O_4) amount as the magnetic nanomaterial caused various negative effects seen in growth of the plant. The level of chlorophyll was magnified at low fluid concentrations of Fe_3O_4 nanoparticles, whereas it was impeded at higher concentrations. Repressing effect was seen on the plantlets growth that resulted in the formation of brown spots on leaves at greater volumes of Fe_3O_4 nanoparticles (Stephan 2004; Laanbroek 1990; Hartley and Lepp 2008). Oxidative stress was the result of excess Fe_3O_4 nanoparticles, which impaired the rate of photosynthesis and a decreased metabolism rate. The oxidative stress caused by the accumulation of Fe_3O_4 fluid in the living plant's tissues (Hartley and Lepp 2008; John 1988; Bencana et al. 1998; Green and Etherington 1977). TiO_2 nanoparticles have a very small size due to which they have a tendency to form a

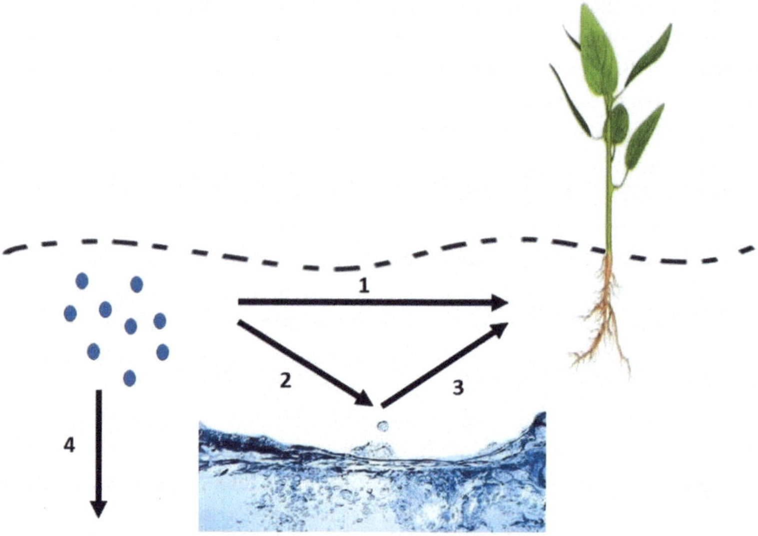

Fig. 10.3 Interaction of nanomaterials in the environment and plants; (1) plant root absorbed nanomaterials directly, (2) nanomaterials combined with water, (3) nanomaterials combined with water and transported to the plant, (4) nanomaterials remained in the soil

covalent bond with no-conjugate natural organic matter, move from one place to another, following the cells' and tissue specific circulation (Garcia et al. 2009; Feizi et al. 2012, 2013; Castiglione et al. 2011). Algal species, *Desmodesmus subspicatus* also showed toxic effects of TiO_2 nanoparticles (Hund-Rinke and Simon 2006). TiO_2 nanoparticles created ROS on contact with organisms or UV irradiation (Khataee et al. 2014; Larue et al. 2012; Dehkourdi and Mosavi; 2013). The phytotoxicity effect on cucumber, maize, carrot, cabbage, and soybean by uncoated and phenanthrene-coated alumina (Al_2O_3) nanoparticles was examined. The results showed that the root elongation was inhibited by uncoated Al_2O_3 nanoparticles at $2mgL^{-1}$ concentrations. The toxicity is shown to be maybe not nano-specified but may be owing to the suspension of nanoparticles of Al_2O_3. ZnO nanoparticles tend to penetrate the cells of the root and impede growth of the seedling (Blamey et al. 1983; Kollmeier et al. 2000; Yamamoto et al. 2001).

10.4.3 *Effect on Animals and Humans*

There are several distinctive properties that make it so promising for nanomaterials in construction. However, massive usage of the building industry can also produce unexpected effects on the human health and environment. Inhalation of nanoparticles during the coating, molding, compounding, and amalgamation process may cause

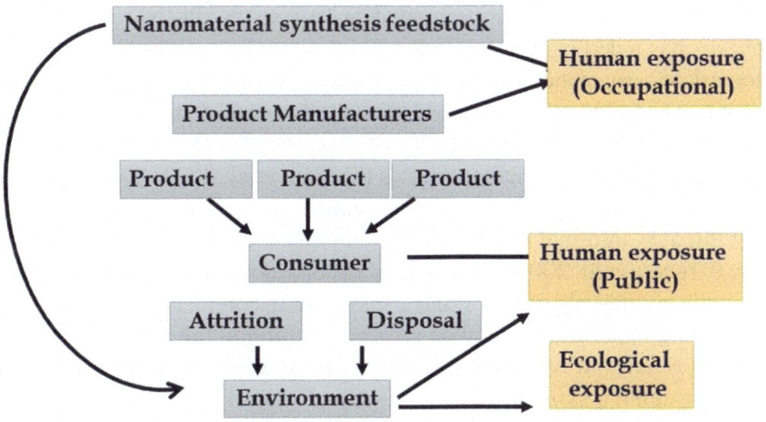

Fig. 10.4 Routes of exposure of nanoparticles

workers to experience many respiratory problems. A risk evaluation questionnaire on the effect of nano-TiO_2 proved that work-related exposure surpassed the normal limit only in the process of packaging. In some cases, nanomaterials exposure may also happen during manufacture and processing before integration into manufactured goods. During aqueous dispersal of CNTs and fullerenes, aerosolized carbon NPs can be produced through sonification, while airborne particles are released during weighing (Rahman et al. 2017). The release of some building nanoparticles can result in activities such as construction, repair, renovation, and demolition (Fig. 10.4).

Figure 10.4 shows the routes of exposure of nanoparticles. Nanomaterial wastes, from industrial processes or building and destruction activities, go through crushing procedure before they are transferred to the disposal unit which may cause their release in the ecosystem. Inhalation is a one of the crucial methods for nanoparticles spread through the body. This has caused an increase in anxiety regarding toxic pulmonary diseases. This is similar to asbestosis, where pleural fibrosis, lung cancer, and mesothelioma are caused by inhaled small crystalline fibers of asbestos (Donaldson and Seaton 2012). Workers were exposed to carcinogenic fibres for many years prior to the initiation of various regulation. This was partly due to the lack of knowledge of the health risks linked with asbestosis. Particles inhalation on building areas has been recognized as a health and safety issue. Studies linked to chronic nasal exposure in male mice were carried out and the results may be comparable with humans. Male mice were subjected to nasal inhalation installation of 1.25, 2.5, or 5 mg/kg TiO_2 nanoparticles for nine months. The outcomes showed a drop in body weight, inflammation and cell permeation, and tumour formation in mouse lung tissue, along with dysfunction of metabolism. Similar experiment was conducted out using nickel hydroxide, subjecting the mice to Ni $(OH)_2$ particles over several time intervals (Gillespie et al. 2010). The findings demonstrated the presence of inflammation of the lungs along with a different reaction varying on the length of the contact.

The rise in the number of polymorphonuclear leukocytes (PMN), lymphocytes and macrophages has aided in an inflammatory reaction (Gillespie et al. 2010).

Nanoparticles can penetrate the skin, gaining entrance into the body. Dermal absorption has been identified via various methods—transcellular, intercellular, and trans-appendageal via follicles of hair or sweat glands. This will also depend on the properties of the specific nanoparticles (Crosera et al. 2009). Dermal absorption is not given much importance in front of studies on inhalation exposure, and the evaluated threat is minimal due to the opinion that they are less permeable via the skin route (Crosera et al. 2009). Polystyrene nanoparticles were spread using vertical diffusion cells at intervals of 0.5, 1 and 2 h in a study conducted using porcine skin. Surface imaging experiments showed an accumulation in the follicular openings of smaller nanoparticles. It is important to carry out studies to determine if the particle will move through the skin at lesion, flexure, and wound sites (Crosera et al. 2009). Experiments have shown that ZnO and TiO_2nanoparticles were incapable of passing via the skin showing them to be safe to cutaneous exposure. Ag nanoparticles were found to enter via the skin but their capability to penetrate is yet to be studied. Their usage in dressings of various medical injuries may cause absorption via the skin, possibly causing damage to the internal organs (Crosera et al. 2009).

Nanoparticles of Ni, Pd, and Co are deemed to be more dangerous due to the high-level release of ions (Filon et al. 2015). Carbon nanotubes (MWNTs and SWNTs) of small sizes create a possible threat as they cause toxicity in the pulmonary system, like fibrosis, inflammation, and epithelioid granulomas in animals (Filon et al. 2015; Karlsson et al. 2008; Park et al. 2007; Reeves et al. 2008). SiO_2 NPs have been reported to exert carcinogenic activity. Exposure to nanosized SiO_2 causes lipid peroxidation and membrane damage on human lung cancer cells and induces tumour necrosis genes in rats (Lin et al. 2006; Donaldson and Seaton 2012; Buchanan et al. 1993; Oravisjärvi et al. 2014). Table 10.2 enlists various health ailments due to the presence of nanoparticles.

10.5 Conclusion

The utilization of nanoparticles in the construction/civil industry presents a number of prospects and challenges. In building, the use of nanoparticles can be used not only to enhance the materials and utilities properties, but also to conserve energy. This is particularly important as residential homes and commercial buildings that provide heating, air conditioning and lighting use a high percentage of all energy consumed. In order to have a greener construction industry, nanomaterials can also take part in the production process. This can happen when we use nanoparticles as a substitute for materials that are dangerous ecological pollutants, like mercury and lead. Prospects for energy savings involve enhanced heat management by using nanoparticles of silica in paint/coating and insulating ceramics that allow energy saving and nano-TiO_2-coated surfaces which are solar-powered and self-cleaning. The usage of Quantum dots and CNTs to increase the performance of energy transfer,

Table 10.2 Health implications of nanoparticles to the human body (Wiemann et al. 2017; Ahamed et al. 2010; Grande and Tucci 2016; Donaldson et al. 2000; Karlsson et al. 2008; Wickrath et al. 2017)

Nanoparticle	Type affected cell/Organ/System
C60 (water-stable colloid)	Absorbed by human keratinocytes; toxicity to human cell lines; peroxidation of lipid, stabilizes proteins
C60 derivatives	Cell death/necrosis; oxidative cytotoxicity; liver accumulation; sarcomas in mice and human cells, gliomas are induced
Carbon nanotubes (CNT)	Cell death/necrosis; cell membrane damage; respiratory functions are inhibited; damage to the mitochondrial DNA; inhibit bacterial clearance from lung tissues; induce granulomas and athereoschlerotic lesions
Zinc oxide nanoparticles (ZnO NP)	Proliferation of the cell
Iron oxide (Fe_3O_4)	Oxidative damage to the DNA
Copper zinc ferrite ($CuZnFe_2O_4$)/Copper oxide (CuO)	Oxidative damage to the DNA
Titanium dioxide (TiO_2)	Damage to DNA; inflammation in lungs; carcinogenesis; cell death; changes in metabolic activities
Silica nanoparticles (SiO_2)	Induces bronchoalveolar carcinoma-derived cells
Silver nanoparticle (Ag NP)	Affects the immune system including the liver, lungs, brain; causes cancerous growth; affects the reproductive organs, vascular system

heating and/or illumination equipment, as well as the integration of graphene and fullerenes to improve systems that store energy like batteries and condensers are further prospects (e.g., solar and wind) (Anikeeva et al. 2009; Ding et al. 2006; Flandrois and Simon 1999; Frackowiak and Beguin 2002).

It is crucial to realize their prospective movement and effects in air, soil, and water life as nanomaterials are devised and used. In order to detect and characterize nanoparticles that may be distributed from or integrated into building supplies, sophisticated analytical resources should be used. They need to measure the amount that could be toxic to the climate. There is also a need to emphasize environmentally friendly life cycle engineering of MNMs in building. Further research and investigation are a necessity to ensure that the workers working on the construction site are safe and that they are minimally exposed to nanoparticles. Work must be undertaken to ensure that the safe methods of design, manufacture, use, and removal and associated recycle, reuse, and remanufacturing programs are sufficient, which can increase the sustainability of both the construction industries and nanotechnology.

References

Abbasi E, Milani M, Fekri Aval S, Kouhi M, Akbarzadeh A, Tayefi Nasrabadi H, Nikasa P, Joo SW, Hanifehpour Y, Nejati-Koshki K, Samiei M (2016) Silver nanoparticles: synthesis methods, bio-applications and properties. Crit Rev Microbiol 42(2):173–180. https://doi.org/10.3109/1040841X.2015.1018131

Aderibigbe BA (2017) Metal-based nanoparticles for the treatment of infectious diseases. Molecules 22:E1370. https://doi.org/10.3390/molecules22081370

Ahamed M, AlSalhi MS, Siddiqui M, AlSalhi PM (2010) Silver nanoparticle applications and human health. Clin Chim Acta 411:1841–1848

AlMatar M, Makky EA, Var I, Koksal F (2017) The role of nanoparticles in the inhibition of multidrug-resistant bacteria and biofilms. Curr Drug Deliv 15:470–484. https://doi.org/10.2174/1567201815666171207163504

Ambika S et al (2015) Green biosynthesis of ZnO nanoparticles using *Vitex negundo* L. extract: spectroscopic investigation of interaction between ZnO nanoparticles and human serum albumin. J Photochem Photobiol B 149:143–148

Anikeeva PO, Halpert JE, Bawendi MG, Bulovic V (2009) Quantum dot light-emitting devices with electroluminescence tunable over the entire visible spectrum. Nano Lett 9:2532–2536

Arefi MR, Javeri MR, Mollaahmadi E (2011) To study the effect of adding Al_2O_3 nanoparticles on the mechanical properties and microstructure of cement mortar. Life Sci J 8:613–617

Bassegoda A, Ivanova K, Ramon E, Tzanov T (2018) Strategies to prevent the occurrence of resistance against antibiotics by using advanced materials. Appl Microbiol Biotechnol 102:2075–2089. https://doi.org/10.1007/s00253-018-8776-0

Bencana M, Moran JF, Iturbe-Ormaetxe I (1998) Independent oxygen free radical generation in plants subjected to environmental stress: toxicity and antioxidant protection. Plant Soil 201(1):137–147

Becher PF (1991) Microstructural design of toughened ceramics. J Am Ceram Soc 74:255–269

Blamey FPC, Edwards DG, Asher CJ (1983) Effects of aluminum, OH: Al and P: Al molar ratios, and ionic strength on soybean root elongation in solution culture. Soil Sci 136(4):197–207

Blecher K, Nasir A, Friedman A (2011) The growing role of nanotechnology in combating infectious disease. Virulence 2(5):395–401

Brown P, Kamat PV (2008) Quantum dot solar cells. Electrophoretic deposition of CdSe_C60 composite films and capture of photogenerated electrons with nC60 cluster shell. J Am Chem Soc 130:8890–8891

Buchanan CM, Gardner RM, Komarek RJ (1993) Aerobic biodegradation of cellulose acetate. J Appl Polym Sci 47:1709–1719

Castiglione MR, Giorgetti L, Geri C, Cremonini R (2011) The effects of nano-TiO_2 on seed germination, development and mitosis of root tip cells of *Vicia narbonensis* L. and *Zea mays* L. J Nanoparticle Res 13(6):2443–2449

Chaipanich A, Nochaiya T, Wongkaew, Tor Kittikul P (2010) Compressive strength and microstructure of carbon nanotubes-fly ash cement composites. Mater Sci Eng 527:1063–1067

Choi O, Deng KK, Kim N-J, Ross L Jr, Surampalli RY, Hu Z (2008) The inhibitory effects of silver nanoparticles, silver ions, and silver chloride colloids on microbial growth. Water Res 42(12):3066–3074

Crosera M, Bovenzi M, Maina G, Adami G, Zanette C, Florio C, Larese FF (2009) Nanoparticle dermal absorption and toxicity: a review of the literature. Int Arch Occup Environ Heal 82:1043–1055

Daniyal Md, Ameer A, Sabih A (2018) Application of nanomaterials in civil engineering. In: Khan ZH (ed) Nanomaterials and their applications. Advanced structured materials. Springer Nature, Singapore Pte Ltd. https://doi.org/10.1007/978-981-10-6214-8_6

De la Rosa G, Lopez-Moreno ML, de Haro D, Botez CE, Peralta-Videa JR, Gardea-Torresdey JL (2013) Effects of ZnO nanoparticles in alfalfa, tomato, and cucumber at the germination stage: root development and X-ray absorption spectroscopy studies. Pure Appl Chem 85(12):2161–2174. https://doi.org/10.1351/pac-con-12-09-05

Dehkordi EH, Mosavi M (2013) Effect of anatase nanoparticles (TiO2) on parsley seed germination (*Petroselinum crispum*) in vitro. Biol Trace Elem Res 155(2):283–286

Diao M, Yao M (2009) Use of zero-valent iron nanoparticles in inactivating microbes. Water Res 43(20):5243–5251

Ding Y, Alias H, Wen D, Williams RA (2006) Heat transfer of aqueous suspensions of carbon nanotubes (CNT Nanofluids). Int J Heat Mass Transf 49:240–250

Donaldson K, Seaton A (2012) A short history of the toxicology of inhaled particles. Part Fibre Toxicol 9:13

Donaldson K, Stone V, Gilmour PS, Brown DM, MacNee W, Malcolm AL, Derwent RG (2000) Ultrafine particles: mechanisms of lung injury. Philos Trans R Soc A Math Phys Eng Sci 358:2741–2749

Doyle ME, Glass KA (2010) Sodium reduction and its effect on food safety, food quality, and human health. Compr Rev Food Sci Food Saf 9(1):44–56

Drexler K (1981) Molecular engineering: an approach to the development of general capabilities for molecular manipulation. Proc Natl Acad Sci 78:5275–5278

Eichert T, Goldbach HE (2008) Equivalent pore radii of hydrophilic foliar uptake routes in stomata and leaf surfaces–further evidence for a stomatal pathway. Physiol Plant 132(4):491–502. https://doi.org/10.1111/j.1399-3054.2007.01023.x

Elliott KC (2011) Nanomaterials and the precautionary principle. Environ Health Perspect 119(6):A240–A245

Fageria NK, Baligar VC, Wright RJ (1990) Iron nutrition of plants: an overview on the chemistry and physiology of its deficiency and toxicity. Pesqui Agropec bras 25(4):553–570

Fang C, Yu X, Yu R, Liu P, Qiao X (2016) Preparation and properties of isocyanate and nanoparticles composite modified asphalt. Constr Build Mater 119:113–118

Feizi H, Kamali M, Jafari L, Rezvani Moghaddam P (2013) Phytotoxicity and stimulatory impacts of nanosized and bulk titanium dioxide on fennel (*Foeniculum vulgare* Mill). Chemosphere 91(4):506–511

Feizi H, Rezvani Moghaddam P, Shahtahmasebi N, Fotovat A (2012) Impact of bulk and nanosized titanium dioxide (TiO2) on wheat seed germination and seedling growth. Biol Trace Elem Res 146(1):101–106

Feynman RP (1960) There's plenty of room at the bottom: an invitation to enter a new field of physics. In: Proceedings of the annual meeting of the American Physical Society, California Institute of Technology, Pasadena, CA, USA, 29 Dec 1959. California Institute of Technology, Pasadena, CA, USA

Filon FL, Mauro M, Adami G, Bovenzi M, Crosera M (2015) Nanoparticles skin absorption: new aspects for a safety profile evaluation. Regul Toxicol Pharmacol 72:310–322

Fiskesjo G (1993) Allium test I: A 2–3 day plant test for toxicity assessment by measuring the mean root growth of onions (*Allium cepa* L.). Environ Toxicol Water Qual 8(4):461–470

Flandrois S, Simon B (1999) Carbon materials for lithium-ion rechargeable batteries. Carbon 37:165–180

Foy CD, Fleming AL (1982) Aluminium of two wheat genotypes related to nitrate reductase activity. J Plant Nutr 5(11):1313–1333. https://doi.org/10.1080/01904168209363064

Frackowiak E, Beguin F (2002) Electrochemical storage of energy in carbon nanotubes and nanostructured carbons. Carbon 40:1775–1787

Garcia JC, Simionato JI, Almeida VDC et al (2009) Evolutive follow-up of the photocatalytic degradation of real textile effluents in TiO2 and TiO2/H2O2 systems and their toxic effects on *Lactuca sativa* seedlings. J Braz Chem Soc 20(9):1589–1597

Ge Z, Gao Z (2008) Applications of nanotechnology and nanomaterials in construction. In: First international conference on construction in developing countries (ICCIDC–I). Advancing and integrating construction education. Research & Practice (Karachi, Pakistan 2008), pp 235–240

Geremias R, Fattorini D, F´avere VTD, Pedrosa RC (2010) Bioaccumulation and toxic effects of copper in common onion *Allium cepa* L. Chem Ecol 26(1):19–26

Giljohann DA et al (2010) Gold nanoparticles for biology and medicine. Angew Chem Int Ed Engl 49:3280–3294

Gillespie PA, Kang GS, Elder A, Gelein R, Chen L, Moreira AL, Koberstein J, Tchou-Wong K-M, Gordon T, Chen LC (2010) Pulmonary response after exposure to inhaled nickel hydroxide nanoparticles: short and long-term studies in mice. Nanotoxicology 4:106–119

Girishkumar G, Rettker M, Underhile M, Binz D, Vinodgopal K, McGinn P, Kamat P (2005) Single-wall carbon nanotube based proton exchange membrane assembly for hydrogen fuel cells. Langmuir 21:8487–8494

Goffredo GB, Citterio B, Biavasco F, Stazi F, Barcelli S, Munafò P (2017) Nanotechnology on wood: The effect of photocatalytic nanocoatings against *Aspergillus niger*. J Cult Heritage 27:125–136

Grande F, Tucci P (2016) Titanium Dioxide Nanoparticles: a risk for human health? Mini Rev Med Chem 16:762–769

Green MS, Etherington JR (1977) Oxidation of ferrous iron by rice (*Oryza sativa* L.) roots: a mechanism for waterlogging tolerance? J Exp Bot 28(3):678–690

Guerrini GL (2012) Photocatalytic performances in a city tunnel in Rome: NOx monitoring results. Constr Build Mater 27:165–175

Hajipour MJ, Fromm KM, Ashkarran AA et al (2012) Antibacterial properties of nanoparticles. Trends Biotechnol 30(10):499–511

Hartley W, Lepp NW (2008) Remediation of arsenic contaminated soils by iron-oxide application, evaluated in terms of plant productivity, arsenic and phytotoxic metal uptake. Sci Total Environ 390(1):35–44

Hauck TS, Ghazani AA, Chan WCW (2008) Assessing the effect of surface chemistry on gold nanorod uptake, toxicity, and gene expression in mammalian cells. Small 4(1):153–159

Hegazy M, Badawi A, El Rehim SA, Kamel W (2013) Influence of copper nanoparticles capped by cationic surfactant as modifier for steel anti-corrosion paints. Egypt J Pet 22:549–556

Hemeg HA (2017) Nanomaterials for alternative antibacterial therapy. Int J Nanomed 12:8211–8225. https://doi.org/10.2147/IJN.S132163

Hood E (2004) Nanotechnology: looking as we leap. Environ Health Perspect 112:A741–A749

Huang FM, Tai KW, Chou MY, Chang YC (2002) Cytotoxicity of resin-, zinc oxide-eugenol-, and calcium hydroxide based root canal sealers on human periodontal ligament cells and permanent V79 cells. Int Endod J 35(2):153–158

Huang X et al (2007) Gold nanoparticles: interesting optical properties and recent applications in cancer diagnostics and therapy. Nanomedicine (Lond) 2:681–693

Huh AJ, Kwon YJ (2011) Nanoantibiotics': a new paradigm for treating infectious diseases using nanomaterials in the antibiotics resistant era. J Control. Release 156(2):128–145

Hund-Rinke K, Simon M (2006) Ecotoxic effect of photocatalytic active nanoparticles (TiO_2) on algae and daphnids. Environ Sci Pollut Res 13(4):225–232

Hussain S, Sastry K (2014) Study of strength properties of concrete by using micro-silica and nano-silica. Int J Res Eng Technol 3:103–108

Ickrath P, Wagner M, Scherzad A, Gehrke T, Burghartz M, Hagen R, Radelo K, Kleinsasser N, Hackenberg S (2017) Time-dependent toxic and genotoxic effects of zinc oxide nanoparticles after long-term and repetitive exposure to human mesenchymal stem cells. Int J Environ Res Public Heal 14:1590

Irie H, Sunada K, Hashimoto K (2004) Recent developments in TiO2 photocatalysis: Novel applications to interior ecology materials and energy saving systems. Electrochem 72:807–812

Iv M et al (2015) Clinical applications of iron oxide nanoparticles for magnetic resonance imaging of brain tumors. Nanomedicine (Lond) 10:993–1018

Jasim B, Thomas R, Mathew J, Radhakrishnan EK (2017) Plant growth and diosgenin enhancement effect of silver nanoparticles in Fenugreek (*Trigonella foenum-graecum* L.). Saudi Pharm J 25(3):443–447. https://doi.org/10.1016/j.jsps.2016.09.012

Jin J, Chen B, Liu L, Liu R, Qian G, Wei H, Zheng J (2019) A study on modified bitumen with metal doped nano-TiO2 pillared montmorillonite. Materials 12:1910
John AR (1988) The iron and molybdenum use efficiencies of plant growth with different energy, carbon and nitrogen sources. New Phytol 109(3):279–287
Kandasamy S, Prema RS (2015) Methods of synthesis of nanoparticles and its applications. J Chem Pharm Res 7(3):278–285
Kang S, Herzberg M, Rodrigues DF, Elimelech M (2008a) Antibacterial effects of carbon nanotubes: size does matter. Langmuir 24:6409–6413
Kang S, Mauter MS, Elimelech M (2008b) Physicochemical determinants of multiwalled carbon nanotube bacterial cytotoxicity. Environ Sci Technol 42:7528–7534
Kang S, Pinault M, Pfefferle LD, Elimelech M (2007) Single walled carbon nanotubes exhibit strong antimicrobial activity. Langmuir 23:8670–8673
Karlsson HL, Cronholm P, Gustafsson J, Möller L (2008) Copper oxide nanoparticles are highly toxic: a comparison between metal oxide nanoparticles and carbon nanotubes. Chem Res Toxicol 21:1726–1732
Kataria S, Jain M, Rastogi A, Živčák M, Brestic M, Liu S, Tripathi DK (2019) Role of nanoparticles on photosynthesis: avenues and applications. Nanomaterials in plants, algae and microorganisms, pp 103–127
Kawashima S, Hou P, Corr DJ, Shah SP (2013) Modification of cement-based materials with nanoparticles. Cem Concr Compos 36:8–15
Khataee AR, Fathinia M, Naseri A et al (2014) Modeling and optimization of simultaneous photocatalysis of three dyes on ceramic-coated TiO2 nanoparticles using chemometrics methods: phyto toxicological assessment during degradation process. Res Chem Intermed 40(3):1283–1302
Khitab A (2012) Materials of construction (Allied Books, Pakistan) Edition: Ist Publisher. Allied Books, 75 Naqi Market, The Mall, Lahore, Pakistan
Khlebtsov N et al (2011) Biodistribution and toxicity of engineered gold nanoparticles: a review of in vitro and in vivo studies. Chem Soc Rev 40:1647–1671
Kollmeier M, Felle HH, Horst WJ (2000) Genotypic differences in aluminum resistance of maize are expressed in the distal part of the transition zone. Is reduced basipetal auxin flow involved in inhibition of root elongation by aluminum? Plant Physiol 122(3):945–956
Koren K et al (2015) Optical sensor nanoparticles in artificial sediments–a new tool to visualize O_2 dynamics around the rhizome and roots of sea grasses. Environ Sci Technol 49:2286–2292
Krishnaraj C, Ramachandran R, Mohan K, Kalaichelvan PT (2012) Optimization for rapid synthesis of silver nanoparticles and its effect on phytopathogenic fungi. Spectrochim Acta A Mol Biomol Spectrosc 93:95–99. https://doi.org/10.1016/j.saa.2012.03.002
Kumar A, Gupta A, Sharma KV, Nasir M (2013) Use of aluminum oxide nanoparticles in wood composites to enhance the heat transfer during hot-pressing. Eur J Wood Wood Prod 71(2):193–198
Kumar A, Vemula PK, Ajayan PM, John G (2008) Silver nanoparticle-embedded antimicrobial paints based on vegetable oil. Nat Mater 7:236–241
Kumar P, Morawska L (2014) Recycling concrete: an undiscovered source of ultrafine particles. Atmos Environ 90:51–58
Kurepa J, Paunesku T, Vogt S et al (2010) Uptake and distribution of ultrasmall anatase TiO2 alizarin red S nanoconjugates in *Arabidopsis thaliana*. Nano Lett 10(7):2296–2302
Laanbroek HJ (1990) Bacterial cycling of minerals that affect plant growth in waterlogged soils: a review. Aquat Bot 38(1):109–125
Larue C, Veronesi G, Flank A, Surble S, Herlin-Boime N, Carriere M (2012) Comparative uptake and impact of TiO2 nanoparticles in wheat and rapeseed. J Toxicol Environ Health A Curr Issues 75(13–15):722–734
Lee J, Mahendra S, Alvarez PJJ (2010) Nanomaterials in the construction industry: a review of their applications and environmental health and safety considerations. ACS Nano 4(7):3580–3590
Lellouche J, Friedman A, Lahmi R, Gedanken A, Banin E (2012) Antibiofilm surface functionalization of catheters by magnesium fluoride nanoparticles. Int J Nanomed 7:1175–1188

Lellouche J, Kahana E, Elias S, Gedanken A, Banin E (2009) Antibiofilm activity of nanosized magnesium fluoride. Biomaterials 30(30):5969–5978
Li GY (2004) Properties of high-volume fly ash concrete incorporating nano-SiO_2. Cem Concr Res 34:1043–1049
Lidén G (2011) The European commission tries to define nanomaterials. Ann Occup Hyg 55(1):1–5
Lin W, Huang YW, Zhou XD, Ma Y (2006) In vitro toxicity of silica nanoparticles in human lung cancer cells. Toxicol Appl Pharmacol 217:252–259
Ma H, Bertsch PM, Glenn TC, Kabengi NJ, Williams PL (2009) Toxicity of manufactured zinc oxide nanoparticles in the nematode *Caenorhabditis elegans*. Environ Toxicol Chem 28(6):1324–1330
Maine MA, Duarte MV, Suñé NL (2001) Cadmium uptake by floating macrophytes. Water Res 35(11):2629–2634
Małgorzata L (2014) Carbon nanotubes influence the compressive strength of cement composites. Tech Trans Civ Eng 1:5–11
Mann S (2006) Nanotechnology and construction. Nanoforum report 30
Mohajerani A, Tanriverdi Y, Nguyen BT, Wong KK, Dissanayake HN, Johnson L, Whitfield D, Thomson G, Alqattan E, Rezaei A (2017) Physico-mechanical properties of asphalt concrete incorporated with encapsulated cigarette butts. Constr Build Mater 153:69–80
Mohajerani A, Vajna J, Ellcock R (2018) Chromated copper arsenate timber: a review of products, leachate studies and recycling. J Clean Prod 179:292–307
Mohamed ASY (2015) Nano-innovation in construction. A new era of sustainability. In: Proceedings of the international conference on environment and civil engineering, Pattaya, Thailand, 24–25 Apr 2015, ICEACE, Pattaya, Thailand
Mohseni E, Ranjbar M, Tsavdaridis K (2015) Durability properties of high-performance concrete incorporating nano-TiO_2 and fly ash. Am J Eng Appl Sci 8:519–526
Momeni S et al (2015) A simple green synthesis of palladium nanoparticles with Sargassum algae and their electrocatalytic activities towards hydrogen peroxide. Appl Biochem Biotechnol 176:1937–1949
Mubaraki M, Ali SIA, Ismail A, Yuso NIM (2016) Rheological evaluation of asphalt cements modified with ASA polymer and Al2O3 nanoparticles. Procedia Eng 143:1276–1284
Nasrollahzadeh M et al (2015) Green synthesis of copper nano-particles using Ginkgo biloba L. leaf extract and their catalytic activity for the Huisgen [3+2] cycloaddition of azides and alkynes at room temperature. J Colloid Interface Sci 457:141–147
Nathan C, Cunningham-Bussel A (2013) Beyond oxidative stress: an immunologist's guide to reactive oxygen Pan species. Nat Rev Immunol 13(5):349–361
Nazari A, Riahi S (2011a) Optimization mechanical properties of Cr_2O_3 nanoparticles binary blended cementitious composite. J Compos Mater 45:943–948
Nazari A, Riahi S (2011b) The effects of Cr2O3 nanoparticles on strength assessments and water permeability of concrete in different curing media. Mater Sci Eng 528:1173–1182
Nazari A, Shadi R (2010) ZrO2 nanoparticles effects on split tensile strength of self compacting concrete. Mater Res 13:485–495
Nazari A, Rafiepour MH, Shadi R (2011) The effects of CuO nanoparticles on properties of self compacting concrete with GGBS as binder. Mater Res 14:307–316
Nazari A, Shadi R (2011) Effects of CuO nanoparticles on microstructure, physical, mechanical and thermal properties of self-compacting cementitious composites. J Mater Sci Technol 27:81–92
Nazari A, Shadi R, Sharin R, Shamekhi SF, Khademno A (2010a) Benefits of Fe_2O_3 nanoparticles in concrete mixing matrix. J Am Sci 6:102–106
Nazari A, Shadi R, Sharin R, Shamekhi SF, Khademno A (2010b) The effects of incorporation Fe2O3 nanoparticles on tensile and flexural strength of concrete. J Am Sci 6:90–93
Nazari A, Shadi R, Shirin R, Seyedeh FS, Khademno A (2010c) An investigation on the strength and workability of cement based concrete performance by using ZrO_2 nanoparticles. J Am Sci 6:29–33
Nejad FM, Azarhoosh A, Hamedi G, Azarhoosh M (2012) Influence of using nonmaterial to reduce the moisture susceptibility of hot mix asphalt. Constr Build Mater 31:384–388

Niroumand H, Zain M, Alhosseini SN (2013) The influence of nano-clays on compressive strength of earth bricks as sustainable materials. Procedia Soc Behav Sci 89:862–865
Nochaiya T, Chaipanich A (2011) Behavior of multi-walled carbon nanotubes on the porosity and microstructure of cement based materials. Appl Surf Sci 257:1941–1945
Oravisjärvi K, Pietikäinen M, Ruuskanen J, Niemi S, Laurén M, Voutilainen A, Keiski RL, Rautio A (2014) Diesel particle composition after exhaust after-treatment of an on-road diesel engine and modeling of deposition into the human lung. J Aerosol Sci 69:32–47
Pan X, Redding JE, Wiley PA, Wen L, McConnell JS, Zhang B (2010) Mutagenicity evaluation of metal oxide nanoparticles by the bacterial reverse mutation assay. Chemosphere 79(1):113–116
Park S, Lee YK, Jung M, Kim KH Chung N, Ahn EK, Lim Y, Lee KH (2007) Cellular toxicity of various inhalable metal nanoparticles on human alveolar epithelial cells. Inhalation Toxicol 19:59–65
Paz Y, Luo Z, Rabenberg L, Heller A (1995) Photooxidative self-cleaning transparent titanium-dioxide films on glass. J Mater Res 10:2842–2848
Pelgrift RY, Friedman AJ (2013) Nanotechnology as a therapeutic tool to combat microbial resistance. Adv Drug Deliv Rev 65(13–14):1803–1815
Peng HI et al (2011) Recent advancements in optical DNA biosensors: exploiting the plasmonic effects of metal nanoparticles. Analyst 136:436–447
Pereira L et al (2015) Metallic nanoparticles: microbial synthesis and unique properties for biotechnological applications, bioavailability and biotransformation. Crit Rev Biotechnol 35:114–128
Potera C (2010) Nanomaterials: transformation of silver nanoparticles in sewage sludge. Environ Health Perspect 118(12):A526–A527
Rahman MT, Hainin MR, Abu Bakar WAW (2017) Use of waste cooking oil, tire rubber powder and palm oil fuel ash in partial replacement of bitumen. Constr Build Mater 150:95–104
Rai M et al (2015) Strategic role of selected noble metal nanoparticles in medicine. Crit Rev Microbiol 1–24
Ramesh M, Palanisamy K, Babu NK (2014) Effects of bulk and nano-titanium dioxide and zinc oxide on physio morphological changes in *Triticum aestivum* Linn. Int J Glob Sci 3(2):415–422
Rana AK, Rana SB, Kumari A, Kiran V (2009) Significance of nanotechnology in construction engineering. Int J Recent Trends Eng 1:46–48
Rao PV et al (2015) Recent advances in nanotechnology-based diagnosis and treatments of diabetes. Curr Drug Metab 16:371–375
Reeves JF, Davies SJ, Dodd NJF, Jha AN (2008) Hydroxyl radicals are associated with titanium dioxide (TiO_2) nanoparticle-induced cytotoxicity and oxidative DNA damage in fish cells. Mutat Res 640:113–122
Rizzello L, Pompa PP (2014) Nanosilver-based antibacterial drugs and devices: mechanisms, methodological drawbacks, and guidelines. Chem Soc Rev 43(5):1501–1518
Royal Society (2004) Nanoscience and nanotechnologies: opportunities and uncertainties. The Royal Society, Science Policy Section, London, England
Ruot B, Plassais A, Olive F, Guillot L, Bonafous L (2009) TiO2-containing cement pastes and mortars: measurements of the photocatalytic efficiency using a rhodamine B-based colourimetric test. Sol Energy 83:1794–1801
Ruparelia JP, Chatterjee AK, Duttagupta SP, Mukherji S (2008) Strain specificity in antimicrobial activity of silver and copper nanoparticles. Acta Biomater 4(3):707–716
Saafi M, Romine P (2005) Nano and micro technology. Concr Int 27:28–34
Sarkar D, Pal M, Sarkar A (2014) Laboratory evaluation of asphalt concrete prepared with over burnt brick aggregate treated by zycosoil. Int J Civ Environ Struct Constr Archit Eng 8:1302–1306
Sato T, Diallo F (2010) Seeding effect of nano-CaCO3 on the hydration of tricalcium silicate. J Transp Res Board 2141:61–67
Schwab F, Zhai G, Kern M, Turner A, Schnoor JL, Wiesner MR (2016) Barriers, pathways and processes for uptake, translocation and accumulation of nanomaterials in plants. Nanotoxicology 10(3):257–278. https://doi.org/10.3109/17435390.2015.1048326

Selid PD et al (2009) Sensing mercury for biomedical and environmental monitoring. Sensors (Basel) 9:5446–5459

Serpone N, Pelizzetti E (1989) Photocatalysis: fundamentals and applications. Wiley, New York

Shaikh FUA, Supit SWM (2014) Mechanical and durability properties of high volume fly ash concrete containing calcium carbonate nanoparticles. Constr Build Mater 70:309–321

Singh P, Kim Y-J, Zhang D, Yang D-C (2016) Biological synthesis of nanoparticles from plants and microorganisms. Trends Biotechnol 34(7). http://dx.doi.org/10.1016/j.tibtech.2016.02.006

Smirnova E, Gusev A, Zaytseva O et al (2012) Uptake and accumulation of multiwalled carbon nanotubes change the morphometric and biochemical characteristics of Onobrychis arenaria seedlings. Front Chem Sci Eng 6(2):132–138

Sondi I, Salopek-Sondi B (2004) Silver nanoparticles as antimicrobial agents: a case study on *E. coli* as a model for Gram-negative bacteria. J Colloid Interface Sci 275(1):177–182

Song GB, Gu HC, Mo YL (2008) Smart aggregates: multifunctional sensors for concrete structures-a tutorial and a review. Smart Mater Struct 17:1–17

Stella WYW, Priscilla TYL, Djuri˘si´c AB, Kenneth MYL (2010) Toxicities of nano zinc oxide to five marine organisms: influences of aggregate size and ion solubility. Anal Bioanal Chem 396(2):609–618

Stephan MK (2004) Iron oxide dissolution and solubility in the presence of siderophores. Aquat Sci 66(1):3–18

Stefanidou M, KazouA (2016) Testing the effectiveness of protective coatings on traditional bricks materials science. Constr Build Mater. Corpus ID: 138561739. https://doi.org/10.1016/J.CONBUILDMAT.2016.02.114

Tani FH, Barrington S (2005) Zinc and copper uptake by plants under two transpiration rates. Part I. Wheat (*Triticum aestivum* L.). Environ Pollut 138(3):538–547

Taniguchi N, Arakawa C, Kobayashi T (1974) On the basic concept of nano-technology. In: Proceedings of the international conference on production engineering, Tokyo, Japan, pp 26–29

Thakkar KN, Mhatre S, Parikh RY (2010) Nanomedicine: nanotechnology. Biol Med 6(2):257–262

Tiwari A, Chowdhury S (2012) An overview of application of nanotechnology in construction materials. Toxicity and assessment of potential occupational and environmental health risks. Crit Rev Toxicol 36:189–217

Vannini C, Domingo G, Onelli E, Prinsi B, Marsoni M, Espen L, Bracale M (2013) Morphological and proteomic responses of *Eruca sativa* exposed to silver nanoparticles or silver nitrate. PLoS ONE 8(7):68752. https://doi.org/10.1371/journal.pone.0068752

Wang S, Lawson R, Ray PC, Yu H (2011) Toxic effects of gold nanoparticles on *Salmonella typhimurium* bacteria. Toxicol Ind Health 27(6):547–554

Wiemann M, Vennemann A, Blaske F, Sperling M, Karst U (2017) Silver nanoparticles in the lung: toxic effects and focal accumulation of silver in remote organs. Nanomaterials 7:441

Yamamoto Y, Kobayashi Y, Matsumoto H (2001) Lipid peroxidation is an early symptom triggered by aluminum, but not the primary cause of elongation inhibition in Pea roots. Plant Physiol 125(1):199–208

Yang K, Ma Y (2010) Computer simulation of the translocation of nanoparticles with different shapes across a lipid bilayer. Nat Nanotechnol 5(8):579–583

Yazdi NA, Arefi MR, Molla Ahmadi E, Nejand BA (2011) To study the effect of adding Fe2O3 nanoparticles on the morphology properties and microstructure of cement mortar. Life Sci J 8:550–554

Ye Q, Zhang ZN, Kong DY, Chen RS (2007) Influence of nano-SiO2 addition on properties of hardened cement paste as compared with silica fume. Constr Build Mater 21:539–545

Zafar H, Ali A, Ali JS, Haq IU, Zia M (2016) Effect of ZnO nanoparticles on *Brassica nigra* seedlings and stem explains growth dynamics and antioxidative response. Front Plant Sci 7:535. https://doi.org/10.3389/fpls.2016.00535

Zahir AA et al (2015) Green synthesis of silver and titanium dioxide nanoparticles using Euphorbia prostrata extract shows shift from apoptosis to G_0/G_1 arrest followed by necrotic cell death in *Leishmania donovani*. Antimicrob Agents Chemother 59:4782–4799

Zhang MH, Li H (2011) Pore structure and chloride permeability of concrete containing nano-particles for pavement. Constr Build Mater 25:608–616
Zhang W, Suhr J, Koratkar N (2006) Carbon nanotube/polycarbonate composites as multifunctional strain sensors. J Nanosci Nanotechnol 6:960–964
Zhenhua L, Huafeng W, Shan H, Yang L, Miao W (2006) Investigations on the preparation and mechanical properties of the nano-alumina reinforced cement composite. Mater Lett 60:356–359
Zhu W, Bartos PJM, Porro A (2004) Application of nanotechnology in construction: a summary of a state-of-the-art report. Mater Struct 37:649–658

Chapter 11
Thermal Insulation of Building Envelope for Ecological Conservation

Mir Firasath Ali and M. M. Vijayalakshmi Natarajan

Abstract Sustainability has been a rising concern for facility projects since the early 1990s. It is conscious that in terms of carbon emissions, energy usage, water use, utilisation of raw materials, waste production and many other factors, construction and operation of buildings have enormous direct and indirect environmental consequences. Increased economy and population will create more and more challenges for the design, construction and operational community to fulfil the new requirements for infrastructure that are affordable, safe, resilient and sustainable while reducing their environmental impact. The building envelope's key feature is to offer shelter, safety, solar and thermal control, control of moisture, control of indoor air quality, access to sunshine, and outdoor views, fire resistance, acoustics, cost-effectiveness, and aesthetics. This study seeks to promote awareness on the energy friendly design strategies which economically minimize operational expenditure, while enhancing comfort of the occupants in the building. This work would allow architects, engineers and builders to choose the most efficient or climate-reactive envelope alternative in order to improve thermal environment efficiency of the construction design being proposed.

Keywords Building envelope · Environmental friendly · Reactive envelope · Sustainable · Thermal comfort

11.1 Introduction

Buildings built and used today, due to exorbitant energy use, and other natural resources, lead to severe environmental problems. The strong relationship between energy usage in buildings and environmental damage comes from a significant depletion of essential environmental resources as energy intensive solutions are pursued to create a building and satisfy its heating, cooling and ventilation & lighting requirements (Owusu and Asumadu-Sarkodie 2016). Sustainability is now a main consideration for prominent stakeholders in the construction industry, given the rising demand

M. F. Ali (✉) · M. M. V. Natarajan
Department of Civil Engineering, Sathyabama Institute of Science and Technology, Chennai, India

J. A. Malik and S. Marathe (eds.), *Ecological and Health Effects of Building Materials*,
https://doi.org/10.1007/978-3-030-76073-1_11

from customers for ecologically sustainable building products and materials (Akadiri et al. 2012). Insulating resources used in residential or commercial buildings are an integral part in the energy efficiency, with a major environmental problem raised by the processing and disposal of a large quantity of insulating materials. In this context, this study proposes to analyse the metrics of alternative insulation used in buildings in a sustainable design process (DOE 2015). One of the most significant criteria for buildings is to provide thermal comfort without unnecessary space conditioning costs. Thermal management in almost any building is therefore an essential feature. In order to assess a building's sustainable energy conscious nature, understanding of heat exchange mechanism, periodic heat-flow in building elements and building envelope is essential (Yüksek and Karadayi 2017).

11.1.1 Heat Exchange Process

Like the human body, the buildings could be treated as a fixed entity and their mechanisms of heat exchange can be studied in the outdoor environment. Different heat exchange mechanisms between a building and the outside world are possible. Heat flows through different building elements like walls, roof, door, windows etc. Heat transmission often occurs by convection and radiation from various surfaces (Ling et al. 2016). Internal heat gain (Qi) heat outputs of human bodies, lamps, motors and appliances can lead to internal heat gain. Solar heat gain (Qs) involves solar radiation on opaque surfaces, but the solar heat gain must be considered separately on the basis of a solar temperature concept via transparent surfaces (Windows). Conduction (Qc) can happen inside or outside through the walls. Ventilation (Qv) heat exchange will take place with air movement in either direction. Heat flow rate and mechanical (Qm) heat removal with any external energy supply. Evaporation (Qe) takes place on the building surface (e.g. a roof pool (or) inside a building, human sweats (or) water from a fountain), and the vapours are removed and a cool effect is caused. The thermal equilibrium condition prevails when $Qi + Qs \pm Qc \pm Qv–Qe = 0$. Building tends to cool down if the sum of this equation is lower than zero and the temperature in the building increases when more than zero (Auliciems and Szokolay 1997).

11.1.2 Periodic Heat Flow in Building Elements

- Climate change creates unbalanced conditions,
- 24 h period—rise in temperature and decrease in temperature,
- During hot times—heat flows from the atmosphere into the building, as some of it is stored,
- The heat flow is reversed at night during the cool time i.e. from building to environment.

$$\text{Decrement factor}(\mu) = (T_{i\text{-max}})/(T_{o\text{-max}})$$

where,

$T_{o\text{-max}}$ is maximum outer surface temperature.

$T_{i\text{-max}}$ is maximum Inner surface temperature.

There are the two variations in external and internal temperatures in thermal regimes (Koenigsberger et al. 2010).

- Outdoor temperature rises in the morning; heat tends to penetrate the wall's outer surface.
- Each wall partition absorbs, in accordance to the specific heat of the wall material, a certain amount of heat for each degree of temperature increase.
- The heat will not be passed on to the next particle until the first particle temperature has risen.
- As shown by the green line, the corresponding rise of the inner surface temperature is thus delayed.
- Before the interior surface temperature is equal, the outside temperature reaches its peak and begins to fall.
- After that, the heat contained in the wall would be partially dissipated outside and only partially inside.
- As the outside air cools, a growing share of this stored heat fluctuates outwards, and the direction of the heat flow reverses fully when the wall temperature falls below the inner temperature.
- Time lag (or phase transition, denoted as Φ) and the decrement factor (or amplitude mitigation, denoted as μ) are the two quantities defined by this periodic change. The second is the ratio between the peak external and internal surface temperature amplitudes derived in accordance with every day average.

If the design allows for efficient daytime lighting, artificial lighting demand of a building can be minimised. In energy-conscious architecture, building materials play a major role too. Day lighting is also a passive solar technology and concludes with a debate on other construction materials (Koenigsberger et al. 2010; Vijayalakshmi et al. 2006).

11.1.3 Building Envelope

The exterior façades of a building like walls, windows, screenings and towers known as the building envelope communicate with the surroundings. The building envelope is mainly known as the building skin. It plays a key part in ensuring that indoor areas are safe, friendly and contributes to the reputation and character of the building within the public sphere. The building envelope needs to balance the need for ventilation and daylight while ensuring thermal safety that suits the site's climate conditions.

The estimation of the operational energy usage and the cost of lifetime of a building is an important factor. The success of the building envelope depends on the choice and combination of the necessary materials and components (Sozer 2010).

11.2 Role of Building Enclosures, Openings and Materials in Thermal Environment

The building envelope functions as a thermal shell that would contribute to energy leakage across all components if built thoughtlessly. Each part must therefore be selected correctly to achieve an effective building. The preference relies upon the location and thus the main goal is to look at the terms of the site. The perfect orientation of the building on a site and the right design layout often play an important part in the efficiency of the building (Gan et al. 2019).

11.2.1 Building Configuration

The heat exchange takes place mainly through the skin of the building between a building and its environment. The heat flow can be regulated by setting the configuration associated with the building that is worthy of climate and use. For instance, in an extremely cold climate, loss of heat from the premises of building to the atmosphere must be reduced (Givoni 1998).

This can be done through (Koch-Nielsen 2013);

a. Use of buffer spaces, for example in favourable weather, sun spaces and balconies can act as sit-outs.
b. The positioning, in the direction of cold winds, of seldom used areas such as storerooms or toilets.
c. Optimise solar radiation exposure, for example, arrangements for large living rooms facing the sun to heat.
d. Proper location of living room, i.e. leeward to escape cold winds, most of the habitable spaces may be held. They can be combined to minimise cold exposure.

The radiation and air movement due to heat flow may be governed by differing the following features of a structure:

11.2.1.1 Surface Area to Volume Ratio (S/V Ratio)

This specifies the degree of heat flow in and out of the building. Higher the S/V ratio, higher the heat gain or heat loss over a certain volume space. On the other hand, it will minimise heat gain/loss by a smaller S/V ratio. For instance, compact building shapes with a minimum S/V ratio can be preferable in cold climates (Kohlhepp and Hagenmeyer 2017).

11.2.1.2 Building Shape

Winds produce pressure differences i.e. positive windward pressures and negative leeward pressures when interrupted by a building. Therefore, the building is surrounded by a modern airflow pattern. Thus, a building's wind pattern may be changed by properly shaping it (Bauman 1988).

11.2.1.3 Buffer Spaces

Shading and catching wind within buffer-spaces in particular balconies, courtyards, atrias and verandahs.

11.2.1.4 Arrangement of Openings

Suitable openings which provide effective ventilation, link high- and low-pressure areas. For receiving or rejecting solar radiation, solid and glazed surfaces must be arranged and orientated (Gulati and Paul 2013).

11.2.2 Building Components

Presence of a structural envelope determines the extent of radiation and wind entering the building. This consisted of elements (A) Roof (B) Walls (C) Floor (D) Exterior colour and texture (E) Fenestration.

Heat flow by means of said elements is marked through resistance, thermal capacity, absorption, transmission and emission of each of these elements. Depending on specific criteria, the resource materials for these elements must be specifically chosen. The heat, density and thermal conductivity are the thermophysical parameters of materials that need to be considered. Whereas the product between the first two defines the material's energy storage capability, and the third specifies the performance of energy flow. Taken together, these three criteria depict time lag (phase shift) and decrement factor. The first is a heat flow delay, while the second is a reduction in heat wave amplitude. According to climate requirements, materials which contribute the intended thermal stock, time delay and amplitude decrease were to be looked for. Surface qualities such as emissivity, reflectivity, absorption and roughness are characterised in colour and texture. These are important for the distribution of light and heat flow. In the case of a painted white ceiling, for instance, the heat transfer in contrast to a dark colour can be decreased by up to 80% (Arnold 2009).

Building components may usually be categorised as opaque (brick wall) and transparent (glazed window). Transparent members permit sunlight in living spaces. An element may also be opened to allow air exchange between the building and its vicinity (such as skylight). Materials like polyurethane foam (PUF) may insulate the walls, floors window, door, etc. By properly insulating them, heat loss or gain from different components of buildings can be minimised and roofs (externally or internally). The introduction of an air cavity on the exterior building envelope is another form of insulation. In cavity walls the air gap prevents heat from being transferred in or out of the building because the air behaves as an inappropriate heat conductor. Variations can be accomplished by the utilisation, modification and usage of various isolation materials in different locations (internal or external). The property of the air gap can be varied in cavity walls by changing its thickness and selecting a ventilated or unventilated air cavity. The efficiency of insulation materials can be adversely affected by water absorption (Mirrahimi et al. 2016).

The heat gain can be varied by the following:

- Component orientation and inclination
- Component Area
- Finishes
- Material characteristics (U-value, lag time, factor of decrement, emissivity, etc.)
- Incoming solar radiation control.

11.3 Sustainable and Energy Efficient Thermal Comfort Techniques

11.3.1 Passive Heating

11.3.1.1 Direct Gain

Direct gain is the most widely used, passive heating technology in cold climatic conditions. It turns out to be a modest and thus commonly used solution. According to said method, direct sunlight through openings or glazed windows is admitted in the living spaces. The sunlight warms up walls and floors that will accumulate the heat within the indoor atmosphere and transmit it. Maximal solar radiation as well as thermal storage mass is the key criteria of a direct gain system. The impacted area of the house has a tendency to get very hot through the day, so the thermal storage mass is made available as uncovered vast walls or floors for the absorption and storage of heat. This particular inhibits the room from being overheated. If it is more and more necessary for space heating, the stored heat is released in the night. The floors and walls used as storage mass should not be covered with carpets and curtains because they hinder the heat flow rate. To be able to inhibit overheating in the summer, sufficient overhangs for shade and open ventilation windows have to be provided (Chiras 2002). A direct gain system therefore comprises the following parts (Johnson 1981):

1. **Glazing**—for transmission and trapping of solar radiation (incoming);
2. **Thermal mass**—towards stocking heat with regard to night usage;
3. **Insulation**—to attenuate losses during evening time;
4. **Ventilation**—Pertaining to cooling during the summer months;
5. **Shading**—to reduce heating up too much during summer.

Reflectors can also be installed outside windows to make the direct benefit system more effective. Heat gain can also be enhanced from Clerestories and skylights. The most popular, simple and efficient approach to heating is direct gain. Some of its drawbacks, however, are overheating, glare as well as deterioration of construction materials due to UV radiation (Tiwari 2002).

Solar energy can be stored in direct recovery systems in the living room floor, walls, ceiling or furnishings if these sections have adequate capacity for night absorption and storage of heat. This capacity is provided by materials (concrete, brick and water) and can be used in direct heat gain use effectively. PCMs, such as salt or wax, are often used to store thermal energy when melting and release heat when solidified. A balanced distribution of mass over the entire living area should be maintained. A thin material extending through a wider area will usually carry out higher than a thick material focused in one section of the room (Yüksek and Karadayi 2017).

Generally, 30% of the storage area must be made available to receive direct sunlight in accordance with per square metre of south-facing glazing area. The storage material requires a thickness of 50–150 mm, and walls should have a thickness of 50–100 mm. With full mortar bedding, the used masonry units should be solid. For the sake of more heat absorption, the storage mass vulnerable to direct sunlight must be dark in colour. Providing thicker rather than thinner storage mass is typically more thermally effective considering optimal thickness. For instance, the storage effect increases with the increase in the thickness of the floor as the storage mass. Increase in the storage effect is insignificant for 100 mm and above thickness. In practice, for thicknesses over 200 mm the performance decreases (Block and Bokalders 2010).

11.3.1.2 Indirect Gain

Thermal wall systems are specifically designed for use in space heating. Actually a wall is situated involving the living area and also the glazing to gain full solar radiation in this way by orienting the southern part of the face associated with the building into the northern hemisphere. Doing this hinders the direct entry of solar radiation to the living spaces, where solar energy is captured, absorbed, stored in, and controlled. The glazing decreases the ambient heat loss. Additionally it is possible to incorporate windows within thermal storage walls to offer illumination, vision plus some direct heating gain. Portable insulation exterior to the glazing façade or in between glazing and the storage wall may be added to minimise heat loss during night times. Usually, shading and reflecting devices are mounted on the external wall (Nayak and Prajapati 2006). In this section, different types of storage walls are discussed.

Trombe Wall

A particular trombe wall is a thermal storage wall manufactured from material, for example concrete, bricks or brick composites, blocks and sand that have high storage capacity. To maximise its absorption, the exterior surface associated with the wall happens to be painted black and is positioned immediately at the rear of the glazing along with some sort of air gap in the middle. The blackened surface absorbs solar radiation and is stored in the wall as sensible heat. The stored heat slowly migrates into the interior of an unvented wall, the place wherein it heats the living space that is adjacent. This wall will offer the living space with sufficient heat during the night if properly built (Mazria 1979).

The heat produced in the air between the glazing and the storage wall is lost through the glass to the outside. Hotter the air, greater is the heat loss in the airspace. This loss of heat can be minimised by venting the top and bottom of the storage wall. These units are referred to as "vented trombe walls." The air between the glazing and the wall is warmed and enters through the upper vent into the living room. Via lower vents, cool room air takes its place, thus creating a natural circulation pattern (thermo-circulation) that involves no mechanical means to move the air.

The impact associated with a thermal storage wall is dependent mostly on the dimensions, kind of material and colour of the outside area of the wall. In Trombe's walls, materials (concrete, brick, and water) with good thermal ability and PCM-phase change materials can be utilized potentially. To absorb solar radiation, the storage mass confronted with direct sunlight needs to have a dark colour. Particular coatings are often added to the exposed surface of storage walls to enhance performance. The coatings have high solar absorptivity and low re-radiation emissivity. The wall inner surface might perhaps be painted or else kept unprocessed. The thermo-circulation part concerning vents ought to be around 2% associated with the wall area, split equally amongst the upper as well as lower vents (Zhang and Li 2019).

Water Wall

Water walls, with the exception that they use water because the thermal storage material, are derived from exactly the same principle as compared to the Trombe wall. The higher specific heat allows water walls for storage of a lot more heat when compared with concrete walls. Actually a water wall can claim to be a thermal storage wall composed of water drums piled in the back of a window. They are externally painted black to make radiation more absorbent. The inner surface could be coated in almost any different colour and may instantly enter into exposure with or insulate the interior space. The heat storage from the water wall is a convective mass, which is very quick compared to a wall of masonry (Wujek and Dagostino 2011).

A large volume of storage offers extended and much more storage capacity, whilst smaller sized products allow for faster distribution. The thumb rule is generally considered as 150 L/m^2 of this water wall facing south to fix the volume of water. The storage mass ratio has a number of heat exchange surfaces including tins, cans, bottles and tubes, bins and drums. Care should be taken to ensure the corrosion-resistant lining of steel and metal containers. Water can also be treated with chemicals that avoid algae. Troughs must be made in the form of safeguard towards container or condensation leakage of water.

The flow of heat through a wall of water is significantly quicker rather than a trunk wall. A check of the heat circulation is required in case heat isn't needed for the building immediately. Through the use of a thin concrete layer or insulating layer, or by supplying air ventilation through vents, this can be accomplished. The rapid heat transfer in the water walls benefits buildings such as schools or government offices working during the day. The water wall glazing is normally filled with insulation at night to minimise heat losses. Using movable overhangs will avoid overheating during the summer (Noseck 2013).

Transwall

Transwall is a semi-transparent thermal storage wall. The solar radiation is partially absorbed and distributed. Direct heating and lighting of living space is caused by the transmitted radiation. The absorbed heat is subsequently transferred to the living room. The heat loss due to the glass is very minimal, since much of the heat is contained in the centre and does not get too hot on the outside of the wall. The system therefore incorporates the desirable characteristics of the trombe wall and direct gain system. There are three main elements in a Standard Transwall segment (Fuchs and McClelland 1979):

- Container made of metal frame parallel glass walls,
- Liquid for thermal storage, typically water,
- A partly absorbent plate mounted parallel to the glass walls on the middle of the transwall.

It is situated directly behind double glazing on the south side of the building (in the north hemisphere). A kind of suppressing agent could possibly be applied to prevent the growth of microorganisms in the storage.

11.3.2 Passive Cooling

The cooling of buildings has created great interest through the use of passive methods. Passive cooling is based on the prevention and/or elimination of heat from entering (or, at least, reducing heat flux) the house. This section addresses the concepts that govern each concept used to cool buildings passively. Ventilation cooling, evaporative cooling are the terms discussed. The applicability of these principles is highly dependent on the climatic conditions in that specific location (Kamal 2012).

11.3.2.1 Ventilation Cooling

The substitution of stagnant air by using fresh air is commonly termed ventilation. Moreover it offers cooling by air movement. The term ventilation should therefore be specified as the external air supply towards inside with regards to air motion and vitiated air replacement. When it comes to hot and humid regions where in fact the outside maximum air temperature will not exceed 28–32 °C, an internal air velocity of one and half to two meter per sec could result in comfort (O'Connor et al. 2016). In the design process of buildings, ensuring adequate ventilation in buildings needs due consideration. In order to create comfortable indoor conditions, a flawed design that results in insufficient ventilation would lead in much higher energy expenditure in the building. Therefore, for various types of occupancy, the ventilation criteria for different seasons should be calculated first. In order to meet the appropriate performance requirements, a ventilation system needs to be then appropriately designed. Ventilation can enhance comfort in a variety of ways. For example, it helps to make people feel cooler in a building by opening up the windows to allow the wind inside and therefore give greater indoor air speed. This technique is called ventilation comfort. Evaporation is the most significant process of the human body's heat loss to maintain thermal comfort in hot environments. As the air around the body is almost filled with moisture, transpiration is harder to evaporate and a feeling of un-comfortableness could be experienced. Extreme humidity and higher temperature combinations are highly oppressive (Nayak and Prajapati 2006).

Perhaps a little air flow close to the body provides easing in these circumstances. A rate of ventilation which must induce required air movement will therefore be desirable. In case of inadequate natural ventilation, running fans inside the building will increase the airflow. Indoor air movement is primarily because of stacking effect (temperature stratification) in addition to wind pressure level. The management of both of these effects will enhance ventilation substantially. The main feature of solar chemistry is the stack effect, for example. The solar powered chimney is employed to rapidly evacuate hot air through the house, thus increasing the cooling down ability associated with the air from all the other open positions (Kleiven 2003).

Wind towers also use wind pressure for cooling. On the top of the terrace, winds are collected and transferred to the interior by wind towers. In addition, Windows may be positioned to make use of the wind pressure and stack effect. In order to cool down the inside mass associated with the building; an indirect type of cooling would be to ventilate the building components exclusively during the night. The cooled mass decreases the rate of indoor temperature rise during the following day and thus offers a cooling effect. This is called night ventilation cooling (Walker 2016).

11.3.2.2 Cross Ventilation

Air movement requirements are typically low during the beginning of summer and later part of the post-monsoon cycles. By providing sufficient cross ventilation through spaces, these can be easily met. The temperature of this interior air and areas accurately matches the surrounding temperature when a building is cross-ventilated during the day. Daytime ventilation also needs to be viewed provided when interior comfort and ease could certainly be enjoyed at the temperature of the outside air (with acceptable indoor speed) (Aman 2017).

11.3.2.3 Wind Tower

For cooling purposes, wind towers are commonly utilized in warm and dried up climates. The particular tower can be intended at higher elevations to "catch" the wind and guide it inside the living space. There may be equal or distinct areas in the air flow passages in the tower. If the wind is mainly in one direction, the tower might perhaps posse's sole opening facing the wind, or could possibly possess openings each of them directions at places with different wind directions. For decades, these technologies have been used for natural ventilation and passive cooling in Western Asian countries (Konya 2013).

A condition for using a wind tower is that winds with a considerably good and constant velocity should be encountered on the location. Depending on the time duration related to day as well as the existence or shortage of wind, a wind tower works in different ways. Altering the temperature and then the denseness regarding the air close to the tower is the main principle of its operation. The density difference produces a draft, dragging air upwards or may be down wards throughout the tower (Malone 2012).

Working at Night The tower area is built to provide high heat storage capacity for the top portion, and also has a large heat transfer surface area. During the day, the tower walls and the internal walls of the air-flow passages collect heat and release it at night, thus warming the tower's cool night air. Warm air rises up to create an upward draft and is exhausted by the openings. The difference in pressure thus created pulls the cool night air into the building through the doors and windows. The tower serves as a chimney in the absence of wind. Further cooling is brought on by night time radiation by using the roof in addition to exterior walls. The cool night air reaches the tower in the presence of wind and drives itself down into the structure. While it is warmed slightly during the process, due to forced circulation, adequate cooling can be achieved. Again, cooling contributes to this process due to nocturnal radiation. Low-rise buildings can easily add wind towers. It must be noted that when cooling is not needed, wind towers may need to be shut down and, thus, such requirements must be included in the design. In order to prevent dust, birds and insects from entering, due consideration must also be given (Sarbu and Sebarchievici 2018).

11.3.2.4 Evaporative Cooling

Evaporative cooling is a passive cooling method in which ambient air is cooled until it is released into the building by evaporating water. The physical theory lies in the use of sensible air heat to evaporate water, thus cooling the air, which in turn cools the living space of the building. At the water–air interface, evaporation occurs. A rise in the proportion of the contact area between water and air increases the evaporation rate and hence the cooling capacity. A cooling effect may be created by the presence of a water source such as a pond, lake or sea near the house, or a fountain in the courtyard. It is also possible to position cisterns or wetted surfaces in the incoming ventilation stream. Usually, such direct systems use little to no auxiliary power, are easy, and can eliminate the need for large water surfaces and significant amounts of air movement. They are, therefore, ideal for hot and dry regions in particular. In these systems, the airflow may be induced mechanically or passively, such as.

It is possible to use evaporative cooling towers that humidify the ambient air. This is direct evaporative cooling. The key drawback of direct systems is that the humidity content of the ventilation air supplied to the interior spaces is increased. Owing to high humidity, high evaporation can result in discomfort (Maheshwari 2001; Amer 2006).

Passive evaporative cooling, however, can also be indirect, the roof can be cooled with a pond, wetted pads or spray, and the ceiling converted into a cooling feature that cools the space below without increasing the indoor humidity through convection and radiation. The evaporation process' efficiency is based on the air and water temperatures, air vapour content and air flow rate over the surface of the water. Evaporation would be improved by the provision of shade and the availability of cool, dry air. Bansal et al. have published a detailed discussion on evaporation. The desert cooler consisting of water, evaporative pads, a fan and a pump is the only evaporative cooling system in North India. It is a direct evaporative cooling system with a hybrid form. The guidelines for use of evaporative cooling have been suggested by Watt (2012);

- The average saturation efficiency of direct evaporative coolers should be 70 percent or more, and cooled air must reach the indoor room without any extra warm-up.
- The cooled air induced indoor air speed must be 1 m/s.
- Before cool air is discharged from room, the room temperature should be reduced to at least 3 °C.
- The cool down area temperature ought to be around four degree centigrade beneath the outside temperature of the dry bulb. The incoming radiant heat must be counteracted.

11.3.2.5 Roof Surface Evaporative Cooling (RSEC)

The solar radiation incident on roofs is very high in a tropical country like India in summer, leading to overheating of rooms below them. By spraying water over sufficient water-retentive materials (e.g., gunny bags) scattered over the roof surface, roof surfaces can be cooled easily and cheaply. It draws much of the necessary latent heat from the surface as the water evaporates, thereby lowering its temperature and reducing heat gain. Besides, the air above the roof is also cooled by evaporation. Via penetration and ventilation, the cool air slides down and reach the living space, providing extra cooling. This is a specific case of the technique of indirect passive evaporative cooling. The sustained wetness of the roof surface is a crucial factor deciding the efficiency of the RSEC system. It is possible to spray the surfaces intermittently, as it is only important to keep them moist (Nayak and Prajapati 2006).

To minimise the cooling load in the summer, the evaporation of water from a roof pond (a large mass of water accumulated on the roof) can also be used. However, the roof needs to be made structurally stable and waterproof to use this cooling technique. Cooling by sprinkling water is more beneficial in contrast, as it gives a greater evaporation surface area without the need for any storage (Bhamare et al. 2019). The following points need to be taken care of in order to implement a roof surface evaporative cooling system (Yannas et al. 2006):

1. Adequate roof waterproofing treatment should be performed.
2. Water-absorbing and water retentive materials must cover the roof. These materials, because of their porosity, serve as a free water surface for evaporation when wet. The durability of such materials is reasonably good, but for fire protection they must be handled.
3. The amount of water required during peak summer is approximately 10 kg/day/m^2 of roof area.
4. Using a water sprayer, the roof must be kept damp during the day. An automatic moisture-sensing system can run or control the sprayer manually. Usually, the sprayer operates at a low water pressure, which can be accomplished by either a water head on the roof of the storage tank or a small water head.

11.4 Sustainable Building Materials

There are several energy efficiency improvement strategies in buildings, and it is the duty of the inhabitants to run them in an energy-conserving manner. However, inhabitants can only run it within the framework of the designers of the building. Finally, it is the designers' duty to provide the owners and tenants with the most energy efficient building. This service is not only economically feasible, but also prevents the building from being unsustainable as a result of high energy costs. In energy-conscious architecture, building materials play an important role. The heat flow rate through different components of a building, it's time lag and amplitude decrement, as well as the building's energy storage capacity are all controlled by the materials used. Therefore, from the perspective of the building's thermal efficiency, the choice of materials is essential. Besides, the materials provide the building with the requisite structural strength. Although traditional building materials are well known to architects, building scientists and consumers, in order to minimise costs and energy usage, it is important to concentrate on alternative materials. It should be remembered that for the production of building materials from their basic raw ingredients a certain amount of energy is consumed. This is known as the materials' embodied energy. The selection of construction materials depends on this aspect (DOE 2015).

11.4.1 Embodied Energy of Building Materials

The energy utilized in the building construction practice needs to be reduced collectively, particularly in the building materials used. The cost of construction is influenced by resources used in the purchasing, manufacturing, and process and recycling of building materials. The energy assessment process consists of three phases: (i) the energy consumption in raw material processing, (ii) the energy consumption in manufacture of finished products, and (iii) the energy used in the manufacturing process for machinery and equipment (Akadiri et al. 2012). The sum of the three is referred to as the intensity of energy. Building materials, based on their energy strength, have been divided into three groups. High-energy materials include products such as aluminium, steel, plastics, glass and cement and are such with energy levels surpassing around 5 GJ/tonne of processed materials. Materials in the medium energy category include those requiring energy inputs of between 0.5 and 5 GJ per tonne of material, including concrete, lime plaster and most types of cement-based, lime-based, fly ash-based and fireclay-based bricks and tiles. The materials of the low energy category include fine and coarse building aggregates, pozzolana soil forms and stabilised soil (Gutowski et al. 2013).

Promoting low-cost, low-energy and medium-energy materials for energy efficiency in building construction is important. These products, however, should be durable as well, need less maintenance and should be recyclable. It should be noted that, while highly energy intensive, materials such as aluminium and steel possibly recycled pretty inexpensively because of energy (Martin et al. 2000). Development Alternatives, New Delhi, has conducted a thorough analysis of the embodied energy of different building materials. The document provides information on various building materials and components at different levels, namely production, processing and manufacturing. Data on material description, technology and resources, environmental effects, statistics of the development and world status of energy data are accessible to designers. The study also provides data on the energy consumed when raw materials, intermediate materials and finished products are mined, manufactured and transported.

11.4.2 Alternative Building Materials

Summary of different sustainable building materials and technologies built to reduce both energy usage and costs.

11.4.2.1 Autoclaved Aerated Concrete (AAC)

In different parts of the world, AAC is acknowledged by different brands as Siporex, Trustone and Environcrete. It is a lightweight precast concrete manufactured by the factory that is obtainable in various sizes and shapes. Thin layer of adhesive can be bonded to AAC blocks, so no mortar is required. Mixtures of OPC, fly-ash or silica sand, lime, water and powder or paste of aluminium are used to make AAC bricks. Millions of very small bubbles of hydrogen extend the mixture to about 5 times its original volume when combined. AAC can be improved by means of traditional carpentry equipment and can be quickly cut. It is a very stable, pollution free, thermally and acoustically insulating, and durable material that resists fire. It needs to be plastered for rain safety, however.

11.4.2.2 Fly Ash

In thermal power plants, fly-ash is a by-product of coal. It consists of non-burnt organic and inorganic matter which could be reused in manufacturing a wide range of construction materials. The characteristics of fly ash find application in manufacturing of stains, solid and hollow bricks, partial cement replacement, and cellular concrete as well as can be used in the manufacturing of paint, distempers etc. A variety of benefits can be accomplished by using fly ash in construction materials.

11.4.2.3 Compressed Earth Blocks

For centuries manual processing of earth blocks has been performed by compacting them in tiny moulds. The method is now automated or industrialized. The earth is made of a mixture of pebbles, sand, silt and clay for blocks. To stabilise earth blocks, approximately 5 percent cement is used. The goods range from solid, cellular and hollow bricks with precise shapes to flooring and paving components. The sun is dry and need not be burned in the compact earth bricks. They are also cost-effective, strong, economical, and easy to produce. A hollow and interlocking soil stabilised block does provide higher thermal isolation than bricks. Fal-G stabilised mud blocks are much stronger than concrete stabilised blocks and absorb less water. Extensive study on this material has been conducted by Development Alternatives, New Delhi and Auroville, Pondicherry. A number of buildings were constructed on compressed blocks of earth at Auroville, Pondicherry.

11.4.2.4 Clay Red Mud Burnt Bricks

These are formed by industrial wastes of aluminium plants in combination with clay, are made of red alumina mud or bauxite. The brick has all the physical characteristics of typical clay bricks. Moreover, the waste management and environmental pollution problems are also solved. Moreover, due to their pleasant colours, they have a strong architectural value as bricks.

11.4.2.5 Lato Blocks

Lato blocks are laterite soil bricks of lime or cement. They are moulded to create solid and better quality blocks which use very less energy when compared to traditional bricks, making them cheaper.

11.4.2.6 Precast Hollow Concrete Blocks

Blocks are constructed from lean cement concrete mixes, extruded by egg laying or static block making machines. They require less amount of cement mortar and thus easy to build in comparison with brick masonry. The thermal safety is enhanced by the cavity in the blocks. Moreover, external or internal plastering may not be needed for the blocks. These can be used as walling or inverted precast t beams as roofing blocks.

11.4.2.7 Bamboo/Timber Mat Based Walls

Such walls consist of a bamboo mat between vertical and horizontal wood/bamboo frames. On both sides, the plastering is achieved with mud or cement. These are simple to build, less costly and common in hilly areas because they can be assembled. However, they do not hold load and need a support system. This improved conventional technology is a key alternative to mitigate damage in case of collapse from the perspective of earthquakes.

11.4.2.8 Rat Trap Bond

The rat trap bond for both English and Flemish Bond is another brick bonding system. It is cost-effective, strong and attractive. The quantity of bricks can be saved by about 25%, while the cost of a wall is about 40%. The rat trap bond is easy to build and has better insulation properties.

11.4.2.9 Composite Ferrocement System

The technique is simple to create and consists of ferrocement (rich mortar reinforced with chicken mesh and welded wire mesh). This reduces the thickness of the wall and makes a greater carpet area. The RCC columns are integrated with precast ferrocement units in trough shape. Units of ferrocement act as a permanent unit for the skin and as an inter-column diagonal stripe. Within coverings it is possible to render mud blocks or any material locally available. This is suitable for seismic conditions.

11.4.2.10 Coconut Fibre and Wood Chips Roofing Sheets

Coconut fibre and wood chips or fibre are submerged in water for two hours and the water is then washed away. They are then mixed with cement, placed on a corrugated mould and held under pressure for 8–10 h. After demoulding before use, the sheets must be cured as well as dried in sunlight.

11.4.2.11 Cement Bonded Fiber Roofing Sheets

This particular is developed in combination with cement as a binder for the manufacture of corrugated or plain roofing sheets of coir waste, coconut pith, wood wool or sisal fibre. These sheets are 50% cheaper than sheets from AC/CGI and use less cement than sheets from AC. They are also lightweight, fire-proof, water-proof and are ideal for sloping roof choices.

11.4.2.12 Micro Concrete Roofing Tiles

Micro cement tiles are composed of a graduated layer of cement mortar built over the inclined roof mould. It is used in roofing pitch systems and is less expensive than ACC/CGI sheets. These tiles are suitable where there are no fired clay tiles and the skeletal system supported by wood is more expensive. When micro concrete roofing tiles are used, the rafter and purlin system cost less. By using ferrocement rafters and purlins, further cost reduction can be carried out.

11.4.2.13 Stone Patti Roofing

Stone Patti roofing is a flat sandstone slab roof system (patties) that rests on steel or sleek RCC beams. The slabs are overlaid for insulation with terracing. This type of roofing is more economical than RCC roofing when (sand) stone sheets are available. The beams are not necessary in locations where large granite stone patties are available, as the Pattis can rest on walls.

11.4.2.14 Precast Brick Arch Panel System

In this method, 50 cm × 50 cm precast brick arches are cast on a base. Side by side, the arches are located above the partly precast joist. Cement concrete covers the haunches between the arches to have a level surface on the top. 30% more such roofs/floors are economical when compared to traditional RCC.

11.4.2.15 Filler Slabs

RCC slabs whereupon filler materials are substituted by bottom half (tension) concrete portions. Filler materials (bricks, cellular concrete blocks, tiles, etc.) are positioned so that structural integrity is not compromised. These are healthy, sound and give ceilings with an aesthetically pleasing pattern. An additional benefit of slabs of filler is that they do not require plastering.

11.4.2.16 Particle Boards

These are made from and bonded by resin from wood waste, cotton stalk and bagasse. They can be used as inserts and they can be used with veneering in panelling, partitioning, false ceiling and furniture as an alternative to timber.

11.5 Conclusion and Recommendation

At minimum 2.35% of the world's energy generation can be able to save by effective passive solar design techniques. Passive design technology includes approaches to collect, reserve, disperse and manage thermal energy by means of different conceptual and physical thermodynamics principles suitable to building materials. These concepts can be converted into attributes inherent in the building design and operation technique through a number of vocabulary frameworks. It creates a structure that is more integrated with its practical environment and gives design opportunity to the architecture and construction profession. The direct gain for heating (office) during sunshine hours is a more convenient passive heating technique whereas the other passive heating principles are used in residential buildings. Dual-glazed systems result in a 9% reduction in heat gain and a 28% decrease in losses compared to single-glazed systems. Double-glazing of the exposed walls should be mandatory to preserve solar radiation within the room with minimal U-value. The combination of evaporative cooling and wind tower is very productive for passive cooling and that can lessen temperature in the range of 12–17 °C. One of most economical methods for cooling a building is evaporative cooling.

Combining Trombe wall, cool roof and thermal isolation as a passive heating/cooling technique will save about 46 and 80% of energy cost respectively in winter and summer. The vast space requirements for thermal energy storage are a major concern in passive design. The building structure can hold approximately 0.23 kWh/°C/ton of its mass. Consequently, the mass of the structure will partially solve the storage issue. The building mass should be focussed within the building's interior to achieve efficient heat storage, e.g. in the inner partition and floors, while the outer walls should be highly isolated. The heat storage inside the building structure, therefore, involves a specific approach to building design.

References

Akadiri PO, Chinyio EA, Olomolaiye PO (2012) Design of a sustainable building: a conceptual framework for implementing sustainability in the building sector. Buildings 2(2):126–152
Aman J (2017) Impact of windows for daylighting on thermal comfort in architecture design studios in Dhaka
Amer EH (2006) Passive options for solar cooling of buildings in arid areas. Energy 31(8–9):1332–1344
Arnold C (2009) Building envelope design guide–introduction. Whole building design guide
Auliciems A, Szokolay SV (1997) Thermal comfort. SL: PLEA
Bauman F, Ernest D, Arens EA (1988) The effects of surrounding buildings on wind pressure distributions and natural ventilation in long building rows. Ashrae Transactions 94
Bhamare DK, Rathod MK, Banerjee J (2019) Passive cooling techniques for building and their applicability in different climatic zones—the state of art. Energy Build 198:467–490
Block M, Bokalders V (2010) The whole building handbook: how to design healthy. Routledge, Efficient and Sustainable Buildings
Chiras DD (2002) The solar house: passive heating and cooling. Chelsea Green Publishing
DOE U (2015) An assessment of energy technologies and research opportunities. Quadrennial Technology Review. United States Department of Energy
Fuchs R, McClelland JF (1979) Passive solar heating of buildings using a transwall structure. Sol Energy 23(2):123–128
Gan W, Cao Y, Jiang W, Li L, Li X (2019) Energy-saving design of building envelope based on multiparameter optimization. Mathematical Problems in Engineering
Givoni B (1998) Climate considerations in building and urban design. Wiley & Sons
Gulati M, Paul A (2013) Principles of energy conscious design
Gutowski TG, Sahni S, Allwood JM, Ashby MF, Worrell E (2013) The energy required to produce materials: constraints on energy-intensity improvements, parameters of demand. Philos Trans Royal Soc Math Phys Eng Sci 371(1986):20120003
Johnson TE (1981) Solar architecture: the direct gain approach. NP, United States
Kamal MA (2012) An overview of passive cooling techniques in buildings: design concepts and architectural interventions. Acta Technica Napocensis: Civil Eng Arch 55(1):84–97
Kleiven T (2003) Natural ventilation in buildings: architectural concepts, consequences and possibilities. Institutt for byggekunst, historie og teknologi
Koch-Nielsen H (2013) Stay cool: a design guide for the built environment in hot climates. Routledge
Koenigsberger OH, Ingersoll TG, Mayhew A, Szokolay SV (2010) Manual of tropical housing and building. Hyderabad
Kohlhepp P, Hagenmeyer V (2017) Technical potential of buildings in Germany as flexible power-to-heat storage for smart-grid operation. Energ Technol 5(7):1084–1104
Konya A (2013) Design primer for hot climates. Elsevier
Ling SJ, Sanny J, Moebs W, Friedman G, Druger SD, Kolakowska A, Anderson D, Bowman D, Demaree D, Ginsberg E, Gasparov L (2016) University physics, vol2
Maheshwari GP, Al-Ragom F, Suri RK (2001) Energy-saving potential of an indirect evaporative cooler. Appl Energy 69(1):69–76
Malone A (2012) The Windcatcher house. Architectural record: building for social change. McGraw-Hill

Martin N, Worrell E, Ruth M, Price L, Elliott RN, Shipley AM, Thorne J (2000) Emerging energy-efficient industrial technologies. Mazria, Edward. Passive solar energy book (1979)
Mirrahimi S, Mohamed MF, Haw LC, Ibrahim NLN, Yusoff WFM, Aflaki A (2016) The effect of building envelope on the thermal comfort and energy saving for high-rise buildings in hot–humid climate. Renew Sustain Energy Rev 53:1508–1519
Mazria E (1979) Passive solar energy book. Rodale Press, Pennsylvania, United States
Nayak JK, Prajapati JA (2006) Handbook on energy conscious buildings. Prepared under the Interactive R & D Project 03(04):03
Noseck RR (2013) The water wall: a passive solar collection and thermal storage device for supplementary radiant heating
O'Connor D, Calautit JKS, Hughes BR (2016) A review of heat recovery technology for passive ventilation applications. Renew Sustain Energy Rev 54:1481–1493
Owusu PA, Asumadu-Sarkodie S (2016) A review of renewable energy sources, sustainability issues and climate change mitigation. Cogent Eng 3(1):1167990
Sarbu I, Sebarchievici C (2018) A comprehensive review of thermal energy storage. Sustainability 10(1):191
Sozer H (2010) Improving energy efficiency through the design of the building envelope. Build Environ 45(12):2581–2593
Szokolay SV, Koenigsberger O (1973) Manual of tropical housing and building. Orient Langman, Bombay
Tiwari GN (2002) Solar energy: fundamentals, design, modelling and applications. Alpha Science Int'l Ltd
Vijayalakshmi MM, Natarajan E, Shanmugasundaram V (2006) Thermal behaviour of building wall elements. J Appl Sci 6(15):3128–3133
Walker A (2016) Natural ventilation: whole building design guide. National Renewable Energy Laboratory, USA
Watt J (2012) Evaporative air conditioning handbook. Springer Science & Business Media
Wujek JB, Dagostino FR (2011) Mechanical and electrical systems in architecture, engineering and construction. Pearson Higher Ed
Yannas S, Erell E, Molina JL (2006) Roof cooling techniques: a design handbook. Earthscan
Yüksek I, Karadayi TT (2017) Energy-efficient building design in the context of building life cycle. Energy Efficient Buildings 93–123
Zhang Z, Li Z (2019) Heat transfer performance of the Trombe wall implanted with heat pipes during daytime in winter. Sci Technol Built Environ 25(7):935–944

Chapter 12
Soil Contamination from Construction Projects

Sirat Sandil and Rabindra Kumar

Abstract Soil plays an essential role in carrying out various ecological functions, acts as a medium for plant growth, and is a base for construction works. Soil contamination is a growing environmental concern worldwide due to the rapid rate of urbanization and industrialization. The problem is pronounced in densely populated countries, where land-use intensity is severe due to construction activities. Procurement of building materials, their manufacture, and processing are environmentally destructive processes, and environmental damage has increased due to an increase in demand for individual living spaces. Further, the construction sector releases a huge amount of hazardous and non-hazardous construction and demolition waste, which negatively impacts the environment through contamination of soil and water. Urban agriculture carried out on contaminated soil leads to an elevated concentration of toxic elements in crops, and consumption of these crops poses a potential health risk to human beings. Construction materials both in existing structures and in new ongoing projects influence the soil on which the construction has been carried out and the soil in its vicinity. In this chapter we elaborate on the different construction and building materials, and how their extraction, manufacture, and usage negatively impacts the soil.

Keywords Construction sector · Construction waste · Demolition waste · Environmental impacts · Soil contamination · Urbanization · Waste disposal

12.1 Introduction

During the last century, the environment has rapidly become contaminated due to uncontrolled industrialization and incessant sprawl of urban areas. The negative impact of human activities on the environment has increased at an alarming rate

S. Sandil (✉)
Cooperative Research Centre of Environmental Sciences, Eötvös Loránd University, Pázmány Péter sétány 1/A, H-1117 Budapest, Hungary

R. Kumar
Faculty of Health Science, University of Pecs, Vörösmarty Mihály Utca 4, 7621 Pecs, Hungary

J. A. Malik and S. Marathe (eds.), *Ecological and Health Effects of Building Materials*,
https://doi.org/10.1007/978-3-030-76073-1_12

especially after the industrial revolution. Environmental contamination and degradation are a serious concern all over the world, especially in countries with a dense population, because they pose major environmental and health issues. An increase in population results in an increase in the needs and wants of the people which produces a degrading effect on the natural resources. Construction activities play an important role in improving the economic status of a society, but they are one of the greatest factors deteriorating the environment. The construction sector globally consumes a major portion of the raw materials and energy which leads to the depletion of natural resources (Singh et al. 2011). Traditional building materials like sand, stone, gravel, timber, cement, and steel are all procured from the natural environment, which results in damage to the environment due to exploitative practices (Pappu et al. 2007). Pollutants arising from construction and demolition activities are myriad and are produced at all stages beginning from the construction materials extraction and manufacturing process and ending with their disposal (Ponnada and Kameswari 2015). Large quantities of wastes are produced by the construction sector in form of construction and demolition waste, industrial by-products from extraction and processing of raw materials, mining and smelting waste, and energy utilization-related wastes. These end-products pose a great challenge for proper disposal due to their high volume and are thus dumped in landfills (Chowdhury et al. 2010). This massive amount of waste poses a significant management problem due to its large volume. Such a huge volume of waste is a problem for the environment, economy, and society (Ginga et al. 2020). The construction sector consumes about 50% of the natural resources and emits about 40% of the global greenhouse gases. The emissions are highest during the construction phase when raw materials are being extracted and processed (Assefa et al. 2007; Hossain and Marsik 2019). Construction and demolition sector and mining and quarrying industry generate the maximum waste as compared to any other sector (Cardoso et al. 2016).

Soil is a natural resource, which provides human beings with raw materials, renewable energy, food, feed, and fiber. It is an essential component of the biosphere because it provides a medium for plant growth, helps in the cycling of nutrients, and maintains the ecosystems and climate (Jie et al. 2002; Harrison and Alloway 2016). It comprises unconsolidated weathered materials derived from the bedrock, decaying organic matter, air, water, and microorganisms (McClintock 2015). Anthropisation or human activities like mining, construction of buildings and roads, and disposal of waste result in deforestation and contamination of soil. Removal of vegetation affects the soil and water quality, increases soil loss through erosion, and rapidly depletes soil fertility by removing nutrient-rich soil. Construction activities destroy the soil at the construction site and contaminate the surrounding soil due to the weight of the structures built on the soil, removal of vegetation, production of wastes and dust, and toxic emissions (Fig. 12.1). Such activities alter the physical, geological, chemical, and biological nature of the soil due to organic and inorganic pollution, and the extent of soil pollution depends on the type of land use (Joimel et al. 2016; Artiola et al. 2019). Interaction between building materials and the environment occurs through the entire life cycle of the construction and has a negative impact (Kobetičová and Černý 2017). Soil removed by excavation, road construction wastes, construction and

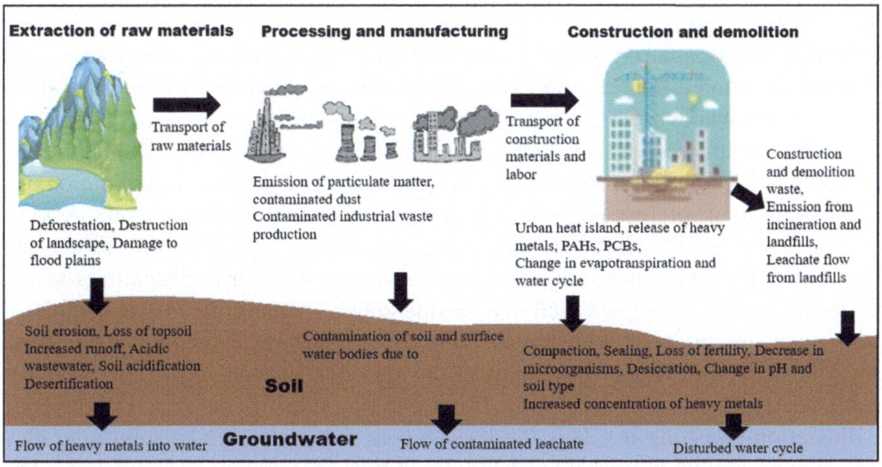

Fig. 12.1 Construction activities related impacts on the soil and groundwater

demolition (C and D) waste, waste rocks, dredge spoils, mine tailings, or mineral waste make up about 97% of the total waste (Cardoso et al. 2016). Soil pollution has a major effect on the sustainable development of the environment and the health of human beings (Yu et al. 2019). To assess the negative impact of construction-related activities on the soil it is essential to consider all stages of the construction process like raw materials, their mining and manufacture, transport, building process, life cycle, demolition, and waste disposal.

12.2 Soil and Its Ecological Importance

Soil is defined as the unconsolidated mineral-rich upper layer of the earth's crust which is an essential part of the ecosystem. Soil develops and serves as a host for various purposes at the interface where there is an interaction between the biosphere, atmosphere, lithosphere, and hydrosphere. Soil is a complex habitat that provides the necessary resources for all living beings and regulates the earth's environment (Singh et al. 2014; Voroney and Heck 2015). Soil is a non-renewable resource, which faces many ecological threats such as organic matter decline, salinization, desertification, and biodiversity loss. Soil pollution still does not generate as much concern as that of air and water pollution and has not received proper attention from citizens and policy makers (Doula and Sarris 2016). Soil is formed from consolidated rock and unconsolidated parent material along with the material deposited by gravity, water, and wind (Olson 2004). Soil formation is a vigorous process depending on five main factors: climate, parent material, topography, time, vegetation, and interaction with living organisms. These factors operate on all soil-forming materials and their relative influence varies with soil type and the geographical location (Harrison and

Strahm 2008). The process of soil formation consists of three stages, overlapping each other: (1) weathering of the parent material; (2) formation of clay and accumulation of organic matter; and (3) translocation of matter and differentiation of horizons. This whole process culminates into a characteristic soil profile, displaying variation of soil properties with depth. Soil forming factors give rise to distinctive soil types depending upon their surrounding environment (Hillel 2008). Examination of a vertical section of soil reveals the occurrence of individual horizontal layers. A well-developed soil profile comprises distinctive layers known as horizons (Brady and Weil 2004). The upper horizon is the A horizon or in popular parlance the topsoil, the zone of major biological activity. In this zone, a largely diverse multitude of microorganisms interacts with plants, animals, and the wastes generated by them. The A horizon is the most fertile part of the soil, rich in organic matter and nutrients. Underlying the A horizon is the B horizon (the subsoil), described as the zone of illuviation (washing in). It is the layer where materials such as migrating clay particles and soluble materials leached from A horizon manage to accumulate. The clay-enriched B horizon contains more clay, but less organic matter and it is thicker than the A horizon. Beneath this layer exists the undifferentiated zone called the C horizon (the parent material). It consists of fragmented and partially weathered rock material, transitioning between the original bedrock below and the soil above (Hillel 2008).

Soil is a dynamic living resource, which needs a minimal environment to carry out its essential functions of conservation, food production, and establishment of a quality environment (Doran and Parkin 1996). Soil produces food, raw materials, and biomass. It plays an important role as a gene pool and habitat and acts as a platform for human functions, landscape, and an archive of heritage. It conserves, filters, and transforms substances, such as nutrients, water, and carbon. It is in fact the largest carbon store in the world (Adriano et al. 1998). Quality of water is directly linked with soils; the soil is a buffering and filtering medium for contaminants. Several physical and chemical characteristics of soil result in clean groundwater for people and animals. Contaminants in soil limit its permeable surface by compaction and sealing, thus affecting its ability to act as a buffering and filtering medium for water. Presence of contaminants also causes extreme changes in soil pH which can drastically affect its retention capacity, triggering the sudden release of contaminants into the groundwater (Doula and Sarris 2016). Soil organic matter (OM) is essential in relation to soil fertility, sustainable agricultural systems, and productivity of crops. The volume of OM in soil depends on the input of organic material, its decomposition rate, the rate at which OM is mineralized, climate, and soil texture. The percentage of OM in clayey soil will be higher than in sandy soil, and in any soil type the OM will be higher for soils with permanent grass cover as compared to soils undergoing continuous arable cropping (Johnston et al. 2009). Soil is essential for construction and building activities. It is directly used to make building materials, such as brick and cement. It is also used to cultivate plants that supply other building materials like timber, wood boards, and insulation fibers. Soils with finer structures are more stable. Clay soils are more stable for construction as compared to sandy soils due to better structure (Joimel et al. 2016).

12.3 Construction Projects and Construction Materials: Extraction, Manufacture, and Transport

The use of raw materials for construction is increasing worldwide because there are lesser people living together in a household and there is a demand for new individual dwellings. Coupled together with the increase in houses, the population requires a certain standard of living which can be enabled by the construction of efficient and extensive roads, airports, and harbors, increased construction of schools, hospitals, shopping malls, office spaces, playgrounds, parks, underground tunnels, etc. (Douglas and Lawson 2003). This leads to an increased spate of building construction. The increase in construction-related activities has a dual impact on the environment. The fast-paced removal of raw materials from the natural environment changes the geomorphology of the natural landscape. Alongside the application of these raw materials and their processed by-products and end-products like concrete in the urban environment changes the urban morphology (Douglas and Lawson 2003). Due to the large number of construction projects happening in the world, construction-related activities have a negative impact on the environment. The operation and maintenance of a completed building produces a significant impact on the surrounding environment. Construction is not an environment-friendly process (Babak 2017). A large amount of energy and raw materials are consumed during the extraction of raw materials, processing and manufacturing, transportation, and construction. Demolition and disposal of wastes is also an energy intensive process (Nautiyal et al. 2015). C and D waste arise from construction activities, partial or complete demolition of constructed spaces like buildings or infrastructures, wastes produced during renovation and maintenance, constructing and maintaining roads, and debris arising during any disaster (Kumbhar et al. 2013; Silva et al. 2017).

The building construction sector is a major consumer of energy and raw materials and has a huge impact on the environment. It emits about 33% of the global carbon dioxide and thus is the second-largest emitter (Zhang et al. 2013). Construction of a building or any other structure requires various raw materials like masonry, sand, timber, and materials prepared through numerous manufacturing and processing techniques. Construction in early times was an ecofriendly process involving biodegradable materials like mud, rocks, and wood, but now it heavily utilizes man-made materials like concrete, bricks, plastics, and glass, which cannot be biodegraded easily (Nautiyal et al. 2015).

Concrete is used in all construction works and is prepared using aggregate, sand, and cement. To increase the strength of concrete, steel bars and fiberglass are used. Another type of concrete uses gypsum plaster with sand and cement. Bricks are important building materials. They are of two types, mud bricks which are air-dried, and fired bricks, which have been treated with fire to increase their durability. Masonry like rocks or stones are highly durable materials used for constructing walls and foundations. They are long-lasting, strong, and heavy and were used for construction in the past (Nautiyal et al. 2015). Cement is a vital building material that is used widely. Cement production factories are a chief source of pollution due to the emission of dust which contains toxic substances like heavy metals (HMs). Dust pollution affects the soil, vegetation, and air in the vicinity of the manufacturing plant. This dust settles on the surrounding soil and is spread further by wind and rain, contaminating the soil. HMs in soil are toxic for the plants and microbes growing in the soil, and the extent of toxicity depends on the metal concentrations. Cement dust settling on soil can negatively impact the soil microorganisms (Isikli et al. 2003; El-Sherbiny et al. 2019; Semhi et al. 2010). Cement dust comprises calcite, dolomite, quartz, and clay minerals, and HMs. HMs in dust were highest near the factory and decreased with distance (Semhi et al. 2010). The dust also contained a significant amount of chromium (Cr) and cadmium (Cd) because of its presence in the machinery (Isikli et al. 2003). Timber is a recyclable and compostable material. It is used for building doors, windows, frames, and furniture. In some places, entire homes are built with timber. They are also used for flooring in cold climates. Timber production causes large-scale deforestation and soil erosion.

Asphalt is used in the construction of roads, bridges, flooring materials, parking lots, and roofing materials. Tiles and ceramics are used as flooring, ceiling, and roofing materials. They are also used on walls. They are lightweight, have a good appearance, are scratch proof, and moisture resistant. They are manufactured using glass and metals (Nautiyal et al. 2015). Marble is highly used for flooring, and its cutting, polishing, and processing produces about 6 million tons of waste only in India. This marble dust is disposed of in riverbeds and reduces soil porosity and permeability, resulting in water logging. The dust remains suspended in the air during the dry season around the manufacturing and disposal site. It increases the soil alkalinity and settles on vegetation affecting their growth (Pappu et al. 2007). Asbestos is a mineral resistant to heat and degradation, used in products like asbestos cement, adhesives, automotive brake, insulation, roofing, and flooring material. Despite being banned in over 50 countries, asbestos-containing products remain intact due to past usage and asbestos production remains steady in other countries like China, Russia, and Brazil. Asbestos is present in dust due to anthropogenic activities which disturb rocks and soil like construction of buildings, roads, and mining (Noonan 2017). Glass is made from sand and silicates, and its manufacture consumes a high amount of energy. It is used in windows and doors, to cover walls, and as partition material. Quartz or silica is a common material found in soil, sand, concrete, granite, masonry, and landscaping materials (Li et al. 2019). Gypsum is used in the construction of wallboards and interiors. Fiber is used in insulation and ceiling system materials. Plastic and polyvinyl chlorides (PVC) are manufactured through polymerization. They are extremely lightweight and have high plasticity. They are used as covering of electrical wires, for carrying water, and in sanitary applications (Nautiyal et al. 2015).

A lot of construction material especially metals are obtained through mining, which causes soil and land pollution. Metals like iron (Fe), aluminum (Al), copper (Cu), lead (Pb), cadmium (Cd), nickel (Ni), zinc (Zn), mercury (Hg), and arsenic (As) are used in the manufacture of a lot of industrial and construction products. HMs like Cd, Cu, Pb, Ni, Zn, Hg, and As are the most hazardous soil pollutants. They are persistent in the soil environment and cannot be degraded. Metals adsorbed on soil can find their way into surface water bodies or groundwater. These further find their way into the food chain through crops grown on contaminated soils or through contaminated drinking water resources (Mulligan and Galvez-Cloutier 2003). The use of metals in the construction sector has increased over the years due to extensive construction activities being carried out. Metals are used to provide structural strength to constructed spaces. They are also used in electrical and sanitary fittings, household appliances. Steel is extensively used in the construction sector due to its durability, strength, and flexibility. Tin (Sn) and aluminum (Al) alloys are also used because they are lightweight and resistant to corrosion (Nautiyal et al. 2015). Natural Cd is found in shale deposits. Sources of Cd in the construction sector are steel plating, Ni–Cd batteries, pigments stabilization, Cd alloys, polyvinyl chloride plastic (PVC), rubber, solder, electroplating, motor oil, and dust produced in cement manufacturing. It could enter the soil through mining waste, disposal of C and D waste on landfills, landfill leachate, leakage from hazardous waste sites, and household waste. Cu is naturally found in sandstones. In the construction sector, it is used in building materials and has industrial applications due to its lightweight, and resistance to corrosion. It is used in roofs, domes, cladding of walls. Sources of Pb are paints, Pb–Zn smelters, batteries, glass, and piping. Sources of Zn are alloys of bronze and brass, rubber, products of galvanization, metal coatings, glass, smelting, and refining of ores, mechanical abrasion of tires, and cement industry (Mulligan and Galvez-Cloutier 2003; Al-Khashman and Shawabkeh 2006). Cr is released by ceramic materials and carbonated samples. Al production requires high energy requirements which put a huge burden on the environment. Alongside, many other pollutants like PAHs, PFCs, CO_2, and SO_2 are also released as by-products of the production process (Diotti et al. 2020).

Despite its economic benefits, mining produces large amounts of useless geologic materials containing a high amounts of toxic metals which are disposed of on-site. Mining activities degrade the soil and water resources and destroy the visual environment. Mining produces two types of waste: (i) mine tailings which are crushed mineral rock from which the desired metal has been extracted, and (ii) mine spoils which are waste rock materials not containing the desired metal. Both mine tailing and spoils containing toxic materials are disposed of in depression or open pits, making the site unstable and susceptible to weathering. Mining operations also give rise to acidic wastewaters when the extraction of metal requires acids, furthering pollution of soil (Artiola et al. 2019; Nguyen et al. 2021). Nguyen et al. (2021) studied soils at five sites near a mining area in Vietnam. They reported that all soils in the vicinity of the mining area had a higher concentration of HMs, and Zn, Cu, and Pb were the dominant HMs at all sites. The type of HM in soil varied with the mining site. Higher Pb was present at the site of clay and rock mining, while Cd and Zn were dominant at the site of clay and bauxite mining.

Pollutants produced in the metallurgical industry cause soil pollution when the waste product (sludge, slag, linings) containing heavy metals are dumped in the landfills. Processing, manufacturing, storage, and delivery of raw materials and end products, and disposal of wastes involves the release of contaminants. These contaminants can spread to wider areas attached to dust and particulate matter, and the area they will spread to depend upon the wind and precipitation factors. Pollutants produced in the metallurgical industry impact the soil, water, and air. Common pollutants are HMs, dust, H_2SO_4, HCl, toxic gases, polycyclic aromatic hydrocarbons (PAHs), and polychlorinated biphenyls (PCBs). The fumes and dust produced during the process also impact soil when they get deposited on the surrounding soil (Sofilić et al. 2013). The rate of pollution will depend on the material being extracted, type of pollutant produced, place of extraction, and the technical procedure applied (Sofilić et al. 2013).

12.3.1 Transportation

Environmental impacts related to transport are manifold: they arise due to vehicle use; production, maintenance, and disposal of the vehicle; and the production of building materials, construction, maintenance, and demolition of the roads, railway tracks, air field, bus, and railway station, and airport. The effect on the environment occurs not only in the vicinity of the road or airport but also at the site from where the raw materials have been procured (Douglas and Lawson 2003). Construction of transport-related infrastructure has additional requirements like the need for office buildings, gas stations, diesel depot, garage, and blocking posts (Babak 2017). Transportation of raw materials and other industrial products to the construction site requires a well-developed road network. All types of construction activities require a large amount of labor, which need to be transported to and from the construction site in vehicles, thus traffic in urban areas contributes significantly to HM pollution (Li et al. 2013). Transport of materials and labor to the construction site has an impact on energy consumption and the environment. Vehicles use energy and emit gases and particulate matter. The fuel consumption will depend on the distance of the construction site from the raw materials procurement site (Cole 2000). Vehicles emit pollutants through the leaking of oil and coolants, corrosion, and wear and tear of parts like tires, brakes, and engines. They release PAHs and Pb by combustion, Cu from the corrosion of brakes and radiators, and Zn from tire dust. The pollutants are spread by wind or water runoff from the road surface which collects the contaminants and transports them to the soil at the sides (Van Bohemen and Van De Laak 2003). Construction of roads has several environmental impacts like deforestation, loss of habitat and habitat fragmentation into small patches, erosion on the road surface, increased sediment runoff during precipitation, pollution of nearby water resources, and slope failure in certain cases (Caliskan 2013). During the construction of roads in a mountainous region in Turkey, Caliskan (2013) reported the destruction of 33–44% of trees by bulldozers.

12.3.2 Energy Requirements

Construction projects and related transportation requires a lot of energy supplied by oil, gas, and coal. The extraction of oil and coal burning for electricity also pollutes the soil. Crude oil drilling uses fluids of high density like sodium chloride, which are disposed of in the oil wells when they are spent. This produces highly saline land and water. Extraction of gas through hydraulic fracturing produces geologic water and sludge with high radioactivity. Combustion of coal for electricity produces fly ash, which is disposed of in ponds or landfills or used as filling material in construction. Depending on the source of the coal, fly ash contains different level of HMs, which can cause soil and groundwater pollution (Artiola et al. 2019). Coal-fired power plants produce coal slag containing HMs like Cd, Cr, Cu, Zn, Au, Hg, and Pb, which leach into the topsoil and contaminate it. Runoff water and erosion help the HMs in infiltrating the soil layers. Liu et al. (2017) studied HM contamination from a coal-fired power plant in Mongolia and reported the HMs concentration to be twice the acceptable concentration. The HMs contamination was also found to increase in the area with increased construction activity, probably due to change in soil pH or reduced absorption by the vegetation. They analyzed the ecological risk posed by the HM contamination and found it to be in the following order: Hg > Cd > As > Ni > b>Cu > Cr > Zn, with the probability of further increase if the construction activities continued unabated. Hg and Cd were high-risk elements while Ni, Pb, Cr, and Cu were medium risk and Zn was a low-risk element.

12.4 Wastes Generated During Construction and Demolition and Their Disposal

The construction and demolition (C and D) industry produces a huge amount of waste every year during the process of construction, demolition, renovation, and maintenance of buildings and infrastructure. Table 12.1 lists the amount of C and D waste produced in some countries. Approximately 10 billion tons of C and D waste is produced worldwide every year which due to its increasing volume poses grave

Table 12.1 Construction and demolition waste produced in different countries

Country	Construction and demolition waste produced per year (tons)	References
India	10–12 million	Ponnada and Kameswari (2015)
Canada	9 million	Yeheyis et al. (2013)
China	2.3 billion	Chen et al. (2020)
USA	700 million	
European Union	857.1 million	Rodríguez-Robles et al. (2015)

environmental impacts (Chen et al. 2020; Ginga et al. 2020). C and D wastes represent one of the biggest waste streams by weight and volume (Diotti et al. 2020). C and D wastes arise from new constructions in process, existing building structures, total or partial demolition of structures, maintenance activities, and road construction. They form a major portion of the municipal solid waste (MSW), generally comprising 20–30% and sometimes up to 50% of the MSW. The majority of the C and D waste is inert, without putrescible materials, since it mainly comprises bricks, concrete, sand, gravel, asphalt, masonry, gypsum, asbestos, and wood. It also contains plastics, metals, insulation materials, glass, cardboard, and paper (Qiang et al. 2015; Yeheyis et al. 2013; Silva et al. 2017). Wastes are generated in the construction phase of a building due to ordering of excessive material, mixing undue amount of materials like concrete or mortar which is ultimately not used, material which is not properly packed and thus spoilt like sand lime bricks, breakage of material during transportation, use of inadequate equipment, improper cutting for stone slabs or tiles which would result in them being wasted (Bossink and Brouwers 1996).

Waste soil during construction originates from the clearance of the building site, during excavation for building the base of buildings or roads, or sanitation works. The waste materials in soil arise from renovation and demolition work and the rubble remains in the upper layers of the soil. Urban soils contain transported soil, remnants of building material like brick, concrete, or paint, organic materials, ash, slag, and any waste released during building construction (Ottesen et al. 2008). Dust released during construction and demolition activities affects the environment and the health of the people living in the vicinity. All suspended and deposited particulate matter of up to 75 μm in size is called dust. Dust arises from silica and asbestos mining, cement manufacturing, marble, and granite processing, from bare soil produced during construction (Li et al. 2019). The production of concrete involves the release of pollutants like heavy metals, organic pollutants, CO, CO_2, SO, NO, and wastewater. Concrete waste is generated by breaking down of foundations, parking areas, driveways, buildings, sidewalks (Asif et al. 2007). Treated wood waste is generated by plywood that has been pressure/creosote-treated or laminated. Untreated wood waste arises from scraps, tops and stumps, and framing material. Timber in demolition waste can be used for the production of wood chips. The waste from the steel industry contains a high amount of iron and iron-containing alloys. This iron finds its way into the groundwater over time (Jhamnani and Singh 2009). The presence of gypsum in wastes releases a high concentration of sulphate (Diotti et al. 2020). Heavy metals are produced through various anthropogenic activities like mining and metallurgy, from constructed roads and residential complexes. Cd is produced in metal production, waste incineration, fossil fuel combustion, and cement manufacturing. Cr is released by metal processing and production industries, and cement manufacturing due to its presence in the rotary lining. Cu is released by wood and piping, while Ni arises from metal processing and electric and electronic waste. Pb arises from paints, old buildings, and metal processing, while Zn is released by metal processing, cement manufacturing, and vehicular emissions (Kasassi et al. 2008).

12.5 Disposal of C and D Wastes

It is difficult to recycle C and D waste at the end of the cycle because of their high chemical contamination level and heterogeneity. It is thus essential to prevent the generation of C and D wastes in the first place and then recycle whatever is produced (Bossink and Brouwers 1996). The disposal of C and D wastes involves an elaborate process of removal of reusable materials. C and D wastes are crushed, metals that can be reused are separated and undesired fractions are removed. C and D waste can be primarily reused as recycled aggregates instead of being disposed of in landfills. After screening, materials defined as recycled aggregates (RAs) are generated. Their composition varies with the waste composition. The RAs are mainly used in the construction of roads, pavement, and drainage, in place of fresh materials. Recycled aggregates are of several types like recycled concrete aggregate, recycled masonry aggregate, mixed recycled aggregate, reclaimed asphalt, and construction and demolition recycled aggregate, depending upon their principal component (Cardoso et al. 2016).

C and D waste are generally disposed of in landfills, and only a small amount is currently recycled. When disposed of in landfills they occupy large areas. Such wastes are heavy and bulky, and thus unsuitable for incineration or composting (Ponnada and Kameswari 2015). Landfills are the cheapest and simplest methods for waste disposal and a major proportion of wastes are dumped in landfills. Landfills are generally situated in existing holes like mines, to reduce costs involved in excavations. When landfill lining with minimum thickness is placed in such areas then there is very little barrier existing between the soil or groundwater and the wastes. Any crack in the lining leads to leakage of the leachate which contaminates the soil and water. Waste degradation in landfills takes a very long time, spanning over 20–30 years and more. During this time the lining of the landfill is susceptible to leakage (Allen 2001). Up to 80% of the demolition waste can be recycled on-site. In contrast, when they are dumped on landfills they are altered physically, mixed with other wastes, and become further contaminated, preventing any reuse or recycling (Cole 2000).

12.6 Effect of Construction Activities on Landscape and Soil Environment

All over the world, the number of people living in urban areas is very high. About 54% of the world's population lived in urban areas in 2014, and this percentage was up to 73% in regions like Europe (Yang and Zhang 2015). The rapid pace of urbanization has augmented the rate of resource consumption and the corresponding environmental degradation. The greatest threat of urbanization is on the soil in the urban areas (Xiang et al. 2020). Cities cover 2% of the land and produce up to 80% of the urban and industrial wastes which affects the urban ecosystem. Urban soils receive a high amount of these wastes which undermines their role in providing ecosystem services, and at this rate could convert the soil from a sink to a source of contaminants (Yang and Zhang 2015). Certain effects of construction activities on the environment are given in Table 12.2.

12.6.1 Impact of Construction Activities on the Landscape

Natural landscape changes on the earth's surface are brought about by the activity of water, ice, or wind. Construction activities can transform the landscape at a faster rate as compared to these natural agents and are thus the most efficient agents changing

Table 12.2 Negative impacts of construction activities on the environment

Stages of construction activity	Effect on the environment	References
Production, manufacture, and transport of building material	• Destruction of natural landscape and soil during extraction of raw materials • Formation of highly contaminated and acidic wastewater during extraction or mining process • Emission of toxic gases during procurement, manufacturing, and transport of raw materials • Emission of dust and particulate matter • Heat pollution • Noise pollution • Destruction of topsoil, which impacts the soil composition and its associated organisms • Increased soil erosion, acidification, and desertification • Increase in runoff and higher chances of flood • Impact on plant growth, habitat, and agriculture	Caliskan (2013) Babak (2017) Burghardt (1994) Allen (2001) Dziri and Hosni (2012) Gangolells et al. (2009) El-Sherbiny et al. (2019)
During construction	• Air pollution due to emission of gases • Emission of particulate matter • Industrial and municipal solid waste formation • Contamination and destruction of soil at the construction site and in the surrounding areas • Destruction of forested areas, water bodies, wetlands	Dziri and Hosni (2012)

(continued)

Table 12.2 (continued)

Stages of construction activity	Effect on the environment	References
During the lifetime of the constructed space, after demolition, and during disposal	• Disturbance of the water cycle due to limited percolation of water in areas covered by buildings or concrete • Change in the temperature of the area due to change in albedo, formation of urban heat island • Emission of greenhouse gases during combustion • C and D wastes comprise a huge fraction of the municipal solid waste • Leachate runoff from landfills, which contaminates soil and groundwater	Allen (2001) Jhamnani and Singh (2009) El-Fadel et al. (1997)

the geology and geomorphology (Douglas and Lawson 2003). Construction activities on steep slopes or on captured flood plains are responsible for economic and livelihood loss (Ferreira et al. 2018). The type and location of urban construction and surrounding land use greatly affect pollutant deposition. Isolated houses with gardens and urban drainage contribute higher pollutant load as compared to residential complexes due to greater garden and road surface area (Ferreira et al. 2018). The urban constructed areas comprise of different land uses like residential buildings, industries, businesses, traffic, parks, and gardens, which have different human activity and thereby a varied impact on the soil, resulting in a mosaic of soil types with dissimilar qualities. Land use and vegetation cover can be an indicator of the degree of pollution and disturbance in an area (Li et al. 2013). In a study on HMs contamination of urban soil in an old industrial city in China, Li et al. (2013), observed that pollution levels varied with the land use type. Industrial lands were the most contaminated followed by construction lands, and then roads. Soil contamination was higher in the vicinity of industrial areas, but pollution could spread to large areas due to wind and precipitation (Fazeli et al. 2019).

Land degradation is the reduced ability of the land to serve in functions like agriculture, construction, transport, etc. (Jie et al. 2002). Deforestation is the removal of large tracts of forests for agriculture, construction, building roads, and mining purposes (Jie et al. 2002). Forested areas are cleared to make space for mining activities, or roads, or to harvest timber needed for construction or as a fuel source. Removal of forest decreases soil stability and makes soil vulnerable to erosion by wind or water. During precipitation, an increased rate of runoff can erode the soil and carry large amounts of sediment downstream. Loss of vegetation leads to soil degradation, desertification, loss of biodiversity due to habitat fragmentation, diminished air quality, and an increase in temperature (Artiola et al. 2019).

12.6.2 Impact of Construction on Soil Environment

Soil composition depends on the initial parent material of the soil, which is acted upon by factors like climate, biota, and topography. Soil is a product of chemical and physical leaching, oxidation, and dissolution of the parent material (Olson 2004). Soil quality is defined by the ability of soil to carry out its essential functions and provide ecosystem services. It is determined by the fertility and the contamination status of the soil (Joimel et al. 2016). Soil is an important natural resource because it acts as a geochemical reservoir for contaminants like heavy metals arising from construction and industrial activities (El-Sherbiny et al. 2019). Any change in the soil quality is dependent on the type of anthropic activities carried out on the soil. Man uses the land to obtain resources and products from it. This usage of the soil environment could be through agriculture, forestry, urbanization, or industrialization, each of which brings about a modification in the natural state of the soil and landscape. The rate of change related to these activities will depend on the duration and intensity of the anthropogenic activity (Joimel et al. 2016). Soil contamination is the accumulation of toxic substances like heavy metals and organic pollutants in the soil (Sharma 2017), at concentrations which would impair the normal functioning of the soil, disrupt vegetation and the biological cycling of the nutrients (Scullion 2006). Soil contamination could also affect the groundwater quality through the percolation of contaminated leachate and pose a danger to aquatic ecosystems, and human health (Scullion 2006).

Soil degradation refers to the loss of the soil's ability to produce plants, which results in food insecurity. Soil degradation occurs through the process of erosion (loss of soil), compaction (decrease in soil space due to mechanical stress), depletion (loss of soil organic matter, fertility, and biological material), and accretion (addition of pollutants to soil, acidification, or alkalinization) (Jie et al. 2002). Soil degradation is a natural process like soil formation which takes place over many years, due to factors like erosion, desiccation, salinization, and compaction of the soil. All these factors are very slow and there is enough time for the regeneration of soil. But a new cause of soil degradation, urbanization is rampant now, which is very fast-paced and does not allow recovery time to the soil. Rapid urbanization is a major cause of soil degradation because of the high population of people living in urban areas (Jie et al. 2002). Anthropogenic activities alter the distribution pattern of materials in the ecosystem. Urban land use can diminish the soil capacity of storing toxic substances like heavy metals and PAHs. It can also impair the ability of the soil to provide ecosystem services. This affects the availability of water, food, and energy to the urban population, and increases their vulnerability to natural hazards (Burghardt 1994).

12.6.3 Formation of Urban Soil

Soil pollution in urban areas is very complex and can pose a serious health risk to human beings. Due to more pollution sources in the urban environment, a large number of pollutants find their way into the soil (Yu et al. 2019). In urban areas, anthropogenic activity is a prominent factor controlling soil quality. Change in land use in urban areas occurs through the construction of residential buildings, schools, playgrounds, etc. The construction of these complexes involves the sealing of the soil surface and sub-surface, construction of underground drainage systems, and laying of pipes for water supply, which gives rise to many watersheds. These activities can alter the percolation of water into the soil and evapo-transpiration, and modify the ability of soil to accumulate toxic substances (Burghardt 1994). Urban soils are comparatively younger due to the mixing of soil layers, and due to the addition of new materials during construction. Modification of urban soil can occur through soil sealing, compaction, excavation and relocation of soil, mixing and filling with other materials (aggregates, ash, building rubble, slag) during construction, contamination (due to construction work, waste disposal, de-icing salts, leakage), and desiccation (lowering of groundwater levels due to reduced water inflow) (Niemelä and Sauerwein 2013). They are excavated, transported, and mixed with other soils on site, causing variation in soil from one place to another, depending on the land use (Li et al. 2013; Meuser 2010). Urban soils are a mixture of anthropic (altered by human influence) soil bodies and unaltered natural soil bodies, together classified as technosols. Soil type is not predetermined, it develops over time depending on the parent material and geogenic and anthropogenic processes. Anthropogenic activities like building construction and demolition, roads, extraction or addition of man-made aggregates (concrete, mortar, asphalt), excavation, and disposal of soil, all impact soil formation and composition (McClintock 2015).

12.7 Effect of Construction on Soil pH, Texture, and Nutrients

Soil pH and texture are dependent on the parent material, land use, and type of anthropogenic activity, and its intensity (Romzaykina et al. 2020). Soils affected by anthropogenic activities display increased pH levels. Kosheleva et al. (2018) reported an increase in pH to 7.5–8.6 in urban areas, caused by the presence of dust, ash, and the dissolution of technogenic materials. Alkalinization is caused by dust deposits from the manufacturing of cement or concrete, processing of marble, and gravel. Disposal of waste and atmospheric deposition from traffic or industries also increases the pH (Joimel et al. 2016). Marble and cement production releases a lot of dust which settles on the surrounding soil and the disposal site, turning the soil alkaline and reduces soil pore size and permeability (Pappu et al. 2007). Urban soils contain a lot of technogenic materials (debris of bricks, rubble, ash, slags, plastics, metals, etc.) other than their original weathered parent material. The presence of sand, gravel, and other demolition wastes in the soil makes the soil more sandy and reduces the fine earth content. Biasioli et al. (2006) reported high pH (up to 8) and high sand content (439–889 g/kg) in urban soils. The high sand content was due to the presence of extraneous coarse materials. Demolition waste like ash, slag, and rubble also change the shape of soil. Soils had a higher fraction of coarse materials and little effect of the parent material. The change in soil texture and shape reduces the water storage capacity of soils and increases the concentration of air-borne contaminants, further enhancing the flow of contaminants into the groundwater. The particles from demolition materials are sharp edged and increase the shear resistance of soil. Their presence hinders the growth of plant roots and occurrence of earthworms. They also prevent the movement of organic substances in the soil, thus limiting the organic matter (OM) to the surface layers. Soils with high OM can bind HMs, preventing their movement into the groundwater. Urban soils also contain a higher amount of black carbon probably contributed by coal fly ash (Burghardt 1994; Yang and Zhang 2015).

Urban activities alter the nutrient content of the soil, especially increasing the P concentration due to added depositions (Joimel et al. 2016). P enters the urban soil through organic manure or sewage sludge and is harvested when plants are cultivated in the soil. But in the case of urban soils, P is not recycled and remains at a very high concentration in the urban environment. This increases the chances of water pollution. Urban environments also alter the carbon and nitrogen content, and soil temperature (Romzaykina et al. 2020).

12.7.1 Physical Degradation of Soil Due to Constructed Spaces

Construction of buildings or roads, airports, or railway tracks causes the sealing and compaction of soil due to the use of heavy machinery and due to the increased weight on soil. Intentional compaction is carried out for building foundations of houses, paving roads, and using heavy machinery for sloping banks on roads or hillsides. Unintentional compaction occurs due to the high traffic burden on soil (Yang and Zhang 2015). Sealing and compaction of soil causes high horizontal and vertical variability in the soil profile depending on the type of anthropogenic influence (Biasioli et al. 2006). Soil compaction causes the physical deterioration of the soil by destroying soil structure, air porosity, reducing the storage capacity of soils, which reduces the movement of water through the soil. It also affects the biological activity in the soil and decreases biomass productivity. Compacted soil restricts the movement of water and air and restricts the growth of roots. Such soils are prone to erosion due to the presence of scarce vegetation (Jie et al. 2002).

Constructed spaces utilize extensive amounts of impermeable substances like concrete, asphalt, and tiles. These materials seal the soil and limit its permeability.

Sealing is a permanent process that impacts the soil properties, limits percolation of water and groundwater recharge, inhibits carbon sequestration, disrupts biogeochemical cycling of nutrients, and causes biodiversity loss. It also results in the formation of urban heat islands because sealed areas absorb increased radiation as compared to vegetation (Niemelä and Sauerwein 2013). Compaction and sealing of soil results in decreased water infiltration and reduced groundwater recharge, increased runoff and increased urban floods, higher pollutant load in surface water bodies due to runoff, formation of urban heat island, increased soil temperature and decreased microbial activity, and weakened plant growth (Yang and Zhang 2015).

Soil erosion is a prominent environmental concern worldwide. Soil erosion depletes the fertile layer of topsoil, degrades the land, and increases siltation in low-lying bodies of water. Sediment loss from construction activities is 95% more than that of forested land. In constructed spaces, soil erosion occurs on bare soils deprived of vegetation, on edges of the road, and on railway tracks. This happens because the soil profile is disturbed, and soil is prone to erosion due to rain or runoff water. Gully erosion in urban areas occurs along construction sites, areas of road construction, and urban drainage (Ferreira et al. 2018).

12.7.2 Impact of Construction Activities on Water in the Urban Environment

Constructed spaces modify the water cycle by altering the soil properties. Soils in urban areas are compacted, which reduces their pore space and water holding capacities. Urban soil surfaces are sealed with concrete or asphalt, and vegetation-free, which hinder water percolation into the soil. This results in an increased amount of run-off. The run-off is collected through artificial drainage channels and transferred to streams or sewers. Reduction in water infiltration leads to decreased groundwater recharge and increased surface runoff. Surface runoff in small catchment areas can lead to flash floods during high-intensity rainfalls. Sealing of the soil surface also reduces the evapo-transpiration rates (Yang and Zhang 2015). Pollutants from the urban soil environment like HMs, PCBs, and PAHs, can make their way to water bodies along with storm water runoff from these systems. Pollutants in water are also derived from landfill leachate, sewage systems, drainage of water from urban gardens, soil erosion, and atmospheric deposition. Urban runoff is a major non-point source of pollution due to its high level of contamination with HMs, PAHs, and P. Runoff water contains a high concentration of HMs collected from rooftops, roads, and paving materials. Road runoffs have significant HMs contribution from vehicular emissions. These pollutants can deteriorate the quality of surface and groundwater thereby impacting drinking water and irrigation water sources. Pollutants and sediment from urban areas also affect the aquatic ecosystems. Sediments eroded by the storm water runoffs in urban areas are 60 times higher than sediment runoff from vegetation-covered surfaces. They enhance turbidity, reduce light penetration, inhibit photosynthesis, decrease dissolved oxygen, and cause eutrophication (Ferreira et al. 2018). Impervious urban surfaces enhance the presence of contaminants like HMs, PAHs, and PCBs in the storm water runoff. These pollutants present in the urban soil could be remobilized due to erosion caused by precipitation. Kartun et al. 2008 studied runoff sediments from storm water traps in Norway. The concentration of Pb, Zn, and Cd in the sediments was 9–675, 51.3–4670, and 0.02–11.1 mg/kg respectively. The PCBs between <0.0004 and 0.704 mg/kg, and the PAHs concentration range was <0.2–80 mg/kg. They reported the origin of runoff sediments to be weathering of buildings, construction and demolition activities, and traffic. A high concentration of contaminants in the urban soil was dispersed into sediments through stormwater runoff from the impervious surfaces.

12.7.3 Impact of the Urban Environment on Soil Microorganisms

Urban sealing decreases soil microbial activity, and the influence on soil microbes depends on the land use type. Soil compaction and trampling may destroy the microbial communities existing in the upper 10 cm layer rich in organic matter. The presence of microbes in the soil will further depend on the N and P content of the soil, as they affect enzyme synthesis and processing of organic matter. Microbial activity is an important indicator of soil contamination levels. Contaminated soils have a diminished microbial mass (Yang and Zhang 2015; Havugimana et al. 2017).

12.7.4 Heavy Metals and Organic Pollutants in Urban Soils

Urban soils are the most important sinks of chemical emissions (Ferreira et al. 2018) and serve as indicators of human exposure to contaminants in the soil environment (Li et al. 2013). Soil contamination in constructed environments occurs due to emissions from residential buildings, factories, transportation, fossil fuels, and disposal of C and D waste (Biasioli et al. 2006). End-products from the construction and demolition sector also contain trace amounts of different contaminants which could contaminate the underlying and surrounding soil and groundwater (Chowdhury et al. 2010). Early soil pollution was mostly due to inorganic pollutants, but there has been an increase in pollution with organic pollutants (polycyclic aromatic hydrocarbons) as well caused by fuel hydrocarbons (Scullion 2006). Pollutants in urban soils are inorganic (heavy metals (HMs) or potentially toxic elements (PTEs)), organic (polychlorinated biphenyls (PCBs), polycyclic aromatic hydrocarbons (PAHs), and phthalate esters), chlor-organic compounds, and radionuclides (Biasioli et al. 2006).

Heavy metal pollution of the soil is a widespread environmental hazard that damages the soil quality, deteriorates water quality, decreases food production, and leads to the production of unhealthy food crops containing a high concentration of HMs. Heavy metal pollution has now become a regional, compound, and non-point source of pollution (Xiang et al. 2020). Heavy metals in urban areas arise from constructed environments, industrial activities, traffic emission, and atmospheric depositions. Soils close to roads, cement factories, are enriched with HMs. HMs are non-biodegradable and thus persistent in soils for several years even if the source of pollution is removed, and difficult to remediate. They are highly toxic even at low concentrations and have a long residence time in soil. They affect plant growth and soil microbes, food web functions, and disrupt biogeochemical cycles (Kasassi et al. 2008; Li et al. 2013; Liu et al. 2017; Nguyen et al. 2021). HMs in urban environments reach human beings through food cultivated in urban polluted soils, through ingestion of soil, through inhalation of dust-laden with HMs, and dermal contact when working in contaminated soils. Factors affecting HMs mobility are pH, organic matter content, and soil texture. HMs display complex spatial distribution in urban soils and the distribution is affected by soil compaction and zoning (Romzaykina et al. 2020).

Heavy metals concentration in urban areas is always greater than in rural areas due to general and diffuse contamination of soils (Biasioli et al. 2006). In Torino, Biasioli et al. (2006) observed high concentrations of Ni (209 mg/kg), Cr (191 mg/kg), Pb (149 mg/kg), and Zn (183 mg/kg) in the urban soils. Pb, Cu, and Zn were associated with traffic sources, Pb with leaded fuel, Cu from brakes, and Zn from tires. Other sources were probably piping, cables, paints, and incinerators. They concluded that large cities strongly influence soil properties and concentrate HM pollutants in the urban soil. Lead, a persistent HM arises from paint, gasoline, waste incineration, and industrial sources. Pavilonis et al. (2020) reported the presence of Pb in soil despite the ban on Pb in fuel and the absence of smelters. They found old houses painted with Pb-based paints to be the most important source of Pb in soil. An increase in population in the area, and renovation and demolition of old buildings caused the release of Pb-containing dust, which settled on the surrounding soil. A positive correlation has been observed between increased Pb concentration in soil and areas with older houses, which also caused an increase in the blood Pb levels of children.

Vegetation in urban areas acts as a surface for the deposition and immobilization of dust particles. Type of land use significantly affects the HMs accumulation in the urban soil (Xie et al. 2019). This was observed by Xia et al. (2011) in Beijing. They analyzed the concentration of Cd, Cr, Cu, Ni, and Pb in soil samples from gardens, residences, business areas, roadside, and schools. All soil had a higher concentration of Cd, Cu, Pb, and Zn due to anthropogenic effects. Among the land uses, the classical garden had the highest HMs concentration due to historical contamination with wood preservatives, pigments, and brassware. The roadside soil had the second-highest concentration of Cd signifying its origin from traffic sources. HMs concentration was higher in lands used for a longer duration, and in lands nearer to the city. Xie et al. 2019 investigated 115 soils from residential areas in Beijing and observed Cu, Cd, Pb, and Zn to be the most common contaminants in urban residential areas. The HMs concentration was significantly correlated with the age of the buildings, population, and distance to the city center, while building height and green cover did not play a significant role.

Building orientation also affects the deposition of contaminants. High-rise buildings transform the direction of wind creating wind shades in closed areas and canyon effects on large open streets. In closed areas decrease in wind speed causes the deposition of dust-containing pollutants, but in open areas, the dust is blown away. Urban soils in a residential area in Moscow were found to accumulate higher HMs concentration as compared to control soils. The soils contained elevated concentrations of Cd (1.8 mg/kg), Zn (190 mg/kg), Cr (98 mg/kg), Pb (50 mg/kg), and Cu (61 mg/kg). The building orientation affected soil contamination in two ways: (i) they acted as protective screens and disrupted the wind pathway, preventing the deposition of contaminants laden dust in the backyards. This led to a decrease in soil concentration of As, Pb, Co, and Sb. The areas near the building had the least concentration of these elements in the soil, and the concentration increased with increasing distance from the building. This protective effect was also dependent on the height of the building (average height 20.7 m). The concentration of Cr, Mo, Ni, and Sn decreased between 1.3 and 1.9 times in the area between high buildings; (ii) densely built high-rise buildings can act as dust traps and increase the contaminants concentration in the inner areas. They sharply decrease the wind speed which leads to dust deposition. This effect caused an increase in the concentration of Cd and Cu in the soils (Kosheleva et al. 2018).

HMs concentration in urban soil is an important geochemical tracer to identify anthropogenic influence. Al-Khashman and Shawabkeh (2006) studied soils near a cement manufacturing plant in Jordan to determine the impact of human influences on HMs presence. They collected soil from the surface at depths of 0–10 and 10–20 cm. They observed high concentrations of Pb, Zn, and Cd in the soils which they attributed to the traffic emission and emission from the cement factory. The metals were also found to be concentrated in the surface layers and decreased with increasing soil depth. Soil contamination was observed in playgrounds in Colombia and Norway by Donado et al. (2021) and Ottesen et al. (2008), respectively. Donado et al. (2021) observed high concentrations of arsenic (As) (26 mg/kg), Ni (10.8 mg/kg), Cu (39 mg/kg), and Pb (89 mg/kg) in the playgrounds. They determined the source of the HMs to be intense construction and industrialization, coal combustion, smelting, and traffic sources near the playground. Ottesen et al. (2008) reported high concentrations of Pb, Cd, Hg, Zn, and PAHs in the surface soil due to the effects of urbanization. The soil at depth of 4–5 m was relatively less contaminated. Higher HMs contamination was observed in soils from old and densely populated areas as compared to the younger settlements. High concentrations of As were seen in playgrounds making use of Cu–Cr–As (CCA) treated wood. Soils have been known to act as geochemical barriers and buffer HMs contamination. Romzaykina et al. (2020) undertook soil studies in Moscow city to determine the HMs concentration in urban soil and the ability of different soils to act as a buffering agent. HMs like Cu, Zn, Pb, Cd, and As were found to be 2–5 times higher than the background values and originated from industrial and traffic sources. Soils were found to be more effective barriers for Ni and Pb as compared to Zn, Cd, and As. Alkaline soils with loamy texture in the city center were better buffers than acidic or loamy sandy soils or soils covered with vegetation.

All PCBs are man-made, there are no natural ones. They were used in buildings in concrete, plaster, joint sealing materials, glue, and paint. High levels of PCBs have been found in building caulking material and in expansion joints of old buildings. They were also used in polyvinyl acetate (PVA) to improve endurance, flexibility, and resistance to erosion in concrete and plaster. Herrick et al. (2007) reported PCBs concentration in buildings with undisturbed caulking to be 10,000–36,200 mg/kg. The corresponding soil PCBs around these buildings ranged from 3.3 to 34 mg/kg. Buildings with the highest PCB concentration contributed the maximum PCB to the soil. The highest PCB concentrations were in the soil near the buildings, and the concentration decreased with increasing distance. They concluded that weathering of caulking could be an important factor contributing to the soil PCB concentration, which are leached as complexes with OM. Andersson et al. (2004) reported a similar finding on PCB emission in building materials in Norway. PCB in the building plaster ranged from undetected to 290 mg/kg, while the corresponding soil contained up to 320 mg/kg. In this study, the PCBs used in the buildings displayed a greater presence in the corresponding soil, contributing to high levels of soil contamination. The PCB level decreased with distance from the building and depended on the type of building use. Offices and industrial buildings had lesser soil PCB while residential buildings and schools were more contaminated. This phenomenon was explained by the renewal of soil and refurbishing of exteriors in the residential areas and schools. The PCB was retained to a greater extent in these soils due to their high OM. Caulking materials and plasters are generally disposed of in landfills, which could cause high levels of PCBs in the leachate, and therefore need to be reconsidered prior to disposal (Andersson et al. 2004; Herrick et al. 2007).

PAHs originate from the combustion of fossil fuels, vehicular emissions, waste incineration, urban wastewater drainage, and manufacture of chemicals, while phthalate esters arise from emissions from construction materials, furniture, plasticizers, industrial processes, and cosmetics. They are attached to particulate matter due to their hydrophobic nature and are deposited on urban soils through long-range transportation and deposition of particulate matter. They are known carcinogens and have low biodegradation in soils (Fazeli et al. 2019; Ferreira et al. 2018). A study was carried out by Fazeli et al. 2019 in an urban area in Tehran to determine the source of PAHs and their co-occurrence with HMs. The anthropogenic fraction in the metals was in the following order: Cd (87.87%) > Cr (40.14%) > Ni (38.64%) > Pb (37.40%) > Co (8.42%) > Cu (7.98%) > Zn (5.57%) > Mn (3.94%). The PAHs in the samples ranged between 0.62 to 3.51 mg/kg, had high molecular weights, and originated from the combustion of petroleum, traffic, and heating systems in houses. The authors reported that both PAHs and metals shared a common origin based on their spatial distribution.

Human activities can increase exposure to natural sources of radiation. Materials used in the construction industry are usually contaminated with natural radioactive materials like radium, thorium, and potassium. Certain raw materials like granite and slate contain a higher content of radionuclides than soils. These materials like granite or marble are used in flooring, and others like kaolin, feldspar, and zircon are used as mixtures. The presence of a significant amount of radioactive materials in buildings can increase radiation exposure. The presence of natural radionuclides in raw building materials is controlled by soil composition and source, density, water content, permeability rate. As compared to undisturbed earth's crust, building materials can increase the gamma exposure due to their geometry. Some building materials and existing buildings can increase the radon generation in closed spaces despite the fact that radon is emitted mainly from the soil (Popovic et al. 1996; Todorovic et al. 2015).

12.7.5 Impact of Construction and Demolition Waste on the Soil During Disposal

Table 12.3 summarizes some of the construction-related effects on soil. Waste during the construction phase arises due to misuse or mishandling of materials or due to the procurement of excessive raw materials which are not utilized ultimately. Demolition

Table 12.3 Negative impacts of construction activities on the soil

Type of land use	Soil degradation process	References
Existing residential buildings	• Destruction of natural vegetation • Increased soil compaction • Loss of soil microbial activity and fertility • Increased stormwater runoff due to sealed soil surface Increase in flash floods • Decreased in water percolation to groundwater aquifers • Increased evapotranspiration	Ferreira et al. (2018) Jie et al. (2002)
Roads, railways, airports, and transport-related structures	• Soil degradation and compaction due to use of heavy machinery • Alteration in water runoff due to the sealing of soil by asphalt or concrete, or recycled aggregates	Babak (2017) Douglas and Lawson (2003)
Mines, smelters	• Soil acidification or alkalization • Biological degradation	Artiola et al. (2019)
Landfills	• Soil compaction • Nutrient loss and fertility reduction • Destruction of naturally thriving microbes • Soil and groundwater pollution from leachate loss	Jhamnani and Singh (2009)

waste is generated through the breaking down and removal of old unusable structures (Ginga et al. 2020). C and D waste is an environmental issue worldwide, due to multiple negative impacts like contamination of soil and water resources. The impact is observed in economic terms too due to the loss of resources like raw materials and energy consumption. It affects the gross domestic product (GDP) of the country and inhibits tourism. It also affects socially by posing potential health risks. Though much of the C and D waste is inert, it contains various toxic substances like heavy metals, organic pollutants, and organic matter, which pose potential health risks. Despite the presence of only a small amount of such pollutants in the C and D waste they are a major concern for the environment due to their susceptibility to microbial activity and release under acidic conditions. These wastes also undergo microbial breakdown in landfills, releasing gas and leachate which further impact the soil (Qiang et al. 2015; Rodríguez-Robles et al. 2015). Heavy metals in C and D waste originates from sources like hazardous building materials (paints containing heavy metals or wood treated with preservatives), heavy metals contaminated soil, and from leaching of toxic elements (Qiang et al. 2015)/(Townsend 2004). The presence of these pollutants in the C and D waste has a negative impact on their re-use. But the removal of these toxic wastes at the site of demolition requires an ample amount of time and effort on part of the management (Qiang et al. 2015). Percolation of water through the wastes in landfills results in leachate carrying dissolved or suspended contaminants, and even high organic matter content in case of new landfills. The composition of the leachate depends on the type of waste and its composition, age, temperature, oxygen availability, and presence of water near the landfill. This leachate migrates into the underlying soil and groundwater and pollutes it, which not only disturbs the natural state of soil but also acts as a potential health risk. Landfill leachate also contains a high amount of metals like Mn, Cu, Pb, Zn, and Cr. Landfill soil contamination with heavy metals occurs due to the leachate flow from untreated waste disposal, and due to incinerated waste as well (Kanmani and Gandhimathi 2013; Dregulo and Bobylev 2020).

12.7.6 Risk Assessment

HMs arising from anthropogenic activities like construction are a prominent source of environmental pollution since they affect the ecology. Determination of HMs in urban soils helps in monitoring pollution (Al-Khashman and Shawabkeh 2006). Risk assessment is done to identify the potential impact of HMs on the soil. It is calculated by the Contamination factor (CF) and Geo-accumulation index (GI). CF is used to study the source and level of contamination in the soil. It is calculated as;

$$CF = \frac{C_i}{B_i} \tag{12.1}$$

where, C_i is the metal concentration in soil and B_i is the background value in absence of any anthropogenic influence. CF < 1 indicates low level of contamination, 1 < CF < 3 indicates moderate contamination, 3 < CF < 6 represents contaminated soil, and CF > 6 denotes highly contaminated soil (El-Sherbiny et al. 2019).

GI helps identify the source of contaminant and the degree of contamination. It is calculated as;

$$GI = \log_2(C_n/1.5B_n) \tag{12.2}$$

where, C_n is the metal concentration in soil, B_n is the background value of the metal in soil, and 1.5 is the constant used for potential variability in reference value. GI was divided into seven classes by Muller: $GI \leq 0$ (unpolluted soil), $0 < GI \leq 1$ (unpolluted to moderately polluted), $1 < GI \leq 2$ (moderately polluted), $2 < GI \leq 3$ (moderately to strongly polluted), $3 < GI \leq 4$ (strongly polluted), $4 < GI \leq 5$ (strongly to extremely polluted), GI > 5 (extremely polluted) (Ihedioha et al. 2017).

12.8 Effect of Construction Activities on Urban Agriculture and Health Risks Due to Contaminated Soil

Urban agricultural projects have been expanding since the beginning of the millennium, from aquaponics, greenhouses, to permaculture, with various alterations in the way food is grown. Urban agriculture plays a positive role in the environment, economics, society, and nutrition, but it is crucial to identify the potential health risks associated with agricultural production in urban settings (Aubry and Manouchehri 2019). Urban soils are influenced by multiple factors. Toxic substances end up in soils as a result of anthropogenic activities. Some substances such as fertilizers and pesticides are deliberately added to soils to improve crop production, while others like industrial and commercial chemicals cause contamination through leakage or spills. Contaminants can also spread through the air and reach soil by precipitation or dust settlement (Turner 2009). Humans in close contact with contaminated materials and soil are at an enhanced risk of potential health issues (Swartjes 2015).

12.8.1 Effect of Dust and Heavy Metals on Plants

Cement manufacture produces a lot of dust which spreads to a wide area due to aeolian activity. This dust pollution affects photosynthesis, respiration, and transpiration in vegetation growing in the vicinity of the manufacturing plant. Dust can settle on the leaf surface and alter growth and stomatal opening, or it can get absorbed into the plant and inhibit essential nutrients uptake. When this dust settles on plants it disrupts the physiological and biochemical processes in plants. It induces the production of reactive oxygen species, resulting in oxidative stress in plants. Cement dust contains calcium hydroxide which denatures the leaf proteins (Semhi et al. 2010). Dziri and Hosni (2012) reported a decrease in essential oil production in pine needles when the pine plants were exposed to cement dust. They concluded that oil production was the most sensitive biochemical process in pine, which could be a result of reduced photosynthesis caused by dust covering the stomata and photosynthetic tissues, or due to an increased rate of senescence. Dust produced by the marble industry also affects the surrounding soil and vegetation, impacting plant growth due to water logging and alkalinity (Pappu et al. 2007).

Background concentrations of HMs in the soil are due to their geogenic origin, but anthropogenic influences increase their concentration, making them harmful for plants (Chibuike and Obiora 2014). These anthropogenic activities include fossil fuel burning, smelting of metals, mining, construction, municipal solid waste disposal, and usage of fertilizers and pesticides (Alloway 1995). In recent years, studies have focused on HMs contamination of agricultural soil and related health risk assessment (Kong et al. 2021). Heavy metals commonly found at contaminated sites are Pb, Fe, Al, Cr, As, Zn, Cd, Cu, Hg, and Ni (Subhashini and Swamy 2016). The effect of HM on the growth and development of plants differs with the HM involved. Elements such as Pb, Cd, Hg, and As, which do not play any beneficial role in plant growth have severe effects on the plants at very low concentrations in the growth medium (Mohnish and Nikhil 2016). Cu, Fe, Al, and Zn are essential components in plant nutrition and development. However, they induce extreme phytotoxicity at high concentrations, significantly affecting plant growth, mineral uptake, and photosynthetic activity (Abdus et al. 2016). Due to alteration in physiological and biochemical activities, plants growing on HMs polluted soils display reduced growth (Chibuike and Obiora 2014). This growth reduction can be attributed to decreased photosynthetic activity, altered mineral nutrition, and reduced activity of some enzymes (Kabata-pendias and Pendias 2001).

Soil is the most common repository for airborne pollutants, but it cannot be considered a permanent sink. Contamination of soil enhances the opportunity for contaminants to be absorbed and recycled into the human food chain through agricultural activity, home gardening, and grazing animals (Swartjes 2015). Several research findings suggest that the presence of HMs in the soil beyond a certain limit results in toxic effects on plants, animals, and soil micro-organisms (Mohnish and Nikhil 2016). The presence of metals and metalloids like As, Cd, Ni, Pb, Cr, Cu, Mo, and Se in the soil produces an antagonistic effect on the yield and nutritional quality of food crops (Ahmad et al. 2018). In order to maintain the ecological harmony of our planet, there is a need to understand the interaction of HMs with plants. HMs produces a negative effect on plants, and plants have an inbuilt defense mechanism to protect against such toxic effects. Mahmood and Islam (2006) researched the effect of Cu, Zn, and Pb on seed germination and seedling growth of barley (*Hordeum vulgare*) rice (*Oryza sativa*), and wheat (*Triticum aestivum*). The inhibitory effect of Cu on seed germination was more pronounced in rice than in either wheat or barley. The roots of the seedlings grown in control or at lower concentrations of Cu, Zn, and Pb were white, long, with abundant root hair, and long lateral roots. But at higher concentrations the root tips of all the seedlings turned brown, roots were hairless, stunted, thick, curled, and with numerous small lateral roots.

12.8.2 Indirect Health Risk from Crops Cultivated in Contaminated Soil

Employment of contaminated soils for agriculture and consumption of the resultant crops poses potential health risks for human beings and increases the probability of contaminants entering the food chain (Rutigliano et al. 2019). Agricultural land can be contaminated by dry and wet deposition of atmospheric pollutants from urban activities like construction and industrial activity (Kabata-pendias and Pendias 2001). Other possible sources of contamination are the disposal of C and D waste on soil, leakage of leachate into soil and groundwater, sewage-derived materials, and agricultural practices such as the addition of fertilizers and pesticides (Liu et al. 2013).

Usage of wastewater for irrigation has increased in urban areas. Though it helps in reducing groundwater extraction for irrigation and contains a high amount of essential nutrients and OM (Zhang and Shen 2017), this practice has several drawbacks (Murtaza et al. 2010). Wastewater contains HMs such as Zn, Cr, Cu, Cd, Ni, Pb, and Hg, which can contaminate the environment and induce severe health risks in human beings (Khan et al. 2008). The usage of untreated wastewater for irrigation can result in soil and groundwater contamination, and cause hardening of soil (Liu et al. 2017). The contamination of agricultural soils with HMs can have negative implications on human health for a very long period because HMs is persistent and non-biodegradable in soil (Khan et al. 2008). Soil quality should be monitored continuously to control the health risks arising from the consumption of crops contaminated with toxic elements (Piekut et al. 2018).

12.8.3 Daily Intake of Metals (DIM) and Health Index Risk (HRI)

The DIM is determined by the following equation (Khan et al. 2008),

$$\text{DIM} = C_{\text{metal}} \times C_{\text{factor}} \times D_{\text{food}}/\text{B.W.} \tag{12.3}$$

where, C_{metal}, C_{factor}, D_{Food}, and B.W. represent the HM concentration in plants, conversion factor, daily intake of vegetables (grams), and average body weight, respectively. A conversion factor of 0.085 was used to convert the fresh weight of the food sample to dry weight.

To evaluate the health risk of HMs, it is important to calculate the level of exposure to that metal by tracing the route of exposure. Health risks arise when food crops enriched with a higher concentration of HMs are consumed by the receptor population (Ávila et al. 2017).

$$\text{HRI} = \text{DIM/RfD} \tag{12.4}$$

where, DIM is the daily intake of metal, and RfD is the reference oral dose, below which there are no observable health effects (Ahmad et al. 2018).

12.8.4 Potential Health Risks from Direct Contact

Exposure of human beings to contaminated soil may cause many health risks varying from nausea, and skin eruptions, to cancer (Swartjes 2015). High concentrations of HMs in the body can affect several organs like the blood, liver, brain, kidneys, and lungs. Long-term exposure to even low levels of HMs can result in neurological and physically degenerative processes (e.g., Parkinson's disease and Alzheimer's disease), and cancer (Brevik et al. 2020). The concentration of possible contaminants in a place depends on the land use and activity carried out there (Turner 2009). Soil despite yielding many essential nutrients can pass on harmful elements through food, and contaminated soil particles can travel thousands of miles and affect people. Direct exposure to soil contaminants can occur through oral ingestion of soil, dermal contact, and inhalation of soil particles. Thus, depending on the route, contaminants enter the body through the mouth, stomach; skin; nose, trachea, and lungs (Swartjes 2015). People working directly with soil on a daily basis like farmers, miners, and construction workers are at a greater risk of health problems. Soil ingestion or geophagy can occur intentionally or accidentally during hand-to-mouth contact particularly in children, or when vegetables or fruits are consumed without washing (Yu et al. 2019; Lindern et al. 2016). Soil ingestion is especially common in children; they are at a greater risk from urban pollutants because they play on the ground and can ingest dust stuck on their hands (Biasioli et al. 2006).

Exposure to the soil through ingestion is controlled by soil ingestion rates, concentrations of HMs, body weight, and the relative bioavailability factor in the human body. Soil ingestion rates have been determined by tracer studies, using typical soil elements such as aluminum, silicon, titanium, and yttrium in feces and urine as indicators (Calabrese et al. 1997). In rare cases, biokinetic models and Pb isotope methodology are used to estimate ingestion rates. From tracer studies, combined soil and dust intake rates range from 31 to 195 mg/day (Swartjes 2015). Skin absorption or penetration can expose a person to chemicals and dangerous pathogens in soil.

Inhalation of soil particles and mineral dust particles has been linked to a range of respiratory problems, ranging from acute inflammatory problems to fibrotic changes. Long term exposure to toxic dust can cause an irreversible respiratory disease, collectively termed as pneumoconiosis. It is caused by extended exposure of lungs to metallic or mineral dust. Inhalation of coal dust causes black lung disorder or coal worker's pneumoconiosis, quartz or silica dust causes silicosis, and asbestos dust results in asbestosis and lung disorders such as bronchitis, emphysema, and mesothelioma (Li et al. 2019; Zosky et al. 2014).

12.8.5 Effect of HMs on Human Beings

HMs also called potentially toxic elements can cause nausea, anorexia, vomiting, gastrointestinal abnormalities, and dermatitis in the human body. They may also affect the immune system and basic physiological processes (Chui et al. 2013). Cd and Pb are the most toxic elements for human health. Pb arises from traffic sources, processing of ores, and pigments, while Cd originates mainly from industries (Galušková et al. 2011). Pb severely affects and damages organs such as the kidney, liver, lungs, reproductive system, central nervous system, urinary system, and immune system and changes the composition of blood (Bansal 2018). Women are more vulnerable to the adverse effects of Cd and have a higher body burden due to increased dietary absorption of Cd in the body. Low-level cumulative exposure has been related to changes in bone metabolism and renal functions (Salt et al. 1995). Workers in close contact with Ni powder are more susceptible to respiratory cancer and nasopharyngeal carcinoma. At low concentrations, Cu acts as a co-factor for various enzymes of redox cycling (Farhan et al. 2016). However, at higher concentrations, Cu disturbs the human metabolism leading to anemia, liver and kidney damage, stomach, and intestinal irritation. Arsenic induces skin, liver, lung, and uterine cancers (Bansal 2018). Certain HMs, their sources, and related health effects are summarized in Table 12.4.

12.9 Mitigation Measures for Construction and Demolition Waste Disposal

Due to the ever-increasing rate of urbanization and re-development, new constructions are taking place which demands more landfill space. C and D wastes are bulky and take up a lot of space in the designated landfills thereby overstraining the landfill capacity. C and D wastes dumped on landfills are hazardous and difficult to remediate. To control the spread of heavy metals the landfill areas need to be stabilized and an adequate plantation cover needs to be provided (Sofilić et al. 2013). Recycling of C and D waste will help in reducing the environmental impact by: (a) reducing depletion of raw materials and natural resources, (b) less pollution due to lesser virgin resources requirement, which will reduce the manufacturing and transportation-related pollution, (c) decrease in energy requirement for all processes which will further reduce environmental damage incurred during generating energy from coal.

Table 12.4 Contaminants and their health effects on human beings

Contaminant	Source	Health effects	References
Lead (Pb)	Mining, Pb based paints, Pb–Zn smelters, solder, glass, piping	Carcinogen, damages the brain and kidneys, lowers IQ, causes miscarriage, anemia, and muscle pain, bone deterioration, and hypertension	Griswold and Ph (2009) Sharma (2017)
Copper (Cu)	Found naturally in sandstones, artificial presence is due to building materials, industrial emissions	Liver toxicity, gastrointestinal distress, headache, irritation of nose and eyes	Mahurpawar (2015)
Chromium (Cr)	Cement factory dust Metal processing industry	Causes nose ulcers, runny nose, and breathing problems such as asthma, cough, shortness of breath	Griswold and Ph (2009) Isikli et al. (2003)
Arsenic (As)	Mining and processing, wood preservatives, paints, pigments, glass, electronics industry	Chronic arsenicosis, nausea, vomiting, liver tumors, and gastrointestinal infections, various cancers, skin, heart, and liver damage, risk of miscarriage	Mohammed Abdul et al. (2015) Sharma (2017)
Zinc (Zn)	Brass and bronze alloys, rubber, tires, glass, metal coatings	Metal fume fever and restlessness	Cooper (2008) Mulligan and Galvez-Cloutier (2003)
Cadmium (Cd)	Water pipes, smelting of Zn, pigments stabilization, metal plating, Ni–Cd batteries, alloys	Carcinogen, causes lung fibrosis, liver and kidney damage, low bone density	Järup (2003) Sharma (2017)
Asbestos/Silica or Quartz/Coal	Direct exposure from asbestos mining, processing, or disposal, products like asbestos cement, textile, adhesive, roofing, flooring material, insulation, indirect exposure from clothes of mineworkers/Silica mining/Coal mining	Cancer (Mesothelioma), Asbestosis/Silicosis/Coal worker's pneumoconiosis	Noonan (2017) Li et al. (2019)
Organic pollutants (PAHs, PCBs)	PAHs from the combustion of fossil fuel, waste, vehicle emission PCBs from old buildings plaster and caulking materials	Carcinogenic Mutagenic	Fazeli et al. (2019) Andersson et al. (2004)

Alongside it will also help in reducing economic costs involved in purchasing, manufacturing, processing, packaging, and transporting raw materials. It will also generate employment for people in the recycling sector (Kumbhar et al. 2013).

To tackle the menace of C and D waste overburden there is a need to implement certain new methods (Yeheyis et al. 2013; Zhang et al. 2013);

a. **Adaptive Reuse**
It involves the process of recycling existing buildings by finding a new use for them. When buildings are structurally strong and well-built but have been discarded then they are altered and repaired and re-used (Cantell and Huxtable 2005).

b. **Regulatory Framework**
There is a need for the implementation of laws that focus on reducing the waste content of C and D waste. Outlawing the practice of landfilling certain waste materials could help retain more space in the landfills. Several countries in Europe have imposed stringent measures to reduce the disposal of C and D waste on landfills. Any demolition activity can be done only with legal permission whereby each demolished component and its disposal method needs to be specified and the demolition wastes need to be disposed of separately. Failure to do so attracts a huge penalty. The disposal of C and D waste is also taxed, while recycling is recommended and free of charge (Ponnada and Kameswari 2015).

c. **Life Cycle Assessment (LCA)**
Waste generation during construction activities occurs during the entire life cycle of the construction process. The quantity and quality of the waste is dependent on the stage of the construction. To address this waste generation and environmental degradation LCA is used as a decision-making tool. LCA methodology helps evaluate the environmental costs of the processes and products related to the entire construction process. Methodologies for determining the LCA were formulated by the International Organization for Standardization (ISO), and ISO14040 deals with the potential environmental impacts of construction-related activities. LCA is called a 'cradle to grave' system because it evaluates the environmental impact of all raw materials and processes involved in the construction like extraction of raw materials, processing, application, demolition products, and their recycling and disposal (Singh et al. 2011). It helps create a defined scope, inventory of the life cycle of products, impact assessment and interpretation of any project. LCA application in the building sector has increased since it is a major consumer of material resources and energy and generates a high amount of environmental impact. The operation stage has the highest environmental impact of 80–90% while the main construction phase has 8–20% impact.
Based on the LCA, the management of C and D waste occurs at three stages of the life cycle: pre-construction, construction, maintenance, and renovation, and post-construction (demolition). At each stage, there is the application of 3 Rs (reduce, reuse, and recycle) to ensure minimal wastage of materials. 3R's help

in conserving raw materials, saving energy, reducing pollution, and reducing the need for landfill spaces or incineration.

d. **Circular Economy**
Circular economy utilizes the idea that there is no end of life of any product and wastes are a useful resource. All materials are constantly reused and recycled at all stages of construction, reducing the wastage of raw materials during construction and minimizing the resultant waste produced (Ginga et al. 2020).

e. **Re-utilization of C and D Waste**
Re-use or recycling of C and D waste is a sustainable and viable process. Recycled products enable less demand for virgin raw materials, which helps in reducing the environmental costs and the expenses incurred during the exploration of raw material and their extraction. Less waste produced also lessens the burden on the landfills. Some examples of recycled wastes are recycled concrete bricks, crushed tiles, and glass in the form of coarse and fine aggregate, and latex paint used as a binding agent in concrete.

12.9.1 Recycling of C and D Waste

C and D wastes need to be managed properly for disposal in accordance with environmental and health considerations (Kumbhar et al. 2013). The handling involves the following steps:

a. **Separation, Storage, and Transportation**
The C and D waste need to be segregated at the site of generation because it is extremely difficult to separate them at the disposal site where they are mixed. This also helps save energy and time. Prior to this, materials like glass, plastic, and wood need to be recycled for reuse. Trucks or tractors are used to transport individual containers to the disposal site.

b. **Recycling for Reuse**
Due to the large volume of C and D wastes they cannot be composted or incinerated. They are dumped on landfills, which are growing relatively scarce due to the increased demand for land for other purposes. Disposal on landfills is not only an eyesore to the public it also has negative impacts like the release of toxic gases and pollution of the underlying soil and groundwater due to leachate. Recycling is therefore important to reduce the impact on land resources for landfilling. It will also help reduce costs involved in the extraction of raw materials, transportation, and energy utilization.

c. **Disposal**
Since C and D wastes are inert, they do not cause pollution if devoid of heavy metals or organic components. They can thus be used for leveling low-lying areas, for building a base for building construction or roads. This would help reduce the need for landfill disposal.

Despite the regulatory measures proposed for the management of C and D waste, the implementation rates are low. LCA has a high variability when applied to real-life scenarios due to the difficulty in accurate measurement of data. This uncertainty is because a large number of raw materials are involved in the construction process, and their life span extends over several hundred years. A common LCA study cannot be applied globally due to the difference in the type of raw materials utilized and due to the difference in the building structure and utilization (Hossain and Marsik 2019).

12.10 Mitigation Measures for Contaminated Soil

People living in urban environments are largely affected by pollutants from the surroundings. Pollutants in the soil can find their way to human beings through inhalation of dust or particulate matter, from the food chain when people consume plants cultivated in contaminated soil, and when children consume soil while playing with it (Galušková et al. 2011). Remediation of contaminated soil is done by removing or reducing the contaminants in the soil through methods like excavation of soil and transport to another place, chemical oxidation, phytoremediation, and thermal desorption (Sharma 2017; Scullion 2006).

These methods have been proposed to tackle the menace of soil contamination in urban areas: Removal of soil and its disposal and treatment in another area, treatment of soil at the site, and containment at the site (Gailey et al. 2020; Sharma 2017; Scullion 2006; Mulligan and Galvez-Cloutier 2003).

a. **Removal and Transport to Another Place**
   In this method, the soil is collected from the contaminated site and transported elsewhere for disposal. The void formed is filled with uncontaminated soil. Despite being the quickest method, it is not useful due to the high transportation costs involved, and due to the distribution of contaminated soil into a larger area.
b. **Chemical Oxidation**
   Chemicals are applied to the soil, either in situ or ex situ. They help in destroying the pollutants, breaking them down into less toxic forms, or rendering them immobile in the soil. The addition of liming material, alkaline biosolids, chelating agents like ethylenediaminetetraacetic acid (EDTA) and nitrilotriacetate (NTA), can help immobilize mine wastes containing heavy metals.
c. **Physical Treatment**
   Soil is treated thermally ex situ at a temperature more than 1000 °C, to destroy the organic pollutants. A high-temperature treatment called vitrification is also carried out for trapping inorganic pollutants in a solid ceramic material. But both these processes destroy the soil, and the resultant products are disposed of in landfills. Vapor extraction and air sparging are used for the extraction of materials like benzene and toluene by volatilizing them. Soil washing is another procedure done ex situ whereby more soluble pollutants are extracted based on their size, density, or surface properties.

d. **Biochemical Processes**
This involves the use of microbes and chemicals to remove heavy metals in soils. Bacteria like *Thiobacillus* sp. under aerobic conditions at sufficient temperature (15–55 °C) can oxidize metal sulfides to produce sulphuric acid, which helps in desorbing metals from the soils through the process of leaching. Another technique involves the use of the fungus *Aspergillus*, which produces citric acid and gluconic acid. These acids act as chelating agents and help in removing Cu. Bacteria like *Bacillus subtilis* and sulfate-reducing bacteria can oxidize Hg and Cd and reduce As and Fe, in presence of sulfur. Biosorption is another process that utilizes bacterial or algal cells for the adsorption of metals into their biomass. This is useful for removing a low concentration of metals in water.

e. **Soil Treatment at Site**
Contaminated soil is covered on-site with a geotextile material, which is then layered with 15 cm of relatively uncontaminated soil. This is done specially for the reconstruction of sensitive areas like playgrounds, parks, schools. But the process of procuring the uncontaminated soil is destructive for other ecosystems.

f. **Containment at Site**
This is carried out with the help of microbes through the process of bioremediation or with the help of both plants and microbes through the process of phytoremediation. Phytoremediation employs plants in ex situ or in situ for cleaning contaminated hazardous waste sites (Subhashini & Swamy 2016). Phytoextraction is the most common method of phytoremediation used for the treatment of heavy metal polluted soil (Mohnish and Nikhil 2016). Plants and microbes associated with plant roots and soil help in breaking down organic pollutants and enable greater uptake of heavy metals. These metals uptake by plants are sequestered in the plant roots or stem and leaves. This process is beneficial because it is the least invasive of all processes and does not damage the soil properties. Containment at the site is also carried out through urban gardening. Urban gardening helps in providing food, reduces storm water run-off, provides habitat for a variety of organisms in the urban environment, thus preventing their displacement, and helps in carbon sequestration. Urban gardening when carried out with the help of soil amendments like biochar, compost or P further help in containing pollutants through absorption or complexation. Application of chelating agents like organic acids also helps in rendering the heavy metals immobile. Certain plants known as hyperaccumulators tend to uptake and store high amounts of heavy metals in their plant parts. Some of them accumulate specific metals while others can efficiently uptake all metals. Plants like *Pteris*, *Trifolium*, *Silene*, *Thlaspi*, *Urtica*, *Chenopodium*, *Alyssim*, and certain species of *Brassica* are efficient hyperaccumulators. Trees like *Salix* and *Populus* are also useful in phytoremediation in areas where the pollutants are present beyond the root zone of other plants.

To ensure that construction and demolition activities cause the least damage to the environment the following issues need to be considered (Cole 2000):

- Protection of the soil and vegetation at the building site and at the site of raw materials excavation.
- Ensuring that the excavated soil is properly disposed of and covered with vegetation.
- Contaminated soil should be treated with an appropriate method according to the degree of soil contamination and should be covered properly to avoid loss of contaminated leachate.
- Prevention of storm water or run-off contamination during construction.
- Construction of green buildings and applying proper maintenance to ensure the long life of all constructed spaces.
- Generating minimum waste during construction by using recycled materials, or materials that can be easily decomposed, are environment-,friendly and locally sourced.
- Conducting demolition when necessary and recycling all components that can be salvaged and reused.
- Minimizing toxic releases to soil and groundwater during all construction and demolition processes and the life cycle of the constructed space.
- And ensuring that individuals living in the surroundings are least impacted by construction-related potential health effects.

12.11 Conclusion

Construction activities are the backbone of every developing economy. To keep up with the rapid pace of globalization construction activities are a requirement. But this industry plays a major role in the destruction of the ecosystem and environment. All processes involved in the construction sector negatively impact the environment especially the soil. Processes like mining for extraction of raw materials lead to soil acidification and desertification. Removal of sand increases the chances of floods and erodes soil. To reduce the impact of the construction processes on the environment, greater emphasis needs to be placed on the reduction of C and D waste production and recycling and re-use of such demolition wastes needs to be encouraged. This would help in reducing the energy costs incurred during incineration and reduce the burden on landfills. There is also a need to formulate stronger laws to ensure that construction companies and individuals are not involved in the misuse of natural resources. Any disobedience to follow the rules should be meted with levying of heavy fines. Soils that have been already contaminated with pollutants arising from the construction can be remediated with the help of chemicals or through bioremediation. Hyperaccumulating plants can uptake a high amounts of metals from the soil and accumulate in their plant parts. Employment of such plants can help in remediating soils without any impairment of the soil parameters. Remediation of contaminated soil is essential and obligatory for the maintenance of the soil environment and its functions. It is also vital for the health of the people living in and near all urban constructed spaces.

References

Abdus M, Kaipiainen E, Mohsin M, Villa A (2016) Effects of contaminated soil on the growth performance of young Salix (*Salix schwerinii* EL Wolf) and the potential for phytoremediation of heavy metals. Effects of contaminated soil on the growth performance of young Salix (*Salix schwerinii* EL) https://doi.org/10.1016/j.jenvman.2016.08.082

Adriano DC, Chlopecka A, Kaplan DI (1998) Role of soil chemistry in soil remediation and ecosystem conservation. Soil Chem Ecosyst Health 52:361–386

Ahmad K, Ashfaq A, Khan ZI et al (2018) Metal accumulation in Raphanus sativus and Brassica rapa: an assessment of potential health risk for inhabitants in Punjab, Pakistan. Environ Sci Pollut Res 25:16676–16685. https://doi.org/10.1007/s11356-018-1868-7

Al-Khashman OA, Shawabkeh RA (2006) Metals distribution in soils around the cement factory in southern Jordan. Environ Pollut 140:387–394. https://doi.org/10.1016/j.envpol.2005.08.023

Allen A (2001) Containment landfills: the myth of sustainability. Eng Geol 60:3–19. https://doi.org/10.1016/S0013-7952(00)00084-3

Alloway BJ (1995) Heavy metals in soils. Springer, Netherlands

Andersson M, Ottesen RT, Volden T (2004) Building materials as a source of PCB pollution in Bergen, Norway. Sci Total Environ 325:139–144. https://doi.org/10.1016/j.scitotenv.2003.11.014

Artiola JF, Walworth JL, Musil SA, Crimmins MA (2019) Soil and land pollution, 3rd edn. Elsevier Inc

Asif M, Muneer T, Kelley R (2007) Life cycle assessment: a case study of a dwelling home in Scotland. Build Environ 42:1391–1394. https://doi.org/10.1016/j.buildenv.2005.11.023

Assefa G, Glaumann M, Malmqvist T et al (2007) Environmental assessment of building properties-where natural and social sciences meet: the case of EcoEffect. Build Environ 42:1458–1464. https://doi.org/10.1016/j.buildenv.2005.12.011

Aubry C, Manouchehri N (2019) Urban agriculture and health: assessing risks and overseeing practices. F Actions Sci Rep 2019:108–111

Ávila PF, Ferreira da Silva E, Candeias C (2017) Health risk assessment through consumption of vegetables rich in heavy metals: the case study of the surrounding villages from Panasqueira mine, Central Portugal. Environ Geochem Health 39:565–589. https://doi.org/10.1007/s10653-016-9834-0

Babak NA (2017) Transport construction negative impact on the environment. Procedia Eng 189:867–873. https://doi.org/10.1016/j.proeng.2017.05.135

Bansal OP (2018) The influence of potentially toxic elements on soil biological and chemical properties, metals in soil—contamination and remediation. IntechOpen, Rijeka. https://doi.org/10.5772/intechopen.81348

Biasioli M, Barberis R, Ajmone-Marsan F (2006) The influence of a large city on some soil properties and metals content. Sci Total Environ 356:154–164. https://doi.org/10.1016/j.scitotenv.2005.04.033

Bossink BAG, Brouwers HJH (1996) Construction waste: quantification and source evaluation. J Constr Eng Manage 122:55–60. https://doi.org/10.1061/(asce)0733-9364(1996)122:1(55)

Brady NC, Weil RR (2004) Elements of the nature and properties of soils. J Chem Inf Model

Brevik EC, Slaughter L, Singh BR et al (2020) Soil and human health: current status and future needs. Air, Soil Water Res 13. https://doi.org/10.1177/1178622120934441

Burghardt W (1994) Soils in urban and industrial environments. Zeitschrift für Pflanzenernährung und Bodenkd 157:205–214. https://doi.org/10.1002/jpln.19941570308

Calabrese EJ, Stanek EJ, Pekow P, Barnes RM (1997) Soil ingestion estimates for children residing on a superfund site. Ecotoxicol Environ Saf 268:258–268

Caliskan E (2013) Environmental impacts of forest road construction on mountainous terrain. J Environ Health Sci Eng 10:1–8

Cantell SF, Huxtable AL (2005) The adaptive reuse of historic industrial buildings: regulation barriers, best practices and case studies. Submitted in partial fulfillment of the requirement for the degree Master of Urban and Regional Planning Virginia Polytechnic Institute and State U

Cardoso R, Silva RV, de Brito J, Dhir R (2016) Use of recycled aggregates from construction and demolition waste in geotechnical applications: a literature review. Waste Manag 49:131–145. https://doi.org/10.1016/j.wasman.2015.12.021

Chen K, Wang J, Yu B, et al (2020) Critical evaluation of construction and demolition waste and associated environmental impacts: a scientometric analysis. J Clean Prod 125071. https://doi.org/10.1016/j.jclepro.2020.125071

Chibuike GU, Obiora SC (2014) Heavy metal polluted soils: effect on plants and bioremediation methods. Appl Environ Soil Sci 2014. https://doi.org/10.1155/2014/752708

Chowdhury R, Paul D, Fry T (2010) A life cycle based environmental impacts assessment of construction materials used in road construction. Resour Conserv Recycl 54:250–255. https://doi.org/10.1016/j.resconrec.2009.08.007

Chui SH, Wong YH, Chio HI et al (2013) Study of heavy metal poisoning in frequent users of chinese medicines in Hong Kong and Macau. Phy Ther Res 27:859–863. https://doi.org/10.1002/ptr.4816

Cole RJ (2000) Building environmental assessment methods: assessing construction practices. Construct Manage Econ 18(8):949–957. https://doi.org/10.1080/014461900446902

Cooper RG (2008) Zinc toxicology following particulate inhalation. Indian J Occup Environ Med 12(1):10 https://doi.org/10.4103/0019-5278.40809

Diotti A, Galvin AP, Piccinali A et al (2020) Chemical and leaching behavior of construction and demolition wastes and recycled aggregates. Sustain 12:1–12. https://doi.org/10.3390/su12241 0326

Donado EP, Oliveira MLS, Gonçalves JO et al (2021) Soil contamination in Colombian playgrounds: effects of vehicles, construction, and traffic. Environ Sci Pollut Res 28:166–176. https://doi.org/10.1007/s11356-020-09965-w

Doran JW, Parkin TB (1996) Quantitative indicators of soil quality: a minimum data set 25–37

Douglas I, Lawson N (2003) Airport construction: materials use and geomorphic change. J Air Transp Manage 9:177–185. https://doi.org/10.1016/S0969-6997(02)00082-0

Doula MK, Sarris A (2016) Soil environment. In: Environment and development. Elsevier, pp 213–286

Dregulo AM, Bobylev NG (2020) Heavy metals and arsenic soil contamination resulting from wastewater sludge urban landfill disposal. Polish J Environ Stud 30:81–89. https://doi.org/10.15244/pjoes/121989

Dziri S, Hosni K (2012) Effects of cement dust on volatile oil constituents and antioxidative metabolism of Aleppo pine (*Pinus halepensis*) needles. Acta Physiol Plant 34:1669–1678. https://doi.org/10.1007/s11738-012-0962-6

El-Fadel M, Findikakis AN, Leckie JO (1997) Environmental impacts of solid waste landfilling. J Environ Manage 50:1–25. https://doi.org/10.1006/jema.1995.0131

El-Sherbiny MM, Ismail AI, El-Hefnawy ME (2019) A preliminary assessment of potential ecological risk and soil contamination by heavy metals around a cement factory, western Saudi Arabia. Open Chem 17:671–684. https://doi.org/10.1515/chem-2019-0059

Farhan M, Khan HY, Oves M et al (2016) Cancer therapy by catechins involves redox cycling of copper ions and generation of reactive oxygen species. Toxins (Basel) 8:1–24. https://doi.org/10.3390/toxins8020037

Fazeli G, Karbassi A, Khoram Kejadian S, Nasrabadi T (2019) Evaluation of urban soil pollution: a combined approach of toxic metals and polycyclic aromatic hydrocarbons (PAHs). Int J Environ Res 13:801–811. https://doi.org/10.1007/s41742-019-00206-8

Ferreira CSS, Walsh RPD, Ferreira AJD (2018) Degradation in urban areas. Curr Opin Environ Sci Heal 5:19–25. https://doi.org/10.1016/j.coesh.2018.04.001
Gailey AD, Schachter AE, Egendorf SP, Mielke HW (2020) Quantifying soil contamination and identifying interventions to limit health risks. Curr Probl Pediatr Adolesc Health Care 50: https://doi.org/10.1016/j.cppeds.2019.100740
Galušková I, Boruvka L, Drábek O (2011) Urban soil contamination by potentially risky elements. Soil Water Res 6:55–60. https://doi.org/10.17221/55/2010-swr
Gangolells M, Casals M, Gassó S et al (2009) A methodology for predicting the severity of environmental impacts related to the construction process of residential buildings. Build Environ 44:558–571. https://doi.org/10.1016/j.buildenv.2008.05.001
Ginga CP, Ongpeng JMC, Daly MKM (2020) Circular economy on construction and demolition waste: a literature review on material recovery and production. Materials (Basel) 13:1–18. https://doi.org/10.3390/ma13132970
Griswold W, Ph D (2009) Human health effects of heavy metals, pp 1–6
Harrison R, Alloway B (2016) Soil pollution and land contamination. Causes, effects and control. Pollution causes Eff Control 352–377
Harrison RB, Strahm BD (2008) Soil formation, pp 3291–3295
Havugimana E, Bhople BS, Kumar A et al (2017) Soil pollution-major sources and types of soil pollutants. Environ Sci Eng 11:53–86
Herrick RF, Lefkowitz DJ, Weymouth GA (2007) Soil contamination from PCB-containing buildings. Environ Health Perspect 115:173–175. https://doi.org/10.1289/ehp.9646
Hillel D (2008) Soil formation. Soil Environ 15–26. https://doi.org/10.1016/b978-0-12-348536-6.50008-3
Hossain Y, Marsik T (2019) Conducting life cycle assessments (LCAs) to determine carbon payback: a case study of a highly energy-efficient house in rural Alaska. Energies 12:25–31. https://doi.org/10.3390/en12091732
Ihedioha JN, Ukoha PO, Ekere NR (2017) Ecological and human health risk assessment of heavy metal contamination in soil of a municipal solid waste dump in Uyo, Nigeria. Environ Geochem Health 39:497–515. https://doi.org/10.1007/s10653-016-9830-4
Isikli B, Demir TA, Ürer SM et al (2003) Effects of chromium exposure from a cement factory. Environ Res 91:113–118. https://doi.org/10.1016/S0013-9351(02)00020-8
Järup L (2003) Hazards of heavy metal contamination. Br Med Bull 68:167–182. https://doi.org/10.1093/bmb/ldg032
Jhamnani B, Singh S (2009) Groundwater contamination due to Bhalswa landfill site in New Delhi. Int J Environ Sci Eng 3:121–125
Jie C, Jing-zhang C, Man-zhi T, Zi-tong G (2002) Soil degradation: a global problem endangering sustainable development. J Geogr Sci 12:243–252. https://doi.org/10.1007/bf02837480
Johnston AE, Poulton PR, Coleman K (2009) Soil organic matter: its importance in sustainable agriculture and carbon dioxide fluxes, 1st edn. Elsevier Inc
Joimel S, Quartet J, Jolivet CC et al (2016) Physico-chemical characteristics of topsoil for contrasted forest, agricultural, urban and industrial land uses in France. Sci Total Environ 545–546:40–47. https://doi.org/10.1016/j.scitotenv.2015.12.035
Kabata-pendias A, Pendias H (2001) Trace elements in soils and plants trace elements in soils and plants. CRC Press, FL
Kanmani S, Gandhimathi R (2013) Assessment of heavy metal contamination in soil due to leachate migration from an open dumping site. Appl Water Sci 3:193–205. https://doi.org/10.1007/s13201-012-0072-z
Kartun M, Ottesen RT, Steinnes E, Volden T (2008) Runoff of particle bound pollutants from urban impervious surfaces studied by analysis of sediments from stormwater traps. Sci Total Environ 396:147–163. https://doi.org/10.1016/j.scitotenv.2008.02.002
Kasassi A, Rakimbei P, Karagiannidis A et al (2008) Soil contamination by heavy metals: measurements from a closed unlined landfill. Bioresour Technol 99:8578–8584. https://doi.org/10.1016/j.biortech.2008.04.010

Khan S, Cao Q, Zheng YM et al (2008) Health risks of heavy metals in contaminated soils and food crops irrigated with wastewater in Beijing, China. 152:686–692. https://doi.org/10.1016/j.envpol.2007.06.056
Kobetičová K, Černý R (2017) Ecotoxicology of building materials: a critical review of recent studies. J Clean Prod 165:500–508. https://doi.org/10.1016/j.jclepro.2017.07.161
Kong F, Chen Y, Huang L et al (2021) Ecotoxicology and environmental safety human health risk visualization of potentially toxic elements in farmland soil: a combined method of source and probability. Ecotoxicol Environ Saf 211:111922. https://doi.org/10.1016/j.ecoenv.2021.111922
Kosheleva NE, Vlasov DV, Kotlyakov ID, Kasimov NS (2018) Contamination of urban soils with heavy metals in Moscow as affected by building development. Sci Total Environ 636:854–863. https://doi.org/10.1016/j.scitotenv.2018.04.308
Kumbhar SA, Gupta A, Desai DB (2013) Recycling and reuse of construction and demolition waste for sustainable development. OIDA Int J Sustain Dev 6:83–92
Li X, Liu L, Wang Y et al (2013) Heavy metal contamination of urban soil in an old industrial city (Shenyang) in Northeast China. Geoderma 192:50–58. https://doi.org/10.1016/j.geoderma.2012.08.011
Li CZ, Zhao Y, Xu X (2019) Investigation of dust exposure and control practices in the construction industry: implications for cleaner production. J Clean Prod 227:810–824. https://doi.org/10.1016/j.jclepro.2019.04.174
Liu X, Song Q, Tang Y et al (2013) Science of the total environment human health risk assessment of heavy metals in soil—vegetable system: a multi-medium analysis. Sci Total Environ 463–464:530–540. https://doi.org/10.1016/j.scitotenv.2013.06.064
Liu D, Quan Y, Ren Z, Wu G (2017) Assessment of heavy metal contamination in soil associated with Chinese coal-fired power plants: a case study in Xilingol, Inner Mongolia. Int J Sustain Dev World Ecol 24:439–443. https://doi.org/10.1080/13504509.2016.1273269
Mahmood T, Islam KR (2006) Response of rice seedlings to copper toxicity and acidity. J Plant Nutr 29:943–957. https://doi.org/10.1080/01904160600651704
Mahurpawar M (2015) Effects of heavy metals on human health. Int J Res-Granthaalayah 2394–3629. ISSN-23500530
McClintock N (2015) A critical physical geography of urban soil contamination. Geoforum 65:69–85. https://doi.org/10.1016/j.geoforum.2015.07.010
Meuser H (2010) Causes of soil contamination in the urban environment. In: Contaminated urban soils, Springer, Dordrecht, pp 29–94
Mohammed Abdul KS, Jayasinghe SS, Chandana EPS et al (2015) Arsenic and human health effects: a review. Environ Toxicol Pharmacol 40:828–846. https://doi.org/10.1016/j.etap.2015.09.016
Mohnish P, Nikhil KAA (2016) Effect of heavy metals on plants: An overview. Int J Appl or Innov Eng Manag
Mulligan CN, Galvez-Cloutier R (2003) Bioremediation of metal contamination. Environ Monit Assess 84:45–60. https://doi.org/10.1023/A:1022874727526
Murtaza G, Ghafoor A, Qadir M et al (2010) Disposal and use of sewage on agricultural lands in Pakistan: a review. Pedosph An Int J 20:23–34. https://doi.org/10.1016/S1002-0160(09)60279-4
Nautiyal H, Shree V, Khurana S et al (2015) Recycling potential of building materials: a review. Environ Footprints Eco-Design Prod Process 31–50. https://doi.org/10.1007/978-981-287-643-0_2
Nguyen MH, Van HT, Thang PQ et al (2021) Level and potential risk assessment of soil contamination by trace metal from mining activities. Soil Sediment Contam 30:92–106. https://doi.org/10.1080/15320383.2020.1811203
Niemelä J, Sauerwein M (2013) Urban soils—characterization, pollution, and relevance in urban ecosystems. Urban Ecol 2:24–32
Noonan CW (2017) Environmental asbestos exposure and risk of mesothelioma. Ann Transl Med 5:1–11. https://doi.org/10.21037/atm.2017.03.74

Olson KR (2004) Factors of soil formation—parent material. Encycl Soils Environ 4:532–535. https://doi.org/10.1016/B0-12-348530-4/00005-9

Ottesen RT, Alexander J, Langedal M et al (2008) Soil pollution in day-care centers and playgrounds in Norway: national action plan for mapping and remediation. Environ Geochem Health 30:623–637. https://doi.org/10.1007/s10653-008-9181-x

Pappu A, Saxena M, Asolekar SR (2007) Solid wastes generation in India and their recycling potential in building materials. Build Environ 42:2311–2320. https://doi.org/10.1016/j.buildenv.2006.04.015

Pavilonis B, Maroko A, Cheng Z (2020) Lead in New York City's soils: population growth, land use, and contamination. Int J Hyg Environ Health 229: https://doi.org/10.1016/j.ijheh.2020.113564

Piekut A, Baranowska R, Marchwińska-Wyrwał E, Ćwieląg-Drabek M, Hajok I, Dziubanek G, Grochowska-Niedworok E (2018) Is the soil quality monitoring an effective tool in consumers' protection of agricultural crops from cadmium soil contamination?—a case of the Silesia region (Poland). Environ Monit Assess 190(1):1–9

Ponnada MR, Kameswari P (2015) Construction and demolition waste management—a review. Int J Adv Sci Technol 84:19–46. https://doi.org/10.14257/ijast.2015.84.03

Popovic D, Djuric G, Todorovic D (1996) Radionuclides in building materials and radon indoor concentrations. Radiat Prot Dosimetry 63:223–225. https://doi.org/10.1093/oxfordjournals.rpd.a031533

Qiang T, Heejong K, Kazuto E et al (2015) Size effect on lysimeter test evaluating the properties of construction and demolition waste leachate. Soils Found 55:720–736. https://doi.org/10.1016/j.sandf.2015.06.005

Rodríguez-Robles D, García-González J, Juan-Valdés A et al (2015) Overview regarding construction and demolition waste in Spain. Environ Technol (United Kingdom) 36:3060–3070. https://doi.org/10.1080/09593330.2014.957247

Romzaykina ON, Vasenev VI, Paltseva A et al (2020) Assessing and mapping urban soils as geochemical barriers for contamination by heavy metal(loid)in Moscow megapolis. J Environ Qual 1–16. https://doi.org/10.1002/jeq2.20142

Rutigliano FA, Marzaioli R, De Crescenzo S, Trifuoggi M (2019) Science of the total environment human health risk from consumption of two common crops grown in polluted soils. Sci Total Environ 691:195–204. https://doi.org/10.1016/j.scitotenv.2019.07.037

Salt DE, Prince RC, Pickering IJ, Raskin I (1995) Mechanisms of cadmium mobility and accumulation in Indian mustard. Plant Physiol 109:1427–1433. https://doi.org/10.1104/pp.109.4.1427

Scullion J (2006) Remediating polluted soils. Naturwissenschaften 93:51–65. https://doi.org/10.1007/s00114-005-0079-5

Semhi K, Al-Khirbash S, Abdalla O et al (2010) Dry atmospheric contribution to the plant-soil system around a cement factory: spatial variations and sources-a case study from oman. Water Air Soil Pollut 205:343–357. https://doi.org/10.1007/s11270-009-0079-8

Sharma RK (2017) Soil pollution–various causes and their remedy 214–236

Silva RV, de Brito J, Dhir RK (2017) Availability and processing of recycled aggregates within the construction and demolition supply chain: a review. J Clean Prod 143:598–614. https://doi.org/10.1016/j.jclepro.2016.12.070

Singh A, Berghorn G, Joshi S, Syal M (2011) Review of life-cycle assessment applications in building construction. J Archit Eng 17:15–23. https://doi.org/10.1061/(asce)ae.1943-5568.0000026

Singh B, Cattle SR, Field DJ (2014) Edaphic soil science, introduction to. Encycl Agric Food Syst 3:35–58. https://doi.org/10.1016/B978-0-444-52512-3.00092-9

Sofilić T, Brnardić I, Šimunić-Mežnarić V, Šorša A (2013) Soil pollution caused by landfilling of nonhazardous waste from steel production processes. Kem u Ind Chem Chem Eng 62:381–388

Subhashini V, Swamy AVVS (2016) Phytoremediation of heavy metals contaminated soils by *Catharanthus roseus*. Int J Sci Res 5(12):726–729

Swartjes FA (2015) Human health risk assessment related to contaminated land: state of the art. Environ Geochem Health 37:651–673. https://doi.org/10.1007/s10653-015-9693-0
Todorovic N, Bikit I, Krmar M et al (2015) Natural radioactivity in raw materials used in building industry in Serbia. Int J Environ Sci Technol 12:705–716. https://doi.org/10.1007/s13762-013-0470-2
Turner AH (2009) Urban agriculture and soil contamination: an introduction to urban gardening. Cent Environ Policy Manage 1–20
Van Bohemen HD, Van De Laak WHJ (2003) The influence of road infrastructure and traffic on soil, water, and air quality. Environ Manage 31:50–68. https://doi.org/10.1007/s00267-002-2802-8
Von Lindern I, Spalinger S, Stifelman ML, Stanek LW (2016) Supplemental material estimating children' s soil/dust ingestion rates through retrospective analyses of blood lead biomonitoring from the Bunker Hill Superfund Site in Idaho 10–15. https://doi.org/10.1289/ehp.1510144
Voroney RP, Heck RJ (2015) The soil habitat, 4th edn. Elsevier Inc
Xia X, Chen X, Liu R, Liu H (2011) Heavy metals in urban soils with various types of land use in Beijing, China. J Hazard Mater 186:2043–2050. https://doi.org/10.1016/j.jhazmat.2010.12.104
Xiang M, Li Y, Yang J et al (2020) Assessment of heavy metal pollution in soil and classification of pollution risk management and control zones in the industrial developed city. Environ Manage 66:1105–1119. https://doi.org/10.1007/s00267-020-01370-w
Xie T, Wang M, Chen W, Uwizeyimana H (2019) Impacts of urbanization and landscape patterns on the accumulation of heavy metals in soils in residential areas in Beijing. J Soils Sediments 19:148–158. https://doi.org/10.1007/s11368-018-2011-6
Yang JL, Zhang GL (2015) Formation, characteristics and eco-environmental implications of urban soils—a review. Soil Sci Plant Nutr 61:30–46. https://doi.org/10.1080/00380768.2015.1035622
Yeheyis M, Hewage K, Alam MS et al (2013) An overview of construction and demolition waste management in Canada: a lifecycle analysis approach to sustainability. Clean Technol Environ Policy 15:81–91. https://doi.org/10.1007/s10098-012-0481-6
Yu H, Li T, Liu Y, Ma L (2019) Spatial distribution of polycyclic aromatic hydrocarbon contamination in urban soil of China. Chemosphere 230:498–509. https://doi.org/10.1016/j.chemosphere.2019.05.006
Zhang Y, Shen Y (2017) Wastewater irrigation: past, present, and future: wastewater irrigation. WIRES Water e1234:10. https://doi.org/10.1002/wat2.1234
Zhang X, Shen L, Zhang L (2013) Life cycle assessment of the air emissions during building construction process: a case study in Hong Kong. Renew Sustain Energy Rev 17:160–169. https://doi.org/10.1016/j.rser.2012.09.024
Zosky GR, Boylen CE, Wong RS, Smirk MN, Gutiérrez L, Woodward RC, Cook A (2014) Variability and consistency in lung inflammatory responses to particles with a geogenic origin. Respirology 19(1):58–66. https://doi.org/10.1111/resp.12150

Chapter 13
Water Pollution from Construction Industry: An Introduction

Keshava Joshi, Lokeshwari Navalgund, and Vinayaka B. Shet

Abstract Water is one of the key natural resources utilized for drinking and other developmental purposes. Water is said to be polluted, when the quality of water is harmful to environment and human health due to unwanted materials entering into the water bodies. Water pollution is a problem that cannot be tolerated even by a construction sector. The pollutants and toxic chemicals generated at the construction sites should be managed well, before discharged into the water bodies. The contaminants like cement, paint, glues, sand, heavy metals, oil, toxic chemicals generated at construction sites enter water bodies due to runoff. Pollutants from construction sites can soak into the groundwater as well, which is more difficult to treat than the surface water. Chemical pollutants especially toxic chemicals, arsenic, lead entering into the water bodies can have a serious human health impact including cancer. Wastewater from the construction sites creates severity to the environment as it can harm or disrupt the entire ecosystem. Managing how much pollution of water can be minimized is a challenging issue to balance between construction business and environment. Hence proper planning is needed to bring the strategies and its implementation in mitigating the water pollution from construction industries.

Keywords Groundwater pollution · Health effects · Strategies · Toxic chemicals · Water bodies

13.1 Introduction

According to sociologist Gideon Sjoberg, the development of the city depends on good environment, fresh climate and water, advanced technology, strong community relation to ensure community steadiness and budget. Construction is an economic

K. Joshi (✉) · L. Navalgund
Department of Chemical Engineering, SDM College of Engineering and Technology (V.T.U Belagavi), Dharwad 580002, Karnataka, India

V. B. Shet
Department of Biotechnology Engineering, NMAM Institute of Technology (V.T.U Belagavi), Nitte, Mangalore 574110, Karnataka, India

J. A. Malik and S. Marathe (eds.), *Ecological and Health Effects of Building Materials*,
https://doi.org/10.1007/978-3-030-76073-1_13

activity and a part of urbanization facilitating infrastructure which is advantageous to humans in nearly new aspects and damaging in the few other aspects (Muhwezi et al. 2012). These construction activities have global environmental apprehensions emphasizing on water pollution, air pollution, destruction of resources etc. (Hussin et al. 2013). In construction industry, there exist the commercial activities covering constructions and modernization of the city sector including the basic sub-sectors like water source, transportation, schools, medical facilities etc. The construction sector has highest commercial activities; maximum through steel, glass, paint and different material manufacturers for the diverse production of infrastructures, import and export cargo, power generation, industries, houses, complexes and so on (Scott et al. 2013).

Construction sector is one of the big sectors for every growth of the country. It's the main source of income of nearly 30% of the population of the world. Today the construction industry has a wide and diverse range of enterprises globally; however the majority of construction activity is still undertaken by local firms (Gunhan and Arditi 2005). Construction industry provides a job opportunity for a huge sector of people, by hiring from other enterprises, obtaining specialized services by subcontractors, designs by separate professional entities. Though construction activities are a vast throughout the globe, the adverse influence of constructions on the atmosphere and human health is a challenging issue (Garetti and Taisch 2012). In India, construction industry contributes nearly 30% of the waste of the country which includes concrete (65%), bricks and tiles (25%), wood (5%), metals (2%), plastic (2%) and other wastes (1%) (Akhtar and Sarmah 2018). The quantity of waste generation is being increased as many structures are getting deteriorated and have to be demolished due to the age factor. The quantity of the waste is also increasing due to increase in natural calamities like earthquakes, cyclones and floods. The stringent laws are to be implemented in minimizing the damages and pollution due to the construction activities (Wang et al. 2004). Many technological changes are needed in the industry with increasing use of environmental friendly materials. There is also a need for skills training and managerial training for enterprises to adapt to such changes.

The understanding and managing the pollution levels at the construction industry with proper planning and monitoring can reduce the adverse effect on the workers, public and environment. The study in construction sector shows that major contribution of pollution is from water (40%), air (25%) and at the landfill sites (35%) (Yeheyis et al. 2013). The proper strategies to be implemented at construction site to enhance the commercial activities in the city with significant positive impact. Hence this chapter covers the water pollution in the construction industry, sources and characteristics of waste water from the construction industry, environmental and health effects, techniques to control water pollution and strategy for sustainable development of the construction industry.

13.2 Water Pollution at Construction Industries

Construction industry creates enormous water pollution which varies due to different stages of construction activities, type of construction and different construction practices and technology on the site. Compared to other industries, the construction industry creates a huge damage on the environment and also there is a need for strategies to be implemented to minimize these impacts of pollution. Due to overpopulation in the different cities across the globe, the construction industry is being expanded and creating a huge market and opportunities, hence there is a need for stringent policies and law for curbing the pollution problems (Gan et al. 2015).

The different types of pollution in any industrial activity are air, water, noise and landfill pollution. Water pollution is one such problem, which cannot be neglected by the construction industry. Every industry should have a precautionary measure to manage harmful waste, as it causes irreversible damage to public health and the surrounding. Construction site induced pollution problems could harmfully distress the environment as well as the economic and community of people (Tzoulas et al. 2007). Bequeathing to the environmental protection agency (EPA), construction activity has brought significant variation on the exterior of a land as vegetation is being cleared for many construction projects (Belayutham et al. 2016). This has resulted in the surrounding environment being heavily polluted. Today water contamination due to the construction industry has brought threat to the environment in the world. Among all the industries, the construction industry generates a large amount of water pollutants killing fish and animals or entire ecosystems living in water bodies and in turn affects human health.

Water contamination is the release of unwanted materials into the water bodies, where they interfere with the natural functioning of the ecosystem and in turn have impact on human health. Construction events regularly comprise the use of sediments, cement, toxic chemicals, heavy metals, wood, plastics, oil, solvents, paints and detergents which enter water bodies if not handled properly (Horvath 2004).The sewage from construction sites is created due to the concrete preparations and pouring it in pipes, the hydrostatic tests conducted, domestic wastage due to workers, solid waste discharged into the sewer lines, seepage in pipelines (Morledge and Jackson 2001). These construction pollutants dumped at sites and with surface runoff can soak into the groundwater and in turn strengthen the urban water pollution. The treatment of groundwater is much harder than surface water and it will have an impact on human health.

13.3 Sources and Characteristics of Water Pollution from Construction Industry

Construction industry is one among the most water intensive industries. Water is a vital resource for mixing the concrete, washing the equipment or wetting the dry surfaces and in all stages of the building process. Water gets polluted due to different construction materials at the surface and beneath the ground close to a structure location.

The probable sources of contamination of water from nearby location include erosion of soil, cement working, stockpiles created, lubricants due to maintenance of vehicles and equipment. The sources of contamination also include waste due to demolition and repair work, renovation, land clearing and earth works, concrete material, packing material, wood works, brick wastes, plastic wastes, waste paints and thinners, hazard and toxic materials, electrical wiring, insulation material, sanitary pieces (US EPA 1998; Tang Soon and Larsen 2003). The main source of water pollution from the construction industry is soil erosion and due to runoff and weather conditions it results in sediments (Issaka and Ashraf 2017). Most of the soil surfaces at construction sites are spilled with oil due to diverse events of vehicles and waste paints and solvents.

It is estimated that annually India is generating around 10 million tons of construction waste (Rao et al. 2014). Waste generated from the construction industry can be of the same size or there can be large variation in the size. The waste generated at Tier-I and II cities are bigger in size and in more quantity because of huge buildings and hence more loss of materials (Barbuta et al. 2015). These toxic materials like lead, mercury, arsenic and huge uncontrolled discharge entering water bodies are a threat to human and environment. The huge construction waste generated is transported by the private sectors and hence ends at their need places or at other construction sites for further use (Ponnada and Kameswari 2015). Hence there is a need for policies and guidelines for developers or a contractor for proper handling of waste generated at the site, its transportation regulations and disposal methods. To overcome the water pollution due to the construction industry, the material of concrete is replaced with green materials, which includes inorganic polymer concrete and its being replaced with actual concrete (Bozkurt and İslamoglu 2013). A huge variability of waste from the construction industry is being reused for acquisition of different concrete material requirements and also to upgrade the strength, durability, hardening and resistance.

The appearances of wastewater have a lot of divergence based on the source from the different activities through different sectors like residential area, commercial area, agricultural area, industrial area, which may comprise physical, chemical, and biological pollutants. The strength and composition of waste water depends on the physical, chemical and biological characteristics and to decide the suitable treatment system before the final discharge.

13.3.1 Physical Characteristics

The general physical characteristic of commercial wastewater includes grayish color, odor and settleable solids. The settleable solids by their characteristics, size and shape can be both in the suspended and dissolved form or they are classified as settleable, suspended, dissolved, volatile or nonvolatile. The solids can be organic material plants, fibers, organisms, food waste etc. and inorganic materials like salts, soap, metals, paper, sand, grit etc. (Kumar et al. 2010). In wastewater, the solids are measured in terms of turbidity. Turbidity is quantified by the extent of light being absorbed or scattered through material in water (Kitchener et al. 2017). Both the magnitude and external features or characters of the solid material effect absorption and scattering. The color in the water is an indicator for water being polluted and it is measured on the platinum cobalt scale as per the APHA standards. The color of wastewater continues to change from grayish to dark grey brownish and eventually to black as it passes through a collection system and also as it approaches the anaerobic conditions (Phukon and Bora 2016).

13.3.1.1 Suspended Solids

The very major cause of water pollution on construction locations is suspended solids as it is the major source of material of construction (Pitt et al. 2007). The soil surface doesn't have any essential component to safeguard it from precipitation and overflow. In absence of vegetation and usage of heavy equipment at the site, the amount of overflow increases and worsens the condition of suspension of soil in water. The machineries working in wet conditions further releases dust elements that are suspended in the shallow water. The construction industry has to take certain steps and precautionary measures to minimize such silt pollution.

13.3.2 Chemical Characteristics

The general municipal or commercial waste water is characterized with 70% organic and 30% inorganic materials. The organic characteristic includes carbohydrates, proteins and fats, which is not found much in construction wastes, while inorganic wastes include heavy metals, alkalinity, sulphur, chlorides, nitrogen, phosphorus, and toxic compounds. Hence a chemical characteristic includes alkalinity, dissolved oxygen (DO), biochemical oxygen demand (BOD) and chemical oxygen demand (COD).

The presence of carbonates and bicarbonates of calcium, magnesium, sodium, potassium, or ammonia leads to alkalinity of the water in construction industry waste. This parameter of alkalinity is a significant value of both normal water and wastewater which is measured in terms of pH using a pH probe. The average pH of wastewater should be in the range of 6–9 to protect the organism and alteration in this value needs treatment before the wastewater is discharged. Dissolved oxygen (DO) is to be maintained in the water for the proper respiration and one of the important parameters for assessing the quality of water. The DO level should be in the range of 5–6 ppm, too low or too high DO can harm aquatic life and affect water quality. DO is required for the inhalation of aerobic bacteria and is present as a free oxygen molecule and it can be measured using a dissolved oxygen probe. The DO level gets affected with change in temperature, the increase in temperature decreases the DO (Vega et al. 1998) and hence more stress on water bodies in summer than the other seasons.

Biochemical Oxygen Demand (BOD) is equal to the quantity of oxygen consumed by microorganisms to oxidize carbon-based matter present in the water at a specific temperature. BOD oxidizes all the organics present in wastewater that are biochemically degradable during 5 days' time period at 20 °C. Chemical Oxygen Demand (COD) measures the content of organic matter of wastewater that is oxidized using a chemical agent $K_2Cr_2O_7$. The COD values of waste water are usually higher than the BOD, as it includes the oxygen demand created by biodegradable as well as non-biodegradable substances. The advantage of COD is that the period of digestion is 3 h more than the 5 days incubation period for BOD measurement. Once the correlation has been studied between COD and BOD measurements, the treatment system design can be controlled. Total organic chemical (TOC) is another method to measure the both organic and inorganic content of the waste water. The waste water discharged from the construction industry has high organic content due to the different materials like elements, toxic materials, detergents, cements, sand etc.

13.3.2.1 Hydrocarbons

The source of hydrocarbons on the construction site is from machineries, paints etc. like petrol, diesel, kerosene and oils, paints. These hydrocarbon spillages at site are captivated into the soil and also from improper mapping of pipe networks at construction sites (Chauhan et al. 2010). These hydrocarbons are originated to be in a dissolved phase and hence the treatment remains more expensive. The common treatment is captivating the hydrocarbons through adsorption onto granular activated carbon containers and recovering back the hydrocarbons through desorption. Spillage of hydrocarbons can be minimized at construction sites by providing designated areas for the vehicle cleaning and for filling the fuel. The floating hydrocarbons are generally removed through the flotation process.

13.3.2.2 High pH

The pH is the one of the sources of wastewater on the construction locations due to the wash of building concrete and equipment at the location. The lime stabilization and reuse of the aggregate concrete at the site also increases the pH (Alyafei et al. 2020). The alkalinity of wastewater is much more harmful than the silt or oil, as the alkalinity of concrete waste water is extremely high in the range of pH 12 to 13. Neutralization is an essential treatment process in many industrial manufacturing environments to meet the standards of discharge of wastewater.

13.3.3 Biological Characteristics

In general waste water has a huge quantity of microorganisms, most of which are innocuous to man and microorganisms range from 500,000 to 5,000,000 per ml. However, few microorganisms are harmful and cause disorders in health are also present in wastewater. These microorganisms decompose complex compounds into stable ones with the help of enzymes. Depending upon the respiration, bacteria are classified as aerobic, anaerobic and facultative bacteria. The number of organisms in wastewater is counted using the standard most probable number (MPN) test. It is observed that, construction industry waste water doesn't have the characteristics of biological waste.

13.4 Environmental and Health Effect of Construction Waste

The technology in the construction industry has brought more beauty and creativity due to human intelligence in transferring the design into reality. At the same time, health, safety and environment should be an integral part of any industrial activity like agriculture, manufacturing or construction industry. People should be trained adequately in designing and implementation of a system with due priority with safety, health and environment (Nigam et al. 2007). The surrounding environment of any activity should be free from pollution, as these areas affect the well-being of the public and in turn the rate of production and environment, as safety, health and environment are interrelated (Luhar and Luhar 2019).

Health and safety are still not given the top priority in the construction industries. Construction industry is regarded as one of the highest environment degradation across the world and has caused serious environmental problems and in turn affects the economic condition of the city. The construction activities are enhancing the environmental degradation along with high energy consumption and depletion of natural resources. Through the different sources and discharge of elements from the construction sites affects the water bodies and has implications on human health due to unsafe drinking water which leads to diarrheal, accounting for 70% of death (Schwarzenbach et al. 2010). The significant discharge of suspended materials and toxic substances in water bodies, chunk the gills of fish and seriously disrupt aquatic ecosystems due to lack of dissolved oxygen (Pandey and Madhuri 2014). The metals like lead, arsenic and mercury are highly toxic leading to the depletion of organisms and in turn affect the human systems. The photosynthetic activity of the plants is hindered in water bodies as the hydrocarbon layers block the entry of the sunlight (Carr and Neary 2008). During Heavy showers sediments, paints, lubricants, fuel, solvent, pesticides etc. enter water bodies leading to the reduction of the oxygen in water and in turn damage the marine life and human immunity level (Jain et al. 2016). The construction activities wastewater also destroys the land fertility in nearby areas and adjacent pavements. The groundwater also gets contaminated, which is a source of drinking water through different heavy metals causing health issues like cancer, when consumed (Mahurpawar 2015).Water pollution from construction industry may cause dangerous like cancer, hormonal imbalance, liver and kidney problems, and damage to DNA and reproductive systems.

13.5 Control of Water Pollution from Construction Sites

The enormous quantity of wastewater is generated at the construction sites, as water being one of the key vital components at different stages of a project work. This significant quantity of wastewater generated at a construction site, frequently desires treatment before being recycled or discharged to the natural environment. The water gets polluted due to sewage produced due to different activities by concrete stirring, curing, pouring pipes, hydrostatic test, domestic sewage of construction workers, abundant solid wastes, seepage failure in drain line. These pollutants discarded from the construction locations are connected with surface overflow, which strengthens urban and groundwater pollution.

Water being a key component of a construction project, it must be properly managed to optimize its consumption and to ensure it does not harm the environment. Most countries today have the national and local regulations and standards for discharging the water into the water bodies.If the water quality doesn't meet the standards it cannot be discharged into a public effluent, hence treatment is required. The waste water discharged from concrete construction activities have high suspended solids and pH value (Al-Jumeily et al. 2018). The recent updates in the law, pushes the construction industry to reuse all the waste water generated at the sites and already few of the industries are achieving zero discharge of waste water.

The general waste water treatment systems consist of pH adjustment, coagulation-flocculation, decantation, flotation, sedimentation etc. This primary treatment system needs more space, relatively complex equipment, more chemical additives etc. In addition, the sludge obtained must be treated and managed as waste before discharge. Because of these difficulties, most of the planned solutions of construction industries intended at reusing the water rather than the treatment of construction industry effluents. There are number of treatment studies shown by different researchers, like 2 stage treatment method of sedimentation and neutralization (Tsimas and Zervaki 2011), coagulation and sedimentation (De Paula et al. 2014), adsorption followed by electrochemical techniques, to minimize the turbidity and COD in the effluents (Alyafei et al. 2020).

Biological treatment is of much significance as it treats wastewater from either domestic, commercial or industrial wastes. The biological treatment process is considered to have a cheaper and safer operating process compared to conventional physical and chemical methods. The aerobic activated sludge process is well practiced across the globe for any commercial waste water. The biological treatment includes both aerobic and anaerobic processes. Aerobic treatment, in presence of oxygen, converts organic matter through microorganisms into carbon dioxide, water and biomass. The anaerobic treatment in the absence of air converts organic matter into methane and carbon dioxide (Kolade 2018). The change in stringent discharge standards by the concerned has forced the application of a different advanced biological treatment processes in current years (Shivaranjani and Thomas 2017).

Today's construction projects are causing enormous environmental pollution. Researchers are concentrating on survey of green materials and its suitability in the construction industry as it minimizes the environmental pollution. Green construction is focusing to facilitate the sustainable development of industry and also to protect the ecosystem (Gupta and Vegelin 2016). Green constructions are trying to attempt towards saving energy, water wastage, reducing material cost etc. In recent years substantial water wastage has been reduced in few of the construction locations due to the technological advancement in different usage of water fixtures, water harvesting, water audits, and leak detection machines.

13.6 Strategy for Sustainable Development of Construction Industry

Accomplishment of sustainable development of the construction industry needs a high standards technology and design with green materials to have a clean and ecofriendly environment. The sustainability in the construction industry is in terms of energy, water, air, green space etc. Environmental Impact assessment (EIA) of construction projects emphases on reducing the adverse impact on the environment examining both positive and negative impact of the project and provides different predictions and options for decision makers (Shah et al. 2010).

The building construction project falls under 8(a) category of EIA notification 2006 (as amended) by the Ministry of Environment and Forests (MoEF). The MoEF has made it obligatory to get environmental clearances for construction projects with area greater than 20,000 m^2 (Gupta et al. 2015). It is required to prepare an EIA report on the basis of a guidance handbook and then submit it to the suitable authority. These impact studies in the construction industry include all factors relevant in having impact on the environment, natural resources and also cost-effective projects. Hence these impact studies require a multidisciplinary approach towards all factors involved and come out with feasibility stages of the project.

The Environmental Protection Agency (EPA) guides in protecting the environment as first priority at the beginning of any construction work. All the contract workers should be aware of the rules and the company should safeguard the environment critically. There are many materials and chemicals used at some stage of construction work and these materials may have an impact on employees and the environment if not taken care in handling them properly. Hence EPA proposes proper execution of the effective pollution hindrance and management procedures safely during the construction work (Petraru and Gavrilescu 2010). Also, energy star programs are generated by EPA along with the energy department for using energy efficient materials and buildings across the world. The research process has been developed to use eco-friendly materials in the construction procedure that can save 250 metric tons of CO_2 emissions yearly. Another proposal of EPA is a recycling program initiated at construction garbage which helps to study the impact on the environment and to recycle all the materials on the site.

Good construction site practice can overcome and minimize pollution. Have proper design systems to reduce the release of contaminants and avoid erosions. Depending on the location and structure of construction the soil stabilization process should be implemented based on local rules and regulations. All the construction materials must be kept safe and secure, to avoid runoff due to rains and other waterways. All the roads and footpaths near the construction site must be kept neat and clean. The regular examination of spillages at the site, use of nontoxic paints, and nonhazardous materials at all possible areas of location at construction sites will reduce the pollution. All these practices at the construction site reduce pollution and implementation of these strategies leads to sustainable development in the construction industry.

13.7 Reduce, Reuse and Recycle of Construction and Demolition of Wastes

Construction industry is trying to minimize or reduce the wastage at the source itself by preserving or optimizing the existing buildings instead of the new construction. Adopt construction methods or technology in the system such that it can be disassembled or reused with different techniques for the better savings in terms of economics and environment. The concept of minimizing the waste (reducc), using items more than once (reuse) and using the product to the new use (recycle) in the construction industry is the need of the hour. Recovering the valuable materials like concrete, rubber, wood, metals, steel, brass from the construction site is a very effective approach to save money and protect the natural resources (Hussin et al. 2013). The construction industry can adopt different technology and design parameters to discharge less waste, avoid too many materials, and incorporate safe and secure storage areas and weather proof conditions.

Currently industry should preserve the good materials and store them in harmless places for further use on the same site or different site. Different materials like bricks, tiles, paint, inert materials, wood, plaster, packing materials, glass, plastics, metals can be reused. Paybacks of reducing disposal of construction materials, 3Rs concept creates services and increases in monetary benefits in recycling industries. The 3Rs concept increases commercial openings within the local community, and minimizes the environmental effects with potential to turn 100 percent of materials back into the construction location (Jain 2012). Hence presenting an effective waste management in the launch of the design systems at construction locations avoids enormous landfilling, thereby improving the recycling process.

13.8 Conclusion

Construction activities are one of the significant sources accountable to devastate the environment and natural ecological units. Any country cannot stop the construction site activities as it is the backbone of the economic status of the particular place. Therefore, there is a need to implement advanced technologies and methods following sustainable construction to bring down the pollution level at the sites. Water being one of the key components of the construction site, can pollute the environment and hence use of recycled water streams in construction areas is the need of the hour. The enormous quantity of water wherever possible can be reused from greywater and rainwater harvested at construction sites. Today energy savings and green buildings are getting popularized in the construction industries in order to save water and energy. Sustainability is being achieved in construction industries with environmental friendly approaches and move towards economic feasibility with comfort and safety of the residents. The changing technology of the construction industries are minimizing the consequence on the environment, improving safety and health of workers, reduction in disposal costs with accomplishment of environmental goals. Hence water pollution in the construction industry can be minimized with more environmentally friendly approaches and awareness, favorable government policies and continuous education for efficient water use.

References

Akhtar A, Sarmah AK (2018) Construction and demolition waste generation and properties of recycled aggregate concrete: a global perspective. J Clean Prod 186:262–281

Al-Jumeily D, Hashim K, Alkaddar R, Al-Tufaily M, Lunn J (2018) Sustainable and environmental friendly ancient reed houses (Inspired by the past to motivate the future). In: 11th international conference on developments in esystems engineering (DeSE). Cambridge UK, pp 214–219

Alyafei A, Al-Kizwini RS, Hashim KS, Yeboah D, Gkantou M, Al Khaddar R, Zubaidi SL (2020) Treatment of effluents of construction industry using a combined filtration-electrocoagulation method. In: IOP conference series: materials science and engineering 888(1)

Barbuta M, Bucur RD, Cimpeanu SM, Paraschiv G, Bucur D, Agroecology C (2015) Wastes in the building materials industry. INTECH Croatia 81–99

Belayutham S, Gonzalez VA, Yiu TW (2016) A cleaner production-pollution prevention-based framework for construction site induced water pollution. J Clean Prod 135:1363–1378

Bozkurt O, Islamoglu M (2013) Comparison of cement-based and polymer-based concrete pipes for analysis of cost assessment. Int J Polym Sci 2013

Carr GM, Neary JP (2008) Water quality for ecosystem and human health. UNEP

Chauhan SK, Sharma S, Shukla A, Gangopadhyay S (2010) Recent trends of the emission characteristics from the road construction industry. Environ Sci Pollut Res 17(9):1493–1501

De Paula HM, De Oliveira Ilha MS, Andrade LS (2014) Concrete plant wastewater treatment process by coagulation combining aluminum sulfate and *Moringa oleifera* powder. J Clean Prod 76:125–130
Gan X, Zuo J, Ye K, Skitmore M, Xiong B (2015) Why sustainable construction? why not? an owner's perspective. Habitat Int 47:61–68
Garetti M, Taisch M (2012) Sustainable manufacturing: trends and research challenges. Prod Plan Control 23(2–3):83–104
Gunhan S, Arditi D (2005) Factors affecting international construction. J Constr Eng Manag 131(3):273–282
Gupta J, Vegelin C (2016) Sustainable development goals and inclusive development. Int Environ AgreemS: Polit, Law Econ 16(3):433–448
Gupta S, Sharma S, Saini RS (2015) Evaluation of stipulated conditions imposed at the time of grant of environment clearance from the perspective of implementation of compliance-a case study for construction projects in India. Chin J Urban Environ Stud 3(01):1550006
Horvath A (2004) Construction materials and the environment. Annu Rev Environ Resour 29:181–204
Hussin JM, Rahman IA, Memon AH (2013) The way forward in sustainable construction: issues and challenges. Int J Adv Appl Sci 2(1):15–24
Issaka S, Ashraf MA (2017) Impact of soil erosion and degradation on water quality: a review. Geol, Ecol, Landsc 1(1):1–11
Jain G, Gupta V, Pandey M (2016) Case study of construction pollution impact on environment. Int J Emer Tech Eng Res (IJETER) 4(6)
Jain M (2012) Economic aspects of construction waste materials in terms of cost savings–a case of Indian construction industry. Int J Sci Res Publ 2(10):1–7
Kitchener BG, Wainwright J, Parsons AJ (2017) A review of the principles of turbidity measurement. Prog Phys Geogr 41(5):620–642
Kolade DM (2018) Review paper on industrial wastewater treatment processes. University of Nigeria
Kumar A, Bisht BS, Joshi VD, Singh AK, Talwar A (2010) Physical, chemical and bacteriological study of water from rivers of Uttarakhand. J Hum Ecol 32(3):169–173
Luhar S, Luhar I (2019) Potential application of E-wastes in construction industry: a review. Constr Build Mater 203:222–240
Mahurpawar M (2015) Effects of heavy metals on human health. Int J Res-Granthaalayah 2394–3629
Morledge R, Jackson F (2001) Reducing environmental pollution caused by construction plants. Environ Manag Health
Muhwezi L, Kiberu F, Kyakula M, Batambuze AO (2012) An assessment of the impact of construction activities on the environment in Uganda: a case study of Iganga municipality. J Constr Eng Proj Manag 2(4):20–24
Nigam NC, Maheshwari AK, Rao NP (2007) Safety and health in chemical industry. Indian J Fertil 3(7):13
Pandey G, Madhuri S (2014) Heavy metals causing toxicity in animals and fishes. Res J Anim Vet Fish Sci 2(2):17–23
Petraru M, Gavrilescu M (2010) Pollution prevention, a key to economic and environmental sustainability. Environ Eng Manag J 9(4):597–614
Phukon J, Bora A (2016) Water characteristics of dibrugarh town protection drain: its impact on aquatic organisms and human health
Pitt R, Clark SE, Lake DW (2007) Construction site erosion and sediment controls: planning, design and performance. DEStech Publications Inc
Ponnada M, Kameswari P (2015) Construction and demolition waste management—a review. Safety 84:19–46
Rao PB, Shivakumar B, Suresh HS (2014) Waste minimization in construction industry. Indian J Appl Res 4(6):23–26
Schwarzenbach RP, Egli T, Hofstetter TB, Von Gunten U, Wehrli B (2010) Global water pollution and human health. Annu Rev Environ Resour 35:109–136

Scott K, Owen A, Barrett J (2013) Estimating emissions associated with future UK consumption patterns. A report for the UK committee on climate change
Shah A, Khan S, Shah MH, Khan R, Jan I (2010) Environmental impact assessment (EIA) of infrastructure development projects in developing countries. OIDA Int J Sustain Dev 1(4):47–54
Shivaranjani SK, Thomas LM (2017) Performance study for treatment of institutional wastewater by activated sludge process. Int J Civ Eng Technol 8
Tang HH, Soon HY, Larsen IB (2003) Solid waste management in Kuching, Sarawak, DANIDA/Sarawak government UEMS project, natural resources and environment board (NREB), Sarawak & Danish International Development Agency (DANIDA)
Tsimas S, Zervaki M (2011) Reuse of waste water from ready-mixed concrete plants. Manag Environ Qual: Int J 22:7–17
Tzoulas K, Korpela K, Venn S, Yli-Pelkonen V, Kazmierczak A, Niemela J, James P (2007) Promoting ecosystem and human health in urban areas using Green Infrastructure: a literature review. Landsc Urban Plan 81(3):167–178
United States Environmental Protection Agency (US EPA) (1998) Characterization of building-related construction and demolition debris in the United States, report no. EPA530-R-98-010, U.S. Environmental Protection Agency Municipal and Industrial Solid Waste Division Office of Solid Waste
Vega M, Pardo R, Barrado E, Deban L (1998) Assessment of seasonal and polluting effects on the quality of river water by exploratory data analysis. Water Res 32(12):3581–3592
Wang SQ, Dulaimi MF, Aguria MY (2004) Risk management framework for construction projects in developing countries. Constr Manag Econ 22(3):237–252
Yeheyis M, Hewage K, Alam MS, Eskicioglu C, Sadiq R (2013) An overview of construction and demolition waste management in Canada: a lifecycle analysis approach to sustainability. Clean Technol Environ Policy 15(1):81–91

Chapter 14
Design and Development of Improved Methods of Curing of Bricks During Manufacturing Process and Construction Work to Save Water, Minimize Pollution and Human Effort

Ramesh Chandra Nayak, Manmatha K. Roul, Payodhar Padhi, and Saroj K. Sarangi

Abstract Curing of bricks during manufacturing process and construction work has a significant impact on the strength, durability and wear resistance of concrete. If the concrete is not properly and adequately cured, it will fail to satisfy the purpose for which it is designed. Minimum 12–24 days curing is required for bricks during the manufacturing process and minimum 7 days water supply is required for masonry wall and concrete. Bricks require a lot of water and time for curing. Maximum use of water is an environmental issue and supply of water by laborers from tube wells to construction sites has an adverse impact on their health. In the present work we have developed two methods for curing bricks and concrete. First method will help for curing of bricks and concrete during the manufacturing process without the requirement of a single drop of water and the second method will help for better supply of water to constructional workplaces with minimum effort.

Keywords Cement concrete · Construction · Curing · Steam · Tube well · Vacuum chamber · Vacuum pump · Water supply

---

R. C. Nayak (✉)
Department of Mechanical Engineering, Synergy Institute of Engineering & Technology, Dhenkanal 759001, Odisha, India

M. K. Roul
Department of Mechanical Engineering, Gandhi Institute for Technological Advancement (GITA), Bhubaneswar 752054, Odisha, India

P. Padhi
Department of Mechanical Engineering, KIST, Bhubaneswar 752050, Odisha, India

S. K. Sarangi
Department of Mechanical Engineering, NIT, Patna 800005, Bihar, India

J. A. Malik and S. Marathe (eds.), *Ecological and Health Effects of Building Materials*,
https://doi.org/10.1007/978-3-030-76073-1_14

14.1 Introduction

India, being a developing nation, provides ample opportunities in the construction sector which includes the construction of smart cities, roads, buildings and technology parks all over the country. The Government of India also introduces several programs like PMAY which is meant for constructing homes for the needy families in the urban areas. Such a demand paves the way for fledgling entrepreneurs to choose the construction sector as their major area of operation. For any kind of construction there is a constant requirement of a workforce which consists of skilled as well as unskilled laborers and the issues related to their health and environment have become prime concerns for the entrepreneurs. Bricks, being an important element in construction work, are required to be produced with a better technique in order to minimize environmental pollution so that the health condition of the workers will be improved. Due to the huge requirement of bricks, a large number of entrepreneurs are setting up brick industries as their first choice of business. Mainly there are two types of bricks—the fly ash brick and the clay bricks. Both the bricks, used largely in construction works, require proper curing after they are manufactured. A lot of water is used for curing cement concrete. As our objective is to save water, we focus on not using a single drop of water for curing. This system will make curing of cement concrete slabs, hume pipes and bricks, etc., which require curing, without water. At present there are water and steam curing processes available. In this process a lot of water is going to be wasted and it requires a period of at least 28 days for curing. In our proposed innovation, curing will be done without a single drop of water and within 24 h with the same strength and by saving a lot of cement. Hence, two benefits are going to be achieved at the same time. In this system, an airtight chamber will be designed in order to accommodate all the items that need to be cured. A vacuum pump will be attached to it. A chemical is going to be used, which generates water vapors for the purpose of curing. Curing will be done within 24 h. The technology is unique. The system of curing is completely new since people are using either water or steam for curing till now. Here, we are using neither water nor steam. This is the uniqueness of our product. A number of works have been done by researchers, where they have worked on different types of curing methods, focused on their working principles and advantages, but not a single work has been done earlier on vacuum curing without the use of water. In this chapter, we have also presented another product that helps for the supply of water for curing of constructional items, like masonry wall, concrete etc. with minimum effort. During construction work curing of bricks and concrete is also a major part. If proper water is not supplied to construction work, then there is a chance of failure of construction.

In this chapter two methods have been described, that is curing of bricks during its manufacturing process by vacuum curing method and curing of construction work by a simple method, where effort required for supplying water to construction work will be negligible. Nguyen et al. (2020) in their work describes how the pure slag destroys the performance of concrete and adding of gypsum changes the performance. Work by Shi et al. (2020a) presents how the performance of concrete changes due to implementation of steam curing methods. They found that due to this steam curing method the mechanical properties of the concrete developed. Mohammed et al. (2020) in their work described the disadvantage about the provision of plastic within concrete and mentioned that microwave curing method improves the chemical and hydrophobic resistivity and improved concert. Shi et al. (2020b) studied the advantages and disadvantages of steam curing methods in this work they indicate that steam curing method is a good method but higher temperature causes harm to the microstructure of the concrete. Yin et al. (2020) studied experimentally, the control of fly ash on the curing features of an epoxy resin and found that due to increasing rate of heat there is enhancement of curing rate takes place. Liu et al. (2020) in their work describe the advantages of steam curing method, highlighting that due to steam curing method, compressive strength and permeability improves and chances of damage decreases. They also represented that curing time is also a factor for steam curing methods. Based on the work of Shi et al. (2020c) on steam curing, it is indicated that in the microstructure of moisturizing products there are clear gradients of porosity and differences on the surface and inside of the concrete treated by steam, and the coefficient of surface absorption is more on the inside. By extending the preheating time higher strength can be accomplished, and the bonded porosity of the surface to atmosphere can be reduced by surface treatment.

Reddy and Hamsalekha (2020) worked on advanced curing method, where they used internet of things (IoT) to develop automatic curing system to create an automatic water-saving treatment mechanism for curing which depends on the moisture available in the concrete and the temperature of the ambient by the use of a humidity sensor. Chen and Gao (2020) studied carbonation curing methods and, in their work, they compared their work with conventional methods. Carbon treatment can effectively improve compressive strength, the carbonation becomes deeper and more homogeneous in the preceding concrete, and the effect of filtration can be reduced on the critical pore diameter through carbon treatment. However, the increase of large pore content which is caused by freeze-thawing can hardly be prevented. Zhang et al. (2013) found that the resin system has a curing time of 5 min with a 95%curing degree, and the processing time of the flakes can be controlled with the studied resin within 13 min under 120 °C with a curing degree of more than 95% and having a few defects. A slight decrease in thermal stability and bending property is exhibited by slides which are manufactured as per the short term curing schedule when it is compared to the cured slides used in the traditional curing schedule which takes more than two hours of curing time. Chang et al. (2020) studied the environmental effect on curing methods, and found that efflorescence crystallization can be inhibited when curing is made by using a cling film seal. Ge et al. (2020) studied the use of recycled clay and found that with pre-hydration time and an increase in RFCBA (recycled fine clay brick aggregate) content, there is a decrease in the initial stagnation flow and there is also an observation of the opposite trend of slum flow loss. It is also found that with an increase in the RFCBA content, there is a decrease in the compressive and bending strength of RFCBA slurry. RFCBA can significantly lag in reducing indoor relative humidity. Moreover, as the porous RFCBA stored water is released, the RFCBA content increases and the dry shrinkage resistance of the RFCBA slurry mixture gets improved. Taki et al. (2020) from their work found that the prepared fired bricks with lower thermal conductivity (0.48 W/m K) indicate better property of insulation. The unique highlight of our study is the potential use of the FA in order to stabilize SS so that the natural clay fired bricks can be sustainably replaced. A future study may investigate the compositional effect of FA on the thermodynamic and thermodynamic properties of bricks. Cheng et al. (2020) studied on laminates by composite methods and its effect; they found that the tensile and the repaired laminates with their comprehensive properties are hardly affected by the curing condition. However, the bonding quality is highly affected due to the adhesive inner voids. Sankar and Das (2019) worked to enhance the strength of bricks by reinforcing their method of work. They made a comparative study between standard composite samples and composite samples with graphene of equal dimensions to find out the effects on the comprehensive strength of the brick by supplementing various amounts of graphene. Dinh and Vinh (2019) worked on solar cure of bricks and found that temperatures above 50 °C can be achieved which is required for curing the concrete bricks. Hence, the feasibility for the use of solar energy could be confirmed for the treatment of concrete bricks depending on the condition of the climate. Li and Zhao (2016) worked on autoclave curing of bricks and they explained that the compressive strength increases due to such a type of curing method.

Zhou and Qu (2011) studied on autoclaved sludge bricks. They could find the produced brick with a comprehensive strength of 20.8 MPa and with a bending strength up to 5.4 MPa. They also discussed the forming and stirring conditions on the mechanical properties of the aseptic sludge bricks due to the effect of curing. Nam Boonruang et al. (2011) worked on soil bricks, present that Fly ash formulations greater than 25% by weight based on soil and from up to 14 days of curing time have proven to be economical mixtures of bearing slabs or brick-type structural elements according to the Thai Industrial Standard (TIS) for structural clay bearing - tile bearing. Therefore, commercial development is very promising. Sadrmomtazi and Haghi (2008) worked on thermal drying of bricks and found that thermal gradients play a role in describing the moisture profiles within a material when the thickness is large. Predictions of temperature and moisture content show that the leading edge dries faster as compared to other sides of the solid. The distributions of drying temperature and moisture content in the porous solid were not uniform due to the forward slack effect during convection drying.

Spraying water on constructional work, like masonry walls, concrete is also known as curing. A Lot of work has been done by a number of researchers; they focused on smart curing for construction and irrigation purposes. Not a single work is there, where minimum effort for lifting of water from tube wells for curing has been described. But our work is a unique work, where minimum effort is required for lifting water from tube wells for constructional and agricultural work. Singh et al. (2020) studied River liquid irrigation with weighty metal load effects soil organic actions and risk aspects and found Potential biological risk aspects (Er) were under little risk and all-inclusive probable ecological jeopardy indices (Ri) were found to be under truncated, reasonable and high-risk categories. Yanala and Pagilla (2020) studied usage of biochar to produce domesticated liquid for irrigation usage and found it commercially accessible. Granular activated carbon (GAC) performed much better than biochar for all the amalgams measured. Abioye et al. (2020) studied an assessment on monitoring and cutting-edge control approaches for meticulous irrigation and found that this assessment aims to support researchers in detecting guidelines and gaps for future study and work in this field. Salahoui et al. (2020) studied an amended tactic to guesstimating the infiltration physiognomies in surface irrigation systems and found that the infiltration function projected by means of the suggested tactic was more precise and rational than the infiltration function projected using the double ring infiltrometer (DRI), and pragmatic (Kostiakov model) approaches. Terence and Purushothaman (2020) studied organized assessment of the Internet of Things in clever farming and found that this lessens man power and upsurges resource exploitation in farming. Kukal and Irmak, (2020) studied Influence of irrigation on interannual unpredictability in the United States agricultural efficiency and found that the demonstration of spatial and chronological dynamic forces in Irrigation-Induced Reduction in Crop Yield Variability (IITV) could support in irrigation-water apportionments and implementation. Zhu et al. (2020) studied founding of agronomic drought loss replicas: A comparison of arithmetic procedures and found that the root mean square fault and the mean absolute fault of the multivariate step by step deterioration were 1.31 times and 1.38 times respectively greater than the root mean square fault and the mean absolute fault from the arbitrary forest model.

Anuradha et al. (2020) studied mathematically about water users connotation for justifiable improvement in agricultural products in rural areas and found that the deterioration equation condition for revenue depends on the size of the agronomic farms and disbursement for cultivation events which should be properly monitored to improve the living conditions of the people living in rural areas. Dehghan et al. (2019) studied the influence of weather change on Agronomy and Irrigation Network and found that the performance of irrigation grids is estimated in rapports of equity and appropriateness indices. Li et al. (2019) found in their study on agronomic water apportionment under ambiguity restructuring of water deficiency risk that the part of water privileges in risk restructuring was more important when the probability dispersal of water scarcity risk was asymmetric. Rajasekaran and Anandamurugan (2019) studied an assessment of remonstrance and implementation of Wireless Sensor Networks in Clever Farming and compared different conventional methods with clever farming in the agronomic domain. Singh (2018) studied an assessment on salinization of agronomic domains due to deprived drainage and provided an outline of various procedures and their appropriateness and restrictions in managing the land salinization and increasing groundwater level complications of irrigated zones. Aleotti et al. (2018) studied A Clever Accuracy-Agronomy Platform for Lined Irrigation Arrangements and found that such systems could help the farmers in various operations of the irrigation domain. Li et al. (2018) studied on Handling irrigation and pollination for the supportable gardening of greenhouse vegetables and found that mitigation procedures for N leaching contamination from greenhouse vegetable grounds should consider guidelines on irrigation and pollination. Prabha et al. (2018) designed and developed a smart irrigation system for farming of chilli based on IOT for improvement in fertilization and irrigation with reduced man power and water supply for reaching higher yields. Rahman et al. (2020) explained the usability of coconut husk. Wang (2020) studied the use and environmental effects of phosphogypsum (PG) in agriculture and found that although use of PG had several benefits, using waste PG could create issues of radiological impact, salt concentration and heavy metal toxicity. The effects of incentive mechanisms of different agricultural models on agricultural technology was discussed by Yu et al. (2020) and they found that ecological agriculture was preferred by most researchers and agricultural information technology management systems had a major role in the development of agriculture. The effects of climate change on the Hamadan-Bahar plain on various sides were studied by Mosavi et al. (2020) and found that climate change had negative effects on the agricultural sector in this region which could be tackled by improved irrigation technologies and by use of an ideal deficit irrigation strategy. Lalehzari and Kerachian (2020) studied a new methodology for distributing groundwater to agricultural lands and found that this strategy increased water productivity, economic efficiency of land and provided highest values of benefit per cost ratio. Water saving by agricultural virtual water trade (VWT) by considering various irrigation factors was studied by Cao et al. (2020) and found that irrigation played an important role in cultivating crops for both the import and export regions without virtual water trade. The simulation of the surface energy balance (SEB) was done by Ishola et al. (2020) who found that soil properties played an important role in finding surface fluxes. The calculation of efficiency of a semi-closed horizontal tubular photo-bioreactor (PBR) for removal of target compounds was done by Vassalle et al. (2020) and they found that this system could be a solid choice for treatment because of its parameters like pH in the closed system, the size of the reactors, high temperatures and the developed specific mixed cultures.

Arrieta-Escobar et al. (2020) studied the importance of 3D printing for improving the understanding of soil functioning and found that the accessibility of additive manufactured soil models could help researchers to conduct experiments for better understanding of soil functioning factors. The use of Machine learning for calculating emission of greenhouse gas from agricultural fields was studied by Hamrani et al. (2020) and they found that the LSTM model could be used for determining these emissions. Zhou et al. (2020) studied an integrated irrigation strategy WSQI and found that this could help researchers by providing a theoretical basis as well as help in improving agricultural production. Boyer (1982) studied the productivity of plants according to the environment and found that by understanding the fundamental mechanisms with help of scientific advances could help to improve productivity. Matson et al. (1997) studied the effect of agricultural intensification on ecosystems and found that although it is having a negative impact, these effects could be reduced as well as the agricultural sector could be improved by using ecologically based management strategies. The assessment of various prospects for improving yields in order to tackle the supply and demand problem due to increased population was studied by Mueller et al. (2012).They found that the demand could be met by utilizing underperforming landscapes and improving its yield by eliminating the overuse of nutrients and proper management of water supply.

14.2 Experimental Setups

Curing of bricks and concrete is an important task for any type of construction work. Generally bricks are prepared separately and need to be cured afterwards. Perfect cured bricks provide better results with regard to their strength. If bricks are not cured properly, failure of construction work may result due to the reduced strength of bricks. There are a number of methods available for curing of bricks such as water curing and steam curing. A number of works have been done by a number of researchers on these methods and it was found that there still exist a number of issues which need to be addressed. It was found that by the water curing method a lot of water is required and a minimum 24 days of curing period is required. Steam curing method requires a skilled laborer and it is a costly method. To overcome such problems, new techniques and methods are required. Nowadays, there is also unemployment which is a great challenge. A large number of entrepreneurs are now choosing the brick manufacturing industry as their first choice. Due to advancement of technology, it is required to apply new methods for curing to save the environment and maximize profit of business. In our work, we have developed two new methods. One of the methods, where a single drop of water is not required for curing of bricks during its manufacturing process and strength of bricks by using our curing method provides better strength than traditional methods. Time required for our curing method is only 24 h, and the other new method has been presented, where lifting water from a tube well is to be done by laborers with minimum effort. This method will help for the curing (Water supply) of concrete and bricks during construction work. So, in this work two experimental set-ups have been described below.

14.2.1 Curing of Bricks During Their Manufacturing Process

In this work, a new way of curing to bricks has been explained, where the chemical calcium sulphate dehydrate ($CaSO_4{\cdot}2H_2O$), vacuum chamber, vacuum pump are the main requirements. This method is known as Vacuum curing which is used in fly ash based brick Industry. Steam curing is the cheapest method till now to improve the strength of concrete. However, steam curing has the defect of causing cracks in reinforced precast concrete members. Further, immediate disposal of these products from the curing plant was also not possible due to rise in temperature and unavailability of skilled labourers. These defects can be overcome with our vacuum technology, where the concrete can be kept in a vacuum chamber with a low cost chemical under some external heat which is used for curing within 24 h. It is very cheap and does not require water at all.

For prototype development it will take 12 months. For scaling up to commercialization model requires 12 months and to reach break—even point for business, it requires 6 months. The profit margin is 60% of the sales. Since we are not going

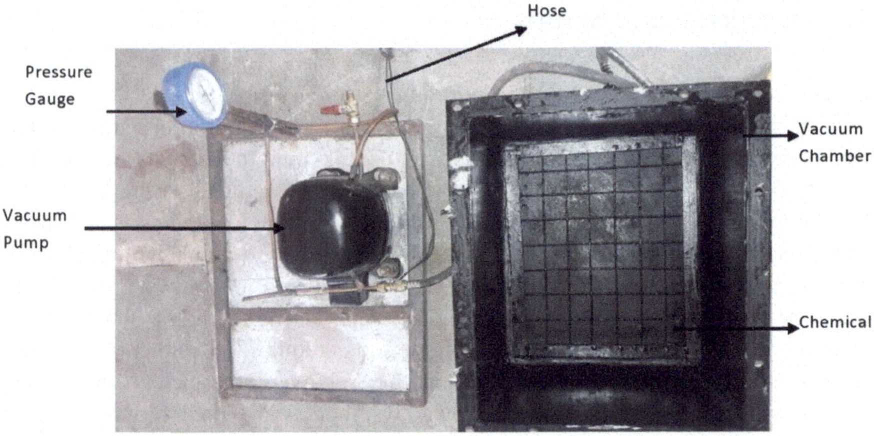

Fig. 14.1 overall arrangements for curing of bricks with chemical and vacuum chamber

to provide any material for construction, it is the responsibility of the customer to provide the materials as per our design for the preparation of the vacuum chamber.

The Fig. 14.1 below shows the overall arrangement of the setup. It consists of a vacuum pump, vacuum chamber, chemical, hose pipes and pressure gauge. Our experimental setup is designed for curing a single fly ash brick. In this system an air tight chamber has been designed to accommodate all the items to be cured, which is known as a vacuum chamber. Vacuum can be created by using a vacuum pump. One end of the vacuum pump hose is connected to the vacuum chamber and the other end is opened to the atmosphere. Vacuum pump is operated by an external electric supply. Proper vacuum creation is observed by the help of a pressure gauge.

In our work we have prepared a 6 in. fly ash brick by using a proper brick manufacturing method, and then kept in the vacuum chamber, a chemical in a container is placed outside the chamber and then the chamber is sealed, after that air from the chamber is extracted by using a vacuum pump. Due to extraction of air from the chamber, it becomes vacuum. The chemical, placed outside the vacuum chamber, is heated with less heat. Reactions from the chemical are sprayed to the vacuum chamber by nano nozzles to penetrate inside the voids of the brick. Curing is to be done for 24 h.

When the chemical calcium sulphate dihydrate ($CaSO_4 \cdot 2H_2O$) is heated, water vapour is produced. This water vapour can easily pass through the capillaries of the concrete. A traditional solution-free epoxy is provided to cover the concrete which acts as a vapour seal to prevent evaporation of water from the concrete. Hydrostatic pressure and osmotic stress can cause waterborne problems, with the latter having a greater impact on concrete. The pressure created by osmosis can surpass other forces and eventually degrade the coating of a floor. The conditions required for generation of osmosis are (i) water, (ii) semi-permeable membrane and (iii) soluble salts. As all these conditions are generally satisfied in concrete, it favours the generation of osmosis pressure.

The difference in the concentration of soluble salt associated with the permeability of concrete to inorganic salts between the upper and lower segments has been recognized as an important contributor to the development of osmotic pressure. The moisture content of concrete can vary from about 4% when the concrete fully cures to about 18% in freshly prepared "green concrete".

14.2.2 Curing of Bricks During Construction Work

In this chapter, a simpler method for curing of bricks and concrete during construction work has also been described. During construction work, water supply to masonry walls and concrete is required in order to improve the strength and accomplish other desired properties for their intended use. During this time water supply to construction parts like masonry wall, plaster, and concrete is very important. It is found that during this process, water is manually taken from the tube well and applied on these places. So, a number of extra labourers are required for this process. Moreover, lifting water and operating a tube well also affect the health of the labourers. So, in order to simplify this method, our designed product will be helpful with minimum effort and maximum efficiency.

The arrangement of water supply for curing bricks and concrete consists of a pair of spur gears, bearings, shaft, handle and flywheel. The system is connected with the tube well plunger rod in an eccentric way. Gears are arranged in simple gear train arrangements. Two different arrangements have separately been used for irrigation purposes. In simple gear train arrangement, the number of teeth on driver and driven gears are of 200 and 40 respectively. Two shafts having length of 640 mm and 600 mm are taken in this system. Driver and driven gears are attached on two shafts. Handle is connected at the end of one shaft and the flywheel is connected at the end of the other shaft. The shaft carries a flywheel at one end and contains a pulley on its other end. That pulley is connected with the plunger rod of the tube well. The shaft which is connected with the handle carries driver gear which has 200 numbers of teeth and the other shaft having flywheel and pulley carries the driven gear of 40 numbers of teeth. Both driver and driven gears are messed with each other. Bearings are provided for smooth rotation of shafts. The experimental set up is done in the workshop and for preparing the setup we have used a welding machine, cutting machine, and gear hobbing machine for generating teeth on gears. Spur gears having teeth of 200 and 40 numbers are manufactured by using gear hobbing machines with improved accuracy. After manufacturing the gears, black oxide finish method was used for preventing rust. The dimension taken for this purpose may be varied for any type of requirement.

A flywheel of 30 kg weight is provided for energy storage and smooth output power deliberation purposes. When the handle is rotated, the driver shaft and driver gear also rotate as the driven gear is meshed with the driver gear, so the driven gear causes to rotate the driven shaft, one end of the driven shaft is attached with a flywheel and the other end is attached with the pulley. Due to the rotation of the pulley the plunger rod connected with the pulley, moves up and down causing it to move up and down the plunger poppet. All such arrangements are provided on a frame made of (40 × 40 × 5) mm angle of mild steel. And water is lifted at the upward direction of the plunger rod.

A timeline has been developed for the product design and development, where time required for product development is 1 month, for scaling up to commercialization model it takes 2 months and for achieving break—even point requires 2 months. The profit margin for business purposes is 40% of the sales.

14.3 Results and Discussion

In this chapter, two experimental set ups have been developed and explained for curing of bricks and concrete. Both these designed models are helpful for construction work. One of the methods is for curing of bricks during its manufacturing process and other is for supply of water after construction work. In the Sect. 14.3.1 below, the curing of bricks without use of water during its manufacturing process has been described. Section 14.3.2, describes an innovative method for the supply of water with minimum effort for curing purposes after the construction work.

14.3.1 Method for Curing of Bricks During Their Manufacturing Process

In this work, a 6-in. fly ash brick before curing has been taken and placed inside a vacuum chamber. A vacuum pump is used to extract air from a vacuum chamber. Chemical is heated externally and products from the chemical are sprayed inside the chamber through a nano sprayer. This process continues for 24 h. After 24 h the chemical spray is stopped. And the brick from the vacuum chamber is taken out. The compressive strength of the brick has been measured by using a Universal testing machine (UTM). It is found that the compressive strength of the brick is more than a brick which is cured by the supply of water. This method of curing produces better finished bricks than traditional methods due to absence of water spray to bricks.

Our system has tremendous demand in the market. It has a B2B (business to business) and B2C (business to consumer) approach. It implies that it can be considered as a selling of service as well as product directly to the consumers. The product will be designed and modified as per the consumer requirement. Construction sector, brick industries, hume pipe industries, railway concrete sleeper and industries producing manhole slabs will be highly benefited by this system. They will place an order for this set up as per their requirement. The product can be customized as per the requirement of the consumers. From a competitor point of view the technology is unique and innovative. The system of curing is completely new since till date for

Table 14.1 Comparison between traditional and designed work

Factors	Traditional methods	Designed work
Curing	Water required	Not a single drop of water
Time required	15–24 days	24 h
Strength	8 MPa	9 MPa
Cost (frequency)	Always	Only initial set up
Environmental Issue	Mercury Pollution	Environment friendly

curing purposes people are using either water or steam. Here, we are using neither water nor steam. This is the uniqueness of our product. Our product will be sold in the market instantly. Every house will come forward and take this from the sales counter. Only we have to promote our product by giving an advertisement in the media and social network etc.

This developed system helps to create maximum opportunities to minimize unemployability. It is a continuous process. Throughout the year there are customers available as construction never stops. Hence, there is a repeated purchase of this product. The idea can be easily implemented because the system will be designed to serve the intended purpose. The material, which is the trade secret, will be provided by us. Although this method is designed for smaller scale of construction, we are in the process of developing the set up for a bigger scale of construction also. Table 14.1 shows the advantages of the newly developed curing method of bricks over the traditional method of curing. It clearly indicates that this new method of curing is helpful for addressing both environmental issues and business purposes.

14.3.2 Method for Curing of Bricks After Constructional Work

We have conducted a survey by taking our design model on construction work of a three storeyed building having 1200 ft^2 constructional area. During the time of construction, there were no facilities for water supply. Most of the construction works are done by using a tube well which was drilled earlier. We found that during construction work extra labourers were required for supply of water from tube wells for masonry and concrete work as operating the tube well for lifting of water is not an easy task. So, we set up our system in that construction area and found that lifting water became very easy for construction purposes. Table 14.2 shows a comparative statement of expenditure for curing during construction of the three storeyed building by manually operated tube well and by our designed model.

From Table 14.2, it can be observed that minimum man power is required for lifting water from tube wells for constructional work using our designed model as

Table 14.2 Comparison between manually operated tube well and designed model

Sl. no	Factors	Manually operated tube well	Designed model
1	Manpower required for water supply	60 labours	10 labours
2	Time required for water lift	200 h	100 h

compared with the traditional manually operated tube well. When our designed model was installed with the tube well at the construction area, it was found that even a lady and handicapped labourer could lift water with minimum effort.

The timeline to develop the product would be as follows. One month of time may be required for product development, one month for scaling up to commercialization of the model, and two months for achieving break—even point. The profit margin would be 40% of sales. There is a large market for this system as this is the best system for supply of water for curing of bricks and concrete. This model can also be used for supply of water for irrigation purposes. There will be no issue for such type of product development as it is an environment friendly system. The purpose of this system is to decrease environmental pollution without the use of electricity, petrol or diesel.

Our designed model is environment—friendly and it produces zero emission, as this system does not need any external source like electric, petrol and diesel for its operation. From a business point of view, our product will be used as there is a large number of construction works carried out all over the world. This is a unique product. Hence, there would be hardly any problem for marketing. The idea can be easily implemented because the raw materials used for this system are easily available in the market.

14.4 Conclusions

In this chapter, the design and development of two products have been described. Both methods are helpful for providing a better environment and better opportunities for business with reduced cost and minimum effort. The conclusions drawn from this analysis are given below.

14.4.1 Method of Curing of Bricks During Their Manufacturing Process

Nowadays, brick manufacturing industries are chosen for the first choice of business, due to huge demand on constructional works. A business and an industry will sustain if proper technologies are adopted. Generally in constructional work, supply of water for curing purposes is a main requirement. Our new technology will provide a better way for curing bricks during its manufacturing process.

(i) The technology is innovative and unique. The system of curing is completely new as people are using either water or steam for curing till today. Here, we are using neither water nor steam for curing purposes. This is the uniqueness of our product.
(ii) Our product will be sold in the market instantly. Every house will come forward and take this product from our sales counter by paying the cost of the product.
(iii) Throughout the year there are customers available as construction never stops. Hence, there is a repeated purchase of this product.
(iv) The idea can be easily implemented because the system will be designed by us. The material, which is the trade secret, will be provided by us. Steps are taken to set up the product for bigger construction projects.
(v) The designed method is environment—friendly.
(vi) Time of curing is reduced and the compressive strength of the bricks and concrete is improved by this method.
(vii) The initial cost of this product may be high but the running cost is very less and it produces better finished products as compared to other traditional methods of curing.

14.4.2 Method of Curing of Bricks After Constructional Work

Water supply for construction work in a number of construction sites is done by lifting water from tube wells or bore wells. Lifting of water from borewells or tube wells requires diesel, petrol, kerosene or electricity for the operation of pumps. In some constructional areas manual workers are engaged for such purposes. Lifting water manually is not an easy task especially when women workers and aged labourers are engaged for such purposes. To overcome such difficulties, our designed product will help a lot. Number of conclusions from this work is enumerated below.

(i) The concept that the increase of mechanical advantage which results in optimum efficiency is relevant right here.
(ii) The idea of energy storage by means of flywheel has been involved here, due to which by means of much less effort the lever of the tube performs smoothly.
(iii) There is no need for electrical supply, petrol or diesel for running this product.
(iv) The price of this product is so low that it can be purchased by poor people.
(v) The weight of this product is also less so that it can be easily taken to the workplaces for any constructional work.
(vi) This method also helps for agriculture purposes as the soil will be no longer affected because irrigation at any time can be made possible by the use of this technique.

(vii) Installation and maintenance of this product is so simple and is not a matter of concern for such a system.
(viii) Since there is no requirement of petrol, diesel or electricity, there will be no emission at all from this system which is a main advantage of such a system.
(ix) Any type of labourers and farmers can utilize this method.
(x) This could boost productivity resulting in the growth of the economy of the country.
(xi) A physically handicapped worker and farmer also can use it.
(xii) Multi activity is viable with the aid of the usage of this machine.
(xiii) A lady worker can use this product without difficulty for the supply of water to construction work and agriculture sites.

References

Abioye EA, Abidin MSZ, Mahmud MSA, Benyamin S, Ishak MHI, Rahman MKIA, Otuoze AO, Onotu P, Ramli MSA (2020) A review on monitoring and advanced control strategies for precision irrigation. Comput Electron Agric 173:105441

Aleotti J, Amoretti M, Nicoli A, Caselli S (2018) A smart precision-agriculture platform for linear irrigation systems. In: 26th international conference on software, telecommunications and computer networks. Soft COM 2018(8555841), pp 401–406

Anuradha B, Packialakshmi S, Nagamani K, Sarala Devi J (2020) Statistical study of water users association for sustainable agriculture in rural development. J Green Eng 10(5):2212–2224

Arrieta-Escobar JA, Derrien D, Ouvrard S, Asadollahi-Yazdi E, Hassan A, Boly V, Tinet AJ, Dignac MF (2020) 3D printing: An emerging opportunity for soil science. Geoderma 378:114588

Boyer JS (1982) Plant productivity and environment. Science 218(4571):443–448

Cao X, Cui S, Shu R, Wu M (2020) Misestimation of water saving in agricultural virtual water trade by not considering the role of irrigation. Agric Water Manag 241:106355

Chang JN, Wang WC, Wang HY, Hu CY (2020) A study of the mechanical properties and appearance of efflorescence in paving bricks under different curing environments. Constr Build Mater 265:120148

Chen T, Gao X (2020) Use of carbonation curing to improve mechanical strength and durability of pervious concrete. ACS Sustain Chem Eng 8(9):3872–3884

Cheng X, Zhang J, Cheng Y, Guo X, Huang W (2020) Effect of curing condition on mechanical properties of scarf-repaired composite laminates. Steel Compos Struct 37(4):419–429

Dehghan Z, Fathian F, Eslamian S (2019) Climate change impact on agriculture and irrigation network. Clim Change Manage 333–354

Dinh Thien B, Vinh Long H (2019) A study of using solar energy to cure concrete bricks. E3S Web of Conferences 97:06041

Ge Z, Feng Y, Zhang H, Xiao J, Sun R, Liu X (2020) Use recycled fine clay brick aggregate as internal curing agent for low water to cement ratio mortar. Constr Build Mater 264:120280

Hamrani A, Akbarzadeh A, Madramootoo CA (2020) Machine learning for predicting greenhouse gas emissions from agricultural soils. Sci Total Environ 741:140338

Ishola KA, Mills G, Fealy RM, Chonchubhair ÓN, Fealy R (2020) Improving a land surface scheme for estimating sensible and latent heat fluxes above grasslands with contrasting soil moisture zones. Agric for Meteorol 294:108151

Kukal MS, Irmak S (2020) Impact of irrigation on interannual variability in United States agricultural productivity. Agric Water Manag 234:106141

Lalehzari R, Kerachian R (2020) Developing a framework for daily common pool groundwater allocation to demands in agricultural regions. Agric Water Manag 241:106278
Li J, Liu H, Wang H, Luo J, Zhang X, Liu Z, Zhang Y, Zhai L, Lei Q, Ren T, Li Y, Bashir MA (2018) Managing irrigation and fertilization for the sustainable cultivation of greenhouse vegetables. Agric Water Manag 210:354–363
Li M, Xu W, Zhu T (2019) Agricultural water allocation under uncertainty: redistribution of water shortage risk. Am J Agr Econ 101(1):134–153
Li PG, Zhao FQ (2016) Autoclaved brick from volume-stability-modified steel slag and low silicon tailings. MATEC Web of Conferences 67:05015
Liu B, Jiang J, Shen S, Zhou F, Shi J, He Z (2020) Effects of curing methods of concrete after steam curing on mechanical strength and permeability. Constr Build Mater 256:119441
Mallikarjuna Reddy DV, Hamsalekha S (2020) Smart concrete curing system. E3S Web of Conferences 184:01086
Matson PA, Parton WJ, Power AG, Swift MJ (1997) Agricultural intensification and ecosystem properties. Science 277(5325):504–509
Mohammed H, Sadique M, Shaw A, Bras A (2020) The influence of incorporating plastic within concrete and the potential use of microwave curing; a review. J Build Eng 32:101824
Mosavi SH, Soltani S, Khalilian S (2020) Coping with climate change in agriculture: evidence from Hamadan-Bahar plain in Iran. Agric Water Manag 241:106332
Mueller ND, Gerber JS, Johnston M, Ray DK, Ramankutty N, Foley JA (2012) Closing yield gaps through nutrient and water management. Nature 490(7419):254–257
Nam Boonruang W, Rawangkul R, Yoda Sudjai W, Khesari J (2011) Investigation of properties of low thermal conductivity pozzolanics soil bricks towards sustainable development. Adv Mater Res 261–263:469–479
Nguyen HV, Nakarai K, Pham KH, Kajita S, Sagawa T (2020) Effects of slag type and curing method on the performance of expansive concrete. Constr Build Mater 262:120422
Prabha R, Sinitambirivoutin E, Passelaigue F, Ramesh MV (2018) Design and development of an iot based smart irrigation and fertilization system for chilli farming. International Conference on Wireless Communications, Signal Processing and Networking, WiSPNET 2018:8538568
Rahman K, Barua S, Biswas P, Hossain MS (2020) Attenuation of heavy metals from runoff using coconut husk adsorbent in porous asphalt pavement. Adv Eng Forum 37:47–58. https://doi.org/10.4028/www.scientific.net/aef.37.47
Rajasekaran T, Anandmurugan (2019) Challenges and applications of wireless sensor networks in smart farming—a survey. Adv Intell Syst Comput 750:353–361
Sadr Momtazi A, Haghi AK (2008) A numerical study on thermal drying of moist brick. Int J Heat Technol 26(1):61–65
Salahoui MK, Jiao X, LuÂ H, Guo W (2020) An improved approach to estimating the infiltration characteristics in surface irrigation systems. PLoS ONE 15(6):e0234480
Sarkar J, Das DK (2019) Enhanced strength in novel nano composites prepared by reinforcing graphene in red soil and fly ash bricks. Int J Miner Metall Mater 26(10):1322–1328
Shi J, Liu B, Wu X, Qin J, Jiang J, He Z (2020a) Evolution of mechanical properties and permeability of concrete during steam curing process. J Build Eng 32:101796
Shi J, Liu B, Wu X, Tan J, Dai J, Ji R (2020b) Effect of steam curing on surface permeability of concrete: multiple transmission media. J Build Eng R 32:101475
Shi J, Liu B, Zhou F, Shen S, Dai J, Ji R, Tan J (2020c) Heat damage of concrete surfaces under steam curing and improvement measures. Constr Build Mater 252:119104
Singh A (2018) Salinization of agricultural lands due to poor drainage: a viewpoint. Ecol Ind 95:127–130
Singh P, Purakayastha TJ, Mitra S, Bhowmik A, Tsang DCW (2020) River water irrigation with heavy metal load influences soil biological activities and risk factors. J Environ Manage E 270:110517

Taki K, Gahlot R, Kumar M (2020) Utilization of fly ash amended sewage sludge as brick for sustainable building material with special emphasis on dimensional effect. J Clean Prod 275:123942

Terence S, Purushothaman G (2020) Systematic review of Internet of Things in smart farming. Trans Emerg Telecommun Technol 31(6):e3958

Vassalle L, Sunyer-Caldú A, Uggetti E, Díez-Montero R, Díaz-Cruz MS, García J, García-Galán MJ (2020) Bioremediation of emerging micro pollutants in irrigation water. The alternative of microalgae-based treatments. J Environ Manage 274:111081

Wang J (2020) Utilization effects and environmental risks of phosphogypsum in agriculture: a review. J Clean Prod 276:123337

Yanala SR, Pagilla KR (2020) Use of biochar to produce reclaimed water for irrigation use. Chemosphere 251:126403

Yin J, Teng Y, Meng X, Ge Y, Zhang X (2020) Influence of fly ash on the curing characteristics of epoxy resin. Polym Polym Compos 28(8–9):579–588

Yu L, Qin H, Xiang PA (2020) Incentive mechanism of different agricultural models to agricultural technology information management system. Sustain Comput Inform Syst 28:100423

Zhang K, Gu Y, Li M, Wang C, Zhang Z (2013) Rapid curing epoxy resin and the properties of carbon fiber/epoxy composite. Fuhe Cailiao Xuebao/acta Materiae Compositae Sinica 30(6):21–27

Zhou H, Chen J, Wang F, Li X, Génard M, Kang S (2020) An integrated irrigation strategy for water-saving and quality-improving of cash crops: theory and practice in China. Agric Water Manag 241:106331

Zhou SJ, Qu JT (2011) Experiment on making autoclaved sludge bricks with the dewatered sludge from water-supply plants. Adv Mater Res 261–263:829–833

Zhu X, Hou C, Xu K, Liu Y (2020) Establishment of agricultural drought loss models: a comparison of statistical methods. Ecol Ind 112:106084

Chapter 15
Embodied Carbon in Construction and Its Ecological Implications

Maheen Javaid, Idrees Yousuf Dar, Zaiema Rouf, Mohmad Younis Dar, and Arshid Jehangir

Abstract In the present scenario, embodied carbon constitutes one of the grave concerns as it shares a substantial amount of greenhouse gas emissions mainly resulting from construction activities. The greenhouse gas emission or carbon impact can be categorically divided into two aspects viz., the operational carbon and embodied carbon (EC). With respect to building life cycle, EC is considered as CO_2 equivalent which is usually linked to the non-operational stage of the building. The overall carbon of the building includes embodied carbon as well as carbon accompanied with the operation (cooling, heating, powering, and other processes). Whereas the considerable amount of the building's carbon is sealed into the materials and structures. Taking embodied carbon into consideration, it can render economic opportunities for carbon savings and lowering of costs against those conventionally addressed through operational savings. Hence, it offers a great chance to lower the carbon impact of the construction industry and increase their carbon savings. Consequently, the embodied carbon emissions that are produced by humans bring about climate change by elevating the temperature of the globe. Various steps and actions have been taken already such as many economic and legislative instruments to mitigate climate change and achieve net zero carbon buildings.

Keywords Climate change · Embodied carbon · Heating · Mitigation · Temperature

15.1 Introduction

The sectors such as building and construction are considered very significant as it appreciably contributes to the economic development of a country (Trinh et al. 2017). The industries associated with the construction consume a large amount of energy resources to extract the huge amount of materials and thus also produce vast quantities of deleterious pollutants that are emitted in the immediate environment (Hammond and Jones 2008a, b). As documented by Chau et al. (2015), out of the world's total

M. Javaid · I. Y. Dar (✉) · Z. Rouf · M. Y. Dar · A. Jehangir
Department of Environmental Science, University of Kashmir, Hazratbal, Srinagar 190006, India

J. A. Malik and S. Marathe (eds.), *Ecological and Health Effects of Building Materials*,
https://doi.org/10.1007/978-3-030-76073-1_15

energy consumption, the building sectors have been reported to consume about 40% energy, 33% emission of greenhouse gases, 30% utilization of the raw materials, generate 25% solid waste, and, 12% of land use. As per the reports, around two-fifths of the worldwide supply of crude rock, stone, and sand whereas one- fourth of the Earth's overall supply of wood in virgin form is consumed every year (Ding 2004; Langston and Langston 2007; Dixit et al. 2010a). It has been reported that the United States from 1975 to 2003 has made considerable utilization of building or construction materials like cement and steel and has witnessed an increase of 57% and 108% respectively (Peters 2010).

As per the evaluation of the United States Geological Survey (USGS), it is estimated that utilization of overall crude materials during the year 2006 was 26 times that of the consumption recorded in 1900 (Matos 1998). As the consequences of immense energy consumption and generation of greenhouse gases or environmental pollution by the construction activities, they are in turn responsible for intensifying the climate change (Ürge-Vorsatz et al. 2007). It is quite evident from the statistics given by the USCB (2010), the building sector in the USA is liable for 40% consumption of the nation's energy, whereas the energy usage by the building sector in the UK comprises over 60% of total utilized energy. In addition, several other researches conducted verified about increasing energy usage of the industries dealing with building activities as well as their tremendous role in emissions of greenhouse gases, destruction of the ecology, and exhaustion of the resources (CICA 2002; Melchert 2007; Zimmerman et al. 2005). Usually estimation of the carbon equivalent is done by the conversion of the specified amount of greenhouse gases (GHG) to an equal amount of carbon dioxide that results in a similar global warming effect (Hong et al. 2014).

In the sphere of the building and construction processes, emissions of GHGs or carbon impacts may be categorized into two classes: operational carbon which is associated with operational energy usage in the survival phase of building, and the second one is embodied carbon which is related to construction materials including effects from extraction of materials, manufacture and transportation along with architecture of buildings and renovations, components of building replaced, consequent demolition and process of disposal (Waldman et al. 2020). During the overall lifecycle of the building, carbon is embodied in each stage that includes the process in which the building materials are extracted and manufactured, construction operations, activities related to the demolition of a building, and other fuel-related activities that happen all through the life expectancy of the building. Furthermore, the generation of waste at a time of various activities of construction, development, maintenance, remodeling, operation stage, and demolition along with the activities of transportation results in the emission of embodied carbons (RICS 2014a, b).Various studies have reported that there is a constant increase of around 23 billion tons annually in the utilization of concrete in the construction processes (Miller et al. 2015), whereas 5% to the global anthropogenicCO2 emissions is solely contributed by cement industries (Flower and Sanjayan 2007). The rapid increase in construction activities has intensified the waste generation and its coupled embodied carbon in different phases of the building lifetime (Ding 2018). Embodied carbon as a significant contributor of GHG emissions, building and infrastructure promptly decarbonize tons of embodied carbon before

2050 in order to attain global GHG reduction goals (IPCC 2018). Therefore, in the field of development it is mandatory to compose cost-effective approaches, methods, and technologies to adapt eco-friendly options.

15.2 Sources of Carbon Emissions in Construction

Generally, the sources of carbon are described concerning the different lifecycle of the building that includes arrangement and planning, design, development and manufacture, installation, experimentation, commissioning, operation, discarding of residue (Gangolells et al. 2009).Moreover, the transformation between various lifecycle phases of building usually leads to significant transportation, which is associated with emissions and should be taken into consideration while evaluating the carbon emissions. According to USEPA (2002) these stages have been classified sequentially into three different phases that is "cradle to the entry gate, the entry gate to exit gate, and exit gate to grave. But on the contrary, Sodagar and Fieldson (2008) have characterized such stages in three different stages viz., the first stage is the initial impact (encompassing the constituents of fabrics in the construction activities), the second stage is the operational impact (operational to maintenance phases), and the third stage is the end of life impact (the demolition activity to waste material).

On the other hand, a building's life cycle is expressed in several different phases that comprises the planning and design phase, the materials and the construction activity phase, the operational phase, the maintenance and replacement phase, and the dismantling and disposal phase as shown in Fig. 15.1 (Ng et al. 2012). In addition to the mentioned categories, carbon emission is further categorized into two subdivisions, operating carbon emissions (OC) and embodied carbon (EC) and Whereas, this OC consists of carbon emissions obtained in the course of lifetime of a building

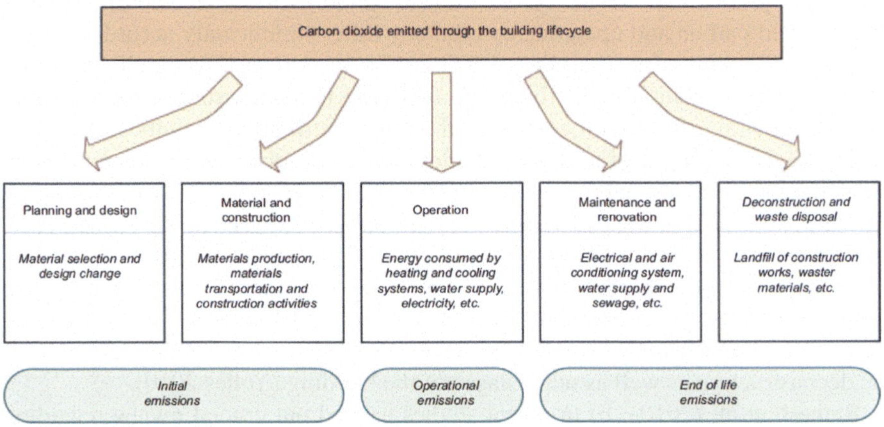

Fig. 15.1 Carbon emissions in the construction (Adapted after: Ng et al. 2012)

and incorporates the carbon emissions experienced in sustaining the indoor building environment by various processes viz., heating, cooling, lighting, and the operation of appliances (Dixit et al. 2012).

EC consists of carbon emissions acquired while manufacturing transportation and construction of building components. In recent times, embodied carbon (EC) became very significant in assessing the building's life-cycle carbon. Embodied carbon is also defined as the carbon footprint of a material. Despite that, generally embodied carbon is defined in several ways which is based on the frontier of the studies and EC's different forms. Hammond and Jones (2011) defined the embodied carbon as "the sum of fuel-related carbon emissions and process-related carbon emissions". Therefore, the carbon released in the course of the construction process of building fabrics owing to fuel utilization as well as concerned chemical processes that are mainly responsible for embodied carbon of that material. On the contrary, some materials such as wood sequester and withdraw carbon in the atmosphere. Consequently, embodied carbon of building elements is appropriately taken as the total emissions related to the fuel consumption and manufacturing related process (Victoria and Perera 2018). It is possible to categorize embodied carbon chiefly into two forms, initial embodied carbon and recurring embodied carbon (Chen et al. 2001; Ramesh et al. 2010a, b). Initial embodied carbon is the emissions related to the mining processes, processing, transport, and building whereas recurring embodied carbon consists of emissions due to building processes such as maintenance, renovation, substitution of building elements, and instruments. Li et al. (2014) divided embodied carbon into two categories, one is direct emissions arising due to assembly operations, and the other is indirect emissions from feedstock extraction, processing, transferring ultimate construction elements to the building site. There are three common meanings based on the chosen device boundary: cradle-to-gate; cradle; incurred in the extraction of the base material, the production, and transportation of final building materials to the construction site. As per the selected system boundary, there are three interpretations related to EC which are cradle-to-gate, cradle-to-site and cradle-to-grave embodied carbon. In the net life cycle carbon of buildings, the appreciable proportion of embodied carbon and operating carbon may differ significantly according to the function and nature of the building (RICS 2012) coupled with factors like climate, location, the orientation of the building, fuel type used, massing of the building, etc. (Nebel et al. 2008).With context to this, the contribution of embodied carbon in the life cycle carbon of traditional buildings was observed to fluctuate between 20% (in case of traditional office and housing buildings) and 80% (for low-energy buildings like washhouses) (RICS 2012; Thormark 2002a, b, 2006; Nemry et al. 2010; Bastos et al. 2014). Approximately 66% of the life cycle greenhouse emissions come from operating processes identified as operational carbon, while 27% greenhouse emissions are because of manufacture and transportation of building elements commonly called embodied carbon. However, below 7% is released from the deconstruction as well as other stages of the building (Yolles 2010).

Ramesh et al. (2010a, b) in recent studies carried out crucial review regarding the evaluation of lifecycle energy of about 73 case studies across the 13 countries. The evaluated results showed that embodied and operating stages contribute about 10–20%, and 80–90% respectively, assumed to be the primary factors of the building's life cycle energy requirement. According to a preliminary examination (Tae et al. 2010), the carbon produced by materials is about 90% or more, while the implications of the one due to equipment is not significant. Hence, this survey showed that only building elements are responsible for EC. Comparatively, the share of embodied carbon is significantly less than operational carbon but EC can become substantial as far as various time frames are concerned. For example, building life span may vary from 25–30 years in various developing countries such as Mainland, China. In this instance, the overall contribution of embodied carbon can elevate much greater, reaching about 50% (Yolles 2010). At the same time, OC has been constantly reduced through the multi-layered efforts which are linked to technology and policy aspects, such as the improvement of ventilation, heating, performance of air-conditioning, adoption of the zero-energy building design, utilization of new and renewable energy, and the introduction of green building certification policies.

A rise in the number of stakeholders have obtained certification in compliance with the green building certification in South Korea, known as G-SEED (Green Standard for Energy and Environmental Design) and around 4958 buildings have been accredited since G-SEED was founded in 2002. The concentration of EC in the life cycle is rising due to these measures (Kim et al. 2009; Dixit et al. 2010b, 2012, 2013; Ibn-Mohammed et al. 2013a, b). Embodied energy is equal to a few years of operating energy for housing built as per traditional standards, except in certain cases like low-energy buildings (Lippke et al. 2004). For the low-energy buildings, the embodied carbon is of great significance (Thormark 2006) mainly due to the reason that during occupation considerably less energy is consumed; extra energy is also needed for the development of enhanced insulation, the utilization of heavier mass elements and deployment of the alternative technologies.During the lifespan of a building, the embodied carbon of a low-energy house is likely to add more to its total life cycle carbon emissions as compared to traditional houses. Hammond and Jones (2008a, b) in an investigation relevant to UK housing building recorded an average of 5.3 GJ perm2 embodied energy and 403 $kgCO^2/m^2$ embodied carbon during 14 case studies focused on primarily UK processes employing an open access inventory of carbon and energy data for a broad range of construction materials. The mean embodied energy is similar to the results of Nässén et al. (2007). Specifically, greater implementation of energy-efficient structures in the building sectors, emission of carbon and consumption of energy from the operational stage of buildings can be reduced to a greater extent (Trabucco 2012). On the contrary to the operational phase CO_2 emissions, which can be curtailed eventually, it is not possible to reverse embodied carbon (Circular Ecology 2014). The modern methods of environmental assessment focus completely on the EC, which means that the knowledge gap has been established (De Wolf and Ochsendorf 2014). As a result, lowering of carbon in the construction industry cannot be done effectively as long as EC is neglected (Ibn-Mohammed et al. 2013a, b). The carbon emissions from the construction sector have been depicted in Fig. 15.1.

15.3 Embodied Carbon Hotspots

Carbon hotspots are defined as "The carbon significant aspects of projects, which are not only carbon intensive but also easily measurable and with high reduction potential" (RICS 2014a, b). The important feature of the carbon is explained on the basis elements of building or processes involved in the production. Emphasizing designing elements of the building that are significant sources of carbon has been demonstrated and suggested to be crucial in limiting emissions of embodied carbon to an appreciable extent. Such elements of the building are scientifically termed as 'Carbon Hotspots'. In general terms, for the building case studies, the floors such as first and top floors, exterior walls, frame, along with roofs mainly designated as hot spots of carbon (Clark 2013; Davies et al. 2014; Yolles 2010). Various studies reported that only those elements of building resulting in about 80% of the embodied can be identified as the hotspots of a specific kind of building. Moreover, these hotspots widely differ for different types of buildings mainly because of different element intensities. Furthermore, the substructure such as external walls, upper floors, frame, and services were constantly reported as carbon hotspots, thus labeled as 'Lead positions'. While ceiling finishes, external doors, floor finishes, internal walls, partitions, wall finishes, roof and windows are designated as carbon hotspots in some of the buildings and were also identified as 'special positions'. However, the rest of elements which comprise the equipment, fittings, furnishings, internal doors, and stairs have not been recognized as carbon hotspots in every assessed building, therefore, these elements were labeled as 'remainder positions'. This suggests that the categories of building elements such as lead positions, and special positions of the building sectors require more attention to attain the maximum possible reduction in embodied carbon (Victoria and Perera 2018).

15.4 Estimation of Carbon Emissions

Embodied carbon can be evaluated within the framework of cradle to grave depicting the whole limiting condition. Such a framework consists of mining of materials from the earth, transport, processing, assembly, product use and ultimately its end of a lifetime (Circular Ecology 2014). In the building sector or construction industry, the documented studies of LCA have led to identification of the significant functions which embodied carbon can exhibit within the building sectors (Heinonen et al. 2011). It is due to this reason that through an LCA, construction and manufacture phases can be estimated and evaluated, along with end-of-life activities. It is evident from various studies that these phases of building may contain entire embodied carbon that is around the emissions of carbon through the operational phase (Pomponi and Moncaster 2017). Life cycle assessment (LCA) is regarded as a valid and efficient environmental impact assessment technique established in the framework of ISO 14,040 standards (ISO 1997). LCA is extensively employed in measuring the embodied carbon of various building fabrics and constituents along with their emission coefficients of carbon, especially of machines and processes mainly involved in the manufacturing and functional phases of the building (Dixit et al. 2012). For estimation of environmental effects, the LCA establishes a comprehensive approach including resulting emissions and energy usage of the buildings (Akbarnezhad and Xiao 2016). A comprehensive research study of LCA provides a significant knowledge of the net consequences of the building. Hence there arises ambiguity in characterizing buildings as "sustainable" in absence of various environmental consequences that need to be taken into account during the expected lifetime. LCAs in association with data related to LCA must be employed in the building industry for estimating buildings as well as helping in design, attaining decision-making, and specification (Waldman et al. 2020). As per the documented reports of Hammond and Jones (2008a, b), that suggested LCA survey must preferably initiated in following manner i.e. the mining of raw materials till end of the production period (mostly cradle-to-grave), and currently, it requires characterization of life cycle from cradle-to-gate, covering every information till the formation of a product. An LCA evaluation is conducted in four stages as per the ISO viz.; Goal, scope and definition, Inventory analysis (LCI), Impact assessment (LCIA), and Interpretation.

(1) **Goal, Scope, and Definition**. This stage is mainly based on the matter and the purposeful utilization of the survey and can differ to a greater extent depending upon the specific project (ISO 2006), that covers the carbon or energy flow, the functional unit, and the system boundary. So far as the LCA implementation in designing a building is concerned, this method emphasizes the components of construction from cradle to site, instrumentation, and discarding of waste. The main aim of the LCA for various plans is to quantify and differentiate between the embodied carbon and the energy consumed during construction and to identify the critical contributors that could help minimize the emissions of the embodied carbon.
(2) **Life Cycle Inventory (LCI)**. For all activities and elements within the device boundary, this step is a data collection exercise for input–output analysis. In the case study, a specific inventory is given.
(3) **Life Cycle Impact Assessment (LCIA)**. LCIA assesses the magnitude of potential environmental impacts by applying the LCI results and provides information for the final interpretation phase. By applying the LCI findings, life cycle impact assessment assesses the severity of possible environmental consequences and results in data for the interpretation stage.
(4) **Interpretation**. Interpretation is the stage in which the findings of the two stages mentioned above are interpreted in accordance with the purpose of the assessment and where analytical delicacy and ambiguity are carried out to evaluate findings.

Currently, there are a number of technical difficulties that should be tackled while carrying out LCA in the building sectors, (Hammond and Winnett 2006; Hammond 2000; Graedel and Allenby 1995; Udo and Heijungs 2007) including the description of the system boundaries, nature of data available and in a manner the outcomes will

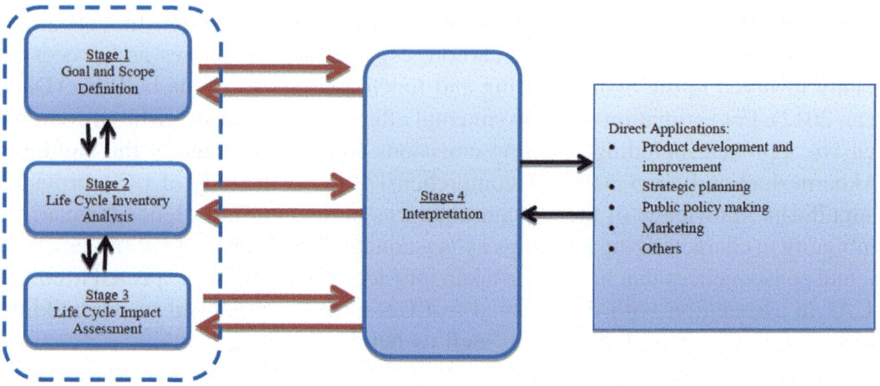

Fig. 15.2 Life cycle assessment framework defined by ISO 14,040 (Adapted after: Haynes 2013)

be standardized (Hammond and Winnett 2006; Hammond 2000; Udo and Heijungs 2007).

The process linked to the goal definition is very essential in the context of the planning phase for conducting LCA. Collecting the data for the purpose of life-cycle inventory may prove to be laborious activity as several companies either have privacy issues or may not have recorded a comprehensive overview which is required for a sound lifetime assessment. The interpretation stages and impact assessment are continuously subjected to modification processes, even though these stages have been coded in the ISO 14,040–14,044 standards methodology recommended by ISO which was launched in 2000, and revised in 2006. The LCA framework defined by ISO 14,040 is shown in Fig. 15.2.

Usually authenticity of LCA depends upon valid life-cycle inventories, which includes requirements of the resources and energy consumption besides emissions to water, air and land in the process of fabrication of a product, operation, or supply of a service (Nisbet et al. 2000). LCA studies are data-based procedures that need time and work dedication (Pomponi et al. 2017). Various measures viz., Waste Reduction Action Program (WRAP 2017), database of the embodied quantity outputs (Böhringer et al. 2018) and the methodology of Royal Institution of Chartered Surveyors (2014) targeted to establish a universal methodology for estimation of embodied carbon and additionally intended to provide combined databases on embodied carbon in buildings. Such measures undertook investigations on a large number of buildings in different areas to provide numerical data on buildings. A range value of 266 and 515 kgCO2-eq/m2 were depicted by 291 organizations so far as the Embodied Carbon Benchmark Study is concerned which took into consideration greater than 1000 entities. The excellence of a research is determined by the accessibility of various kinds of data viz., Operation data, environmental data on evaluated processes; system data on transportation of resources, and energy or products as well as to show comparison between various end products. Environmental product declarations (EPDs) serve as the source of environmental data on the market for construction goods and are increasingly being used during the final stages of building design for environmental performance evaluation of buildings and product comparison for procurement decisions. A hindrance of data is an obstacle in characterization and buying lower embodied carbon goods (Waldman et al. 2020).

Therefore in certain cases, the scope of the analysis therefore differs as per the limits and specifications defined by the researcher. These factors thus induce subjectivity and ambiguity in the LCA research, suggesting a lack of systematic study that is difficult to draw conclusions (Kayaçetin and Tanyer 2018). In EC assessment, there exists ambiguity since a single value must be anticipated within the range of prescribed values. In addition, variability due to the type and amount of resources used in a building that could change depending on the design, type of construction, condition of the site, and characteristics of the owner. A probabilistic analysis may be done to deal with ambiguity and instability (Kang et al. 2015). In other words, probabilistic analysis is necessary for a meaningful estimation of EC where uncertainty and variability are inherent, and a statistical characteristics analysis must take precedence to retain value for the input variables in simulation. It is important to acknowledge that the outputs from LCA are a product of the accuracy of the inputs, and a number of (consistently applied) assumptions are made about processes and supply chains. It is, however, very effective in drawing comparisons between different design options such as material type or elemental build-up which can then inform the overall design.

15.5 Mitigation Strategies

Importance of mitigation mechanisms for embodied carbon is well known among researchers and government organizations of many developed countries. Till now a significant research was carried out to explore different approaches in order to alleviate embodied carbon of buildings. Such approaches can be mostly categorized in the following way (1) Utilization of Low-carbon materials; (2) better architecture; (3) local procuring of materials and minimizing transport (4) reuse and recycling of materials (5) Utilization of prefabricated elements and (6) Law Enforcement.

15.5.1 Utilization of Low-Carbon Materials

The environmental impact caused by materials can be assessed during the complete life period, i.e. from cradle to grave. Thus it becomes important to use low environmental impact building materials in designing which can lead to betterment into the system. Use of materials with low carbon and low embodied energy in building architecture is vital in the creation of a building environment with sustainability. This issue aims designers to anticipate such decisions somewhat critical in determining or selecting a building's possible material with regards to its embodied impact (Ünalan et al. 2016). Designers mostly choose from less available options of different materials which are made procurable for each structural and non-structural component in a building only in view of their performance against technical needs (Ahmadian et al. 2016a). The material chosen following screening based on technical efficiency necessities could have significantly distinct embodied carbon repercussions for the buildings (Hammond and Jones 2006 and 2008a, b). As per reports that distinct materials constitute different embodied carbon dioxide quantities and thus have many environmental consequences. As documented,CO_2 emissions declined by about 30% after substituting old fabrics with low EC fabrics, especially in buildings, thus indicating significance of opting fabrics in decreasing the carbon emissions of buildings González and Navarro (2006). Studies show that this method involves consumption of natural substances such as bamboo, hemp-lime composites and timber. Similarly, Reddy (2009) worked on the application of stabilized mud blocks as an option against load bearing brickwork and reported around 50% decline in embodied costs. While considering use of other options of building elements against conventional materials for construction of a 28 floored housing in Hong Kong and estimation of their related embodied carbon reductions showed a 34.8% decline. Salazar and Meil (2009) evaluated the carbon implications of a 'wood-intensive' house in contrast to a structure based on brick cladding in Canada and indicated that about 20 tCO2e were released for the former whereas 72 tCO2e in the case of the latter resulting in great changes. Many works in Japan, Spain and Sweden conducted by Gerilla et al. 2007 concluded that less energy consumption in comparison to steel and concrete. Similarly Vukotic et al. (2010) showed that school buildings mainly built of timber had fewer repercussions as compared to frames made of steel. However it was claimed that "instead of encouraging debate about which type of material is 'better' than any other", the finest option is possible by selecting material in different circumstances. Overall, the advantages of using wood are quite prominent. It can be noticed that in certain related works, utilization of fabrics possessing lower embodied energy and embodied carbon may consist of conventional materials. Similar observation was reported in the work conducted by You et al. (2011) in which they demonstrated 4.2% Carbon dioxide decline while utilizing steel–concrete frames as compared to masonry-concrete structures implying significance design. However, while taking into account earths' supplies of wood and its depletion as a whole system implying that presently huge amounts of wood are continuously spent rather than replenished, most of it is used as fuel in third world nations (Hammond and Jones 2010). Hence, it is wrong to consider wood having a negative global warming potential because finally a great proportion of this is reduced to ashes or added to land-fill, thus generating 0.0036 kg carbon dioxide and 1.47 kg carbon dioxide per kg of wood respectively. According to the SwisOekoInventa record, neutralizing its non-permanent impacts on carbon dioxide equilibrium (Peuportier 2001). Consequently, wood should be recycled and reused to a possible extent in the same way as other building materials in order to conserve the environment. In addition to comparing the embodied carbon of fabrics, choosing within the available traditional options for use in a construction, literature proposes two effective schemes to alleviate the embodied carbon of buildings. These comprise decreasing the embodied carbon of available materials by increasing the amount of recycled waste or byproduct constituents in their structure (Habert and Roussel 2009) and designing new low-carbon materials (Davidovits and Davidovics 1991).

15.5.2 Better Architecture

Better architecture plans and suitable options during the designing phase in collaboration with methods like design for demolishing have been designated as critical approaches for EC decline and abatement. It is well documented in the input and output study of the Irish building industries that better design showed a 20% decline in indirect releases as well as direct releases by 20% and 1.6% respectively amounting 3.43 $MtCO_2e$ (Acquaye and Duffy 2011). Similarly, evaluation of restoration of tall cemented buildings in Hong Kong reported about the importance of design. It was further claimed that the effective option is to consider 15–30% of the existing structural and non-structural building materials as it can lessen the CO2 footprint by 17.3% (Chau et al. 2012). Häkkinen et al. (2015) stressed significance and they suggested an orderly and progressive continuation throughout various stages and phases of a design so that precise assessment of GHG emissions and attainment of low-carbon buildings is possible.

15.5.3 Local Procuring of Materials and Minimizing Transport

From the available literature it is obvious that transportation being a major contributor has immense consequences because it adds to embodied carbon of constructions (Gonzalez and Navarro 2006; Yan et al. 2010). It has been found that key elements influencing emissions due to transportation consists of the amount of material to be supplied, transportation distance, dimensions of material and means of transport (Ahmadian et al. 2014, 2016b). Broadly fabrics are divided into different types viz., Made-To-Order (MTO), Engineered-To-Order (ETO), Assembled-To-Order (ATO), and Made-To-Stock (MTS), and Assembled-To-Order (ATO) products; all having considerably characteristics supply chain systems and thus characteristics emissions rates (Olhager 2003). Many researchers suggested that the decline of EC is mainly because of greater consumption of local materials and thus will minimize transportation effects (Gustavsson et al. 2010; Asdrubali 2015; Chou and Yeh 2015). In a comprehensive evaluation regarding production of stone conducted in compliance with PAS2050 recommendations, it was stated that considering the type of stone and its source, about 2% - 84% of the embodied carbon of stones procured from foreign lands could be saved by using stones of UK origin.

15.5.4 Material Reuse and Recycling

While taking into consideration carbon emissions consequences during adoption of end –of- life approach, the cradle to grave embodied carbon of constructions could be decreased. The main approaches to manage buildings at the end of their service life consist of "demolition and landfilling", reuse and recycling of fabrics (Xiao et al. 2016; Akbarnezhad et al. 2013). Tam (2009) mentioned that among the various strategies recycling of construction waste proved to be excellent techniques to amend its environmental consequences. Recycling of concrete has been considered as a better option for decreasing carbon emissions and charges involved in transporting, discarding waste at far-off landfill sites, decreasing the availability of landfill area and furnishing a viable source of options (Akbarnezhad and Nadoushani, 2014; Chowdhury et al. 2010; Marinkovic et al. 2010). In fact, more recycling components in building material generates highly energy-efficient and less wasteful materials, unconventional techniques and subsequently new construction methods, collectively helping in decreasing the total energy usage for all new buildings (RCIS 2010). WRAP (2010) emphasized the significance of utilizing secondary or recycled components for building along with tackling wood wastes in the construction industry. Various researches have shown that in the waste stream every recycled aggregate was reused, and in 2009 replaced about 25% of primary aggregates (MPA 2010). From a recycling perspective it was reported that recycling concrete showed minimal effect on net embodied energy of a building (Harris and Elliot 1997). It was shown that while keeping else things similar, the highest CO_2 reserves within the industry sector are hopefully to be accomplished because of addition of extra cementitious substances (Tyrer et al. 2010), like fly ash (Pedersen et al. 2008) or ground granulated blast-furnace slag (O'Rourke et al. 2009). An exclusive cementitious binder formed on magnesium oxide is prepared, while the manufacturing industry declares that about half of the CO_2 is generated by the product in comparison to Portland cement.

15.5.5 Refurbishment

Refurbishment involves the energy efficiency of current buildings by addition of insulative materials or substituting previous systems with much energy-efficient ones, like low energy lighting or improved boilers. Refurbishment means inhabitants use less energy day-to-day, but on the other hand it also means more embodied carbon as new materials are being added to the building while older equipment is being thrown away. These changes can prolong the lifespan of the building while they can also help to avoid the embodied carbon required to substitute an old building with a new one.

15.5.6 Utilization of Prefabricated Elements

It is a strategy mainly related to highly effective building developments but it is quite evident from the available literature targeting off-site construction while prefabrication was considered separately. In most works, only estimation of emission savings was also done. Su's approach was supported with the investigation conducted by Mao et al. (2013) which indicated that semi-prefabrication led to release 3.2% less in comparison to traditional construction. Off-site manufacturing in association with other approaches (such as utilization of low embodied carbon) were also assessed (Monahan and Powell 2011).

15.5.7 Law Enforcement

As expected, the application as well as amendment of rules and regulations by statutory law appeared to be a commonly mentioned approach for EC alleviation (Blengini and Di Carlo 2010; Dakwale et al. 2011; Giesekam et al. 2014). This approach in certain studies mostly aimed to assist other mitigation approaches, such as extensive consumption of low embodied energy/ embodied carbon fabrics, while other policies have a greater potential. For instance it was reported in China and Japan that a 50% CO_2 decline could be attained via the influence of policies on designing and construction methods Dhakal (2010).

15.6 Ecological Implications

Protection of environment protection is an area of concern developing as well as developed countries and (Tse 2001).In the natural environment building is not an eco-friendly activity (Li et al. 2010). Levin (1997) pointed out that construction processes have immense both indirect and direct effects on the ecosystem. Shen et al. (2005) pointed out that construction is a chief source of environmental degradation with respect to other sectors. In the sphere of construction, predominant environmental consequences evaluated till present are carbon emissions due to energy utilization in operation of building, contributing around 40 and 32% of US and global annual energy production respectively (USDE 2012). The environmental consequences of construction are usually categorized into two classes: operational embodied operational (Ibn-Mohammed et al. 2013a, b). These were also described in the year 2012 with the help of European Standards EN 15,804 and EN 15,978 (CEN 2011, 2019) providing a description regarding overview of the life-cycle stage implications of a construction project, building and product. Since embodied consequences were not prominent in contrast to operational impacts, mitigation has paid much attention to the latter (Szalay, 2007). In view of the fact regarding the literature of the standards, interest in research as well as industry related to the buildings' embodied impacts has increased fastly (WGBC 2019). But the complications and differences in the data, coupled with disparities in temporal and spatial limits, implied that results remain unclear (Moncaster et al. 2018). In recent time International Energy Agency (IEA) project Annex 57 evaluated the data of about 80 individual LCA'S of organizations (Moncaster et al. 2019). In the majority of the case studies assessed, the product stage showed significant impact. The construction activity is a major producer of CO_2 emissions and about one –fourth of total CO_2 emissions are due to energy consumption in buildings in the world (Metz et al. 2007). Moreover 5% is due to the manufacture of cement, a chief construction component (Worrell et al. 2001). In the process of manufacturing, installation, and transportation of a building materials like concrete and steel need huge amounts of energy, even though depicting a small portion of the total cost in the building. When one ton of concrete consisting of the minimum proportion of embodied energy is multiplied with the maximum proportion of concrete consumed results in concrete material possessing a huge quantity of carbon in the world. The value EE of concrete is highest i.e. 12.5 MJ/kg EE followed by steel i.e. 10.5 MJ/kg EE and the lowest is that of wood with a value of 2.00 MJ/kg EE. (Hsu 2010). It is well documented that the embodied energy value of each building material differs to a large extent, especially of concrete most probably due to the reason that production of cement requires vast fossil fuel and energy, forming it a major producer of CO_2 emissions causing global warming (Shams et al. 2011). As far as steel and concrete is concerned their respective embodied energy showed significant environmental consequences. Recently, an IPCC report clearly demonstrated that a difference in the temperature i.e. between 1.5 °C and 2 °C of global warming significantly escalates the threats of devastating climate change (IPCC 2018). It is predicted that global warming will result in destruction of ecosystems. Such rapid increases in global temperatures are showing profound effects on the environment like increase in sea levels as a result of thermal expansion of the ocean and melting of glaciers on a wider scale (Lu et al. 2007). The deleterious impact will be long lasting and in certain cases permanent. In fact, the impact of global climate change presently occupies a central position in mass awareness and different countries have taken initiatives to tackle the challenges. Some of the measures introduced were the Framework Convention on Climate Change (1992), Kyoto Protocol (December 1997) and Copenhagen Accord (2009) with a main goal to achieve emission reduction targets. The Kyoto Protocol identifies carbon dioxide, methane, nitrous oxide, sulfur hexafluoride and hydrofluorocarbons as greenhouse gases (ISO 2006; WRI/WBCSD 2004). Carbon dioxide is the main vital human induced gas, responsible for 80% of the increase in the global warming phenomena (Borges 2011; IPCC 2007). Presently "The scientific evidence is now surprising as climate change presents very serious global risks, and it demands an urgent global attention" (Stern 2007). In view of these facts, it is assumed that cutting off the energy requirements and subsequent carbon emissions as a result of buildings is a crucial step as far as climate regulation by the government is concerned. So, we should positively increase the speed and possibilities of decarbonisation measures, in collaboration with all construction value chains to accomplish the magnitude of transformation required.

15.7 Conclusion

Currently there is a deep concern over the significance of embodied carbon in the construction industry. As great quantities of lifetime carbon of building is trapped in the material and structures therefore tackling embodied carbon may also have cost-effective carbon mitigation opportunities savings. Research scientists and government bodies from several industrialized nations addressed the need for embodied carbon mitigation. However numerous strategies have been used in mitigating the operational emissions but mostly little attention was paid to embodied emissions. The contribution of embodied carbon is on the rise while operating carbon is constantly reduced to approach zero energy or zero carbon building. This increase in the proportion of embodied carbon resulted in the measures necessary to reduce embodied emissions. Hence, many attempts are also being made to minimize carbon emissions over the entire life of buildings. In order to attain more valid results through these efforts, the carbon emission status must be clearly defined and a target based on this status must be created and it is essential to clearly designate the carbon emission status and establish a goal based on this status.Implementation of LCA and assessment of embodied energy and carbon emission can be useful in determination and tackling the environmental consequences in the development stage thus promoting sustainability in the building industry. More research is needed on the empirical findings of embodied energy and carbon emission evaluation. Due to the value of embodied carbon and the use of low-carbon buildings, it is important to improve the EC computing system with robust inventory data and genuine methodology. More opportunities for creativity and collaboration in new methods and products are emerging as industry increasingly seeks to tackle embodied carbon and minimize impact.

References

Acquaye A, Duffy AP, Basu B (2011) Stochastic hybrid embodied CO_2-eq analysis: an application to the Irish apartment building sector. Energy Build 43:1295–1303

Accord Copenhagen (2009) Draft decision-/CP. 15. In Conference of the Parties to the L NFCC. Fifteenth Session, Copenhagen, Vol. 7, p.18

Ahmadian FFA, Akbarnezhad A, Rashidi T, Waller T (2016a) BIM-enabled sustainability assessment of material supply decisions. Eng Constr Archit Manag

Ahmadian FFA, Akbarnezhad A, Rashidi TH, Waller ST (2016b) Accounting for transport times in planning off-site shipment of construction materials. J Constr Eng Manage 142

Ahmadian FFA, Akbarnezhad A, Rashidi T, Waller S (2014) Importance of planning for the transport stage in procurement of construction materials. In: Proceedings of the 31st international symposium on automation and robotics in construction and mining (ISARC 2014). Sydney, Australia, pp 466–473. 9–11 July 2014

Akbarnezhad A, Moussavi Nodoushani ZS (2014) Estimating the costs, energy use and carbon emissions of concrete recycling using building information modelling. In: Proceedings of the 31st international symposiumon automation and robotics in construction and mining (ISARC 2014). Sydney, Australia, pp 385–392. 9–11 July 2014

Akbarnezhad A, Ong KCG, Zhang MH, Tam CT (2013) Acid treatment technique for determining the mortar content of recycled concrete aggregates. J Test Eval 41:441–450

Akbarnezhad A, Xiao J (2016) Estimation and minimization of embodied carbon of buildings: a review analysis. Energy 32(9):1593–1602

Asdrubali F, D'Alessandro F, Schiavoni S (2015) A review of unconventional sustainable building insulation materials. Sustain Mater Technol 4:1–17

Bastos J, Batterman SA, Freire F (2014) Life-cycle energy and greenhouse gas analysis of three building types in a residential area in Lisbon. Energy Build 69:344–353

Blengini GA, Di Carlo T (2010) Energy-saving policies and low-energy residential buildings: an LCA case study to support decision makers in Piedmont (Italy). Int J Life Cycle Assess 5:652–665

Böhringer C, Carbone JC, Rutherford TF (2018) Embodied carbon tariffs. Scand J Econ 120(1):183–210

Borges AV (2011) Present day carbon dioxide fluxes in the coastal ocean and possible feedback under global change. In: da Silva Duarte PM, Santana Casiano JM (eds) Oceans and the atmospheric carbon content. Springer, Dordrecht, The Netherlands, pp 47–77

CEN (2011) Sustainability of construction works. Assessment of environmental performance of buildings. Calculation method. Brussels: Comité Européen de Normalisation 15978

CEN (2019) Sustainability of construction works. Environmental product declarations.Core rules for the product category of construction products. Brussels: Comité Européen de Normalisation 15804

Chau CK, Hui WK, Ng WY, Powell G (2012) Assessment of CO_2 emissions reduction in high-rise concrete office buildings using different material use options. Resour Conserv Recycl 61:22–34

Chau CK, Leung TM, Ng WY (2015) A review on life cycle assessment, life cycle energy assessment and life cycle carbon emissions assessment on buildings. Appl Energy 143:395–413

Chen T, Burnett J, Chau C (2001) Analysis of embodied energy use in the residential building of Hong Kong. Energy 26:323–340

Chou JS, Yeh KC (2015) Life cycle carbon dioxide emissions simulation and environmental cost analysis for building construction. J Clean Prod 1:137–147

Chowdhury R, Paul D, Fry TA (2010) Life cycle based environmental impacts assessment of construction materials used in road construction. Resour Conserv Recycl 54:250–255

CICA (Confederation of International Contractors' Associations) (2002) Industry as a Partner for Sustainable Development. CICA Paris, France

Circular Ecology (2014) Carbon Footprint, LCA, Embodied Energy and Sustainability Experts. See http://www.circularecology.com. Accessed 25 May 2014

Clark DH (2013) What colour is your building? measuring and reducing the energy and carbon footprint of buildings, UK: RIBA Enterprises Limited

Dakwale VA, Ralegaonkar RV, Mandavgane S (2011) Improving environmental performance of buildings through increased energy efficiency: a review. Sustain Urban Areas 1:211–218

Davidovits J, Davidovics M (1991) Geopolymer—ultra-high temperature tooling material for the manufacture of advanced composites. In: Proceedings of the 36th international SAMPE symposium and exhibition, Book 1 and 2. San Diego, CA, USA, pp 1939–1949

Davies PJ, Emmitt S, Firth SK (2014) Challenges for capturing and assessing initial embodied energy: a contractor's perspective. Constr Manag Econ 32:290–308

De Wolf C, Ochsendorf J (2014) Participating in an embodied carbon database. Connecting structural material quantities with environmental impact. Struct Eng 92(2):30–33

Dhakal S (2010) GHG emissions from urbanization and opportunities for urban carbon mitigation. Curr Opin Environ Sustain 2:277–283

Ding GKC (2004) The development of a multi-criteria approach for the measurement of sustainable performance for built projects and facilities, PhD thesis, University of Technology, Sydney

Dixit MK, Culp CH, Fernandez-Solis JL (2013) System boundary for embodied energy in buildings: a conceptual model for definition. Renew Sustain Energy Rev 21:153–164

Dixit MK, Fernandez-Solis JL, Lavy S, Culp CH (2010a) Identification of parameters for embodied energy measurement: a literature review. Energy Build 42:1238–1247

Dixit MK, Fernández-Solís JL, Lavy S, Culp CH (2010b) Protocol for embodied parameters, TG66-Special Track 18th CIB world building congress. Salford, United Kingdom, p 188
Dixit MK, Fernandez-Solis JL, Lavy S, Culp CH (2012) Need for an embodied energy measurement protocol for buildings: a review paper. Renew Sustain Energy Rev 16:3730–3743
Ding GKC (2018) Embodied carbon in construction, maintenance and demolition in buildings. Embodied Carbon Build 217–245
Flower DJ, Sanjayan JD (2007) Greenhouse gas emissions due to concrete manufacture. Int J Life Cycle Assess 12(5):282
Gangolells M, Casals M, Gasso S et al (2009) A methodology for predicting the severity of environmental impacts related to the construction process of residential buildings. Build Environ 44(3):558–571
Gerilla GP, Teknomo K, Hokao K (2007) An environmental assessment of wood and steel reinforced concrete housing construction. Build Environ 42(7):2778–2784
Giesekam J, Barrett J, Taylor P, Owen A (2014) The greenhouse gas emissions and mitigation options for materials used in UK construction. Energy Build 78:202–214
Gonzalez MJ, Navarro JG (2006) Assessment of the decrease of CO_2 emissions in the construction field through the selection of materials: Practical case study of three houses of low environmental impact. Build Environ 41:902–909
Graedel TE, Allenby BR (1995) Industrial ecology. Prentice-Hall, Englewood Cliffs, NJ
Gustavsson L, Joelsson A, Sathre R (2010) Life cycle primary energy use and carbon emission of an eight-storey wood-framed apartment building. Energy Build 42:230–242
Habert G, Roussel N (2009) Study of two concrete mix-design strategies to reach carbon mitigation objectives. CemConcr Compos 31:397–402
Hakkinen T, Kuittinen M, Ruuska A, Jung N (2015) Reducing embodied carbon during the design process of buildings. J Build Eng 4:1–13
Hammond G, Jones C (2008a) Embodied energy and carbon in construction materials. Proc Inst Civ Eng Energy 161:87–98
Hammond GP, Jones CI (2008b) Embodied energy and carbon in construction materials. ProcInstCivEng Energy 161:87–98
Hammond G, Jones C (2011) A BSRIA guide embodied the inventory of carbon and energy (ICE). BSRIA, UK
Hammond G, Jones CI (2006) Inventory of carbon and energy (ICE), 1.5a Beta ed. United Overseas Bank Ltd. London, UK
Hammond GP (2000) Energy and the environment. In: Warhurst A (ed) Towards a collaborative environment research agenda: challenges for business and society. Macmillan, Basingstoke, pp 139–178
Hammond GP, Jones CI (2010) Embodied carbon: the concealed impact of 99 residential construction. Global Warming-Green Energy & Technology, pp 367–384
Hammond GP, Winnett AB (2006) Interdisciplinary perspectives on environmental appraisal and valuation techniques. Proc Inst Civ Eng, Waste Resour Manag 159(3):117–130
Harris DJ, Elliot CJ (1997) Energy accounting for recycled building components. In: Proceedings of the second international buildings and the environment conference. Paris, France, pp 485–492
Heinonen J, Seinajoki A, Junnila S (2011) A longitudinal study on the carbon emissions of a new residential development. Sustainability 3:1170–1189
Hong T, Ji C, Jang M, Park H (2014) Assessment model for energy consumption and greenhouse gas emissions during building construction. J Manag Eng 30:226–235
Haynes R (2013) Embodied energy calculations within life cycle analysis of residential buildings, pp 1–16
Hsu SL (2010) Life cycle assessment of materials and construction in commercial structures: variability and limitations (Doctoral dissertation, Massachusetts Institute of Technology)
Ibn-Mohammed T, Greenough A, Taylor S, Ozawa-Meida L, Acquaye A (2013a) Operational versus embodied emissions in buildings—a review of current trends. Energy Build 66:232–245. https://doi.org/10.1016/j.enbuild.2013.07.026

Ibn-Mohammed T, Greenough R, Ozawa-Medida S, Acquaye A (2013b) Operational versus embodied emissions in buildings: a review of current trends. Energy Build 66:232–245

International Energy Agency. FAQs: Energy Efficiency. http://www.iea.org/aboutus/faqs/energyefficiency/

IPCC (2018) Summary for policymakers—Global warming of 1.5°C. 2018. Intergovernmental Panel on Climate Change (IPCC). https://www.ipcc.ch/sr15/chapter/summary-for-policy-makers/

IPCC (Intergovernmental Panel on Climate Change) (2007) The physical science basis, intergovernmental panel on climate change (IPCC) fourth assessment report: climate change 2007. Cambridge University Press, Cambridge, UK

ISO 14040 (2006) Environmental management: life cycle assessment principles and framework. International Standards Organisation, Paris, France

ISO International Standard 14040 (1997) Environmental management—life cycle assessment–principles and framework. Geneva: International Organization for Standardization (ISO)

Kang G, Kim T, Kim YW, Cho H, Kang KI (2015) Statistical analysis of embodied carbon emission for building construction. Energy Build 105:326–333

Kayaçetin NC, Tanyer AM (2018) Analysis of embodied carbon in buildings supported by a data validation system. In: Embodied carbon in buildings. Springer, Cham pp 143–164

Kim J, Choi K, Kang J, Yang G, Lee S (2009) Research on the elements for developing zero energy buildings: the importance of embodied energy in lifecycle perspective. J Spring Conf Archit Inst Korea 29:669–672

Langston YL, Langston CA (2007) Building energy and cost performance: an analysis of 30 Melbourne case studies. Aust J Constr Econ Build 7(1):1–18

Levin H (1997) Systematic evaluation and assessment of building environmental performance (SEABEP), paper for presentation to "buildings and environment". Paris

Li X, Yang F, Zhu Y, Gao Y (2014) An assessment framework for analyzing the embodied carbon impacts of residential buildings in China. Energy Build 85:400–409

Li X, Zhu Y, Zhang Z (2010) An LCA-based environmental impact assessment model for construction processes. Build Environ 45(3):766–775

Lippke B, Wilson J, Perez-Garcia J, Bowyer J, Meil J (2004) CORRIM: Life-cycle environmental performance of renewable building materials. For Prod J 54(6):8–19

Lu J, Vecchi GA, Reichler T (2007) Expansion of the hadley cell under global warming. Geophys Res Lett 34:1–5

Mao C, Shen Q, Shen L, Tang L (2013) Comparative study of greenhouse gas emissions between off-site prefabrication and conventional construction methods: two case studies of residential projects. Energy Build 66:165–176

Marinkovic S, Radonjanin V, Malesev M, Ignjatovic I (2010) Comparative environmental assessment of natural and recycled aggregate concrete. Waste Manag 30:2255–2264

Matos G, Wagner L (1998) Consumption of materials in the United States, 1900–1995. Annu Rev Energy Env 23:107–122

Melchert L (2007) The dutch sustainable building policy: a model for developing countries? Build Environ 42(2):893–901

Metz B, Davidson OR, Bosch PR, Dave R, Meyer LA (2007) Contribution of working group III to the fourth assessment report of the intergovernmental panel on climate change. IPCC Fourth Assessment Report (AR4)

Miller D, Doh JH, Mulvey M (2015) Concrete slab comparison and embodied energy optimisation for alternate design and construction techniques. Constr Build Mater 80:329–338

Monahan J, Powell JC (2011) An embodied carbon and energy analysis of modern methods of construction in housing: a case study using a life cycle assessment framework. Energy Build 43:179–188

Moncaster AM, Rasmussen FN, Malmqvist T, Wiberg AH, Birgisdottir H (2019) Widening understanding of low embodied impact buildings: Results and recommendations from 80 multi-national quantitative and qualitative case studies. J Clean Prod 235:378–393

Moncaster AM, Pomponi F, Symons KE, Guthrie PM (2018) Why method matters: temporal, spatial and physical variations in LCA and their impact on choice of structural system. Energy Build 173:389–398. https://doi.org/10.1016/j.enbuild.2018.05.039
MPA (Mineral Products Association) (2010) Concrete Industry Sustainability Performance Report. MPA, The Concrete Centre, Camberley, UK
Nässén J, Holmberg J, Wadeskog A, Nyman M (2007) Direct and indirect energy use and carbon emissions in the production phase of buildings: an input–output analysis. Energy 32(9):1593–1602
Nebel B, Alcorn A, Wittstock B (2008) Life cycle assessment: adopting and adapting overseas LCA data and methodologies for building materials in New Zealand; Fraunhofer: Rotorua, New Zealand
Nemry F, Uihlein A, Colodel CM, Wetzel C, Braune A, Wittstock B, Hasan I, Kreißig J, Gallon N, Niemeier S et al (2010) Options to reduce the environmental impacts of residential buildings in the European union—potential and costs. Energy Build 42:976–984
Ng ST, Wong JM, Skitmore S, Veronika A (2012) Carbon dioxide reduction in the building life cycle: a critical review. Proc Inst Civ Eng-Eng Sustain 165(4):281–292
Nisbet M, Van Geem MG, Gajda J, Marceau M (2000) Environmental life cycle inventory of portland cement concrete SN. 2137, Portland Cement Association, Skokie, IL
O'Rourke B, McNally C, Richardson MG (2009) Development of calcium sulphate–GGBS–portland cement binders. Constr Build Mater 23(1):340–346
Olhager J (2003) Strategic positioning of the order penetration point. Int J Prod Econ 85:319–329
Pedersen KH, Jensen AD, Skjøth-Rasmussen MS, Dam-Johansen K (2008) A review of the interference of carbon containing fly ash with air entrainment in concrete. Prog Energy Combust Sci 34(2):135–154
Peters GP (2010) Carbon footprints and embodied carbon at multiple scales. Curr Opin Environ Sustain 2(4):245–250
Peuportier BLP (2001) Life cycle assessment applied to the comparative evaluation of single family houses in the French context. Energy Build 33(5):443–450
Pomponi F, Moncaster AM (2016) Embodied carbon mitigation and reduction in the built environment—what does the evidence say? J Environ Manage 181:687–700
Pomponi FD, Amico B, Moncaster AM (2017) A method to facilitate uncertainty analysis in LCAs of buildings. Energies 10:524
Ramesh T, Prakash R, Shukla K (2010a) Life cycle energy analysis of buildings: an overview. Energy Build 42(10):1592–1600
Ramesh T, Prakash R, Shukla KK (2010b) Life cycle energy analysis of buildings: an overview. Energy Build 42:1592–1600
Reddy BV (2009) Sustainable materials for low carbon buildings. Int J Low-Carbon Technol 4:175–181
RICS (2010) The future of UK housebuilding. Research report. RICS, London, UK
RICS (2012) Methodology to calculate embodied carbon of materials; RICS information paper, IP 32/2012; RICS, Surveyor Court, Westwood Business Park: Coventry, UK
RICS (2014a) Methodology to calculate embodied carbon, 1st edn. RICS, London
RICS (2014b) Methodology to calculate embodied carbon, RICS guidance note, global, 1st edn, Parliament Square, London, SW1P 3AD, UK
Salazar J, Meil J (2009) Prospects for carbon-neutral housing: the influence of greater wood use on the carbon footprint of a single-family residence. J Cleaner Prod 17(17):1563–1571
Shams S, Mahmud K, Al-Amin M (2011) A comparative analysis of building materials for sustainable construction with emphasis on CO2 reduction. Int J Environ Sustain Dev 10(4):364–374
Shen LY, Lu WS, Yao H, Wu DH (2005) A computer-based scoring method for measuring the environmental performance of construction activities. Autom Constr 14(3):297–309
Sodagar B, Fieldson R (2008) Towards a low carbon construction practice. Constr Inf Q 10(3):101–108
Stern N (2007) The economics of climate change: the stern review, Cabinet Office HM Treasury

Szalay AZ (2007) What is missing from the concept of the new European building directive? Build Environ 42:1761–1769. https://doi.org/10.1016/j.buildenv.2005.12.003
Tam VWY (2009) Comparing the implementation of concrete recycling in the Australian and Japanese construction industries. J Clean Prod 17(7):688–702
Tae S, Baek, C and Shin, S (2011) Life cycle CO_2 evaluation on reinforced concrete structures with high-strength concrete. Environ Impact Assess Revi 31(3):253–260
Thormark C (2002a) A low energy building in a life cycle-its embodied energy, energy need for operation and recycling potential. Build Environ 37(4):429–435
Thormark C (2002b) A low energy building in a life cycle—its embodied energy, energy need for operation and recycling potential. Build Environ 37:429–435
Thormark C (2006) The effect of material choice on the total energy need and recycling potential of a building. Build Environ 41:1019–1026
Trabucco D (2012) Life cycle energy analysis of tall buildings: design principles. In: Proceedings of 652 CTBUH 9th world congress. Shanghai, China, pp 447–453
Trinh TMKH, Doh JH, Hou L (2017) An overview of building lifecycle embodied carbon emissions research. In: The proceedings of the 22nd international conference on advancement of construction management and real estate, Swinburne University of Technology
Tse RYC (2001) The implementation of EMS in construction firms: case study in Hong Kong. JEAPM 3(2):177–194
Tyrer M, Cheeseman CR, Greaves R, Claisse PA, Ganjian E, Kay M, Churchman-Davies J (2010) Potential for carbon dioxide reduction from cement industry through increased use of industrial pozzolans. Adv Appl Ceram 109(5):275–279
Udo De Haes HA, Heijungs R (2007) Life-cycle assessment for energy analysis and management. Appl Energy 84(7–8):817–827
Unalan B, Tanrivermis H, Bulbul M, Celani A, Ciaramella A (2016) Impact of embodied carbon in the life cycle of buildings on climate change for a sustainable future. Int J Hous Sci Appl 40(1):61–71
USDE (United States Department of Energy) (2012) Building energy data book: 1.1 buildings sector energy consumption. Washington, DC
Urge-Vorsatz D, Harvey LD, Mirasyedi S, Levine MD (2007) Mitigating CO_2 emissions from energy use in the world's buildings. Build Res Inf 35(4):379–398
USCB (US Census Bureau) (2010) Guide to data sources, definition: NAICS 23, Construction, US Census Bureau. http://www.census.gov/epcd/naics02/def/NDEF23.HTMBusinessDictionary.com
USEPA (2002) Tool for the reduction and assessment of chemical and other environmental impacts (TRACI): User's guide and system documentation, USEPA, Washington, DC, USA
Victoria M, Perera S (2018) Carbon and cost hotspots: an embodied carbon management approach during early stages of design. Springer International Publishing AG
Vukotic L, Fenner RA, Symons K (2010) Assessing embodied energy of building structural elements. Proc Inst Civ Eng-Eng Sustain 163:147–158
Waldman B, Huang M, Simonen K (2020) Embodied carbon in construction materials: a framework for quantifying data quality in EPDs. Build Cities 1(1):12–24
WGBC (2019) Bringing embodied carbon upfront: coordinated action for the building and construction sector to tackle embodied carbon. London: World Green Building Council (WGBC). https://www.worldgbc.org/bringing-embodied-carbon-upfront-reportwebform
Worrell E, Price L, Martin N, Hendriks C, Media LO (2001) Carbon dioxide emissions from the global cement industry. Annu Rev Energy Env 26(1):303–329
WRAP (2017) Embodied carbon database: share and compare embodied carbon data. https://www.ecdb.wrap.org.uk. Analyze embodied carbon data. https://www.carbondeqocom/database/graph
WRAP (Waste and Resources Action Programme) (2010) Environmental benefits of recycling–2010 Update. Final report WRAP, UK

WRI/WBCSD (World Resources Institute/World Business Councilfor Sustainable Development) (2004) The greenhouse gas protocol: a corporate accounting and reporting standard. WRI/WBCSD, Geneva, Switzerland

Xiao J, Ma ZM, Ding T (2016) Reclamation chain of waste concrete: a case study of Shanghai. Waste Manag 48:334–343

Yan H, Shen QP, Fan LCH, Wang YW, Zhang L (2010) Greenhouse gas emissions in building construction: a case study of one peking in Hong Kong. Build Environ 45:949–955

Yolles H (2010) Embodied carbon—sustainable offices. South West of England Regional Development Agency (SWRDA), UK

You F, Hu D, Zhang H, Guo Z, Zhao Y, Wang B et al (2011) Carbon emissions in the life cycle of urban building systems in China—a case study of residential buildings. Ecol Complex 8:201–212

Zimmerman M, Althaus HJ, Haas A (2005) Benchmarks for sustainable construction–a contribution to develop a standard. Energy Build 37:1147

Chapter 16
Human Health Hazards Associated with Asbestos in Building Materials

Alessandro F. Gualtieri, Magdalena Lassinantti Gualtieri, Valentina Scognamiglio, and Dario Di Giuseppe

Abstract Asbestos is a fibrous natural material that possesses outstanding technological properties exploited since the time of its discovery for the manufacture of various building materials. Unfortunately, as known since the mid-1950s, both humans and animals exposed to asbestos fibres may develop a number of lethal respiratory diseases. Consequently, international medical and health organizations have classified asbestos as a human carcinogen and many countries worldwide have banned its use. Besides a short historical chronicle, this chapter provides a classification of asbestos minerals, applications in building materials, as well as its toxicity and pathogenicity mechanisms. The global asbestos issue and its use as a building material today will be the core of the chapter. In addition, a section is dedicated to the description of the reclaim, disposal and recycling of asbestos containing materials and a description of the substitutes of asbestos used today in building materials.

Keywords ACBM · Asbestos · Asbestos substitutes · Disposal · Pathogenicity · Reclamation · Recycling · Toxicity

16.1 Introduction

The word *asbestos* refers to a family of mineral fibres known and used for millennia (Dilek and Newcomb 2003; Ross and Nolan 2003). It was discovered that anthophyllite asbestos, one of the five amphibole asbestos minerals, was utilized to manufacture fireproof pottery and ceramics in Northern Finland (Lapland) during the Stone Age, some like 7,000–10,000 years ago and more commonly during the Early Metal Age, ca. 2,000 B.C. to 300 A.D. (Gualtieri 2017). Chrysotile, the only serpentine asbestos mineral, was utilized for the first time on the island of Cyprus about 5,000 years ago

A. F. Gualtieri · V. Scognamiglio · D. Di Giuseppe (✉)
Department of Chemical and Geological Sciences, University of Modena and Reggio Emilia, Modena, Italy
e-mail: dario.digiuseppe@unimore.it

M. Lassinantti Gualtieri
Department of Engineering "Enzo Ferrari", University of Modena and Reggio Emilia, Modena, Italy

J. A. Malik and S. Marathe (eds.), *Ecological and Health Effects of Building Materials*,
https://doi.org/10.1007/978-3-030-76073-1_16

for weaving clothes (Gualtieri 2017 and references therein). Greeks and Romans knew asbestos and praised this magic fibre in many different ancient scripts. In the first century A.D. the Roman philosopher Pliny the Elder wrote about asbestos in his *Naturalis Historia* calling it as *linumvivum* (living linen) (Rosselli 2014). Over time, asbestos continued to attract noblemen, alchemists, and magicians from Western Europe to the Far East. Legend has it that Charlemagne, the Emperor of the Sacred Roman Empire from 800, possessed a fireproof tablecloth very likely made of asbestos that he was using to impress his guests by cleaning it into the fire (Rosselli 2014). In the sixteenth century, Georg Agricola provided the first scientific explanations to the magical properties of asbestos (Alleman and Mossman 1997).

The industrial age of asbestos dates back to about 1850 when manufacturing sites opened in Germany and the United Kingdom. Mining activity of white (chrysotile) asbestos in Quebec (Canada) began in 1878 in the Thetford district. Since then, asbestos rapidly became an invaluable resource and every-day life commodity all over the world (Gualtieri 2017). At that time, in countries like the Russian Empire where asbestos from the Ural deposits had been extensively exploited since 1884 (Shcherbakov et al. 2001) and used for the manufacture of various building materials, it was truly considered a marvel material. Asbestos was such a wonder that in 1908 Aleksandr Aleksandrovič Bogdanov, in his fiction book Краснаязвезда (The red star), narrates that the Martians weaved their clothes using *"fibrous minerals of the asbestos type"*.

In 1915, the Russian Empire was second only to Canada as far as the production of asbestos in the world.The period across the nineteenth and twentieth century saw the birth of other asbestos mine districts all over the world. Among them, the South African mines soon became of paramount importance as two different amphibole asbestos species were mined there: crocidolite and amosite (A=asbestos M=mines O=of S=South Africa) whose exploitation began in 1893 (Beukes and Dryer 1986) andin 1907 (Bowles 1955), respectively. Even Italy had its own natural source of asbestos from the mining district of Balangero and Corio, ca. 20 km northwest of Torino, where the exploitation of chrysotile began in 1918 and ended in 1990. Primo Levi, one of the most famous Italian writers of the twentieth century, also worked for some time at the Balangero chrysotile mine. In one of his books, Levi reports a fragment of his infernal experience there: "*There was asbestos everywhere, like gray snow: if you left a book on a table for a few hours and then removed it, you would find its negative profile.*" (Levi 1975). Figure 16.1 is an image of the abandoned Balangero mine which is now a reclaimed superfund site of national interest.

The outstanding technological properties of commercial asbestos have been extensively exploited at industrial scale since the beginning of the twentieth century. It is possible to claim that asbestos minerals have been utilized to create more than 3,000 different asbestos containing materials (ACMs) used in practical and industrial applications (Gualtieri 2012). Among the asbestos minerals, chrysotile is by far the most exploited one. The asbestos-cement industry is the largest user of chrysotile fibres (about 85% of all applications) and it is estimated that about 95% of mining activity regards chrysotile asbestos (Ross et al. 2008).

Fig. 16.1 The reclaimed Balangero mine; an asbestos remediation Italian superfund site

The state of the art of the definitions of asbestos, pinpointing the gaps in this sphere of knowledge, and classification of asbestos minerals are described in Sect. 16.2 with the goal to deliver a synthetic clear picture of the complex area. Another goal of this chapter will be to describe what we know today about the toxicity and pathogenicity effects of asbestos (Sect. 16.3) within an uncertain and conflicting global scenario (Sect. 16.5).

As described in a dedicated paragraph of this chapter (Sect. 16.4), a number of building materials were or are actually made of composite mixtures including asbestos. ACMs used in buildings are classified as *loose* (or *friable*) and *compact*. *Loose* or *friable* asbestos building materials are mechanically crushed or pulverized with little effort with fibres (usually 70–95 wt.% of the product) that can be easily released into the surrounding environment. *Compact* asbestos building materials are made of a cement or polymeric material added with asbestos fibres (about 4–15 wt. %). Asbestos fibres are well fixed to the matrix and are released only if the material is damaged by mechanical tools. Common examples of *friable* asbestos in building materials are: suspended ceilings and floors, coal stoves, fireproofing spray and fire door interiors, insulating boards/panels, acoustical panels and finishes, lagging like steam pipes, boilers, pipework, asbestos blanket or asbestos paper tape, anti-vibration gaiters, ducts, walls and soundproofing or decorative spray coatings (Gualtieri 2012, 2013). Selected examples of *compact* asbestos in building materials are: masonry fillers, mortars, planar or corrugated roofing (by far the most common ACMs), shingles, vinyl asbestos, pipes, and water tanks (Gualtieri 2012, 2013). Figure 16.2 portrays selected examples of asbestos containing building materials (ACBMs): corrugated cement-asbestos roofing of a building in an urban industrial

Fig. 16.2 Examples of asbestos containing building materials (ACBMs) **a** corrugated cement-asbestos roofing of a building in an urban industrial area; **b** bent corrugated cement-asbestos roofing of a building in a rural industrial area; **c** a grain silo in a rural area. Original pictures taken by A.F.G

area (a); bent corrugated cement-asbestos roofing of a building in a rural industrial area (b); a grain silo in a rural area (c).

One of the objectives of this chapter is to deliver an updated classification of ACMs and discuss the problems related to their reclamation and disposal, with special attention to the differences in the existing directives and laws (Sect. 16.6). A further objective of the chapter is to report an updated list of asbestos substitutes that are used nowadays in building materials with an unbiased analysis of their pros and cons, especially in terms of toxicity/pathogenicity effects.

16.2 Classification of Asbestos Minerals

Although most of the outstanding technological properties and health hazards of asbestos are known, there is no consensus to date on a single definition of this term (Mossman and Gualtieri 2020). Depending on the context (commercial, regulatory, mineralogical, etc.), there are many ways to define the word "asbestos". Asbestos is often used as a generic term to identify minerals that can be mechanically ground to generate thin flexible fibre bundles of single fibres (Case et al. 2011). According to the commercial definition, asbestos are mineral fibres that possess exceptional properties (like heat resistance, mechanical strength and many more) that make them valuable materials for industrial purposes (Niklinski et al. 2004). As concerns the regulatory framework, asbestos is a group of fibrous minerals with length >5 μm and aspect ratio (length/width) ≥3 (National Institute for Occupational Safety and Health 2019). Unfortunately, the incomplete and inadequate definition of asbestos led to general confusion in scientific, social, health, and legal frameworks (Mossman and Gualtieri 2020). In this chapter, we will refer to the general definition adopted by international health and regulatory agencies (such as International Agency for Research on Cancer and the International Labour Organization) for which chrysotile and five fibrous amphiboles form the family of "asbestos" minerals (Gualtieri 2017). The amphibole asbestos species are amosite (asbestos grunerite), crocidolite (asbestos riebeckite), fibrous actinolite, fibrous anthophyllite and fibrous tremolite (Gualtieri 2012) (Table 16.1).

Asbestos minerals are silicates classified based on their crystalline chemistry and structural features (Ballirano et al. 2017). Amphiboles are double chain silicates with Si/O=4:11. These chains are linked to a layer of octahedral sites: $M(1)$, $M(2)$, $M(3)$ are regular octahedral cavities and $M(4)$ is a large and distorted 6- to eightfold cavity (Fig. 16.3a). In addition, there is an even larger 10- to 12-fold A site (Fig. 16.3b). OH^- groups occur in the interiors of the rings in the double chains (Fig. 16.3a). The structure of amphiboles has the general formula $A_{0-1}B_2C_5T_8O_{22}W_2$ (Hawthorne et al. 2007). A=Ca^{2+}, K^+, Na^+, Li^+, with 6 to 12-fold coordination (i.e., A site); B=Ca^{2+}, Mg^{2+}, Na^+, Fe^{2+}, Mn^{2+}, Li^+, with distorted eightfold coordination (i.e.,

Table 16.1 The ideal chemical composition and crystal symmetry of the six asbestos minerals (adapted from: Gualtieri 2012)

Commercial term	Mineral species	Idealized chemical formula	Space group	References
Amphibole asbestos				
Amosite/brown asbestos	Grunerite	$(Fe^{2+},Mg)_7Si_8O_{22}(OH)_2$	*C2/m*	Pollastri et al. (2017a)
Crocidolite/blue asbestos	Riebeckite	$Na_2(Fe^{2+},Mg)_3Fe_2{}^{3+}Si_8O_{22}(OH)_2$	*C2/m*	Pacella et al. (2019)
Actinolite asbestos	Actinolite	$Ca_2(Mg,Fe^{2+})_5Si_8O_{22}(OH)_2$	*C2/m*	Pollastri et al. (2017b)
Anthophyllite asbestos	Anthophyllite	$(Mg, Fe^{2+})_7Si_8O_{22}(OH)_2$	*Pnma*	Pollastri et al. (2017a)
Tremolite asbestos	Tremolite	$Ca_2Mg_5Si_8O_{22}(OH)_2$	*C2/m*	Giacobbe et al. (2018)
Serpentine asbestos				
White asbestos	Chrysotile	$Mg_3(OH)_4Si_2O_5$	*Cc*	Pollastri et al. (2016)

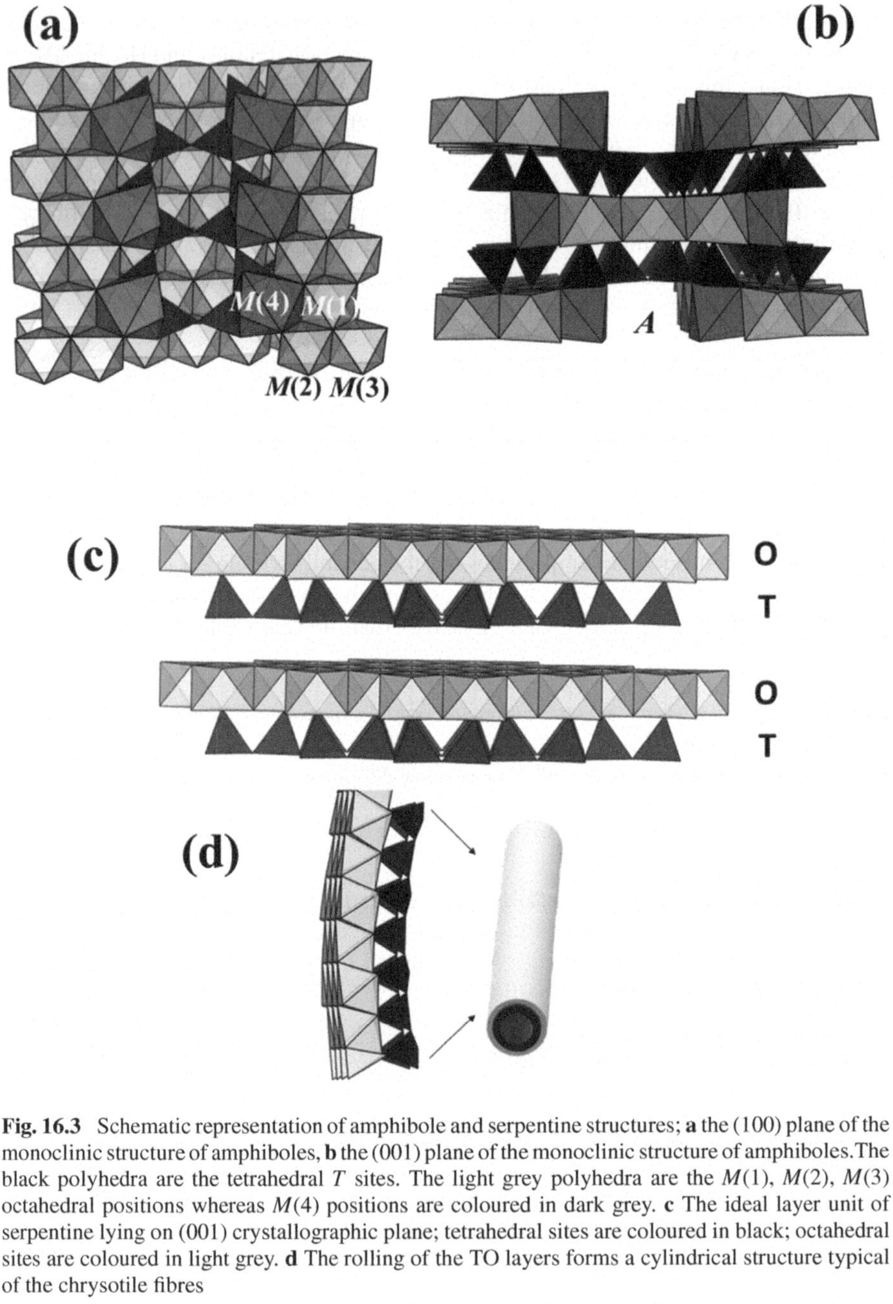

Fig. 16.3 Schematic representation of amphibole and serpentine structures; **a** the (100) plane of the monoclinic structure of amphiboles, **b** the (001) plane of the monoclinic structure of amphiboles. The black polyhedra are the tetrahedral *T* sites. The light grey polyhedra are the *M*(1), *M*(2), *M*(3) octahedral positions whereas *M*(4) positions are coloured in dark grey. **c** The ideal layer unit of serpentine lying on (001) crystallographic plane; tetrahedral sites are coloured in black; octahedral sites are coloured in light grey. **d** The rolling of the TO layers forms a cylindrical structure typical of the chrysotile fibres

the $M(4)$ sites); C=Mg^{2+}, Fe^{3+}, Fe^{2+}, Al^{3+}, Mn^{3+}, Mn^{2+}, Ti^{4+}, Li^{+}, with regular sixfold coordination(i.e., $M(1)$, $M(2)$ and $M(3)$ sites); T=Si^{4+}, Al^{3+}, Ti^{4+}, at the tetrahedral sites $T(1)$ and $T(2)$ running along the chains; W=OH^{-}, Cl^{-}, F^{-}, O^{2-}. These minerals preferentially crystallize along the *c*-axis and mono-dimensional growth determines their fibrous crystal habit (Gualtieri 2012). Amphiboles crystallize in different space groups (Hawthorne et al. 2007) but, apart from anthophyllite, the asbestos varieties are monoclinic with space group *C2/m* (Table 16.1). Idealized chemical formulas of amphibole asbestos are shown in Table 16.1. For detailed information regarding the classification, crystal chemistry and structural characteristics of amphiboles, the reader can refer to the work of Hawthorne et al. (2007).

The serpentine group of silicate minerals includes the fibrous polymorph chrysotile, and lamellar antigorite and lizardite (Ballirano et al. 2017). To a first approximation, the structure of serpentine species is characterized by units of tetrahedral (T) sheets centred by Si and octahedral (O) sheets centred by Mg with T/O=1:1 (Fig. 16.3c). Because the size of an ideal T sheet (b=9.10 Å) is smaller than the size of an ideal O sheet (b=9.43 Å), a mismatch between the T and O sheets occurs inducing a differential strain (Pollastri et al. 2016). To compensate for the size differences and strain of the sheets, structure distortions occur in the different polymorphs of serpentine (Ballirano et al. 2017). Concerning chrysotile, the rolling of the TO layer releases the strain and forms a cylindrical lattice (Pollastri et al. 2016). The curvature of the lattice propagates along a preferred axis leading to the formation of the tubular structure typical of chrysotile fibres (Fig. 16.3d). The general chemical formula of serpentine is $Mg_3(OH)_4Si_2O_5$. Ionic substitutions are usually limited in chrysotile compared to other serpentine minerals (Ballirano et al. 2017). The most common substitution occurs between Fe^{2+} and Mg^{2+}in the octahedral site (Ballirano et al. 2017). Moreover, Al^{3+} can replace Si^{4+} in the T sheet and Fe^{3+}can replace Mg^{2+} in the O sheet (Pollastri et al. 2016; Ballirano et al. 2017). Gualtieri et al. (2019a) recently found that Cr, Ni, Mn and V can replace Mg in the O sheet. The concentration of this group of metals in chrysotile is highly variable and depends on the geological origin (Bloise et al. 2016): 2,044 and 2,064 mg/kg for Italian chrysotile samples from Valmalenco and Balangero, respectively (Bloise et al. 2016); 1,704 mg/kg for a chrysotile sample from Quebec, Canada (Bloise et al. 2016); 13,473 mg/kg for a commercial sample of chrysotile from Orenburg, Russia (Di Giuseppe et al. 2021).

The peculiar structural features of asbestos minerals give them exceptional physical and chemical properties that building material manufacturers have found incredibly useful (Gualtieri 2012). The main properties of the asbestos fibres are: high tensile strength, non-flammable, sound isolation, low thermal conductivity, chemical resistance, high surface area, thermal stability and thermal resistance (Gualtieri 2012). Although all types of asbestos have these properties, amphibole asbestos and chrysotile are different. In particular, amphibole asbestos fibres are resistant to any type of chemical attack, while chrysotile fibres dissolve quickly in an acidic environment (Gualtieri et al. 2018a). Concerning chrysotile, the low pH induces the replacement of Mg^{2+} for H^{+} or H_3O^{+} and leads to the breakdown of the octahedral layer resulting in a form of amorphous silica (Gualtieri et al. 2018a, 2019b). In contrast, the substitution of octahedral cations for H^{+} in amphibole species occurs

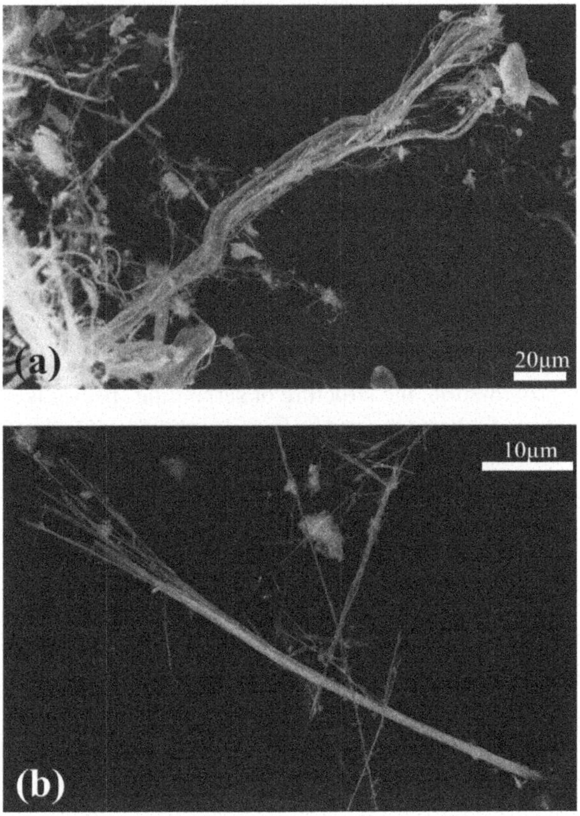

Fig. 16.4 Electron micrograph images of asbestos fibres. **a** Image of chrysotile fibre bundles from Balangero (Italy). Chrysotile forms aggregates of very long, thin and curvilinear fibres that show a high degree of flexibility. **b** Micrograph images of UICC standard crocidolite from South Africa. Crocidolite fibres show the typical columnar and straight aspect of amphibole asbestos

without major structural changes (Gualtieri et al. 2018a). The other main difference between these two types of asbestos concerns their crystalline habit. As displayed in Fig. 16.4, chrysotile is characterised by long, very thin and curled fibres (Pollastri et al. 2016) whereas amphibole asbestos fibres commonly exhibit a quite rigid and straight columnar aspect (Belluso et al. 2017).

16.3 Toxicity and Health Effects of Asbestos

We are aware of the potential risks of asbestos to human health since the end of the nineteenth century (Alleman and Mossman 1997) and literature from the 1930 and 1940s reported health problems affecting workers exposed to asbestos (Niklinski et al. 2004). However, the first unequivocal evidence of asbestos carcinogenicity was delivered in the mid-1950s by Sir Richard Doll whose pioneering epidemiological studies correlated lung cancer among asbestos workers to asbestos exposure (Doll 1955). Later, many more scientific evidences were found to prove that exposure to asbestos is linked to the development of respiratory diseases such as lung cancer, malignant mesothelioma (MM) and asbestosis (pulmonary fibrosis affecting asbestos workers) (Mossman and Gualtieri 2020). The International Agency for Research on Cancer (IARC), after carefully evaluating available data (e.g., from epidemiological, in vivo and in vitro studies), determined in 2012 that all asbestos types (chrysotile and amphibole asbestos) are carcinogenic for a number of target organs including lung, pleura, peritoneum and larynx (International Agency for Research on Cancer 2012). Hence, chrysotile and amphibole asbestos are now included in Group 1 "carcinogen for humans" (International Agency for Research on Cancer 2012) and classified as Category 1A carcinogens (European Chemicals Agency 2015). According to the latest data available in the literature, asbestos causes approximately 255,000 deaths per year, of which 233,000 are related to occupational exposure (Furuya et al. 2018). In particular, MM is estimated to cause the deaths of 13,883 people annually in Asia, 3,354 in Africa and 2,794 in Europe (Odgerel 2017).

Exposure to asbestos fibres occurs through inhalation, mainly in the workplace but also in the vicinity of natural geologic occurrence of asbestos or inside buildings with ACM (Kamp 2009; Gualtieri 2020). According to the World Health Organization criteria, regulated asbestos fibres are longer than 5 μm, thinner than 3 μm and with an aspect ratio (length/width) $\geq$3 and can be airborne and inhaled (World Health Organization1997). Once released into the air, these fibres easily penetrate the upper airways (i.e., nasal and oral cavities) and travel along the airflow pathway (Gualtieri et al. 2017). The fate of a fibre in the respiratory tract depends on its aerodynamic diameter D_{ae} (Gualtieri et al. 2017): when D_{ae}>5 μm, a fibre is filtered in the upper respiratory tract where it is cleared (Bustamante-Marin and Ostrowski 2017); when $3 \leq D_{ae} \leq 15$ μm, a fibre reaches the laryngeal/bronchial tracts; when $D_{ae} \approx 2$–3 and <0.2, a fibre reaches the lower alveolar respiratory section. The airways are covered with a mucus layer that is constantly moved by the ciliated epithelium from the lower to upper airways (Bustamante-Marin and Ostrowski 2017). Mucociliary escalator traps and carries particles to the pharynx where they can be expelled by coughing or swallowing (Bustamante-Marin and Ostrowski 2017). Particles that can bypass the mucociliary escalator and reach the alveolar spaces are cleared by alveolar macrophages (AMs) through phagocytosis (Oberdörster 1993). Phagocytosis involves several steps, from recognizing the exogenous particle (via a receptor-mediated mechanism) and engulfing it in a phagosome (i.e., a membrane-bound vacuole) to the maturation of phagosome into a phagolysosome (i.e., a new vacuole created by the fusion of phagosome with lysosome) and the dissolution of the internalized particle in the phagolysosome acidic environment (Rosales and Uribe-Querol 2017). Unfortunately, phagocytosis fails when AMs attempt to engulf fibres longer than the AM diameter (Fig. 16.5a), leading to frustrated phagocytosis (Donaldson et al. 2010). Toxicological studies on lung tissues from rodents and human subjects suggest that the fibre length threshold value for inducing frustrated phagocytosis is 5 μm (Mossman 2018; Roggli and Green 2019). Fibres that are not biodurable (i.e., with biodurability defined as the resistance of a particle to biochemical decomposition) such as chrysotile (Gualtieri 2018; Gualtieri et al. 2018a), are dissolved by the acidic environment of the phagolysosome, broken into small fragments and easily

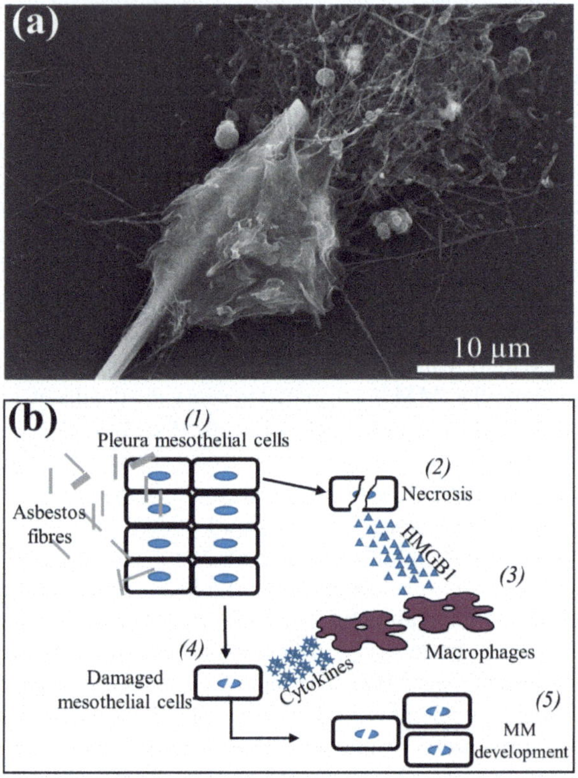

Fig. 16.5 **a** Electron micrograph showing an alveolar macrophage attempting (without success) to phagocytose a long asbestos fibre. **b** Sketch illustration of the chain process responsible for the onset of MM. (*1*) Thin and long asbestos fibres can reach the pleural surface. (*2*) In the pleural cavity, asbestos fibres induce frustrated phagocytosis and chronic inflammation that prompt programmed necrosis of mesothelial cells. (*3*) Necrotic cells release the proinflammatory protein HMGB1 which stimulates macrophages to secrete monocyte-derived cytokines. (*4*) Cytokines stimulate the survival of cells with damaged DNA and lead to the pathway of carcinogenesis of MM (*5*). (adapted from: Carbone and Yang 2012, 2017)

engulfed by the AMs (Bernstein et al. 2013; 2014, 2015). On the contrary, biodurable fibres, such as amphibole asbestos and fibrous zeolites (Gualtieri et al. 2018a), are designed to induce chronic frustrated phagocytosis (Donaldson et al. 2010).

Literature data shows that frustrated phagocytosis is pivotal in asbestos-induced toxicity (Carbone et al. 2019). During frustrated phagocytosis, in the attempt to clear the fibrous particles, AMs trigger respiratory bursts with production of reactive oxygen species (ROS). NADPH oxidase enzyme helps the macrophages to reduce O_2 to a superoxide free radical ($O_2{\bullet}^-$) that rapidly reacts with itself to form hydrogen peroxide (H_2O_2). $O_2{\bullet}^-$ and H_2O_2 via iron-catalysed reaction or myeloperoxidase-catalyzed oxidation, produce hydroxyl radicals (HO•) and hypochlorite (ClO^-) (Babior 1984; Dahlgren and Karlsson 1999; Thomas 2017). In addition to ROS, frustrated phagocytosis may be modulated by highly reacting nitrogen species (RNS) (Shukla et al. 2003). Active sites on the surface of the asbestos fibres also produce ROS (Turci et al. 2017). For example, O_2 reacts with Fe^{+2} exposed at the fibre surface generating H_2O_2 and HO• through Haber–Weiss/Fenton cycles (Turci et al. 2017; Gualtieri et al. 2019c). Prolonged production of ROS/RNS overcomes the antioxidant cell defence and induces alteration of the cell membrane, cell injury and DNA damage (Carbone and Yang 2012; Mossman 2018). The combined cyto/genotoxic activity of asbestos induces chronic inflammation and leads to the growth of mutated lung cells prone to cancer development (Carbone and Yang 2017; Gualtieri 2018; Mossman 2018; Carbone et al. 2019).

The size of a fibre is also determinant in inducing MM (Donaldson et al. 2010). The small diameter of asbestos fibres makes it possible for them to reach the pleural cavity (Donaldson et al. 2010; Gualtieri et al. 2017). Short fibres (<5 μm) are easily cleared from the pleura as they pass through the stomata (cavities in the parietal side of the pleura) and end up in the lymphatic capillary system (Donaldson et al. 2010). Conversely, long fibres (>5 μm) cannot leave the cavity, because they are too long to pass through the stomata (Donaldson et al. 2010). According to Carbone and Yang (2012), the asbestos fibres retained in the pleura cause chronic inflammation (via frustrated phagocytosis and ROS/RNS generation) and induce programmed necrosis of pleural mesothelial cells (Fig. 16.5b). The proinflammatory protein HMGB1 released from necrotic cells stimulates AMs to secrete cytokines like IL-1β and TNF-α prompting the proliferation of DNA-damaged cells responsible for the onset of MM (Carbone and Yang 2017; Carbone et al. 2019).

Carcinogenesis is a complex multistep process in which genetic susceptibility plays a primary role (Carbone et al. 2007; 2019). Recent studies demonstrated that subjects with germline mutations of the *BAP1* gene are prone to develop MM after asbestos exposure (Carbone et al. 2019). *BAP1* modulates several cellular functions like DNA repair and apoptosis (i.e., programmed cell death). Mesothelial cells with germline *BAP1* mutations are unable to adequately repair DNA damage and fail apoptosis (Carbone and Yang 2012). It follows that, if exposed to asbestos fibres, the mesothelial cells mutated in *BAP1* easily undergo neoplastic processes (Carbone and Yang 2012; Carbone et al. 2019).

Although all six asbestos species revealed carcinogenic effects (International Agency for Research on Cancer 2012), a part of the scientific community questions the toxicity of chrysotile (see Sect. 16.5) due to its low biopersistence (i.e., the resistance of a particle to biochemical decomposition and mechanical clearance) with respect to amphibole (Camus 2001; Bernstein et al. 2008; 2013). However, as stressed by IARC in Monograph 100c, low biopersistence is not synonymous with low toxicity and chrysotile toxicity must be related to other chemical/physical parameters (International Agency for Research on Cancer 2012). In this context, the recent toxicity model proposed by Gualtieri et al. (2019a) suggests that chrysotile induces toxicity in the same way as nanoparticles do (Studer 2010). During phagolysosome-induced dissolution, chrysotile fibres release the metals (e.g., Fe, Mn, Cr and Ni) stored at the octahedral site (Gualtieri et al. 2018; Gualtieri et al. 2019a). These metals when released into lung tissues prompt ROS formation (Gualtieri et al. 2019c). Low biodurability of chrysotile allows a fast release of these redox-active metals and induces an acute cytotoxicity (Gualtieri et al. 2018, 2019a). Furthermore, recent in vitro tests have shown that the amorphous silica fibre produced by the dissolution of chrysotile possesses moderate cyto/genotoxicity against monocytes and mesothelial cells (Gualtieri et al. 2019b).

16.4 The Use of Asbestos in Building Materials

Several building materials contain asbestos. Asbestos can be found in both private (farms, industrial sites, houses, etc.) and public (hospitals, railway stations, schools, etc.) buildings. There are basically two different ACBMs, friable and compact, which differ in asbestos content and nature of the asbestos rich composite material (see Becklake et al. 2007; D'Orsi 2007; Gualtieri 2012). A working definition of a friable ACBM is that it can be ground or pulverized by hand when it is dry, releasing free fibre bundles of single fibres. The content of asbestos in friable ACBMs is very high (from 70 to 95 wt.%) and for this reason, friable ACBMs should be considered very hazardous. A working definition of a compact ACBM is that the asbestos fibres are strongly anchored to a matrix and not easily released unless the material is processed using mechanical tools. They are thus composite materials where the asbestos fibres are intimately adhered to a cement or polymeric matrix. The content of asbestos in compact ACBMs is generally from 1 to 20 wt.%. Because of the lower content of asbestos and the resistance to release fibres, compact ACBMs should be considered much less hazardous than friable ACBMs.

Chrysotile has been the most used form of asbestos in the past and is the only asbestos mineral used today. Ross et al. (2008) reported that about 95% of asbestos mining commercial activities regard chrysotile. Table 16.2 contains a list of the most common friable and compact ACBMs (Gualtieri 2012). The table is not meant to be comprehensive. Figure 16.6 depicts a few selected examples of ACBMs: a flat tile with about 5–10 wt.% of chrysotile (compact) (a); a gasket made of chrysotile (friable) (b); a fragment of corrugated slate with about 10 wt.% of chrysotile (compact) (c); a fragment of vinyl asbestos floor with about 2 wt.% of chrysotile (compact) (d).

The presence of asbestos in building materials prompts long term exposure of the population to this contaminant and represents a great health concern. The progressive decomposition of ACMs causes the release of airborne fibresand subsequent possible exposure to this hazardous dust (Obmiński 2020). Degradation of these building materials is chiefly due to use damage. For indoor materials, the major deterioration factors are: natural ageing and wear, anthropic action, shrinkage and thermal contraction/expansion. For outdoor materials, the major deterioration factors that lead to the decomposition of the matrix and release of the asbestos fibres are: mechanical and chemical action of atmospheric agents (e.g., acid rains, hailstorms, etc.), mosses and lichens colonization, natural ageing and wear, anthropic action, shrinkage and thermal expansion in areas subjected to high thermal excursion (Obmiński 2020).

Obviously, the nature of the ACBM (friable or compact) plays a key role although it is mainly outdoor products that suffer degradation due to weathering. The mechanism of releasing asbestos fibres from the materials is well described in Spurny (1989) and Obmiński (2020).

Table 16.2 The most common friable and compact asbestos containing building materials (ACBMs)

Friable (loose) ACBMs	Compact ACBMs
Acoustic panels	Asphalt
Anti-vibration gaiters	Bonding cement and adhesives for floor tiles
Boilers, rope seals on boiler access room and flue	Caulk
Cavities and partitions of floors and ceilings	Ceiling tiles
Ceilings	Chimney flues and tops
Cork boards	Fire refractory bricks and cements
Coverings	Floor tiles
Ducts	Glassbestos
Expansion and compound joints	Guttering and drain pipes
Filled mastics	Insulating blocks and seals
Fire door interiors	Linoleum
Fireproofing blankets, boards and sprays	Masonry filler
Flat/flexible boards	Mastics
Floors	Mortars
Frames of windows and doors	Pipes
Gaskets	Putties
Insulating components of various nature and forms	Roof felt
Joint and patching composites	Roof planar or corrugated cement-asbestos slates
Lagging	Rubber
Panels to lift shafts	Shingles
Pipework	Sidings
Range hoods	Transit wallboards
Roof flashing	Vinyl asbestos and vinyl sheet flooring
Steam pipes	Wall boards, cladding and papers
Textured coatings and paints	Water tanks
Walls	

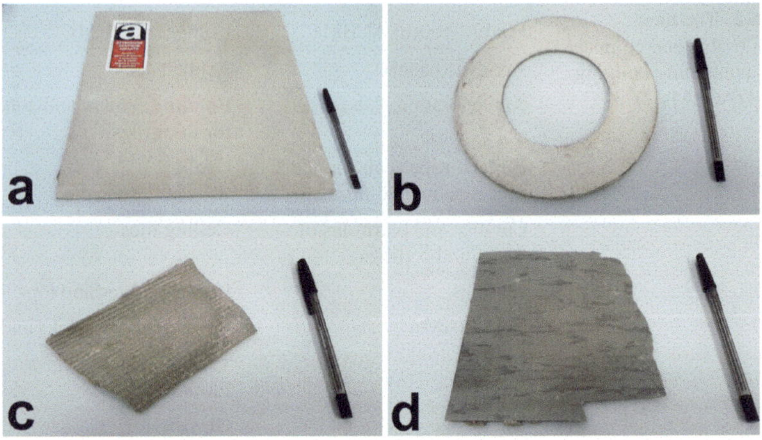

Fig. 16.6 Examples of ACBMs: **a** a flat tile; **b** gasket; **c** fragment of corrugated slate; **d** fragment of vinyl asbestos floor. Original pictures taken by A.F.G (pen included as scale marker)

16.5 The Actual Global Asbestos Issue

To describe the role of asbestos in the actual global market of building materials, one should start from the time when the scientific evidence of the carcinogenicity of asbestos minerals was shared among the global scientific, institutional and social communities. Since the mid-late 80's, following the compelling requests of most of the scientific community, the national associations of the asbestos victims, the health and environment protection groups, the workers' unions and many political forces, a number of countries worldwide (namely European countries) began to limit or ban asbestos in building materials.

In 1983, the European Union (EU) started to deliver issues against asbestos. The European Union Council directive 76/769/EEC (see European Council 1976) has been the framework directive to successfully limit the use of asbestos, through several amendments over the years beginning with the council directive 83/478/EEC (fifth amendment) which puts limitations to the commercialization and use of crocidolite (European Council 1983). A couple of years later, the other five asbestos fibres (i.e., chrysotile and the other four amphibole asbestos species) were banned from certain products such as toys, materials applied by spraying, paints, etc. (85/610/EEC, seventh amendment; see European Council 1985). In 1991, the total ban extended to include amphibole asbestos whereas chrysotile was only partially banned (European Council 1991). Finally, with the 1999/77/ECdirective (European Council 1999), manufacture, marketing and use of asbestos fibres and of products containing them was prohibited by 2005. Nevertheless, ACMs in use before the beginning of 2005 are allowed to remain in use if their degradation level is limited. The regulation indeed reports that "Member States may, for reasons of protection of human health, restrict, prohibit or make subject to specific conditions, the use of such articles before they are disposed of or reach the end of their service life." In this context, the Italian law banning the extraction and use of asbestos and ACMs (Italian Ministry of Health 1994) includes a paragraph that obliges building owners to notify the local authorities (AUSL = Azienda Unità Sanitaria Locale) if friable asbestos is found on their premises. Furthermore, removal on their own expense may be required if considered necessary by authority. Today the asbestos ban is incorporated in the EC regulation 1907/2006 regarding the Registration, Evaluation, Authorisation and Restriction of Chemicals (REACH), without further limitations on the use of ACMs installed before 2005 (European Parliament 2006). The United States was the most important world user of asbestos during the major part of the twentieth century (Virta 2006). In 1989, most asbestos-containing materials were banned in a document delivered by the United States Environmental Protection Agency (USEPA). Prior to this act, only some spray-applied surfacing materials and some friable asbestos-containing materials were banned beginning from 1973. However, in 1991, following a decision of the Fifth Circuit Court of Appeals of New Orleans, the 1989 regulation was amended to ban new uses of asbestos and some specific asbestos-containing products (United States Environmental Protection Agency 1994).

Today, raw bulk chrysotile is imported for the fabrication of diaphragms for use in chlorine and sodium hydroxide production, whereas a limited number of chrysotile asbestos containing products are imported for use in chemical, automotive and oil industries. These products are now facing a Toxic Substance Control Act (TSCA) risk evaluation (United States Environmental Protection Agency 2016). Following the final rule issued by USEPA in 2019, any other uses of asbestos that began prior to 1989 and therefore not prohibited by the 1989 partial ban, cannot be commercialized again without USEPA review (United States Environmental Protection Agency 2019). Asbestos-containing building materials are covered by this rule (e.g., roof coatings, cement products, millboard, roofing felt, vinyl-asbestos floor tiles and others).

Although in the last two decades asbestos was subjected to intensive multidisciplinary studies, the causes of toxicity and pathogenicity as well as the relationship between fibres' exposure and the development of respiratory diseases are still unclear. In this scenario, two blocks have risen in the global scenario facing each other with no concession or negotiation (the global chrysotile or asbestos issue): one block considers all the regulated asbestos fibres as toxic and carcinogenic and supports their ban; the other side recognizes that only the five amphibole asbestos species are toxic and carcinogenic while chrysotile is carcinogenic but with a low to null toxicity and low potential for producing mesothelioma (see for example Bernstein et al. 2013; Camus 2001; Liddell et al. 1997; McDonald et al. 1997) if compared to amphibole asbestos. With this assumption, these countries promote the safe ("controlled"; LaDou et al. 2010) use of chrysotile. The chrysotile defence is based upon the assumption that lung diseases due to chrysotile exposure should be attributed to the fact that chrysotile can be contaminated by amphibole asbestos. This "amphibole hypothesis" (Mossman 1993; Stayner et al. 1996) is supported by the biodurability paradigm (Hogson and Darnton 2000; Berman and Crump 2008; Gualtieri et al. 2018a). Chrysotile has a low biodurability (Gualtieri et al. 2018a) and promptly dissolves in the intracellular macrophage environment (see Sect. 16.3), during phagocytosis, whereas amphiboles are biodurable and are kept in the lungs for decades (Giacobbe et al. 2021; Gualtieri et al. 2018a; Oze and Salt 2010; Bernstein et al. 2008; Hume and Rimstidt 1992). Although the defence of chrysotile based on the "amphibole hypothesis" is rejected by most of the scientific community, there are relatively few epidemiological studies on this issue to systematically compare the effects of the different fibre types and this somehow leaves the issue open to debate (Gualtieri 2017; 2018).

The global chrysotile litigation determines a *status quo* situation with just 67 out of 196 countries (34%) in the world that have banned asbestos whereas a "controlled use" of chrysotile is permitted in the remaining (66%) countries (International Ban Asbestos Secretariat 2021a; Zen et al. 2013). Uncertain contradicting policies result in bizarre outcomes. Above all, the example of China should be addressed: although China is among the countries allowing a "controlled use" of asbestos, the building materials used to build the 2008 Olympic venues were asbestos-free because the Chinese government assumed that some foreign athletes would not participate if they had to eventually get in contact with asbestos (Frank et al. 2014).

In any case, the global situation reveals that the chrysotile asbestos market is active, although the production of asbestos worldwide is markedly decreasing. The 2019 asbestos (chrysotile) trade data (United States Geological Service 2020) shows that the top producing countries (t/year) in decreasing order are: Russian Federation (750,000), Republic of Kazakhstan (200,000), People's Republic of China (125,000), Federative Republic of Brazil (15,000) and Republic of Zimbabwe (2,500), (total world production is 1,100,000 t/year). It should be said that in Brazil there is actually only one producer which temporarily suspended its activity. In Zimbabwe, asbestos is mined from the tailings of mines closed in 2007 (United States Geological Service 2020).

The 2018 asbestos top five users (t/year) in decreasing order are (International Ban Asbestos Secretariat 2021b): Republic of India (308,000), People's Republic of China (288,000), Russian Federation (234,000), Federative Republic of Brazil (120,000), and Republic of Indonesia (114,000).

Besides the uncertainties of the current scientific models, circumventing the global chrysotile issue will not be an easy task due to the political pressure in favour of chrysotile of the world's economy leading countries like India and Russia. In those countries, chrysotile asbestos is a primary strategic raw material whose economic importance prevails over health concern. The case of India is exemplary. India has very limited asbestos mining activity but is one of the top importers of asbestos in the world for the production of building materials with low taxes for asbestos containing products and higher taxes for the corresponding products made with asbestos substitutes (Frank et al. 2014). The situation is getting even more (intentionally) confused as other countries of the leading global economy like the United States adopt contradictory behaviour. As described above, in the United States chrysotile asbestos is not actually banned and, under the mandate of President Trump, the new issues delivered by USEPA have the potential to even increase its utilization in the future and not its conclusive ban (Landrigan and Lemen 2019).

In this scenario, it is not surprising that chrysotile asbestos has not been included so far in the category of hazardous substances concerned by the Rotterdam convention. Consequently, ACBMs are freely distributed worldwide without any labelling. This situation originates enormous problems and eventually international incidents: consider for example that an ACBM produced in a country like China (where chrysotile asbestos is safely used) can be anonymously exported to a country like Italy where import, export, and use of ACBMs are banned since 1992 (Gualtieri 2012).

16.6 Asbestos Reclamation, Disposal and Recycling

Despite the ban of asbestos in the EU, ACBMs are still found in the living environment as current legislations do not necessarily oblige the removal of installations undertaken prior to 2005 (REACH, EC regulation 1907/2006; see European Parliament 2006). As explained in Sect. 16.4, most of those materials may be found in buildings both indoor and outdoor and represent a health concern only if the asbestos fibres become airborne, for example if damaged, friable, or located in places where they are exposed to vibration, movements or air currents. Hence, a risk assessment by certified companies is performed before deciding which actions are needed to be taken. This is done by examining the conditions of the material, environmental factors that could determine fibre release and eventually also by measuring the fibre concentration in the indoor air (Italian Ministry of Health 1994). Workers performing demolition, maintenance and refurbishing activities have the highest risk of being exposed to airborne fibres. Hence, these actions are regulated by laws assuring that necessary measures are taken to avoid exposure for both workers and the population. The 2009/148/EC directive is the roadmap for the European Union to protect workers against asbestos exposure exceeding 0.1 fibres/cm^3 on 8h long exposure time, the only activity allowed being treatment and disposal of asbestos (European Parliament 2009). Whenever workers are or may be exposed to asbestos fibres, the employer is obliged to guarantee that the exposure limit is not exceeded through risk assessment, safe work practice, protective equipment, training, etc. The employer is also required to provide a work plan to the local health/environment authority. In Italy, companies authorized to the disposal and reclamation of sites contaminated by asbestos must be listed in a national register and report their activities to the local authority "Azienda Unità Sanitaria Locale", AUSL (see Italian Law n. 257 1992).

Regulations and technical methodologies for the reclamation of ACBMs foresee the following reclamation methods (the choice of method being determined by the nature of the material, the state of preservation as well as environmental factors) (Italian Ministry of Health 1994, 1999);

Encapsulation: This method prevents the dispersion of fibres into the environment by spraying specially designed products that either isolates the material from the environment by forming a protective coating or fixes the fibres to the matrix by penetrating into the pores of the material. Although being economical and fast, this method requires periodical controls and maintenance of the encapsulation to assure its integrity over time.

Isolation: This method is mainly used to reclaim cement-asbestos roofs in relatively good conditions and consists of placing a new lightweight metal cover over the asbestos-cement slates. Before proceeding with the new coverage, the slates are subjected to preventive encapsulation. As for encapsulation, periodical controls and maintenance are required.

Removal: This method becomes mandatory when the ACBM is very deteriorated and the risk of release of fibres into the surrounding environment is high. Prior to removal, the ACBM is encapsulated in order to limit fibre dispersion during handling. Any dust that is formed during these operations must be aspirated. Although being a final solution, the produced hazardous waste must be dealt with.

During the reclamation activity, employers are invariably obliged to provide the workers with protective clothing including gloves and shoes. Respiratory protection must also be used and the choice of device depends on the degree of air pollution (Italian Ministry of Health 1994, 1999). Suitable preparation of the working place is particularly important for indoor reclamation sites (Italian Ministry of Health 1994, 1999). Firstly, all movable items must be removed whereas those that cannot be moved should be sealed by plastic wrapping and adhesive tape including, window- and door-frames, sockets, radiators, etc. Filters in ventilation, heating and air conditioning systems must be removed and treated as asbestos-containing waste. Floors and walls must also be covered by polyethylene sheets and a temporary power supply system should be installed. An air extraction system equipped with efficient filters must be installed so that a negative pressure is kept inside the reclamation area with respect to the surroundings. This system has the dual advantage of avoiding fibre leakage and reducing the concentration of asbestos fibres inside the work area.

According to Regulation (EC) 1272/2008 (see European Parliament 2008a), asbestos is classified as a carcinogenic substance (see Sect. 16.3) with hazard statement codes H350 and H372 (i.e., may cause cancer and damage to organs, respectively). The 2008/98/EC directive, ruling the waste management in general (defining terms and basic management codes), states that a waste containing more or equal to 0.1% of a category 1A carcinogenic substance with hazard statement code H350 must be classified as carcinogenic waste with hazardous property HP7 (European Parliament 2008b). As a consequence of the waste directive, all of the asbestos-containing wastes listed in the European waste catalogue (delivered as Commission issue2000/532/EC and modified in 2014 by the 2014/955/UE issue) are classified as hazardous (European Commission 2014). Waste producers and handlers in the member states must refer to a six-digit code, the first two ones identifying the source generating the waste, when producing documentation required under waste legislations. ACMs from buildings are identified by the codes 17.01.05* (asbestos containing materials) or 17.06.01* (asbestos containing materials for insulation). It should also be mentioned that asbestos-contaminated materials produced during reclamation is identified by the code 15.02.02* (asbestos contaminated absorbents, filters, protective clothing and more). Paglietti et al. (2016) recently analysed the European laws on the management of wastes underlined the importance of correct classification of asbestos-containing waste in order to protect workers involved in waste handling and guarantee disposal in landfills of suitable category according to current regulations (1999/31/EC; 2003/33/EC) (Paglietti et al. 2016).

ACMs are today disposed of in controlled landfills even though viable technical solutions exist that transform the hazardous fibres into harmless compounds (Plescia et al. 2003; Paolini et al. 2019; Gualtieri, 2013). The landfill option is certainly in conflict with the EU directives (European Parliament 2008b) that lists a priority order for the waste management: (1) prevention; (2) preparing for reuse; (3) recycling; (4) different recovery plans; (5) disposal. In addition, the following statements can be found in the 2012/2065 (INI) document of the European Parliament: "delivering asbestos waste to landfills would not appear to be the safest way of definitively eliminating the release of asbestos fibres into the environment (particularly into air and groundwater) and therefore it would be far preferable to opt for asbestos inertisation plants"; "whereas creating landfills for asbestos waste is only a temporary solution to the problem, which in this way is left to be dealt with by future generations, as asbestos fibres are virtually indestructible over time" (European Parliament 2013). It is thus rather clear that the European waste management policy regarding ACMs is navigating towards recycling rather than disposal.

The 2008/98/EC directive reports the novel term "*end-of-waste*" that, under certain conditions, indicates how and when a waste ceases to be so and becomes a secondary raw material following a proper well-defined conversion process (European Parliament 2008b). However, every state of the European community may independently establish objective criteria to assess when a waste ceases to be so (article 6(4)). Since 2004, the crystal-chemical conversion process of ACMs through chemical, mechanical or thermal conversion and recycling of the product of transformation has been legalized (Italian Ministry of Environment and Protection of Land 2004).

There has been an intense research activity regarding the ACMs inertisation processes and possible recycling options of the transformation product in Europe under the last two decades. Developed processes for the total destruction of the crystal-chemical structure of asbestos can roughly be divided into three groups: thermal; chemical and mechano-chemical (Plescia et al. 2003; Paolini et al. 2019; Gualtieri 2013). Thermal treatment processes can either bring on melting of the entire system, i.e., *vitrification* or result in recrystallization of the hazardous fibres in harmless phases, i.e., *ceramization*. The lower energy consumption in ceramization processes render them economically more competitive. In some processes, vitrification is followed by a controlled recrystallization that leads to the formation of a glass–ceramic. Other ones modify the reaction path and the final properties of the product through the addition of other inorganic materials such as clay and fly ash. Chemical methods utilize chemical agents, usually strongly alkaline ones, to destroy the asbestos fibres. The various methods using mechanical action to completely convert asbestos minerals into non-hazardous material belong to mechano-chemical processes.

The by far most studied method is thermal treatment and a couple of solutions have actually been launched on an industrial scale (Gualtieri 2013; Tomassetti et al.

2020). The development of many of these processes has been coupled with research on possible recycling opportunities in the building sector, such as partial replacement of Portland cement and aluminosilicate raw materials in concretes and geopolymers, respectively; flux in ceramic tile production (Gualtieri 2013) and frits to manufacture stoneware slabs (Ligabue et al. 2020). Nevertheless, existing designs of treatment facilities have not yet been able to compete with landfill options even in a country like Italy where such solutions are allowed by current legislation.

16.7 Substitutes of Asbestos in Building Materials

With the introduction of asbestos bans in numerous countries over 30 years ago, continuous progress has been made in the development of asbestos substitutes. The most common asbestos substitute materials are basically grouped into two macro-categories: man-made and natural fibres (Table 16.3; Gualtieri 2012). Man-made fibres can be classified as organic (e.g., carbon, cellulose, polyolefins, polyester and polyvinyl) or inorganic. The latter are divided into two classes: *man-made mineral fibres* (MMMF) and *man-made vitreous fibres* (MMVF). The family of natural fibres include natural inorganic fibres (e.g., erionite, sepiolite, wollastonite), and natural organic fibres (e.g., cotton, wool, hemp). The fibrous variety of erionite is deemed to have even greater toxicological and pathogenic properties than asbestos and is known to be a potent human carcinogen causing the MM epidemics that affected some villages in Cappadocia (Turkey), where it was used as a building material (Dogan 2003; Carbone et al. 2007). Cases of MM related to erionite exposure have been reported for Mexico and the United States (Carbone et al. 2019). For this reason, fibrous erionite has recently been included in Group 1 as carcinogen for humans (International Agency for Research on Cancer 2012).

Man-made organic fibres are industrially produced by high-temperature processing of monomeric organic compounds, such as polyalkylene, polyester, polyamides, cellulose, and carbon (Singh and Wagner 2011). These materials are widely utilized as a substitute for asbestos to improve various technological properties like mechanical strength to the bulk materials (Gualtieri 2012).

Nowadays, MMVF are the most common asbestos substitutes in the global market (Gualtieri 2012). MMVF include amorphous silicate-based glass fibres, refractory ceramic fibres (RCF), rock-wool and many more (Table 16.3). These materials are composed of SiO_2, CaO, Na_2O, K_2O, MgO and Al_2O_3 (Singh and Wagner 2011). Glass wool is manufactured starting from sand, soda-ash, limestone and recycled glass; rock wool is produced by melting a mixture of various slags and basaltic rocks; slag wool is obtained from iron or copper production waste derived from the metallurgical industry; RCF is produced by melting a mixture of kaolin clay, alumina and silica (International Agency for Research on Cancer 2002). The European 67/548/EEC directive classifies MMVFs on the basis of their chemical composition. The main uses of MMVFs are in commercial and residential thermal and acoustic insulation, e.g., wall, roof, floor and pipe insulation (Gualtieri et al. 2009),

Table 16.3 General classification of asbestos substitute materials (adapted from Gualtieri 2012; Park 2018)

Fibres type	IARC classification
MAN-MADE FIBRES	
Organic	
Carbon fibres	–
Cellulose fibres	–
p-aramid	Group 3
Polyalkylene fibres	–
Polyester	–
Polyvinyl alcohol (PVA)	Group 3
Polyvinyl chloride (PVC)	Group 3
Inorganic	
Man-made mineral fibres (MMMF)	
PCW	Group 2B
SiC whiskers	Group 2A
SiN whiskers	–
Man-made vitreous fibres (MMVF)	
Glass wool	Group 2B
Refractory ceramic fibres	Group 2B
Rock wool	Group 2B
Slag wool	Group 2B
NATURAL FIBRES	
Inorganic (Minerals)	
Attapulgite, fibre length >5 μm	Group 2B
Attapulgite, fibre length <5 μm	Group 3
Erionite	Group 1
Sepiolite	Group 3
Wollastonite	Group 3
Organic (Vegetable, Animal)	
Cotton	–
Flax	–
Jute	–
Hemp	–
Sisal	–
Silky	–
Tendinous	–
Wool	–

but also as reinforcing material of vehicles and appliances (International Agency for Research on Cancer 2002).

The MMMF category includes several polycrystalline fibres such as PCW (polycrystalline wools), SiC (silicon carbide) whiskers and SiN (silicon nitride) whiskers (Table 16.3). They mainly consist of silicates and mineral oxides manufactured from molten material by spinning/blowing or by sol–gel mediated synthesis (Harrison et al. 2015). Due to their excellent heat resistance, these fibres are extensively used in high-temperature applications (e.g., fire protection, furnace coatings) (Harrison et al. 2015). For example, PCW and SiC whiskers resist temperatures up to 1300 °C and 2000 °C, respectively. Moreover, SiC whiskers have mechanical strengths useful for the aerospace industry (Usukawa 2018), while PCW can be used in chemically aggressive industrial processes (Brown and Harrison 2014).

The IARC has evaluated several asbestos substitutes (Park, 2018; Table 16.3): Attapulgite fibres >5 μm are included in Group 2B (i.e., possibly carcinogenic to humans) whereas shorter attapulgite fibres (<5 μm) are classified as Group 3 (not classified as far as carcinogenicity in humans is concerned) together with sepiolite and wollastonite (International Agency for Research on Cancer 1997). Furthermore, also organic man-made fibres, such as carbon, cellulose, polyester, rayon, polyvinyl alcohol (PVA) and *p*-aramid, have raised concern about their safety for human health (International Agency for Research on Cancer 1997; Cullen et al. 2002; Inthavong et al. 2013; Singh and Bhalla 2017). According to IARC, PVA and *p*-aramid fibres are less respirable than chrysotile but more biopersistent (Gualtieri, 2012). Despite the fact that p-aramid fibres are classified in Group 3 (International Agency for Research on Cancer 1997), studies conducted by Friendmann et al. (1990) show that these fibres cause adverse pulmonary effects (Gualtieri 2012). As regards the MMVF category, RCF and glass/rock/slag wool are included in Group 2B (Park 2018). Although the evidence of carcinogenicity of RCF is limited, in vivo tests have shown that exposure to these fibres is linked to the onset of lung cancer and MM (Park 2018). Consequently, RCF is classified into Group 2B (IARC 2002). Literature data demonstrated the carcinogenicity of MMMF using in vivo animal models (Singh and Wagner 2011). In particular, PCW are classified as ceramic fibres and hence possible human carcinogens (Group 2B) by IARC in (1988) (Cannizzaro et al. 2019). According to IARC, SiC whiskers induce bio-chemical adverse effects similar to those observed for asbestos wherefore they are included in Group 2A as probably carcinogenic to humans (International Agency for Research on Cancer 2014).

Health risks can also arise from exposure to asbestos-contaminated natural raw materials utilized in building materials (International Agency for Research on Cancer 2012). Noteworthy in this regard are the health problems experienced by the employees of the vermiculite plant in Libby, Montana, United States,where about 5.8 Mt of raw vermiculite containing fibrous amphiboles were extracted and processed between 1924 and 1990 (Horton et al. 2008; Larson et al. 2010). A cohort study of workers exposed to Libby's vermiculite revealed excess lung cancer and asbestosis morbidity compared to the mean mortality in Montana and the United States (Larson et al. 2010). Since then, asbestos fibres have been identified in several raw materials considered as non-hazardous such as talc (Dyer 2019; Tran et al. 2019) and feldspar (Gualtieri et al. 2018b). In this context, the first case of a commercial product of brucite contaminated with chrysotile asbestos was recently documented by our group. This particular sample was imported to Italy from China in 2018 and destined for the glass and ceramic industries. Thanks to an established analytical protocol (Gualtieri et al. 2018b), a significant amount of respirable chrysotile fibres was detected. The results imply that commercial raw brucite currently marketed should be checked for the presence of asbestos to avoid situations similar to those of commercial talc and feldspar. The case of asbestos-contaminated brucite from China is one example of problems created by different asbestos regulations in trading partner countries (see Sect. 16.5).

16.8 Conclusions

In this chapter the following topics were presented:

- classification and definitions of asbestos minerals;
- application of commercial asbestos in building materials;
- toxicity and pathogenicity mechanisms of asbestos with a view on the global asbestos issue;
- past and current uses of asbestos in building materials;
- potential exposure to asbestos during, e.g., restoration and remediation of old buildings;
- current regulations regarding restoration and remediation of buildings containing asbestos, including waste management and recycling;
- substitutes of asbestos in building materials.

The classification and, sometimes conflicting, definitions of asbestos minerals have been revised and updated with respect to the existing literature. The issues raised by the absence of an universally shared definition of asbestos have been highlighted.

Regarding the applications of asbestos in building materials, it was evidenced that the extensive use of asbestos at an industrial scale started in the twentieth century, and prompted the manufacture of more than 3000 different asbestos-containing materials. The largest economic impact has been, and continues to be in some part of the world, the fibre-reinforced cement industry. Nevertheless, a clear decreasing trend is observed in the global use of asbestos even in countries like China and Brazil where the production and consumption of these fibres are among the highest in the world. Considering the development of asbestos policies in countries like Italy where, e.g., chrysotile was mined, processed and consumed, the global decreasing trend is expected to be decisive for future bans. Unfortunately, the ceased use of these toxic materials will not put an immediate end on human exposure.

Potential occupational and non-occupational exposure to asbestos has been illustrated. If looking at the European countries where the extraction and use of asbestos fibres are banned since a couple of decades, ACBM are still found in the living environment mainly due to their long service-life and the lack of firm legislation that forces building owners to remediate contaminated sites.

The chapter also revised the current regulations regarding the restoration and remediation of old buildings containing asbestos with a special attention to the problem of waste management and recycling that should be universally followed. An immediate solution to the asbestos-problem would include more restrictive legislations on remediation coupled with the adoption of existing technologies capable of transforming ACBM into non-hazardous compounds that may eventually be re-entered in the construction sector as secondary raw material.

The classification of synthetic fibres, in particular for use in building materials made of fibre-reinforced cement has been revised. It is clear that the availability of asbestos-substitutes is decisive for the decreased reliance of asbestos.

References

Alleman JE, Mossman BT (1997) Asbestos revisited. Sci Am 1:70–75

Babior BM (1984) The respiratory burst of phagocytes. J Clin Invest 73:599–601

Ballirano P, Bloise A, Gualtieri AF et al (2017) The crystal structure of mineral fibres. In: Gualtieri AF (ed) Mineral fibres: crystal chemistry, chemical-physical properties, biological interaction and toxicity. European Mineralogical Union, London, pp 17–64

Becklake MR, Bagatin E, Neder, JA (2007) Asbestos-related diseases of the lungs and pleura: uses, trends and management over the last century. Int J Tuberc Lung D 11:356–369

Belluso E, Cavallo A, Halterman D (2017) Crystal habit of mineral fibres. In: Gualtieri AF (ed) Mineral fibres: crystal chemistry, chemical-physical properties, biological interaction and toxicity. European Mineralogical Union, London, pp 65–109

Berman DW, Crump KS (2008) A meta-analysis of asbestos-related cancer risk that addresses fiber size and mineral type. Crit Rev Toxicol 38:49–73

Bernstein DM, Donaldson K, Decker U et al (2008) A biopersistence study following exposure to chrysotile asbestos alone or in combination with fine particles. Inhal Toxicol 20:1009–1028

Bernstein DM, Dunnigan J, Hesterberg T et al (2013) Health risk of chrysotile revisited. Crit Rev Toxicol 43:154–183

Bernstein DM, Rogers R, Sepulveda R et al (2014) Evaluation of the deposition, translocation and pathological response of brake dust with and without added chrysotile in comparison to crocidolite asbestos following short-term inhalation: Interim results. Toxicol Appl Pharm 276:28–46

Bernstein DM, Rogers RA, Sepulveda R et al (2015) Evaluation of the fate and pathological response in the lung and pleura of brake dust alone and in combination with added chrysotile compared to crocidolite asbestos following short-term inhalation exposure. Toxicol Appl Pharm 283:20–34

Beukes NJ, Dreyer CJB (1986) Amosite asbestos deposits of the Penge area. In: Anhaeusser CR, Maske S (eds) Mineral deposits of Southern Africa, vol I. Geological Society of South Africa, Johannesburg, pp 901–910

Bloise A, Barca D, Gualtieri AF et al (2016) Trace elements in hazardous mineral fibres. Environ Pollut 216:314–323

Bowles O (1955) The asbestos industry. US Bureau of Mines, Washington DC

Brown TP, Harrison PTC (2014) Crystalline silica in heated man-made vitreous fibres: a review. Regul Toxicol Pharmacol 68:152–159

Bustamante-Marin XM, Ostrowski LE (2017) Cilia and mucociliary clearance. CSH Perspect Biol 9:a028241

Camus M (2001) A ban on asbestos must be based on a comparative risk assessment. CMAJ 164:491–494

Cannizzaro A, Angelosanto F, Barrese E et al (2019) Bile solubility of high temperature insulation wools in simulated lung fluids. J Occup Med Toxicol 14:15

Carbone M, Adusumilli PS, Alexander Jr HR et al (2019) Mesothelioma: scientific clues for prevention, diagnosis, and therapy. CA: Cancer J Clin 69:402–429

Carbone M, Emri S, Dogan AU et al (2007) A mesothelioma epidemic in Cappadocia: scientific developments and unexpected social outcomes. Nat Rev Cancer 7:147

Carbone M, Yang H (2012) Molecular pathways: targeting mechanisms of asbestos and erionite carcinogenesis in mesothelioma. Clin Cancer Res 18:598–604

Carbone M, Yang H (2017) Biological activities of asbestos and other mineral fibres. In: Gualtieri AF (ed) Mineral fibres: crystal chemistry, chemical-physical properties, biological interaction and toxicity. European Mineralogical Union, London, pp 435–445

Case BW, Abraham JL, Meeker G et al (2011) Applying definitions of "asbestos" to environmental and "low-dose" exposure levels and health effects, particularly malignant mesothelioma. J Toxicol Env Heal B 14:3–39

Commission E (2014) 2014/955/EU: commission decision of 18 December 2014 amending decision 2000/532/EC on the list of waste pursuant to directive 2008/98/EC of the European parliament and of the council. Off J Eur Union L370:44–86

Council E (1991) Commission directive 91/659/EEC of 3 December 1991 adapting to technical progress annex I to council directive 76/769/EEC on the approximation of the laws, regulations and administrative provisions of the member states relating to restrictions on the marketing and use of certain dangerous substances and preparations (asbestos). Off J Eur Comm L363:36–38

Council E (1999) Commission directive 1999/77/EC of 26 July 1999 adapting to technical progress for the sixth time annex I to council directive 76/769/EEC on the approximation of the laws, regulations and administrative provisions of the member states relating to restrictions on the marketing and use of certain dangerous substances and preparations (asbestos). Off J Eur Comm L207:18–20

Cullen RT, Miller BG, Jones AD et al (2002) Toxicity of cellulose fibres. Annoccuphyg 46:81–84

D'Orsi F (2007) Amianto valutazione, controllo, bonifica. EPC Libri, Rome

Dahlgren C, Karlsson A (1999) Respiratory burst in human neutrophils. J Immunol Methods 232:3–14

Di Giuseppe D, Zoboli A, Nodari L et al (2021) Characterization and assessment of the potential toxicity/pathogenicity of Russian commercial chrysotile. Am Mineral (in press)

Dilek Y, Newcomb S (2003) Ophiolite concept and the evolution of geological thought. The Geological Society of America, Boulder

Dogan AU (2003) Mesothelioma in Cappadocian villages. Indoor Built Environ 12:367–375

Doll R (1955) Mortality from lung cancer in asbestos workers. Brit J Ind Med 12:81–86

Donaldson K, Murphy FA, Duffin R, Poland CA (2010) Asbestos, carbon nanotubes and the pleural mesothelium: a review of the hypothesis regarding the role of long fibre retention in the parietal pleura, inflammation and mesothelioma. Part Fibre Tox 7:5

Dyer O (2019) Johnson & Johnson recalls its baby powder after FDA finds asbestos in the sample. BMJ 367:l6118

European Chemicals Agency (2015) Annex, V.I. of regulation (EC) No 1272/2008 (CLP Regulation). https://echa.europa.eu/it/information-on-chemicals/cl-inventory-database. Accessed 02 Jan 2021

European Council (1976) Council directive 76/769/EEC on the approximation of the laws, regulations and administrative provisions of the member states relating to restrictions on the marketing and use of certain dangerous substances and preparations. Off J Eur Comm L262:201–203

European Council (1983) Council directive 83/478/EEC of 19 September 1983 amending for the fifth time (asbestos) directive 76/769/EEC on the approximation of the laws, regulations and administrative provisions of the member states relating to restrictions on the marketing and use of certain dangerous substances and preparations. Off J Eur Comm L263:33–36

European Council (1985) Council directive 85/610/EEC of 20 December 1985 amending for the seventh time (asbestos) directive 76/769/EEC on the approximation of the laws, regulations and administrative provisions of the member states relating to restrictions on the marketing and use of certain dangerous substances and preparations. Off J Eur Comm L375:1–2

European Parliament (2006) Regulation (EC) No 1907/2006 of the European parliament and of the council of 18 December 2006 concerning the registration, evaluation, authorisation and restriction of chemicals (REACH), establishing a European chemicals agency, amending directive 1999/45/EC and repealing council regulation (EEC) No 793/93 and commission regulation (EC) No 1488/94 as well as council directive 76/769/EEC and commission directives 91/155/EEC, 93/67/EEC, 93/105/EC and 2000/21/EC. http://data.europa.eu/eli/reg/2006/1907/2009-06-27. Accessed 6 Jan 2021

European Parliament (2008a) Regulation (EC) No 1272/2008 of the European parliament and of the council of 16 December 2008 on classification, labelling and packaging of substances and mixtures, amending and repealing Directives 67/548/EEC and 1999/45/EC, and amending Regulation (EC) No 1907/2006. Off J Eur Union L353:1–1355

European Parliament (2008b) Directive 2008/98/EC of the European parliament and of the council of 19 November 2008 on waste and repealing certain directives. Off J Eur Union L312:3–30

European Parliament (2009) Directive 2009/148/EC of the European parliament and of the council of 30 November 2009 on the protection of workers from the risks related to exposure to asbestos at work. Off J Eur Union L330:28–36

European Parliament (2013) European parliament resolution of 14 March 2013 on asbestos related occupational health threats and prospects for abolishing all existing asbestos (2012/2065(INI)). Off J Eur Union C36:102–110

Frank AL, Joshi TK (2014) The global spread of asbestos. Ann Glob Health 80:257–262

Friedmann AC, Fiel SB, Radecki PD, Lev-Toaff AS (1990) Computed tomography of benign pleural and pulmonary parenchymal abnormalities related to asbestos exposure. Semin Ultrasound CT MR 11:393–408

Furuya S, Chimed-Ochir O, Takahashi K et al (2018) Global asbestos disaster. Int J Environ Res 15:1000

Giacobbe C, Di Giuseppe D, Zoboli A et al (2021) Crystal structure determination of a lifelong biopersistent asbestos fibre using single-crystal synchrotron X-ray micro-diffraction. IUCrJ 8:76–86

Gualtieri AF (2012) Mineral fibre-based building materials and their health hazards. In: Pacheco-Torgal F, Jalali S, Fucic A (eds) Toxicity of building materials. Woodhead, Cambridge, pp 166–195

Gualtieri AF (2013) Recycling asbestos-containing material (ACM) from construction and demolition waste (CDW). In: Pacheco-Torgal F, Tam VWY, Labrincha JA, Ding Y, de Brito J (eds) Handbook of recycled concrete and demolition waste. Woodhead, Cambridge, pp 500–525

Gualtieri AF (2017) Introduction. In: Gualtieri AF (ed) Mineral fibres: crystal chemistry, chemical-physical properties, biological interaction and toxicity. European Mineralogical Union, London, pp 1–15

Gualtieri AF (2018) Towards a quantitative model to predict the toxicity/pathogenicity potential of mineral fibers. Toxicol Appl Pharmacol 361:89–98

Gualtieri AF (2020) Naturally occurring asbestos: a global health concern? state of the art and open issues. Environ Eng Geosci 26:3–8

Gualtieri AF, Foresti E, Lesci IG et al (2009a) The thermal transformation of man made vitreous fibers(MMVF) and safe recycling as secondary raw materials (SRM). J Hazard Mater 162:1494–1506

Gualtieri AF, Lusvardi G, Zoboli A et al (2019b) Biodurability and release of metals during the dissolution of chrysotile, crocidolite and fibrous erionite. Environ Res 171:550–557

Gualtieri AF, Lusvardi G, Pedone A et al (2019c) Structure model and toxicity of the product of bio dissolution of chrysotile asbestos in the lungs. Chem Res Toxicol 32:2063–2077

Gualtieri AF, Andreozzi GB, Tomatis M, Turci F (2019d) Iron from a geochemical viewpoint. Understanding toxicity/pathogenicity mechanisms in iron-bearing minerals with a special attention to mineral fibers. Free Radical Bio Med 133:21–37
Gualtieri AF, Mossman BT, Roggli VL (2017) Towards a general model for predicting the toxicity and pathogenicity of minerals fibres. In: Gualtieri AF (ed) Mineral fibres: crystal chemistry, chemical-physical properties, biological interaction and toxicity. European Mineralogical Union, London, pp 501–526
Gualtieri AF, Pollastri S, Bursi Gandolfi N, Lassinantti Gualtieri M (2018a) In vitro acellular dissolution of mineral fibres: a comparative study. Sci Rep 8:7071
Gualtieri AF, Bursi Gandolfi N, Pollastri S et al (2018b) Assessment of the potential hazard represented by natural raw materials containing mineral fibres-The case of the feldspar from Orani, Sardinia (Italy). J Hazard Mater 350:76–87
Harrison P, Holmes P, Bevan R et al (2015) Regulatory risk assessment approaches for synthetic mineral fibres. RegulToxicolPharmacol 73:425–441
Hawthorn FC, Oberti R, Della Ventura G, Motta A (2007) Amphiboles: crystal chemistry, occurrence, and health issues. Rev Mineral Geochem 67:1–54
Hodgson JT, Darnton A (2000) The quantitative risks of mesothelioma and lung cancer in relation to asbestos exposure. Ann Occup Hyg 44:565–601
Horton DK, Bove F, Kapil V (2008) Select mortality and cancer incidence among residents in various US communities that received asbestos-contaminated vermiculite ore from Libby, Montana. Inhal Toxicol 20:767–775
Hume LA, Rimstidt JD (1992) The biodurability of chrysotile asbestos. Ame Min 77:1125–1128
International Agency for Research on Cancer (1988) Man made mineral fibres. IARC Monogr Eval Carcinog Risks Hum 43:39–171
International Agency for Research on Cancer (1997) Silica, some silicates, coal dust and *para*-aramid fibrils. IARC Monogr Eval Carcinog Risks Hum 38:1–506
International Agency for Research on Cancer (2002) Man-made vitreous fibres. IARC MonogrEval-Carcinog Risks Hum 81:1–418
International Agency for Research on Cancer (2012) Asbestos (chrysotile, amosite, crocidolite, tremolite, actinolite, and anthophyllite). IARC MonogrEvalCarcinog Risks Hum 100C:219–309
International Agency for Research on Cancer (2014) Silicon carbide. IARC MonogrEvalCarcinog Risks Hum 111:243–306
International Ban Asbestos Secretariat (2021a) Current asbestos bans. http://www.ibasecretariat.org/alpha_ban_list.php. Accessed 6 Jan 2021
International Ban Asbestos Secretariat (2021b) USGS asbestos trade data. http://ibasecretariat.org. Accessed 6 Jan 2021
Inthavong K, Mouritz AP, Dong J, Tu JY (2013) Inhalation and deposition of carbon and glass composite fibre in the respiratory airway. J Aerosol Sci 65:58–68
Italian Law n. 257 (1992) Norme relative alla cessazione dell'impiego dell'amianto. https://www.gazzettaufficiale.it/eli/id/1992/04/13/092G0295/sg. Accessed 12 Feb 2021
Italian Ministry of Environment and Protection of Land (2004) Decreto 29 luglio 2004, n. 248. Regolamento relativo alla determinazione e disciplina delle attività di recupero dei prodotti e beni di amianto e contenenti amianto. https://www.gazzettaufficiale.it/eli/gu/2004/10/05/234/sg/pdf. Accessed 6 Jan 2021
Italian Ministry of Health (1994) Normative e metodologie tecniche di applicazione dell'art.6, comma 3, e dell'art.12, comma 2, della legge 27 marzo 1992, n. 257, relativa alla cessazione dell'impiego dell'amianto. https://www.gazzettaufficiale.it/eli/id/1994/09/20/094A5917/sg. Accessed 6 Jan 2021
Italian Ministry of Health (1999) Ampliamento delle normative e delle metodologie tecniche per gli interventi di bonifica, ivi compresi quelli per rendere innocuo l'amianto, previsti dall'art. 5, comma 1, lettera f), della legge 27 marzo 1992, n. 257, recante norme relative alla cessazione dell'impiego dell'amianto. https://www.gazzettaufficiale.it/eli/gu/1999/10/22/249/sg/pdf. Accessed 6 Jan 2021

Kamp DW (2009) Asbestos-induced lung diseases: an update. Transl Res 153:143–152

LaDou J, Castleman B, Frank A et al (2010) The case for a global ban on asbestos. Environ Health Perspect 118:897–901

Landrigan PJ, Lemen RA (2019) A most reckless proposal—a plan to continue asbestos use in the United States. N Engl J Med 381:598–600

Larson TC, Antao VC, Bove FJ (2010) Vermiculite worker mortality: estimated effects of occupational exposure to Libby amphibole. J Occup Environ Med 52:555–560

Levi P (1975) Il Sistema periodico. Einaudi, Torino

Liddell FDK, McDonald AD, McDonald JC (1997) The 1891–1920 birth cohort of Quebec chrysotile miners and millers: development from 1904 and mortality to 1992. Ann Occup Hyg 41:13–36

Ligabue ML, Gualtieri AF, Lassinantti Gualtieri M et al (2020) Recycling of thermally treated cement-asbestos for the production of porcelain stoneware slabs. J Clean Prod 247:119084

McDonald AD, Case BW, Churg A et al (1997) Mesothelioma in Quebec chrysotile miners and millers: epidemiology and aetiology. Ann Occup Hyg 41:707–719

Mossman BT (1993) Mechanisms of asbestos carcinogenesis and toxicity: the amphibole hypothesis revisited. Br J Ind Med 50:673–676

Mossman BT (2018) Mechanistic *in vitro* studies: What they have told us about carcinogenic properties of elongated mineral particles (EMPs). Toxicol Appl Pharm 361:62–67

Mossman BT, Gualtieri AF (2020) Lung cancer: mechanisms of carcinogenesis by asbestos. In: Anttila S, Boffetta P (eds) Occupational cancers. Springer, Heidelberg, pp 239–256

Niklinski J, Niklinska W, Chyczewska E et al (2004) The epidemiology of asbestos-related diseases. Lung Cancer 45:S7–S15

National Institute for Occupational Safety and Health (2019) Asbestos and other fibers by PCM, In: NIOSH manual of analytical methods (NMAM). https://www.cdc.gov/niosh/nmam/pdf/7400.pdf. Accessed 17 Dec 2020

Oberdörster G (1993) Lung dosimetry: pulmonary clearance of inhaled particles. Aerosol Sci Tech 18:279–289

Obmiński A (2020) Asbestos in buildings and its destruction. Constr Build Mater 249:118685

Odgerel CO, Takahashi K, Sorahan T et al (2017) Estimation of the global burden of mesothelioma deaths from incomplete national mortality data. Occup Environ Med 74:851–858

Oze C, Solat KL (2010) Biodurability of chrysotile and tremolite asbestos in simulated lung and gastric fluids. Am Min 95:825–831

Pacella A, Andreozzi GB, Nodari L et al (2019) Chemical and structural characterization of UICC crocidolite fibres from Koegas Mine, Northern Cape (South Africa). Period Mineral 88:297–306

Paglietti F, Malinconico S, Conestabiledella Staffa B et al (2016) Classification and management of asbestos-containing waste: European legislation and the Italian experience. Waste Manage 50:130–150

Paolini V, Tomassetti L, Segreto M et al (2019) Asbestos treatment technologies. J Mater Cycles Waste 21:205–226

Park SH (2018) Types and health hazards of fibrous materials used as asbestos substitutes. SH@W 9:3601364

Plescia P, Gizzi D, Benedetti S et al (2003) Mechanochemical treatment for recycling asbestos-containing waste. Waste Manage 23:209–218

Pollastri S, Perchiazzi N, Gigli L et al (2017) The crystal structure of mineral fibres. 2. Amosite fibrousAnthophyllite Period Miner 86:55–65

Pollastri S, Gigli L, Ferretti P et al (2017) The crystal structure of mineral fibres. 3. Actinolite Asbestos Period Miner 86:89–98

Pollastri S, Perchiazzi N, Lezzerini M et al (2016) The crystal structure of mineral fibres. 1. Chrysotile Period Miner 85:249–259

Roggli VL, Green CL (2019) Dimensions of elongated mineral particles: a study of more than 570 fibers from more than 90 cases with implications for pathogenicity and classification as asbestiform versus cleavage fragments. Ultrastruct Pathol 43:1–5

Rosales C, Uribe-Querol E (2017) Phagocytosis: a fundamental process in immunity. Bio Med Res Int 2017:9042851
Ross M, Langer AM, Nord GL et al (2008) The mineral nature of asbestos. Regul Toxicol Pharmacol 52:S26–S30
Ross M, Nolan RP (2003) History of asbestos discovery and use and asbestos-related disease in context with the occurrence of asbestos within ophiolite complexes. Geol Soc Am Spec Pap 373:447–470
Rosselli M (2014) The asbestos lie: the past and present of an industrial catastrophe. European Trade Union Institute, Brussels
Shcherbakov SV, Kashansky S, Domnin SG et al (2001) The health effects of mining and milling chrysotile: the Russian experience. In: Nolan RP, Langer AM, Ross M, Wicks FJ, Martin RF (eds) The health effects of chrysotile asbestos: contribution of science to risk management decisions. The Canadian Mineralogist, Ottawa, pp 187–198
Shukla A, Gulumian M, Hei TK et al (2003) Multiple roles of oxidants in the pathogenesis of asbestos-induced diseases. Free Radical Bio Med 34:1117–1129
Singh Z, Bhalla S (2017) Toxicity of synthetic fibres & health. Adv Res Tex Eng 2:1–5
Singh J, Wagner A (2011) Man-made mineral fibres. In: Vernon ER, Cohrssen B (eds) Patty's industrial hygiene, 6th edn. Wiley, Hoboken, pp 333–372
Spurny KR (1989) On the release of asbestos fibers from weathered and corroded asbestos cement products. Envir Res 48:100–116
Stayner LT, Dankovic DA, Lemen RA (1996) Occupational exposure to chrysotile asbestos and cancer risk: a review of the amphibole hypothesis. Am J Public Health 86:179–186
Studer AM, Limbach LK, Van Duc L et al (2010) Nanoparticle cytotoxicity depends on intracellular solubility: comparison of stabilized copper metal and degradable copper oxide nanoparticles. Toxicol Lett 197:169–174
Thomas DC (2017) The phagocyte respiratory burst: historical perspectives and recent advances. Immunol Lett 192:88–96
Tomassetti L, Di Giuseppe D, Zoboli A et al (2020) Emission of fibres and atmospheric pollutants from the thermal treatment of asbestos containing waste (ACW). J Clean Prod 268:122179
Tran TH, Steffen JE, Clancy KM et al (2019) Talc, asbestos, and epidemiology: corporate influence and scientific incognizance. Epidemiology 30:783–788
Turci F, Tomatis M, Pacella A, Gualtieri AF (2017) Surface and bulk properties of mineral fibres relevant to toxicity. In: Gualtieri AF (ed) Mineral fibres: crystal chemistry, chemical-physical properties, biological interaction and toxicity. European Mineralogical Union, London, pp 171–214
United States Environmental Protection Agency (1994) Technical amendment in response to court decision on asbestos; manufacture, importation, processing and distribution prohibitions. Fed Regist 59:33208–33211
United States Environmental Protection Agency (2016) Designation of ten chemical substances for initial risk evaluations under the toxic substances control act. Fed Regist 81:91927–91929
United States Environmental Protection Agency (2019) Restrictions on discontinuous uses of asbestos; significant new use rule. Fed Regist 84:17345–17360
United States Geological Service (2020) Asbestos statistics and information. https://pubs.usgs.gov/periodicals/mcs2020/mcs2020-asbestos.pdf. Accessed 6 Jan 2021
Usukawa R, Ishikawa T (2018) High-performance SiC-polycrystalline fiber with smooth surface. Ceram Int 1:165–174
Virta RL (2006) Worldwide asbestos supply and consumption trends from 1900 through 2003. US Geological Survey Circular 1298. http://pubs.usgs.gov/circ/2006/1298. Accessed 6 Jan 2021
World Health Organization (1997) Determination of airborne fibre number concentrations; a recommended method, by phase contrast optical microscopy (membrane filter method). https://apps.who.int/iris/bitstream/handle/10665/41904/9241544961.pdf. Accessed 6 Jan 2021
Zen S, Ahamad I, Gopal Rampal R et al (2013) Use of asbestos building materials in Malaysia: legislative measures, the management, and recommendations for a ban on use. Int J Occup Med Environ Health 19:169–178

Chapter 17
Development of Eco-efficient Geopolymer Masonry for Sustainability

Radhakrishna and K. Venugopal

Abstract Geopolymer has been considered a popular alternative to traditional cement. Materials which are rich in silica and alumina can be used along with traditional aggregates to prepare geopolymer mortar and masonry units. This chapter deals with the development of geopolymer masonry units such as brick (GPB), solid block (GPSB) and hollow block (GPHB). All these products were tested for dimensionality, water absorption, strength, modulus of elasticity, etc. They possessed better properties compared to traditional masonry units. The same units were used for making masonry prisms, and wallets. The prisms and wallets were loaded with axial and eccentric loads and tested for compression. The modulus of elasticity was also determined along with the crack pattern. Results indicated that the prisms and wallets performed exceedingly well compared to the traditional masonry structures. Moreover, no traditional cement is required in any stage of construction. The traditional water curing and thermal input is avoided. Based on this study, geopolymer masonry units are strongly recommended for the structural masonry without compromising properties.

Keywords Geopolymer · Sustainability · Eco-efficient · Masonry · Masonry units · Brick · Block · Hollow block · Prism · Wallet

17.1 Introduction

Geopolymers are the innovative, low carbon, energy efficient and eco-friendly green material for the sustainable environment (Dao et al. 2019; Zuo and Ye 2021; Bhogayata et al. 2020; Verma and Dev 2021). It is formed by the reaction of solid alumino-silicates with high concentration alkaline hydroxide or silicate solution

---

Radhakrishna (✉)
Department of Civil Engineering, RV College of Engineering, Bengaluru, Affiliated to Visvesvaraya Technological University, Mysuru Road, Bengaluru, Karnataka, India
e-mail: radhakrishna@rvce.edu.in

K. Venugopal
Department of Civil Engineering, SEA College of Engineering, Bengaluru, Affiliated to Visvesvaraya Technological University, Bengaluru, Karnataka, India

J. A. Malik and S. Marathe (eds.), *Ecological and Health Effects of Building Materials*,
https://doi.org/10.1007/978-3-030-76073-1_17

(Ren and Zhang 2019). This network of alumina-silicate materials was entitled as "geopolymers" by a French Scientist Prof. Davidovits (Norton and Provis 2020). Construction industry being one of the largest contributors for carbon emissions, research on sustainable material like geopolymer composites has gained momentum (Rahman and Al-Ameri 2021). Geopolymer products can reduce the emission of green-house gases which are responsible for global warming unlike cement products. Geopolymer has already shifted from emerging technology to adopted structural material for various purposes. Even though pre-processing and procurement of raw material is tedious, it is expanding its application and adaptability to meet the market demands (Norton and Provis 2020). Geopolymers possess higher mechanical properties, bond strength, resistance to heat, chemicals, alkalis and reduces corrosion compared to OPC composites (Bhogayata et al. 2020; Sakthidoss and Senniappan 2020; Rehman and Sglavo 2020; Nuruddin et al. 2016). With the increase in awareness about environmental impact and sustainable products, Research and Development activities are focused towards replacement of eco-detrimental Cement products by eco-friendly Geopolymer Composites (Ridha et al. 2018). Geopolymers/Alkali activated products are the third generation materials after lime and cement, have proved their low Carbon emission over OPC products to carry forward a sustainable environment for the next generation (Sikandar et al. 2019).

Masonry is a systematic arrangement of masonry units for a specific purpose. Masonry units are made from clay, mortar, concrete, calcium silicate etc. Bricks, solid blocks and hollow blocks are popular among masonry units. Solid and hollow blocks are manufactured using traditional concrete. Blocks are relatively larger in size than brick. Blocks are normally used to construct partition and load bearing walls. Hollow concrete blocks have one or more hollow cavities. The cavities act as good thermal insulators. To construct the wall, bricks/blocks are placed one at a time and held together firmly with fresh mortar which are timely tested and used. Though traditional masonry units are popular, they are neither eco-friendly nor sustainable, as they consume cement or soil or heat energy. Geopolymers are relatively new material without any traditional cement. The binders for making geopolymers are Class F fly ash, GGBS or any material which are dominant in silica and alumina. Fly ash is a marginal material from thermal power plants. The health and ecological challenges of disposal of fly ash can be minimized by using this by-product as building material. The technology used in making geopolymer masonry is the same as traditional masonry units except in basic ingredients. The aggregates used are the same but instead of water, an alkaline solution made out of sodium silicate and sodium hydroxide is used to mix the ingredients. Geopolymer mortar can be effectively used to make sustainable bricks/blocks which can make effective structural masonry (Venugopal and Radhakrishna 2016a, b; Radhakrishna et al. 2015; Radhakrishna and Venugopal 2020).

17.2 Bricks and Blocks

Sodium hydroxide flakes were added to the water depending on the required concentration to make sodium hydroxide solution. After preparing the sodium hydroxide solution, it is allowed to cool and sodium silicate was added to the same mix in required proportion. The alkaline solution was prepared 24 h earlier to the casting of bricks/blocks. The alkaline solution of 8 molar concentrations was prepared.

The geopolymer bricks (GPB), geopolymer solid blocks (GPSB) and geopolymer hollow blocks (GPHB) were mechanically cast using Class F fly ash and GGBS as binders. Alkaline solution was used as fluid, M-sand as fine aggregates. The proportions of these materials used in casting are as follows.

- Binders—80% Class F fly ash, 20% GGBS
- Binder: aggregates—1:1
- Ratio of alkaline solution to binders—0.2

The casting of Geopolymer bricks and blocks was done mechanically in the casting yard of a brick factory. The binders and aggregates were first added to the mixer and dry mixed for 5 min before the addition of alkaline solution. GPB, GPSB and GPHB were cast in an 8 MPa compression machine; pan mixer and belt conveyor are as shown in Fig. 17.1. The schematic representation of the typical masonry units and dimensions are as shown in Fig. 17.2.

The bricks and blocks were cast, kept in ambient curing conditions at room temperature in open air without any water or external curing agents. The finished geopolymer bricks and blocks are shown in Fig. 17.3a–c.

17.2.1 Tests on Geopolymer Masonry Units

Following tests were conducted

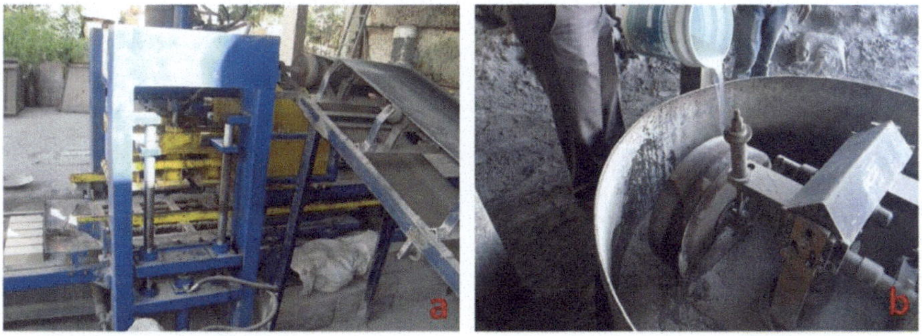

Fig. 17.1 Casting process

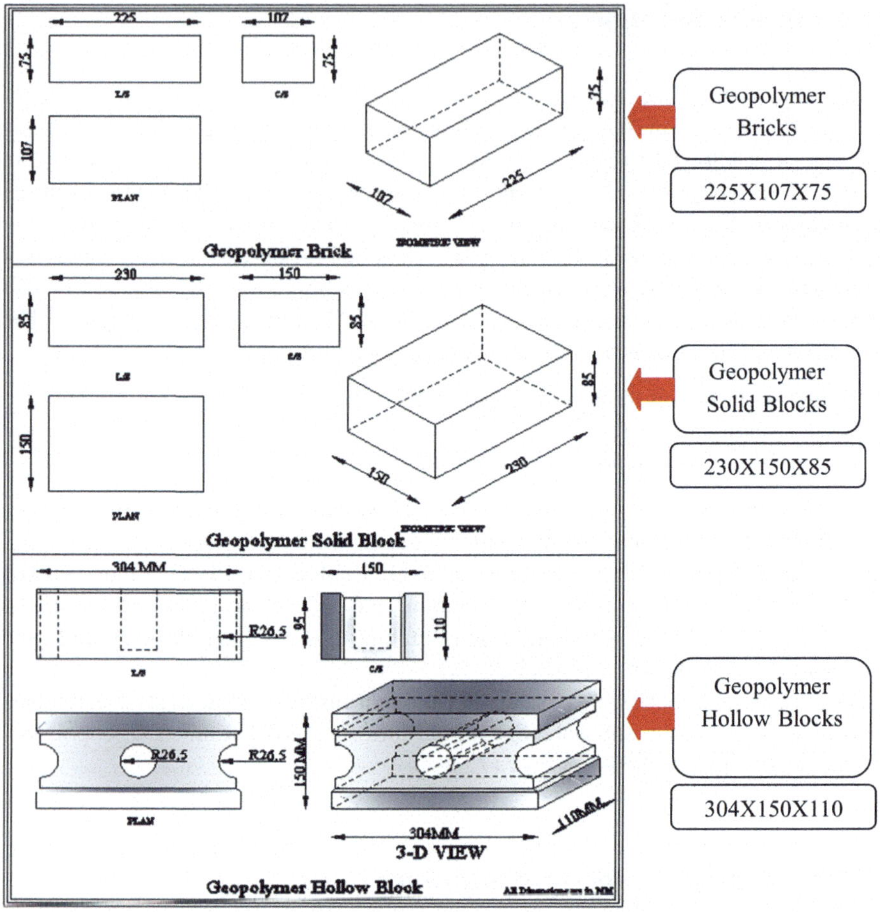

Fig. 17.2 Schematic representation showing the dimensions of the brick/block

- Dimensionality
- Density
- Water absorption
- Initial rate of water absorption
- Compressive strength
- Modulus of elasticity
- Microscopic Analysis
- X-ray Diffraction Analysis
- Flexural strength.

Fig. 17.3 **a** Geopolymer bricks, **b** geopolymer solid block, **c** geopolymer hollow block

17.2.1.1 Dimensionality Test

This test was conducted according to IS 1077-1992 (Standard 1997). The maximum deviation of the length of individual bricks should not be more than ±5 mm, the maximum deviation of width and height should not be more than ±3 mm. Twenty bricks were selected at random lined on a level floor successively in a straight line as shown in the Fig. 17.4a, b. The overall dimensions of the bricks were measured using steel tape.

The average variation of deviation of the length, breadth and depth of brick/blocks is shown in Table 17.1. It shows that the variation in the size of the bricks and block is

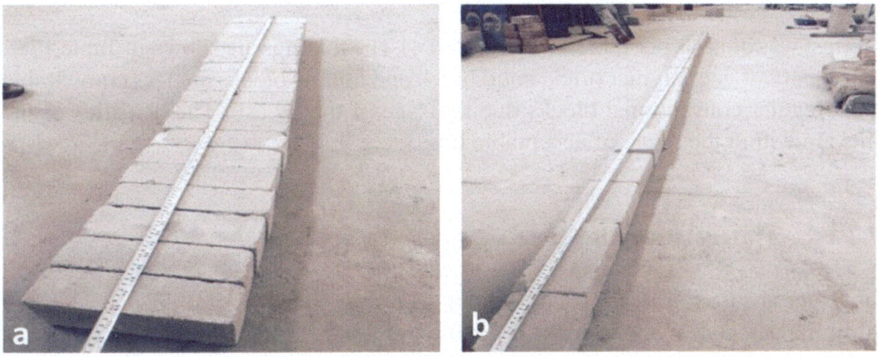

Fig. 17.4 Dimensionality test: **a** along breadth, **b** along length

Table 17.1 Results of dimensionality test

Brick/block	Dimensions measured along	Size (mm)	Measurements (mm)	Average measurements (mm)	Variation in measurements (mm)	IS 1077:1992
Geopolymer brick	Length (L)	225	4560	228.0	+3.0	±5
	Breadth (W)	107	2157	107.85	+0.85	±3
	Height (H)	75	1509	75.45	+0.45	±3
Geopolymer solid block	Length (L)	230	4614	230.70	+0.75	±5
	Breadth (W)	150	3012	150.60	+0.60	±3
	Height (H)	85	1724.4	86.24	+1.24	±3
Geopolymer hollow 'block	Length (L)	304	6103	305.15	+1.15	±5
	Breadth (W)	150	3015	150.75	+0.75	±3
	Height (H)	110	2221	111.05	+0.05	±3

Table 17.2 Results of density test

Type of block	Density (kg/m^3)	IS 2185:2008 (kg/m^3)
Geopolymer brick	1800	1800–2000
Geopolymer solid block	1810	
Geopolymer hollow block	1750	

within the range of codal limits. Bricks/blocks satisfied the specifications for length, width and height.

17.2.1.2 Density Test

This test was done as per IS 2185-2008 (Part 4).The findings are shown in Table 17.2. The density of geopolymer brick, solid block and hollow blocks are less comparable to the regular conventional blocks due to usage of the fly ash. The densities of the units are within the range of acceptable limits.

17.2.1.3 Water Absorption Test

The water absorption test was carried out according to IS 3495: 1992-Part 2. They fall in the category of up to class brick of strength 12.5 MPa as per code and it is represented in Table 17.3.

Table 17.3 Results of water absorption Test

Type of block	Water absorption (%)
Geopolymer brick	8.5
Geopolymer solid block	8.3
Geopolymer hollow block	9.1

Table 17.4 Results of initial rate of absorption test

Type of block	Initial rate of absorption IRA ($kg/m^2/min$)
Geopolymer brick (GPB)	3.0
Geopolymer solid block (GPSB)	2.7
Geopolymer hollow block (GPHB)	2.5

17.2.1.4 Initial Rate of Water Absorption (IRA) Test

This test was conducted according to ASTM C67-1299 (ASTM 2009). The results are indicated in Table 17.4. Initial rate of water absorption of GPB, GPSB and GPHB is less than the regular conventional blocks; these values are within codal limits.

17.2.1.5 Compressive Strength Test

This test was conducted according to IS 1077-1992 (Standard 1997). The compressive strength development of brick/blocks with age is represented in Fig. 17.5. IS 1077-1992 specifies minimum compressive strength of conventional bricks at 28 days as 3.5 MPa, whereas the compressive strength of the Geopolymer masonry units is

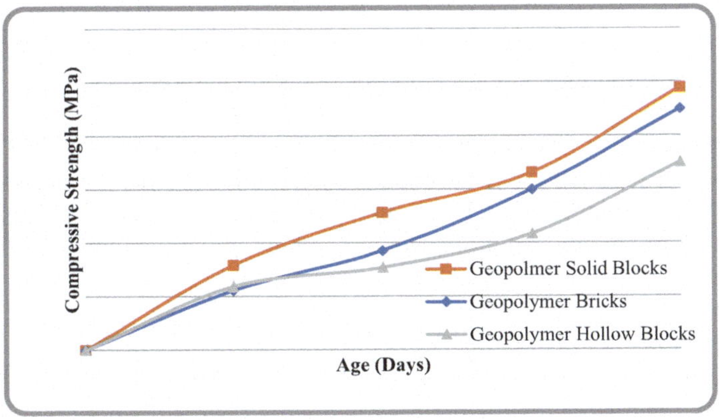

Fig. 17.5 Compression strength of bricks and blocks

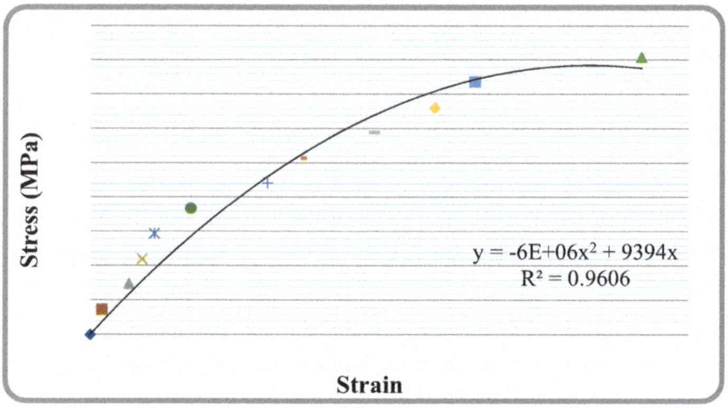

Fig. 17.6 Stress-strain curve for 8M NaOH bricks/blocks

more than the minimum strength and these were satisfied by the code provision. Also noted that the geopolymer brick and the solid blocks are having more strength compared to the geopolymer hollow blocks due to the voids or hollow portions of the blocks.

17.2.1.6 Modulus of Elasticity Test

The elastic modulus (E) is a very important characteristic in determining how the entire masonry will behave. It is particularly helpful for design calculations and assessment of deformation characteristics. Five specimens were tested to find modulus of elasticity. Initial tangent modulus of 8M NaOH bricks/blocks was found out to be 9394 MPa which is higher than that of traditional clay bricks manufactured in India (Rao 1986). A typical stress-strain curve for the geopolymer brick/block represented in Fig. 17.6.

17.2.1.7 Scanning Electron Microscope (SEM) Analysis

Microscopic image of 8M NaOH brick/blocks is as shown in Fig. 17.7. It indicates the presence of unreacted fly ash particles and aluminosilicate gel phases and the un-reacted fly ash particles of less than 10 μm size. It is due to the use of low concentration of alkaline solution. The use of high molarity solution will develop higher strength.

Fig. 17.7 Scanning electron microscope image of 8M NaOH brick

17.2.1.8 X-RAY Diffraction (XRD) Analysis

X-ray diffraction is a method used to obtain the internal lattice of crystalline materials and to obtain information on unit cell dimensions from the information generated by X-ray. XRD test on materials of 8M NaOH bricks/blocks was conducted. The X-ray diffractogram (XRD) is shown in Fig. 17.8. The diffractogram indicates the presence of crystalline phases of quartz and mullet. The base hump in the graph indicates the presence of amorphous silica or reactive silica. The corresponding material composition of the brick/blocks is shown in Table 17.5.

17.2.1.9 Flexural Strength

This test was conducted according to IS 4860-1968 as shown in Fig. 17.9. Flexural strength of the 8M NaOH Bricks and blocks are shown in Table 17.6. The flexural strength of the GPB at 28 days as per IS 4860-1968 is minimum of 10% that of the compressive strength. The flexural strength of GPB, GPSB and GPHB are 1.36, 1.55 and 1.79 MPa respectively. Flexural strength of geopolymer bricks/blocks is much higher than the regular conventional bricks and also noted that the flexural strength of the hollow block is much higher than that of the solid block and the brick because of the increase in thickness of the hollow blocks.

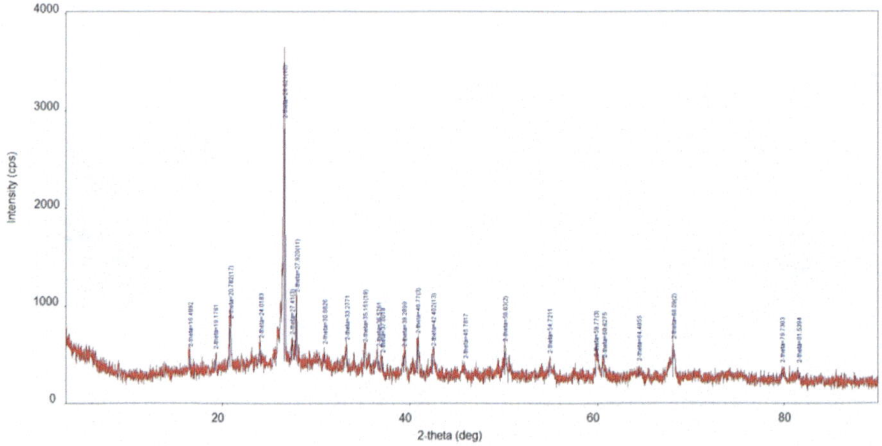

Fig. 17.8 X-ray diffraction results of 8M NaOH bricks/blocks

Table 17.5 Material composition of the brick/block

Phase name	Content (%)
Quartz	69.0
Mullite	12.2
Gossypol acetic acid	0.0
Vaterite	18.0

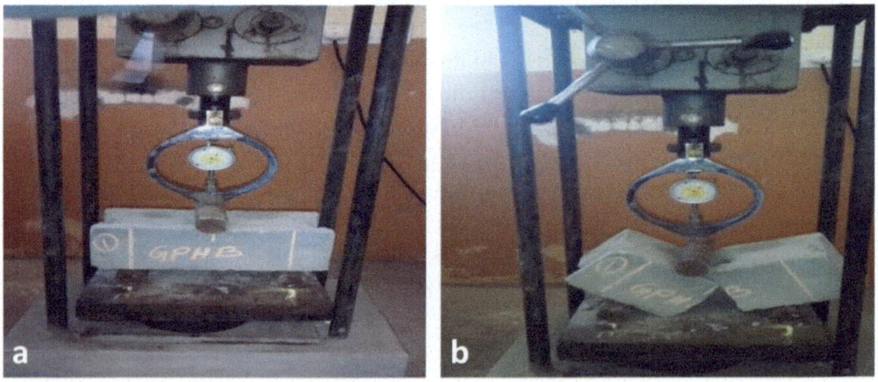

Fig. 17.9 Flexural strength test setup: **a** test setup, **b** failure pattern

Table 17.6 Results of flexural strength of the bricks/blocks

Type of block	Average flexural strength (MPa)
Geopolymer brick	1.36
Geopolymer solid block	1.55
Geopolymer hollow block	1.79

Cement mortar and geopolymer mortar was used for casting masonry prisms and wallettes. The mortars were cast as per the recommendations of IS 2250-1981. The binders and inert filler materials were first mixed dry till a homogenous mixture was obtained. The water/alkaline solution was added and mixed for 5–10 min to obtain the uniform mix of cement mortar/geopolymer mortar.

17.3 Geopolymer Masonry Prism

One of the major forces that masonry has to deal with is compression. Therefore, it is important to know the behavior of masonry in compression. Masonry efficiency in compression is an important parameter to determine the permissible stresses required for the design of masonry structures. Masonry efficiency was determined by a prism test as specified by IS 1905-1987. As per the code, the h/t ratio of the prism is to be in the range 2–5.

Stack bonded prisms are the prisms cast by keeping masonry units, one above the other with mortar placed in between them. Five brick thick stack bonded prisms were cast using 8M NaOH bricks and 12.5 mm cement mortar/Geopolymer mortar and the 4brick thick and 3 brick thick stack bonded prisms were cast using 8M NaOH bricks and 7.5, 10 and 12.5 mm cement mortar joints. Five-block, four-block and Three-block thick stack bonded solid and hollow block prisms were also cast using 8M NaOH blocks and 7.5, 10 and 12.5 mm cement mortar joints. The stack bonded prisms are as shown in Fig. 17.10.

Unrendered is the portion not covered by plaster or stucco for all faces. To study the effect of mortar joint thickness on masonry efficiency, stack bonded prisms were

Fig. 17.10 Stack bonded prisms

Fig. 17.11 Unrendered stack bonded prisms

cast using 8M NaOH bricks and 7.5, 10 and 12.5 mm cement mortar/Geopolymer mortar. The bricks were immersed in water before casting to ensure the saturated condition of the brick so that it will not absorb water from the mortar. The unrendered prisms are as shown in Fig. 17.11. Prisms which were cast using cement mortar were cured in water while the prisms cast using geopolymer mortar were kept in open air for curing. Prisms were cured for 28 days before the test.

17.3.1 Testing of Brick Prisms

The testing arrangements of brick prisms are shown in Fig. 17.12. Prisms were kept on the loading frame and leveled using the level tube. Steel plates were provided on top of the specimen to ensure uniform distribution of loads and also to act as filler materials. The centerline of the prisms was marked. Two metal studs were fixed on both the sides of the center line along the vertical direction. Then the displacement measuring demes gauge was fixed on these studs. Loading was done at the rate of 350 kN/min by a hydraulic jack of 500 kN capacities. For axial loading, the jack was placed exactly in the center of the bearing surface of the prism and the eccentric loading; the jack is placed from the center from a distance of 1/6th from the center. Proving ring of 500 kN capacity was placed to measure load at regular intervals. Deformation was noted at regular intervals of loading until the failure of the specimen occurred. The compressive stress calculated as 0.25 times the compressive strength was compared with the values given in Table 21.8 of IS 1905-1987.

Fig. 17.12 Testing arrangement of stack bonded prisms

Variations of masonry efficiency in mortar joint thickness for prisms cast using cement mortar after applying the correction factor given in IS 1905-1987 are as shown in Table 17.7. Permissible stresses for masonry are based on the values of basic compressive stress. Basic compressive stress is given in the code IS 1905-1987

Table 17.7 Basic compressive stress of stack bonded geopolymer brick prisms

h/t ratio	Mortar joint thickness (mm)	Average Compressive strength (f′m)	Basic compressive stress = 0.25 × f′m
3.97	12.5	3.00	0.75
3.88	10.0	3.05	0.76
3.78	7.5	3.25	0.81
3.15	12.5	2.72	0.68
3.08	10.0	3.01	0.75
3.01	7.5	3.20	0.80
2.33	12.5	2.55	0.64
2.28	10.0	2.71	0.67
2.24	7.5	2.95	0.74

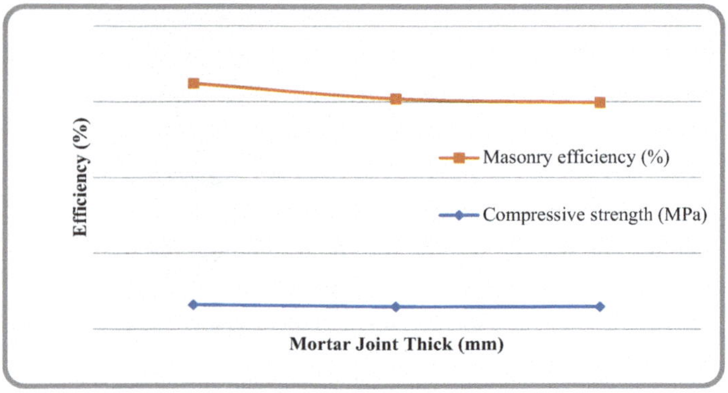

Fig. 17.13 Stack bonded 5 brick thick geopolymer prism with cement mortar joints

for burnt clay bricks. These values were compared with that of geopolymer bricks obtained from the prism test. Basic compressive stress for the geopolymer brick prisms is 10 MPa, the Crushing strength and M2 grade mortar as per Table 8 of the code after applying suitable corrections is 0.6 MPa. The basic compressive stress of prism test can be calculated as Basic Compressive stress = 0.25X f'm Where f'm is Compressive strength of prism

$$\text{Basic compressive stress} = 0.25 \times f'_m$$

It was noticed that the masonry efficiency of prisms cast using Geopolymer Bricks and cement mortar decreased with the increment in mortar joint thickness as shown in Fig. 17.13. Basic compressive stress by prism tests is given in Table 17.7. It was found that the basic compressive stress of all the prisms was greater than that given in Table 8 of the code IS 1905-1987.

Stress and strain were recorded at various intervals. The typical variation was plotted as shown in Fig. 17.14 after normalizing the values. Modulus of elasticity of the prisms increased slightly with the decrease in mortar joint thickness. Initial tangent modulus for prisms with 12.5 mm mortar thickness was 7005, 7111 MPa for prisms with 10 mm mortar thickness and 7314 MPa for prisms with 7.5 mm mortar thickness. There was no much difference in modulus of elasticity with various h/t ratios of the prisms. The variation of stress with strain is almost linear. The behavior is like typical burnt brick masonry prisms.

The vertical cracks and vertical splitting of narrow faces originated from the top of a brick. The cracks propagate further down. It also noticed that the bottom most brick was crushed to a considerable extent. Typical failure patterns of the prisms are as shown in Fig. 17.15a–d.

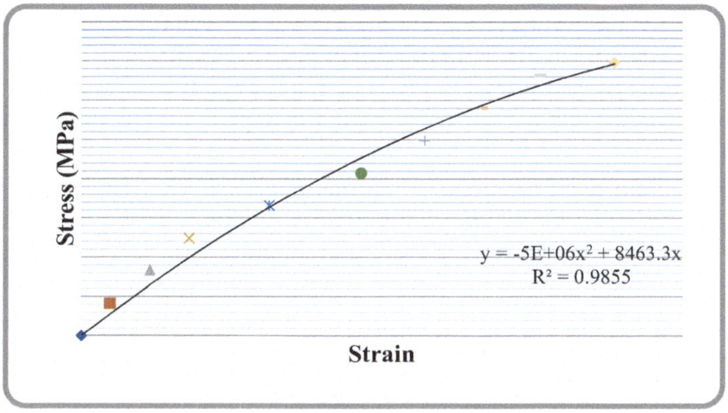

Fig. 17.14 Stress-strain curve for 10 mm thick joint cement mortar prism with h/t = 3.08

Fig. 17.15 Failure pattern of brick prisms

17.3.2 Testing of Geopolymer Solid Block Prisms

The test was done as per IS 1905-1987. The testing arrangements and method of rendered and unrendered solid block prisms are the same as geopolymer brick prisms and was discussed in the previous section. This study was made on geopolymer solid block prisms cast using cement mortar. It was noticed that masonry efficiency, decreasing with the increment in mortar joint thickness as shown in Fig. 17.16. Basic compressive stress by prism tests is given in Table 17.8. It was found that the basic compressive stress of all the prisms was greater than that given in Table 8 of the code IS 1905-1987.

Stress and strain were recorded at various intervals. The typical variation was plotted as shown in Fig. 17.17 after normalizing the values. It was observed that the modulus of elasticity of the prisms increased slightly with the decrease in mortar joint thickness. Initial tangent modulus for prisms with 12.5 mm mortar thickness was 5831 and 6313 MPa for prisms with 10 mm mortar thickness and 8471 MPa for prisms with 7.5 mm mortar thickness. There was no much difference in modulus of elasticity with different h/t ratios of the prisms. The variation of stress-strain is almost linear. The behavior is like typical burnt brick masonry prisms.

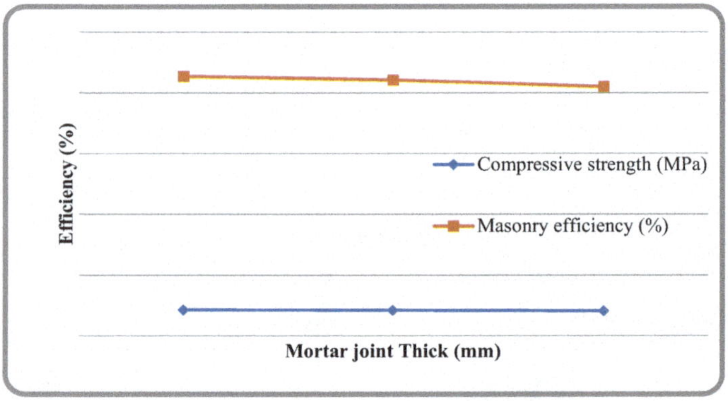

Fig. 17.16 Stack bonded 5 brick thick geopolymer solid block prism with cement mortar joints

Table 17.8 Basic compressive stress of prisms

Mortar joint thickness (mm)	h/t ratio	Average compressive strength (f′m)	Basic compressive stress = 0.25 × f′m
12.5	3.17	4.11	1.02
10.0	3.10	4.22	1.05
7.5	3.03	4.28	1.07
12.5	2.51	3.89	0.97
10.0	2.46	3.87	0.96
7.5	2.41	4.06	1.01
12.5	1.86	3.73	0.93
10.0	1.83	3.76	0.94
7.5	1.80	3.89	0.97

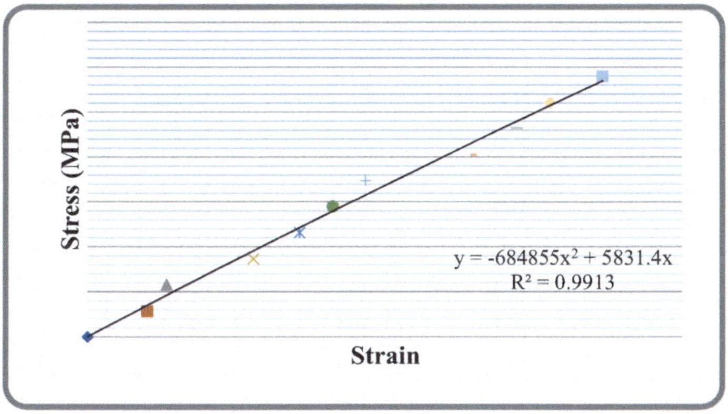

Fig. 17.17 normalized stress-strain curve for 12.5 mm thick joint cement mortar solid block prism with h/t = 3.17

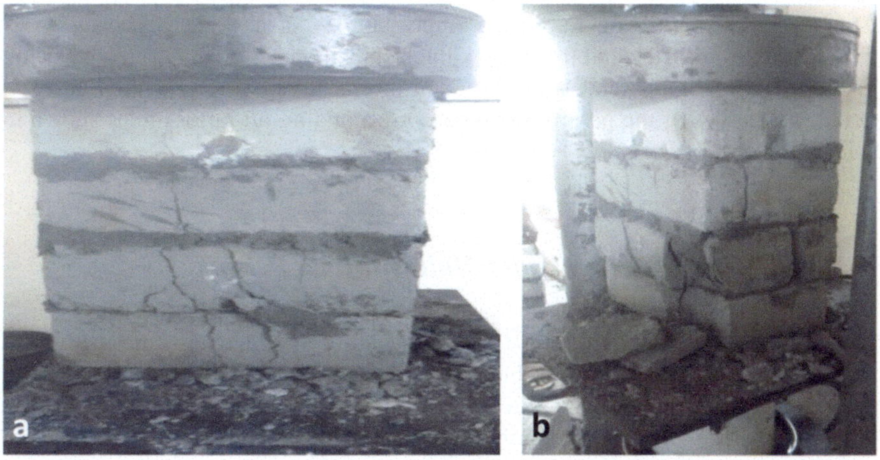

Fig. 17.18 .

The vertical cracks are developed at the bottom of the block and it propagates till the top of the prisms and bottom most corners of the blocks where spalling had taken place. Typical failure patterns of the prisms are as shown in Fig. 17.18a, b.

17.4 Geopolymer Hollow Block Prisms (GPHB)

Casting procedure of 8M NaOH geopolymer hollow block prisms is the same as geopolymer brick prisms and this has been discussed in the previous section. Masonry efficiency is found as specified by IS 1905-1987. As per the code, the h/t ratio of the

Table 17.9 Mortar thicknesses and h/t ratios of stack bonded hollow block prisms

Mortar thickness (mm)	Prisms dimensions			h/t ratios
	H (mm)	B (mm)	T (mm)	
12.5	600	304	150	4.00
10	590	304	150	3.93
7.5	580	304	150	3.86
12.5	477.5	304	150	3.18
10	470	304	150	3.13
7.5	462.5	304	150	3.08
12.5	355	304	150	2.36
10	350	304	150	2.33
7.5	345	304	150	2.30

prism must be between 2 and 5. Rendered and unrendered stack bonded prisms were cast to determine their behavior under compression.

Five-brick, four-brick and three-brick thick stack bonded prisms were cast using 8M NaOH hollow block and 10 mm cement mortar joints. The prism dimensions, mortar thicknesses of the prisms and their corresponding h/t ratios are as shown in Table 17.9.

17.4.1 Testing of Geopolymer Hollow Block Prisms

The test was done as per IS 1905-1987. The testing arrangements and method of rendered and unrendered hollow block prisms are as same as geopolymer brick prisms and was discussed in the previous 5.1.2. Test setup is as shown in Fig. 17.19.

This study was made on geopolymer hollow block prisms cast using cement mortar. The influence of mortar joint thickness and effect on the unrendered masonry efficiency of stack bonded prisms is discussed in this section.

It was noticed that masonry efficiency, decreasing with the increment in mortar joint thickness as shown in Fig. 17.20. Basic compressive stress by prism tests is given in Table 17.10. It was found that the basic compressive stress of all the prisms was greater than that given in Table 8 of the code IS 1905-1987.

Stress and strain were recorded at various intervals. The typical variation was plotted as shown in Fig. 17.21 after normalizing the values. Modulus of elasticity of the prisms increased slightly with the decrease in mortar joint thickness. Initial tangent modulus for prisms with 12.5 mm mortar thickness was 5831 and 6313 MPa for prisms with 10 mm mortar thickness and 8471 MPa for prisms with 7.5 mm mortar thickness. There was no much difference in modulus of elasticity with different h/t ratios of the prisms.

Fig. 17.19 Testing arrangement of geopolymer hollow block stack bonded prisms

Fig. 17.20 Stack bonded 5 brick thick geopolymer hollow block prism with cement mortar joints

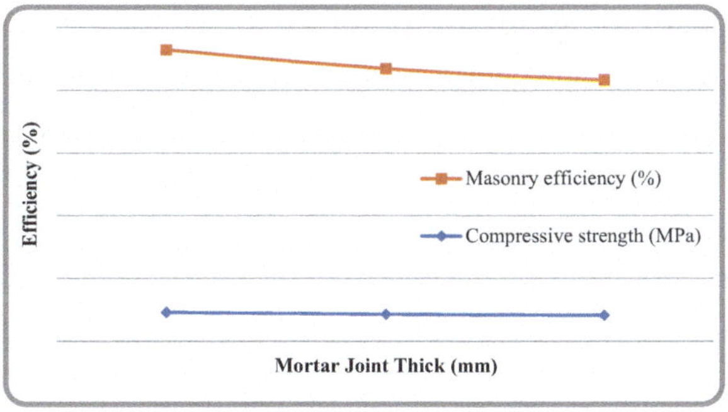

Table 17.10 Basic compressive stress of stack bonded prisms

h/t ratio	Mortar joint thickness (mm)	Average compressive strength (f′m)	Basic compressive stress = 0.25 × f′m
4.00	12.5	4.17	1.04
3.93	10.0	4.35	1.08
3.86	7.5	4.65	1.16
3.18	12.5	4.05	1.01
3.13	10.0	4.35	1.08
3.08	7.5	4.49	1.12
2.36	12.5	3.69	0.92
2.33	10.0	3.94	0.98
2.30	7.5	4.11	1.02

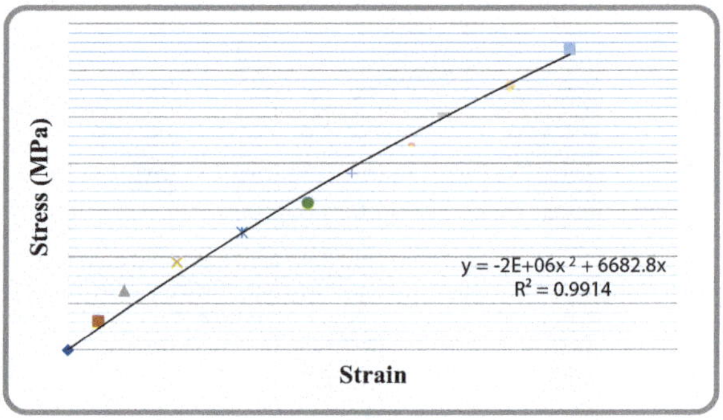

Fig. 17.21 Normalized stress-strain curve for 12.5 mm thick joint cement mortar hollow block prism with h/t = 4.0

The vertical cracks are developed at the center of the block and it continues till the bottom. Typical failure patterns of the prisms are as shown in Fig. 17.22a, b.

17.5 Wallettes

A brick/block wallette may be defined as the composite continuum of bricks/blocks and cement mortar joints. The details are as shown below.

- Bricks—8M NaOH bricks.
- Mortar—1:6 cement: River sand mortar of type M2 as per IS 1905-1987.
- Bed joint—10 mm thick mortar.
- Head joint—10 mm thick mortar.
- Concrete capping on top—75 mm thick.
- Wallette dimensions—(h × b × t) = 1105 × 1165 × 107 mm.
- h/t ratio = 10.32.

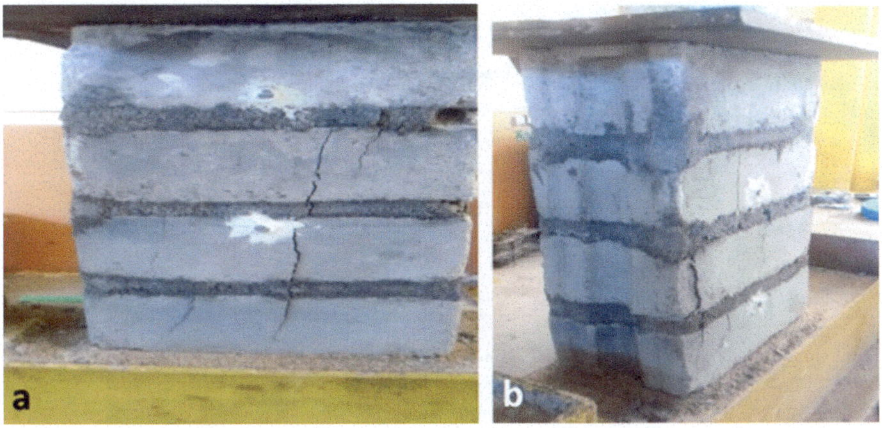

Fig. 17.22 Failure pattern for the unrendered geopolymer block prisms

The schematic representation of the typical wallette showing the dimensions is as shown in Fig. 17.23. The actual arrangements of units of geopolymer brick wallettes is shown in Fig. 17.24.

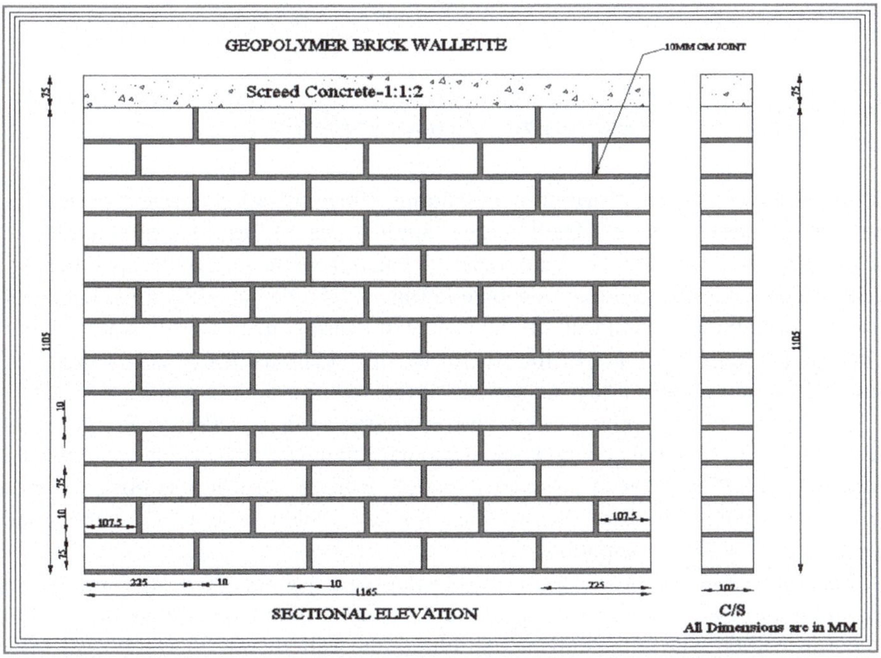

Fig. 17.23 Representation of typical geopolymer brick wallette

Fig. 17.24 Stretcher bonded geopolymer brick wallette kept for curing

The eccentricity adopted for wallette testing was the same as that adopted for English bonded prism, i.e., e/d = 1/6.

where e = eccentricity, d = bearing width = 107 mm.

From the above relation, eccentricity = 107/6 which is approximately equal to 18 mm.

17.5.1 Testing of Geopolymer Bricks Wallettes

To make sure of uniform distribution of loading, 100 mm thick concrete capping was done on the brick wallettes. The load was applied at a uniform rate by a hydraulic jack of 1000 KN capacity. The load was distributed from the jack to the specimen by means of a ladder arrangement as shown in Fig. 17.25a. For axial loading, center of loading assemblage was placed on the center of bearing area of brick wallette. For eccentric loading, the center of the ladder arrangement was placed at a distance of 18 mm from the center of bearing surface of the wallette as shown in Fig. 17.25b.

The results of the axial loaded and eccentric loaded wallettes are given in Table 17.11. It was noticed that the masonry efficiency of eccentrically loaded wallette was 83% of that of an axially loaded wallette. Also it was observed that the compressive strength of the stretcher bonded wallette was 65% of the stack bonded prism of the same parameters.

Stress and strain were recorded at various intervals. The variation was plotted, and the initial tangent modulus of the wallette was higher in axial loading compared to eccentric loading. The initial tangent modulus of the wallette was 3528 and 2791 MPa for axial and eccentric loading respectively. Figure 17.26a and b show the stress-strain curve for axially loaded wallette and eccentrically loaded wallette respectively.

Fig. 17.25 Ladder arrangement for geopolymer brick wallette: **a** axial loaded brick wallette, **b** eccentric loaded

Table 17.11 Results of stretcher bonded geopolymer brick wallette

Type of loading	Wallette No	Load at initial crack (KN)	Failure load (KN)	Compressive strength (MPa)	Avg. compressive strength (MPa)
Axial loading	1	240.0	267.0	2.14	1.99
	2	200.5	229.8	1.84	
Eccentric loading	1	200.0	218.6	1.75	1.66
	2	170.0	197.2	1.58	

It was noticed that vertical cracks originated from the head joints on top of the wallette and it propagated, till three fourth of the height from top of the wallette. It also noticed that the width of the crack is increased up to 3 mm. Spalling of bed mortar joints have taken place in the eccentrically loaded wallettes. The failure patterns of axially loaded wallette and eccentrically loaded wallette are shown in Fig. 17.27a and b respectively. This behavior is similar to masonry wallettes.

17.5.2 Solid Block Wallets

In this research, stretcher bonded solid block masonry wallettes were cast using GPS and cement mortar to check the effect of eccentric compression on masonry efficiency. Four wallets were cast in the study, two wallettes each of axial compression and eccentric compression. The casting details are the same as geopolymer brick wallettes and as explained in the previous section and some of them are shown below.

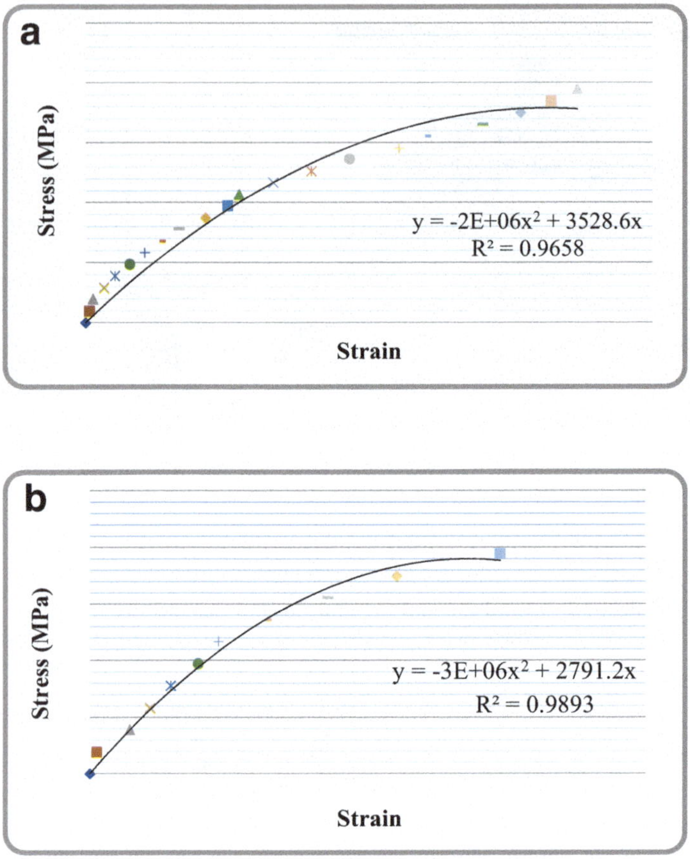

Fig. 17.26 **a** Stress-strain curve for axially loaded wallette. **b** Stress-strain curve for eccentrically loaded wallette

- Solid Block—8M NaOH blocks
- Wallette dimensions—(h × b × t) = 1034 × 1190 × 150 mm.
- h/t ratio = 6.90

The schematic representation of the typical solid block wallette showing the dimensions are as shown in Fig. 17.28a.

The wallettes cast was cured for 28 days after casting. Wallettes kept for curing are as shown in Fig. 17.28b.

The eccentricity adopted for wallette testing was the same as that adopted for English bonded prism, i.e., e/d = 1/6. Bearing width = 150 mm.

From the above relation, eccentricity = 150/6 which is approximately equal to 25 mm.

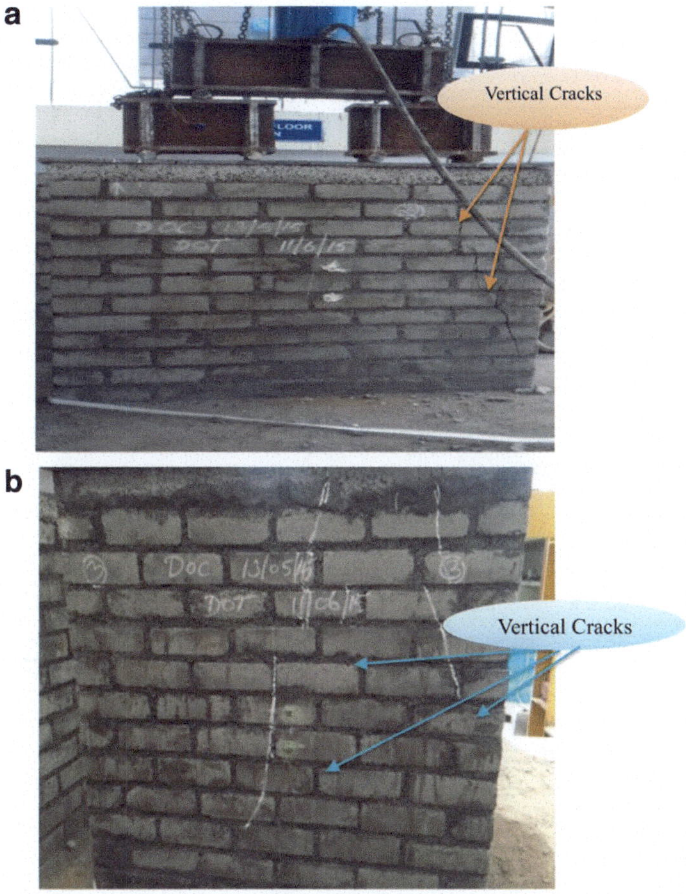

Fig. 17.27 **a** Failure pattern of axially loaded wallette, **b** Failure pattern of eccentrically loaded wallette

17.5.3 Testing of Geopolymer Soil Block Wallets

The testing arrangement of the solid block wallette was the same as that adopted for the geopolymer brick wallette and was discussed in the previous section. For eccentric loading, the center of the ladder arrangement was placed at a distance of 25 mm from the center of the bearing surface of the wallette.

The results of the axial loaded and eccentric loaded wallets are given in Table 17.12. It was noticed that the masonry efficiency of eccentrically loaded wallette was 90% of that of an axially loaded wallette. It was also noticed that the compressive strength of the stretcher bonded wallette was 50% of the stack bonded prism of the same parameters.

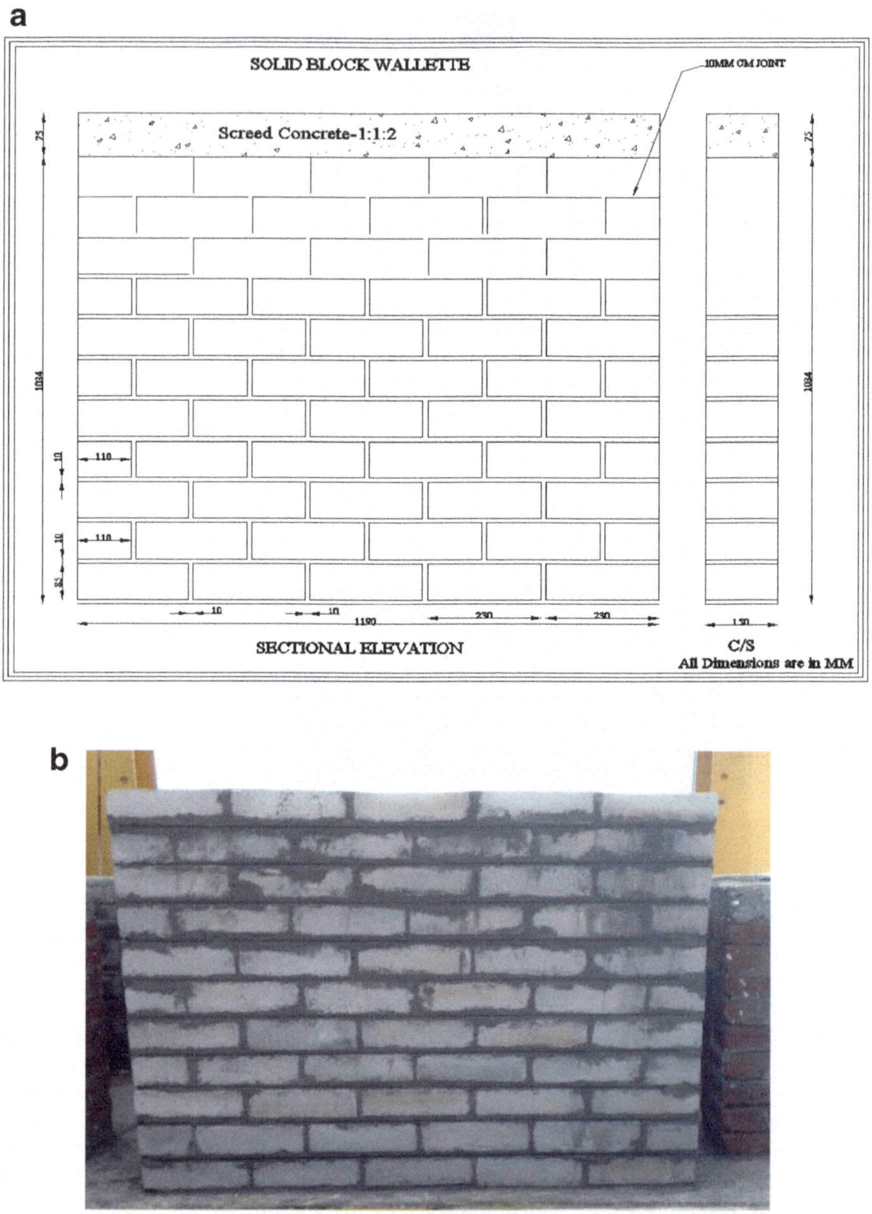

Fig. 17.28 **a** Schematic representation of typical geopolymer solid block wallette. **b** Stretcher bonded solid block wallette kept for curing

Table 17.12 Test results of stretcher bonded geopolymer solid block wallette

Type of loading	Wallette No	Load at initial crack (KN)	Failure load (KN)	Compressive strength (MPa)	Avg. compressive strength (MPa)
Axial loading	1	300	395	2.21	2.07
	2	275	345	1.93	
Eccentric loading	1	290	340	1.90	1.87
	2	265	330	1.84	

Stress and strain were recorded at various intervals. The variation was plotted, and the initial tangent modulus of the wallette was higher in axial loading compared to eccentric loading. The initial tangent modulus of the wallette was 3551 and 2787 MPa for axial and eccentric loading respectively. Normalized stress-strain curves for the same conditions are shown in Fig. 17.29a and b respectively.

It was noticed that vertical cracks were originated on top of the wallettes and it continued, till 2/3rd of the height from top. Spalling of bed mortar joints has taken place in the eccentrically loaded wallettes. The failure patterns of axially loaded wallette and eccentrically loaded wallette is shown in Fig. 17.30a and b respectively. This behavior is similar to any masonry wallettes.

17.5.4 Geopolymer Hollow Block Wallettes

In this research, stretcher bonded hollow block masonry wallettes were cast using GPHB and cement mortar to check the effect of eccentric compression on masonry efficiency. Four wallets were cast in the study, two wallettes each of axial compression and eccentric compression. The casting details are the same as geopolymer brick wallettes and as explained in the previous section and some of them are shown below.

- Hollow Block—8M NaOH blocks
- Wallette dimensions—(h $\times$ b $\times$ t) = 1080 mm $\times$ 1246 mm $\times$ 150 mm.
- h/t ratio = 7.2

The schematic representation of the typical hollow block wallette showing the dimensions is as shown in Fig. 17.31.

The geopolymer hollow block wallettes were cured for 28 days after casting. Wallettes kept for curing are as shown in Fig. 17.32a, b.

The eccentricity adopted for wallette testing was the same as that adopted for English bonded prism, i.e., e/d = 1/6.

where e = eccentricity, d = bearing width = 150 mm.

From the above relation, eccentricity = 150/6 which is approximately equal to 25 mm.

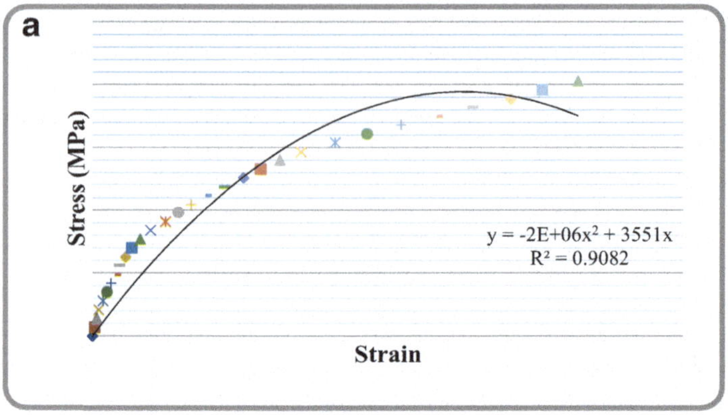

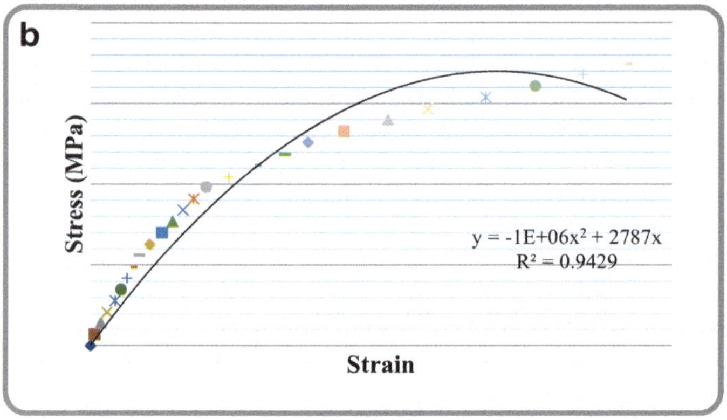

Fig. 17.29 **a** Stress-strain curve for axially loaded wallette. **b** Stress-strain curve for eccentrically loaded wallette

17.5.5 Testing of Geopolymer Hollow Block Wallettes

The testing arrangement for the hollow block wallette was the same as that adopted for the geopolymer brick wallette and was discussed in previous Sect. 5.7.1.1. The ladder arrangement for the geopolymer hollow block wallette is as shown in Fig. 17.33a. For axial loading, the center of loading assemblage was placed on the center of the bearing area of brick wallette. For eccentric loading, the center of the ladder arrangement was placed at a distance of 25 mm from the center of bearing surface of the wallette as shown in Fig. 17.33b.

The results of the axial loaded and eccentric loaded wallettes are given in Table 17.13. It was noticed that the masonry efficiency of eccentrically loaded wallette was 84% of that of an axially loaded wallette. It was noticed that the compressive strength of the stretcher bonded wallette was 53% of the stack bonded prism of the same parameters.

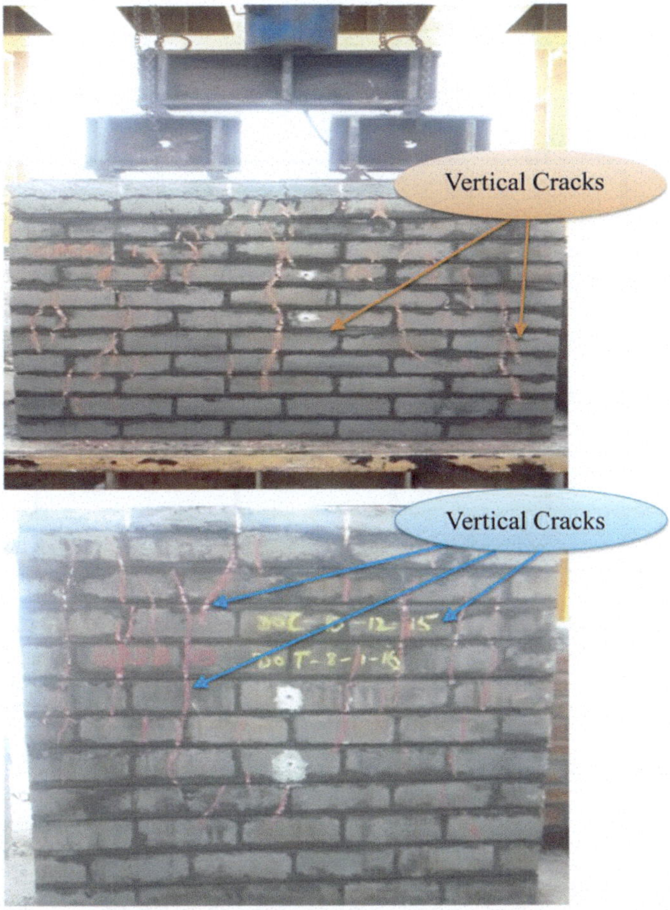

Fig. 17.30 Failure pattern: **a** axially loaded wallette, **b** eccentrically loaded wallette

Stress and strain were recorded at various intervals. The variation was plotted, and the initial tangent modulus of the wallette was higher in axial loading than eccentric loading. The initial tangent modulus of the wallette was 5834 and 2048 MPa for axial loading and eccentric loading respectively. Stress-strain curves for axially loaded wallette and eccentrically loaded wallette are shown in Fig. 17.34a and b respectively.

It was noticed that vertical cracks were originated at the top of the wallettes and it continued, till 2/3rd of the height from the top and the crack width was increased up to 3 mm. It was observed that spalling of bed mortar joints has taken place in the eccentrically loaded wallettes. The failure patterns of axially loaded wallette and eccentrically loaded wallette is shown in Fig. 17.35a and b respectively. This behavior is similar to any masonry wallettes.

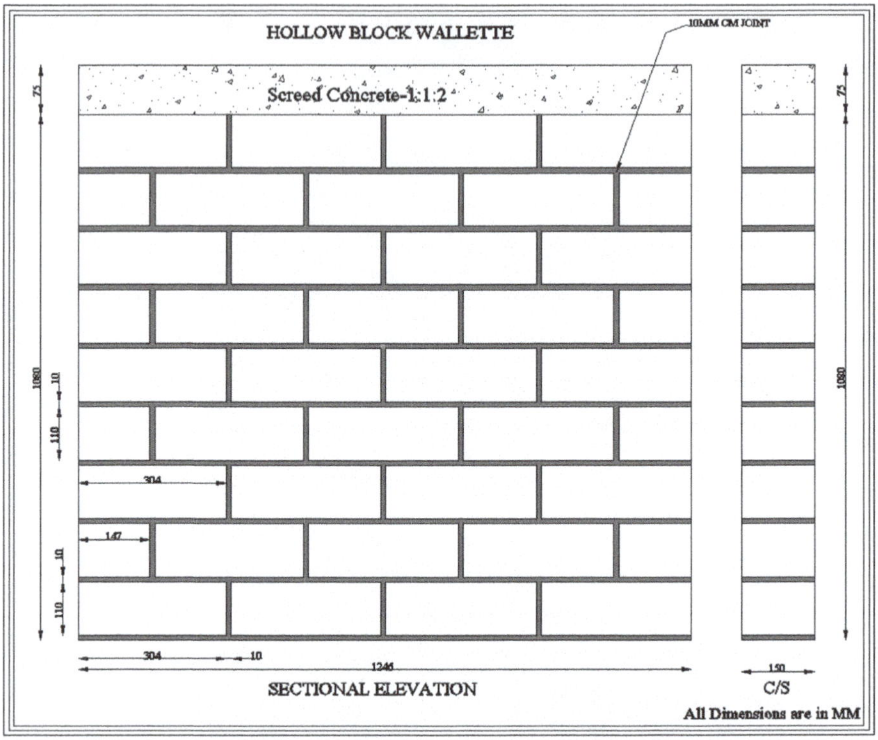

Fig. 17.31 Representation of typical geopolymer hollow block wallette

Fig. 17.32 Stretcher bonded hollow block wallette kept for curing: **a** front view, **b** top view

Fig. 17.33 Ladder arrangement for geopolymer hollow block wallette testing: **a** axial loaded wallette, **b** eccentric loaded

Table 17.13 Test results of stretcher bonded geopolymer hollow block wallettes

Type of loading	Wallette No	Load at initial crack (KN)	Failure load (KN)	Compressive strength (MPa)	Avg. compressive strength (MPa)
Axial loading	1	300	425	2.27	2.31
	2	380	442	2.36	
Eccentric loading	1	270	375	2.01	1.95
	2	245	355	1.89	

17.6 Conclusions

The following are the conclusions based on the discussion:

Fly ash and slag which are by-products of industry can be effectively used in making eco-efficient masonry units without compromising the properties. This addresses the challenges of disposing the by-products/marginal materials. The masonry units also depicted enhanced properties over conventional ones. They can be recommended to use in making structural masonry. Hence geopolymer masonry units proved to be eco-friendly and sustainable alternative building materials.

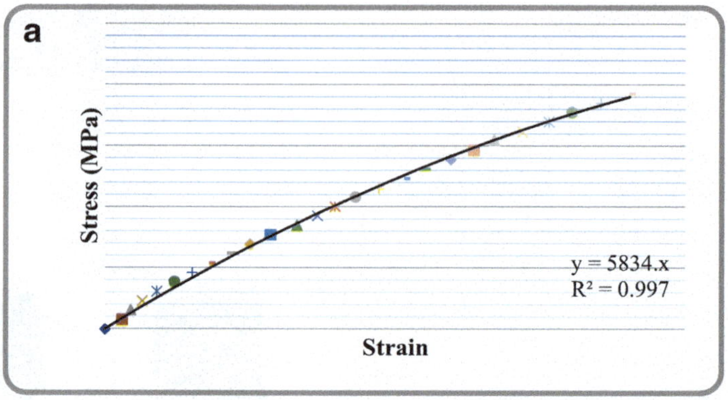

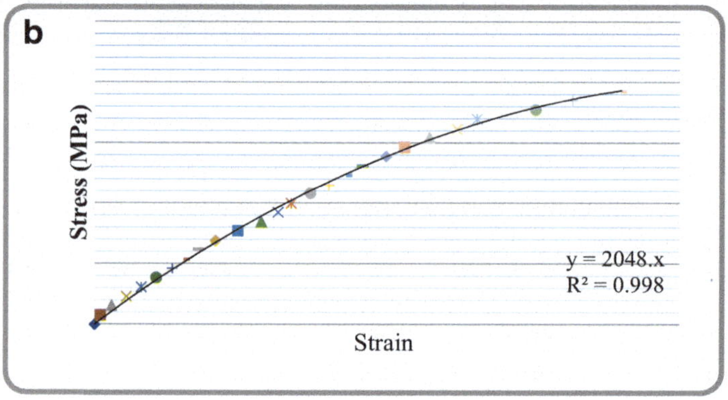

Fig. 17.34 **a** Stress-strain curve for axially loaded wallette. **b** Stress-strain curve for eccentrically loaded wallette

Fig. 17.35 Failure pattern: **a** axially loaded wallette, **b** eccentrically loaded wallette

References

ASTM (2009) ASTM C67-09. Standard test methods for sampling and testing brick and structural clay tile. American Society for Testing and Materials

Bhogayata A, Dave SV, Arora NK (2020) Utilization of expanded clay aggregates in sustainable lightweight geopolymer concrete. J Mater Cycles Waste Manag 22:1780–1792

Dao DV, Trinh SH, Ly HB, Pham BT (2019) Prediction of compressive strength of geopolymer concrete using entirely steel slag aggregates: novel hybrid artificial intelligence approaches. Appl Sci 9(6):1113

IS 1905-1987: Code of practice for structural use of unreinforced masonry. Bureau of Indian Standards, New Delhi

IS 2185 (Part 1)-2005: Concrete masonry units—Specifications Bureau of Indian Standards, New Delhi

IS 2250-1981: Code of practice for preparation and use of masonry mortars. Bureau of Indian Standards, New Delhi

IS 3495 (parts 1 to 4)-1992: Methods of tests of burnt clay building bricks. Bureau of Indian Standards, New Delhi

IS 4860-1968: Specification for acid resistant brick. Bureau of Indian Standards, New Delhi

Norton MG, Provis JL (2020) 1000 at 1000: geopolymer technology—the current state of the art. J Mater Sci 55:13486–13489

Nuruddin MF, Malkawi AB, Fauzi A, Mohammed BS, Almattarneh HM (2016) Geopolymer concrete for structural use: recent findings and limitations. In: IOP conference series: materials science and engineering, vol 133, No. 1. IOP Publishing, p 012021

Radhakrishna VK, Sasalatti V, Venumadhav T (2015) Study of geopolymer masonry as sustainable building material. J Environ Res Dev 9(3A):925–932

Venugopal K, Radhakrishna (2020) experimental study on behavior of geopolymer structural masonry. Indian Concr J 94(7):59–65

Rahman SK, Al-Ameri R (2021) A newly developed self-compacting geopolymer concrete under ambient condition. Constr Build Mater 267:

Rao NKS (1986) Structural masonry: properties and behavior. Department of Civil Engineering, IISc, Bangalore

Rehman AU, Sglavo VM (2020) 3D printing of geopolymer-based concrete for building applications, concrete for building applications. Rapid Prototyp J

Ren X, Zhang L (2019) Experimental study of geopolymer concrete produced from waste concrete. J Mater Civ Eng 31(7):04019114

Ridha S, Abd Hamid AI, Setiawan RA, Shahari AR (2018) Influence of sulfuric and hydrochloric acid on the resistance of geopolymer cement with nano-silica additive for oil well cement application. Int J Struct Integr. https://doi.org/10.1108/IJSI-11-2017-0066

Sakthidoss DD, Senniappan T (2020) A study on high strength geopolymer concrete with alumina-silica materials using manufacturing sand. Silicon 12:635–746

Sikandar MA, Jo BW, Baloch Z, Khan MA (2019) Properties of chemically synthesized nano-geopolymer cement based self-compacting geopolymer concrete (SCGC). J Wuhan Univ Technol Mater Sci Ed 34(1):98–106

Standard I (1997) IS 1077: 1992, Common burnt clay building bricks–specifications. Bureau of Indian Standards, New Delhi

Venugopal K, Radhakrishna (2016a) Structural behavior of geopolymer masonry. Indian J Sci Technol 9(25):12–26

Venugopal K, Radhakrishna (2016b) Sustainable units for structural masonry. Indian J Sci Technol 9(25):1–7

Verma M, Dev N (2021) Effect of SNF-based superplasticizer on physical, mechanical and thermal properties of the geopolymer concrete. Silicon 1–11

Zuo Y, Ye G (2021) GeoMicro 3D: A novel numerical model for stimulating the reaction process and microstructure formation of alkali activated slag. Cem Concr Res 141:

Chapter 18
Utilization of Waste Brick Powder for Manufacturing Green Bricks and Cementitious Materials

Anwar Khitab, Riaz Akhtar Khan, Muhammad Saqib Riaz, Kashif Bashir, Seemab Tayyab, and Raja Bilal Nasar Khan

Abstract Clayey bricks and cementitious materials present an important class of building materials. Apart from the benefits, they also pose serious environmental threats. Clayey bricks involve huge utilization of fertile land. Cementitious composites are one of the major sources of greenhouse gas emissions. One of the ways to minimize the soil depletion in the case of clayey bricks and the greenhouse emission in cementitious composites is the replacement of the hazardous materials by certain green materials. One such eco-friendly material is the Waste Brick Powder (WBP). It has been reported that WBP consists of a reasonable amount of the minerals necessary for clayey bricks. On the other hand, it has certain compounds, which make WBP a good pozzolanic material for cementitious composites. Pozzolanic materials replace cement, and are therefore, environmentally beneficial and cost effective.

Keywords Brick powder · Cementitious composites · Clayey bricks · Environmental hazards · Recycling · Waste management

18.1 Introduction

Clayey bricks present one of the oldest building materials. The oldest mud bricks can be found in the ruins of many ancient civilizations like Egyptian and Indus Valley. Bricks offer many advantages: They are light in weight, easy to transport, require no dressing, and have adequate strength. As fertile clay is available abundantly, they are also cheap. According to a recent report, 282 billion bricks are being manufactured in only three South Asian Countries, India, Bangladesh and Nepal (Eil et al. 2020).

Clay is the main ingredient of the bricks. Owing to huge production of the bricks all over the world, the fertile clay is rapidly depleting and there is a serious need to conserve the clay for the coming generations. The researchers over the last few decades are seriously working to minimize the amount of clay in the manufacturing of this important building block. Some significant efforts are described as follows:

A. Khitab (✉) · R. A. Khan · M. S. Riaz · K. Bashir · S. Tayyab · R. B. N. Khan
Department of Civil Engineering, Mirpur University of Science and Technology (MUST), Mirpur 10250, AJ&K, Pakistan
e-mail: anwar.ce@must.edu.pk

J. A. Malik and S. Marathe (eds.), *Ecological and Health Effects of Building Materials*,
https://doi.org/10.1007/978-3-030-76073-1_18

Kazmi et al. studied the effect of sugarcane bagasse ash (SBA) and rice husk ash (RHA) by partially replacing clay in the bricks (Kazmi et al. 2016a, b): They have reported light weight bricks with good durability and slightly lesser mechanical strength. Munir et al. investigated the partial replacement of clay by waste marble powder (Munir et al. 2018). Their study revealed that the replacement led to lighter bricks with reduced linear shrinkage and mechanical strength. Javed et al. partially replaced the clay content by lime and bentonite (Javed et al. 2020). They have reported lower unit weight with enhanced strength and reduced thermal conductivity. Kadir et al. investigated the partial replacement of clay by Palm oil waste (Kadir et al. 2017). They have reported that in the range of 1–10% partial replacement, bricks with permissible shrinkage, and mechanical strength can be produced.

Cementitious materials gained immense popularity after the invention of Ordinary Portland Cement by Joseph Aspdin in 1824 (Khitab 2012). According to a recent report, 4.1 billion metric ton of cement is produced all over the globe (Raynor de Best 2020). Manufacturing of cement and cementitious composites is nevertheless an environmentally hazardous process and one of the major sources of greenhouse gas emissions (Ahmed et al. 2020). Additionally, the products like mortars, plasters and concrete are responsible for rapid depletion of natural resources like sand and rocks. Research on partial replacement of the ingredients of the cementitious products has gained rapid momentum in the past few decades and several useful studies are available on the topic (Jain et al. 2015; Jalil et al. 2019; Khan and Khitab 2020).

This chapter particularly addresses the partial replacement of clay in clayey bricks and cement in cementitious materials by a single material: Waste Brick Powder (WBP). It is suitable for clay replacement as it contains the mineralogical composition similar to that of clay. It is equally useful for cement replacement in cementitious composites owing to its pozzolanic characteristics. An overall green building can be constructed using one material that partially replaces both the clay and the cement. This chapter discusses the details to acquire the material, its characterization, the content of replacement, characteristics of the finished products and its comparison with the control materials. Due to the problems arising from the building materials, recycling of the waste is a viable solution. This in turn can preserve the environment and reduce the depletion of natural resources.

18.2 Clayey Bricks

18.2.1 Clay

Clay is one of the most abundant natural minerals containing materials on earth. However, for brick manufacturing, clay must possess some specific properties and characteristics. Such clays must have plasticity, which permits them to be shaped or molded when mixed with water. Also, when subjected to appropriate temperatures, the clay particles must fuse together. The clay used for making ordinary bricks should consist of alumina and silica in varying proportions, mixed with other substances like lime, magnesia, and iron etc. Good clay used for making bricks should contain 20–35% alumina, 50–60% silica, 5–10% lime, and 5–10% iron oxide. The proportions of alkalis and Magnesia should preferably be less than 7 and 1% respectively (Khitab and Anwar 2016). Alumina imparts plasticity; silica provides heat resistance and preserves the form of brick even at very high temperatures; lime acts as a binding agent to tie the clay particles together, and iron oxide provides it a reddish color as well as strength. The excess of alkalis makes the clay unsuitable for burning and an excess of magnesia imparts a bluish black color to the brick. Additionally, the presence of vegetation and organic matter is also harmful as these organic matters get burnt during the manufacturing process leaving voids inside the brick.

18.2.2 Waste Brick Powder

Waste brick powder is a lavish material that is generated as a waste at construction sites and brick kilns. It goes into dumping and pollutes our environment. Researchers around the globe are working to use wastes in construction materials to solve economic and environmental problems. Use of waste material may impart high or same structural properties (Ayaz Khan et al. 2018). Main source of brick waste powder is brick kilns; approximately 59 billion bricks are produced in Pakistan yearly in 12,000 brick kilns. During the colonial era, Bull's Trench Kiln technology for brick manufacturing was introduced in British India. Till date, such technology is in use in most parts of India and Pakistan. Being an old technology, where controlling parameters are not strict, a huge quantity of brick dust (powder) is generated as waste (Riaz et al. 2019). Structures are demolished after completion of their life span around the globe, which produces waste material and waste brick waste powder is one of them. Particles of different sizes are present in waste brick powder from coarse to finer. Specific gravity of waste brick powder is 2.42, and density is 1.251 g/cm^3. The chemical and mineralogical compositions of waste brick powder and ordinary clay are given in the literatures (Brînduş-Simuţ et al. 2018).

Looking at Table 18.1, it is evident that WBP can possibly be a substitute for fresh clay as it principally contains the same compounds as required for the fresh clay (for brick manufacturing) in almost desired quantities. Similarly, for a pozzolanic material, the cumulative amount of SiO_2, Fe_2O_3 and Al_2O_3 should be more than 70% (Sutter et al. 2013). The sum of the percentage as given in Table 18.2 is 78.3%, which makes WBP a pozzolanic material for cementitious composites. The pozzolanic character of WBP has also been confirmed by several other researchers: According to Brînduş-Simuţ et al. 356 mg of $Ca(OH)_2$ were consumed by one gram of WBP in the modified Chapelle test (Brînduş-Simuţ et al. 2018).

Table 18.1 Chemical composition of WBP

Compound	Percentage	
	WBP	Clay for bricks
SiO_2	55.5	50–60
Al_2O_3	17.0	20–35
Fe_2O_3	5.8	5–10
CaO	2.1	5–10
MgO	2.4	<1
K_2O	2.8	<7
Na_2O	0.7	<7
P_2O_5	0.2	–
TiO_2	0.8	–
SO_3	1.9	–

Table 18.2 Mineralogical composition of WBP

Compound	Percentage
SiO_2	26.2
Fe_2O_3	2.3
$NaAlSi_3O_8$	13.0
$KAlSi_3O_8$ (Microcline)	3.6
$KAlSi_3O_8$ (Orthoclase)	3.5
$KAl_2(AlSi_3O_{10})(OH)_2$	12.5
$K0._{65}Al_2(Al0._{65}Si_3._{35}O_{10})(OH)_2$	3.8
$CaMgSi_2O_6$	4.4
$Ca_2MgSi_2O_7$	2.8
Amorphous	27.8

Modified after: Brînduş-Simuţ et al. (2018)

18.3 Use of WBP for Bricks

The XRD patterns of both clay and WBP were reported in literature (Riaz et al. 2019). It can be seen that both the materials resemble in mineralogical composition except the diminution of Calcite in WBP. As clayey bricks are burnt to a temperature of about 900 °C, the major part of calcite gets decomposed, leaving its minor traces in the burnt units. Hence, partial replacement of clay can be successfully adapted. It shall lead to saving in the fertile clay and large reserves of clay can be preserved for the coming generations. Also, the landfill for dumping waste materials can be reserved. The minor quantities of lime in WBP are favorable (<5%), as the sole function of lime in clay is to help silica fuse at lower temperatures and its excess quantity de-shapes the finished product. According to Riaz et al. the partial replacement of clay by WBP in bricks should be limited to 10%, as the WBP is porous in nature: On one hand it reduces the density, but on the other hand, it also induces pores, which are although beneficial for insulation, reduces the strength and enhances water absorption (Riaz et al. 2019). The SEM micrographs reported in the literatures indicate that the clay particles possess denser structure, which increases the dead weight and strength of the finished product. On the other hand, the WBP has a less dense structure, which is beneficial, where light weight and insulation are required. The extent of partial replacement of clay by WBP thus depends on the required properties of the finished units. Nevertheless, the partial replacement from 5 to 10% of clay by WBP results in lighter bricks with improved insulation characteristics and acceptable strength requirements as per ASTM standards (Riaz et al. 2019). But for light-weight partitions, where structural characteristics are less important, higher replacements are possible and advantageous. One additional advantage with WBP inclusion is the lesser appearance of efflorescence on the surface of the bricks: The obvious reason behind that is the lesser to zerolime content in WBP, which is the primary source of efflorescence. Apart from WBP, several other replacement materials like Rice Husk Ash and Sugarcane Bagasse Ash were also found to lessen the effects of efflorescence (Kazmi et al. 2016a, 2017).

18.4 Use of WBP for Cementitious Composites

Pozzolanas are powdered siliceous, aluminous or silico-luminous materials, which in the presence of moisture react with $Ca(OH)_2$ and form compounds that possess cementing characteristics (Bahadori 2015). Looking at Table 18.1, the WBP is mainly a silico-Aluminous material with high content of SiO_2. The pozzolanic behavior may be assessed either by measuring the consumption of $Ca(OH)_2$, the strength activity index or by the development of the strength (Donatello et al. 2010). The main factors affecting the pozzolanic reactivity include the particle size, shape, and phase of the material under study (Black 2016): The behavior is pronounced if glassy or zeolitic phases are present. According to Rogers (2011), the materials used for brick manufacturing are in fact non-pozzolanic (Rogers 2011): Clay is composed of (50–60)% Silica, (20–30)% Alumina, other oxides and carbonates. Clay is non pozzolanic but when it is heated in the presence of lime during brick manufacturing it gains pozzolanic character: According Brînduş-Simuţ et al. (2018), WBP contains about 27% amorphous content (Brînduş-Simuţ et al. 2018), as referred in Table 18.2. Uddin et al. (2004) added WBP and explained the mechanism as how WBP contributes to the strength of the concrete and reported that WBP is pozzolana. Lime in Portland cement reacts with water and makes hydraulic compounds(Uddin 2004). Calcium-Silicate-Hydrate gel (CSH), and Calcium hydroxide (Portlandite). Portlandite In the presence of moisture, further reacts with Pozzolana and forms strong adhesives like Calcium Silicate Hydrate gel. The governing chemical reactions are shown as equations (Eqs. 18.1 and 18.2):

$$\text{Portland cement} + H_2O \rightarrow CSH + Ca(OH)_2 + CO_2 \tag{18.1}$$

$$\text{Pozzolana} + Ca(OH)_2 + H_2O \rightarrow \text{CHS (adhesive)} \qquad (18.2)$$

The emission of CO_2 as demonstrated in Eq. 18.1 is a greenhouse gas and is nevertheless hazardous. Portland cement reacts faster with water and provides early strength of concrete and pozzolana in the presence of lime reacts with water slowly, which although affects the early age strength, but contributes to the strength of concrete a few days later.

Following is the summary of the important works carried out, while using WBP as partial replacement of cement in cementitious composites. Li et al. (2020) recycled waste brick powder based on paste substitution method to make mortars and studied workability and resistance to sulfate, water and shrinkage (Li et al. 2020). Waste brick powder increased the superplasticizer requirement, but also enhanced the resistance to sulfate, water and shrinkage and reduced cement content. The authors have reported that 20% WBP by volume of the mortar can reduce 33% of the cement content without any undesirable effects.

Kırgız et al. (2016) worked on microstructure observation, chemical characterization of cement paste and monitored strength gaining up to 90 days of curing (Kırgız 2016a). Cement mortar and cement paste were prepared using WBP up to 35% and 65% respectively by weight of Portland cement. Several specimens were prepared and cured at 19–21 °C and relative humidity was kept at 98% until testing. Substitutions up to 35% by weight of WBP improved compressive and flexural strength development. According to the authors, WBP did not densify the microstructure of the cement paste. Rather an increase in calcium oxide (CaO) and iron oxide (Fe_2O_3) contents was observed. Ayaz Khan et al. (2018) replaced 5, 10, 15 and 20% of cement by WBP in plain cement concrete and performed tests for workability and strength of concrete, using w/c ratio of 0.55 (Ayaz Khan et al. 2018). They have reported that the replacement increased workability, compressive and split tensile strengths, which showed that the cement can be successfully replaced by WBP. Kırgız et al. (2016) worked on the partial replacement of cement and fine aggregates by WB and marble powder respectively, in concrete(Kırgız 2016b). SEM monitoring for hydration growth of cement paste, rheology for fresh state of concrete, measured strength gains and the mechanical properties of 7, 14, 28 and 90 days showed an improvement due to the partial replacement (Kırgız 2016b).

Heidari and Hasanpour (2013) studied the mineralogical and chemical composition of WBP and reported that the waste material possesses pozzolanic properties and therefore, can replace cement in concrete. They have reported that WBP possesses pozzolanic character and the concrete with 10% partial replacement of cement by WBP produces mechanical strength almost equal to that of the control specimens and that the strength further decreases with increase in the replacement level. Sharma et al. (2017) worked on replacement of cement by WBP and marble powder in concrete and performed a test of workability and strength (Sharma et al. 2017). They have reported an increase in workability and compressive strength up to 10% replacement of cement by WBP. Dubey and Porwal (2014) mixed WBP and fly ash as admixtures in concrete and performed tests on the strength of concrete (Dubey and Porwal 2014).

Their results showed that 20% cost of cement could be saved without affecting the strength of concrete. Bharti and Patel (2014) replaced cement by 10% waste brick powder and 15% silica fume in concrete at 0.50 w/c ratio and performed tests of compressive and tensile strengths (Bharti and Patel 2014). Both compressive and tensile strength of concrete were improved by replacement of cement by waste brick powder.

Harbi et al. (2017) used kaolin dust, glass powder, WBP and metakaolin to fill in voids between sand particles in mortar and performed tests such as compressive strength, tensile strength, water absorption, porosity, and sulphate resistance (Harbi et al. 2017): The authors have reported an enhancement of physico-mechanical properties and durability. Kirgiz (2014) prepared paste and mortar using blended cements, which were prepared by using 5% gypsum and 6, 20, 21 and 35% WBP for CEMI42.5 N cement clinker and grinding in ball mill for 30 min (Kirgiz 2014). Strength gaining mechanisms of blended cements were studied experimentally. Chemical compositions of pastes, compressive strength and flexural strength of mortars were experimentally determined at 7, 28 and 90 days, which showed increasing contents of silica in blended cement pastes and an improved strength gain mechanism up to 35% replacement. Letelier et al. (2017) replaced cement by 0, 5, 10 and 15% of WBP in concrete and performed tests for strength of concrete after 28 days. Results reported by Letelier et al. (2017) showed that cement can be replaced in concrete up to 15% without loss of strength.

Ma et al. (2020) used WBP as supplementary cementitious material (SCM). Mortar specimens were prepared by using various fineness levels and replacing cement with various percentages of WBP. They have reported that the filler effect and pozzolanic activity both refine the pore network of the cementitious material. With the incorporation of WBP, the water requirement increases and the slump decreases which increases the resistance to drying shrinkage. Pozzolanic activity increases with decreasing particle size of WBP. WBP with smaller particle size as compared to cement increases the compressive strength up to 15% as compared to control specimen; adding appropriate WBP ratio with smaller particle size decreases the water absorption of the finished product. The SEM images of the paste with various WBP replacement ratios (28d, 50 mm/10 mm/2 mm) are repotrted by Ma et al. (2020).

Lin et al. (2010) investigated the properties of WBP blended cement at replacement ratios (0–50) %. WBP consisted of SiO_2 (63.21%), Al_2O_3 (16.41%), Fe_2O_3 (6.05%), Na_2O (1.19%), K_2O (2.83%) and MgO (1.11%), and had a pozzolanic activity index of 107%. Experimental results show that 10, 20, 30, 40 and 50% of cement can be replaced by WBP with an increase in setting time of cement. Initially, strength development was slow in WBP blended cement but after some time strength development was significant. Various reaction products of pozzolanic paste are Calcium-Silicate-Hydrate (CSH), Calcium Aluminate Hydrate (CAH) and Calcium Aluminium Silicate Hydrate (CASH), resulting in utilization of $Ca(OH)_2$ during the late ages of curing. Change in the properties of WBP blended pastes increased with increase of blended ratio. Pores of CSH gels and CAH are filled via pozzolanic reaction, which is responsible for densification and subsequently enhances the gel ratio and degree of hydration. They have reported that the WBP can be successfully employed as SCM.

18.5 Conclusions

Past studies on the replacement of clay by waste brick powder (WBP) in clayey bricks are limited; although WBP is actually burnt clay and mineralogically similar to the fresh clay, except having reduced amount of Calcite. WBP is more porous than the clay and therefore, its inclusion in clayey bricks induces pores, which reduce density, and enhance thermal insulation. On the other hand, the mechanical strength also reduces; Smaller replacement levels (<10% by mass of clay) are recommended, where strength and durability are of prime importance. However, larger replacement levels can be used, where higher insulation is required and lower strength is not a concern. WBP produces lesser to nil effects of efflorescence: The efflorescence is reduced as the replacement level is increased.

WBP is a pozzolanic material for cementitious composites. Its pozzolanic character has been confirmed by various researchers, using strength activity index and strength development of various materials, including cement paste, mortar and concrete. A partial replacement as high as 35% by mass of cement has been successfully examined. The resulting products are reported to exhibit higher mechanical strength, resistance to salts, water and shrinkage. Secondary hydration reactions, consuming $Ca(OH)_2$ content to form Calcium-Silicate-Hydrate (CSH), Calcium Aluminate Hydrate (CAH) and Calcium Aluminium Silicate Hydrate (CASH) have been reported, which densify the material's structure and impart strength and durability properties. Nevertheless, pozzolanic property is dependent on the particle size and finer particles lead to higher pozzolanic character.

Recycling of WBP in clayey bricks as well as cementitious materials as partial replacement of clay and cement respectively is beneficial in terms of environmental impacts and cost effectiveness. Construction activities are on the rise around the globe. Huge consumption of fertile clay and cement are posing serious threats to the environment; also landfill spaces are limited. Recycling is a viable solution and recycling WBP in bricks and cementitious composites will lead to lesser utilization of the natural raw materials. Additionally, it will lessen the emission of greenhouse gases, reduce the cost of the structures, and produce the finished products with enhanced properties.

References

Ahmed S, Khitab A, Mehmood K, Tayyab S (2020) Green non-load bearing concrete blocks incorporating industrial wastes. SN Appl Sci 2:266. https://doi.org/10.1007/s42452-020-2043-6

Ayaz Khan MN, Liaqat N, Ahmed I, Basit A, Umar M, Khan MA (2018) Effect of brick dust on strength and workability of concrete. IOP Conf Ser Mater Sci Eng 414:012005. https://doi.org/10.1088/1757-899X/414/1/012005

Bahadori A (2015) Engineering and technical fundamentals of lining. Essentials of coating, painting, and lining for the oil, gas and petrochemical industries. Elsevier, pp 227–308

Bharti RK, Patel RD (2014) A study on low performance concrete using mineral admixtures (Brick Kiln Dust and Silica Fume). Int J Sci Res Dev 2:217–221

Black L (2016) Low clinker cement as a sustainable construction material. Sustainability of construction materials. Elsevier, pp 415–457
Brînduş-Simuţ J, Vyšvařil M, Bayer P et al (2018) Effect of particle size of waste brick powder on the properties of alkaline activated materials. IOP Conf Ser Mater Sci Eng 379: https://doi.org/10.1088/1757-899X/379/1/012019
Donatello S, Tyrer M, Cheeseman CR (2010) Comparison of test methods to assess pozzolanic activity. Cem Concr Compos 32:121–127. https://doi.org/10.1016/j.cemconcomp.2009.10.008
Dubey S, Porwal R (2014) Studies on effect of cement concrete strength by using waste materials like fly ash and brick kiln dust. Int J Eng Res Technol 3:42–57
Eil A, Li J, Baral P, Saikawa E (2020) Dirty stacks, high stakes: an overview of the brick sector in South Asia. Washington, DC 20433
Harbi R, Derabla R, Nafa Z (2017) Improvement of the properties of a mortar with 5% of kaolin fillers in sand combined with metakaolin, brick waste and glass powder in cement. Constr Build Mater 152:632–641. https://doi.org/10.1016/j.conbuildmat.2017.07.062
Heidari A, Hasanpour B (2013) Effects Of waste bricks powder of Gachsaran Company as a pozzolanic material in concrete. Asian J Civ Eng (Building Housing) 14:755–763
Jain N, Garg M, Minocha AK (2015) Green concrete from sustainable recycled coarse aggregates: mechanical and durability properties. J Waste Manag 2015:1–8. https://doi.org/10.1155/2015/281043
Jalil A, Khitab A, Ishtiaq H et al (2019) Evaluation of steel industrial slag as partial replacement of cement in concrete. Civ Eng J 5:181. https://doi.org/10.28991/cej-2019-03091236
Javed U, Khushnood RA, Memon SA et al (2020) Sustainable incorporation of lime-bentonite clay composite for production of eco-friendly bricks. J Clean Prod 263: https://doi.org/10.1016/j.jclepro.2020.121469
Kadir AA, Sarani NA, Abdullah MMAB et al (2017) Study on fired clay bricks by replacing clay with palm oil waste: effects on physical and mechanical properties. IOP Conf Ser Mater Sci Eng 209:012037. https://doi.org/10.1088/1757-899X/209/1/012037
Kazmi SMS, Abbas S, Munir MJ, Khitab A (2016a) Exploratory study on the effect of waste rice husk and sugarcane bagasse ashes in burnt clay bricks. J Build Eng 7. https://doi.org/10.1016/j.jobe.2016.08.001
Kazmi SMS, Abbas S, Saleem MA et al (2016b) Manufacturing of sustainable clay bricks: utilization of waste sugarcane bagasse and rice husk ashes. Constr Build Mater 120:29–41. https://doi.org/10.1016/j.conbuildmat.2016.05.084
Kazmi SMS, Munir MJ, Abbas S et al (2017) Development of lighter and eco-friendly burnt clay bricks incorporating sugarcane bagasse ash. Pakistan J Eng Appl Sci 21:244–256
Khan RBN, Khitab A (2020) Enhancing physical, mechanical and thermal properties of rubberized concrete. Eng Technol Q Rev 3:33–45. https://doi.org/10.5281/zenodo.3852541
Khitab A (2012) Materials of construction, 1st edn. Allied Books, Lahore, Pakistan
Khitab A, Anwar W (2016) Classical building materials. Advanced research on nanotechnology for civil engineering applications, pp 1–27
Kirgiz MS (2014) Effects of blended-cement paste chemical composition changes on some strength gains of blended-mortars. Sci World J 2014:1–11. https://doi.org/10.1155/2014/625350
Kırgız MS (2016a) Strength gain mechanism for green mortar substituted marble powder and brick powder for Portland cement. Eur J Environ Civ Eng 20:s38–s63. https://doi.org/10.1080/19648189.2016.1246691
Kırgız MS (2016b) Fresh and hardened properties of green binder concrete containing marble powder and brick powder. Eur J Environ Civ Eng 20:s64–s101. https://doi.org/10.1080/19648189.2016.1246692
Letelier V, Tarela E, Moriconi G (2017) Mechanical properties of concrete with recycled aggregates and waste brick powder as cement replacement. Procedia Eng 171:627–632. https://doi.org/10.1016/j.proeng.2017.01.396

Li LG, Lin ZH, Chen GM, Kwan AKH (2020) Reutilizing clay brick dust as paste substitution to produce environment-friendly durable mortar. J Clean Prod 274: https://doi.org/10.1016/j.jclepro.2020.122787

Lin K-L, Chen B-Y, Chiou C-S, Cheng An (2010) Waste brick's potential for use as a pozzolan in blended Portland cement. Waste Manag Res 28:647–652. https://doi.org/10.1177/0734242X09355853

Ma Z, Liu M, Duan Z et al (2020) Effects of active waste powder obtained from C&D waste on the micro properties and water permeability of concrete. J Clean Prod 257: https://doi.org/10.1016/j.jclepro.2020.120518

Munir MJ, Abbas S, Nehdi ML et al (2018) Development of eco-friendly fired clay bricks incorporating recycled marble powder. J Mater Civ Eng 30. https://doi.org/10.1061/(ASCE)MT.1943-5533.0002259

Raynor de Best (2020) World and US cement production 2010–2019

Riaz MH, Khitab A, Ahmed S (2019) Evaluation of sustainable clay bricks incorporating brick kiln dust. J Build Eng 24:100725. https://doi.org/10.1016/J.JOBE.2019.02.017

Rogers SB (2011) Evaluation and testing of brick dust as a pozzolanic additive to lime mortars for architectural conservation. University of Pennsylvania

Sharma A, Panchal S, Khan MM, Bharti AA (2017) Innovative use of brick powder and marble dust as a mineral admixture in concrete. Int J Civ Eng Technol 8:987–990

Sutter LL, Hooton RD, Schlorholtz S (2013) Methods for evaluating fly ash for use in highway concrete. Transp Res Board

Uddin MK (2004) Use of brick dust in concrete as mineral admixture and partial replacement of cement. J Civ Eng 32:69–78

Chapter 19
Health Impacts of Construction Workers: A Short Introduction

Summia Rehman, Ishfaq Ahmad Sheergojri, Ishfaq Ul Rehman, Tajamul Islam, Subzar Ahmad Nanda, and Rayees Ahmad Rather

Abstract Construction workers are the key and indispensable contributors to every construction project. For many years, the construction industry has been documented, having a great possibility of occupational illnesses, injuries and accidents. Such threats of construction workers can result in social problems and low productivity. Accordingly, construction workers' well-being ought to be extremely self-addressed to boost competency and potency. The factors like rising temperature and warmth stress risks within the dynamic climate state of affairs may doubtlessly have an effect on the workers globally, particularly those with strenuous work in tropical settings. Filth generated by construction activities deeply affects air quality and adversely affects the health of people residing near to construction sites. All through this book chapter, the authors aim to investigate the factors that have an effect on construction workers, continuing health supported by a scientific review of revealed research within the vicinity of construction. Conclusions from the study can give solid evidence of the causes and results of construction workers' future health.

Keywords Construction · Low productivity · Social problems · Unsafe conditions · Workers' health

19.1 Introduction

Construction activity is one of the main threats around the world as a tough, significant, and labor-intensive industry (Mahmoudi et al. 2014). Substantial injuries and illnesses forever cause the decline of craft efficiency, work eagerness and an increased delinquency rate. During this case, the labor potency and productivity enhancements arise from the hope that a healthier hand needs less input, produces a lot of output of higher excellence and performs better (Ødegaard and Roos 2014). On the other hand, the development trade additionally may have a harmful result on the psychological welfare of workers because of excess activity demands in manufacturing

---

S. Rehman · I. A. Sheergojri · I. U. Rehman (✉) · T. Islam · S. A. Nanda · R. A. Rather
Department of Botany, University of Kashmir, Srinagar 190006, J&K, India

J. A. Malik and S. Marathe (eds.), *Ecological and Health Effects of Building Materials*,
https://doi.org/10.1007/978-3-030-76073-1_19

projects that may additionally have an undesirable pressure on individual and structure performance (Love et al. 2011). For that reason, construction workers' potency and yield is improved by focusing on the mental state and physical health (Yorio and Wachter 2014; Martin et al. 2016).

With the intensive employees that the development business employs, health and caring problems within the housing industry became necessary, since the business continues to be the one among the most casualty and accident rates (Zhou et al. 2015). Consequently, wellbeing and safety management within the housing industry ought to be extremely increased; thereby serving construction organizations to bring about their H&S objectives. On the other hand, several studies are conducted to boost the security presentation of construction projects, a very few studies have centered on the health organization facet. The relationship between health managing and work outcomes has not been fully examined (Kwon and Adler 2014; Oswald et al. 2015). Moreover, the effect of an SNC on labor presentation conjointly ought to be known so as to assist construction organizations to cut back work pressure, improve work potency, productivity and improve the social capital (Thomson et al. 2017).

The pivotal role of workers is central to almost all types of construction activities. No description of luxuries of this world would be complete without accepting the importance of the occupational/working class. While it is true that working class is making construction of physical infrastructure and allied things possible, it is largely at the cost of their precious health. The construction industry in its multifarious nature has consumed a lot of precious human lives of occupational groups. Workers, most often while working for a living, eventually become diseased due to working in bad working conditions. The construction activities have been flourishing at the cost of working or occupational groups who are at the greater risk of working places and environment. The unsafe working conditions and other equally hazardous situations have left the working class at the receiving end.

The working class is becoming an ever growing victim of unsafe working places in two ways. The first category of disadvantageous health related problems are accidental or visible ones. This category would include physical injuries, accidents, accidental deaths and other daily issues. Thanks to a number of pro-labour constitutional safeguards these immediate hazards are being taken into account with some immediate measures of relief. However the hidden invisible and the long term impacts of construction activities affecting the precious lives of the working class are not being taken into account. The long term impact on human health of occupational groups or simply termed as occupational human health has not been fully understood. The lack of will to accept the long term impacts of unsafe working conditions is adding insult to the injuries.

When operating in unsafe building environments, there are few advanced risks that are implemented to protect a stable and healthy working environment. The gruesome picture demands ample attention. The welfare and health condition of construction employees require empathetic support of the research community and pro-labour government policies. On the academic front over the last few years multiple efforts have created a foundational ground to discuss the long term health impact of the construction industry with special reference to the working class. It has been found that due to several constitutional measures all across the world construction industries seem to be bound to cater and address the immediate health impacts of the working class. Number of construction industries are making an account of Injuries, deaths, accidents and providing some relief to the class. However the long term impacts are not the concerns of construction corporations. Over the recent few years a handful of studies have been able to examine and investigate the long term effects of construction works on the workers' health. These studies, founded on the data generated by a number of agencies like National Institute for Occupational Safety and Health (NIOSH 2018) and Occupational Safety and Health Administration (OSHA 2019), sensitizes the issue by exploring the long term toll of human health of the laboring class. These studies are predominantly the struggles of investigators and specialists addressing the security issues (Zuluaga et al. 2016). Their overall emphasis is to develop the safe climate and surroundings of the construction manufacturing units (Pandit et al. 2019). They desire to highlight the longstanding effects of construction work on workers' health which have hardly been cited. Such studies have been able to establish the long term consequences of toiling in harmful, harsh, unhygienic working places and how the current safety measures are meant only to address the immediate effects (Sacks et al. 2009). By coining terms like occupational health, occupational disease, chronic diseases and a ratification of the same by the world health organization, studies focusing on long term implications of inhumane working conditions express primarily elimination of hazards as a pertinent need.

A few studies, quite recent in date, have been able to locate the influence of climate variation on the workers' health. The increasing global temperature is affecting the working class associated with the construction industry by developing a heat stress. Workers with least knowledge of increasing global temperature have become greatly vulnerable. On a regular basis they are facing troubles, accidents and diseases which they rarely heard of in the past. The thrilling heat due to varying climate, sideways with additional factors upsurge the susceptibility of construction workers to heat strain (Xiang et al. 2014). The growing heat stress is becoming a global phenomenon. There has been higher gratitude for high temperature hazards to agricultural staff because of perceptibly greater heat-related mortality rates within the industry (Jackson and Rosenberg 2010). Studies have documented a high frequency of warmth stress connected symptoms among North American nation migrant farm staff (32%) (Fleischer et al. 2013), Japanese forest staff (32%) (Maeda et al. 2006), Costa Rican sugar cane staff (Crowe et al. 2013), and South African horticulture staff (Mathee et al. 2010).

Heat strain considered as a direct outcome of environment change poses a considerable risk to construction workers globally. More people are facing it. One study shows in the U.S. construction workers are 13 folds more expected to perish from a heat-related illness (HRI) related to workers in different trades, and inside the industry, road construction workers and roofers face predominantly a high threat of HRIs (Gubernot et al. 2015). It is being estimated that the extreme heat on account of greater global change in climate would result in substantial enhancement in the susceptibility of construction workers to high temperature stress. The growing heat stress would have a huge impact not only on the workers' health, it is expected it would even hamper the growth of the global construction industry. Under such circumstances distinctive consideration necessities to be compensated to the threats faced by the worldwide construction staff from professional heat stress.

The growing climatic concerns and its impacts on workers' health are becoming alarming. The emission of greenhouse gasses makes the scientific community vocal in saying climate change is happening with a universal trend of growing temperatures (Change 2007). Consequently, the quantity of maximum scorching days is estimated to last longer with additional regularity and strength within the future (Meehl and Tebaldi 2004). And there are square measure studies that show a positive correlation between scorching temperature and injury claims (Xiang et al. 2014). And there are studies that show a positive co relationship between increasing temperature and wound claims (Xiang et al. 2014). The increasing heat waves or heat stress had direct bearings on the workforce by causing more work-related complaints and damages in temperate climatic conditions. While heat waves of heat stress affect the workforce in all types of working environment it is outdoor working conditions that have become gruesome. Largely outdoor industry workers during heat waves have witnessed more fatalities and brutal experiences of being victims of growing heat stress. (Maeda et al. 2006; Mathee et al. 2010; Crowe et al. 2013; Fleischer et al. 2013), with the bulk displaying that heat pressure levels outdo the suggested criteria of Yankee Conference of Governmental Industrial Hygienists (National Institute for activity Safety and Health (NIOSH) 1986).

Almost all outside employees are one or other getting affected by the heat stress. However amid unsafe working conditions the section of outside workers undertaking very intensive and physical actions suffer more. They face a relatively great threat of heat-associated illnesses and damages during exceptionally scorching weather. What adds more to their problems is the lack of precautionary measures that are not satisfactorily adopted. The picture of outdoor workers affected by heat stress is more pathetic in a number of middle-low revenue countries (Kjellstrom and Lemke 2009). Heat stress is causing greater challenges for outdoor workers and the construction industry seems to be least aware of it. The available data proves that heat stress is consuming more lives and inflicting more injuries. It is believed that due to heat stress increases the number of potential injuries as workers suffer with less attention and management, wet palms, fogged-up protection glasses, and unintentional contact with scorching surfaces. Kjellstrom (2009), Kjellstrom and Lemke (2009) argue that due to climate change, contact to dangerous heat in the factory can upsurge the risk of work-associated damages and calamities, chiefly for those operating outside or close to high temperature sources.

It is being explored to what extent employees are affected by high temperature influences i.e. when ambient heats remain greater over numerous uninterrupted days. It is probable that it may have consequences on an employee's level of tiredness and negotiation of wellbeing and protection. The climate resulting in heat stress is however affecting men more as compared to the women. This perhaps owes to the fact that men are involved in more intensive physical work in outdoor environmental conditions. More importantly it has been found that heat stress impacts more middle-aged employees (35–54 years), tradespersons, transitional fabrication and transportation employees and workers.

19.2 Work-Related Illnesses and Injuries

Heat strain considerably upsurges the amount of the 'shocking injuries', 'injuries, musculoskeletal and connective tissue ailments, slashes, and amputations'. The maximum common damages and criticisms are 'body stressing', 'striking objects with a portion of the body', trips, falls and slides of a person'. Grievance claims through high temperature waves for 'being smashed by moving objects', 'chemicals and other constituents', electricity, and 'heat, and other ecological factors' have also grown. Mechanism-specific evaluation shows that employees are smashed by moving objects enhanced by 9.7% during high temperature waves. Central temperature advancement and sunstrokes can result in lethargy, fatigue, laziness, and failure of attentiveness which might be responsible for these damages. The noteworthy intensification of wounds due to 'interaction with harmful chemicals' during high temperature waves might be due to employees dropping the use of protective kits like personal protective equipment (PPE) under situations of life-threatening hotness (Park et al. 2009), or the amplified incorporation frequency of liquid substances through the skin in advanced ambient temperatures (Cherrie et al. 2004). Grown-up male laborers are at larger risk of heat-associated deaths. Similarly agronomic laborers die more due to heat strain (Petitti et al. 2013). Grown-up and elderly males and females cannot tolerate heat and are more expected to be affected from physiological high temperature anxiety than younger persons (Blatteis 2012). In the course of heat waves more persons become exposed to chemical constituents (Park et al. 2009). Outside male employees and tradespersons, persons employed in 'forestry and fishing' farming, and water, electricity, and gas manufacturing units are at greater threat of damage during heat waves. Work-related burns, cuts, amputations, and heat ailments are health consequences meaningfully related with thrilling heat.

19.3 Musculoskeletal Ailments as the Major Occupational Disorders

One of the core concerns of studies on long term consequences of working in unsafe and safety-less working environments was to provide a list of diseases affecting the working class. The long-term health impacts commonly found prevalent in the construction workers are MSDs, audible damage, breathing exposures, skin ailment, and psychosocial symbols. Spielholz et al. (2006) terms Musculoskeletal Disorders as the major work-related ailment in the production industry. MSDs Symptoms include body pain largely found in lower back, neck, shoulders, joints, hands and other kinds of ache (Wang et al. 2015). MSDS are the outcome of a set of factors and may often include Bodily exposures, lifting heavy loads, uncooperative lifting, recurrent back bending, greater vibrations, life-threatening temperature (Wang et al. 2015). Manual material handling, especially the mechanism of lifting, oftenleads to MSDs (Valero et al. 2016). MSDs are commonly found in the construction workers associated with masons, carpenters, brick makers and heavy material controllers (Valero et al. 2016).

19.4 Respiratory Diseases

Respiratory diseases constitute another portion of occupational diseases which are taking a toll on occupant groups. The large scale dust production during construction, with a total suspended particulate matter (TSP), fine particulate matter and coarse particulate matter causes respiratory diseases (Paschalidou et al. 2016; Yan et al. 2018). The particulate matter (PM) has opposing properties on human healthiness. Studies show that it causes many respiratory diseases (Le et al. 2017; Sanyal et al. 2018; Kowalska et al. 2019). Continued contact to air pollutants cause tumors (Kim et al. 2018). Respiratory diseases in occupational groups are common due to longer exposure to dust, especially lead. Oliver et al. (2001) has found dust a main reason leading to respiratory disorders. Asbestos, similarly, are potential reasons for respiratory diseases (Everatt et al. 2007). Likewise other forms of dust such as silica dirt, solvent and isocyanate vapors, and other poisonous constituents cause main respiratory ailments. In a major study it has been found that Lead dust is an active agent in promoting occupational diseases like damaging of respiratory organs (Levin and Goldberg 2000). Working with activities generating different types of smoke and glasses also causes respiratory problems (Bakke et al. 2001). Synthetic paints used for the purpose of polishing, finishing and painting are significantly causing respiratory problems (Kaukiainen et al. 2008). Paintings leading to respiratory problems have been described by Park et al. (2016) and (Meijer et al. 2001). Beaudry et al. (2013) believes that tunnel workers are more susceptible to respiratory disorders due to production of large quantities of dust and deficiency of qualitative air to breathe.

19.5 Hearing Problems

Loss of hearing power is one of the occupational diseases which have been trending. Working in high sound creating construction sites causes loss of hearing. The high sound affects the working groups gradually and makes them hear sounds only at high volume or they stop hearing. The hearing disabilities due to noise, dust and other hazardous conditions of the workplace are rendering most of the workers as long term impacts of constructional noises. In a study it has been found that between

60 and 70% of the building employees suffer everyday contact to a noisy condition all over their functioning day (Fernández et al. 2009). The noise at construction sites plays a destructive role. According to Li et al. (2016) such noise comes from machines, comprising air compressor appliances, rebar cutting, digger engines, and bending machines. There are a number of studies that show the harmful effects of noise like working conditions. Noise due to constructional activities had long term impacts of hearing (Lee et al. 2019). Frequent exposure to heavy noise created by Construction machinery and equipment substantially adds to the hearing problems (Kwon et al. 2018). It has been established that those working construction apparatus or working near the equipment are more likely prone to hearing disorders. Likewise men working as ironworkers, carpenters and masons suffer with hearing diseases (Seixas et al. 2005).

19.6 Skin Diseases

Skin diseases of divergent scales with potential to affect the patients majorly are largely found in occupational groups. It has been reported that Occupational skin diseases (OSD) are one of the hazardous outcomes of construction manufacturing units (Bock et al. 2003). Timmerman, et al. (2014) argues that construction workers are more prone to OSD. It has been found that long-term outdoor working increases the risks of OSD. Varghese et al. (2019) opines that heat and cold exposure of workers increases chances for increasing skin diseases. Likewise long exposure and working with hazardous chemicals such as epoxy resin, chrome iron, cobalt chloride, potassium dichromate etc. causes skin diseases (Bock et al. 2003).

19.7 Psychological Diseases

Construction workers have also been found vulnerable to the number of psychological and social problems. Being under the burden of heavy weights, lifting like activities, their mental health deteriorates. As their High psychological demands fall apart they descend to depression. According to Boschman et al. (2012) the little job gratification, low community support, greater work-associated stresses including post-traumatic stress often takes a toll on their health. Most common symptoms comprise high necessity for rescue after work (14%; 25%), distress (5%; 7%), dejection (18%; 20%), and post-traumatic stress disorder (11%; 7%) (Boschman et al. 2013). High workload with meager income also damages their mental peace (Maqsoom et al. 2018).

19.8 Recommendations

Central to aforementioned studies is the concern that working class or occupational groups are at the receiving end in the construction industry. Their precious lives are in peril due to heat stress, unsafe working conditions and working behaviors. The precious working force and its protection is a greater possibility. Indication displays that construction employees in Japan are not bodily defied by heat strain if actual deterrence procedures are taken on construction places (Morioka et al. 2006). Similarly the heat strain organization strategy of Construction, Energy Union and Forestry and Mining needs to be implemented and extended. To avoid heat stress work must be regulated, delayed or shifted to shadowy intervals. Likewise work in morning or evening hours may protect workers from getting exposed to heat stress. Role of technology in eradicating the heat stress has also been found crucial especially by making workers to work under air conditioner conditions, proper sops and work protocols prevents workers from a number of long term health impacts. Additional to field tools to display heat pressure conditions (Jackson and Rosenberg 2010), enforceable permissible guidelines may be looked-for to make sure the employment of heat stress regulation and prevention values such as California Code of Regulations, Title 8, Section 3395 (2014), mostly in industries with a huge percentage of temporally engaged non-union workers. Workers need to be regulated as the age, experience according to regulations received by enforcing agencies. For example it has been stated that compared with older workers, fresh workers are more susceptible to work-related wounds due to undertaking more vigorous jobs and absence of security training and services (US for Disease Control and Prevention 2003; Xiang et al. 2014). By regulating their working hours, a precious part of human capital would be saved.

Consequently, age-linked dissimilarities in the susceptibility to heat-related injury or accident should be taken into consideration when evolving new heat prevention strategies and regulations and service procedures. This highlights the importance of continued instructions in the use of PPE during high temperature waves among those unprotected to chemical constituents (Park et al. 2009). While construction industry counts the immediate health hazardous to some extent there however seems least awareness among occupational groups about the long term implications of the work they are engaged in. Many chronic health problems prevalent in the working class are actually an outcome of long term impacts of the nature of their job. From all these studies it is obvious that impacts of unsafe working conditions and behaviors would eventually deteriorate the health of occupational groups. The several pollutants generated at the construction sites spoil workers. They had least information about the chemicals, dust, gas and solvents that made them vulnerable to a number of diseases. Nearly all kinds of lasting health influences are due to insecure performances in dangerous atmospheres. MSDs, respiratory exposures, hearing loss skin disease, and psychosocial symptoms, cardiovascular diseases and other long term impacts are something that could be managed and mitigated to a greater extent.

The unsafe working behavior and less pro-working class conditions have been the basis for a number of risk factors. Consequently their alteration would ensure safety and long-term health of workers. Construction industry is required to restrain long term impacts like MSDs, respiratory exposure, hearing loss, skin diseases, and psychological symptoms. The need of the hour is therefore to expect the production units to place sufficient care to workers' health. Furthermore, it is predicted that the outcomes from this chapter will offer the industry more evidence about how to offer a safe atmosphere and how to safeguard safe worker performances on construction sites.

Acknowledgements The authors are highly thankful to the Department of Botany, University of Kashmir Srinagar – 190006, J&K, India, for providing the facility and also thankful to those who made their literature available for compiling the chapter.

References

Bakke B, Stewart P, Ulvestad B, Eduard W (2001) Dust and gas exposure in tunnel construction work. AIHAJ-Am Ind Hyg Assoc 62(4):457–465

Beaudry C, Lavoué J, Sauvé JF, Bégin D, SenhajiRhazi M, Perrault G, Gérin M (2013) Occupational exposure to silica in construction workers: a literature-based exposure database. J Occup Environ Hyg 10(2):71–77

Blatteis CM (2012) Age-dependent changes in temperature regulation–a mini review. Gerontology 58(4):289–295

Bock M, Schmidt A, Bruckner T, Diepgen TL (2003) Occupational skin disease in the construction industry. Br J Dermatol 149(6):1165–1171

Boschman JS, van der Molen HF, Sluiter JK, Frings-Dresen MH (2012) Musculoskeletal disorders among construction workers: a one-year follow-up study. BMC Musculoskelet Disord 13(1):1–20

Boschman JS, Van der Molen HF, Sluiter JK, Frings-Dresen MHW (2013) Psychosocial work environment and mental health among construction workers. Appl Ergon 44(5):748–755

Change IPOC (2007) Climate change 2007: the physical science basis. Agenda 6(07):333

Cherrie JW, Semple S, Brouwer D (2004) Gloves and dermal exposure to chemicals: proposals for evaluating workplace effectiveness. Ann Occup Hyg 48(7):607–615

Crowe J, Wesseling C, Solano BR, Umaña MP, Ramírez AR, Kjellstrom T, Nilsson M (2013) Heat exposure in sugarcane harvesters in Costa Rica. Am J Ind Med 56(10):1157–1164

Everatt RP, Smolianskienė G, Tossavainen A, Cicėnas S, Jankauskas R (2007) Occupational asbestos exposure among respiratory cancer patients in Lithuania. Am J Ind Med 50(6):455–463

Fernández MD, Quintana S, Chavarría N, Ballesteros JA (2009) Noise exposure of workers of the construction sector. Appl Acoust 70(5):753–760

Fleischer NL, Tiesman HM, Sumitani J, Miz T, Amarnath KK, Bayakly AR, Murphy MW (2013) Public health impact of heat-related illness among migrant farmworkers. Am J Prev Med 44(3):199–206

Gubernot DM, Anderson GB, Hunting KL (2015) Characterizing occupational heat-related mortality in the United States 2000–2010: an analysis using the census of fatal occupational injuries database. Am J Ind Med 58(2):203–211

Jackson LL, Rosenberg HR (2010) Preventing heat-related illness among agricultural workers. J Agromed 15(3):200–215

Kaukiainen A, Martikainen R, Riala R, Reijula K, Tammilehto L (2008) Work tasks chemical exposure and respiratory health in construction painting. Am J Ind Med 51(1):1–8

Kim HB, Shim JY, Park B, Lee YJ (2018) Long-term exposure to air pollutants and cancer mortality: a meta-analysis of cohort studies. Int J Environ Res Public Health 15(11):2608

Kjellstrom T (2009) Climate change direct heat exposure, health and well-being in low and middle-income countries. Global Health Action 2

Kjellstrom HI, Lemke B (2009) Workplace heat stress health and productivity–an increasing challenge for low and middle-income countries during climate change. Glob Health Action 2(1):2047

Kowalska M, Skrzypek M, Kowalski M, Cyrus J, Ewa N, Czech E (2019) The relationship between daily concentration of fine particulate matter in ambient air and exacerbation of respiratory diseases in Silesian agglomeration Poland. Int J Environ Res Public Health 16(7):1131

Kwon N, Song K, Lee HS, Kim J, Park M (2018) Construction noise risk assessment model focusing on construction equipment. J Constr Eng Manag 144(6):04018034

Kwon SW, Adler PS (2014) Social capital: maturation of a field of research. Acad Manag Rev 39(4):412–422

Le Blond JS, Woskie S, Horwell CJ, Williamson BJ (2017) Particulate matter produced during commercial sugarcane harvesting and processing: a respiratory health hazard. Atmos Environ 149:34-46

Lee SC, Kim JH, Hong JY (2019) Characterizing perceived aspects of adverse impact of noise on construction managers on construction sites. Build Environ 152:17–27

Levin SM, Goldberg M (2000) Clinical evaluation and management of lead-exposed construction workers. Am J Ind Med 37(1):23–43

Li H, Wang R, Cao S, Chen Y, Huang W (2016) A method for low-frequency noise suppression based on mathematical morphology in microseismic monitoring. Geophysics 81(3):V159–V167

Love PE, Davis PR, Chevis R, Edwards DJ (2011) Risk/reward compensation model for civil engineering infrastructure alliance projects. J Constr Eng Manag 137(2):127–136

Maeda T, Kaneko SY, Ohta M, Tanaka K, Sasaki A, Fukushima T (2006) Risk factors for heatstroke among Japanese forestry workers. J Occup Health 48(4):223–229

Mahmoudi S, Ghasemi F, Fam M, Soleimani E (2014) Framework for continuous assessment and improvement of occupational health and safety issues in construction companies. Saf Health Work 5(3):125–130

Maqsoom A, Mughees A, Khan AK, Arif MI (2018) Labor productivity in construction industry: investigating the influence of internal psychosocial stressors. In: ICCREM 2018: analysis of real estate and the construction industry, pp 10–17. American Society of Civil Engineers, Reston VA

Martin A, Karanika-Murray M, Biron C, Sanderson K (2016) The psychosocial work environment, employee mental health and organizational interventions: improving research and practice by taking a multilevel approach. Stress Health 32(3):201–215

Mathee A, Oba J, Rose A (2010) Climate change impacts on working people (the HOTHAPS initiative): findings of the South African pilot study. Glob Health Action 3(1):5612

Meehl GA, Tebaldi C (2004) More intense, more frequent, and longer lasting heat waves in the 21st century. Science 305(5686):994–997

Meijer E, Kromhout H, Heederik D (2001) Respiratory effects of exposure to low levels of concrete dust containing crystalline silica. Am J Ind Med 40(2):133–140

Morioka I, Miyai N, Miyashita K (2006) Hot environment and health problems of outdoor workers at a construction site. Ind Health 44(3):474–480

Ødegaard F, Roos P (2014) Measuring the contribution of workers' health and psychosocial work-environment on production efficiency. Prod Oper Manag 23(12):2191–2208

Oliver LC, Miracle-McMahill H, Littman AB, Oakes JM, GaitaJr RR (2001) Respiratory symptoms and lung function in workers in heavy and highway construction: a cross-sectional study. Am J Ind Med 40(1):73–86

Oswald AJ, Proto E, Sgroi D (2015) Happiness and productivity. J Law Econ 33(4):789–822

Pandit B, Albert A, Patil Y, Al-Bayati AJ (2019) Impact of safety climate on hazard recognition and safety risk perception. Saf Sci 113:44–53

Park H, Park HD, Jang JK (2016) Exposure characteristics of construction painters to organic solvents. Saf Health Work 7(1):63–71

Park HR, Jeong GO, Lee SL, Kim JY, Kang SA, Park KY, Ryou HJ (2009) Workers intake too much salt from dishes of eating out and food service cafeterias: direct chemical analysis of sodium content. Nurs Res Pract 3(4):328

Paschalidou AK, Kassomenos PA, Kelessis A (2016) Tracking the association between metro-railway construction works and PM levels in an urban Mediterranean environment. Sci Total Environ 568:1326–1332

Petitti DB, Harlan SL, Chowell-Puente G, Ruddell D (2013) Occupation and environmental heat-associated deaths in Maricopa County, Arizona: a case-control study. PLoS ONE 8(5):e62596

Sacks R, Rozenfeld O, Rosenfeld Y (2009) Spatial and temporal exposure to safety hazards in construction. J Constr Eng Manag 135(8):726–736

Sanyal S, Rochereau T, Maesano CN, Com-Ruelle L, Annesi-Maesano I (2018) Long-term effect of outdoor air pollution on mortality and morbidity: a 12-year follow-up study for metropolitan France. Int J Environ Res Public Health 15(11):2487

Seixas NS, Goldman B, Sheppard L, Neitzel R, Norton S, Kujawa SG (2005) Prospective noise induced changes to hearing among construction industry apprentices. Occup Environ Med 62(5):309–317

Spielholz P, Davis G, Griffith J (2006) Physical risk factors and controls for musculoskeletal disorders in construction trades. J Constr Eng Manag 132(10):1059–1068

Thomson DS, Carter K, Grant F (2017) Unpacking cohort social ties: the appropriateness of perceived social capital to graduate early career performance in construction project teams. Int J Constr Educ Res 13(4):299–316

Timmerman JG, Heederik D, Spee T, Smit LA (2014) Skin symptoms in the construction industry: occurrence and determinants. Am J Ind Med 57(6):660–668

US Department of Health and Human Services (1986) National Institute for Occupational Safety and Health-NIOSH

Valero E, Sivanathan A, Bosché F, Abdel-Wahab M (2016) Musculoskeletal disorders in construction: a review and a novel system for activity tracking with body area network. Appl Ergon 54:120–130

Varghese BM, Barnett AG, Hansen AL, Bi P, Hanson-Easey S, Heyworth JS, Pisaniello DL (2019) The effects of ambient temperatures on the risk of work-related injuries and illnesses: evidence from Adelaide Australia 2003–2013. Environ Res 170:101–109

Wang D, Dai F, Ning X (2015) Risk assessment of work-related musculoskeletal disorders in construction: state-of-the-art review. J Constr Eng Manag 141(6):04015008

Xiang J, Bi P, Pisaniello D, Hansen A (2014) Health impacts of workplace heat exposure: an epidemiological review. Ind Health 52(2):91–101

Yan D, Zhang T, Su J, Zhao LL, Wang H, Fang XM, Yu LY (2018) Structural variation in the bacterial community associated with airborne particulate matter in Beijing China during hazy and non hazy days. Appl Environ Microbiol 84(9)

Yorio PL, Wachter JK (2014) The impact of human performance focused safety and health management practices on injury and illness rates: do size and industry matter? Saf Sci 62:157–167

Zhou Z, Goh YM, Li Q (2015) Overview and analysis of safety management studies in the construction industry. Saf Sci 72:337–350

Zuluaga CM, Namian M, Albert A (2016) Impact of training methods on hazard recognition and risk perception in construction. In: Construction research congress, pp 2861–2871

Chapter 20
The Benefits of Eco-efficient Plasters for Occupant's Health—A Case Study

Maria Idália Gomes, José Lima, Tânia Santos, João Gomes, and Paulina Faria

Abstract The health and comfort of building inhabitants are significantly affected by indoor air properties. Currently, there is sufficient scientific evidence associating discomfort and unpleasant indoor environment, reported by building occupants, with construction materials used inside those buildings. Hygienic and human-toxicological aspects need to be further studied in buildings to guarantee the existence of pleasant and comfortable built environments, but mainly healthy ones. Plasters, coating the surface of indoor walls and ceilings, can perform an important role on indoor conditions. In this chapter, the contribution of different plasters to the interior comfort, namely regarding the ability to regulate relative humidity by its hygroscopic capacity, is analyzed. The drying shrinkage, bulk density and mechanical performance are also compared to ensure that all the mortars can perform well when used as plasters. The analyzed plasters are made of earth, without and with low content of air lime and gypsum addition, as well as lime, gypsum and cement. It is shown that earth plasters have a more active effect on the hygrothermal balance when compared to air lime, gypsum and cement plasters, and that the addition of low binder content to earth plasters seems to be negative.

Keywords Adsorption · Air lime · Cement · Earth · Gypsum · Hygroscopic capacity · Mechanical performance · Plasters

---

M. I. Gomes (✉)
CERIS and Department Civil Engineering, ISEL, Polytechnic Institute of Lisbon, 1959-007 Lisbon, Portugal
e-mail: idalia.gomes@isel.pt

J. Lima
Faculty of Architecture, University of Lisbon, 1349-055 Lisbon, Portugal

T. Santos · P. Faria
CERIS and Department Civil Engineering, FCT, NOVA University of Lisbon, 2829-516 Caparica, Portugal

J. Gomes
CERENA and Department Chemical Engineering, ISEL, Polytechnic Institute of Lisbon, 1959-007 Lisbon, Portugal

J. A. Malik and S. Marathe (eds.), *Ecological and Health Effects of Building Materials*,
https://doi.org/10.1007/978-3-030-76073-1_20

20.1 Introduction

Throughout history, humanity has noticed that polluted air can be harmful to health and wellbeing indoors, due to the emissions from indoor sources, as well as too humid or too dry environments. The energy crisis of the 70's of last century resulted in more airtight buildings, having low air intakes, thus increasing occupants' exposure to pollutants and lowering the quality of the indoor air. The Sick Building Syndrome (SBS) identified by the World Health Organization (WHO 1999), in the 70s of the past century is characterized by several symptoms and diseases due to inadequate ventilation. Low indoor air quality (IAQ) is recognized as one of the worst threats to human health, as humans spend around 80–90% of their lifetime inside buildings, and the indoor air can be more contaminated than outdoors (WHO 2010; EEA 2013; Al Horr et al. 2016). According to an environmental report issued by OCDE (2012), by 2030 indoor air pollution will be one of the major casualties caused by environmental issues, resulting in 2.3 million deaths. Although there are several sources of indoor pollutants, indoor air quality is also influenced from numerous factors that are bound to affect health and wellbeing of occupants, such as the building conservation state, ventilation air renovation ratio, temperature and relative humidity, pollutant emission rate, existence of indoor sources, maintenance and cleaning, outdoor air quality, number of occupants and their activities indoors. As referred, air temperature clearly affects indoor air quality: if occupants are subjected to conditions such as hot/dry as well as cold/humid, these conditions are not favorable to the human respiratory system and could induce lung infections, as well health problems in occupants having asthma. The excess heat also affects, negatively, persons with health conditions such as cardiovascular disease, diabetes, Alzheimer and epilepsy (Ormandy and Ezratty 2012). Low humidity promotes dermatological conditions, namely dry skin and associated diseases, dryness of eyes and nose, and vocal problems. These negative effects are also increased by high temperature with high humidity, which are also favorable to microbial growth, such as fungi and bacteria (Reinikainen and Jaakkola 2003; CCRSA 2008; FPP 2015). Microbial growth is also responsible for the emission of spores, cells, microparticles and volatile organic compounds (VOCs), which negatively affect indoor air quality (APA 2010; US-EPA 2010). VOCs emitted from microbial agents also result in malodorous environments. International standard ISO 7730 (2005) defined the optimal range for relative humidity of 30–60%. However, relative humidity lower than 50% inhibits growth of fungi, mites and bacteria (Pegas et al. 2011). Therefore, the range of 30–50% of relative humidity, usually recommended for housing (US-EPA 2010).

Nevertheless, according to WHO (2009), fungi do not appear for relative humidity lower than 75%, if the air temperature is within the range 5–40 °C, whereas the development of mites requires a relative humidity in the range 45–50% (WHO 2009).

It is also important to consider that a high relative humidity also influences the chemical degradation of the materials, contributing to increase the degradation of indoor air quality, and these concentrations can still increase with inadequate ventilation (Bornehag et al. 2005). Furthermore, high relative humidity can contribute to decreasing mechanical performance of building products and elements.

Hygienic and human-toxicological aspects are currently being studied in the built environment, to guarantee the health and comfort of the indoor environments.

Plasters, coating the surface of indoor walls and ceilings, can perform an important role in indoor conditions. When plasters have a hygroscopic capacity, i.e., ability to absorb and release the moisture, they can make an active contribution in the regulation of the RH of the indoor environments where they are applied. It has been shown that earth plasters have a more active effect on the hygrothermal balance when compared to other plasters.

The interest in earth plastering mortars in the scientific community has been growing. Some studies were carried out about the behavior of earth plasters with addition of binders, such as lime and gypsum, and other types of additions, such as geopolymers and enzymes (Rescic et al. 2021).

According to the authors' knowledge, there are few studies analyzing the mechanical and hygroscopic characteristics of earth-based plasters compared to current binder plasters. The same for the ability to regulate IAQ by plasters' contribution to a healthy indoor environment.

To prove this statement, different types of plaster will be analyzed, consisting of earthen plasters, without and with low content of air lime and gypsum addition, as well as lime, gypsum and cement-based plasters.

The present study intends to analyze the contribution of each plaster to indoor comfort, namely regarding the ability to regulate RH through its hygroscopic capacity, promoting the comfort of building occupants. In addition, the drying shrinkage, bulk density and mechanical strengths will also be compared to ensure that the mortars have mechanical performance adequate for plastering.

20.2 Materials, Mortars Composition and Test Methods

20.2.1 Materials and Mortars Composition

In this study, eleven different mortars were analyzed, consisting of: earth plasters ($E_{1\_f}S$; $E_{1\_m}S$; $E_{2\_f1c}S$; $E_{3\_f}S\_Fib$); earthen plasters with the addition of low contents of air lime ($E_{1\_m}S + CL$; $E_{4\_c}S + CL_p$); earthen plaster with the addition of a low content of gypsum ($E_{1\_m}S + G$); air lime plaster ($CL\__{f}S$); gypsum plasters ($G\__{f}S$ and G_m); and cement plaster (C_m).

The mortar composition and fresh state characterization are presented in Table 20.1. The reddish clayish earth (E_1) is from Algarve, the southernmost region of Portugal; it was studied (Lima and Faria 2016; Lima et al. 2016). The remaining clayish earths (E_2, E_3 and E_4) were characterized by Santos et al. (2020), wherein: E_2 is a clayish earth, also reddish, from the center of Portugal; E_3 is a pre-mixed earth

Table 20.1 Mortars composition and fresh state characterization

Mortar's name	Volumetric proportions									Weight proportions		Wet density[c] (kg/m^3)	Consistency[d] (mm)
	E (%)	$_f$S (%)	$_m$S (%)	$_c$S (%)	CL (%)	G (%)	C (%)	Volumetric ratio	Water[a] (%)	Mass ratio	Water[b] (%)		
E$_1$_$_f$S	25	75	–	–	–	–	–	1:3 (E$_1$:$_f$S)	24.6	1:3.42 (E$_1$:$_f$S)	16.9	2018.8 ± 9.0	173.2 ± 1.2
E$_1$_$_m$S	25	–	75	–	–	–	–	1:3 (E$_1$:$_m$S)	19.6	1:3.63 (E$_1$:$_m$S)	12.8	2130.7 ± 0.0	173.2 ± 0.0
E$_2$_f1$_c$S	18.2	54.5	–	27.3	–	–	–	1:3:1.5 (E$_2$:$_{f1}$S:$_c$S)	11.7	1:3.04:1.78 (E$_2$:$_f$S:$_c$S)	9.5	1555.9 ± 56.7	136.0 ± 19.0
E$_3$_$_f$S_F$_{ib}$	Pre-mixed earth product							nk	20	nk	14.7	2033.3 ± 19.0	125.0 ± 8.0
E$_1$_$_m$S + CL	25	–	75	–	5.0[(e)]	–	–	1:3 + 5%CL (E$_1$:$_m$S + 5%CL)	24.6	1:19.43:70.45 (CL:E$_1$:$_m$S)	15.9	2052.1 ± 0.0	176.0 ± 0.0
E$_4$_$_c$S + CLp	Pre-mixed earth product							nk	–	nk	20	1988.8 ± 20.3	153.0 ± 1.0
CL_$_f$S	–	75	–	–	25	–	–	1:3 (CL:$_f$S)	25.4	1:12.88 (CL:$_f$S)	20.9	1958.4 ± 10.3	169.6 ± 3.4
E$_1$_$_m$S + G	25	–	75	–	–	5.0[(e)]	–	1:3 + 5%G (E$_1$:$_m$S + 5%G)	20.4	1:10.08:36.54 (G:E$_1$:$_m$S)	13.1	2073.9 ± 8.9	168.8 ± 4.8
G_$_f$S	–	75	–	–	–	25	–	1:3 (G:$_f$S)	24	1:6.90 (G:$_f$S)	18.6	1936.3 ± 3.0	175.9 ± 3.5
Gm	Pre-mixed mortar, produced using Gp product							nk	nk	nk	43	1578.9 ± 50.3	–

(continued)

Table 20.1 (continued)

Mortar's name	Volumetric proportions									Weight proportions		Wet density[c] (kg/m^3)	Consistency[d] (mm)
	E (%)	$_f$S (%)	$_m$S (%)	$_c$S (%)	CL (%)	G (%)	C (%)	Volumetric ratio	Water[a] (%)	Mass ratio	Water[b] (%)		
Cm	Pre-mixed mortar, produced using Cp product							nk	22	nk	14	1896.4 ± 3.5	138.0 ± 14.0

Notation: E_1—Illitic clayish earth; E_2 and E_3—reddish clayish earth; E_4—yellow clayish earth; $_f$S—fine sand (washed); $_m$S—medium sand (not washed); $_c$S—coarse sand (washed); CL—Calcitic air lime (powder); CLp—calcitic air lime (putty); G—Gypsum; Gp—pre-mixed product; Cp—pre-mixed product; nk—not known

[a]Volume of water added, considering the total volume of mortar dry components without additions

[b]Mass of water added, considering the total mass of mortar dry components including additions

[c]Fresh state density

[d]Flow table consistency

[e]Volume of binder added, considering the total volume of mortar dry components;

mortar product from Embarro company, formulated with a reddish clayish earth from the region of E_1, sand and vegetable fibers, and E_4 is a yellowish clayish earth supplied by Sorgila company, located in the center of Portugal. Illite is the predominant mineral in all clay soils (E_1, E_2, E_3 and E_4), but E_2 also contains kaolinite minerals.

The sands used are all siliceous, namely fine sand ($_fS$ and $_{f1}S$), medium sand ($_mS$)—used unwashed—and coarse sand ($_cS$). The hydrated calcitic air lime (CL) used, is a commercial product designated as "H100" from Musical company (Lhoist group), distributed in a form of powder. CLp is a calcitic air lime hydrated with abundant water and used as a putty, provided by a builder (Aldeias de Pedra company). The hemi hydrated gypsum (G) designated "Gesso Estuque" and the pre-mixed product (G_p) designated "PROJECT 2010" are commercial products from Sival company, both supplied in powder. The mortar C_p is a commercial pre-mixed product provided by SecilArgamassas company, designated "RHP Manual Interior", provided in powder.

The distribution curves for the raw materials particle size were characterized by EN 1015-1 (1998); these results are available in Santos et al. (2020) and Lima et al. (2016). The gypsum-based product (G_p) was characterized regarding the distribution of the particle size nor the product date sheet presents that information. According to the producer, the particle size of the cement-based product (C_p) is lower than 1.2 mm, according to EN 1015-1 (1998).

Table 20.2 presents the loose bulk density for the raw materials that are part of the mortar´s formulation. They were determined in accordance with EN 1097-3 (2002), considering the average of three samples for each material.

20.2.2 Mortars Characterization and Preparation, Characterization in Raw and Specimen Obtention

Eleven studied mortars are formulated as explained below:

- $E_1\_{_f}S$ was formulated considering a volume ratio of 1:3, respectively of clayish earth E_1 and fine sand ($_fS$)
- $E_1\_{_m}S$ was formulated considering a volume ratio of 1:3, respectively of clayish earth E_1 and medium sand ($_mS$)
- $E_2\_{_{f1c}}S$ was formulated considering a volume ratio of 1:3:1.5, respectively of clayish earth E_2, fine sand ($_{f1}S$) and coarse sand ($_cS$)
- $E_3\_{_f}S\_F_{ib}$ was a pre-mixed product of earth materials, composed of the clay earth E_3, fine sand ($_fS$) and cut straw fibers (proportions for each constituent are unknown)
- $E_1\_{_m}S + CL$ and $E_1\_{_m}S + G$ have the same base formulation of $E_1\_{_m}S$ but with 5% volumetric addition of hydrated calcitic air lime (CL) and hemi hydrated gypsum (G), respectively

Table 20.2 Materials, pre-mixed products loose bulk density (kg/m^3)

E_1	E_2	E_3	E_4	${}_{f}S$	${}_{f1}S$	${}_{m}S$	${}_{c}S$	CL	LP	G	Gp	Cp
1317.0 ± 1.8	1359.8 ± 8.1	1396.7 ± 5.5	1201.5 ± 4.2	1500.0 ± 2.0	1384.2 ± 4.3	1591.8 ± 0.6	1606.2 ± 4.9	350.7 ± 1.7	1374.6 ± 19.4	652.1 ± 4.7	810.9 ± 11.0	1502.0 ± 1.2

- $E_{4\_c}S + CL_p$ has in its composition the clayish earth E_4, coarse sand ($_cS$), limestone powder (LP) and some addition of hydrated calciticair lime putty (CL_P); however, the content of each constituent are unknown as the mortar was delivered already pre-mixed
- $CL\_{_f}S$ and $G\_{_f}S$ were formulated considering a volume ratio of 1:3, respectively of the corresponding binder, hydrated calcitic air lime (CL) or hemi hydrated gypsum (G), and fine sand ($_fS$)
- Gm and Cm mortars were prepared based on the pre-mixed products G_p and C_p, respectively, requiring only water to mix; the proportions of constituents and other components are not known.

To produce the mortars containing the clayish earth E_1 the German standard DIN 18947 (2018) for earth plasters was followed, except the preparation of the mortar with E_1 and G addition which followed the EN 1015-2 (1998). In all these mortars the volume of water added to the mortars' mixture was the minimum required to achieve a flow table consistency in a range defined by the DIN 18947 (2018) and to ensure adequate workability.

All other mortars preparation tried to mimic, as much as possible, the procedure usually executed on construction sites. The mortars were prepared by means of a mixer blade system. First, the dry materials were put in a vase and water was added in order to obtain a first mixture, during a period of about 8 min. Then, the mortar adhering to the walls of the vase was removed and put together with the remaining mortar, and a second mixture was carried out for 3 min. To achieve a better connection between earth and the hydrated calciticair lime putty the $E_{4\_c}S + CLp$ mortar was prepared one day prior to the production of the specimens. All other mortars were prepared on the same day of the molding of specimens. Pre-mixed mortars, $E_{3\_f}S\_F_{ib}$, $E_{4\_c}S + CL_p$ and C_m, were prepared only by addition of the water content indicated by their producer. It should be noticed that Gm is made with a pre-mixed material (G_p product), and the water content was not indicated by the manufacturer. Therefore, the water content of Gm mortar was, then, defined by an experienced craftsman in order to assure an acceptable workability. The same procedure was followed for the preparation of mortar $E_{2\_f1c}S$.

The obtained mortars were characterized as raw (Table 20.1) in terms of wet density, according to standard EN 1015-6 (1998) and flow table consistency according to standard EN 1015-3 (1999). Table 20.1 allows to see that flow table consistency varies significantly between the different binder mortars, what can thus influence the hardened state properties of the mortars and plasters, and particularly its porous structure, also depending on their type of curing: just drying for the earth plasters, by carbonation for the plasters with air lime, hydrating for the hydraulic plasters.

Each plastering mortar resulted in different types of specimens which were produced in metallic molds:

- for carrying out tests of bulk density, linear shrinkage, elasticity dynamic modulus and flexural and compressive strength, 6 prismatic specimens having dimensions 40 mm × 40 mm × 160 mm, molded as two layers mechanically compacted at 20 strokes/layer and leveled manually and were finally demolded when dried, after 7 days, at least.

- for carrying out adsorption and desorption tests, 3 planar specimens having dimensions 200 mm × 500 mm × 15 mm were compacted manually and leveled; the metallic mold ensures that adsorption/desorption cycle occurs only in the top exposed surface; in the Gm planar specimens a gypsum finishing coat designated "Massa de acabamento", provided by Sival, having a thickness of 1 mm, was applied; this application was done 24 h after these mortars were applied in the planar specimens.

Specimens were placed in environmental conditions of 20 ± 2 °C and 65 ± 5% RH prior to the execution of the characterization tests.

The specimens' age of each plastering mortar for all the tests and the type of specimen is shown in Table 20.3. The aging process for prismatic specimens of mortar CL_fS included a period of accelerated carbonation comprising 30 days in a CO_2 rich confined environment. The specimens' carbonation was confirmed through a phenolphthalein test, carried out on the fracture surface of the specimens, immediately after the flexural strength test. The specimens of mortar $E1_{\_m}S + CL$ were tested after an aging of 1.5 months. The specimens' carbonation was not confirmed after flexural strength test due to the mortar's reddish color. A longer aging period of 4.5 months was allowed for the specimens of mortar $E_{4\_c}S + CLp$ to compensate for the lime's slow carbonation reaction.

Table 20.3 Number, age and type of specimens for each performed test

Test	Testing age (months)				Minimum number of specimens	Type of specimen
	$E_{1\_f}S$ $E_{1\_m}S$ $E_{1\_m}S + CL$ $E_{1\_m}S + G$ $G_{\_f}S$	$CL_{\_f}S$	$E_{2\_flc}S$ $E_{3\_f}S\_F_{ib}$ Gm Cm	$E_{4\_c}S + CL_p$		
Linear shrinkage	When demolding				6	Prismatic
Bulk density, dynamic modulus of elasticity, flexural and compressive strength	1.5	2	2	4.5	6	Prismatic
Sorption and desorption	1.5	6.5	4	6.5	3	Planar on metallic mold

20.2.3 *Test Methods*

Mortars were assessed in terms of linear drying shrinkage, dry bulk density, dynamic modulus of elasticity, flexural and compressive strength and water vapor adsorption and desorption capacity. First tests characterize the mortars' physical and mechanical performance while the adsorption/desorption test allows to assess the contribution of plasters to balance indoor RH. Thus, there is a clear distinction between "mortar" and "plaster", depending on the characterization test performed.

20.2.3.1 Linear Shrinkage, Bulk Density, Dynamic Modulus of Elasticity and Flexural and Compressive Strength

Linear shrinkage was determined on the prismatic specimens, in accordance with DIN 18947 (2018), by measuring the difference of the linear geometrical length of mortar between the raw and hardened state. The dry bulk density was determined geometrically in accordance with standard EN 1015-10/A1 (1999), by measuring the ratio between the dry mass and the volume of each specimen. The dynamic elasticity modulus (Ed) was determined according to standard EN 14146 (1414), by using Zeus XRM equipment. Flexural (FStr) and compressive (CStr) strengths were determined according to standard EN 1015-11 (1999), by using a ZwickRoell Z050 equipment, using load cells of 2 kN and a velocity of 0.2 mm/min for flexural strength and 50 kN and a velocity of 0.7 mm/min for compressive strength. 6 halves of the prismatic specimens, after subjected to the FStr test, having about 80 mm long, were used in the determination of the CStr for each mortar, having a compressive area of 40 mm × 40 mm, as specified by standard EN 1015-11 (1999).

20.2.3.2 Adsorption and Desorption

According to DIN 18947 (2018), absorption capacity was determined with some complements based on (Santos et al. 2020). In the climatic chamber at 23 °C and 50% relative humidity the specimens were placed until they reached constant mass (that is, mass less than 2%). The adsorption phase of the plasters it was achieved at the expense of an increase to 80% of relative humidity inside the climatic chamber, keeping the temperature at 23 °C. By weighing the specimens after 1, 3, 6, 12 and 24 h, the water vapor gained (in g/m^2) by the plasters was determined. The absorption test must end at 12 h by DIN 18947 (2018). Nonetheless, to better understand the adsorption effect on plasters, this test was extended to 24 h. The standard also reported a first weighing at 30 min, but this was not carried out, the first weighing was performed at 1 h after the test started. It was considered that if the first weighing was at 30 min it would result in a negative effect and destabilize the climatic chamber, since 30 min is considered a very short period of time to stabilize the climate chamber.

Although DIN 18947 (2018) just define the analysis of the adsorption capacity of plasters, the reverse process was used to determine the desorption capacity, since it is considered important to evaluate the water release (Schroeder 2018; Maddison et al. 2009; Veiga et al. 2010). To evaluate the decrease of water vapor content, relative humidity was decreased to 50%. This parameter was also measured in g/m^2, from 1 up to 24 h in the same time periods defined in the previous test.

20.3 Results and Discussion

20.3.1 Dry Bulk Density and Linear Shrinkage

The dry bulk density of each mortar is presented in Fig. 20.1, and is related to the loose bulk density of the materials that comprise its composition (Table 20.2). Loose bulk density is equivalent in all raw materials and pre-mixed mortar products, exception made too hydrated calcitic air lime (CL), hemi hydrated gypsum (G) and gypsum-based pre-mixed product (Gp); these materials exhibit low loose bulk density and, consequently, $CL\_{}_{f}S$, $G\_{}_{f}S$ and Gm mortars, present low dry bulk density. However, although Table 20.1 shows that flow table consistency varies significantly between the different binder mortars, the differences in terms of dry bulk density are not very high, considering the different types of hardening the mortars have.

Observing in Fig. 20.1 the dry bulk density, and taking into account DIN 18947 (2018), the mortars $E_{1}\_{}_{m}S$ and $E_{3}\_{}_{f}S\_F_{ib}$ fall within bulk density class 2.0 (from 1.81 kg/dm^3 till 2.00 kg/dm^3), and $E_{1}\_{}_{f}S$, $E_{2}\_{}_{flc}S$ can be classified as class 1.8

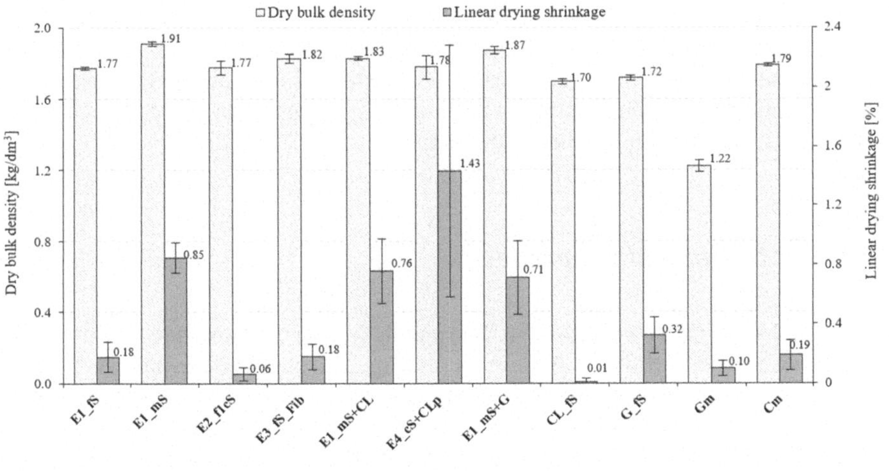

Fig. 20.1 Dry bulk density and linear drying shrinkage

(which is considered between 1.61 and 1.80 kg/dm^3); these classification are only applicable for unstabilized earth plasters.

All the remaining earth mortars that are stabilized with lime or gypsum, namely E_1_mS + CL, E_4_cS + CLp, E_1_mS + G and the cement mortar (Cm)exhibit high values for dry bulk density (between 1.78 and 1.87 kg/dm^3).

Regarding linear drying shrinkage, all mortars presented in this study show significantly low linear drying shrinkage (Fig. 20.1). Unstabilized earth mortars, E_1_fS, E_1_mS, E_2_flcS, E_3_fS_F_{ib}, exhibit very low linear shrinkage.

According to Röhlen and Ziegert (2011), earth mortars present a dry bulk density around 1400–1800 kg/m^3. All earth-based mortars analyzed in the present study have a dry bulk density in this range of values. In addition, the mortars with sand with different particle size distribution and with the addition of fibers and air lime show higher results. The high dry bulk density can be justified by a better packaging obtained with the different grains of sand and the air lime; the influence of the fibers cannot be compared because of the use of a different type of earth.

The low linear shrinkage found, may be due to the low swelling characteristic of the clay mineral that is present in the earths used, namely illite (which is present in E_1_fS, E_1_mS, E_2_flcS, E_3_fS_Fib) and kaolinite (also presented in E_2_FlcS). The mortars E_1_mS + CL and E_4_cS + CLp, that are stabilized with lime, present the highest value for linear shrinkage (excluding the mortar E_1_mS). The observed result may be due to the use of air lime (powder and putty), known by shrinking during carbonation. It is important to highlight the high standard deviation value in E_4_cS + CLp mortar, leading to a maximum value of 2.3% for linear shrinkage.

However, the earth mortars exhibit linear shrinkage less than 3%, taking into account DIN 18947 (2018) and the NZS (1998). Röhlen and Ziegert (2011) also refer that shrinkage in earth mortars should not be more than 2%. The average values of all mortars were found to be within the requirement (Fig. 20.1). For cement mortars, Röhlen and Ziegert (2011) refer that shrinkage can be 0.09%. In the present study, the cement-based pre-mixed mortar presents linear shrinkage of 0.19%, showing that the range can vary with the mortar formulation, which is unknown in the case of the pre-mixed mortar. The mortars based on lime (CL_fS), gypsum (G_fS and G_m) and cement (C_m) also exhibit low linear shrinkage.

20.3.2 Mechanical Properties

Figure 20.2 shows the compressive (CStr) and flexural (FStr) strengths and dynamic modulus of elasticity (Ed) for all mortars (average and standard deviation).

Observing flexural and compressive strength and after analyzing the DIN 18947 (2018) standard, the minimum mechanical strength values defined in the class S-I (CStr $\geq$ 1.0 N/mm^2 and FStr $\geq$ 0.3 N/mm^2), for all the earth mortars (with and without the addition of binder), have not been reached; the exception was the mortar E_1_mS + G that achieved both conditions (CStr and FStr), and E_2_flcS, that reached class S-I in CStr. Nevertheless, it is possible to notice that two other mortars, E_3_fS_F_{ib}

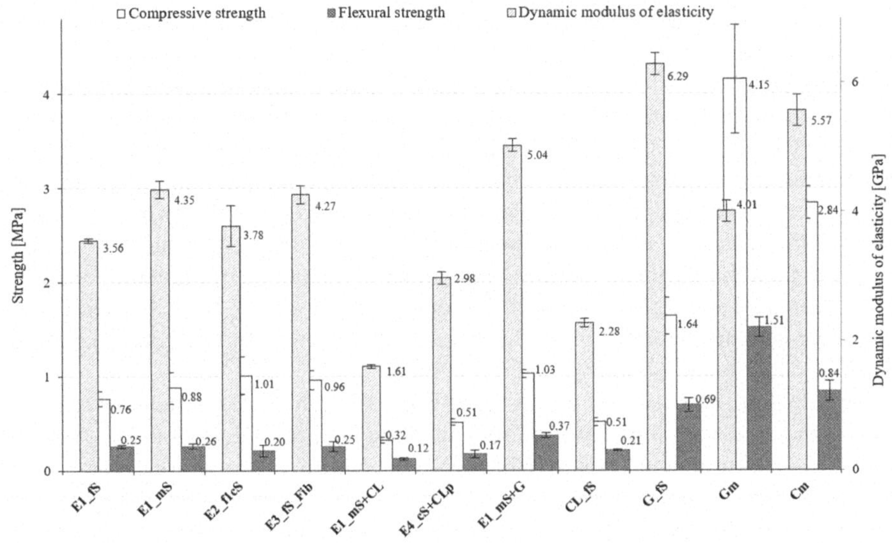

Fig. 20.2 Compressive and flexural strength and dynamic modulus of elasticity

and $E_{1\_m}S$, almost achieved the minimum values of mechanical resistance, CStr and FStr, defined for class S-I; and $E_{1\_f}S$ almost reached the minimum value for FStr.

Different classes for compressive strength at 28 days, for plastering mortars, are defined in the standard EN 998-1 (2010). These are: CS I values between 0.4 and 2.5 N/mm^2; CS II values between 1.5 and 5.0 N/mm^2; CS III values between 3.5 and 7.5 N/mm^2 and CS IV for values $\geq$6 N/mm^2. All earth mortars analyzed (with and without binder) are in the CSI class, except for $E_{1\_m}S + CL$ (0.32 N/mm^2). This earth mortar where powder hydrated air lime was added, did not reach the minimum limit to be classified. $G\_{}_fS$, Cm and Gm fit within the CS II class. According to this standard (EN 998-1 2010), these studied mortars meet the requirements for interior and exterior plasters and renders, respectively. However, it should be noted that these were not tested for 28 days. Röhlen and Ziegert (2011) refer that earth plasters must be made with earth mortars that have a compressive strength classified as CS II, namely values between 1 and 3 N/mm^2 for earth mortars ($E_{2\_f1c}S$ and $E_{1\_m}S + G$ fall into this condition) and 3 N/mm^2 for gypsum plaster mortars (just Gm fits in this condition).

EN 13279-1 (2008) defines a range of values for gypsum plasters, namely for CStr of 2–6 N/mm^2 and aFStr of 1–2 N/mm^2. Is possible to observe that Gm mortar is in these limits but $G\_{}_fS$ stays outside this range. The mortars that stand out from the others by the higher values in mechanical resistance are Gm, Cm and $G\_{}_fS$, both in CStr and FStr strengths, what could be expected in comparison to all the other non-hydraulic mortars.

According to Schroeder (2018), the compressive strength (according to EN 1015-11 (1999)) of earth mortars applied in secondary spaces must be greater than 0.5 MPa.

In the present study, all the earth mortars supersede the defined limit, exception made to $E_{1\_m}S + CL$ (Fig. 20.2).

Note should be made that the lowest values in mechanical resistance, CStr and FStr, are found in the earth mortars with lime addition and in the lime mortar, namely $E_{1\_m}S + CL$, $E_{4\_c}S + CLp$, $CL_{\_f}S$. The later presents the highest mechanical resistance in this mortar group and, as previously mentioned, its specimens' carbonation was confirmed through phenolphthalein test. Among the two earth mortars with lime addition the lowest mechanical resistance (CStr and FStr) is presented by mortar $E_{1\_m}S + CL$. This might be related to the shorter aging period of these mortar specimens or related to a higher lime addition content on mortar $E_{4\_c}S + CLp$. However, once the constituent proportions of the latter are unknown no further conclusions are possible.

Nevertheless, is possible to assess the effect over mechanical resistance (CStr and FStr) of the small addition of lime present in mortar $E_{1\_m}S + CL$ by comparison with mortar $E_{1\_m}S$, formulated exactly with the same constituent proportions (apart from the lime addition). In this comparison is clear that the small amount of lime addition reduced the mechanical resistance of the earth mortar by more than half. These results are not likely to be justifiable only by the shorter aging period of mortar $E_{1\_m}S +$ CL, furthermore, considering the low mechanical resistance presented by the lime mortar $CL_{\_f}S$, formulated with the same sand and volumetric proportions of the earth mortar $E_{1\_m}S$, and having its carbonation confirmed through phenolphthalein test, as previously mentioned. Therefore, a limited statement can be made pointing that the low addition of air lime tends to significantly decrease the mechanical strength of earth mortars.

Similar results were obtained by Santos et al. (2017) and Gomes et al. (2018) for contents from 5 to 15%: after 60 days, the addition of 5% of air lime in an illitic earth mortar decreased its mechanical strengths (Santos et al. 2017); after 90 days, the addition of 5–15% of air lime in a kaolinitic earth mortar, Gomes et al., (2018) obtained similar results. This should be justified by the weakness of the structure defined by the lime; interrupting the bonds between the clayish lamellas, in turn make the overall structure weaker in comparison with earth mortars without lime addition. However, considering that the carbonation of the lime may not be complete, some increase of strength can yet occur with aging.

Houben and Guillaud (1994) reported that the compressive strength of earth-air lime mortars tends to increase with the age of the mortar, easily reaching values of 2–5 N/mm^2. They also refer that additions of 2–6% of lime tend to increase compressive strength and, for larger additions, this strength tends to fall. As previously mentioned, that behavior was not observed for mortar$E_{1_m}S$ + CL with the addition of 5% of lime (CL) in volume (Table 20.1) which presented a CS of 0.32 N/mm^2 after an aging of 42 days (Fig. 20.2), value very low compared to those reported by the researchers. Also, the pre-mixed mortar $E_{4\_c}S + CLp$, obtained a value of 0.51 N/mm^2 (Fig. 20.2) for CStr, the value is also considerably lower when compared to the indicated by the authors, although, in this mortar the percentage of air lime is unknown, so it is not possible to effectively assess the effect of the lime addition.

Regarding the dynamic modulus of elasticity in earth mortars Röhlen and Ziegert (2011) refer that values lie, typically, within the range of 450–3000 N/mm^2. In this case, just $E_{1\_m}S + CL$, $E_{4\_c}S + CLp$ are within these limits (Fig. 20.2), the remaining mortars present higher values. These results may be due to the clay type or the mortar formulation, leading to a higher deformability of the mortars analyzed.

Low values of Ed in mortars may be advantageous if these mortars are applied in substrates with low mechanical properties. The mechanical characteristics of the mortars must not exceed those of the substrate on which they are applied, to guarantee compatibility between the mortar and the substrate in the long-term. Otherwise, premature anomalies and detachment of the mortar may occur due to this lack of compatibility between the substrate and the mortar. Veiga et al. (2010) defined general requirements for the application of plastering mortars on old buildings; one of the characteristics presented was Ed values, in a range between 2000 and 5000 N/mm^2. Observing these ranges, $E_{1\_m}S + G$ is at the upper limit and $E_{1\_m}S + CL$ does not fit in the limits presented by the authors (Fig. 20.2). All the remaining earth mortars are within the limits shown.

20.3.3 Hygroscopic Properties

Considering DIN 18947(2018), all earth plasters without the addition of binder, $E_{1\_f}S$, $E_{1\_m}S$, $E_{2\_flc}S$, $E_{3\_f}S\_F_{ib}$ (Table 20.3), obtained adsorption values above the lower limit of the water adsorption class WS-III (adsorption water vapor adsorption greater than 60.0 g/m^2 after 12 h at a temperature of 23 °C and a relative humidity of 80%), the higher class defined in the standard. However, and although the classes are only for unstabilized plasters, $E_{1\_m}S + CL$ and $E_{1\_m}S + G$ plasters were very close to reach the WS-III class, with values of 57.7 g/m^2 and 56.3 g/m^2, respectively. Considering the standard, they meet the class WS-II with a water vapor ≥47.5 g/m^2.

The higher adsorption observed in the earth plasters group in comparison with the binder-based plasters group is clearly associated with the presence of the clayey material in the mortar's formulation as well as its mineralogy. Among the three main clay groups with significant available in the nature to allow being consider as plastering materials, namely, montmorillonitic, illitic and kaolinitic clays, the first is characterized by having very high hygroscopicity and shrinkage, while the latter is known for its low hygroscopicity and shrinkage. In turn, the illitic clays, the one prevalent in the earth mortars assessed in this study, present an average condition of hygroscopicity and shrinkage (Lima et al. 2020).

Comparing the adsorption among the four unstabilized earth plasters it is possible to conclude that plaster $E_{2\_flc}S$, whose prevalent clay minerals are illite and kaolinite, presents the lowest adsorption. This result can be associated with the lower hygroscopicity of kaolinite minerals, as well as with the lower clayey material content of this mortar formulation (Table 20.1).

The higher water vapor adsorption capacity of unstabilized earth plasters is even more evident when looking at the behavior of mortar $E_{3\_f}S\_F_{ib}$: after 12 h, it reached a value higher than 78.3 g/m^2 of water vapor adsorption (31% higher than the 60 g/m^2 limit defined in class WS-III of the DIN 18947 (2018)). The behavior of this mortar becomes more surprising at 24 h, when water vapor adsorption is 103.9 g/m^2 with a tendency to continue the increase; therefore, this mortar could still increase its adsorption capacity for a significantly longer period of time. As this mortar contains fibers in its formulation, it may be necessary to have some attention, since the fibers provide an increase in adsorbed water (Ashour et al. 2011; Gomes et al. 2018) but can also provide an enhanced growth of mold. That will be most probable if the indoor environment is not well ventilated, or if it is in prolonged conditions of high relative humidity (Gomes et al. 2019).

The results obtained with the earth mortars are extremely important: the water vapor adsorption capacity allows the earth-based plasters to act as a moisture buffer, and thus positively contribute to balance the relative humidity of the indoor environment, reducing its peaks. Through this passive process the health and comfort of the buildings' occupants are promoted.

Surprisingly, $E_{4\_C}S + CLp$ plaster, despite being based on earth and with addition of lime putty lime, presents a weak adsorption; even after 24 h only absorbed about 24 g/m^2, value that does not reach the minimum classification by DIN 18947 (2018), that is class WS-I with a water vapor $\geq$35 g/m^2 after 12 h; nevertheless, the classification of this standard is applied only for unstabilized earth plasters, as previously mentioned. It is reported in the literature that the addition of air lime decreases hygroscopicity in earth plasters, as it appears that the airline network blocks the clay structure, creating a new structure between the clay lamellar structure and inhibiting the hygroscopic characteristics of the clay; the performance of this mixture becomes representative of the air lime and loses the dynamic behavior of the clay (Gomes et al. 2016; Santos et al. 2020). However, observing Fig. 20.3, it appears that $E1_{\_m}S + CL$ mortar, where lime was also added, but hydrated in powder form, presents quite high values for water adsorption; in this case the lime structure does not seem to interrupt the clay matrix's hygroscopic and dynamic behavior. The difference in behavior of these two plasters may be in the lime type (one was hydrated to become a powder and the other was hydrated with abundant water to turn into a putty) and particularly its content (the added percentage). However, the percentage of lime putty added in $E_{4\_C}S + CLp$ mortar is not known; therefore, it is impossible to be sure that the air lime content in $E_{4\_C}S + CLp$ mortar is higher than the one in $E1_{\_m}S + CL$ mortar. Another difference between these two plasters is the main material—the earth—as it may behave differently with lime.

The Gm, $G_{\_f}S$ and $CL_{\_f}S$ plasters present low adsorption capacity compared with the unstabilized earth plasters, being respectively 9 g/m^2, 12 g/m^2 and 21 g/m^2. However, the Cm plaster absorbed 39 g/m^2, higher than the $E_{4\_C}S + CLp$ plaster adsorption. This proves the good ability to regulate the relative humidity of the interior environments of earth mortars compared to mortars based on other types of binder mortars.

All plasters showed a good performance with respect to desorption since they desorbed almost the total water vapor they initially adsorbed. The mortars that, after 24 h desorption, still retained the highest values of moisture (although low) were

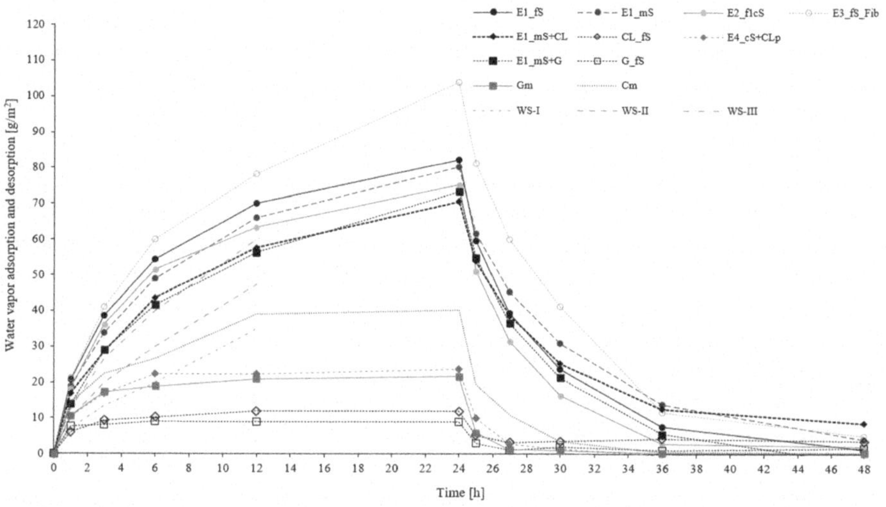

Fig. 20.3 Water vapor adsorption and desorption curves and class limits (WS-I, WS-II and WS-III) defined by DIN 18947 (2018)

$\mathrm{E1\_}_{m}\mathrm{S} + \mathrm{CL}$ and $\mathrm{E}_{3}\mathrm{\_}_{f}\mathrm{S\_F}_{ib}$, with 8.7 and 5.1 g/m^2 of water vapor, respectively. However, if the test was carried out for a longer period of time, the plasters would probably reach a desorption value similar to the initial values, as they show a downward trend. But they would take a longer period to achieve it. The hysteresis presented by these mortars may be related to the addition of air lime and fibers: the air lime may have blocked the mortar matrix and the fibers may retain water vapor inside, making the desorption more difficult. The $\mathrm{E}_{4}\mathrm{\_}_{C}\mathrm{S} + \mathrm{CLp}$, Cm and Gm plasters desorbed all the water vapor they had adsorbed for the same period: after performing the desorption, these plasters present a similar water vapor content, close to zero (similar to the one at 0 h).

The results observed in this study are quite similar to other studies carried out on adsorption and desorption in earth plasters (Minke 2006; Maddison et al. 2009; Maskell et al. 2018). Although slightly different values are observed in the results, the studies are unanimous in considering that earth plasters have a higher water vapor adsorption when compared with other plasters, namely air lime, gypsum and cement plaster.

The adsorption and desorption results obtained in the present study by the different plasters analyzed confirmed the capacity of clayey plasters to absorb and desorb water vapor more quickly and in greater quantities than other building products, such as lime, gypsum or cement plasters, as observed by other researchers (Minke 2006; Morton 2008). That shows that, although all the earths used for plasters can be different, with diverse types and different contents of clays, silts and sands, the results are now conclusive.

The capacity of earth plasters to absorb water vapor can strongly contribute to create healthy environments inside the buildings, through moderate regulation of the relative humidity of the indoor air, and it may even reduce the peak of high humidity that may be promoted by cooking food or water vapor from the baths or, in turn, the peak of low humidity associated with the continuous heating of buildings (Morton 2008).

By analysis of a psychrometric diagram (Moret-Rodrigues et al. 2009), it appears that the thermal comfort inside the buildings is favored by the regulation of relative humidity, since the temperature increases with the decrease in relative humidity. The perception of thermal comfort in indoor spaces can be improved through the regulation of relative humidity, taking into account that a high relative humidity leads to an increase of the thermal conductivity of the air and moderates the evaporation of the skin; consequently, it can increase the sensation of discomfort associated with the perception of cold or heat (Moret-Rodrigues et al. 2009).

The mineralogical, chemical and microstructural composition of the clayey materials may allow them to absorb pollutants from the indoor air, although this mechanism is not well known yet. According to Minke (2006) and Morton (2008), the clay soil can absorb and bind pollutants dissolved in water. This mechanism may be related to the hygroscopic capacity of mortars, due to the fact that pollutants may be dissolved in moisture and, in turn, be absorbed by the plaster as water vapor. Lamble et al. (2011) and Darling et al. (2012) also refer that earth plasters can contribute to improve the quality of the indoor air, since the clay can act as a passive removal material, decreasing the internal ozone concentrations and, therefore, reducing the likelihood of an ozone reaction with other building materials inside buildings. Earth plasters are known for not releasing toxins compounds to the indoor environment, at least if they have no additions in their formulation.

20.4 Conclusion

Currently, there is a growing concern with the environment and the indoor air quality in buildings. Around the world, researchers are looking for eco-efficient building materials and products that are not harmful to human health. The building materials used inside buildings can influence positively or negatively the health and comfort of the inhabitants. Since plasters can have a significant interior area coating walls and ceilings, they should contribute to a healthy indoor environment.

In the present study eleven plasters based on different mortars were produced and samples were tested: two earth plasters with the same earth and added sand, just varying the particle size of the sand; two earthen plasters with the same base formulation as one of the previous but with 5% volumetric addition of hydrated calcitic air lime and hemi hydrated gypsum, respectively; a ready-mixed earthen plaster with another type of earth, coarse sand, limestone powder and addition of calcitic air lime putty (characteristics and proportions of each constituent are not known); an earth plaster with another earth and a higher content of added sand; a premixed earth mortar known for including cut straw fibers and having earth from the same region as the first ones (proportions of each constituent are not known);

powder hydrated air lime and gypsum hemi-hydrated mortars with the same volume ratio of binder and sand, respectively; two premixed gypsum and cement mortars, respectively(the proportions of constituents and other components are not known).

It is important that plasters perform well, not only to contribute to indoor comfort and health but also to be durable and visually interesting, namely without cracks. The results from the present study have shown that all the mortars evaluated have a low linear drying shrinkage, including the earthen mortars. Regarding the mechanical strength, the study discussed more deeply earthen mortars because they are considered low strength mortars. For the tested earth mortars, the results of flexural and compressive strength indicated that they can be suitable to use in indoor spaces. However, an improvement in mechanical performance of earth plasters may be achieved by a reinforcement placed in corners within the mortar layer. Furthermore, an increase in mechanical strength can be achieved through the formulation of mortars with different earths, with diverse types of clays and different contents of the fractions of clays, silts and sands and a good particle size distribution of added sand, which predictably will also lead to changes in terms of the hygroscopic capacity.

Contrary to what is reported in some of the literature, the stabilization of earth mortars with low contents of binders does not always contribute to increasing their mechanical performance. It seems that the addition of low contents of air lime decreases the mechanical strength of the earthen mortars. On the contrary, the addition of low contents of gypsum seems to increase the mechanical strength of the earthen mortars.

Concerning hygroscopicity, unstabilized earth plasters present the highest water vapor absorption capacity, while air lime and gypsum plasters have the lowest. Therefore, this study shows the ability of earth plasters to contribute to the regulation and balance of indoor air humidity; thus, earth plasters can influence in a positive way the healthiness of the indoor environment, when compared to other plasters, such as based in lime, gypsum or cement. Earth as a building material acts as buffer protection against significant variations in humidity, contributing to balance the relative humidity of the indoor environments in buildings, promoting comfort and health of occupants. This capacity in an earth plaster comes from the exchange of water vapor with air, releasing moisture when the air is drier and adsorbing it when the air is more humid. It is important to note that this capacity for hygrothermal rebalancing depends on factors such as the clay type of the earth with which the plaster is produced, the clay proportion in the plaster, and possible stabilization with mineral binder or other added products. Another aspect that may influence the adsorption and desorption capacity are finish systems that can be applied on some plasters, such as paints. Furthermore, the effect of consecutive cycles of water vapor adsorption and desorption should also be studied because they can lead to a decrease of plasters hygroscopic capacity. Finally, as a contribution to indoor air quality, not only the plasters cannot release pollutants but, if possible, they should be optimized in order to capture pollutants present indoor.

The main conclusions regarding the eleven mortars and regarding the tests carried out are as follows:

- as it was expected, earth mortars have lower mechanical resistance when compared to gypsum and cement mortars, still these are appropriate to use in indoor spaces; the addition of lime in an earth mortar, although low, did not seem to be mechanically advantageous, unlike gypsum (in low quantities) seems to increase mechanical strength of the earthen mortars
- earth plaster exhibit greatest advantages when compared to the others studied plasters, validating their ability to regulate the relative humidity inside the buildings; they present excellent capacity for adsorption and desorption; when the absorption test was interrupted, at 24 h, all earth plasters (with the exception of the earth plaster stabilized with lime), would continue to absorb water
- When applied on ceilings and walls, the high capacity of earth plasters to capture water vapor contributes to indoor comfort in buildings, leading to very positive consequences on indoor air quality, namely a healthier environment while, at the same time, contributes to energy savings as a way of ensuring comfort.

Therefore, being an area of interest and due to the importance for the indoor air quality field, more research is still needed, namely for earth plasters. That research is justified by the fact that earth is a natural building material and earth plasters present low CO_2 emissions in its manufacture and application, and low incorporated energy. All these factors are important for the sustainability of the building and the planet, as well as for the health of building users.

Acknowledgements This study was supported by Operational Programme Competitiveness and Internationalization (COMPETE 2020), by FEDER funds through the Regional Operational Programs of Lisbon and of Center (Lisbon 2020, Center 2020) and by national funds through the Portuguese Foundation for Science and Technology (FCT) in the scope of the project SAICT/23349/2016: INDEEd—Indoor air quality regulation through the usage of eco-efficient mortars. The authors acknowledge the FCT for the support given to CERIS, through the strategic projects UIDB/04625/2020 and for Tânia Santos PhD fellowship SFRH/BD/147428/2019. Acknowledgements are due to Aldeias de Pedra, Embarro, SecilArgamassas and Civil Companies for providing the materials and pre-mixed mortars.

References

Al Horr Y, Mohammed M, Katafygiotis M, Mazroei A, Kaushik A, Elsarrag E (2016) Impact of indoor environmental quality on occupant well-being and comfort: a review of the literature. Int J Sustain Built Environ 5:1–11. https://doi.org/10.1016/j.ijsbe.2016.03.006

APA (2010) Indoor air quality, a technical guide (in Portuguese). Agência Portuguesa doAmbiente, Amadora

Ashour T, Georg H, Wu W (2011) An experimental investigation on equilibrium moisture content of earth plaster with natural reinforcement fibres for straw bale buildings. Appl Therm Eng 31(2–3):293–303. https://doi.org/10.1016/j.applthermaleng.2010.09.009

Bornehag CG, Sundell J, Hägerhed-Engman L, Sigsgaard T (2005) Association between ventilation rates in 390 Swedish homes and allergic symptoms in children. Indoor Air 15(4):275–280. https://doi.org/10.1111/j.1600-0668.2005.00372.x

CCRSA (2008) Opinion on risk assessment on indoor air quality, an opinion produced by the SCHER (Scientific Committee on Health and Environmental Risks) of the European Commission
Darling EK, Cros CJ, Wargocki P (2012) Impacts of a clay plaster on indoor air quality assessed using chemical and sensory measurements. Build Environ 57:370–376. https://doi.org/10.1016/j.buildenv.2012.06.004
DIN 18947(2018) Earth plasters. Terms and definitions, requirements, test methods (in German). DIN, Berlin
EEA (2013) Indoor air quality, European Environment Agency. https://www.eea.europa.eu/signals/signals-2013/articles/indoor-air-quality. Accessed 5 Dec 2020
EN 1015-1 (1998) Methods of test for mortar for masonry. Part 1: Determination of particle size distribution (by sieve analysis), A1:2006. CEN, Brussels
EN 1015-10 (1999) Methods of test for mortars for masonry. Part 10: Determination of dry bulk density of hardened mortar, A1: 2006. CEN, Brussels
EN 1015-11 (1999) Methods of test for mortar for masonry. Part 11: Determination of flexural and compressive strength of hardened mortar, A1: 2006. CEN, Brussels
EN 1015-2 (1998) Methods of test for mortar for masonry. Part 2: Bulk sampling of mortars and preparation of test mortars. CEN, Brussels
EN 1015-3 (1999) Methods of test for mortars for masonry. Part 3: Determination of consistency of fresh mortars, A1:2004/A2:2006. CEN, Brussels
EN 1015-6 (1998) Methods of test for mortars for masonry. Part 6: Determination of bulk density of fresh mortars. CEN, Brussels
EN 1097-3 (2002) Tests for mechanical and physical properties of aggregates. Part 3: Determination of loose bulk density and voids. CEN, Brussels
EN 13279-1 (2008) Gypsum binders and gypsum plasters. Part 1: Definitions and requirements. CEN, Brussels
EN 14146 (2006)Natural stone test methods. Determination of the dynamic modulus of elasticity (by measuring the fundamental resonance frequency). CEN, Brussels
EN 998-1 (2010) Specification for mortar for masonry. Part 1: Rendering and plastering mortar. CEN, Brussels
FPP (2015) Environment and indoor air quality (in Portuguese). http://www.fundacaoportuguesadopulmao.org/ambiente_interior.html. Accessed 1 Sept 2015
Gomes MI, Faria P, Gonçalves TD (2019) Rammed earth walls repair by earth-based mortars: the adequacy to assess effectiveness. Constr Build Mater 205:213–231. https://doi.org/10.1016/j.conbuildmat.2019.01.222
Gomes MI, Gonçalves TD, Faria P (2016) Hydric behavior of earth materials and the effects of their stabilization with cement or lime: study on repair mortars for historical rammed earth structures. J Mater Civil Eng 28:04016041. https://doi.org/10.1061/(ASCE)MT.1943-5533.0001536
Gomes MI, Faria P, Gonçalves TD (2018) Earth-based mortars for repair and protection of rammed earth walls. Stabilization with mineral binders and fibers. J Clean Prod 172:2401–2414. https://doi.org/10.1016/j.jclepro.2017.11.170
Houben H, Guillaud H (1994) Earth construction: a comprehensive guide. Intermediate Technology Publications, London
ISO 7730 (2005) Ergonomics of the thermal environment. Analytical determination and interpretation of thermal comfort using calculation of the PMV and PPD indices and local thermal comfort criteria. ISO, Geneva
Lamble SP, Corsi RL, Morrison GC (2011) Ozone deposition velocities, reaction probabilities and product yields for green building materials. Atmos Environ 45(38):6965–6972. https://doi.org/10.1016/j.atmosenv.2011.09.025
Lima J, Faria P (2016) Eco-efficient earthen plasters: the influence of the addition of natural fibers. In: Fangueiro R, Rana S (ed) Natural fibres: advances in science and technology towards industrial applications. Springer, Netherlands, Dordrecht, pp 315–327. https://doi.org/10.1007/978-94-017-7515-1_24

Lima J, Faria P, Santos Silva A (2016) Earthen plasters based on illitic soils from Barrocal region of Algarve: contributions for building performance and sustainability. Key Eng Mat 678:64–77. https://doi.org/10.4028/www.scientific.net/KEM.678.64

Lima J, Faria P, Santos Silva A (2020) Earth plasters: the influence of clay mineralogy in the plasters' properties. Int J ArchitHerit 14(7):948–963. https://doi.org/10.1080/15583058.2020.1727064

Maddison M, Mauring T, Kirsimäe K, Mander Ü (2009) The humidity buffer capacity of clay-sand plaster filled with phytomass from treatment wetlands. Build Environ 44:1864–1868. https://doi.org/10.1016/j.buildenv.2008.12.008

Maskell D, Thomson A, Walker P, Lemke M (2018) Determination of optimal plaster thickness for moisture buffering of indoor air. Build Environ 130:143–150. https://doi.org/10.1016/j.buildenv.2017.11.045

Minke G (2006) Building with earth. Design and technology of a sustainable architecture. Birkhäuser—Publishers for Architecture, Berlin

Moret-Rodrigues A, Canha-Piedade A, Braga A (2009) Thermal of buildings (in Portuguese), 1st edn. Alfragide, Portugal

Morton T (2008) Earth masonry. Design and construction guidelines. IHS BRE Press, UK

NZS 4298 (1998) Materials and workmanship for earth buildings. NZS, New Zealand

OCDE (2012) OECD environmental outlook to 2050: the consequences of inaction. Organization for economic co-operation and development

Ormandy D, Ezratty V (2012) Health and thermal comfort: from WHO guidance to housing strategies. Energy Policy 49:116–121. https://doi.org/10.1016/j.enpol.2011.09.003

Pegas PN, Alves CA, Evtyugina MG, Nunes T, Cerqueira M, Franchi M, Pio CA, Almeida SM, Freitas MC (2011) Indoor air quality in elementary schools of Lisbon in Spring. Environ Geochem Health 33(5):455–468. https://doi.org/10.1007/s10653-010-9345-3

Reinikainen L, Jaakkola K (2003) Significance of humidity and temperature on skin and upper airway symptoms. Indoor Air 13(4):344–52. https://doi.org/10.1111/j.1600-0668.2003.00155.x

Rescic S, Mattone M, Fratini F, Luvidi L (2021) Earthen plasters stabilized through sustainable additives: an experimental campaign. Sustainability 13(3): 1090. https://doi.org/10.3390/su13031090

Röhlen U, Ziegert C (2011) Earth building practice—planning, design, building, 1st edn. Bauwerk, Beuth Verlag GmbH, Berlin

Santos T, Nunes L, Faria P (2017) Production of eco-efficient earth-based plasters: influence of composition on physical performance and bio-susceptibility. J Clean Prod 167:55–67. https://doi.org/10.1016/j.jclepro.2017.08.131

Santos T, Gomes MI, Santos Silva A, Ferraz E, Faria P (2020) Comparison of mineralogical, mechanical and hygroscopic characteristics of earthen, gypsum and cement-based plasters. Constr Build Mater 254: https://doi.org/10.1016/j.conbuildmat.2020.119222

Schroeder H (2018) The new DIN standards in earth building—the current situation in Germany. J Civil Eng Arch 12:113–120

US-EPA (2010) A brief guide to mold, moisture, and your home. United States Environmental Protection Agency

Veiga R, Fragata A, Velosa A, Magalhães A, Margalla G (2010) Lime-based mortars: viability for use as substitution renders in historical buildings. Int J Archit Herit 4:177–195. https://doi.org/10.1080/15583050902914678

WHO (1999) Strategic approaches to indoor air policy-making. World Health Organization, Copenhagen, Denmark

WHO (2009) WHO guidelines for indoor air quality: dampness and mould. World Health Organization, Copenhagen, Denmark

WHO (2010) WHO guidelines for indoor air quality: selected pollutants. World Health Organization, Copenhagen, Denmark

Chapter 21
RETRACTED CHAPTER: Occupational Health Problems of Construction Workers

Manoj Kumar Karnena, Madhavi Konni, and Vara Saritha

The Editor of Ecological and Health Effects of Building Materials has retracted this chapter. After publication, concerns were raised that the chapter shares a high degree of similarity with Chapter 4 in [1]. Additionally, the contents of Fig. 21.1 are identical to those of Fig. 4.1 in [1]. This chapter is therefore redundant.

Author Manoj Kumar Karnena does not agree to this retraction. Authors Madhavi Konni and Vara Saritha have not responded to any correspondence from the editor or publisher about this retraction.

[1] Lingard, H. and Turner, M. (2019). Construction Workers' Health. In Integrating Work Health and Safety into Construction Project Management (eds H. Lingard and R. Wakefield). https://doi.org/10.1002/9781119159933. ch4

M. K. Karnena (✉) · V. Saritha
Department of Environmental Science, GITAM Institute of Science, GITAM (Deemed to be) University, Visakhapatnam 530045, India

M. Konni
Department of Basic Science and Humanities, Dadi Institute of Engineering and Technology, Visakhapatnam, India

J. A. Malik and S. Marathe (eds.), *Ecological and Health Effects of Building Materials*,
https://doi.org/10.1007/978-3-030-76073-1_21

RETRACTED CHAPTER

RETRACTED CHAPTER

Chapter 22
Impact of Construction Material on Environment

Sukanya Mehra, Mandeep Singh, Geetika Sharma, Shiv Kumar, Navishi, and Pooja Chadha

Abstract Increasing utilization of crude materials by the building units result in the diminution of natural resources as well as raising the ecological shocks including CO_2 secretions all over the surroundings. Being prominently used, steel and concrete industries are dominating among the construction industries leading to symbolic damage to the environment. Thus this article determines various problems of hostile ecological affects due to the employment of the major building materials (steel and concrete) in the building industries. To diminish these ecological affects, there are two approaches which have been also focused in the present article such as reducing consumption and material selection to reduce impacts.

Keywords Concrete · Environment · Negative impact · Steel · Consumption reduction · Material selection

22.1 Introduction

The major environmental damages like tsunami, floods, wildfires, global warming, rise of sea level, ozone layer depletion which leads to increase in human illnesses and soil contamination are the result of rise in anthropogenic activities. The major part of these environmental problems has been contributed by the construction industries. Due to building and construction activities, three billion tons of raw materials have been consumed. The yearly construction output of the UK requires 170 million tons of primary material and products, 125 million of quarry products and 70 million tons of secondary recycled and reclaimed products. Around 6 million tons of energy is consumed and 23 million tons of CO_2 is released to manufacture and deliver these products. Steel and concrete are the prime materials which are used in construction. Steel (32 MJ/kg) and concrete (1.9 MJ/kg) require high embodied energy resulting in huge CO_2 emission. Steel generates a huge amount of carbon dioxide and along with this; the processing of construction material and the transport of these materials also leads to the release of tremendous amounts of CO_2. Because of the high usage of

S. Mehra (✉) · M. Singh · G. Sharma · S. Kumar · Navishi · P. Chadha
Department of Zoology, Guru Nanak Dev University, Amritsar, India

J. A. Malik and S. Marathe (eds.), *Ecological and Health Effects of Building Materials*,
https://doi.org/10.1007/978-3-030-76073-1_22

these construction materials, they overpower the impact of other sources. All around the world, because of increasing consumption of steel and concrete and embodied energy, the depletion of resources is reaching its peak. This chapter provides an insight about the environmental hazards which are faced due to uncontrolled usage of these construction materials. Though the main goal is to minimize these impacts but with the rise in usage of construction materials these goals do not look promising. The main goal is to minimize these hazardous and unfavourable environmental impacts. On the basis of this statement, the problem is devised to estimate the unfavourable environmental impacts which are caused as a result of consumption of steel and concrete and to define the important methods to mitigate these impacts. The unfavourable environmental impacts can be mitigated to some extent by minimizing the consumption of construction material or by minimizing the impacts caused by these materials.

22.2 Impact of Steel Industry on Environment

The increased global warming post-industrialization is a big problem and resulted by growing working industries viz iron and steel industry and different types of processes involved in production in these industries that consume either artificial or natural energy. The iron and steel industry are the world's most important and biggest manufacturing energy consuming industries. In both developing and developed nations, a plethora of manpower works in these industries and has the highest share in the world's economy (Cortez 2009; Vatani et al. 2016). The manufacturing of steel consumes high energy, because most of the manufacturing proceeds at high temperatures. The steel is used in a wide variety of products as it has a strong set of properties that makes it appropriate for plenty of applications at a cheap price. The steel is also used in the sectors such as mechanical engineering, construction, and ships building to everyday utilization things, and also in construction projects such as road, bridge or rail.

Globally, the iron and steel industry constitutes about 7% CO_2 emission of total CO_2 production by human activity, making it one of the leading industrial emitters of CO_2 (An et al. 2018). CO_2 is major greenhouse gas that is emitted in steel industry. From 2017–2018 each one ton of steel production accounts for on an average 1.83 tons of CO_2 emission (Sun et al. 2019). About 7–9% of CO_2 emission occurs due to the burning of fossil fuels' burning in the steel industry and it is considered as the major environmental problem (Xu et al. 2016). Thus, the lowering of ecological impact of steel manufacturing mainly focuses on decrease of energy consumption. In steel production process, particular matter such as dust is produced that comes from steel shops, coke ovens, ore piles, blast furnaces, and sinter plants. The dust accumulated was used either or slightly for landfilling. There are some other emissions also occurs from steel industry such as organic compounds, inorganic compounds (H_2S, CO, SO_x, NO_x, O_3), and heavy metals (lead, nickel, mercury, zinc, cadmium, copper, etc.) (Yoro and Daramola 2020). The CO_2 emission differs as per process chosen for production of steel (Ceccolini et al. 2017). Iron making, coke production, sinter production, alloying, steelmaking, shaping and casting, and finishing are fundamental processes involved in production of steel (Mars Sudiro 1994). The casting operations are primary sources of blast furnace emissions and but these are also caused by drilling and plugging the taphole. Sometimes, the taphole clogged and oxygen lance is used to open taphole that can cause heavy emissions. Moreover, the contact of air above the surface of the melted iron and slag generates particulate emissions and mainly the iron oxides, magnesium oxide, and carbonaceous compounds are generated as particulates (Salonitis et al. 2016; Dehghani et al. 2018).

In tropical areas, due to high ambient temperatures, the workers serving in the steel industry are liable to thermally traumatic environments that cause heat-related illnesses and limit workers' productivity. Hence, workers are more apparent to excessive heat stress and its adverse effects (Chen et al. 2003). Studies have shown that heat stress inhibits DNA repair processes and in addition acts as a DNA damaging agent (Kantidze et al. 2016; Krishnamurthy et al. 2017). Oxidative stress, one of the main reasons for DNA damage, has been reported to be caused by heat stress in some animal and human studies (Bridge et al. 2003; Kantidze et al. 2016). Consequently, the oxidative stress within cell changed cellular redox environment that stimulate protein degradation and cell death (Purschke et al. 2010), affects sperm quality and risk of infertility increases (Harrouk et al. 2000; Durai Rajanayagam et al. 2015). It was reported that workers had over-expression of HSP70 levels and with huge DNA damage, when exposed to high heat conditions (Pelham 1984; Xiao et al. 2002). During steel production, the process of iron making involves physical conversion and chemical reduction of iron oxides in the blast furnace (BF) into fluidic iron called "hot metal" with "liquid slag" and "liquid iron". This hot molten metal runs into the channels and it is one of the biggest sources of radiant heat. The impure liquid slag settled at the bottom of the furnace is also very hot and hard to remove. Thus, many workers engaged in disposal of molten slag were susceptible to radiant heat released from the process heat generated. In coke making, coke is formed by heating coal at high temperatures in absence of oxygen that removes volatile components. Coal was carbonized in slot ovens at the battery area and the process releases heat at 1100 °C in an oxygen deficient environment. Especially while loading and unloading of coal and coke from the furnace, extreme radiant heat is generated. In addition, the radiant heat was also generated when coke was cooled for subsequent processes in the cooling bed area (Krishnamurthy et al. 2017).

Globally over the past decades, the increased ambient outdoor air pollution resulted in acute and chronic health effects in the population (Dockery et al. 1993; Pope et al. 1995; Brook et al. 2010; Cakmak et al. 2011; Franklin et al. 2015). Specifically, the chronic exposure to increased air pollution significantly affects all-cause mortality (Dockery et al. 1993), as well as cardiovascular and pulmonary morbidity and mortality (Brook et al. 2010; Franklin et al. 2015). But, recently the acute effects on human health of air pollution have also been reported (Cakmak et al. 2011, 2014; Franklin et al. 2015) and more specifically heart related morbidity and death, including increased risks of disturbance in cardiac cycle (Cakmak et al. 2014).

In the United States, iron and steel industries generally produced the coke using the "byproduct" coke process. The "byproduct" coke process may release volatile organic compounds (VOCs) such as methane, toluene, benzene, xylene, butylene, butane, ethane, hydrogen cyanide, propane, butane, and propylene. These VOCs are released to the environment through leaks in manufacturing system parts like in the coke oven doors, lids, standpipes, and may be within the plant itself; and also, when the coke is being removed, or "pushed," from the oven (Mars Sudiro 1994). Coke oven emission associated with steel making industry consists a mixture of coal tar, coal tar pitch, naphthylamine, chromium (Cr), benzene, volatiles, xylene, arsenic, chromium, beryllium, and toluene that show carcinogenicity in humans (Liberti et al. 2006). The coke oven emissions are collective mixtures of solid, liquid, and gases phases. It comprises approx. 40 PAHs, as well as other PICs; notably formaldehyde (HCHO), acrolein, aliphatic aldehydes, ammonia (NH_3), carbon monoxide (CO), nitrogen oxides (NO_X), phenol, cadmium (Cd), arsenic (As), and mercury (Hg). In coke production, 635 kg of coke was produced from a metric ton of coal, with up to 90 kg large coke particulates (coke breeze), 27.5–34 l of coke oven gas tar, 7–9 kg of $(NH_4)_2SO_4$, 55–135 lt. of ammonia liquor, and 8–12.5 lt. of light oil. Up to 35% of the initial coal charge is emitted as gases and vapours. Coke production releases a mixture of toxic gases and gas tar that comprises H_2, CH_4, ethane, CO, CO_2, pyridine, tar acids, ammonia (NH_3), creosote oil, acetylene (C_2H_2), and Hydrogen sulfide (H_2S), etc. Moreover, light oil fraction may be the result of the extraction of (C_6H_6), toluene (C_7H_8), xylene (C_8H_{10}), benzene, and solvent naphtha. Some coke oven emissions are the benzene-soluble fraction of the particulate matter (PM) generated during the production of coke that are reported to contain human carcinogens. These emissions comprise a highly toxic mixture of gases and aerosols (Valero 2014).

Open-hearth furnaces are one of several types of furnace; the pig iron is burnt to produce steel by removing excess carbon and other impurities. The burning emitted pollutants like sulphur oxides (SO_x), carbon monoxide (CO), and nitrogen oxides (NO_x). There are three basic processes that determine the dust generation in an open-hearth furnace are combustion, mechanical impact of furnace atmosphere and charge, and the chemical processes. During the initial rise of heat and in the process of refining, due to the flow of gases in the furnace, the fine particles get charged in the furnace working chamber. Moreover, increased dust generation can be observed from molten metal during chemical processing specifically in the intensive refining process. The particles are thrown to the surface of the molten metal by the rising carbon monoxide (CO) bubbles that then remain suspended in furnace gases, which increases the dust load (Schino 2018a).

In scrap based mini-mills, the electric furnace, finishing and rolling contribute 45%, 36%, and 16%, respectively to greenhouse gas emission. The main emission of electric furnaces alongside CO and CO_2 is dust and dust particles smaller than 10 microns contribute 60% of total dust particles (Schino 2018a). The dust consists of heavy metals such as zinc and lead besides nickel (Ni), chromium (Cr), mercury (Hg) arsenic (Ar), and cadmium (Cd) because polluted scrap is used in these mills. Sometimes traces of dioxins, furans, and hexachlorobenzene are also emitted. The emission of NO_x and CO also occurs due to gradual heating and cooling. Depending upon the scrap grade used the emission from tapping includes majorly sulphur oxides, iron oxides, and other metallic oxides. The polyaromatic hydrocarbons (PAHs), one of the most harmful environmental contaminants due to their high persistence, bioaccumulation, and toxicity, are also emitted from the iron and steel industry and are major sources for PAHs. Protection gas contains polycyclic aromatic hydrocarbons. Moreover, coke ovens are main point sources for emission of PAHs. Several PAHs have mutagenic and/or carcinogenic properties. The major source for air pollution is also the release of dibenzofurans (PCDDs/Fs) and polychlorinated dibenzo-p-dioxins mainly from electric arc furnaces. These emissions from iron and steel industries had a significant impact on the environment and human health. The increased global warming, climate change, melting of glaciers, increasing ocean level and the amount of exposure to chronic diseases, etc. was associated with industrial emissions (Wiley et al. 2011).

China is the top producer of iron-steel, followed by Japan and the U.S. The total of 2.1% of world's production of iron-steel has been shared by Turkey in the year 2016 with 34.7 million tons of production (Schino 2018b). China, the largest steel and iron utilising and producing country, contributes enormously to air pollution. The generation of NO_x occurs when nitrogen from solid fossil fuel reacts with atmospheric oxygen at relatively high temperature. It is also emitted from the heating furnace of various steel and iron industries, during the sintering and pelleting processes. The SO_2 is mainly emitted during the sintering and pelleting process that contributes 84% to the total emission. High iron ore and sulphur content found in solid fuel distributed desulphurization facilities occurring in sintering machines and low desulfurization efficiency are the responsible factors for the emission of SO_2 (Wu et al. 2015).

In China's, the steel and iron industry, is known well for the continuous emission from sintering machines that includes approximately 98.0% of PCDD/Fs, 49.4% of NO_x, 69.6% of VOCs, 22.5% of TSP, 72.4% of SO_2, and 24.0% of PM 2.5 during the year 2012. These primary air pollutant emissions are mainly derived from the basis of information specifically conveyed on the production technique development and various equipment scales, removal of pollutant situations, and the comparative rise of emission standards in the iron and steel industry of China. During the year 2013, China's steel industry emitted apps. 20% of SO_2 and 27% of particulate matter in China contributes significantly to various global as well as regional environmental consequences. In 2015, China's steel production was about 804 million metric tonnes that was 49.5% of the total world production (Wu et al. 2015). The major contribution of air pollution is the iron and steel industries and it is confirmed with the high values of iron emission from these industries. Slag that is the byproduct from the iron and steel industry is disposed of in huge amounts from these industries. This slag is not fully utilized and consequently it is converted into a dump that surrounds the plant site therefore encouraging the valuable land. The slag from steel and iron industry is majorly composed of calcium silicon and iron along with traces of toxic elements like chromium and slags (Chaurand et al. 2007). Blast furnace slag gives off sulphur dioxide when present in fresh and it gives off hydrogen sulfide and sulfuric acid in the presence of water therefore this blast furnace slag is considered unfriendly to the environment (Kachhap 2010). Another major problem in the world is contamination of soil due to pollutants released from the steel production as it can cause chronic consequences on human life (Ene et al. 2011). Rapid advancement and innovation in the steel and iron industry has been the enormous source of various hazardous pollutants such as heavy and toxic metals chromium, cadmium, copper, iron, mercury, nickel, lead etc. Smelting and tailings that get deposited in soils in the region of industrial activities causes pollution that lead to determination of soil quality. Large amount of solid waste consisting of toxic heavy metals is usually produced from iron steel manufacturing industries (Ene et al. 2009; Ene and Pantelica 2010). The release of these toxic heavy metals by weathering from dumps and landfills lead to pollution of surrounding soil and along with this also pollute the surface and groundwater. Studies have stated that the continuous emission of various constituents obtained from basic oxygen furnace slag was inconsequential with respect to the environmental impact (Sofilić et al. 2013).

Human health is affected with advancements in industrialization and also with increasing environmental pollution (Yağdı et al. 2000). Air pollution for which the major contributors are Industries is off-putting aesthetically and at some time it can be hazardous for humans and vegetation (Onder et al. 2007). Industrial and environmental pollution air pollution affects human health as well as the vegetation. Industrial emission along with the use of fossil fuel commonly releases heavy metals into the air. Raipur and Raigarh (Chhattisgarh states) have heavy air pollution due to the particulate matter released from the steel and sponge iron industry. The ambient air quality in Raipur is getting degraded due to heavy metal and SPM released from the steel industry. The environmental degradation caused by steel and iron industry occurs due to the emission of SO_2, NO_x, H_2S, PAH etc. in the air and organic matter, cyanide, Ammonia, oil, metals, suspended solids etc. into the water and slag, sludge, sulphur compound, salts into the soil (Tiwari et al. 2016).

22.3 Impact of Concrete on Environment

The concrete is used as the structural material composed of particulate substances called aggregate. The aggregates are hard and chemically inert that includes sand, gravel which are bonded together with different compositions of cement and water. Concrete is the necessary component used in building material today because of its strength and its durability. Every construction nearby is possible because of the concrete. One of the main components of concrete is cement that is more often interchangeably used with concrete. Cement is used as a binding for tying different components of concrete together that harden when mixed with water. During ancient times, the bonding material used was clay. The Egyptians developed a substance called binder using lime and gypsum that resemble cements in its properties. Concrete is characterized on the basis of the method of its production and type of cementing agent; whereas when metal or steel components are embedded with concrete material for making structurally strong components are called reinforced concrete, or ferro-concrete. Concrete that has been hardened onto embedded metal is called reinforced concrete, or ferroconcrete. For long spans of time and heavy stress sustainability we rely on tensile strength and comprehensive strength of concrete. Large size structures, sky touching skyscrapers are formed from these reinforced cement and concrete structures. Although wooden constructed sculptures are also in trend, when it comes to durability and mass appeal for big buildings, everyone looks up to concrete buildings. However, with the impressive strength and permeance of the concrete industry, come some severe repercussions that cannot be disregarded (Babor et al. 2009).

The building industry is unrivalled consumers of the world's material extraction; about 24% of the global extractions are utilized for construction. That is extracted from natural as well as manmade resources. But these ultimately affect the natural ecosystem in many diverse ways (Bildirici 2019). Apart from depletion these extractions lead to some severe impact on our environment such as:

- Ecosystem disruption,
- Damage to landscape,
- Damage to flora either during the manufacturing or transport of raw material for construction industry,
- Health care damage by contamination of environment during production, processing, maintenance and demolition of building materials,
- Different emissions from the building materials cause contamination in soil, water and air.

22.4 Raw Material for Concrete Production

Typical concrete mixture consists of Portland cement, sand, crushed stone or gravel and water. Cement plays one of the important factors as it is the combining agent during manufacturing of concrete and provides structural shape to the layout of buildings. Across all the factors, cement is one of the factors that require a special manufacturing unit. Sand used for the construction purpose extracted from sites near the river bank as they contain porosity and are considered suitable for the construction purpose. For crushed stone, explosives are used in stone mines for the crushing and digging of the gravels from the large part of stones that further can be used in the building materials. Cement industry became one of the important parts of the construction industry. For concrete production, cement becomes the relevant part. The characteristic and structural component of concrete is set on by the ratio of water to cement. The proportion of water is kept low, as the water content is lower than the concrete but the amount of water is adjustable in a way to make missile cement paste (Babor et al. 2009). For durability, the relation between cement and aggregate is to be maintained, as for strong concrete the relative ratio of aggregate is to be kept low.

Resources for the production of cement are in multitude. For minerals like calcium, limestone, marlstone and chalk are utilized. And for silicon sources they make use of clay, sand and shale. Ore forms of iron and aluminum can be provided, but recycled metals can also be used. For sulfur- and calcium-based mineral requirements, gypsum is added, approximately 5% by weight. To make cement, limestone that is actually calcium carbonate is heated to an extensive temperature of about 10,000 °C, along with other component materials such as calcium components and silicates. The end product of this intensive heat comes out to be calcium oxide, silicon oxide and carbon dioxide. The further combined product of these oxides comes out to be di and tri-calcium Silicate that is further converted to a 'clinker' that is formed from grinding of these final products into fine powdered form. For preventing flash setting of cement, gypsum is added to the clinker and after pulverizing the final mixture the end product comes out as cement that is bedrock for the composition of concrete. According to the environmental research report of the University of British Columbia, for mass production of one ton of finished cement, it takes around 1,455–1,597 kg of raw materials (Miller and Moore 2020).

22.5 Environmental Impact of Manufacturing of Raw Material

In the process of production of cement, depletion of natural resources like limestone, malt stone and clay occurs. Substances that can impact our environment are also emitted such as clinker, CO_2, flying ash etc. (Oh et al. 2014). One of the reasons for soil erosion and ecosystem destruction is extraction of natural aggregate. The wastewater and sludge excreted from concrete plants affect the water ecosystem of nearby localities and cause harmful impacts on flora and fauna (Cucchiella et al. 2014). Radon gas that is radioactive in nature is released by some of the aggregates from concrete production. However, some of the natural stones and minerals used in concrete production also release uranium that ultimately gets converted into radon. Concrete has several impacts on the environment; during the constructional process diverse punctuations are also observed. Different operative machinery and limestone quarries are visible from long distances. Also, the onsite disturbances like dust, noise, gases and vibrations may permanently disfigure the local environment. There are several cement production companies that use equipment and machinery for reducing different macro and micro emissions that take place during quarrying and manufacturing of cement. Equipment to trap and separate exhaust gases are also increasingly used but their effectiveness varies with the type of component to be trapped. Annually about 7–10% of carbon dioxide is added to the atmosphere due to cement production (Mehta 2001). One of the leading hydraulic cement used is Portland cement that is responsible for the production of immense amounts of greenhouse gases. In the tally of industries that produce greenhouse gases, coal-powered production of electricity tops the list and after that the cement manufacturing sector becomes the next biggest emitter of greenhouse gases. Even for the production of 1 ton of Portland cement, it requires about 4 GJ energy. In this way it becomes one of the most energy intensive processes of all industrial manufacturing. Portland cement clinker manufacture adds a huge volume of carbon dioxide into the atmosphere. Intensive amounts of CO_2 are produced during production of cement, as different effective units use fossil fuel as a source for combustion that is used to operate the rotary kiln. Another source for carbon dioxide during cement kiln is calcination limestone into lime. These are the two major sources of carbon dioxide emissions. Combining these two sources, the relevant amount of CO_2 is released. Worldwide estimate of CO_2 emission by cement production is 1.6×10^9 ton, which is a leading factor of carbon dioxide production among major anthropogenic activities (Malhotra 1999; Mehta 2001). Along with combustion, land use conversions, usage of fossil fuel, cement production become one of the major contributors to the CO_2 emitter. Cement and concrete industries come up with one of the major contributors of air pollutants; that CO_2 becomes the major component of the emissions. It's estimated total particulate dust emissions of 360 pounds per ton of cement produced, the majority of that results from the cement production handling raw materials, grinding cement clinker, and packaging or loading finished cement, that is grinded to a very fine powder.

For extraction of raw material such as limestone and clay, indiscriminate damage is created to the forest and consequently results in soil erosion. Ordinary concrete typically contains about 12% cement and 80% aggregate by mass. Every year the rate of consumption of raw material for the concrete industry is extensively increasing. One of the major concerns regarding cement and concrete production is energy consumption. For every ton of cement production, it takes about 1700–1800 kWh utilized in direct fuel usage for mining and transportation of raw material. Consequently, the emission of different air pollutants such as dust and toxic oxides such as CO_2, nitrous oxide and Sulphur are mainly due to the reliance of this industry on coal as an energy source. Furthermore, for the production of electricity, coal is utilized that subsequently adds to emission of air pollutants in the atmosphere (Liu et al. 2020). Operations like mining, processing and transport involve large amounts of energy consumptions that consequently results in disturbances in the ecosystem via affecting the ecology of the forest and riverbeds. About trillions of litres of fresh water are used every year by the concrete industry for various purposes. For creating the subtle concrete mixture different ingredients are incorporated into it such as cement, water, aggregates and numerous chemical and mineral admixtures that involve massive energy inputs. All these operations are exclusively energy-intensive (Bildirici 2019; Sivakrishna et al. 2020).

22.6 Environmental Concerns Due to Construction Industry

Ramifications of the concrete industry are assessed via analysing the impact of manufacturing of raw products for the concrete industry. The negative influence of the cement industry is illustrated on different components of atmosphere, hydrosphere and lithosphere. The major offender is different emissions during the manufacturing processes and an additional colossal amount of carbon dioxide in the atmosphere. Along with carbon dioxide, other toxic oxides are also released in the atmosphere such as Sulphur dioxide, nitrogen dioxide etc. The emissions of fly ash come out as a new problem that contaminates the atmosphere. In addition to air pollution, the concrete industry also results in water pollution. Discarded water from the batch plant is released to the ponds in nearby locality that results in a setting of discharge and solid waste in the peripheral region of the pond that imparts hazardous impact on the aquatic ecosystem. Along with the extensive consumption of water, the concrete industry pollutes the water bodies by discharge of the left-out products either during manufacturing or during any construction products. By burning out some of the components such as cement kiln fuel and fly ash from concrete mixture, the concrete industry helps to reduce some of the solid waste problem as the left-out products are used in landfills. But after any of the construction or demolition waste of any construction process, concrete becomes the final left out product of the process while most of it is used as a highway substrate or as a clean fill around buildings. Similarly, we cannot ignore the dust pollutant and disturbance created to the pre-referral working or construction site either in construction or during transport of construction material. As the dust particles from the concrete industry also influence the environmental factors during their transport process. Throughout the conveniences process of construction material sluggish amounts of dust, waste, sand, gravel is released. These chunks directly or indirectly affect the environment. As we observe that the flora near the construction site is dull and are desiccated that halt the growth of nearby herbs and shrubs. The plantation nearby the construction unit is affected by different ways. As the canopy of plants present the peripheral region are more exposed to the dust and macro emissions from concrete industry, that results in covering of the stomatal surface of leave and haling of the respiration process in plants and ultimately affect the metabolism and affecting miscellaneous functions of plant. The left-out cement from construction events enhance the alkalinity of the soil and hence indirectly affect the root absorption properties. Water quality of construction sites is also affected by the construction events that add to the detrimental impact on flora of the locality. Chunks left out from the concrete industry also affect the environment (Guggemos and Horvath 2005; Gheibi et al. 2018).

22.7 Health Concerns Regarding Construction Material

According to the studies, the pollution and harmful gases create more industrial regions rather than outdoor city pollution. Construction industry is a growing commerce sector that is based on huge investments and to fulfil the return from this sector many of the illegal working takes place. As they do not follow the norms and guidelines for the manufacturing units, construction site and health care terms. These businesses do not follow the healthcare guidelines that result in different health negligence for the construction workers that may cost their lives. And cause some of the severe life-threatening diseases such as cancer, asthma, allergic reactions and many more are in total. Building materials play a very important role to create an ecologically and sustainably healthy environment but at what cost should it be done?

The most exposed part of our body is skin, while working with wet concrete the alkalinity of the cement causes skin problems. The only way to protect you is to use boots and rubber gloves. But mostly there is a shortage of safety equipment for the workers. Although concrete is one of the inert building materials, for the enhancement of different characteristics like setting time, plasticity, strength, water-content and pumpability, several chemicals are added to it. One of the problems with concrete floors and walls are moisture leaden mold and mildew growth that can create grievous health issue in some of the individuals (Babor et al. 2009). Certain construction material contains toxicants—chemicals synthesized or concentrated during the manufacturing of raw material as they are become relatively toxic to human health even at low concentration, while some of them results in some ill effect during chronic exposure that may results in carcinogenic activity, otherwise impact the health of the respiratory, neurological, endocrine or other systems of the human body.

One of the health concerns regarding the construction industry is skin and lung diseases whereas Cement and dust are one of the reasons for that. However another combination effect is high alkalinity and abrasiveness of wet cement. Due to alkalinity of wet cement, it may cause chemical burns and skin irritation (Poupon et al. 2005; Leon-Kabamba et al. 2018). Silica dust is one of the prominent components of air-pollutants near the building sites. Without proper protection, over several standing years, adverse effects were observed on lungs that lead to several diseases like silicosis that is associated with chronic wheezing, arthritis, cancer and reduced life expectancy. These dust particles may cause disablement in the normal functioning of respiratory systems specially lungs such as lung function impairment, chronic obstructive lung disease, restrictive lung disease, pneumoconiosis. Similarly, carcinoma of the lungs, stomach and colon is also observed. In addition to accumulations in the alveolar region, these dust particles also affect the nasopharyngeal region and also cause desiccation in the throat region that may affect the functioning of thyroid gland affect the Some studies have shown that during long term exposure to cement and dust, it may enter in all the essential organs of body and affects different tissues of major organs like heart, liver spleen and observable results are also obtained in muscles bone and hair that ultimately affect micro-structure and physiological performance of the body (van Thienen and Spee 2008; Miller and Moore 2020). It's is also important to note that construction workers may carry danger to of several diseases for his family and expose them to many occupational diseases, the term used for this is para-occupational or secondary exposure, as when a family member comes into contact with dust particles, cement dust, silicon particles or asbestos like secondary dust particles, they are placed at risk for developing respiratory issues as well. Indirectly they came in contact with these particulate matters and became passive consumers of these construction by-products. Construction work is extremely laborious work and as it often requires long hours of hard manual labor, that too in harsh weather conditions that can exhaust the labor worker both physically as well as mentally. Hence results in ill effect on the health of workers and may cause multiple health issues in them.

22.8 Strategies to Mitigate the Environmental Impact of Steel and Concrete

Reduction in consumption of steel and concrete by recycling and reuse through implementing latest technology and policies are the potential ways to mitigate the ecological impacts of building materials.

22.8.1 *Recycling of Steel*

Steel is one of the most recycled materials which get downgraded because structural steel is the lowest grade of steel and so steel is called the "enviro steel". Material can be reused to reduce the consumption of construction material. Renovation and construction of buildings take place as steel structures do not serve for a greater service life and hence, they need to be reused. From recycling institutes, it is estimated that 90% of the construction steel is being recycled (Gorgolewski 2006).

About 90% of the heavy voluminous products like steel beams are highly recycled and 80% of the products like reinforcing bars are recycled with very low recycling rates because of the constraints in the separation of concrete from steel while recycling. Reinforced bars' 4 and 10% of the sections are produced out of the total crude steel produced in the world. Products like beams are highly recycled while the case is not the same for the reinforced bars (Gorgolewski 2006).

22.8.2 *Reuse of Steel*

Reuse of steel can be done to simplify deconstruction or disassemble the steel building before their life ends. It has many advantages and an important one is to reduce pollution by preventing the generation of CO_2 and other environmental troubles which are created during the process of manufacturing construction steel (Pongiglione and Calderini 2014).

22.8.3 *Reuse and Recycling of Concrete*

Though developed concrete does not match the virgin one, it is utilized as a base material for bridges, roads, pavements etc. It can be used for landfills. Reuse of concrete is difficult due to its density and deficiency in modification. It can be used as a filling material in large constructions like footings, foundations, concrete slabs, etc. (Kim and Choi 2012).

22.8.4 *Selection of Material to Reduce Environmental Impact of Steel and Concrete*

The most prominent technique used by designers for selection of material across all the industries is life cycle assessment [LCA] technique. Another important technique is the Eco indicator system which is devised in the Netherlands; both these techniques are used by the designers in the construction industries. Building for environmental and economic sustainability is one of the LCA tools. It requires users to allot relative values to various categories like global warming, acidification etc. (Hu et al. 2014).

Another important tool is ATHENA which in contrast to BEES is a product comparison tool. It is an assessment tool for buildings.in the database of ATHENA various types of assemblies which are used in construction are present. BEES are more useful than ATHENA as BEES have a database of various materials with their life cycle cost and life cycle impact (Hu et al. 2014).

22.9 Conclusion

As a result of huge use of steel and concrete, it is apparent that with this continued course of usage of steel and concrete will lead to major environmental impacts all around the world. The hard environmental impact is of the steel with the highest emission per unit weight and it also has a very high embodied energy. Comparing steel and concrete, the performance of concrete is much better than steel per unit weight. New materials which have less environmental impact and are more environmentally friendly will help to minimize these environmental hazards caused by these materials which are more harmful globally.

References

An R, Yu B, Li R, Wei YM (2018) Potential of energy savings and CO_2 emission reduction in China's iron and steel industry. Appl Energy 226:862–880. https://doi.org/10.1016/j.apenergy.2018.06.044

Babor D, Plian D, Judele L (2009) Environment impact of concrete. Bul Institutului Politeh din Iași Sect Constr Arhit 55:27

Bildirici ME (2019) Cement production, environmental pollution, and economic growth: evidence from China and USA. Clean Technol Environ Policy 21:783–793. https://doi.org/10.1007/s10098-019-01667-3

Bridge AJ, Pebernard S, Ducreux A, Nicoulaz AL, Iggo R (2003) Induction of an interferon response by RNAi vectors in mammalian cells. Nat Genet 34:263–264. https://doi.org/10.1038/ng1173

Brook RD, Rajagopalan S, Pope CA, Brook JR, Bhatnagar A, Diez-Roux AV, Holguin F, Hong Y, Luepker RV, Mittleman MA, Peters A, Siscovick D, Smith SC, Whitsel L, Kaufman JD (2010) Particulate matter air pollution and cardiovascular disease: an update to the scientific statement from the american heart association. Circulation 121:2331–2378. https://doi.org/10.1161/CIR.0b013e3181dbece1

Cakmak S, Dales R, Leech J, Liu L (2011) The influence of air pollution on cardiovascular and pulmonary function and exercise capacity: Canadian Health Measures Survey (CHMS). Environ Res 111:1309–1312. https://doi.org/10.1016/j.envres.2011.09.016

Cakmak S, Kauri L, Shutt R, Liu L, Green MS, Mulholland M, Stieb D, Dales R (2014) The association between ambient air quality and cardiac rate and rhythm in ambulatory subjects. Environ Int 73:365–371. https://doi.org/10.1016/j.envint.2014.08.015

Ceccolini R, Martini U, Mengaroni S, Neri S, Rinaldi S, Di Schino A (2017) Effect of slag control on steel process optimization and on the environment. J Mater Environ Sci 8:2403–2408

Chaurand P, Rose J, Briois V, Olivi L, Hazemann JL, Proux O, Domas J, Bottero JY (2007) Environmental impacts of steel slag reused in road construction: a crystallographic and molecular (XANES) approach. J Hazard Mater 139:537–542. https://doi.org/10.1016/j.jhazmat.2006.02.060

Chen ML, Chen CJ, Yeh WY, Huang JW, Mao IF (2003) Heat stress evaluation and worker fatigue in a steel plant. Am Ind Hyg Assoc J 64:352–359. https://doi.org/10.1080/15428110308984827

Cortez OD (2009) Heat stress assessment among workers in a Nicaraguan sugarcane farm. Glob Health Action 2:2069. https://doi.org/10.3402/gha.v2i0.2069

Cucchiella F, D'Adamo I, Gastaldi M (2014) Sustainable management of waste-to-energy facilities. Renew Sustain Energy Rev 33:719–728. https://doi.org/10.1016/j.rser.2014.02.015

Dehghani F, Omidi F, Rafizadeh O, Barati Chamgordani S, Gharibi V, Sotoudeh Manesh A (2018) Occupational health risk assessment of volatile organic compounds emitted from the coke production unit of a steel plant. Int J Occup Saf Ergon 26:227–232. https://doi.org/10.1080/10803548.2018.1443593

Dockery DW, Pope CA, Xu X, Spengler JD, Ware JH, Fay ME, Ferris BG, Speizer FE (1993) An association between air pollution and mortality in six U.S. Cities. N Engl J Med 329:1753–1759. https://doi.org/10.1056/nejm199312093292401

Durai Rajanayagam D, Agarwal A, Ong C (2015) Causes, effects and molecular mechanisms of testicular heat stress. Reprod Biomed Online 30:14–27. https://doi.org/10.1016/j.rbmo.2014.09.018

Ene A, Pantelica A (2010) Study of transfer of minor elements during ironmaking by neutron activation analysis. Radiochim Acta 98:53–57. https://doi.org/10.1524/ract.2010.1685

Ene A, Popescu IV, Ghisa V (2009) Study of transfer efficiencies of minor elements during steelmaking by neutron activation technique. Rom Reports Phys 61:165–171

Ene A, Pantelică A, Freitas C, Boşneagă A (2011) EDXRF and INAA analysis of soils in the vicinity of a metallurgical plant. Rom J Phys 56:993–1000

Franklin BA, Brook R, Arden Pope C (2015) Air pollution and cardiovascular disease. Curr Probl Cardiol 40:207–238. https://doi.org/10.1016/j.cpcardiol.2015.01.003

Gheibi M, Karrabi M, Shakerian M, Mirahmadi M (2018) Life cycle assessment of concrete production with a focus on air pollutants and the desired risk parameters using genetic algorithms. J Environ Health Sci Eng 16:89–98. https://doi.org/10.1007/s40201-018-0302-x

Gorgolewski M (2006) The implications of reuse and recycling for the design of steel buildings. Can J Civ Eng 33:489–496. https://doi.org/10.1139/L06-006

Guggemos AA, Horvath A (2005) Comparison of environmental effects of steel- and concrete-framed buildings. J Infrastruct Syst 11:93–101. https://doi.org/10.1061/(asce)1076-0342(2005)11:2(93)

Harrouk W, Codrington A, Vinson R, Robaire B, Hales BF (2000) Paternal exposure to cyclophosphamide induces DNA damage and alters the expression of DNA repair genes in the rat preimplantation embryo. Mutat Res DNA Repair 461:229–241. https://doi.org/10.1016/S0921-8777(00)00053-7

Hu JY, Gao F, Wang ZH, Gong XZ (2014) Life cycle assessment of steel production. In: Materials science forum. Trans Tech Publications Ltd., pp 102–105

Kachhap S (2010) Waste management in mining and allied industries. National Institute of Technology, Rourkela

Kantidze OL, Velichko AK, Luzhin AV, Razin SV (2016) Heat stress-induced DNA damage. Acta Naturae (англоязычная версия) 8(2):29

Kim YJ, Choi YW (2012) Utilization of waste concrete powder as a substitution material for cement. Constr Build Mater 30:500–504. https://doi.org/10.1016/j.conbuildmat.2011.11.042

Krishnamurthy M, Ramalingam P, Perumal K, Kamalakannan LP, Chinnadurai J, Shanmugam R, Srinivasan K, Venugopal V (2017) Occupational heat stress impacts on health and productivity in a steel industry in Southern India. Saf Health Work 8:99–104. https://doi.org/10.1016/j.shaw.2016.08.005

Leon-Kabamba N, Kakoma SJB, Oscar-Luboya N, Pascal-Kimba M, Banza-Lubamba NC, Ngatu NR, Nemery B (2018) Occupational cement dermatitis and cement burns. In: Occupational and environmental skin disorders: epidemiology, current knowledge and perspectives for novel therapies. Springer Singapore, pp 63–69

Liberti L, Notarnicola M, Primerano R, Zannetti P (2006) Air pollution from a large steel factory: polycyclic aromatic hydrocarbon emissions from coke-oven batteries. J Air Waste Manag Assoc 56:255–260. https://doi.org/10.1080/10473289.2006.10464461

Liu Z, Ciais P, Deng Z, Davis SJ, Zheng B, Wang Y, Cui D, Zhu B, Dou X, Ke P, Sun T, Guo R, Zhong H, Boucher O, Bréon FM, Lu C, Guo R, Xue J, Boucher E, Tanaka K, Chevallier F (2020) Carbon monitor, a near-real-time daily dataset of global CO_2 emission from fossil fuel and cement production. Sci Data 7:1–12. https://doi.org/10.1038/s41597-020-00708-7

Malhotra VM (1999) Making concrete "Greener" with fly ash. Concr Int 21:61–66

Mars Sudiro PJ (1994) Pollution prevention in the integrated iron and steel industry and its potential role in MACT standards development, 94-TA28. 02. In: US EPA, Washington, DC

Mehta PK (2001) Reducing the environmental impact of concrete. Concr Int 23:61–66

Miller SA, Moore FC (2020) Climate and health damages from global concrete production. Nat Clim Chang 10:439–443. https://doi.org/10.1038/s41558-020-0733-0

Oh DY, Noguchi T, Kitagaki R, Park WJ (2014) CO_2 emission reduction by reuse of building material waste in the Japanese cement industry. Renew Sustain Energy Rev 38:796–810. https://doi.org/10.1016/j.rser.2014.07.036

Onder S, Dursun S, Gezgin S, Demirbas A (2007) Determination of heavy metal pollution in grass and soil of city centre green areas (Konya, Turkey). Polish J Environ Stud 16

Pelham HR (1984) Hsp70 accelerates the recovery of nucleolar morphology after heat shock. EMBO J 3:3095–3100. https://doi.org/10.1002/j.1460-2075.1984.tb02264.x

Pongiglione M, Calderini C (2014) Material savings through structural steel reuse: a case study in Genoa. Resour Conserv Recycl 86:87–92. https://doi.org/10.1016/j.resconrec.2014.02.011

Pope CA, Thun MJ, Namboodiri MM, Dockery DW, Evans JS, Speizer FE, Heath CW (1995) Particulate air pollution as a predictor of mortality in a prospective study of U.S. adults. Am J Respir Crit Care Med 151:669–674. https://doi.org/10.1164/ajrccm/151.3_pt_1.669

Poupon M, Caye N, Duteil F, Pannier M (2005) Cement burns: retrospective study of 18 cases and review of the literature. Burns 31:910–914. https://doi.org/10.1016/j.burns.2005.04.007

Purschke M, Laubach HJ, Anderson RR, Manstein D (2010) Thermal injury causes DNA damage and lethality in unheated surrounding cells: active thermal bystander effect. J Invest Dermatol 130:86–92. https://doi.org/10.1038/jid.2009.205

Salonitis K, Zeng B, Mehrabi HA, Jolly M (2016) The challenges for energy efficient casting processes. In: Procedia CIRP. Elsevier B.V., pp 24–29

Schino D (2018a) Survey on environmental impact and circular economy aspects related to steel production in Europe. J Mater Environ Sci 9:1630–1635

Schino A Di (2018b) Environmental Impact of the Steel Industry. In: Handbook of environmental materials management. Springer International Publishing, pp 1–21

Sivakrishna A, Adesina A, Awoyera PO, Kumar KR (2020) Green concrete: a review of recent developments. In: Materials today: proceedings. Elsevier Ltd, pp 54–58

Sofilić T, Brnardić I, Šimunić-Meţnarić V, Šorša A (2013) Soil pollution caused by landfilling of nonhazardous waste soil pollution caused by landfilling of nonhazardous waste from steel production processes. Kem Ind 62:381–388

Sun W, Zhou Y, Lv J, Wu J (2019) Assessment of multi-air emissions: case of particulate matter (dust), SO_2, NO_x and CO_2 from the iron and steel industry of China. J Clean Prod 232:350–358. https://doi.org/10.1016/j.jclepro.2019.05.400

Tiwari MK, Bajpai S, Dewangan UK (2016) Air and leaching pollution scenario by iron and steel plants, in central India. Elixir Pollut 101:44011–44017

Valero D (2014) Air pollutant emissions. In: Fundamentals of air pollution. Elsevier, pp 787–827

van Thienen G, Spee T (2008) Health effects of construction materials and construction products. J Appl Occup Sci 14:2–23

Vatani J, Golbabaei F, Dehghan SF, Yousefi A (2016) Applicability of universal thermal climate index (UTCI) in occupational heat stress assessment: a case study in brick industries. Ind Health 54:14–19. https://doi.org/10.2486/indhealth.2015-0069

Wiley DE, Ho MT, Bustamante A (2011) Assessment of opportunities for CO_2 capture at iron and steel mills: an Australian perspective. In: Energy procedia. Elsevier Ltd., pp 2654–2661

Wu X, Zhao L, Zhang Y, Zheng C, Gao X, Cen K (2015) Primary air pollutant emissions and future prediction of iron and steel industry in China. Aerosol Air Qual Res 15:1422–1432. https://doi.org/10.4209/aaqr.2015.01.0029

Xiao C, Chen S, Li J, Hai T, Lu Q, Sun E, Wang R, Tanguay RM, Wu T (2002) Association of HSP70 and genotoxic damage in lymphocytes of workers exposed to coke-oven emission. Cell Stress Chaperones 7:396–402. https://doi.org/10.1379/1466-1268(2002)007%3c0396:AOHAGD%3e2.0.CO;2

Xu W, Wan B, Zhu T, Shao M (2016) CO_2 emissions from China's iron and steel industry. J Clean Prod 139:1504–1511. https://doi.org/10.1016/j.jclepro.2016.08.107

Yağdı K, Kaçar O, Azkan N (2000) Heavy metal contamination in soils and its effects in agriculture. Ondokuz Mayıs Üniversitesi, Ziraat Fakültesi Derg 15:109–115

Yoro KO, Daramola MO (2020) CO_2 emission sources, greenhouse gases, and the global warming effect. In: Advances in carbon capture. Elsevier, pp 3–28

Chapter 23
Ecological Impacts of Land Conversion on Wildlife Conservation: A Case of Construction Sector in Tanzania

Cosmas Benedict Mabalika Haule

Abstract Establishment of protected areas is globally employed as the mechanism for conservation of biological diversity. However, the protected areas are subject to land conversion. The land conversions taking place in the protected areas can be illegal or ones graced by authorities mandated to conserve the biodiversity. In Tanzania, the changing socio-economic forces and population are behind the growing construction projects in the wildlife habitats. In most cases, the development initiatives such as the construction projects lead to downgrading, downsizing and degazettement of the protected areas. Various studies have reported notable impacts of construction projects on wildlife conservation in protected areas. The obvious identified ecological impacts of construction projects include habitat loss, habitat fragmentation, and blockage of migratory corridors and loss of dispersal areas, increased poaching rates, pollution, and disturbance to wildlife. The Undertakings of environmental impacts assessments on proposed construction projects are vital not only for enhancement of preparedness among those responsible for wildlife conservation but also provision of insights on how to mitigate the negative impacts of those projects. In that case, it is imperative necessary to document the growing concerns in biodiversity conservation so as to reduce the ecological impacts of the increasing construction works.

Keywords Construction · Ecological impacts · Land conversion · Protected areas · Wildlife conservation

23.1 Introduction

Establishment of protected areas is one the global strategy to maintain biological diversity including wildlife. The protected areas offer ecosystem services which support the conservation of biodiversity (Ingram et al. 2012). In line with global efforts to maintain the biological diversity, Tanzania has designated about 40.5%

C. B. M. Haule (✉)
Department of Geography, The Open University of Tanzania, Dar es Salaam, Tanzania
e-mail: cosmas.haule@out.ac.tz

J. A. Malik and S. Marathe (eds.), *Ecological and Health Effects of Building Materials*,
https://doi.org/10.1007/978-3-030-76073-1_23

(383,000 km^2) of its land for conservation. Besides, the country has designated 33% (279,000 km^2) of its land surface as protected areas (Kideghesho et al. 2013; USAID 2012). In that respect, Tanzania is among the leading countries which value biodiversity conservation globally and Africa in particular.

According to Wildlife Conservation Act No. 5 of 2009 (URT 2009), Protected area means an area of land or sea specifically protected for maintenance of biological diversity, natural or associated cultural resources and managed through legal or other effective means.

In ideal conditions, the protected areas are taken to be undisturbed areas on the landscape set aside for conservation purpose (Mascia and Pailler 2010). This assumption is based on the fact that most of the protected areas are located not only in the remote areas but also in marginal lands that do not attract much attention for land conversion (Joppa and Pfaff 2010).

However, Bühne et al. (2020) study in West Africa revealed that the protected areas as for other land uses are prone to pressure for land conversion. Similar observations are reported by Hermann et al. (2016) study in Brazil who found low efficiency of protected areas in preventing conversion. In most cases, the land conversion involves but not limited to change in existing land uses form and extent. Furthermore, the study (Bühne et al. 2020) found that protected areas not only experience diverse kinds and rates of land conversion but also differ in their effectiveness to prevent land conversion in the protected areas.

Even though all kinds of protected areas are prone to land conversion, it is expected that higher potential of land conversion will be found where there is weak protection status in terms of resource allocation and legal protection. Similarly, protection areas with sufficient funding, enough political and local communities support, and trained staff are likely to effectively conserve biodiversity (Chardonnet 2019).

In Tanzania, USAID (2012) study identified land conversion as the major threat to biodiversity. Despite the restrictions made on development of protected areas, the construction projects are among the emerging forces which drive land conversion. Tanzania has a notable number of construction projects in her protected areas which obviously convert the land. This come despite the country ratifying a number of international conventions and protocols that promote conservation of wildlife conservation.

Basically, all construction works have an impact on wildlife conservation. However, the extent and magnitude of the ecological impacts of construction works will depend on the location of the proposed development, timing of the construction works, and kind of species that occupy the particular area as their habitat. In that case, this chapter draws cases from both legal and illegal land conversion to illustrate the ecological impacts of construction works on wildlife.

23.2 Forms of Land Conversion in Protected Areas

There are various forms of land conversion in protected areas, namely, legal and illegal conversion. The illegal forms of land conversion are committed when there is unauthorized entry and carrying out any activity without the discretion of the authorities mandated to manage the protected areas. The most reported form of illegal conversion in protected areas is the deforestation brought by either agriculture or logging for timber and firewood. The deforestation of protected areas especially those set for wildlife is not the only form of illegal land conversion. In Tanzania, there are reported cases of illegal mining in Amani Nature Reserve, illegal livestock keeping in Malagarasi-Moyowosi and Ihefu wetlands, and illegal construction in some of the protected areas.

Another form of land conversion is legal conversion which involves legal authorization of individuals and/or organizations to carry out a number of activities which result in change in land use of the protected areas. In this form of land conversion, there is a legal requirement of conducting Environmental Impact Assessment (EIA) for all proposed development projects in the protected areas. However, the International Union for Conservation of Nature (IUCN) (2019) reported inappropriateness of EIA for the Stiegler's Gorge hydropower project in the Selous Game Reserve. The recommendations provided by the EIA for the project are questionable as they did not comply with IUCN guidance on impact assessment and International Finance Corporation (IFC) performance standards. The negligence of required procedures for the EIA mighty be behind the unnecessary loggerheads among stakeholders in wildlife conservation when it comes to establishment of various development initiatives in the protected areas.

23.3 Forces Behind Land Conversion in Protected Areas

Land conversion in protected areas is driven by a number of forces. The first major cause of land conversion is the fact that protected areas have socially constructed governance regimes (Mascia and Pailler 2010). In most cases, the protected areas respond to existing social demands by changing the existing land tenures. From time to time, the governments, including that of Tanzania have changed land tenures of the protected areas by downgrading, downsizing and/or degazettement. Zimmerer et al. (2004) cite the degazettement of Ruvu Game Reserve following heavy encroachment and degradation of the protected area by smallholder farmers and pastoralists. In some instances, local communities living adjacent to protected areas have been evicted and their settlement areas destroyed to give space for expansion of the wildlife habitats. This is illustrated by the eviction of Maasai pastoralists from both Serengeti and Mkomazi National Park (Kideghesho et al. 2013).

Close related to social pressures is the openings provided by prevailing policy and regulations. For instance, article six, section two of Land Act No. 4 of 1999 provide that the President has authority to declare and revoke the status of the reserved lands including protected areas (URT 2013).The power bestowed to authorized officer such as the Commissioner offer a permission for land conversion be conducted in the protected area. The typical examples of the land conversion based on existing policy and regulations include the establishment of the Mkuju River Uranium project and the construction of Julius Nyerere Hydropower plant (formerly known as Stiegler's Gorge) in the Selous Game Reserve (SGR). The construction of the two projects involved alteration of boundaries for one of the famous UNESCO World Heritage Sites, the SGR. The conflicting interests in the project between international organizations with interest in wildlife conservation and the national endeavor to develop its resources have made the decision to continue with and ultimately commencement of construction of hydropower generation plants to take more than five decades.

Another social pressure in the context of the construction sector in Tanzania which exacerbated the conversion of land in protected areas is the country's development vision. The main goal of the Tanzania Development Vision (TDV) is to be a middle income by 2025 (URT 1999).One of the major strategies set for realization of a strong and competitive economy target is the development of infrastructure which will support the national need to industrialize. In that case, the development of Stiegler's Gorge for hydropower generation became inevitable not only to meet energy demand which is sufficient, affordable and reliable power for industrialization but also to ensure better utilization of local resources. The project is expected to add 2115 MW in the national power grid. Besides, with reliable power, the project is strategically initiated to support the Standard Gauge Railway project under construction. In that context, the hydropower project in Selous Game Reserve and Stiegler's Gorge in particular took off despite the international NGOs interests on its valuable ecosystem services (WWF 2017).

In addition to social pressure, the rapid growing population implies increasing pressure in the protected areas. The rapid growth of population is behind the encroachment of protected areas. The growing population is associated with an increase in demand for food and settlements. The increased demand for food and settlements imply increasing forest clearance for establishment of new fields and settlements. As far as land is a limited resource, then, most of the clearance is expected to be undertaken in the protected areas. On a number of occasions, agro-pastoralists have encroached and settled in the protected areas. From time to time, the government has been obliged to change the protected area boundaries in order to accommodate the demand for settlements and grazing land for the agro-pastoralists. The most affected part of protected areas that is affected by this kind of land conversion is the wildlife corridors. This comes as most of them are either already heavily degraded and/or lack legal status as protected areas.

Moreover, the most cited reason for land conversion is poverty. The state of deprivation which the societies living close to protected areas face are responsible for the conversion of land. For instance, lack of affordable energy force local communities living close to the protected areas to engage in the charcoal making which lead to deforestation which threaten the wildlife habitats.

23.4 Ecological Impacts of Construction Works on Wildlife Conservation

The nature of development in the construction industry is one among the many activities that have direct and indirect impact on wildlife conservation. The construction activities that have impact on wildlife conservation include establishment of settlements, industrial development, infrastructure development such as construction of roads, railways, power lines, and dams. The major ecological impacts of construction work on conservation can be categorized into four major groups, namely, habitat loss, habitat fragmentation, habitat degradation, and mortality of animals (Bennett et al. 2011; Gunson et al. 2016; Jackson and Griffin 2000; Kideghesho et al. 2006; Özdemirel et al. 2016).

Construction of infrastructures such as roads, railways and dams are associated with large land clearance and leveling. The process of land clearance and levelling involves tree cutting, excavation of earth's surface, and fill of other areas (Başkaya et al. 2011). For instance, the on-going construction of Julius Nyerere Hydropower in Selous Game Reserve involves building of diversion channels to allow main concrete dams to be built, construction of tunnels that will lead water to the turbines responsible for power generation, construction of power house and staff quarters. All these activities will not only lead to loss of original habitat used by wildlife in the Stiegler's Gorge but also add materials that change the quality of the wildlife habitats.

Similarly, construction of highways, not only open the remote located protected areas but also have been reported to be responsible for habitat fragmentation (Bennett et al. 2011; Jackson and Griffin 2000; Kideghesho et al. 2006; Gunson et al. 2016). The linear infrastructures such as roads and railways are blamed for breaking apart habitat and creating isolated and patch habitats which reduce connectivity and put species at risk of extinction. According to Özdemirel et al., (2016) the requirements for large home ranges make the large mammals most affected species by habitat fragmentation. In Tanzania, Kideghesho et al. (2013) reported habitat fragmentation of the Selous Game Reserve and Magombera Forest Reserve due to the construction of Tanzania Zambia Railway (TAZARA) in the 1970s.

Apart from highway construction, lack of habitat connectivity can come from establishment of settlements in the wildlife corridors. In most cases, the houses built in wildlife corridors block the animal movements and their migratory routes to the dispersal areas which are very important for animal's species feeding, water, shelter and breeding places. The typical example where animal species abundance and distribution have been disturbed in Tanzania is the settlements established in the Kwakuchinja wildlife corridor which connect Tarangire National Park and Lake Manyara National Park.

Moreover, it should be taken into mind that any construction project might result in habitat degradation. In most cases, the construction works such as dams, roads, or railways involve the consumption of varieties of resources. In that case, large construction projects such as construction of Julius Nyerere Hydropower Project, Standard Gauge Railway, and other roads passing in the protected areas are associated with generation of wastes. The presence of waste water, oil spills, fuels, dust, glass, nylon, and paper in the construction site lead to pollution which reduce the quality of habitats used by wildlife. Similarly, activities like blasting lead to noise pollution which make some animals avoid a particular habitat.

Another common ecological impact related to construction activities is the direct mortality of animals. Koemle et al. (2018) term the highways construction as the single greatest threat to wildlife as they increase their exposure to wildlife-vehicle collision. Similar observation is shared by Gunson et al. (2016) who reported that amphibians and reptiles are the most affected species. In Tanzania, Tanzania Zambia Highway which has 50 km passing through Mikumi National Park has been reported to be responsible for a number of animals killed in the park. The increased number of animal-vehicle collisions in the highway despite placement of speed bumps has led to consideration of using alternative roads which pass through Kilosa town. The experiences gained from animals knocked by vehicles in the Mikumi National Park has sparked a strong resistance to the construction of a proposed 53 km stretch of unpaved road passing through Serengeti National Park (Fyumagwa et al. 2013; Kideghesho et al. 2013).

In addition to animal-vehicle collision, increased numbers of animals killed might come from increased poaching. This comes as construction of road networks increases the accessibility of the very remote areas and places them close to markets of wildlife products. In that case, the opening of remote areas is expected to increase the incidence of illegal hunting of animals. However, this claim is disputed due to the fact that the improved transportation networks are likely to add security in the protected areas. This is true as improved road networks make the surveillance provided by game scouts and visiting tourists much easier. In addition, in most cases, poachers prefer the inaccessible area which provides easy escape when tracked by protected areas security staff. Moreover, when we take the case of Selous Game Reserve, the construction of Mwalimu Nyerere Hydropower plant at Stiegler's Gorge is associated with upgrading part of the reserve into a newly established national park. The increased status of the area will not only overcome the long blamed vastness of the game reserve but also its failure to contain poaching.

23.5 Conclusion

This chapter has demonstrated that establishment of the protected areas is fundamental in the context of wildlife conservation. Furthermore, the chapter has found that the presence of protected areas though necessary measures to maintain biodiversity, is not sufficient to prevent land conversion. Since, there are a number of driving forces which put land conversion such as construction projects inevitable in our protected areas. Besides, it has been demonstrated that despite the ecological impacts of construction works being obvious, the rate and extent of the ecological impacts experienced by a particular protected area will vary depending on its legal conservation status and the prevailing social pressures in a particular area. In that case, despite its importance, protected areas should not be taken as a panacea to biodiversity loss. Similar inadequacy of protected area in conserving biodiversity especially wildlife is also reported by Kideghesho et al. (2006). In order to minimize the ecological impacts of the construction works, this chapter calls for thorough undertaking of studies such as the environmental impact assessment prior to implementation of any proposed construction project.

References

Başkaya S, Başkaya E, Sari A (2011) The principal negative environmental impacts of small hydropower plants in Turkey. Afr J Agric Res 6(14):3284–3290

Bennett VJ, Smith WP, Betts MG (2011) Toward understanding the ecological impact of transportation corridors. Gen Tech Rep PNW-GTR-846, Portland, pp 3–11

Bühne HS, Wegmann M, Durant SM, Ransom C, de Ornellas P, Grange S, Beatty H, Pettorelli N (2020) Protection status and national socio-economic context shape land conversion in and around a key transboundary protected area complex in West Africa. Remote Sens Ecol Conserv 3(4):190–201. https://doi.org/10.1002/rse2.47

Chardonnet B (2019) Africa is changing: should its protected areas evolve? reconfiguring the protected areas in Africa. France-IUCN Partnership. Nature and Development. https://papaco.org/wp-content/uploads/2019/03/etudesAP_configAP_EN.pdf. Accessed 7 Jan 2021

Fyumagwa R, Gereta E, Hassan S, Kideghesho JR, Kohi EM, Keyyu J, Magige F, Mfunda IM, Mwakatobe A, Ntalwila J, Nyahongo JW, Runyoro V, Røskaff E (2013) Roads as a threat to the Serengeti ecosystem. Conserv Biol 00(0):1–4. https://doi.org/10.1111/cobi.12116

Gunson K, Seburn D, Kintsch J, Crowley J (2016) Best management practices for mitigating the effects of roads on amphibians and reptiles species at risk in Ontario. Ontario Ministry of Natural Resources and Forestry, Queen Printerfor Ontario, Ontario, pp. 7–8

Hermann JM, Lang M, Gonçalves J, Hasenack H (2016) Forest–grassland biodiversity hotspot under siege: Land conversion counteracts nature conservation. Ecosyst Health Sustain 2(6):e01224. https://doi.org/10.1002/ehs2.1224

Ingram JC, Redford KH,Watson JEM (2012) Applying ecosystem services approaches for biodiversity conservation: benefits and challenges, S.A.P.I.EN.S 5.1 http://journals.openedition.org/sapiens/1459. Accessed 15 Feb 2021

International Union for Conservation of Nature (IUCN) (2019) Technical review of the environmental impact assessment for the Rufiji hydropower project in Selous Game Reserve, Tanzania. IUCN, Gland, Switzerland, 23 p

Jackson SD, Griffin CR (2000) A strategy for mitigating highway impacts on wildlife. In: Messmer TA, West B (eds) Wildlife and highways: seeking solutions to an ecological and socio-economic dilemma. The Wildlife Society, Amherst, pp 143–159

Joppa L, Pfaff A (2010) Reassessing the forest impacts of protection: the challenge of non random location and corrective method. Ecol Econ Rev 1185:135–149

Kideghesho JR, Nyahongo JW, Hassan SN, Tarimo TC, Mbije NE (2006) Factors and ecological impacts of wildlife habitat destruction in the Serengeti ecosystem in Northern Tanzania. AJEAM_RAGEE 11:917–932

Kideghesho JR, Rija AA, Mwanende KA, Selemani I (2013) Emerging issues and challenges in conservation of biodiversity in the rangelands of Tanzania. Nat Conserv 6:1–29. https://doi.org/10.3897/natureconservation.6.5407

Koemle D, Zinngrebe Y, Yu X (2018) Highway construction and wildlife populations: Evidence from Austria. Land Use Policy 73:447–457. https://doi.org/10.1016/landusepol.2018.02.021

Mascia MB, Pailler S (2010) Protected area downgrading, downsizing, and degazettement (PADDD) and its conservation implications. Conserv Lett 00:1–12. https://doi.org/10.1111/j.1755-263X.2010.00147.x
Özdemirel B, Turak AS, Bilgin CC (2016) Impact of large scale dam construction on movement corridors of mammals in Artvin, North –eastern Turkey. Appl Ecol Environ Res 14(3):489–507. https://doi.org/10.15666/aeer/1403-489507
United Republic of Tanzania (URT) (1999) The Tanzania development vision 2025. Planning Commission, Dar es Salaam, 31 p
United Republic of Tanzania (URT) (2009) The wildlife conservation act no 5 of 2009. Ministry of Natural Resources and Tourism, Dar es Salaam
United Republic of Tanzania (URT) (2013) Chapter 113 act no. 4 of 1999 Land Act (Principal Legislation), Revised Edition. Ministry of Lands,Dar es Salaam, pp11–12
USAID (2012) Tanzania environmental threats and opportunities assessment. The United States agency for international development (USAID). November 2012. http://www.brucebyersconsulting.com/wp-content/uploads/2013/01/Tanzania-Environmental-Threats-and-Opportunities-Assessment-2012.pdf. Accessed 7 Jan 2021
WWF (2017) The true cost of power: the facts and risks of building Stiegler's Gorge Hydropower dam in Selous Game Reserve, Tanzania. World Wide Fund for Nature, Gland, Switzerland
Zimmerer KS, Gatt RE, Buck MV (2004) Globalization and multi-spatial trends in the coverage of protected area conservation (1980–2000). Ambio 33:520–529

Chapter 24
Perception of Construction Workers on Psychophysical Health and Safety Issues: A Qualitative Investigation

Jaya Bharti and Megha Singh

Abstract Research from the National Institute for Occupational Safety and Health (NIOSH) suggests that construction workers face more risk than workers in other industries. Based on previous studies, it can be said that construction workers have to face more work related injuries and accidents. The purpose of the present research is to assess the physical, environmental and emotional well-being among construction workers across gender. The sample consisted of 100 construction workers. The Health Interview Schedule was used to assess the physical, emotional and environmental wellbeing of construction workers. Content analysis was done for presenting the appropriate result. It was found that both women and men engaged in construction work have different problems, where on the one hand, men have to work harder and get less salary due to which they continue to struggle with many physical problems. At the same time, women do not get equal pay as men nor do they get free time from home work even after working hard. The findings of the study serve as the roadmap for furthering research into physical, environmental and emotional wellbeing of the construction workers and would serve as the base for creating an improved physical, environmental and emotional wellbeing for construction workers.

Keywords Construction workers · Emotional health · Environmental health · Physical wellbeing · Unskilled workers

24.1 Introduction

The progress of any country depends on the economic development of that country; the economic development of that country shows how many possibilities of development are available in that country (Paruchuru et al. 2020). Veeramani (2019) concise that GDP growth in India has followed an inverted U-shaped curve, accelerating

J. Bharti (✉)
Department of Psychology, ANDNNMM (CSJM University), Kanpur, India

M. Singh
Department of Psychology, University of Lucknow, Lucknow, India

J. A. Malik and S. Marathe (eds.), *Ecological and Health Effects of Building Materials*,
https://doi.org/10.1007/978-3-030-76073-1_24

from a low of 5.5% in 2012–13 to a peak of 8.2% in 2016–17 and then decelerated to 6.8% in 2018–19.

Construction industry contributes to the increasing economy of our country with reference to its gross domestic product. Construction has emerged as a foundation in the industrial development of our country. According to the Government of India, the construction sector contributes 9% to GDP as well as employs 44 million people. It is the second largest employer in India as per 2017 data and it is also reported that by 2025 the construction market will emerge as the third largest power and market in the world.

In construction industries, the people who work as unskilled labourers are usually known by the name of construction labourers. Construction workers are labourers who leave different areas, villages and sites in search of a good job, bread or to meet some money and basic needs. To fulfil their basic needs, they leave their ancestral village and take up construction work. They are the most exploited ones among the other labourers even after five decades of independence (Kumar 2013). Service in construction sites can be considered into three groups "Management and technical workforce", "Skilled workforce" and "Semi-skilled and unskilled workforce" (Vitharana et al. 2015).

People with higher educational qualifications and knowledge who deliver innovative ideas (Odusami et al. 2007) and usually work as site managers who instruct the construction workers in designing and managing the project comes under the management and technical workforce (Vitharana et al. 2015; Odusami et al. 2007).

People who present themselves as a professional in construction activity have extensive knowledge and experience related to construction activities. Such people have been identified as skilled workers or semi-skilled workers. On the other hand, there are also some unskilled workers who have little knowledge related to construction work (Vitharana et al. 2015).

Thus, a construction worker is the person who works in construction sites and requires a lot of physical labour. Thus, the area of unskilled construction workers are majorly at stake because they usually face safety issues while working in their sites. This labour force is strained to exertion in very unacceptable circumstances with no safety, sanitation and health care in our country. Yadav (2015) has found in her study that construction workers are mostly the migrants, unskilled, illiterate and socially backward individuals who belong to poor socio-economic status. Their lifestyle is very limited and does not possess any infrastructure. Construction workers working in construction sites face more accidents and risks due to poor construction planning, unsafe design, inadequate safe training, and neglect of site regulations (Saeed 2017). Peyton and Rubio (1991) have found that accidents in the construction sites occur due to two reasons—unsafe conditions and unsafe acts. Thus, the government of India has established a law to ensure the safety of the unskilled labour. The act talks about the work-related wellbeing, health of people working there and working circumstances of the labourers. It states that every employer shall ensure that the condition of his workplace is free from all types of hazards. The welfare of the employees working there is fully taken care of. There is no need to fear them from any risk or accident etc. They will be provided a completely safe environment which will be healthy as well as risk free.

Undoubtedly, construction industries are the best industries in the setup of the Indian economy, and are also a major means of providing employment to the people of India. It increases the employment rate and gives people a better opportunity to improve their skill as well as lifestyle (Khahroet al. 2016). While there is a positive side in the construction industry, on the contrary, work related risks, mortality and accidents due to mishaps and terrible conditions, and injuries have made this industry one of the dangerous industries, due to which people experience fear (Camino et al. 2008; Cheng et al. 2010). Working day and night in such dangerous places and carrying a lot of load on their back and shoulders, hanging on the ladder for hours affects their physical health negatively. This not only hampers their physical strength but they also get some problems of backache, pain on their knees and shoulders.

Studies have shown that the least attention is paid to health-related accidents and risks, while the emphasis is more on the causes of work-related absenteeism (Monnery 1998). The physical illness may take a long time to be identified within the worker's body. Most of the people working as construction workers experience occupational health related problems. The health related constraints not only affect the physical health of individuals but also affect their mental health due to strenuous working hours. At times they have to stand on the ladder for hours or are hung on building sites at a great length. This not only leads them towards physical health related issues but their job also hampers their mental health.

According to Kahn and Fawcett (2004) "Mental Health refers to an individual's ability to negotiate the daily challenges and social interactions of life without experiencing undue emotional and behavioural incapacity; Mental Health is more than just absence of mental disorders."

Physical and mental health is the base of well-being of individuals. Physical fitness can be understood as proper physical activity, proper exercise, proper work and better physical health (Nazeer et al. 2016). These things are completely missing in the construction workers. They can hardly afford two meals per day so they can afford a nutritious diet with fruits and vegetables. This further affects the wellbeing of the individuals. Well-being is associated with mental health and physical health, that is, integrating the mind and body. It shows a holistic approach towards prevention of diseases found in our body and development of health (Russo and Terraneo 2020). The World Health Organisation (WHO) defines health as a state of being full of happiness, prosperity and success (WHO 1948). Studies show that good and sound sleep for long hours, physical activity with good physical health and substance etc. use are good predictors of mental wellbeing (Ridner et al. 2016).

Emotional wellbeing can be seen as a positive, optimistic, supportive mental health factor. Emotional wellbeing focuses on a person's life satisfaction, his happiness, his good moments, his personal experiences. Similarly, Hedonic tradition has also stated that welfare also includes happiness, hope, pleasant feelings which play an important role in strengthening a person's life (Ojedokun and Idemudia 2014). Emotional well-being can be seen as a cognitive appraisal for which our direct thinking, problem solving, creativity and perception play a direct role (Lamers et al. 2011). Stress is a system that creates an imbalance between mind and body. Due to stress, a person feels uncontrolled and uncomfortable (Jahoda 1982; Warr et al. 1982; Gottfredson and Duffy 2008). Context free wellbeing and job related wellbeing are positively correlated as well as they are related to the demands of the organization's work, their difficulties and specifications, feelings, stress and the attitude towards work etc. (Fisher et al. 2003).

Environmental wellbeing with respect to environment refers to a person's sense of belonging and social inclusion. Environmental welfare inspires us to live a life that is positive and optimistic for us and our family. Environment wellbeing informs that the more positive our environment is, the better our and those around us will develop. The negative environment has a pessimistic and negative impact on us and our society.

24.2 Methods

The purpose of the present research paper was to explore the perceptions of psychophysical health and safety among construction workers. Data were collected using an interview schedule. In this study, purposeful sampling procedures were used. According to Singleton et al. (1998), cited in De Vos (2002), this type of sample is entirely based on the will of the researcher. This sample is also known by the names of selective sampling, decision, and subject samples, which are types of non-probability samples, in which researchers are confident that people who have been selected as samples in their studies are a good representative of population or they have the characteristic of representation. The researcher deliberately selected some construction sites and construction workers for the purpose of his study. The researcher selected those who had full understanding of the problem and purpose of research with full confidence. There were 50 male construction workers and 50 female construction workers. Their age range varied from 25 to 45 years. Ex-post facto research with exploratory orientation was used. The interviews were recorded after taking participants prior consent. The language of the schedule was Hindi. Thus, the effort was to use the closest possible English translation so that the meaning of the content remains the same and does not lose their significance.

Construction sites were visited before starting the data gathering process. Construction workers were assimilated at the construction sites and the research objectives and their benefits were discussed in detail. At the same time, aspects of maintaining confidentiality were well explained, followed by introductory sessions with construction workers so that they could not put any questions inside themselves and give their answers without hesitation.

The researcher also made sure that none of the questions in the interview were likely to pose any threat to construction workers, or that construction workers experienced any kind of problem. It was also kept in mind that no question should have any different meaning. The researcher did not hide any objective of the study. He presented all the objectives to the construction workers. The researcher made these trips to delve deeper into the physical, emotional, and environmental conditions of construction workers throughout the study.

24.3 Results

With reference to results of the interview schedule the researcher tried to explore three dimensions of well-being i.e. physical, emotional and environmental. The questions that were included in the interview schedule were designed very thoughtfully, so that it could extract the meaning from the depth and make more statements in fewer words without losing its meaning. The questions used in the schedule were open ended, so that it could meet the target of the current study.

24.3.1 Part I: Perceived Health—Physical

In the first part (physical well-being) the researcher tried to explore the physical complaints by the construction workers. According to Bharti (2020), "Physical well-being is an important component of construction workers. Physical well-being refers to physical functioning, fatigue, sleep quality, health problems and self-care behaviours such as exercise, nutrition, recreational activities, rest and sleep. Physical well-being can be influenced by caregivers' health status and health-related behaviours. Construction workers' health status is initially similar to that of the normal population. Eventually, construction workers report their own health problems, mainly in the form of musculoskeletal pain, gastro-intestinal diseases, chronic back injury/pain, and respiratory problems". In the present research this dimension is loaded with few issues like:

- Physical complaints by the construction workers
- Perceived appetite of construction workers
- Complaints about pain
- Physical deterioration experienced by construction labourers.

The most frightening symptom of musculoskeletal disorders is pain, this pain is extremely unbearable. Table 24.1 clearly reflects that most of the male workers (60%) experience musculoskeletal pain followed by chronic back injury. As the workers pull heavy weights at their back and whole days are working in the same position, they are unable to take care of their body and thus, experience musculoskeletal pains. 60% males reported that—"*Fatigue persists throughout the day; there is a lot of back pain, widespread pain, aching or stiffness of the entire body, burning sensations and so much unbearable pain that it is not tolerable… But work has to be done otherwise what we will eat and how we are going to feed our family…we do not have enough money for treatment…. Thus, spend our own money for all treatment*

Table 24.1 Physical complaints by the construction workers

Categories	Percentage	
	M (%)	F
Musculoskeletal pain	60	20%
Chronic back injury/pain	20	30%
Gastro-intestinal diseases	4	50%
Respiratory problems	10	–
Work-related injuries	6	–

fees". This highlights the dismal state of these construction workers that despite their poor health, they are unable to take any rest which worsens their health conditions.

In contrast, the female construction workers are affected by the gastro-intestinal disease. 50% females in their responses said that "*Me and my other colleagues suffered from constipation, irritable bowel, hemorrhoids, anal fissure, perianal abscess, anal fistulas, perianal infections, diverticular disease, colitis, colon polyps, and cancer. When I showed my symptoms to the government doctor the doctor said that this problem has occurred due to being engaged in construction work or having a wrong lifestyle*". Even the female construction workers work day and night too hard to earn a living. Even the pregnant women were seen working in the construction areas without having any break in between. At times they faint in between and lose their child.

The next question was related to their appetite. Researchers tried to explore whether these workers feel hungry or not.

Almost 60% of the males reported that they feel hungry. Some of them said that "*You feel hungry after doing so much work. After eating the food I go to sleep*", "*When I feel hungry, I cook for myself and eat all the food*". This initiates that after a long day they do not have enough strength to do other work and while eating the food, they move towards their bed.

However, 100% of the females feel hungry but they were not able to satisfy their hunger (Table 24.2). As there is not enough food for them and they are more concerned about the hunger of their children. One of them discussed that "*I feel very hungry, but I am unable to eat even after wishing. The amount of work I do, I do not get enough money to feed the entire family. If I eat the whole chunk what would my children and family members eat?*".

This shows their impoverished state where despite working the whole day they do not even save a handful of food to feed themselves and their entire family. This is the

Table 24.2 Perceived appetite of construction workers

Categories	Percentage	
	M (%)	F
Yes	60	–
No	20	–
Yes, but not able to satisfy hunger	20	100%

Table 24.3 Complaints about pain

Categories	Total	
	M (%)	F (%)
Pain related to whole body	60	10
Pain related to lower part of body	20	70
Pain related to upper part of body	20	20

reason why despite being unhealthy, they do not take even a day's rest and continue to earn as most of them are daily wage earners.

Table 24.3, discussed the pains that these construction workers go through while working hard daily. While 60% of the male construction workers feel pain related to the whole body, 70% of the female workers feel pain related to their lower body parts. Some of the verbatims include—"*There is pain in the whole body. Whenever I work for long hours in a standing posture or work from sitting, prolonged sitting or lying down, doing a new physical activity etc. creates unbearable pain. I feel like leaving the job but due to constraints of my family responsibility I am unable to do so.*"

One of the female construction labourers said that—"*I experience pain in the lower back and legs which is caused by overwork…. The pain is such that no medicine works. Even the medicine given by the doctor has no effect on my body*", "*At times I eat 4-5 painkillers to bear my pain but it is also of no use. Even in such a severe pain I have to go for work again to earn a livelihood*". These narratives clearly reflects that some strict laws need to be generated so that even after working for such a long day with a tedious task they are able to feed their children and hospitals need to charge only a basic amount from these workers.

The next question that was asked from them was "*Have you faced any physical deterioration within your body after joining this job?*".

Functional fitness is the ability by which a normal human being can easily do all his work, all work in his workplace without stopping, without tiredness, and he does not even feel tired and in doing this task he maintains full interest (Kostic et al. 2011). Both males and females reported that they have experienced physical deterioration after joining this job (Table 24.4). They are unable to move properly, they have muscular pain in their body and at times feel like lying down on bed and do nothing. Males reported that "*Because of working so hard and because of continuous work, physical ability has decreased, does not even have strength to do easy work at home, the body is getting weak before time*".

Table 24.4 Physical deterioration experienced by construction labourers

Categories	Percentage	
	M	F
Yes	100%	100%
No	–	–

Females are facing problems worse than males as they have to complete household tasks before and after going to work. They reported, "*I have to do the tedious jobs both at work and at home. Sometimes it feels like running away from these places, but do not have a choice*".

24.3.2 Part II: Perceived Health—Emotional

According to Bharti (2020), "Emotional well-being refers to a state of mental well-being in which a person realizes his/her abilities, can cope with the normal stresses of life, work productively and fruitfully, and in creating a future for the family and society. Emotional well-being describes one's psychological and social well-being. Our mental health reflects our eating habits, levels of physical activity, substance use behaviours and how we think, feel and face difficult situations. We face mental health every day. A person's mental health is just as important for their physical health, and mental health conditions are as real as physical illnesses. It is important to keep this in mind during your interactions and interactions with others". The second part deals with the emotional well-being of these construction workers. It includes four questions that focus on the mental health issues like:

- Cause of mental health issues
- Impact of working conditions on mental health
- Satisfaction with work
- Satisfaction with salary.

Table 24.5 on mental health issues brought forth four major categories for mental health issues. Almost 70% of the male respondents face work related mental health issues. They have safety issues related to the places where they work. The conditions are not very safe and at times safety issues are completely missing in the construction sites where they work. Despite the various laws and regulations made for the safety of construction labourers most of the industries are not following the safety rules which hampers the mental and physical health of their employees. At times, their life is also at stake. Some of the verbatims by male respondents clearly brought forth the dangers and stress these people face in their lives. One of them said "*Mental problems are faced by us most of the time. We are most worried when we go to work.*

Table 24.5 Cause of mental health issues

Categories	Percentage	
	M	F
Related to work	70%	–
Related to family	00	60%
Related to health	10%	–
Related to future	20%	40%

Table 24.6 Impact of working conditions on mental health

Categories	Percentage	
	M	F
Anxiety	–	10%
Week memory	–	70%
Deterioration in overall mental health	20%	20%
Week concentration	20%	–
Problems related to emotion	60%	–

I wonder whether I will come back home alive or not." Another male reiterated "*If something happens to me, what will happen to our parents, what will happen to the children? There is not much protection provided in workplaces where I work. Even I do not receive a good amount of money. If something happens to me, my family will not even get a two times meal. I do not receive enough money to save for their future*".

While males are more worried related to their work, females have issues related to family. 60% of the females are concerned about their family and their safety. They reported "*We are too much concerned about our family. We are needy so are forced to work in such unsafe conditions. Our children will follow our path and will be engaged in the same profession. There is neither much money nor safety in this work. I have worked like this my whole life but will not be able to see my children working in this environment. This anxiety always persists in my mind*".

It is very important to do our work to maintain health. The more work we do, the more we keep ourselves busy, the better our mental health gets, but if the working conditions are not favourable, then it adversely affects our mental health, due to which we get many types of mental You may also have to face diseases like—anxiety, depression, stress and Tension etc. The table on impact of working conditions on mental health brought forth five significant responses (Table 24.6). 60% male construction workers predominantly reported in the category of problems related to emotion; "*Due to too much work, I am no longer able to control my anger. I get angry at small things, feel sad at small things, my own emotions are out of my control, I start worrying very quickly, I feel like I have lost control of myself somewhere. I start screaming too in difficult situations*". On the other side 70% female construction workers reported weak memory with narrative like "*Memory has become very weak due to working. Many hours have to be worked, no time is available for us and I have been engaged in this work for many years. There is no time to think for ourselves in this work. We work 15 to 16 hours out of 24 hours due to which our memory is getting weaker now. We are not able to remember things which happened in the evening and morning, forget a lot of things, problems occur when we make a mistake in our place of work, and Wages are cut*".

All the respondents reported that they are not satisfied with their work as the amount of hard work required is too much in comparison to the amount of money they receive. Besides that the working conditions do also affect their mental and physical health. Some of the verbatim of males are—"*No, not satisfied with work at all. Satisfaction is attained only when everyday needs are easily met. We do not even get two times of meal properly so what about our other basic needs. If we are unable to meet our needs, where will the satisfaction come from?*".

Females also gave similar statements where they pointed that "*I do not think any person doing this work is satisfied, and will not get satisfaction either. I am paid only a menial amount so I am unable to meet the basic needs of my children. I am unable to send them to schools, unable to provide them with food and clothing. I am also unable to take them to a good doctor when they are ill as I am unable to meet their fees*".

This response clearly reflects that despite working for the whole day and risking their life the satisfaction level of the employees are completely missing. In contrast, they have physical and mental health issues. A lot of dissatisfaction with work was found among both male and female construction workers (Table 24.7). As most of the people working in different fields seem to be dissatisfied with their work but continue to do it as they receive a good amount of salary to satisfy their daily needs. Thus, a pertinent question appears to the researcher's mind was whether they were satisfied with the salary they received?

Table 24.8 shows that only a menial amount of people were satisfied with their salary. 90% of the males and 80% of the females were not satisfied with their salaries. They have many complaints with their employers "*I am not satisfied with the salary I receive. The amount of money I receive ends in a day or two. I have to work extra hard for my family so that if I am unable to go to work any day I have some savings so that I can feed my family*".

Another respondent said "*I am ashamed to share with my friends and family about the money that I receive. It disturbs my mind and even does not feed my family properly*".

Table 24.7 Satisfaction with work

Categories	Percentage	
	M	F
Yes	–	–
No	100%	100%

Table 24.8 Satisfaction with salary

Categories	Percentage	
	M (%)	F (%)
Yes	10	20
No	90	80

24.3.3 Part III: Perceived Health—Environmental

Wellbeing with respect to environment reflects the degree to which one experiences a sense of belongingness and social inclusion. Part III of the interview schedule deals with the environmental well-being of the construction workers. They were asked about the safety issues, uneven health conditions due to the environment in which they work and the dangers in the construction sites. This dimension takes up issues like:

- Perceived safety in the job environment
- Safety equipments available in the construction sites
- Physical damage due to unsafe working conditions
- Risks at construction sites.

The environment around a person reflects the thinking of a particular place, its environment and the people living there, as well as it creates a unique picture of personality development, because the more pessimistic and uncooperative our environment is, it creates equal disappointment in our personality.

Both males and females have safety issues in their job environment. They accepted that their life is at stake while working on the construction sites (Table 24.9). At times even when there are safety measures they are at risk of losing their lives. They reported "*There is risk of life in the construction sites where they work. Almost in 15 days, one or two accidents definitely took place. At times we do not have any safety equipment in the places where we work*".

Table 24.10 brought forth 5 categories which show that 40% of the males receive safety shoes while working in the construction sites while 40% of females report that they receive safety hats while working in their job. Most of them emphasized that they do receive only one or two items for their security while most of the items

Table 24.9 Perceived safety in the job environment

Categories	Percentage	
	M	F
Yes	–	–
No	100%	100%

Table 24.10 Safety equipments available at the construction sites

Categories	Percentage	
	M (%)	F
Safety hat	20	40%
Safety belt	20	–
Safety shoes	40	20%
Safety gloves	10	20%
Safety vest jacket	10	20%

that they receive are very old and broken and thus, has to be borrowed from others. Apart from a few equipments they do not receive anything else. This information needs to be used very cautiously as this indicates that people stake their life to earn a livelihood and brokers even do not provide them with proper safety equipment which is necessary to save their life.

Next question deals with the physical damage that is being experienced by the construction workers while working in the construction sites.

Table 24.11 brought forth 5 categories which show that 60% of the males reported that construction sites are unsafe and harmful to health. Work in construction is related to physical ability. Construction workers mostly perform tasks that involve heavy and powerful machinery, use such powerful machines and climb to heights which presents a very frightening picture in front of us. Unfortunately, due to the inherent nature of the work, male construction workers are at serious risk of injuries at work such as the loss of broken bones and sprains due to slipping and falling on a ladder with narrative "*Due to foot slipping while working, there was a sprain problem in the leg, due to which I could not work for 1 month. After that it happened two or three times and my legs got stretched. Bone of my hand was also injured because of these type of working circumstances*". On the other hand 50% of females report that most of the accident are related to ladder or a ladder's collapse at the construction sites with verbatim like "*The hand was broken due to slipping from the ladder, and also suffered a lot of injuries. The doctor had to apply plaster, the work was stopped, and the earning was also stopped. Every day there are some injuries*".

Table 24.11 Physical damage due to unsafe working conditions

Categories	Percentage	
	M	F
Damage of hand and leg—ladder accident/ladder collapse	20%	50%
Broken bones, sprains, and other injuries—slip and falls	60%	–
Electrocution—electrical equipment	10%	10%
Serious accidents—crane or hoist	–	20%
Accidents by heavy machines	10%	20%

Table 24.12 Risks at construction sites

Categories	Percentage	
	M (%)	F
Gas leaks, fires, and explosions	10	–
Elevator shaft accidents	40	–
Forklift accidents	10	20%
Falling objects	20	80%
Repetitive motion injuries	20	–

Table 24.12 brings forth 5 major categories of the various risks that exist while working in the construction sites. Most of the males (40%) face elevator shaft accidents while most of the females (80%) experience falling objects accidents. Some of the verbatim by male respondents are; "*Many times we have seen accidents from the elevator with our own eyes. Feel very scared to work on it. Whenever any goods have to be taken upstairs with the elevator, it seems whether I will come back down or not, or I will go from above*".

Females reported "*In construction work, there is a fear of falling heavy loads from the top, and while working, the heavy loads fall from the top, which often results in losing their lives*".

24.4 Implications and Suggestions

The findings strongly suggest the need to provide intervention in the form of psychosocial support to these construction workers. Need for counselling services was essentially felt. Counselling services by trained mental health professionals should be enforced in all construction sites, both public and private. It is also clear to the researcher that there is a need for occupational safety technicians and inspectors/observers who are responsible for areas related to construction. These professionals will be responsible for visiting construction projects as well as the health and safety of workers. Thus, the minimum requirements regarding accident prevention can be met and the mental and physical problems of workers can be minimized.

The present research has focused only on the construction workers of Lucknow. Other areas and cities which are popular for construction could also be taken up for further research. Interventional studies may be done to target population (life style, continuous working hour, support system etc.) and its effect can be assessed on quality of life of construction workers. The present research has primarily focused on the construction workers' psychophysical health and safety. Therefore other similar conditions could also be explored. A comparative study could also be taken up between construction sites workers and other employees (as building surveyor, building service engineer, site manager, trade and craft workers etc.) in construction sites. Effect of intervention programs on these children could also be studied.

24.5 Conclusion

The results clearly reflect that the construction workers go through a lot of stresses, strains and health related issues. Despite working the whole day, they hardly receive two meals a day. They suffer from various diseases and health risk problems. Poor working conditions and lack of financial incentives demotivate them to work under this area.

Thus, the governmental policies and acts need to be modified and their strict implementation is required. Most of the acts or policies focus on the physical aspects of the construction labourers and their psychological health is completely ignored. If the construction workers are being provided with both mental and physical security then they will be motivated to work in this area. As there are major requirements of such workers we need to enhance their motivation without affecting their health. There are various safety management issues for such workers and their life is at stake.

More research is required in this area to find out ways to improve their safety and give them psychological relief. There is a need to find policies which are strictly laid to help them physically, emotionally and financially. The paper concluded that the unique perception of construction workers on psychophysical health and safety issues. Further studies can be conducted on such psychophysical health and safety issues that can be adopted to improve the health and safety regarding mental and physical aspects. There is a requirement that employers can always check legislation and draw up proper health and safety issues to the employer's workplace and employees.

References

Bharti J (2020) Psychological health among caregivers of person with mental illness. Mukta Shabd J 9(12):1141–1146

Camino LMA, Ritzel DO, Fontaneda I, González OJ (2008) Construction industry accidents in Spain. J Safety Res 39(5):497–507

Cheng CW, Lin CC, Leu SS (2010) Use of association rules to explore cause–effect relationships in occupational accidents in the Taiwan construction industry. Saf Sci 48(4):436–444

De Vos AS (2002) Intervention research. In: De Vos AS, Strydom H, Fouché CB, Delport CSL (ed) Research at grass roots: for the social science and human service professions, 2nd edn. Van Schaik, Pretoria

Fisher J, Katz LA, Miller K, Thatcher A (2003) South Africa at work: applying psychology to organisations. Witwatersrand University Press, Johannesburg

Gottfredson GD, Duffy RD (2008) Using a theory of vocational personalities and work environments to explore subjective well-being. J Career Assess 16(1):44–59

Jahoda M (1982) Employment and unemployment: a social-psychological analysis. Cambridge University, Department of Social and Political Science, Cambridge University Press

Kahn AP, Fawcett J (2004) The encyclopedia of psychiatric disorders, vol III. Mental Health

Khahro SH, Ali TH, Talpur MAH, Akhund MA, Siddique FH (2016) Occupational health problems in construction industry: a case study. Sindh Univ Res Jour (Sci Ser) 48(4):737–740

Kostic R, Pantelic S, Uzunovic S, Djuraskovic R (2011) A comparative analysis of the indicators of the functional fitness of the elderly. Facta Univ Ser Phys Educ Sport 9(2):161–171

Kumar D (2013) Inimitable issues of construction workers: case study. British J Econ Finance Manage Sci 7(2):42–53

Lamers SMA, Westerhof GJ, Bohlmeijer ET, ten Klooster PM, Keyes CLM (2011) Evaluating the psychometric properties of the mental health continuum-short form (MHC-SF). J Clin Psychol 67(1):99–110

Monnery N (1998) The costs of accidents and work related ill-health to a cheque clearing department of a financial services organisation. Saf Sci 31(1):59–69

Nazeer MT, Ilyas M, Butt JI, Gill SA, Shoaib V, Fayyaz M, Jabbar A (2016) Impact of fitness awareness on physical fitness and exercise activities of visitors in public parks. Sci Int (Lahore) 28(5):99–102

Odusami KT, Oyediran OS, Oseni AO (2007) Training needs of construction site managers. Emirates J Eng Res 12(1):73–81

Ojedokun O, Idemudia ES (2014) Psycho-socio-emotional well-being of workers in a high-stress occupation: are men and women really so different? Gender Behav 12(3):5824–5838

Paruchuru M, Mavuri S, Jyothsna M (2020) Challenges for economic growth in India—a critique. J Crit Rev 7(7):169–175

Peyton R, Rubio T (1991) Construction safety practices and principles. Van Nostrand Reinhold, New York. *[hardcover1991-ISBN: 9780442237424; Publisher: Van Nostrand Reinhold]*

Ridner SL, Newton KS, Staten RR, Crawford TN, Hall LA (2016) Predictors of well-being among college students. J Am Coll Heal 64(2):116–124

Russo C, Terraneo M (2020) Mental well-being among workers: a cross-national analysis of job insecurity impact on the workforce. Soc Indic Res 152(2):421–442

Saeed YS (2017) Safety management in construction projects. J Univ Duhok 20(1):546–560

Singleton R, Straits BC, Straits MM, McAllistair RJ (1998) Approaches to social research. Oxford University Press, New York

Veeramani A (2019) Policy reforms for reversing slowdown and accelerating GDP growth in the Foundation for Economic Growth and Welfare. Working paper no 1/2019. https://egrowfoundation.org/research/policy-reforms-forreversing-slowdown-and-accelerating-gdp-growth

Vitharana VHP, Subashi De Silva GHMJ, De Silva S (2015) Health hazards, risk and safety practices in construction sites—a review study. Engineer 48(3):35–44

Warr P, Cook J, Wall T (1982) Scales for the measurement of some work attitudes and aspects of psychological well being. J Occup Psychol 52:129–148

WHO (1948) Constitution of the World Health Organization. World Health Organization, Geneva, Switzerland

Yadav M (2015) Economic conditions of house construction workers. Int J Res Sci Innov 2(7):57–59

Chapter 25
Effect of Different Building Materials on Indoor Radon/Thoron and Associated Health Hazards

Bhupender Singh, Maneesha Garg, and Krishan Kant

Abstract Diverse building materials are utilized for construction all over the world. These emit radioactive pollutants and are liable for continuous exposure of ionizing radiations (radon/thoron) to the indoor environment. As per World Health Organization, radon is recognized as the second significant cause of lung disease subsequent to smoking. Thus, it is of much importance to measure the level of these radioactive gases in dwellings which are constructed using different building materials. The purpose of this chapter is to analyze the impact of different building materials on indoor radon and thoron levels and to find out radon prone regions as it has associated health hazards. In this chapter, indoor radon (^{222}Rn), thoron (^{220}Rn) activity and radon/thoron progeny level were simultaneously measured in 150 houses of twenty locals of district Palwal, Haryana, India. Passive detectors (pinhole twin-cup dosimeter and direct radon/thoron progeny sensors) were utilized for time-integrated monitoring of the exposure period of four months. The variation of radon and thoron gases was observed such as H2 (mud house) > H1 (cemented house) > H3 (traditional house) > H4 (modern house) and H2 (mud house) > H1 (cemented house) > H4 (modern house) > H3 (traditional house) respectively. Similar trends were observed in case of progeny.

Keywords Building materials · DRPS/DTPS · Effective dose · Indoor environment · Pinhole dosimeter · Progeny · Radon/thoron

25.1 Introduction

The primordial radionuclides like ^{238}U and ^{232}Th in soil, rocks, and water are accountable for the presence of ^{222}Rn and ^{220}Rn in indoor and outdoor environments. A radionuclide decays into its daughter products and releases energy as alpha, beta or gamma particles. Then the newly formed nuclei become the head of the decay chain

B. Singh (✉) · M. Garg
J.C. Bose US&T, YMCA, Faridabad 121006, Haryana, India

K. Kant
Aggarwal College Ballabgarh, Faridabad 121004, Haryana, India

J. A. Malik and S. Marathe (eds.), *Ecological and Health Effects of Building Materials*,
https://doi.org/10.1007/978-3-030-76073-1_25

and further decay. Radon and Thoron are the daughter nuclides of ^{238}U and ^{232}Th decay chain. Uranium-238 decay series indicates that the primordial radionuclide uranium-238 is unstable radionuclide and therefore decays into thorium-234 with the release of an alpha particle of energy 4.2 meV to attain stability. But thorium-234 is also unstable and further decay into uranium-234 by emitting beta particles. Uranium-234 further decays into thorium-230 by emitting an alpha particle. Thorium-230 decays into radium-226 by emitting an alpha particle. This radium isotope (^{226}Ra) is a generator of Radon-222 after releasing an alpha particle. Similarly, the thorium-232 decay series indicates that thorium-232 primordial radionuclide is also unstable and goes to radioactive decay. After three alpha decays it reaches Thoron (Radon-220). In both, the decay series only Radon and thoron radionuclides are found in gaseous form. Thus, radon and thoron further decay to attain stability. Decay products of radon are listed as short-lived and long-lived progeny. The short-lived decay products of Radon are ^{218}Po, ^{214}Pb, ^{214}Bi, ^{214}Po, and long-lived decay products are ^{210}Pb, ^{210}Bi, ^{210}Po. Thoron short-lived decay products are ^{216}Po, ^{212}Pb, ^{212}Bi, ^{212}Po. Uranium-238 decay series stops at lead-206 which is a stable nuclide. In this decay series 8 alpha, 6 beta, and associated gamma energy are released. Thorium-232 decay series stops at lead-208 which is a stable nuclide. In this decay series 6 alpha, 4 beta, and associated gamma energy are released. In the periodic table Radon comes in between metals and nonmetals. Thus it is metalloid and has properties of both. It is present diagonally in the periodic table. It is an inert gas and chemically non-reactive. The chemical symbol of Radon is Rn with atomic weight 222 and atomic number 86. It has protons and electrons, 86 protons and 86 electrons in its atomic structure. As the electro-negativity decreases with increase in the atomic number in the column of the periodic table, Radon has low electro-negativity compared to other noble gases. The solubility increases with an increase in atomic number therefore Radon is more soluble than other inert gases. Its solubility is more in organic liquids as compared to water. Also, it was observed that heat vaporization increases and ionization energy decreases with increase in atomic number.

Radon and thoron are inert, carcinogenic and alpha emitter gases. Radon and thoron reach the air viz., indoor environment through many paths. The soil under or nearer the dwellings and construction material are the major sources of the radon gas to the indoor environment whereas only building materials are major sources for thoron gas to the indoor environment. Radon and thoron gas reached the environment from rocks and soil grains through two fundamental processes. The first and the second process is emanation and exhalation from the material grain and the matrix using many transport mechanisms. The process through which radon atoms liberate from the solid mineral grains to the air filled pores is known as emanation. Consequently, transportation of radon gas from the pores of the air to the atmosphere is called exhalation. The transportation of radon in soil pores is significantly due to advection brought out by pressure driven flow of soil gas and diffusion brought out by concentration gradient. The molecular weight of radon is almost 8 times that of air so travels closer to the ground and the progeny can be accumulated as solid radioactive offshoots on water, vegetation and surface of soil. The diffusion of the radon in air is due to Brownian motion. It is well understood that molecular diffusion and advection are responsible for Radon/Thoron transportation through emanation and exhalation processes.

Radon gas reaching the dwellings can accumulate for a longer time (half-life is 3.8 days) and can travel up to 3 m in the air. Thoron gas cannot migrate up to longer distances because of its shorter half-life (55.6s). Firstly, Radon/Thoron gases reach the indoor environment from building materials, soil nearer, or under the ground surface. They diffuse to the indoor environment and decay by emission of alpha particles. The decay products of Radon/Thoron are the isotopes of polonium, bismuth, and lead. These decay products are the combination of coarse and fine fractions of polonium, bismuth, and lead. Most of these progenies are positively charged particles and less are neutral particles. These charged particles can attach with the air vapors and trace gases present in the atmosphere and form clusters and further deposits to surfaces (Fares et al. 2011). Otherwise, these charged particles can interact with aerosol and sulfur dioxide and then become neutral and deposit to the surface. This deposition is mainly governed by the gravity force and Brownian diffusion. Thus, it can occur either by transport towards the surface or by precipitation of aerosol particles on the surface.

The pressure difference, moisture, porosity of the medium, permeability, and temperature has a significant influence on radon concentration in soil. Radon and thoron are transported to the indoor environment through cracks in surface soil, joints of walls and floors of dwellings, pores in buildings materials, etc. The source of radon into the indoor environment is through water. Household activities such as showering, laundering, dishwashing and other activities are the ways to transport radon to an indoor environment. Parameters like temperature, atmospheric pressure, ventilation condition, moisture content, permeability of soil, emanation coefficient, etc. affect the concentration of radon in the environment (Janik et al. 2015).

The World Health Organization reported that over exposure of radon may be considered as a key factor of lung diseases for non-smokers (WHO 2009). During breathing, they may enter the human body and can damage the bronchial epithelial living cells. Secretion and Bronchial stem cells in airways are the primary target cells for lung cancer induction due to the exposure of radon. The deposited daughters of the radionuclide decay and emit ionizing particles such as alpha, beta, and gamma. The densely ionizing alpha radiation causes the potential damage to the deoxyribonucleic acid (DNA) of lung cells. This will cause the initiation of the chain of events leading to lung cancer. However beta particles and gamma rays have lower biological effectiveness and longer range thus have negligible effects on lung tissues. Health hazards from radon are relatively smaller than what is expected from its progeny–^{218}Po, ^{214}Po, ^{214}Pb, and ^{214}Bi due to their longer half-life. The major contribution (>50%) of total dose of ionizing radiation received by the general public is due to radon, thoron, and their progeny. The results of occupational investigations as well as residential studies indicate the human carcinogenicity of radon (IARC 2011). Thirteen case controls studies in Europe and seven case control studies in North America indicated that increase in indoor radon gas is associated with lung cancer risk (Darby et al. 2005). It is predicted by Elío et al. (2018) that household radon can considerably increase the risk of lung diseases. Thus, due to its radiological impact on humans it is very important to simultaneously measure radon, thoron and their progeny levels in indoor regions.

Various investigations were carried out earlier for simultaneous estimation of radon and thoron gases by passive methods in India and other countries (Singh et al. 2019a). Solid state nuclear track detectors based passive equipment such as pinhole dosimeters, Raduet dosimeters, direct progeny sensors were used for these measurements worldwide. In the present investigation, LR-115 type II film based pinhole dosimeters and direct progeny sensors were used to register the alpha tracks formed due to decay of radon/thoron and their progeny. In the past many studies have focused on radon measurements using passive methods. However, thoron had been often neglected and considered as an interrupting factor in the measurement of radon. This may not be entirely correct in the context of India which has regions consisting of thorium rich soil and high background radiation areas (HBRAs). In addition, contrary to the gas measurement, limited investigations were done for direct measurement of decay products. The estimation of decay products concentration is usually calculated from gas concentrations using the equilibrium factor approach. But, for estimating thoron decay products concentration this may not be an appropriate method. Moreover, since the inhalation doses are dominantly because of the decay products of radon and thoron, and not due to the gases, it is important to measure the decay products directly. Thoron along with its progeny is also dominant contributor to annual effective dose due to inhalation as reported in many investigations. Moreover in Indian scenario, the thoron contribution for inhalation dosimetry has also been acknowledged. The exponential decay in thoron concentration from the surface of the wall has been observed. However, a quantification of dose contribution either from thoron or its progeny alone suggests that the majority of dose will come from thoron progeny (~98%) with a very little contribution (~2%) from thoron gas. Therefore in this study, thoron gas and its progeny have been monitored separately using a radon-thoron discriminating dosimeter (Sahoo et al. 2013) and Direct Thoron Progeny Sensor (Mishra and Mayya 2008) respectively and both quantity have been used for dose calculations. Hence the estimated dose will be dominated by the measured value of thoron progeny which is more or less uniform in dwellings and gives reliable results. However, for the completeness of dose, we have added the marginal contribution by thoron gas too.

The aim of this chapter is to compare the impact of different building materials on radon/thoron levels. Secondly, the data will be compared with the safety limits recommended by various agencies like WHO, UNSCEAR, ICRP, etc. For the present work, the area chosen for investigation is under reported and from a geographical point of view it is very important to understand the effect of radiation in this region. This study is conducted under a major project provided by the Board of Research in Nuclear Sciences (BRNS), Bhabha Atomic Research Center (BARC), Mumbai, Government of India. The project will cover the radiation measurement in soil, water, and air of this region. The data provided by this paper will be helpful for the researchers to understand the effect of building materials on radiations. This study is a part of seasonal monitoring (summer, rainy and winter seasons) of indoor radon/thoron and their progeny level in villages of district Palwal, Haryana, India.

25.2 Geology of the Study Area

The study area of the present investigation is located in the southern part of the state of Haryana of Northern India. The latitude and longitude extend from 27°50′29″N to 28°12′30″N and from 77°17′47″E to 77°22′47″E. The area includes 282 villages and the city region of Palwal. It is bounded by districts Gurugram and Faridabad in north-west, by district Mewat in west, and by state Uttar Pradesh in east. About 270 km^2 are irrigated by surface water sources and 770 km^2 by ground water sources. The entire study region has almost flat plains. The soil of this region is tropical and brown. Organic contents in soil are in the range of 0.2 to 0.4% and pH of soil lies in between 6.5 and 8.5. The underground water sources such as borewells, hand-pumps, tap water, etc. and surface water sources such as river Yamuna, Gurugram and Agra canals are present in this region. Tropical and brown soil is present in the major part of district Palwal. Geo-morphological information of the district indicates that organic content in soil varies from 0.2 to 0.4% and in Hathin block it varies from 0.41 to 0.75%. The pH of soil varies from 6.5 to 8.7. The entire study region has almost flat plains. In Palwal, 770 km^2 are irrigated by borewell and 270 km^2 by canals. The sand and gravel are major water-bearing formations. Hydro-geological information of the district describes the study region engrossed by the Indo-Gangetic alluvial plain of the Quaternary age. The main underground water horizon made up of alluvium comprises gravel, kankar, and sands silt. This hydro-geological, geo-morphological, and geological information of this region is based on a report of the Central Ground Water Board (CGWB 2013), government of India.

25.3 Materials and Methods

25.3.1 Preliminary Survey of the Study Area

For impactful study and systematic investigation, a survey has been done before the deployment of detectors. Outdoor gamma level was measured during the preliminary survey to categorize the region into zones to get the zone-wise distribution of radionuclides. During this survey gamma level is measured at a height of 1 m from the ground to avoid any lead or interference by decay products generated in air. The progeny of radon gas (^{214}Pb, ^{214}Bi and ^{210}Pb) and thoron (^{212}Pb, ^{212}Bi and ^{208}Tl) present in the soil are the main sources of gamma radiations. Therefore, quantification of outdoor gamma level is also performed to explore any correlation with parent nuclides such as radon and thoron gases. The Geiger Muller counter based Survey meter (Polimaster PM/1405, Garmin Instrument, Republic of Belarus) was used to measure outdoor gamma level at one meter height from the earth surface. Survey meter incorporates a large energy compensated Geiger Muller tube for precise measurement of the ambient equivalent dose rate of the gamma radiation in the range from background level to 100 mSv/h (10 R/h). It has a gamma energy response from

0.05 to 3 meV and can be used for dose rate measurement varying from 0.01 to 130 mSv h^{-1} suggesting suitability for environmental gamma surveys. It has a calibration accuracy of $\pm(20 + 1/H)$ % where H is the dose rate in μSv h^{-1}. Also, the different types of residential houses in the study region were observed which further categorized based on construction building materials.

25.3.2 Categorization of Investigated Houses

The building material of houses, underground and surrounding surface soil, water used in houses are major contributors to indoor radon level while only building materials are a major source of indoor thoron level as it has a very short half-life. Several building materials such as gypsum, black cement, white cement, stone dust, bricks, marble, tiles, granite, and POP were tested by researchers to find out the level of radon concentration, radon mass exhalation rate, and thoron surface exhalation rates. A wide variation was observed in the level of these radioactive elements in these building materials. It indicated that the different building materials have different impacts on these radioactive elements therefore a systematic investigation was required for it. Therefore the dwellings of the study region were categorized into four types viz H1, H2, H3 and H4 on the basis of building materials used for construction as shown in Fig. 25.1a, b, c, and d respectively.

The first category (H1) included the dwellings having roofs made up of girder and stone slab, walls of houses made up of fired bricks covered with plaster layer and floor covered with cement-plaster layer. Generally, these types of dwellings are present in most of the study region area and nearby areas of similar geological conditions.

The second category (H2) included the dwellings having thatched roofs, walls of houses made up of fired bricks covered with a layer of mixture of clay soil and cow/buffalo dung and open ground floor with coats of the same mixture used on walls. However, the quantity of such dwellings is less compared to the H1 category but peoples all over study regions used these dwellings in present time also. Therefore, this category was also a point of interest for monitoring the radionuclide elements.

The third category (H3) included the dwellings having roof made up of concrete and beam and the roof structure standing on the columns of beam, walls of houses made up of fired bricks and columns at corner and middle of walls which covered with plaster layer and floor covered with cement-plaster layer. These types of dwellings are replacing the category of H1 in the present scenario. Dwellings which were made up of H1 category building materials when damaged over a long period of time are replaced by H3 category. Thus, this category was also a point of interest for measurements of radionuclide pollutants level.

The fourth category (H4) is named as modern houses. It includes the dwellings having roof made up of concrete and beam and the roof structure stand on the columns of beam same as of third category, walls of houses made up of fired bricks and columns at corner and middle of walls which covered with plaster layer and floor covered with

Fig. 25.1 Interior view of investigate houses **a** houses of type H1, **b** houses of type H2, **c** houses of type H3, and **d** houses of type H4 of district Palwal, Southern Haryana, India

cement-plaster layer, also the walls (up to 3 to 4 feet height) and floor covered with tiles, marble, stones etc.

However, there were two more categories viz mud houses and Haveli (tradition in some regions). Mud houses were made up of thatched roofs, walls of houses made up of clay pieces with a layer of mixture of clay soil and cow/buffalo dung and open ground floor with coats of the same mixture used on walls. However, the quantity of such dwellings is very less compared to other residential dwellings. Haveli were made up of small fired brick walls with an open or covered floor. Stone pillars were part of the attraction in this category and had a roof made up of wooden pieces and stone slab. But these were very few in quantity and not to be used for residential or work purposes by the public in current time therefore neglected for investigation.

25.3.3 Pin-Hole Based Dosimeter and Deposition Based Direct Progeny Sensors (DRPS/DTPS)

Pin-hole dosimeter relies upon radon/thoron isolation technique. The length and the radius of the two compartments of the pin-hole dosimeter are 4.1 cm and 3.1 cm respectively. The segregation of the two compartments is done with the help of the disc having thickness 2 mm and 4 pin-holes of diameter 1 mm. The front compartment is a radon + thoron chamber and the rear one is of radon. The complete dimension of the dosimeter is chosen in such a way that the thoron entry into the rear compartment is prohibited. The paper of thickness 0.56 μm is placed at the entry face. The air containing both radon and thoron enters into the front chamber and subsequently the air containing only radon diffuses to the rear chamber through pin-holes. Inner surface of the dosimeter compartments and central disc is coated by metallic substances such as nickel to form a neutral electric field inside the compartment volume. It helps to uniform deposition of charged progeny throughout the inner surface of the dosimeter. Solid-State Nuclear Track Detectors (SSNTDs) are insulating solids widely used for passive measurements. These detectors include plastics, inorganic crystals, glasses, etc. Cellulose nitrate (CN 85, LR-115), allyl diglycol carbonate (CR-39), bisphenol-A polycarbonate (Makrofol, Lexan), etc. are used as SSNTDs. To measure the activity of radionuclides in the air or in powder samples the SSNTDs are widely used. The alpha track etch technique is one of the most widely used techniques to register the tracks created from ionizing radiations (alpha particles). When the radionuclides such as radon and thoron decay they emit alpha particles which can be detected by using SSNTDs films. The alpha particles when passed through the passive detectors release their energies and leave the tracks in detector films. The registration of tracks (latent tracks) in a given SSNTD depends on the orientation, energy, etc. of ionizing particles. These tracks cannot be visualized through a scanning electron microscope and optical transmission microscope because the size of latent tracks is very small (diameter in the range of 1–10 nm). Therefore, by using suitable chemical etchant or reagent the size of these tracks can be enlarged. Thus, fully developed tracks can be visualized or counted by a transmission optical microscope also or by using a spark counter. SSNTDs are easy to handle, are unaffected by humidity, store data up to many years, have low cost, used for time integrating measurements, etc. The type-II LR-115 film (Kodak Path, France) having 12 μm thick cellulose nitrate on a 100 μm thick polyester base has been used to record the tracks generated by alpha particles due to radon and thoron gases inside the chamber. The track recording efficiency of these detectors ranges from 1.7 to 4.8 meV.

DRPS (deposition based direct radon progeny sensors) and DTPS (deposition based direct thoron progeny sensors) were used for measurement of radon/thoron progeny levels in the indoor dwellings. A combination of LR-115 film and a suitable absorber has been used in DRPS and DTPS. DRPS comprises LR-115 film and an absorber of 37 μm thickness (25 μm Mylar sheet and 12 μm cellulose nitrate). The detecting efficiency of DRPS is upto alpha particles of energy 7.67 meV emitted from Polonium-214. DTPS comprises LR-115 film and an absorber of 50 μm thick Mylar sheet. The detecting efficiency of DTPS is upto alpha particles of energy 8.78 meV emitted from Polonium-212. The number of tracks estimated by DRPS and DTPS are used for the calculation of EEC (Equilibrium Equivalent Concentration) using the suitable sensitivity factor. The minimum detection limit for DTPS is 0.1 Bq m^{-3}, whereas that for DRPS is 1.0 Bq m^{-3} which arises due to intrinsic background track density. Intrinsic background tracks are those tracks which are registered on the detector films during transit, manufactured, or packaging period. Detectors having a size of 2.5 cm × 2.5 cm were loaded in DRPS and DTPS were used for the present investigation. The pin holes twin cup dosimeter has been calibrated against standard radon and thoron sources (Model RN 1025 and TH 1025, Pylon, Canada) in a 0.5 m^3 calibration chamber available at Bhabha Atomic Research Centre (BARC), Mumbai, India. Relative humidity controls from 10 to 99% and temperature from 20° to 50 °C in the calibration chamber. DRPS/DTPS were calibrated with active Working level monitors from Tracer lab, Grab-filter-paper sampling and alpha-counting at BARC, Mumbai India.

Dosimeters along with DTPS/DRPS were deployed based on the weight factor assigned to each category of houses after regression analysis. Detectors were deployed for a period of four months (July–October) during the rainy season in district Palwal, Haryana, India according to the standard protocol of Bhabha Atomic Research Center, Mumbai. Detectors were collected back from the dosimeters and progeny sensors on completion of the monitoring time. During this exposure period, tracks are registered on the detectors. These tracks cannot be visualized through a transmission optical microscope because the size of latent tracks is very small (diameter in the range of 1–10 nm). Therefore, by using suitable chemical etchant or reagent the size of these tracks can be enlarged. Thus, fully developed tracks can be visualized or counted by an optical transmission microscope or spark counter. In this study, a constant etching bathtub (model PSI-CTB1) is used for the etching of detectors. It has three compartments (tub), a temperature controller, a timer, a heater coil, and a pump for the circulation of water. The first calibration of the equipment has been carried out by the manufacturer. However, we also calibrated it for bulk etch removal rate of unexposed detector films. A solution of 2.5 N NaOH is prepared and filled in all three compartments of the tub. The temperature is set at 60 °C temperature and after 25 min the solution in all compartments are checked by the thermometer to ensure the temperature of the solution. The detectors are marked by punching at the corner and loaded in a cartridge and put into the compartments. The timer is set for 90 min at this temperature. After the completion of etching time, the cartridges are removed from the compartments and washed with flowing tap water. The detectors are then washed in distilled water and dried for one hour and now the detectors are ready for spark counting.

After the etching process, the next step is to find out the number of tracks on detector films registered due to the alpha particles. Spark counter (we used model PSI-SC1) is used for this purpose. In the spark counter, the thin etched track detector (about 8–10 μm thick) is placed between two electrodes forming a capacitor. The bottom electrode is a thick conductive electrode, commonly made of brass. The thin LR-115 detector is placed on this electrode. The aluminized Mylar is placed on the detector such that the aluminized surface faces the detector as well as the thin electrode. A heavy weight is placed on the top to ensure good contact between the electrodes, the detector, and the aluminized film. These track densities were used to estimate the radon/thoron activity and progeny concentration.

The radon (C_R) and thoron (C_T) gas level were calculated from the number of tracks per unit area observed in exposed LR-115 detector and given by Eqs. (25.1) and (25.2) respectively;

$$C_R\left(\mathrm{Bq\,m^{-3}}\right) = \frac{(T_1 - B)}{d \cdot K_R} C_R \tag{25.1}$$

$$C_T\left(\mathrm{Bq\,m^{-3}}\right) = \frac{\left(T_2 - d \cdot C_R \cdot K_{R'} - B\right)}{d \cdot K_T} C_R \tag{25.2}$$

where T_1 track density (rear chamber), B is the number of tracks per unit area raised from background, d is monitoring days, K_R has the value 0.017 ± 0.002 tr cm^{-2} d^{-1} Bq m^{-3} (rear chamber calibration factor), T_2 track density (front chamber), $K_{R'}$ has the value $0.0172 \pm 0.0\,02$ tr cm^{-2} d^{-1} Bq m^{-3} (front chamber calibration factor for radon) and K_T has the value 0.010 ± 0.001 tr cm^{-2} d^{-1} Bq m^{-3} (front chamber calibration factor for thoron).

The EETC is calculated from Eq. (25.3) and the EERC from Eq. (25.4);

$$EETC\left(\mathrm{Bq\,m^{-3}}\right) = \frac{(T_T - B)}{S_T} \tag{25.3}$$

$$EERC\left(\mathrm{Bq\,m^{-3}}\right) = \frac{(T_{Rn} - B)}{S_R} \tag{25.4}$$

where T_T is the track density in DTPS, S_T and S_R are sensitivity factors for thoron and radon progeny respectively, T_{Rn} track density from radon progeny in DRPS;

$$T_{Rn}\left(Bqm^{-3}\right) = T_{DTPS} - (\eta_{RT}/\eta_{TT})T_{DRPS} \tag{25.5}$$

where T_{DRPD} and T_{DTPS} are total number of tracks in DRPS and DTPS respectively, η_{RT} (0.01 ± 0.0004) and η_{TT} (0.083 ± 0.0004) track registration efficiency for thoron progeny in DRPS and DTPS respectively.

Total annual effective dose (AED_{Rn+Th}) due to inhalation was estimated. It is the sum of annual effective dose calculated from measured concentration of radon (C_{Rn}) along its progeny that is EERC (equilibrium equivalent radon concentration) AED_{Rn} and calculated from thoron (C_{Th}) along its progeny that is EETC (equilibrium equivalent thoron concentration) AED_{Th}.

Annual effective dose due to inhalation of radon and its progeny is calculated from equation

$$AED_{Rn}(mSv\ y^{-1}) = (C_{Rn} \times FC_{Rn} + EERC \times FC_{EERC}) \times 8750 \times O_F \times 10^{-6} \tag{25.6}$$

Annual effective dose due to inhalation of thoron and its progeny is calculated from equation

$$AED_{Th}(mSv\,y^{-1}) = (C_{Th} \times FC_{Th} + EETC \times FC_{EETC}) \times 8750 \times O_F \times 10^{-6} \tag{25.7}$$

where FC_{Rn} (0.17 nSv Bq^{-1} h^{-1} m^3) and FC_{Th} (0.11 nSv Bq^{-1} h^{-1} m^3) are dose conversion factors for radon and thoron concentration respectively, FC_{EERC} (9 nSv Bq^{-1} h^{-1} m^3) and FC_{EETC} (40 nSv Bq^{-1} h^{-1} m^3) are dose conversion factors for radon progeny concentrations and standard occupancy factor (O_F) is 0.8 for 1 year exposure period.

25.4 Results and Discussion

25.4.1 Distribution of Radionuclides

A heterogeneous distribution of radon, thoron, and their progeny concentrations was observed in this study. Literature review revealed that the distribution of radon and thoron are affected by local geology of testing sites, environmental parameters such as moisture contents, atmospheric pressure, temperature difference inside and outside of the houses, radon emanation factor of soil under the surface of floors, etc. The measured radon and thoron activities vary from 4 to 175.1 Bq m^{-3} with an average of 28.2 ± 2.1 Bq m^{-3} and from 2.1 to 195.2 Bq m^{-3} with an average of 29.5 ± 2.8 Bq m^{-3} respectively. The overall thoron concentration was found higher as compared to radon level in district Palwal, Haryana, India. It is due to thorium rich soil in the earth's crust of India. The average radon level observed here is less than the world average value reported for indoor dwellings of 40 Bq m^{-3} (Singh et al. 2019a) and also less than the indoor radon reference level of 100 Bq m^{-3} (WHO 2009) and 200 Bq m^{-3} of ICRP (2014). Radon was found higher than 100 Bq m^{-3} in three dwellings and it can be attributed to natural geology of location, more exhalation of radon from joints of walls or cracks, etc. Therefore, further measurements are required at these locations to ensure the reasons for high values of radon. The average thoron level was observed at 29.5 ± 2.8 Bq m^{-3} which is nearly 3 times higher than the worldwide average value reported for dwellings of 10 Bq m^{-3}. However this will not significantly affect the total dose contribution as thoron has negligible effect on the total dose.

The average radon and thoron concentrations are found highest in mud houses (type H2) which can be attributed to excessive emission of these radioactive gases from the open floor (mud surface). The walls of these houses were made up of fired bricks covered with a layer of clay soil and cow/buffalo dung therefore continuous emission of radon and thoron gases from cracks and opening of walls are responsible for elevated levels. Also, the dimensions of these types of dwellings in comparison with other categories of houses is generally small resulting in improper ventilation and raised the level of radon and thoron. It is also in agreement with results of other investigations (Sannappa and Ningappa 2014; Suman et al. 2020). Radon concentration found second highest in cemented houses (type H1). In cemented houses walls are of fired bricks covered with cement plaster and floor is also cemented; this reduces the emission of radon and thoron gases from earth's crust due to low permeability of cement plaster. The average radon concentration is found lowest in modern houses (H4). However use of enhanced materials like marble, granite, tiles, stones etc. on walls and floors of the dwellings is expected for higher radon emission inside the houses. But this trend was not observed in our study. Thus the results in the present investigation contradict the results of other investigations in the case of elevated radon levels in modern houses (Singh et al. 2019a). Most probable reason for low levels of radon in modern houses is the proper ventilation conditions of rooms having two doors and a window which leads to air exchange between rooms and the outdoor atmosphere. The variation of radon and thoron gases was observed such as H2 (mud house) > H1 (cemented house) > H3 (traditional house) > H4 (modern house) and H2 (mud house) > H1 (cemented house) > H4 (modern house) > H3 (traditional house) respectively.

The variation of EERC is from 1.1 to 41.4 Bq m^{-3} and has an average of 9.1 $\pm$ 0.02 Bq m^{-3} and EETC is from 0.2 to 6.8 Bq m^{-3} and has an average of 1.2 $\pm$ 0.01 Bq m^{-3}. As per the ICRP, limits of average EERC and EETC are 2–50 Bq m^{-3} and 0.04–2 Bq m^{-3} respectively and the present data confirms that the progeny concentration is within the limits in this study area. The average values of EERC (9.1 $\pm$ 0.02) and EETC (1.2 $\pm$ 0.01) in the present study were found below the world average values of 15 Bq m^{-3} for EERC and 0.5 Bq m^{-3} for EETC. In the present investigations, thoron concentration was found higher than radon concentration but reversed in case of progeny. This can be attributed to the different effects of environmental parameters such as temperature and pressure gradient, moisture content, and ventilation conditions, etc. on the gases and their solid decay products. The overall concentration of radon progeny was found higher than the thoron progeny. It is presumably due to higher deposition velocities of radon progeny than the thoron progeny on detector surfaces. Mean values of decay products of radon and thoron gases are found highest in mud houses (type H2) similar to the results observed for radon and thoron gases. The variation of decay products of radon and thoron was observed such as H2 (mud house) > H1 (cemented house) > H3 (traditional house) > H4 (modern house) and H2 (mud house) > H1 (cemented house) > H4 (modern house) > H3 (traditional house) respectively same as observed for radon and thoron gases.

The box whisker plots of measured indoor radon, thoron, radon progeny (EERC), and thoron progeny (EETC) are shown in Fig. 25.2. The upper and lower whisker represents maximum and minimum concentration value respectively. The top line

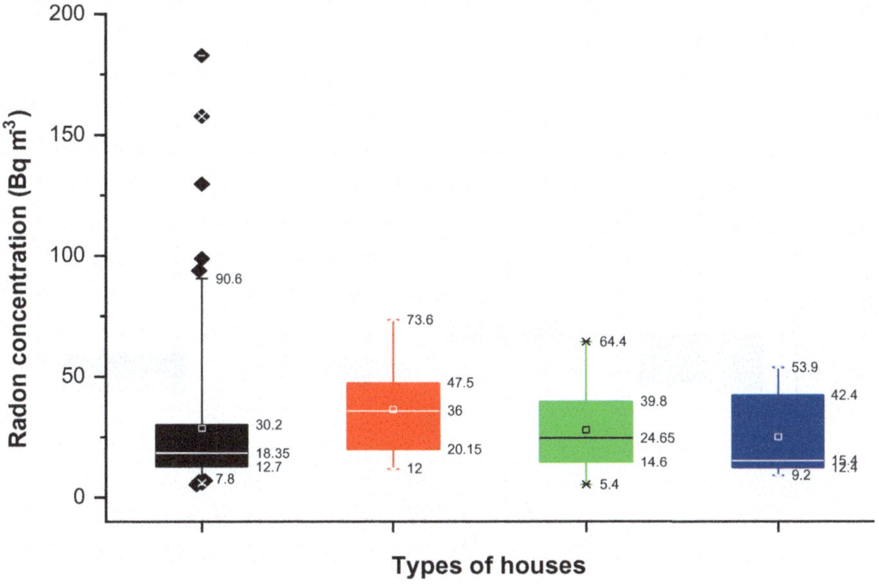

(a)

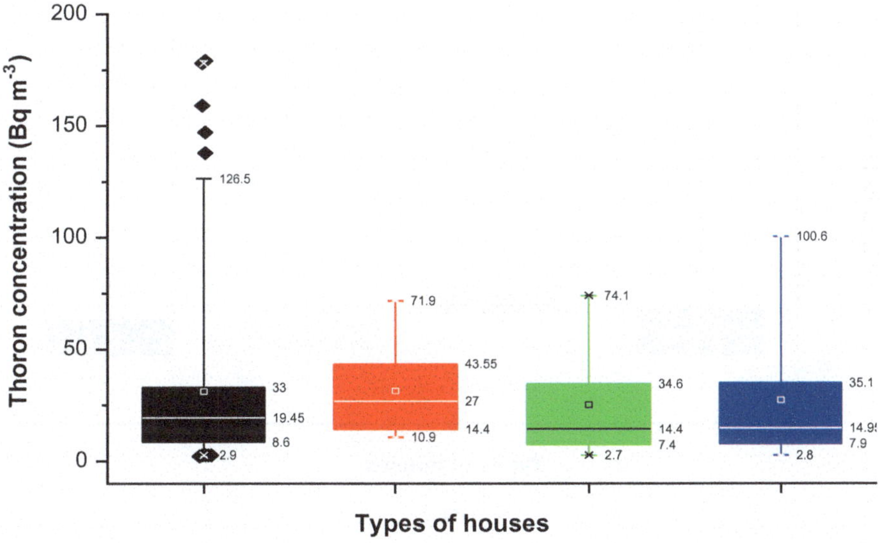

(b)

Fig. 25.2 Box-whisker plots of measured **a** radon, **b** thoron, **c** EERC, and **d** EETC in different type of houses of district Palwal, Southern Haryana, India (for H1 type represents by black colour plot, for H2 type represents by red colour plot, for H3 type represents by green colour plot, and for H4 type represents by blue colour plot)

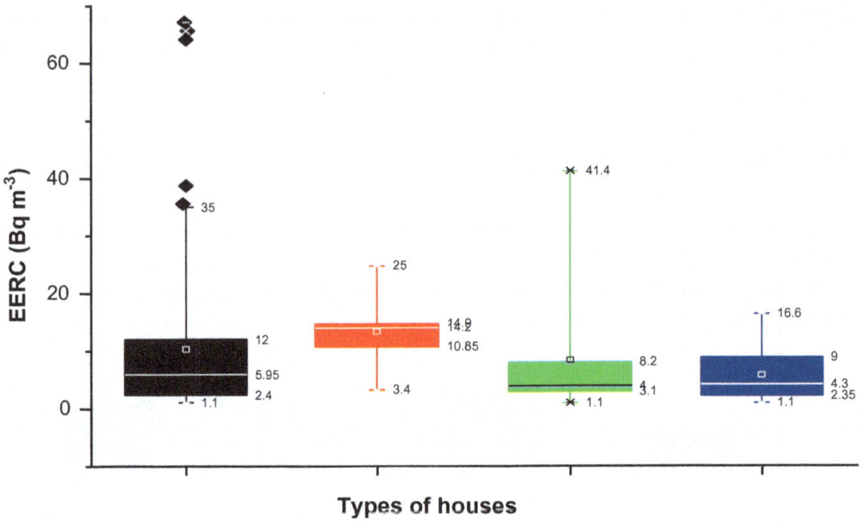

(c)

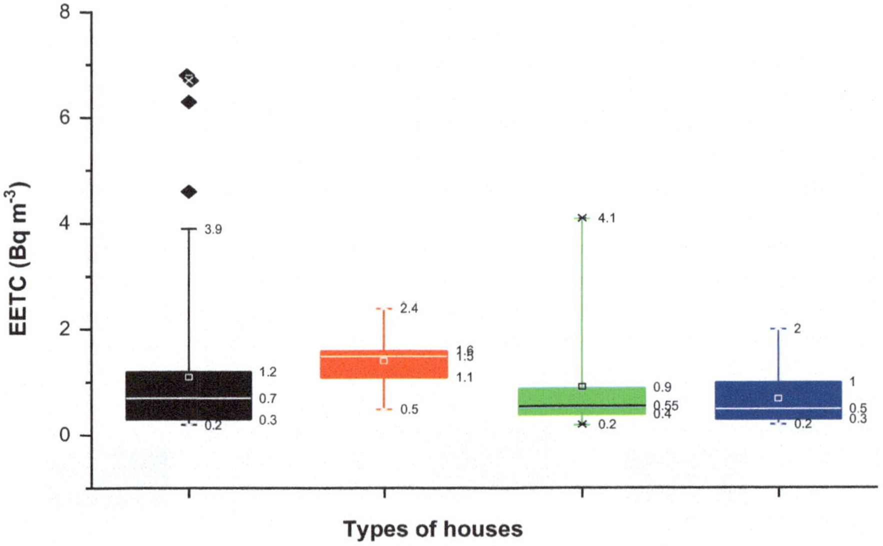

(d)

Fig. 25.2 (continued)

of the box indicates the third quartile that is 75 percentile, the middle line indicates the second quartile or median that is 50 percentile, and the lower line represents the first quartile that is 25 percentile. Outliers (represented by dots above or below the maximum or minimum values in box-whisker plot) are values which lie 1.5 times greater or lower than the top or below lines of the box. The minimum, maximum, mean, first quartile, median, and third quartile values for radon and thoron and their progeny concentration for different types of houses are shown in Table 25.1.

Table 25.1 Variation of radon, thoron and their decay products (EERC and EETC) concentration in different type of investigated houses for rainy season of district Palwal, Haryana, India

Parameters	Statistical parameter	House category			
		Type H1	Type H2	Type H3	Type H4
Radon (Bq m^{-3})	Min	5.2	11.9	5.4	9.2
	Max	90.6	73.5	64.4	53.9
	Mean	47.4	36.6	27.9	25.2
	1st quartile	12.7	20.1	14.6	12.4
	2nd quartile (median)	18.3	36.0	24.6	15.4
	3rd quartile	30.2	47.5	39.8	42.3
Thoron (Bq m^{-3})	Min	2.9	10.9	2.7	2.8
	Max	126.5	71.9	74.1	100.6
	Mean	31.1	31.6	25.1	27.3
	1st quartile	8.6	14.4	7.4	7.9
	2nd quartile (median)	19.4	27.0	14.4	14.9
	3rd quartile	33.0	43.5	34.6	35.0
EERC (Bq m^{-3})	Min	1.1	3.4	1.1	1.1
	Max	35	25	41.4	16.6
	Mean	10.3	13.0	8.2	5.9
	1st quartile	2.4	10.8	3.1	2.3
	2nd quartile (median)	5.9	12.0	4.0	4.3
	3rd quartile	12.0	13.0	8.4	9.0
EETC (Bq m^{-3})	Min	0.2	0.5	0.2	0.2
	Max	3.9	2.4	4.1	2.0
	Mean	1.1	1.5	0.7	0.7
	1st quartile	0.3	1.1	0.4	0.3
	2nd quartile (median)	0.7	1.4	0.6	0.5
	3rd quartile	1.2	1.6	0.7	1.0

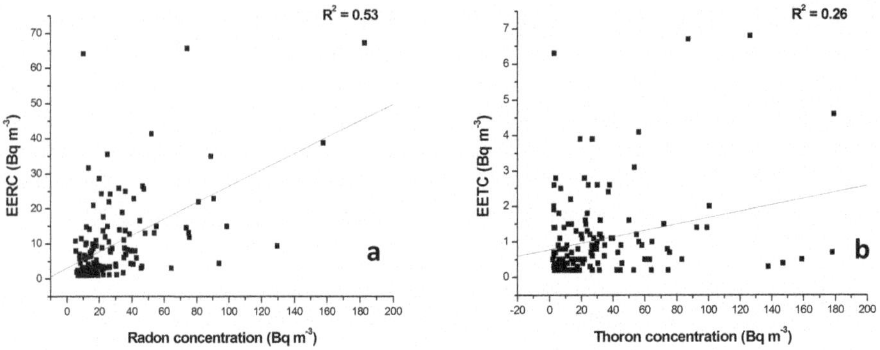

Fig. 25.3 Scatter plot along with Pearson's correlation (R^2) coefficient between **a** radon with EERC and **b** thoron with EETC

25.4.2 Correlation Among Gases and Their Progeny

The scatter plot along with Pearson's correlation coefficient (R^2) between radon and EERC and between thoron and EETC are shown in Fig. 25.3. Radon and EERC have moderate positive correlation ($R^2 = 0.53$) whereas thoron and EETC found to have a weak positive correlation with ($R^2 = 0.26$). Radon and thoron are the gases whereas their decay products are heavy metals and isotopes of lead, bismuth, and lead. Also, the half-life of parent nuclei that is radon and thoron is less compared to their daughter products. Therefore, solid decay products stay for a longer period in the environment compared to the parent nuclides. This weak correlation between thoron and EETC is attributed to the strong influence of moisture contents and ventilation conditions in dwellings of the study region. Also, the environmental parameters affect differently both the parent nuclides (gases) and their daughter nuclides (solid).

25.4.3 Frequency Distribution of Radon, Thoron, and Their Progeny

Frequency distribution of radon, thoron, EERC, and ETC in dwellings of study region is shown in Fig. 25.4. It indicates that 87% of dwellings have radon concentration below the value 50 Bq m^{-3}, 80% of dwellings have thoron concentration below the value 50 Bq m^{-3}, 97% of dwellings have radon progeny concentration below the value 50 Bq m^{-3}, and 97% of dwellings have thoron progeny concentration below the value 4.5 Bq m^{-3}. Three locations in case of radon and 7 locations in case of thoron exceed the value of 100 Bq m^{-3}. Since, the contribution of thoron itself is negligible towards the total annual effective dose due to inhalation of thoron along its progeny therefore it indicates that the thoron is not hazardous in dwellings of the study region. This heterogeneous distribution is most likely due to topography,

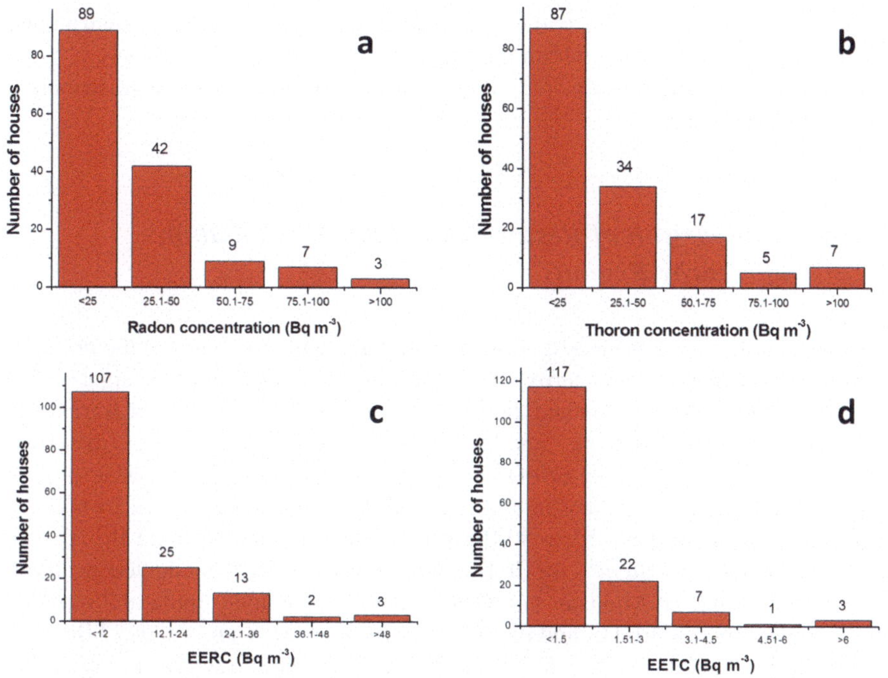

Fig. 25.4 Frequency distribution of measured **a** radon, **b** thoron, **c** EERC, and **d** EETC concentration in houses of district Palwal, Southern Haryana, India

diverse geological location of investigated houses, different building materials used in construction of investigated dwellings, influence of environmental factors such as moisture contents, temperature gradient in dwellings, etc.

25.4.4 Annual Effective Dose Due to Inhalation

Annual effective dose due to inhalation of radon and its progeny is found in the range of 0.07 to 1.11 mSv y^{-1} with an average of 0.29 $\pm$ 0.01 mSv y^{-1}. Annual effective dose due to inhalation of thoron and its progeny is found in the range of 0.03 to 0.38 mSv y^{-1} with an average of 0.15 $\pm$ 0.01 mSv y^{-1}. The estimated total annual effective inhalation dose due to radon, thoron, and their decay products varies from 0.1 to 1.1 mSv y^{-1} with an average of 0.4 $\pm$ 0.01 mSv y^{-1}. Singh et al. (2015) reported that the total annual effective dose rate due to inhalation in Tosham region of Haryana, India varies from 1.33 to 2.44 mSv y^{-1}. Singh et al. (2019a) reported that total AED_{Rn+Th} in the region of district Faridabad of Haryana, India varied from 0.15 to 0.45 mSv y^{-1}. Kumar et al. (2020) reported that total AED_{Rn+Th} in the region of district Dadri of Uttar Pradesh, India varies from 0.29 to 2.06 mSv y^{-1}. Thus, results of the present study are similar to the results of nearby regions. Estimated total annual effective dose was observed within the prescribed safe limits of 3–10 mSv y^{-1} and 10 mSv y^{-1}. It concludes that no radiological hazards are associated with distribution of these radionuclides.

25.4.5 Comparison of Results with Other Investigations of Nearby Regions

The results of the present investigation are compared with outcomes of the previous investigations carried out in nearby regions of India. Sannappa and Ningappa (2014) reported that the indoor radon and thoron in the nearby granite region of Karnataka, India varies from 16–170 Bq m^{-3} and 18–300 Bq m^{-3} respectively. Singh et al. (2015) conducted an indoor investigation of radon, thoron, EERC, and ETC in Tosham region of Haryana, India in rainy season and reported that their values varies from 37 to 80 Bq m^{-3} for radon, from 53 to 80 Bq m^{-3} for thoron, from 12 to 23 Bq m^{-3} for EERC, and from 2 to 7 Bq m^{-3} for EETC. Bangotra et al. (2019) conducted an indoor study in the houses of Muktsar and Mansa districts of Punjab, India and reported that the radon, thoron, EETC, and EERC varies from 19–88 Bq m^{-3}, 22–77 Bq m^{-3}, 11–50 Bq m^{-3}, and 0.7–7 Bq m^{-3} respectively. Singh et al. (2019a) reported that radon, thoron, EERC, and ETC varies from 5.3–128.8 Bq m^{-3}, 9–183.6 Bq m^{-3}, 1.1–18.9 Bq m^{-3}, and 0.1–1.9 Bq m^{-3} respectively in the dwellings of district Faridabad of southern Haryana, India. Kumar et al. (2020) reported that radon, thoron, EERC, and EETC in the nearby region of national capital power station, district Dadri of state Uttar Pradesh of India, varies from 9.7–64.9 Bq m^{-3}, 34–90 Bq m^{-3}, 3.3–27.2 Bq m^{-3}, and 0.3–1 Bq m^{-3} respectively. Thus, it concluded that the results of the present investigation are comparable with results of nearby regions of India.

25.4.6 Seasonal Comparison of Results of Present Investigation

The outcomes of summer season (exposure period of 4 months) for the same study region are published (Singh et al. 2019b) and in the present chapter, results of measurements have been performed for the 2nd season that is the rainy season (exposure period of 4 months) are reported. Thus, the inter-comparison will provide the seasonal variation of radioactive gases and their solid decay products. The average values of measured indoor radon and thoron were found 28.2 $\pm$ 2.1 and 29.5 $\pm$ 2.8 Bq m^{-3} in the present investigation i.e. rainy season and 28.6 $\pm$ 0.03 and 30 $\pm$ 0.04 Bq m^{-3} in summer season respectively. No significant change in the average values of indoor radon and thoron is observed. Moreover, the lifestyle of peoples of the present study region is comparatively different in both seasons. In the summer season people mostly use fans, coolers, air-conditioners, etc. in their homes than during the rainy season. Thus, ventilation conditions are different in both seasons which lead to comparatively high levels of radon and thoron in the summer season. But it contradicts our study. However, significant change observed in distribution of radon and thoron dwelling wise. In the rainy season radon and thoron levels are found maximum in mud houses and minimum in modern houses but it is completely opposite for summer season. Due to temperature gradients inside and outside the dwellings of mud in the rainy season, these gases accumulate up to longer time inside the dwellings.

25.5 Conclusions

The measured average indoor thoron gas concentration (29.5 ± 2.8 Bq m^{-3}) is about 3 times higher than the world average value of 10 Bq m^{-3}. The average values for radon concentration, EERC, and EETC found to be 28.2 ± 2.1, 9.1 ± 0.02, and 1.2 ± 0.01 respectively were found below the world average values of 40, for radon, 15 Bq m^{-3} for EERC, and 0.5 Bq m^{-3} for EETC. The overall level of thoron gas was found to be higher than radon but results were reversed in case of EERC and EETC.

Overall no significant change in average values of radon and thoron concentration is observed compared with summer season outcomes. However, dwelling wise comparison showed that results are completely opposite for mud and modern houses. The variation of radon and thoron gases was observed such as H2 (mud house) > H1 (cemented house) > H3 (traditional house) > H4 (modern house) and H2 (mud house) > H1 (cemented house) > H4 (modern house) > H3 (traditional house) respectively. Similar trends were observed in case of progeny. The higher concentration of gases and their solid decay products were found in dwellings of zone 2 than zone 1.

Radon and EERC have moderate positive correlation with Pearson's correlation coefficient ($R^2 = 0.53$) whereas thoron and EETC found to be weak positive correlation with Pearson's correlation coefficient ($R^2 = 0.26$).

Frequency distribution of radon, thoron, EERC, and EETC shows that the radionuclides in the indoor region are widely distributed. This heterogeneous distribution is presumably due to topography, different geological location of investigated houses, different building materials used in construction of investigated dwellings, influence of environmental factors such as moisture contents, temperature gradient in dwellings, etc.

The estimated total AED_{Rn+Th} varies from 0.1 to 1.1 mSv y^{-1} with an average of 0.4 ± 0.01 mSv y^{-1}. The measured concentration of radon, thoron, EERC, and EETC are found within prescribed limits of WHO, ICRP, and UNSCEAR. It indicates that no radiological hazards are associated with radon, thoron, and their progeny in the present study region.

Acknowledgements This work is supported by the Board of Research in Nuclear Sciences, Government of India, through Project No. (36(4)/14/2014-BRNS/36018) dated 26/02/2016. Authors are also thankful to the people of district Palwal for their cooperation during field work.

References

Bangotra P, Mehra R, Jakhu R, Pandit P, Prasad M (2019) Quantification of an alpha flux based radiological dose from seasonal exposure to ^{222}Rn, ^{220}Rn and their different EEC species. Sci Rep 9(1):1–15. https://doi.org/10.1038/s41598-019-38871-6

CGWB (2013) Ground water information booklet of district Palwal, Haryana. Central Ground Water Board, Ministry of water resources, Government of India, North Western region, Chandigarh

Darby S, Hill D, Auvinen A, Barros-Dios JM, Baysson H, Bochicchio F, Heid I (2005) Radon in homes and risk of lung cancer: collaborative analysis of individual data from 13 European case-control studies. BMJ 330(7485):223. https://doi.org/10.1136/bmj.38308.477650.63

Elío J, Crowley Q, Scanlon R, Hodgson J, Zgaga L (2018) Estimation of residential radon exposure and definition of Radon Priority Areas based on expected lung cancer incidence. Environ Int 114:69–76. https://doi.org/10.1016/j.envint.2018.02.025

Fares S, Yassene AA, Ashour A, Abu-Assy MK, Abd El-Rahman M (2011) Natural radioactivity and the resulting radiation doses in some kinds of commercially marble collected from different quarries and factories in Egypt. Nat Sci 3(10):895. https://doi.org/10.4236/ns.2011.310115

IARC (International Agency for Research on Cancer) (2011) Monographs on the evaluation of carcinogenic risks to humans, vol 78, Ionizing radiation, Part 2: Some internally deposited radionuclides. Lyon, France

ICRP (International Commission on Radiological Protection) (2014) Radiological protection against radon exposure. ICRP Publication 126. Ann ICRP 43(3)

Janik M, Omori Y, Yonehara H (2015) Influence of humidity on radon and thoron exhalation rates from building materials. Appl Radiat Isot 95:102–107. https://doi.org/10.1016/j.apradiso.2014.10.007

Kumar M, Kumar P, Agrawal A, Sahoo BK (2020) Measurements of ^{222}Rn, ^{220}Rn and their progeny concentrations indoors around a coal/gas based power plant and estimation of annual inhalation dose to the public. J Radioanal Nucl Chem 326:65–74. https://doi.org/10.1007/s10967-020-07289-0

Mishra R, Mayya YS (2008) Study of a deposition-based direct thoron progeny sensor (DTPS) technique for estimating equilibrium equivalent thoron concentration (EETC) in an indoor environment. Radiat Meas 43(8):1408–1416. https://doi.org/10.1016/j.radmeas.2008.03.002

Sahoo BK, Sapra BK, Kanse SD, Gaware JJ, Mayya YS (2013) A new pin-hole discriminated ^{222}Rn/^{220}Rn passive measurement device with a single entry face. Radiat Meas 58:52–60. https://doi.org/10.1016/j.radmeas.2013.08.003

Sannappa J, Ningappa C (2014) Indoor concentration of radon, thoron and their progeny around granite regions in the state of Karnataka, India. Radiat Prot Dosim 158(4):406–411. https://doi.org/10.1093/rpd/nct243

Singh B, Kant K, Garg M, Singh A, Sahoo BK, Sapra BK (2019a) A study of seasonal variations of radon, thoron and their progeny levels in different types of dwellings in Faridabad district, Southern Haryana, India. J Radioanal Nucl Chem 320(3):841–857. https://doi.org/10.1007/s10967-019-06544-3

Singh B, Kant K, Garg M, Singh A, Sahoo BK, Sapra BK (2019b) Radiological impact of radon and thoron levels in dwellings measured using solid state nuclear track detectors. In: AIP conference proceedings, vol 2142, no 1, AIP Publishing LLC, p 120002. https://doi.org/10.1063/1.5122498

Singh P, Singh P, Singh S, Sahoo BK, Sapra BK, Bajwa BS (2015) A study of indoor radon, thoron and their progeny measurement in Tosham region Haryana, India. J Radiat Res Appl Sci 8(2):226–233. https://doi.org/10.1016/j.jrras.2015.01.008

Suman G, Reddy KVK, Reddy MS, Reddy CG, Reddy PY (2020) Indoor radon and thoron in the vicinity of proposed uranium mining site: a case study at Dasarlapally village, Telangana State, India. Radiat Prot Dosim 189:1–8. https://doi.org/10.1093/rpd/ncaa032

WHO (2009) WHO handbook on indoor radon: a public health perspective. World Health Organization, Geneva

Chapter 26
Sustainable Techniques for Building Waste Disposal

Tarun Kumar Kumawat, Vishnu Sharma, Varsha Kumawat, Manish Biyani, Anjali Pandit, and Agrima Bhatt

Abstract The building industry plays an important role in setting up the infrastructure needed for socio-economic sustainability. Owing to rapid urbanization, the substantial development of the construction industry has led to the formation of building material waste that has a negative impact on the environment, such as pollution of soil, air, and water. Building material (BM) waste consists primarily of inert and non-biodegradable materials such as concrete, plaster, metal, masonry, non-ferrous metal, paper and cardboard, mortar, bricks, roofing tiles, glass, paints, pipes, electrical fixtures, wood, plastics, etc. These waste products contain large concentrations of toxic materials that have a detrimental effect on the atmosphere and the health of humans. During the building process, various hazardous compounds are released into the environment. The conventional approach to waste disposal has long been to deposit BM waste in sanitary landfills, but this would not be possible in the years to come. It is necessary to control and manage the production of BM waste. Sustainable management of waste produced from construction is becoming increasingly compulsory to protect public health, minimize environmental burden and preserve existing natural resources. Extensive research has been dedicated to encouraging the safe management and disposal of waste building materials. This chapter addresses the sustainability of waste from construction materials and describes the related management and disposal techniques for the protection of natural resources and the environment, such as reduction, reuse, recycling, and incineration.

Keywords Environment · Hazardous · Waste · Sustainable · Building · Bioremediation

T. K. Kumawat (✉) · V. Sharma · A. Pandit · A. Bhatt
Department of Biotechnology, Biyani Girls College, Jaipur 302039, Rajasthan, India

V. Kumawat
Naturilk Organic & Dairy Foods Pvt. Ltd., Jaipur 302012, Rajasthan, India

M. Biyani
Department of Bioscience and Biotechnology, Japan Advanced Institute of Science and Technology, Ishikawa, Japan

J. A. Malik and S. Marathe (eds.), *Ecological and Health Effects of Building Materials*,
https://doi.org/10.1007/978-3-030-76073-1_26

26.1 Introduction

Building material waste is defined as the solid waste generated from the construction sector, including construction activities, material procurement, finishing, renovations, reconstruction, and demolition (Ajayi et al. 2017; Jamal et al. 2021). Construction activities have been on the rise to accomplish extensive property development, better public transport, and enhanced infrastructure for increasing urbanization and lead to more abstraction of natural raw materials, which results in a significant increase in the production of building material waste (Shooshtarian et al. 2020a, b). Building material waste is a highly heterogeneous material mixture that usually includes metals, concrete, asphalt, mortars, stone, gypsum wallboard, timber, plastic roofing, and cardboard (El-Haggar 2007). At least 30% of the total solid waste produced around the globe accounts for building material waste. Building material waste is caused by unnecessarily organized supplies or by unskilled laborer's mishandling of materials (Ginga et al. 2020).

Building material waste production has reached 3 billion tonnes worldwide, with India, China, and the USA contributing more than 2 billion tonnes (Akhtar and Sarmah 2018). Building material waste is categorized between non-inert and inert waste, where the non-inert waste is dumped at landfills and inert waste is usually disposed of in urban fillings as reclamation products (Poon et al. 2013). A significant factor of building waste is non-inert waste (timber and wood) (Wu et al. 2019a). Building material waste account for a significant part of the solid waste taking up landfills on a global scale and contribute to environmental issues. Limited landfill areas, water pollution, energy consumption and greenhouse gas emissions, waste materials produced from building demolition have become a major challenge to sustainable development (Bribián et al. 2011; Ding et al. 2016a).

Building material waste dumped into forests, streams, and vacant lots causes erosion, pollutes wells, water levels, and surface waters, attracts pests, creates fire hazards, and detracts from the beauty of natural areas. Building material waste containing toxic contaminants such as asbestos, heavy metals, persistent organic compounds, and volatile organic compounds (VOCs) is also much harder to disposal (Esin and Cosgun 2007). Every year, a huge volume of building waste is produced all over the world. Many investigations have demonstrated that waste management fees are an efficient solution that can minimize waste generation and increase the rate of landfill diversion (Wang et al. 2019).

The building industry accumulates a substantial volume of pollution that has a substantial effect on the atmosphere and the flow of energy (Sozer and Sozen 2020). Waste from the building sector faces enormous difficulties for sustainable waste management. This is not only due to the vast quantities of waste produced during building activities but also to the inherent toxins present in these materials (Amaral et al. 2020; Raskovic et al. 2020). In this chapter, we are proposed to discuss building material waste, their impact on the environment, human health, and various means and ways for the management and safe disposal of waste building material.

26.2 Sources and Causes of Building Material Waste

Building waste typically refers to debris created by practices such as construction, destruction, and reconstruction, and describes what involves building waste from country to country (Hoang et al. 2020). There are many interrelated sources of building waste existing (Ekanayake and Ofori 2004). Based on the particular building methods used, the forms and composition of onsite waste generated are extremely variable (Osmani 2011). The building material waste is produced during the project from the pre-construction stage, construction stage, and the final finishing stage. In general, the building waste in the industry can be categorized in many divisions as shown in Fig. 26.1.

The major components of the building waste include cement concretes, bricks, steel, stone, wood/timber, and cement plaster, whereas the minor components include iron plastic conduits, pipes, glass, tiles, etc. (Shrivastava and Chini 2005).

Excessive cement mix or leftover mortar after work of building operation is overdue to rejection caused by the change in plan or incorrect quality of work, etc. Concrete is considered as waste in 2 forms: reinforced (building structural elements) concrete and non-reinforced (foundations) concrete (Ponnada and Kameswari 2015). Concrete, substances that make it an affordable material and easy to manufacture anywhere, is found in organic granules, cement, and water. About 12% cement, 80% bulk aggregates, and 8% water is used in conventional concrete (Barbuta et al. 2015).

During demolition, large quantities of bricks and brickwork mixed with cement, mortar, or lime are produced as trash. During the demolition of old buildings, the stone is produced as waste material. Metal waste is produced from tubes, transmission lines, and light sheet material used throughout the air vents, cords, and plumbing fixtures, and structural concrete. Various waste materials include plastic, glass, paper, etc. (Ponnada and Kameswari 2015). A fraction of demolition waste was analyzed by Briere et al. (2014) and found that 52.8% of the most efficient waste was classified as

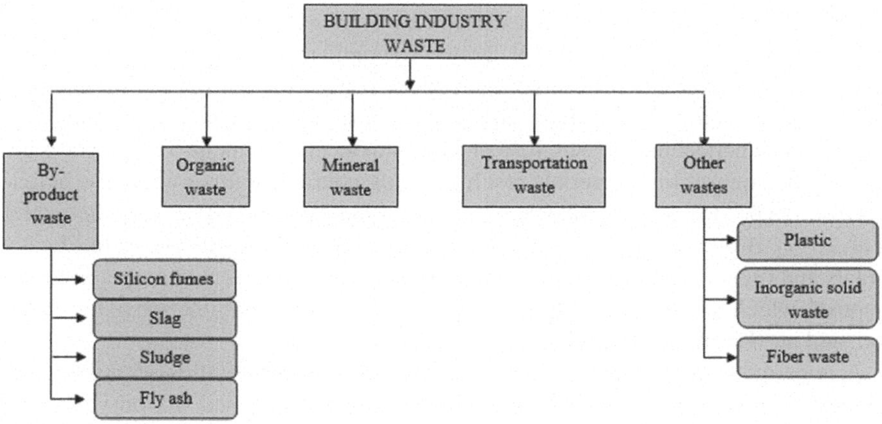

Fig. 26.1 Building material waste

masonry, 26.4% reinforced concrete, 9.3% mixed inert waste, and the other 11.5%. The key causes of waste in construction were categorized into six categories: design source, procurement, material handling, operation or service, residuals, and other sources (Gavilan and Bernold 1994; Bossink and Brouwers 1996; El-Haggar 2007).

26.3 Impact of Building Material Waste

Buildings are liable for the exploitation of resources and emission of pollutants in the environment (Feng and Lam 2021). As a result, the building sector has thus become a global target for reducing environmental impacts and curbing the depletion of resources (Hossain and Thomas 2019). In both urban and dynamically developing cities in the world, construction companies are booming (Hasan et al. 2021). Rapid urbanization has boosted unlimited construction in both developed and developing nations. Building material waste has been extensively produced which resulting in catastrophic and tragic effects in terms of economic standards and environmental conservation on urban health and longevity (Aslam et al. 2020). Improper management of building material waste affects our environment due to water, soil, and air pollution and have hazardous impacts on our ecosystem's flora and fauna, it is also responsible for social along with public health issues which can cause health problems, improper working and human safety and can have economic impacts because of loss of raw materials, resources and use of fuel supply during waste transportation (Asgari et al. 2017).

26.3.1 Impact on Environment

Traditionally, the building industry is environmentally unfriendly. Soil pollution, water contamination, and landscape erosion are among the environmental effects of building material waste (Fadiya et al. 2014). The rapid increase in building construction activities has resulted in large amounts of waste generation causing environmental change and pollution (Rani 2017). The construction industry consumes 35% of generated energy and releases 40% of CO_2 into our environment. This sector is the largest consumer of raw materials resulting from natural resources and these activities result in producing material waste that harms the environment (Luangcharoenrat et al. 2019). Building material waste comprising of hazardous substances like heavy metals and organic pollutants could result in leaching if not treated properly before disposal, which will be toxic to the atmosphere and cause contamination to the water, soil, and air (Huang et al. 2017).

A huge quantity of waste is produced due to building activities accounting for 20–30% of total solid waste of which 70–80% is concrete and masonry, thereby causing detrimental effects to both human life and the environment (Gupta 2018). Due to the continuous increase of building material waste, there is a shortage of landfill space which in turn causes pollution in the atmosphere because of illegal dumping in the world i.e., China, Malaysia, Hong Kong, and Israel (Yu et al. 2013). Building material waste landfilling's not only consumes limited landfilling resources but also contributes to an increase in energy consumption, greenhouse gas emissions, cause public health issues, and environmental contamination (Chen et al. 2017a, b; Mah et al. 2018; Zhang et al. 2019). These wastes are contributing to the global warming phenomena leading to an increase in temperature and weather extremes, further causing heat waves and poor air quality (Marzouk and Azab 2014).

Building material waste contributes to air pollution and water pollution mostly occurring from the transportation of waste, greenhouse gas emissions, and leachates released from the wastes in landfill areas (Yahya and Halim Boussabaine 2006). Sulfates present in gypsum drywall in building material debris are mainly responsible for causing harm to the environment and the excess of chemicals above the limit in water can cause a cathartic effect on humans (Jang and Townsend 2001). Improper disposal of building material waste limits any land reclamation job which could have taken place on large surfaces along with contributing to the contamination of air with toxic gases such as CO_2 and CH_4, two of the main factors responsible for the greenhouse effect at the planetary level (Iacoboaea et al. 2010). Other chief pollutants in construction activities are mostly particulate matter 10 (PM 10) and 2.5 (PM 2.5), N_2 and SO, and volatile organic compounds (VOCs) which attributes to 23% of air pollution present in cities, 50% of climate change through greenhouse gas emission, 40% of pollution in drinking water, 50% of landfill pollution and 50% of Ozone depletion (Jain et al. 2016).

26.3.2 Impact on Public Health

Building material waste is a major part of industrial waste. This waste is heterogeneous and comprises a large degree of building materials, but even considerable amounts of hazardous chemicals are included (Trankler et al. 1996). The building material waste has been disposed of in the sanitary landfills. Due to improper disposal of waste, heavy metals (HMs) such as Cd, As, Pb, Ni, Cr, Cu, and Zn gets leached into the water bodies, which poses a grave threat to the ecosystem and public health and can cause health problems like ulcers, liver damage, diarrhea, cancer, respiratory disorders, and cardiovascular disease, through unintended utilization of the heavy metals (Yu et al. 2018). The main constituents of building materials such as calcite and gypsum are responsible for causing respiratory tract infection and inflammation in the mucus membranes of the eyes, and their inhalation in form of these ultra-fine particles e.g. PM 2.5, could potentially lead to toxic respiratory effects (Oliveira et al. 2019).

Informal workers i.e., recyclers were also found to be suffering majorly from body pain due to lifting of heavy loads, back pain due to constant bending down, long working hours, and improper working conditions (Gutberlet and Baeder 2008). Dust is the fourth highest emission in percentage resulting from building activities impacting the environment directly by contributing to the direct visual, image, and health of the ecosystem and public health (Estevez et al. 2003). Some of the additional risks of handling building material waste involve asbestos removal whose fibers upon inhalation could be the causative agent for cancer and other lung diseases (Duan and Li 2016). The improper management of the building material waste affects the community's health and causes various adverse health effects and disorders associated with pollution, such as respiratory illness, birth defects, cancer, etc. (Aboginije et al. 2020).

26.3.3 Impact on Economy

The built environment is the core of every economy, providing the infrastructure needed to improve growth, but the way it consumes natural resources makes it responsible for some of the most serious changes in the local and global environment (Dania et al. 2008). Globally, the building industry is known as an economic investment, and its association with economic development is well asserted as many developing and developed nations have learned the significance of it at a socio-economic level necessary for employment and sustainable development of a country. It is vibrant for the progress of a nation since it provides necessary infrastructures and materials for activities like commerce, services, and utilities (Hasmori et al. 2020).

Building material waste is a major economic concern since the raw materials involved in the building elements are not further recycled, which is not economically and environmentally feasible (Marrero et al. 2017). Proper recycling of building material waste can also boost up the economy by creating a variety of jobs in the field of recycling and salvaging of building material waste which also helps in the business sense by creating a powerful social image, improving production efficiency and their profits, and hence improving product quality and overall environmental performance (Jain 2012; Oyenuga and Bhamidimarri 2015).

26.4 Traditional Disposal Strategies for Building Material Waste

The traditional method for waste management has often been the dumping of waste construction materials in landfills, but this would not be feasible in the years to come (Ginga et al. 2020). Maximum building material waste is transported for disposal to landfills that consume vast land resources and posing safety issues because of the excessive accumulation. In the metropolitan lifestyle, the growth in construction and urbanization technology has led to a significant quantity of constructive wastes. The maximum part of constructive wastes represents waste by reconstruction after demolition. Improper disposal of such solid wastes generates unsanitary and soil barren conditions or turns to environmental pollution (Luangcharoenrat et al. 2019). It is essential to reduce the dumping of building material waste at landfills (Berge et al. 2018).

26.5 Sustainable Technologies for the Disposal of Building Material Waste

Sustainable construction is an integrative and holistic building process that seeks to restore harmony between the natural and the built environment. Appropriate construction waste management, reduction, reuse, and recycling methods are important for each nation (Lei et al. 2019). Five key groups consist of the building waste management hierarchy (El-Haggar 2007);

1. Reduce,
2. Reuse,
3. Recycle,
4. Recover, and
5. Disposal

United Kingdom, North America, Europe, and various parts of Asia accepted the 3R Principle, which is to reduce, reuse, and re-process waste (Shen et al. 2002; Sakai et al. 2011; Allwood et al. 2011; Hasmori et al. 2020). Coventry and Guthrie (1998) stated remarkable reasons to prefer all such traditional approaches like minimizing the risk of environmental pollution and uncertainty to human health. This report was also found economically in support as can reduce project costs, expanded business support (Shen et al. 2002; Dania et al. 2007).

26.5.1 Reduce, Reuse, Recycle, Recover (4R) Strategy

In 4R approaches, waste reduction is the primary method related to reducing building material waste production at its local stage. The reduction could be accomplished by diminishing or withdrawing unspecified activities at running projects, or by preserving current buildings instead of constructing new ones. Though, it can also be managed by improving the dimension of modern construction or can build new construction with substantiality to prolong life. The reduction in building trash is recognized as the most effective and efficient method but several factors including design changes, poor material handling, lack of capability among the laborers, weather, etc. arise the building material wastes. Besides, the reduction of building material wastes not only manages the generation of waste but also helps to reduce the running cost of disposal and recycling. Reduction in building material waste saves the landfilling premises and lowers the environmental negative impacts (Kralj and Markic 2008; Osmani et al. 2008). Reduce, Reuse, Recycle are the strategies

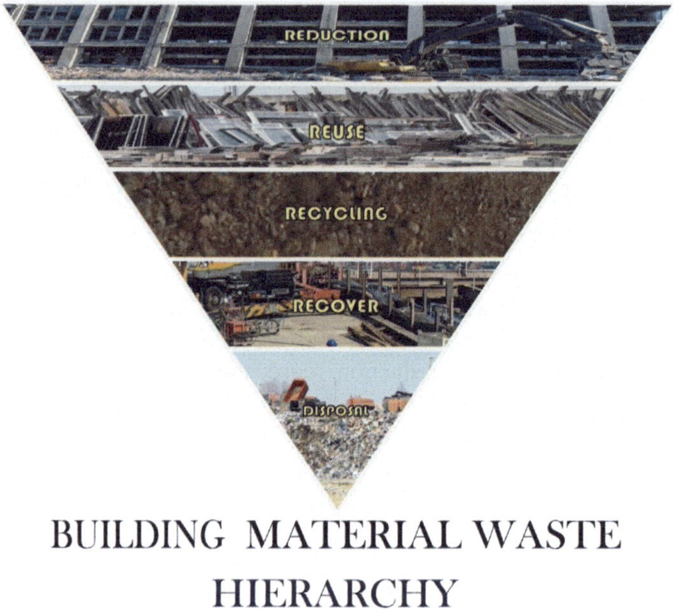

Fig. 26.2 4Rs rule of building material waste management hierarchy

to reduce the consequences of building material waste production (Yu et al. 2021) (Fig. 26.2).

The reuse and recycling of building waste eliminate the demand for raw materials and resources, reducing environmental pollution, aesthetic impacts, and disruption to natural habitats accordingly (Craighill and Powell 1997). Reusing building material waste is also the right sustainable approach for environmental management. The reuse approach is a good alternative if they satisfy some criteria like dimensions and quality, to manage and delaying final disposal or recycling of building material waste production. It is a simple repairing, refurbishing, or effortless recovery of used appliances from a building material waste manufacturer. Hence, reusable waste can be marketed at their production site or by auction. Even, a lot of raw building material debris including gravel and aggregate products, Concrete, Clean wood, Plastics, Insulation materials, iron materials, etc. can be used as a resource.

In Recycle strategy, all used, re-used, or unused matters are considered as waste and processed into valuable new products. Building materials enterprises (BMEs) and waste recycling enterprises (WREs) are two primary players in the recycling of building waste, and their waste recycling decisions influence the development of the waste recycling industry (He and Yuan 2020). Before recycling, these building wastes are screened on-site or off-site and will be transported to the processing center and re-manufactured into new or same products. Waste screening is based on laboratory research or scientific approach and durability evaluation. For example, concrete and gravel can be recycled into aggregated concrete products; Wood can be into furniture;

Metals like steel, copper, etc. can be separately recycled into new valuable products. Bricks and cement chips, sand, stone breaks, etc. can be refurbished as tiles. Hence, a large quantity of building material waste is left after typical management and is dumped in the landfills. It is referred to as waste disposals (Dania et al. 2008). Waste glass from the building sector is one of the major problems for effective waste treatment and reduction. The recycled glass has been developed to be used in structural buildings (Adekomaya and Majozi 2021). Bakchan and Faust (2019) estimated the production of construction waste (CW) from institutional building ventures and also quantified the benefits of CW recycling as saving trees, water, energy, and greenhouse gas pollution in the atmosphere.

26.6 Building Material Waste Management: Global Best Practices and Plan

The disposal and management of building waste have become crucial challenges all over the world (Chen et al. 2021). Building material waste includes large volumes of dangerous materials that adversely impact public health and the environment (Nadeem et al. 2021). The strategy of building material waste management is minimizing waste at its root source (Ferguson et al. 1995). The building material waste management is ongoing in U.K, France, Denmark, Germany, U.S.A, Japan, etc. In these developed countries, around 94% of architects, engineers, consultants, and owners are interested in green buildings or sustainable buildings. In China, the Chinese government is founded the "Polluter Pays Principle" under environmental law to promote sustainable building and integrated construction waste management (Chen et al. 2017a, b).

Many nations decrease construction material waste by implementing new laws and creating attentiveness. Singapore, Japan, and European nations are at the front for handling and reusing construction waste. In Japan, 20 subdivisions of 'building material by-products' are systematically administered rendering to categories. The core principle in managing building material waste in Japan is to reduce and recover the produced waste at the building site. Singapore emphasizes developing principles for green buildings to mitigate the production of building material waste from the source (Lei et al. 2019). Though, recycling of building material waste was initiated during World War-II in Germany and was to tackle disposing of the large quantity of demolished waste. At present, studies throughout the world have demonstrated possibilities for the reuse or recycling of building material waste. In India, Central Building Research Institute (CBRI), Roorkee, and Central Road Research Institute (CRRI), New Delhi are in focus to work on recycling of aggregates (Ponnada and Kameswari 2015).

Building material waste, if not adequately handled, has harmed the environment. Efficient disposal of building material waste is also of key significance for sustainable development. The stakeholders in the building industry need to regulate and handle the production of construction waste. There are several efforts have been made to recycle and process building material waste for use in the manufacturing of different construction materials (Wahi et al. 2016). For the application of suitable management techniques, construction waste composition analysis is always important (Wu et al. 2019b). Ding et al. (2016b) developed a system dynamics (SD) based environmental efficiency simulation for the reduction and control of building material waste during the construction process in Shenzhen, China. The findings of the simulation demonstrate that source reduction is an efficient measure of waste reduction that can minimize 27.05% of overall waste generation. Implementation of construction waste reduction may decrease 53.77% of landfills. Ding et al. (2018) built a two-stage device dynamics (SD) model for environmental benefit assessment of construction waste reduction using Vensim tools. This model included the subsystem for building waste reduction management, the waste generation, and recycling subsystem, and the subsystem for environmental profit evaluation. The findings of the simulation highlight that reduction management will minimize waste generation by 40.63%.

Building information modeling (BIM) design validation is an important way of reducing the amount of construction waste. BIM may be practiced as a less costly, simulated, and technological framework to encourage designers to analyze numerous design alternatives, or contractors to assess various construction schemes (Lu et al. 2017). The construction waste is produced largely due to incorrect design and unforeseen changes in the design and construction processes (Won et al. 2016). Shi and Xu 2021 developed a Building Information Modeling (BIM) CDW information framework using Revit software to reduce the CDW based on the 3R principle, which overcomes the barriers to the implementation of BIM and illustrates the information directly applicable to CDW.

Construction Waste Reduction (CWR) refers to reducing the quantity of hazardous waste produced during construction projects and facilitating the sustainability of the construction industry. This research seeks to enhance the awareness of the essential steps to be engaged in the application of CWR in construction by building stakeholders (Liu et al. 2020). In the course of the construction phase, the construction sector has become extra involved in pushing towards the introduction of a creative approach for minimizing waste and environmental impacts (EIs). Life Cycle Assessment (LCA) is an internationally recognized approach for the environmental impact assessment or concern of buildings (Khasreen et al. 2009; Jalaei et al. 2019).

The utilization of Recycled Concrete Aggregate (RCA) in cement concrete as a partial or complete substitute for Natural Aggregate (NA) was proposed by Shaban et al. (2021) to handle the tremendous amount of building material waste. Recycled Aggregates (RA) resultant from construction material waste may be used along with Natural Aggregates (NA). Ali et al. (2021) observed increased concrete strength after assessment of the cumulative effect of Recycled Aggregates Concrete (RAC) and Styrene-Butadiene Rubber (SBR) Latex on concrete workability. Ultimately, recycling waste concrete for fine aggregate processing would conserve landfill spaces and protect natural sand and thereby reduce the carbon footprint of buildings (Soni and Shukla 2021). There is an emerging movement in the global building sector to follow a "zero-waste" objective at the site level, but little is understood about it (Lu et al.

2021). Bilal et al. (2016) proposed Big Data architecture for building waste analytics, which can store and process vast quantities of data scalably and efficiently using a commodity server cluster. Bakshan et al. (2017) used a Bayesian Network study to identify causal behavioral determinants for enhancing construction waste management (CWM) practices. Chi et al. (2020) analyzed the success of LEED (Leadership in Energy and Environmental Design) approved projects in the United States and China for the minimization of construction waste. Minimization of building waste is an important environmental priority of green building ranking systems.

26.7 Conclusion

After increased urbanization and construction activities, the issue of building material waste has become more severe over the last two decades. There are myriad approaches in which building material waste and pollution can have economic, environmental, and social impacts. The treatment of building waste worldwide has been committed to growing research efforts. Priority must be given to improving waste management methods for the decrease of building waste. The reuse of building material waste can prevent the production of building waste. The methods should be developed concerning reduce, reuse, recycling, and adequate disposal of building waste, thereby delivering environmental and economic benefits through the execution of waste management. Today, sustainable development includes the smart implementation of environmentally-friendly green resources and creative concepts.

Acknowledgements We thank the Director, Research & Development, Biyani Group of Colleges, Jaipur for support and encouragement.

References

Aboginije A, Aigbavboa C, Thwala W, Samuel S (2020) Determining the impact of construction and demolition waste reduction practices on green building projects in Gauteng Province, South Africa. In: Proceedings of the international conference on industrial engineering and operations management 2428–2437

Adekomaya O, Majozi T (2021) Mitigating environmental impact of waste glass materials: review of the existing reclamation options and future outlook. Environ Sci Pollut Res 28(9):10488–10502

Ajayi SO, Oyedele LO, Akinade OO, Bilal M, Alaka HA, Owolabi HA (2017) Optimising material procurement for construction waste minimization: an exploration of success factors. Sustain Mater Technol 11:38–46

Akhtar A, Sarmah AK (2018) Construction and demolition waste generation and properties of recycled aggregate concrete: a global perspective. J Clean Prod 186:262–281

Ali U, Shahid S, Ali S (2021) Combined effects of styrene-butadiene rubber (SBR) latex and recycled aggregates on compressive strength of concrete. J Rubber Res 1–4

Allwood JM, Ashby MF, Gutowski TG, Worrell E (2011) Material efficiency: a white paper. Resour Conserv Recycl 55(3):362–381

Amaral REC, Brito J, Buckman M, Drake E, Ilatova E, Rice P, Sabbagh C, Voronkin S, Abraham YS (2020) Waste management and operational energy for sustainable buildings: a review. Sustainability (Switzerland) 12(13):5337
Asgari A, Ghorbanian T, Yousefi N, Dadashzadeh D, Khalili F, Bagheri A, Raei M, Mahvi AH (2017) Quality and quantity of construction and demolition waste in Tehran. J Environ Heal Sci Eng 15(1):14
Aslam MS, Huang B, Cui L (2020) Review of construction and demolition waste management in China and USA. J Environ Manage 264:110445
Bakchan A, Faust KM (2019) Construction waste generation estimates of institutional building projects: leveraging waste hauling tickets. Waste Manag 87:301–312
Bakshan A, Srour I, Chehab G, El-Fadel M, Karaziwan J (2017) Behavioral determinants towards enhancing construction waste management: a Bayesian network analysis. Resour Conserv Recycl 117:274–284
Barbuta M, Bucur RD, Cimpeanu SM, Paraschiv G, Bucur D (2015) Wastes in building materials industry. Agroecology 1(1):81–99
Berge ND, Batarseh ES, Reinhart DR, Stegmann R, Bolyard SC, Lu W (2018) Landfill operation. In: Cossu R, Stegmann R (eds) Solid waste landfilling: concepts, processes, technology. Elsevier, pp 845–866
Bilal M, Oyedele LO, Akinade OO, Ajayi SO, Alaka HA, Owolabi HA, Qadir J, Pasha M, Bello SA (2016) Big data architecture for construction waste analytics (CWA): a conceptual framework. J Build Eng 6:144–156
Bossink BAG, Brouwers HJH (1996) Construction waste: quantification and source evaluation. J Constr Eng Manag 122(1):55–60
Bribián IZ, Capilla AV, Usón AA (2011) Life cycle assessment of building materials: comparative analysis of energy and environmental impacts and evaluation of the eco-efficiency improvement potential. Build Environ 46(5):1133–1140
Briere R, Fresnet AF, Tardivel Y, Baverel O (2014) Environmental approach of the end-of-life of buildings. In: AvniR conference, Lille, France
Chen J, Shen L, Song X, Shi Q, Li S (2017a) An empirical study on the CO_2 emissions in the Chinese construction industry. J Clean Prod 168:645–654
Chen Z, Liu P, Yang Y (2017b) Sustainable building and integrated construction waste management challenges and strategies for China. In: Proceedings of the 2017 3rd international conference on humanities and social science research (ICHSSR 2017), Atlantis Press, pp 328–333
Chen Q, Cheng Q, Wang J, Du M, Zhou L, Liu Y (2021) Identification and evaluation of urban construction waste with VHR remote sensing using multi-feature analysis and a hierarchical segmentation method. Remote Sens 13(1):158
Chi B, Lu W, Ye M, Bao Z, Zhang X (2020) Construction waste minimization in green building: a comparative analysis of LEED-NC 2009 certified projects in the US and China. J Clean Prod 256:120749
Coventry S, Guthrie P (1998) Waste minimisation and recycling in construction–design manual. London, Construction Industry Research and Information Association, London
Craighill AL, Powell JC (1997) Using environmental economics in decision making and policy formulation for sustainable construction waste management. Stud Environ Sci 71:859–867
Dania AA, Kehinde JO, Bala K (2007) A study of construction material waste management practices by construction firms in Nigeria. Proc 3rd Scottish Conf Postgrad Res Built Nat Environ 121–129
Dania AA, Kehinde J, Bala K (2008) A study of construction material waste management practices by construction firms in Nigeria. In: Proceedings of the 3rd Scottish conference for postgraduate researchers of the built and natural environment, Glasgow, pp 121–129
Ding Z, Wang Y, Zou PXW (2016a) An agent based environmental impact assessment of building demolition waste management: conventional versus green management. J Clean Prod 133:1136–1153
Ding Z, Yi G, Tam VWY, Huang T (2016b) A system dynamics-based environmental performance simulation of construction waste reduction management in China. Waste Manag 51:130–141

Ding Z, Zhu M, Tam VWY, Yi G, Tran CNN (2018) A system dynamics-based environmental benefit assessment model of construction waste reduction management at the design and construction stages. J Clean Prod 176:676–692
Duan H, Li J (2016) Construction and demolition waste management: China's lessons. Waste Manag Res 34(5):397–398
Ekanayake LL, Ofori G (2004) Building waste assessment score: design-based tool. Build Environ 39(7):851–861
El-Haggar S (2007) Sustainability of construction and demolition waste management. Sustain Ind Design Waste Manage 261–292
Esin T, Cosgun N (2007) A study conducted to reduce construction waste generation in Turkey. Build Environ 42(4):1667–1674
Estevez B, Aguado A, Josa A (2003) Environmental impact of concrete recycling, coming from construction and demolition waste. Deconstruct Mater Reuse 3:165–175
Fadiya OO, Georgakis P, Chinyio E (2014) Quantitative analysis of the sources of construction waste. J Constr Eng 2014:1–9
Feng K (Emma), Lam C-Y (Anson) (2021) An overview of circular economy in China: how the current challenges shape the plans for the future. Chinese Econ 1–17
Ferguson J, Kermode N, Nash CL, Sketch WAJ, Huxford, RP (1995) Managing and minimizing construction waste: a practical guide. Thomas Telford Ltd
Gavilan RM, Bernold LE (1994) Source evaluation of solid waste in building construction. J Constr Eng Manag 120(3):536–552
Ginga CP, Ongpeng JMC, Daly MKM (2020) Circular economy on construction and demolition waste: a literature review on material recovery and production. Materials (Basel) 13(13):2970
Gupta S (2018) The impact of C & D waste on Indian environment: a critical review. Civ Eng Res J 5(2):1–7
Gutberlet J, Baeder AM (2008) Informal recycling and occupational health in Santo André, Brazil. Int J Environ Health Res 18(1):1–15
Hasan KMF, Horváth PG, Alpár T (2021) Lignocellulosic fiber cement compatibility: a state of the art review. J Nat Fibers 1–26
Hasmori MF, Zin AFM, Nagapan S, Deraman R, Abas N, Yunus R, Klufallah M (2020) The on-site waste minimization practices for construction waste. In: IOP conference series: materials science and engineering, vol 713, no 1, IOP Publishing, p 012038
He L, Yuan H (2020) Investigation of construction waste recycling decisions by considering consumers' quality perceptions. J Clean Prod 259:120928
Hoang NH, Ishigaki T, Kubota R, Tong TK, Nguyen TT, Nguyen HG, Yamada M, Kawamoto K (2020) Waste generation, composition, and handling in building-related construction and demolition in Hanoi, Vietnam. Waste Manag 117:32–41
Hossain MU, Thomas Ng S (2019) Influence of waste materials on buildings' life cycle environmental impacts: adopting resource recovery principle. Resour Conserv Recycl 142:10–23
Huang S, Zhao X, Sun Y, Ma J, Gao X, Xie T, Xu D, Yu Y, Zhao Y (2017) Pollution of hazardous substances in industrial construction and demolition wastes and their multi-path risk within an abandoned pesticide manufacturing plant. Front Environ Sci Eng 11(1):12
Iacoboaea C, Luca O, Aldea M, Sercaianu M (2010) Main issues related to the construction and demolition waste management in Romania. WIT Trans Ecol Environ 533–544
Jain G, Gupta V, Pandey M (2016) Case study of construction pollution impact on environment. Int J Emerg Technol Eng Res 4(6):4–7
Jain M (2012) Economic aspects of construction waste materials in terms of cost savings—a case of Indian construction industry. Int J Sci Res Publ 2(1):2250–3153
Jalaei F, Zoghi M, Khoshand A (2019) Life cycle environmental impact assessment to manage and optimize construction waste using Building Information Modeling (BIM). Int J Constr Manag 1–18
Jamal S, Hegazy Y, Desouki S (2021) A case study of the sustainable management of construction waste of the Grand Egyptian Museum. Int Design J 11(1):59–67

Jang YC, Townsend T (2001) Sulfate leaching from recovered construction and demolition debris fines. Adv Environ Res 5(3):203–217

Khasreen MM, Banfill PFG, Menzies GF (2009) Life-cycle assessment and the environmental impact of buildings: a review. Sustainability 1(3):674–701

Kralj D, Markic M (2008) Building materials reuse and recycle. WSEAS Trans Environ Dev 4:409–418

Lei J, Huang B, Huang Y (2019) Life cycle thinking for sustainable development in the building industry. In: Life cycle sustainability assessment for decision-making: methodologies and case studies, Elsevier, pp 125–138

Liu J, Yi Y, Wang X (2020) Exploring factors influencing construction waste reduction: a structural equation modeling approach. J Clean Prod 276:123185

Lu W, Bao Z, Lee WMW, Chi B, Wang J (2021) An analytical framework of "zero waste construction site": two case studies of Shenzhen, China. Waste Manag 121:343–353

Lu W, Webster C, Chen K, Zhang X, Chen X (2017) Computational building information modelling for construction waste management: moving from rhetoric to reality. Renew Sustain Energy Rev 68:587–595

Luangcharoenrat C, Intrachooto S, Peansupap V, Sutthinarakorn W (2019) Factors influencing construction waste generation in building construction: Thailand's perspective. Sustain 11(13):3638

Mah CM, Fujiwara T, Ho CS (2018) Environmental impacts of construction and demolition waste management alternatives. Chem Eng Trans 63:343–348

Marrero M, Puerto M, Rivero-Camacho C, Freire-Guerrero A, Solís-Guzmán J (2017) Assessing the economic impact and ecological footprint of construction and demolition waste during the urbanization of rural land. Resour Conserv Recycl 117:160–174

Marzouk M, Azab S (2014) Environmental and economic impact assessment of construction and demolition waste disposal using system dynamics. Resour Conserv Recycl 82:41–49

Nadeem A, Khamatova A, Hossain MA, Leung HY (2021) Construction and demolition waste management on construction sites in Kazakhstan. In: Ahmed SM, Hampton P, Azhar S, Saul DA (eds) Collaboration and integration in construction, engineering, management and technology. Advances in science, technology & innovation, IEREK interdisciplinary series for sustainable development, Springer, Cham, pp 57–62

Oliveira MLS, Izquierdo M, Querol X, Lieberman RN, Saikia BK, Silva LFO (2019) Nanoparticles from construction wastes: a problem to health and the environment. J Clean Prod 219:236–243

Osmani M (2011) Construction waste. In: Waste. Academic Press, pp 207–218

Osmani M, Glass J, Price ADF (2008) Architects' perspectives on construction waste reduction by design. Waste Manag 28(7):1147–1158

Oyenuga AA, Bhamidimarri R (2015) Economic viability of construction and demolition waste management in terms of cost savings—a case of UK construction industry. Int J Sci Eng Investig 4(43):16–23

Ponnada MR, Kameswari P (2015) Construction and demolition waste management—a review. Int J Adv Sci Technol 84:19–46

Poon CS, Yu ATW, Wong A, Yip R (2013) Quantifying the impact of construction waste charging scheme on construction waste management in Hong Kong. J Constr Eng Manag 139(5):466–479

Rani HA (2017) The impact of construction waste to the environmental on project development in Aceh. Int J Manag Inf Technol Eng 5(4):1–8

Raskovic M, Ragossnig AM, Kondracki K, Ragossnig-Angst M (2020) Clean construction and demolition waste material cycles through optimised pre-demolition waste audit documentation: a review on building material assessment tools. Waste Manag Res 38(9):923–941

Sakai S, Yoshida H, Hirai Y, Asari M, Takigami H, Takahashi S, Tomoda K, Peeler MV, Wejchert J, Schmid-Unterseh T, Douvan AR, Hathaway R, Hylander LD, Fischer C, Oh GJ, Jinhui L, Chi NK (2011) International comparative study of 3R and waste management policy developments. J Mater Cycles Waste Manag 13(2):86–102

Shaban WM, Elbaz K, Yang J, Thomas BS, Shen X, Li L, Du Y, Xie J, Li L (2021) Effect of pozzolan slurries on recycled aggregate concrete: mechanical and durability performance. Constr Build Mater 276:121940
Shen L, Tam V, Tam CHS (2002) Material wastage in construction activities—a Hong Kong survey author. Creat a Sustain Constr Ind Dev Ctries Copyr 125–131
Shi Y, Xu J (2021) BIM-based information system for econo-enviro-friendly end-of-life disposal of construction and demolition waste. Autom Constr 125:103611
Shooshtarian S, Caldera S, Maqsood T, Ryley T (2020a) Using recycled construction and demolition waste products: a review of stakeholders' perceptions, decisions, and motivations. Recycling 5(4):31
Shooshtarian S, Maqsood T, Wong PSP, Yang RJ, Khalfan M (2020b) Review of waste strategy documents in Australia: analysis of strategies for construction and demolition waste. Int J Environ Technol Manag 23(1):1–21
Shrivastava S, Chini A (2005) Construction materials and C&D waste in India. Int J Environ Res 72–76
Soni N, Shukla DK (2021) Analytical study on mechanical properties of concrete containing crushed recycled coarse aggregate as an alternative of natural sand. Constr Build Mater 266:120595
Sozer H, Sozen H (2020) Waste capacity and its environmental impact of a residential district during its life cycle. Energy Rep 6:286–296
Trankler JOV, Walker I, Dohmann M (1996) Environmental impact of demolition waste—an overview on 10 years of research and experience. Waste Manag 16(1–3):21–26
Wahi N, Joseph C, Tawie R, Ikau R (2016) Critical review on construction waste control practices: legislative and waste management perspective. Procedia—Soc Behav Sci 224:276–283
Wang J, Wu H, Tam VWY, Zuo J (2019) Considering life-cycle environmental impacts and society's willingness for optimizing construction and demolition waste management fee: an empirical study of China. J Clean Prod 206:1004–1014
Won J, Cheng JCP, Lee G (2016) Quantification of construction waste prevented by BIM-based design validation: case studies in South Korea. Waste Manag 49:170–180
Wu H, Zuo J, Zillante G, Wang J, Yuan H (2019a) Status quo and future directions of construction and demolition waste research: a critical review. J Clean Prod 240:118163
Wu Z, Yu ATW, Poon CS (2019b) An off-site snapshot methodology for estimating building construction waste composition—a case study of Hong Kong. Environ Impact Assess Rev 77:128–135
Yahya K, Halim Boussabaine A (2006) Eco-costing of construction waste. Manag Environ Qual An Int J 17(1):6–19
Yu ATW, Poon CS, Wong A, Yip R, Jaillon L (2013) Impact of construction waste disposal charging scheme on work practices at construction sites in Hong Kong. Waste Manag 33(1):138–146
Yu D, Duan H, Song Q, Li X, Zhang H, Zhang H, Liu Y, Shen W, Wang J (2018) Characterizing the environmental impact of metals in construction and demolition waste. Environ Sci Pollut Res 25(14):13823–13832
Yu KH, Zhang Y, Li D, Montenegro-Marin CE, Kumar PM (2021) Environmental planning based on reduce, reuse, recycle and recover using artificial intelligence. Environ Impact Assess Rev 86:106492
Zhang C, Xu T, Feng H, Chen S (2019) Greenhouse gas emissions from landfills: a review and bibliometric analysis. Sustain 11(8):2282

Chapter 27
Impact of Textile Product Emissions: Toxicological Considerations in Assessing Indoor Air Quality and Human Health

Mansoor Ahmad Bhat, Fatma Nur Eraslan, Kadir Gedik, and Eftade O. Gaga

Abstract The presence of an increasing number of chemicals has been detected in nearly all human life aspects. This affects human development, including physical and mental health and well-being. Considering that a normal person spends a substantial part of his/her life indoors, it is important to pay more attention to indoor air quality. Textiles used in interior spaces like beds and seat coverings, curtains, carpets, interior decoration, thermal insulation, moisture defence are believed to be one reason for the deterioration of indoor air quality. Poisonous chemicals are used in these textiles during dyeing, printing, and finishing procedures to reveal an obligatory useful feature to the fabric, like making cotton fabrics wrinkle-free, flame retardant, water repellent, waterproof, antistatic, antibacterial, etc. These chemicals may be released in indoor spaces due to air movement influences, temperature rise, and friction. Thus, the rigorous practice of these materials within the indoor spaces might disturb indoor air quality and cause health complications. The scientific literature concerning the possible antagonistic health consequences of chemical substances in the textile industry is mostly related to human exposure; evidence regarding consumers' exposure is considerably limited. This review reflects evidence concerning the toxicological influences of textile emissions on indoor air quality and human health.

Keywords Carpets · Chemicals · Clothes · Dyes · Emissions · Microplastics · Nanoparticles · Nanoplastics · Skin allergies · Textile industry

27.1 Introduction

The textile industry is spread worldwide, making about 1 trillion dollars, adds 7% of the overall world exports and hires about 35 million workers worldwide (Desire and

M. A. Bhat · F. N. Eraslan · K. Gedik · E. O. Gaga (✉)
Department of Environmental Engineering, Faculty of Engineering, Eskisehir Technical University, Eskisehir, Turkey

M. A. Bhat
e-mail: mansoorahmadbhat@eskisehir.edu.tr

J. A. Malik and S. Marathe (eds.), *Ecological and Health Effects of Building Materials*,
https://doi.org/10.1007/978-3-030-76073-1_27

Narula 2018). Regardless of its indisputable significance, this manufacturing industry is one of the main worldwide polluters, due to excessive quantities of fuels and chemicals used (Bhatia 2017). The particular stress is enlisted on the massive consumption of water in its manufacturing chain's numerous processes, like laundry, bleaching, dyeing, among others (Hossain et al. 2018). The textile sector is accountable for a vast list of environmental complications. The air pollution formed encompasses, for instance, the discharge of particulate matter and dust, oxides of nitrogen and sulfur and volatile organic compounds (VOCs). The leftovers of textile materials and yarns and abandoned packaging comprise the main solid waste. On the other hand, the textile sludge discloses complications associated with excess volumes and undesirable composition, often donating elevated organic matter loads, micronutrients, heavy metals and pathogenic microorganisms (Bhatia 2017). The textile industry is categorized into three fundamental types: cellulose fibres cotton, rayon, linen, ramie, hemp and lyocell, protein fibres wool, angora, mohair, cashmere and silk and synthetic fibres polyester, nylon, spandex, acetate, acrylic, ingeo and polypropylene. The kinds of dyes and chemicals employed in the textile sector vary contingent on the materials manufactured. Reactive dyes remazol, procion MX and cibacron F, direct dyes congo red, direct yellow 50 and direct brown 116, naphthol dyes fast yellow GC, fast scarlet R and fast blue B and indigo dyes indigo white, tyrian purple and indigo carmine are several dyes used to dye cellulose fibres (Lorimer et al. 2001). Protein fibres are dyed using acid dyes azo dyes, triarylmethane dyes and anthraquinone dyes and lanaset dyes Blue 5G and Bordeaux B (Schmidt et al. 2003). Additional dyes, like disperse dyes; disperse yellow 218 and disperse navy 35, basic dyes; "basic orange 37 and basic red 1" and direct dyes, are used to dye synthetic fibres (Burkinshaw 1995). China is the primary textile manufacturing and exporting nation in the world. In 2019, China was the top-ranked worldwide textile exporter with a worth of nearly 120 billion US dollars, next to China, the European Union (EU) (28 countries), with a worth of 66 billion dollars. India, the USA, and Turkey would be subsequent, with a value of 17, 13 and 12 billion dollars, respectively (Statista 2020).

Textiles are the basic fabrics used in buildings' general interior spaces, including chairs, curtains, carpets, upholstery, and wall coverings. In the indoor environment, they form the greatest area of all surface areas, combining wide flooring areas, ramparts, ceiling equipment, and fittings. The different types of textile products used in indoor environments are shown in (Fig. 27.1). Mainly interior engineers choose textiles for indoor applications since they are the quickest medium that can be modified in household fittings and decorations. They are an easy, fast method to modify a particular area's shape and build a novel one; when consumers feel tired and choose to refresh themselves. Textiles often have a part to play in arranging the illumination of a certain location by curtains. Very new applications of interior textile materials have needed improvements to the practical and decorative features of textiles by adding chemical compounds containing poisonous constituents that have been proven to have harmful effects on indoor air quality and human health. There is a common usage of carpet as a floor covering. A variety of chemical ingredients are used in most of the industrially manufactured carpets. During the manufacturing of carpet fibre, additives are impregnated or applied externally as topical treatments

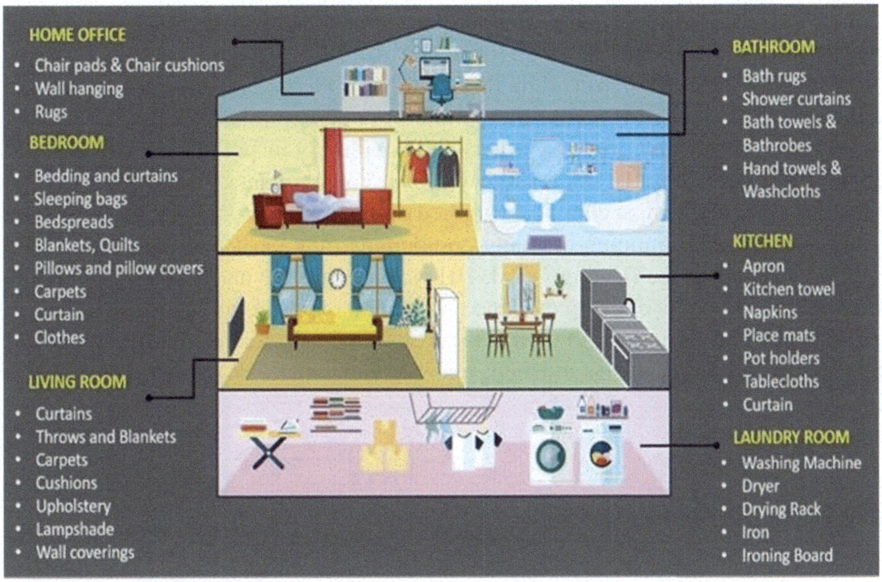

Fig. 27.1 Different types of textile products used in indoor environments

for the finished product. Any of these compounds was meant to eliminate bed bugs, bacteria, moulds and fungi. Using chemical agents to carpets contributes to possible human exposure to toxic chemicals in the household and other indoor conditions. Typically, a conventional carpet is finished of synthetic fabrics that have either been tufted or woven to support. Almost all of the carpet being used in domestic, public and industrial settings comprises an artificial layer, typically nylon 6 or nylon 6,6 pile, tufted into the main backing with a water-based adhesive coating. The tufted stack loop is set into a foundation of latex styrene-butadiene (SBR) sandwiched between the main and secondary supports normally made of woven polypropylene or jute (Dietert and Hedge 1996). The carpet is then treated in a drying oven to absorb more of the water. Based on the dye bath and the inclusion of substances, like stain-resistant and soil retardant compounds, the overall chemical properties of finished carpet items can differ. As per Dietert and Hedge (1996) the novel synthetic carpet would emit VOCs into the indoor environment when introduced first.

The chemicals present in different products of textiles used can get released from them and can have harmful impacts on human health and indoor air quality. The skin is a broad shield organ which defends the body against possible dangers heat, cold, chemicals, mechanical forces, etc. It protects the body's modesty, whereas the clothing structure creates an additional coating(s) of a shield to improve the wearer's visual, thermophysiological and sensory comfort. But, direct touch and contacts between textiles and skin can trigger reactions, including injury or illnesses. General studies on chemical health impacts in textiles involve inflammatory skin problems. Disperse dyes employed to dye synthetic fabrics have been documented to be the main widespread textile allergy source, and contact allergic reaction to disperse dyes is a medically significant issue (Coman et al. 2014; Malinauskiene et al. 2013; Ryberg et al. 2006, 2009). Yet, skin allergy may not be the only concern for human well-being. This is well reported that humans are susceptible to pollutants mostly via nutrition (meals and drinking water) and breathing (air pollution). Still, dermal sensitivity cannot be fully eliminated with any contaminants. In terms of dermal contamination, while most chemicals applied during textile processing processes are rinsed out, residual amounts of many compounds may exist and may be emitted during consumer usage (Luongo et al. 2014). Based on the above, this paper's key purpose was to review existing evidence on human exposure to chemicals found in textile goods used indoors and their effects on indoor air quality. Within the paper, gaps in the literature are established, and recommendations are given for future research.

27.2 Textile Processing

Textile wet processing, including dyeing, printing, and finishing, is used in nearly all textile products to generate colours, designs, and distinct presentation features. These procedures have long been criticized for their negative environmental influences as dyes, and different chemicals are employed. Though the textile industry has effectively recycled and cut down waste, textile wet processing's net environmental consequence is still a concern: as colouring and printing are usually used in most textiles. The textile industry uses input materials like cotton or wool and spinning them into yarn, which is subsequently used to generate a cloth (EPA 1997). Altogether the methods implicated in switching the raw material into a final item: developing, producing, manufacturing, and distributing textiles—are incorporated in the textile sector, which uses several kinds of fabrics, with two main groups, natural and synthetic. Natural fabrics arise naturally from animals and plants, whereas synthetic fabrics are being developed and shaped in a laboratory and human-made (EPA 1997).

The standard textile processing industry comprises sizing, desizing, scouring, bleaching, mercerizing, and dyeing methods (EPA 1997). The steps can be visualized by the process flow chart as given in (Fig. 27.2).

Sizing is the primary preparation step, in which sizing compounds such as starch, polyvinyl alcohol (PVA) and carboxymethyl cellulose are used to give intensity to the fibres and reduce braking.

Desizing is employed to eradicate sizing constituents before weaving.

Scouring eradicates scums from the fibres through alkali solution (usually sodium hydroxide) to breakdown natural oils, fats, waxes and surfactants, and blend and suspend contaminations in the scouring bath.

Bleaching is the phase applied to eradicate the fibres' undesirable color employing chemicals like sodium hypochlorite and hydrogen peroxide.

Fig. 27.2 Typical textile processing steps involved in different textile industries

Mercerising is a constant chemical procedure applied to intensify dye-intensity, lustre and fibre appearance. A rigorous alkaline solution is used in this stage, and an acid solution rinses the fibres before the dyeing step.

Dyeing is the technique of applying colour to the fabrics, which typically entails significant quantities of water in the dye bath and during the rinsing procedure. Various chemicals such as metals, salts, surfactants, organic processing aids, sulphide and formaldehyde (HCHO) might be applied to enhance the adsorption of dye on the fibres, depending on the dyeing process.

The textile manufacturing sector typically incorporates a wide variety of compounds (EPA 1997), like:

Detergents and caustic: To extract dirt, dust, oils, and waxes, detergent and caustic are being used. To maximize whiteness and brightness, bleaching is used.

Size agents that are applied to enhance weaving.

Oils, which are applied to make spinning and knitting easier.

Latex and glues that are employed in the form of binders.

Dyes, fixers, and many in-organics are used to supply the consumer needs with a dazzling variety of colours.

A wide range of distinct compounds is employed, like softeners, stain release agents, and wetting agents.

Most of these contaminants become a portion of the finished item while the others are separated from the cloth, and are removed in the garment run-off.

27.3 Indoor Air Quality and Health Issues

As it occurs with different manufacturing processes, the textile industry also has ecological complications. Different ingredients used in the textile industry can create both air quality and health issues. Among the various compounds whose existence has been revealed in textile wastewater, dyes are between the maximum critical toxins (Brillas and Martínez-Huitle 2015; Mohamed et al. 2016; Zarea et al. 2018). Global ecological issues related to the textile industry are mostly linked with water contamination triggered by the release of unprocessed waste and those due to the use of possibly noxious materials, particularly throughout processing (Khan and Malik 2014; Pattnaik et al. 2018). The technical works concerning the possible harmful health impacts of chemical constituents in the textile industry are primarily correlated to human exposure throughout textile manufacturing. Therefore, cases of physical risks related to fabric and garment production comprise fire threat, sound, temperature, moisture, unsafe equipment, dust and injurious chemicals. However, the evidence about the exposure of customers is minimal (KEMI 2014). Several activities are included in the textile and clothing manufacturing, going from handling raw materials to final steps such as bleaching, printing, dyeing, impregnating, coating, plasticizing, etc. As an outcome of these activities, the primary chemical contaminants are dyes, which comprise hazardous amines, metals, pentachlorophenol, chlorine bleaching, halogen carriers, HCHO, biocides, fire retardants, and softeners (Brigden et al. 2012). Chemical varnishes might be applied on textiles matting, upholstery, drapery textiles for antistatic, antimicrobial, stain-resistant, wrinkle-resistant, or flame-resistant goals. Yet, these procedures consume compounds that disturb air quality and use huge quantities of water and energy. However, the textile industrial sector claims that these processes are harmless (Building Green 1994); numerous customers are worried about air quality and health problems associated with such chemical uses. The different chemicals used in the textiles products and their impacts on indoor air quality and human health are mentioned below.

27.4 Flame Retardants

The growing trend in health-related issues and the serious concern to improve quality of life contributed to the development of studies into the impact of indoor environments on people's health, particularly the quality of indoor air. Humans themselves were the only cause of pollutants inside a closed area, as per pioneers. Different construction materials are now used to design and complete works within homes, and are a big cause of indoor air emissions; they directly affect the quality of air in enclosed spaces. Barker (1975) wrote a paper on fibre and cloth additives, assuming that altogether fibres and textiles encompassed detectable quantities of toxins and additives. It was reported that although the toxicity concentrations of the fibres were, in over-all conditions, very low, at that time large amounts of chemical additives were found in textiles treated for fire resistance or, for example, for oil and water repellency. Indoor textiles may have fire-resistant properties by treating them with flame retardants. These flame-retardant characteristics are obtained by the use of hazardous chemical compounds that are either employed as additives or react with textiles. The flame-retardant surface feature is designed exclusively for cellulose, wool, and human-made fibres (Horrocks 1986). This coating is essential for upholstery fabrics, carpets/rugs, wall coating textiles and curtains in the furniture sector. Polybrominated diphenyl ethers (PBDEs) have performed a significant part in the flame retardants commonly used in the textile industry. PBDEs are a class of brominated flame retardants (BFRs) applied to plastics, polyurethane foam, fabrics and electrical devices to safeguard humans from fires by decreasing the flammability of easily flammable substances (Lorber 2007). PBDEs have become common contaminants in recent years (Guigueno and Fernie 2017; Malarvannan et al. 2015), and has been found in the tissues of the humans (He et al. 2018; Ma et al. 2017; Schuhmacher et al. 2009, 2013). Several researchers have revealed that, as for other persistent organic pollutants, consumption of food is the chief pathway of human exposure to PBDEs (Domingo et al. 2008; Domingo 2012; Linares et al. 2015; Lorber 2007; Perelló et al. 2009). Yet, there is very little evidence on exposure in humans to PBDEs via the skin (Chen et al. 2009) calculated the amounts of different BFRs in kids dolls bought from South China, including PBDEs. "Increased exposures mainly contributed via the mouthing pathway, inhalation, dermal contact and oral ingestion less significant exposure-related paths with toys" (Chen et al. 2009). Repeated PBDE exposures may include abnormal growth of the brain and thyroid, fetal abnormalities, hearing problems, deferred adolescence, and potentially malignancy. PBDEs are already found in household dust, leaves, food and human tissues because of increased rates of production and environmental persistence and bioaccumulation of PBDEs (Bakker et al. 2008; Domingo et al. 2008; Domingo 2012).

Concerning human sensitivity to BFRs by garments, it is essential to mention that clothing encompasses nearly 85% of human skin and can serve as an environmental pollutant shield. Clothing, though, maybe a possible cause of exposure to such contaminants, in particular. Clothing plays an important part for the sorbent of indoor semivolatile organic compounds (SVOCs); however acts as a source for outdoors via washing was recently investigated Saini et al. (2016a). "Phthalates, BFRs and organophosphate esters (OPEs) were also measured". Clothing has been shown to serve as an effective conveyor of soluble SVOCs from indoors to outdoors by air accumulation and discharge throughout washing. The drying of laundry may also result in the discharge of chemicals that electric dryers produce. Such results have consequences for prospective skin exposure. A consequent analysis carried out by the similar scientific community Saini et al. (2016b) analysed the deposition of phthalates and BFRs in cotton and rayon by deploying these fabrics for 28 days indoors in 20 homes and 5 offices and evaluating uptake over 56 days. The findings indicated that all gas-and particle-phase chemicals accumulate in cotton. It was suggested that this broad sorptive potential might have consequences for clothes as a chemical sink for SVOCs indoors, and likewise for human exposure Schecter et al. (2009) calculated the quantities of these pollutants in residential dryer lint in Germany and the USA to explain the amounts of PBDEs in lint. The median U.S. overall PBDE concentrations were ten times greater than the median German concentrations, and the mean U.S. concentrations were two times greater than the mean German concentrations. US concentrations varied from 321 to 3073 ng g^{-1} median: 803 ng g^{-1}, mean: 1138 ng g^{-1} and German concentrations ranged from 330 to 2069 ng g^{-1} median: 71 ng g^{-1}, mean: 361 ng g^{-1}. PBDE concentration of lint was observed in every sample; it was observed that, after being laundered prior to drying, clothing could be a basis of PBDE pollution of dryer lint and might act as a measure of indoor exposure to such contaminants. "The authors proposed that the lint origins of PBDEs may be extracted mostly from electronic parts of the dryer and dust accumulation on garments as well." The Stapleton et al. (2005) assessed that concentrations of PBDE in lint are lower than the dust. According to Lorber (2007); Stapleton et al. (2008) "PBDE lint exposure can occur by hand-to-mouth touch or dermal absorption." Normal household inhabitants are likely to be subject to less lint PBDEs than dust, but susceptibility to PBDE from lint will also add to the body pressure of industrial laundry staff and skilled household workers exposed to significant volumes of lint.

In the textile sectors processing techniques, "decabromodiphenyl ether" (Deca-BDE, fully brominated PBDE congener) is becoming progressively crucial, particularly in the manufacture of yarn and synthetic fibres used for fabrics, carpets, and curtains. BFR prevents the textile from fire without altering the fabrics' structure, tone, or design. In the textile industry, Deca-BDE is the largely extensively used BFR and is widely used as a stimulator in conjunction with antimony trioxide (Sb_2O_3). According to Guzzella et al. (2008) "it is applied to the process baths used in finishing processes such as piece-dyeing post-treatment, mattress garment coating, velour and flat furniture cloth coating, or carpet back-coating." Deca-BDE is emitted into the ambient air by pollution from the processing of Deca-BDE goods and the commodities themselves' use and disposal. As a result of these releases, Deca-BDE levels have increased in the environment (Hale et al. 2006). Studies like (De Boer et al. 2003; Yogui and Sericano 2009) have found that elevated Deca-BDE levels exist in much developed and urbanised environments. The research illustrated the association among polybrominated diphenyl ether (PBDE) levels in the deposits and the Antwerp textile sector, where Deca-BDE has been consumed (De Boer et al. 2003). There has been increasing attention in BFR and Deca-BDE's environmental effects, especially in last (10–18) years (Alcock et al. 2011; Eljarrat et al. 2008; Kemmlein et al. 2003; Law et al. 2006; Ward et al. 2008). In particular, Deca-BDE is regarded as quite persistent and is not listed as hazardous or bio-accumulative (Alcock et al. 2011; Chen and Hale 2010; Guzzella et al. 2008). However, current analysis has shown that there is a good risk of Deca-BDE being split into extra persistent, bio-accumulative, harmful and additional mobile environmental brominated goods. As a result of Deca-BDE's continuous use, its levels in the environment can rise with time. Consequently, the concern in strategies for limiting Deca-BDE pollution in last few years has increased (Alcock et al. 2011; Covaci et al. 2011; Eljarrat et al. 2007, 2008; Guzzella et al. 2008; Kemmlein et al. 2009; Law et al. 2006; Ross et al. 2009; Shaw et al. 2012; Yogui and Sericano 2009).

27.5 Trace Elements

Dyes, containing different chemicals of well-known toxicity, are the major environmental contaminants in textiles. In textile fabrics, various harmful trace elements can be detected, as they are commonly used in different textile processes. Moreover, raw textile supplies may contain some trace elements as well. "Metals in textile products and clothing are used for many reasons, such as metal complex dye (cobalt, copper, chromium, lead), pigments, mordant (chromium), catalyst in synthetic fabrics manufacture (antimony oxide), synergists of flame retardants (Sb_2O_3), antimicrobials (nanoparticles of silver, titanium oxide, and zinc oxide), as well as like water repellents, and odour-preventive agents" (Derden and Huybrechts 2013; Muenhor et al. 2010; Simoncic and Tomsic 2010; Stefaniak et al. 2014; Wöhrle et al. 2012). Association among trace elements and fabric might pose a significant ecological concern for the garment manufacturing sector. However, such harmful trace elements in apparel might pose a health threat to customers. An in-depth analysis of scientific literature reveals that human metal consumption rarely results in disease and extremely infrequently in death. Nevertheless, persistent contact to small amounts of harmful elements like "As, Cd, Hg and Pb" is associated with a variety of negative impacts, among others (García-Esquinas et al. 2015; Jaishankar et al. 2014; Rodríguez-Barranco et al. 2014; Roy et al. 2011). Besides, the different metals important for humans, like "Cu, Co, Fe, Mn, Mo, or Zn," could also be harmful at elevated exposure levels (J L Domingo 1993; Lucas et al. 2015). Heavy metals such as "arsenic (As), lead (Pb), mercury (Hg), chromium (Cr), cadmium (Cd), zinc (Zn), cobalt (Co), nickel (Ni) and copper (Cu)," which may result in significant health and environmental impacts, are regarded as dangerous properties but are commonly used to the manufacture of textile dyes colour pigments (Mathur et al. 2006). During processing, dyeing and printing activities, these metals may infiltrate into textile fabrics or occur in textile structures or through protective agents used during storage (Halimoon and Yin 2010).

Quantities of these elements are already measured in different textile fabrics depending on the possible health risks due to metals exposure. Concentration levels of six metals "Cu, Cd, Zn, Mn, Fe and Ni" were measured in different fabric samples obtained in Turkey (Tuzen et al. 2008). These metals' values varied from 0.10 to 0.25 μg/g for Cd and 3.55 to 34.3 μg/g for Fe, for the lowest and highest metals, accordingly. In the assessed samples, Cu and Cd concentrations were greater than that of the limit values provided by OEKO-TEX (2018). Apart from this, "17 trace elements in 16 textile samples" of various origins were calculated Rezić and Steffan (2007). The findings (minimum-maximum μg/mL) in sweat extracts were as follows: Al (0.11–1.58), Cd (0.02–0.05), Cr (0.01–0.32), Cu (0.05–1.95), Mn (0.01–2.17) and Ni (0.05–0.10). While the levels of further elements were either lower or higher than detection limits. "In cotton and polyester samples, Zn and Cd were found, Cr was found in flax, silk and polyester samples, Cu was found in silk samples, and As was found in silk and polyester samples." The levels of 28 trace elements "Al, As, Ba, Be, Bi, Ca, Cd, Co, Cr, Cu, Fe, K, Hg, Mg, Mn, Mo, Na, Ni, Pb, Sc, Si, Se, Sn,

Sm, Sr, Tl, V and Zn" were calculated in raw textile supplies "cotton, flax, hemp and wool" by the same authors in the subsequent investigation Rezić et al. (2011). These elements' values varied between the < detection limit for different elements and "1170.2 mg/g for K in cotton." However, they varied among < the limit of detection for different elements and "86.6 mg/g for Mg;" in hemp, among < the limit of detection for different elements and "540 mg/g for Ca," and respectively among < the limit of detection for different elements and "660 mg/g for Ca in the wool." The amounts of "Sb, As, Pb, Cd, Cr, Co, Cu, Ni and Hg" in polyamide raw supplies (pellets) and textiles applied in sports T-shirts have been calculated Matoso and Cadore (2012). For Cr in black cloths, the highest amounts of trace elements were detected. However, the extractable quality was smaller than the "Oeko-Tex Standard 100:2017" Standards' proposed limits OEKO-TEX (2018).

It is precisely understood that ingestion and inhalation are the key routes by which trace elements are able to enter the human body. Human exposure to metals by skin contact may additionally be a non-negligible route for certain elements and some exposure situations. Based on this hypothesis, few studies have recently centred on evaluating the quantities of certain trace elements in garments and analysing customers' possible health threats. The Al, As, B, Ba, Be, Bi, Cd, Co, Cr, Cu, Fa, Hg, Mg, Mn, Mo, Ni, Pb, Sb, Sc, Se, Sm, Sn, Sr, Tl, V and Zn concentrations in different skin-contact dresses were calculated Rovira et al. (2015). The studied samples were manufactured of "cotton, polyamide, polyester, spandex, and viscose," categorised by fabric, brand, and eco-labelled categories "T-shirts, blouses, underwear, baby pyjamas and bodies." Elevated concentrations of Cr were found in dark polyamide clothing (605 mg/kg), Sb in polyester clothing (141 mg/kg), and Cu in certain green cotton fabrics (around 280 mg/kg). Intriguingly, in "eco" clothes, low levels of Al and Sr were observed, although no major variations in branded and unbranded clothing items were noted. Furthermore, in EU countries' clothing, Al and Sc amounts were greater than in clothing manufactured outside the EU. The skin interaction exposure to 28 trace elements and the consequent human health hazards by examining the amounts of these elements in 37 skin-contact garments T-shirts, blouses, socks, baby pyjamas and bodies was determined by Rovira et al. (2017b) in their other study. They also performed migration experiments to create more accurate risk assessments by assessing the composition of the same 28 elements in artificial sweat. Trace element dermal sensitivity was measured for women and men and children (having the ages lesser than one year) and related health risks were evaluated. Zn levels (186–5749 mg/kg) were high in zinc pyrithione t-shirts and in polyester, and black polyamide manufactures high Sb, and Cr levels were found, respectively. In contrast, a scanning electron microscope found traces and aggregate of "Ag and Ti particles" in different garments (Rovira et al. 2017b). All the samples evaluated in that analysis met the Oeko-Tex norm parameters (OEKO-TEX 2018). The Oeko-Tex Norm is a globally standardised method of inspection and certification in all phases of manufacturing for clothing raw materials, intermediate and finished goods. Certification refers to several human-ecological characteristics, like hazardous compounds banned or controlled by regulation, contaminants considered to be detrimental to health but not legally barred, and criteria that are used as a safety precaution to protect health. 4 polyester samples surpassed the TOX-Proof standard extractable Sb limit, set at "1.0 mg/kg". Concerning health risks, the mean hazard quotients (HQs) for Sb were "0.44, 0.40 and 0.13" for polyester clothing, and for "adult males, adult females and children <1 year of age," collectively, with single polyester T-shirts hitting a value of 1.2 (Rovira et al. 2017b) inferred, however, that these components may have harmful consequences on human skin.

The amounts of 28 trace elements Ag, Al, As, B, Ba, Be, Bi, Cd, Co, Cr, Cu, Fa, Hg, Mg, Mn, Mo, Ni, Pb, Sb, Sc, Se, Sm, Sn, Sr, Ti, Tl, V and Zn were evaluated by Rovira et al. (2017a) in 78 samples of housewares "towels, bedding and pyjamas" and also their possible health implications throughout their continuous usage. Samples were divided into three groups depending on the fabrics: 100% cotton, cotton + synthetic and synthetic. In almost all samples, "As, Be, Cd, Sc, Se, Sm and Tl" demonstrated concentrations under their corresponding limit of detections (LODs). On the other hand, "Hg, Mo and V" were found just in 2–4 objects. The maximum average values were "Mg 142 mg/kg), Cu 32.8 mg/kg, Sb 26.9 mg/kg, Al 14.7 mg/kg, Fe 12.9 mg/kg and Ti 10.9 mg/kg," respectively. Nevertheless, the maximum levels for Cu in a black sample "100% cotton" and Mn in a brown sample "50% lyocell + 50% cotton" after an independent examination of the tests, with levels of "1065 and 889 mg/kg," accordingly. Polyester materials contained elevated Sb levels, following the findings of their earlier works (Rovira et al. 2015, 2017b), while Ti levels were too raised in specimens of synthetic fibre. And from the other side, the textile colour was the main problem due to the elevated concentrations of Cr, particularly in black polyamide clothing, and Cu in coloured "blue, green, red, and brown cotton clothing." The overall HQ for nearly all trace elements was just <0.01 concerning human health risks. The exception was Sb whose skin exposure settings were "0.4 and >1," accordingly, owing to use of bedclothes/pyjamas and towels. Compared to other every day practices, the use of towels by hand-to-mouth influence was the very important act contributing to dermal exposure of maximum trace elements (Rovira et al. 2017a).

The concentrations of Ag, Al, As, Ba, Be, Bi, Ca, Cd, Co, Cr, Cu, Fe, Hg, K, Li, Mg, Mn, Na, Ni, Pb, Sb, Se, Sr, Ti, V and Zn in 120 samples (63 kinds of cotton, 44 nylons and 13 polyesters) of various brands and colours of women's undergarments produced in 14 countries were calculated in a similar theme to Rovira et al. (2015, 2017a, b) by Nguyen and Saleh (2017). Cotton samples were analysed to be abundant in Al, Fe and Zn, while nylon substrates comprised elevated Cr, Cu and Al levels. Polyester fibres, in particular, had greater Ni and Fe amounts relative to cotton or nylon. The countries with the largest amounts of trace elements present in all garments were China, Egypt and India. In 35% of the samples, chromium surpassed the Oeko-Tex levels, while Pb and Ni did so in 14% and 5% of the analysed products, accordingly. They advised that customers be aware of the possible health threats involved with these metals' quality in their garments.

27.6 Aromatic Amines

The most significant kind of synthetic dyes is azo dyes. While different aromatic amines are being used in the production of azo dyes as substrates (Freeman 2013), Aromatic amines could have mutagenic and carcinogenic properties along with other toxic potentials as well as the capacity for allergens (Brüschweiler et al. 2014; Brüschweiler and Merlot 2017; Platzek 2010). Moreover, dermal, systemic and bacterial biotransformation of azo dyes, can—in turn—release aromatic aromatic amines (Stingley et al. 2010). Accordingly, aromatic amines which can emit one of "22 known carcinogenic" aromatic amines are presently prohibited from garment cloths in the EU. However, there are still many limitations in the understanding of the possible risk of non-regulated aromatic amines. The incidence of dermatological issues created by consumer exposure to dyes on clothes was examined Hatch and Maibach (1985). Allergic contact dermatitis can be caused by 31 dyes, primarily spread through anthraquinone or azo structures. They subsequently studied the textile dye dermatitis literature printed in the previous decade and recognized four novel dye allergens (Hatch and Maibach 1995). In 2000, research on the incidence of allergic contact dermatitis (ACD) and the number of ACD cases identified in the population was summarised Hatch and Maibach (2000). However, most of the experiments were done in Europe, principally in Italy. Besides altogether, the experiments were done by applying dye directly on the skin, often scattered dye, of uncertain purity rather than a dyed cloth. In comparison to colour textile ACD, the term textile-dye ACD was then introduced Hatch et al. (2003b), Since more complex factors are involved in the above case, such as dye particles shifted or emitted from fabrics to the skin, the dyes' transpiration power. It was correspondingly confirmed that dyes that were positive for a patient's patch examination were rarely present in the cloth accused of causing skin issues (Hatch et al. 2003a). This suggests that the judgment and treatment of colored-textile allergic contact dermatitis needs additional research.

It is possible to biotransform Azo dyes into colourless aromatic amines which are more readily absorbed by the skin. Many of these may be less harmful than the actual dye (Collier et al. 1993), while some are possibly carcinogenic, including arylamines and free radicals, (Chung 1983; Mason et al. 1977; Nakayama et al. 1983). Thousands of azo dyes are accessible, and over 500 comprise aromatic amines that are possibly carcinogenic (Platzek et al. 1999). A cancer risk evaluation by the Netherlands National Institute for Public Health and the Environment identified carcinogenic aromatic amines in azo coloured goods, like fabric toys, bedding and wrist straps (Zeilmaker et al. 2000). Upholstery textiles must be made to a standard appropriate for their final use. To retain the beauty and to last the lifespan of the commodity, they have some aftercare needs. The expectations set by the customer define internal apparel efficiency specifications. To fulfil colour fastness criteria for textiles not to fade, manufacturers in this area need to understand textile form and dye operation. The tolerance of dyed/printed fabrics to change in colour before and after a certain element is colour fastness. Fastness to; "damp and dry rubbing/crocking, light and detergent washing" is the main common to indoor textiles. To increase the colour speed, the colouring must have a constant aromatic configuration such as azo dyes and the inclusion of textile auxiliary materials necessary to the colouring process, such as catalysts, dyes and dispersants (Chakraborty 2011). The incidence of bladder cancer, splenic sarcomas, hepatocellular carcinoma, cell defects are associated with experience with some azo dyes (Ventura-Camargo and Marin-Morales 2013).

Azo dyes are already confirmed to be mutagenic (Busk and Ahlborg 1982; Esancy et al. 1990) and can cause certain forms of cancer (Zdzislaw et al. 1991) found that after the long-lasting experience azo dyes painters got cancer of the bladder. These polymers are predominantly metabolised on the intestine and liver and contain free, possibly cancer-causing and mutagenic aromatic amines (Gingell et al. 1971). One hundred fifty-three items of coloured or black clothing were collected to examine the incidence of azo-dye cleavage items (Brüschweiler et al. 2014) T-shirts, underwear, sports clothes, scarves and clothes. Direct interaction with the skin, potential connection with sweat, and delicate skin contact were the most relevant sampling requirements. The 22 toxicologically significant non-regulated aromatic amines were analyzed. 8 of these 22 aromatic amines were measured in 17% of the samples, whereas several of the concerned aromatic amines were identified at >30 mg/kg in 9% of the samples. Finally, 8% of the samples find these aromatic amines between 5 and 30 mg/kg, the maximum calculated concentration of 622 mg/kg textile. The researchers presumed that several aromatic amines isolated from the 470 textile azo dyes have a huge toxicity knowledge gap. Therefore it is important to fill these gaps regularly and scientifically. In the latest report (Brüschweiler and Merlot 2017) this discrepancy was explored by researchers examining the genotoxicity of 397 non-regulated aromatic amines that can theoretically be emitted from the 470 recognized textile azo dyes. Thirty-six mutagenic aromatic amines have been detected, through freely searchable datasets. Also, 40 separate aromatic amines discovered to be mutagenic and theoretically emitted as cleavage substances were also reported from roughly 180 parent azo dyes used in textile garments. On the basis of these findings, the researchers assumed not only that the individual aromatic amine exposure, but also the cumulative exposure in textiles to multiple mutagenic aromatic amines, should be viewed as a full exposure and health risk evaluation, taking into consideration that the mutagenic characteristics of aromatic amines could be of much greater significance than previously anticipated Brüschweiler and Merlot (2017); Nguyen and Saleh (2017) also analyzed the concentrations of azo dyes and aromatic amines in women under apparel in the same line. For the possible discharge of aromatic amines to the skin, 120 samples of women's underwear of various colours, fabric systems, sources, geographical locations of manufacturing and brand names were assessed. In 74 samples, lower concentrations of aromatic amines were found, but 18 samples had elevated concentrations of aromatic amines than those suggested by the EU and China. The researchers noted the significance of studying aromatic amines in clothes.

27.7 Quinoline, Bisphenols, Benzothiazoles and Benzotriazoles

Investigations by numerous scientific, academic institutions have found that construction materials and components used to finish work within the building are largely responsible for the degradation of the indoor air quality. The compounds used for a wide variety of uses include quinoline and derivatives, bisphenols (BPAs), benzothiazoles (BTH), and benzotriazoles (BTRs), including clothes textiles. Quinoline, a heterocyclic aromatic organic compound and its components, is commonly employed in the textile sector to produce dyes. Few of them can be irritants to the skin or can be responsible for cancer in humans. Certain compounds are already identified in textiles (Luongo et al. 2014, 2016a; Yang et al. 2013). In 31 textile samples of various styles, quinoline levels and ten quinoline derivatives were evaluated: including T-shirts and denim to skirts. They embodied numerous colours, fabrics, labels, production nations, and prices (Luongo et al. 2014). Quinolines were found in 29 of the 31 specimensanalysed, and quinoline representing up to about 50% of the overall quinolines. Reflecting that the skin is exposed to a wide area of fabric and the possible health effects of these chemicals, the maximum amounts were 1.9 mg in a single material, an effect that cannot be reduced. Ten quinoline chemicals were again found in fabrics made of cotton, polyamide >70% and polyester in a corresponding analysis performed by the similar investigation team Luongo et al. (2016a). At values between 0.06 and 6.2 μg/g, quinoline was found in all samples. Since quinoline and isoquinoline are known as hazardous chemicals, the researchers emphasized the significance of gathering evidence from this possible origin for everyday human skin exposure. For BTHs and BTRs (Avagyan et al. 2015) screening amounts of 11 derivatives in 26 textile compounds, including products for infants, children and adolescents of various garment fabrics, colours and goods made in 14 countries have been tested. In the textile samples, eight of the 11 examined chemicals could be identified, revealing that garments might be a human exposure pathway to BTHs and BTRs. These are cytotoxic agents that may function as an endocrine disruptor, which may have allergic and irritating effects of subcutaneous sensitizers - between further harmful impacts (Ginsberg et al. 2011; Oda et al. 2008). The maximum proportion of BTH in the assay was "8.3 mg" of this compound. However, it was similarly seen 22 μg/g in a baby body finished from organic cotton labelled with Nordic Ecolabel/Svanenmärkt. Generally, BTH concentration was much greater than BTR concentrations.

The total emissions of BTH and quinolines in the residential wastewater were (0.5 and 0.24 g) correspondingly after one wash 5 kg of polyester cloth (Luongo et al. 2016b). These findings indicated that clothing is a possible cause of human exposure to BTHs, BTRs and quinolines, noting that significant concentrations of these compounds existed in the garments even after ten times of laundering. The presence of benzothiazole benzotriazole and seven main derivatives in an overall of 79 textile samples with raw fabric and baby apparel "blankets, diapers and clothing" has been detected lately (Liu et al. 2016). Depending on textile form, e.g. cotton, polyester, and nylon, places of origin, and colours, the levels of BTHs and BTRs were analysed. The most commonly observed chemical was BTH. BTR levels were increased in some textiles with the maximum BTR level (14,000 ng/g) in an infant's printed bodysuit. In the examined fabrics, the mean average amounts of BTR were greater than those of BTH. Dermal sensitivity to these contaminants has also been measured in infants. The exposure amounts of BTH and BTRs from fabrics up to 3740 pg/kg·bw/day of a thematic graphic imprint in a bodysuit's chest section were determined to have been strong from the use of socks 244-395 pg/kg·bw/day. To assess the frequency of BPAs, including bisphenol A (BPA) and bisphenol S (BPS), benzophenones, bisphenol A diglycidyl ethers (BADGEs) and novolac glycidyl ethers (NOGEs), 77 textiles and baby clothing items were also analysed by the same researchers Xue et al. (2017). Different textile varieties, e.g. cotton, polyester, nylon, colours and places of origin were acquired. Raw fabrics, disposable diapers, blankets, and clothes sold for children <1 year of age were included. The findings revealed that 82% and 53% of textile samples had BPA and BPS and that the average concentrations were 366 and 15 ng/g. On the other hand, benzophenone-3 (BP3) was detected at an average level of 11.3 ng/g in 70% of the samples. Lastly, the prevalent chemical was BFDGE, with an average level of 13.6 ng/g, between the 11 BADGEs and NOGEs was studied. Textile dermal exposure doses for babies aged 6–12 months varied from 201 pg/kg·bw/day to 248 pg/kg·bw/day for newborns whereas BP3 exposure doses varied from 6.17 to 7.62 pg/kg·bw/day, respectively. The dosage of dermal BPA exposure from certain textiles to newborns was as strong as 7280 pg/kg·bw/day. Between the garments examined, maximum BPA sensitivity in babies was socks. The researchers suggested that additional research on the origins and concentrations of exposure of compounds existing in textiles is required (Xue et al. 2017; Liu et al. 2016).

27.8 Phthalates

Phthalates, a broad variety of organic compounds, are employed primarily to improve softness and durability as plasticizers in plastics, particularly polyvinyl chloride (PVC). One of the applications of phthalates in the manufacturing sector is PVC prints; plasticized PVC may result in prolonged phases of straight contact with the skin. It must be taken into consideration that the end-user of fabrics with printed PVC are kids who, in their functional status, are the most susceptible category to these destructive chemicals (Martínez et al. 2018). The research (Pedersen and Hartmann 2004), performed with the analysis of phthalate material in baby clothes was a prime example. In all the clothing evaluated, phthalates were identified "19 samples," with concentrations "sum of all phthalates" ranging from "1.4 mg/kg to 200,000 mg/kg—about 20% of the sample weight." Di-2-ethylhexyl phthalate (DEHP), diisononyl phthalate (DINP) and dihexyl phthalate (DHP) were the phthalates present in elevated quantities. However, phthalate "primarily DEHP and DINP" concentrations in nylon sheets, crib mattresses and diaper-changing mats above 0.1% by weight were identified (Negev et al. 2018), the normal provided by "EU Commission Regulation-EC No 552/2009." A research published by Li et al. (2015), produced similar results "concentrations >0.1%" indicating that the chemicals present in the fabrics should be assessed if the compounds should be followed up.

Comparison to other floor coverings, carpets may retain larger concentrations of dust and surface-sorbed substances. Carpets can function as an indoor chemistry facilitator by "storing" compounds for upcoming reactions. Such compounds in carpets might be accessible only under particular or temporary conditions for contact with short-lived indoor oxidants (Alwarda et al. 2018). Oxidants react at hundreds of times higher to surface-sorbed chemicals than in gas-phase chemistry (Alwarda et al. 2018; Ham and Raymond Wells 2009; Shu and Morrison 2011), products and yields for reactions can vary between the two (Waring and Siegel 2013). Acid-base interaction also influences reaction sites on carpets (Ongwandee and Morrison 2008). Due to esters' presence in carpet materials and adhesives, more attention should be given to the possibility of hydrolysis reactions. Due to esters' involvement in carpet fabrics and adhesives, greater consideration should be paid to hydrolysis. Moisture and the existence of an alkaline base, like concrete, may facilitate ester hydrolysis, resulting in smaller, more volatile species being produced. DEHP hydrolysis is believed to be the main indoor cause of 2-ethyl-1-hexanol (2-EH) (Wakayama et al. 2019), and even at comparatively low levels is an irritant. As per Chino et al. (2013) when the carpet was fixed to floors with the elevated water level, the 2-EH emissions increased using a phthalate containing adhesive. It is no longer popular to use PVC—which contains the plasticizer DEHP as a carpet adhesive. DEHP from several other sources can nevertheless be found in carpet dust (Abb et al. 2009; Bornehag et al. 2005). The prominent microbial and corrosion risks related to humidity illustrate the importance of moisture problems in the indoor environment.

27.9 Volatile Organic Compounds

There is a well-known understanding that outdoor air pollution can have dangerous effects on human health. Scientific proof has already shown that indoor air quality is more extremely contaminated and has a higher health impact than outdoor air. The possible explanation is that humans spend much of their period indoors inside firmly closed building structures. Thus, because of the high risk of exposure to toxins indoors than outside, more people are at higher risk of health issues. Individuals who are subjected for long durations to indoor air emissions are extra prone to the impacts of indoor air contamination. According to Yu and Crump (1998) carpets serve as a primary basis for the indoor VOCs. The primary term applies to contaminants found in the product before they are manufactured and then discharged indoors and thus primary pollutants from many of the building materials are present. Several experiments have shown that "hundreds of VOCs and SVOCs" are released from carpets, underlays and glues (Cox et al. 2002; Hodgson 2000; Koutsoyiannis et al. 2008; Schaeffer et al. 1996; Sollinger et al. 1993, 1994; Wilke et al. 2004). 4-phenylcyclohexene (4-PCH), the cause of the latest carpet scent, aromatic compounds (styrene, benzene, toluene, xylene) and HCHO are several of the known VOCs

(Hodgson et al. 1993; Koutsoyiannis et al. 2008). The main pollution from carpets may affect total indoor VOC levels (Dietert and Hedge 1996). Comparison to other indoor construction materials, they may negatively lead to sensory assessments of indoor spaces (Sakr et al. 2006). Carpet studies record pollution sources or levels of unique or total VOCs (TVOCs) originating from carpet piles or backs and glues ranging in a degree above many orders; different studies record ranges of pollution factors ranging 10–10000 $\mu g\ m^{-2}\ h^{-1}$ (Guo et al. 2004; Koutsoyiannis et al. 2008).

Several SVOCs are also used in carpet production. For instance, several chemicals likefluorinated soil retardants including per- and polyfluoroalkyl substances (PFAS) (Kissa 2001; Prevedouros et al. 2006), triclosan-like antimicrobials (Petersen 2016), plus phthalate plasticizers which might be either existing in the dust or can get released from the industrial products where PVC was used (Langer et al. 2010). As toxins, flame retardants of organohalogen and organophosphorus are found in carpets manufactured from recycled polyurethane foam (Curtis et al. 2015). PFAS is probably the most researched of these SVOCs, and associations among the existence or number of carpet in indoors and PFAS levels in the dust are being identified (Gewurtz et al. 2009; Kubwabo et al. 2005). Although PFAS is not currently used in the manufacturing of novel carpets, however, the turnover of manufactured carpets either in use or present in stores will take years, if not decades. Apart from the chemicals present in the carpet's fabrics, the usage of carpet will also lead to the release of the chemicals into indoors which are applied to it after selling. For instance, the chemicals used for cleaning and pest control. In one notable case, repeated family treatment of an aftermarket stain-protector has been shown to result in increased perfluorohexane sulfonate (PFHxS) levels in the carpet, dust, and blood serum of inhabitants (Beesoon et al. 2012).

By reactive accumulation, sorption/desorption, and particle deposition procedures, carpet surfaces disturb indoor chemistry. Indoor air composition can be altered by the absorption and desorption of VOCs and SVOCs by attenuating the air pollutants overall pollution concentrations and extending exposure to the case following re-emission (Singer et al. 2004; Won et al. 2000, 2001). A broad variety of SVOCs can be easily sorb on the garments or to carpets (Morrison et al. 2015; Saini et al. 2016a). Carpets are found to be a good sink for chemicals like nicotine and phenanthrene with low volatility (Liu et al. 2016; Van Loy et al. 2001), organophosphorus flame retardants (Liu et al. 2016), both phthalates and adipates (Uhde et al. 2019). These SVOCs will then progressively be re-emitted from carpets to indoor environments for several or more years, even for the rest of the carpet's life.

HCHO is listed as human carcinogenic (Group 1) by the International Organization for Research on Cancer (IARC 2012). Textiles and clothing are being processed with HCHO-releasing chemicals and adhesives for a long time to strengthen anti-creasing features (Aldag et al. 2017). The strong press chemical coating was centred on urea-formaldehyde (UF) resin and melamine/formaldehyde resin in the 1950s and 1960s, which leaked large quantities of HCHO into clothing 5000–12000 ppm: 0.5–1.2% (De Groot et al. 2010b). This increased HCHO level has led to several contact dermatitis cases being identified De Groot et al. (2010a, b). Strong press chemical coating is currently based on enhanced dimethylol dihydroxyethyleneurea, which contains minimal HCHO (De Groot et al. 2010b). Among all the VOCs, HCHO is commonly used in manufacturing construction materials and various home products. As a by-product of burning and other natural processes, it can also be found outdoors. HCHO is being used in the fabric polishing to improve wrinkle and flame tolerance, dye and ink penetration, and reliability with other finishing constituents, such as emulsifiers and whitening agents (Niculescu et al. 2012). It has some beneficial uses concerning textiles but is a harmful air pollutant emitted indoors from many industrial goods like permanent press materials and separation by fibreglass (Kelly et al. 1999). It is a highly harmful substance; individual HCHO indoor experiences can cause inflammation of the skin, breathing complications, cough, eye, nose and throat burning sensations (Niculescu et al. 2012).

This compound contributes permanent press characteristics to garments and draperies linked to textiles (USGAO 2010). Different works have measured concentrations of HCHO in garments. Novick et al. (2013) examined 20 fabric objects finding HCHO in just 3 of them. Interestingly, the levels of 2 of these three products identified 3172 and 1391 ppm were 40 times greater than those specified by international textile regulations. Washing and drying methods decreased HCHO levels by between 26 and 72%, as per Novick et al. (2013). Research involving 180 textile products was completed by the USGAO (2010). Thirty-five out of 180 free HCHO samples were found, with 10 of them exceeding the governing norm with concentrations between 75.4 and 206.1 ppm (75 ppm in non-baby direct skin clothing contact). However, Piccinini et al. (2007) examined 221 clothing and linen samples. HCHO concentrations were reported to be below 30 ppm in 89% of them, while the concentrations were below 75 ppm in 97% of them. Just three products, with the highest concentration of 163 ppm, surpassed the 100 ppm limit. The difference was observed as per the place of origin and the stores where the products were bought (Piccinini et al. 2007).

27.10 Nano-Materials and Nanoparticles

Due to its large potential and financial impact, nanotechnology is a focus of research in various countries. However, the presence of such nano-materials can lead to harmful influences on human health and the environment because of the irregularities and anomalies in form, complexity, and chemical characteristics. The textile sector is already a major user of nanotechnology, with a tremendous amount of nanotextiles on the market and various consumer products, including nano-materials. Nanotextiles are considered to be traditional textiles incorporating nano-materials. These innovative constituents provide diverse advanced features such as flame retardant, self-cleaning, dirt-repellent, water repellent, ultraviolet radiation safety or antibacterial properties (Almeida and Ramos 2017; El-Naggar et al. 2018; Elsayed et al. 2020; Gadkari et al. 2020; Ibrahim 2015; Radetić 2013; Sundarrajan et al. 2010; Xue et al. 2020). Nano-coating and nano-finishing boost the potential use of fabric materials in various fields. In evolving practical and high-performance textiles, the use of nanofibers and nanocomposite-based coating materials has highlighted a big potential (Bashari et al. 2018; Faccini et al. 2012; Lund et al. 2018; Riaz et al. 2018; Silva et al. 2019).

Usage of metal nanoparticles has brought a massive boom in different manufacturing sectors during the last decade. Increased development and use of engineered nanoparticles (ENPs) and engineered nano-materials have contributed to the potential to manufacture new engineered nano-materials (ENMs). Between these, metal ENPs are being used increasingly in the textile manufacturing sector (Montazer et al. 2014; Som et al. 2011; Yetisen et al. 2016). The silver-ENP antimicrobial activity and titania-ENP UV absorption are strong illustrations of this application (Lombi et al. 2014; Von Goetz et al. 2013; Windler et al. 2013). Nevertheless, skin exposure to these nano-objects and their aggregates and agglomerates (NOAA) may lead from the mobilisation and migration of ENPs from textiles into human sweat (Von Goetz et al. 2013). Silver (Ag) is among the most commonly employed components of metal nanoparticles. Nanoscale Ag is among the most frequently reported applications of nano-Ag in fabrics. Nevertheless, Ag's appearance in fabrics remains mostly unidentified as the marking of items is inadequate (Lombi et al. 2014). Corresponding to this it has been stated that for Ag-NOAA, its possible dermal exposure through fabrics is roughly proportional to the main source of Ag-ENP, which is a nutritional supplement (Von Goetz et al. 2013). Because of the rising significance of metallic nanoparticles, particularly in the textile industry, interest has also grown in parallel with their environmental and human health consequences.

One of the most commercialized uses for this nano-material is the processing of fabrics with silver nanoparticles, although there are questions about product safety (Arora et al. 2008; Jura et al. 2013; Korani et al. 2011; Samberg et al. 2010; Trop et al. 2006; Vlachou et al. 2007; Wijnhoven et al. 2009). Samberg et al. (2010) observed that unwashed silver nanoparticles induced the focal swelling and localization of silver nanoparticles on the surface and the upper stratum corneum structures of porcine skin to increase significantly in the pro-inflammatory cytokine release regions. Normal human primary keratinocytes were presented in vitro to polyvinyl propylene-capped silver nanoparticles (Jura et al. 2013) and reductions in cell viability, metabolism, proliferation, and migratory behaviour were reported. There are also questions about using Ag nanoparticles in wound gauze when the substance comes into connection with vulnerable skin. Arora et al. (2008) subjected human skin carcinoma cells to Ag nanoparticles in vitro and reported alterations in cellular proliferation and oxidative stress-proof. A temporary spike in silver plasma and urine concentrations through item use was recorded when person skin injuries were cured with Ag comprising surgical instruments (Trop et al. 2006; Vlachou et al. 2007). Besides, Trop et al. (2006) stated that hepatotoxicity and argyria greyish discolouration of the skin formed in people with burn wounds cured with a Ag comprising cured bandage. Skin contact to Ag nanoparticles can also raise a wide variety of detrimental health issues, and further study is worthwhile. Strong bonds among fabric and Ag nanoparticles remain important to safeguard long-term longevity of antimicrobial care. The procedure method is a critical indicator of Ag preservation in fabrics and the long-term effectiveness of the microbial growth management chemical. In industrial practice, throughout also called masterbatch or after manufacturing, antimicrobials such as Ag may be applied to textile fibres also called as finishing. Masterbatch manufacturing is employed for artificial fabrics and adds the Ag during development into the thickness of the fibre polymer threads. On the other hand, finishing is employed for natural and synthetic cloth forms and implies covering the Ag on threads by dipping the thread or fibre during production in a wet bath (Stefaniak et al. 2014).

Research of textiles incorporating Ag nanoparticles had largely focused on researching the escape of Ag into the environment (Benn et al. 2010; Lee et al. 2003; Liu et al. 2012; Perelshtein et al. 2008). More lately, studies have focused on the release of Ag into body fluid and the possibility of contamination from interaction with customer textile items (Quadros et al. 2013; Von Goetz et al. 2013). These findings indicate that Ag was mainly emitted to artificial sweat in ionic form in textiles at differing amounts, based on the substance being studied. The factors affecting Ag exposure from textile consumer goods are, however, little understood. Stefaniak et al. (2014) proposed that the item itself and/or the body fluid's biochemical characteristics that come in contact with the garment throughout use may contain important exposure factors. Explanations of product-related considerations include the fibre (synthetic or natural) form and Ag usage during the manufacturing process. Sweat and saliva are biological fluids that can contact textiles; they have distinct properties due to their particular biological functions. Composition, pH, and temperature are examples of important physiological variables in biological fluids.

27.11 Micro and Nano-Plastics

Micro nano(plastics), origins, fate and consequences have attracted growing attention from the scientific society, the general public and policymakers <5 mm in size of plastic particles and fibres, generally mentioned as microplastics (Anderson et al. 2015). Microplastics below 1000 nm might be known as nanoplastics with certain studies showing that this size group could be the very harmful type of plastic waste in water ecosystems (Koelmans et al. 2015). As 0.19 million tonnes of microfibers from the manufacture and regular use of textile materials, especially residential garment washing, have been projected to reach the marine ecosystem on an annual basis alone (Eunomia 2016) usage continues to increase, which figure appears set to grow more. Fibres, derived from textiles, are one widely known form of microplastic found in the environmental samples. Many substances in our everyday life are formed of fibres, either synthetic or natural-furniture, fabric, etc. In the marine ecosystem, "both natural textile fibres—wool, linen, and cotton and synthetic textile fibres—polyester, polyamide" are present (Mathalon and Hill 2014; Remy et al. 2015). Since textiles are believed to be major sources of synthetic fibre, based on our awareness of fibres amounts in the atmosphere and their paths, it is important to resolve synthetic textile concerns. Use of synthetic fibres in textiles and clothing grew from an annual output of 2.1 million tonnes of synthetic fibres in 1950 to an annual production of about 50 million tonnes of synthetic fibres in 2010 (Aizenshtein 2012; Essel et al.

2015). Synthetic fibres make up about 60% of overall world production of fibres, and polyamide (nylon) and polyester (PET) predominate. A comprehensive report suggested that more than 700,000 fibres might be emitted from laundering 6 kg of clothes (Napper and Thompson 2016). Lower levels have been documented in a previous report; outgoing laundry water might comprise between "100–300" fibres each litre (Browne et al. 2011). The same author Browne et al. (2011) published a report and concluded that 1900 fibres are released per garment after wash. While as found (Almroth et al. 2018), knit fabrics emitted "9 ± 7" fibres per 100 cm^2 of material, on average. An average of 1177 ± 135, 932 ± 59, and 1210 ± 96 fibres per 100 cm^2 of fabric is released by fleece and microfleece. How closely the wool is knitted into the cloth is correlated with the degree of shedding. As shown by the knitting gauge, more closely knitted yarn leads in more fibres on the similar region of fabric, leading to more loss of fibre. This has ramifications for the garment industry, where attempts are tried to manufacture textiles that shed fewer and are thus more ecologically conscious.

Ageing of fabric, i.e. by use and time passing, could influence the amount of microplastic fibres produced throughout laundering. A newly released study conducted by Patagonia (outdoor clothes manufacturers) showed that old clothes currently shed higher fibre masses than new clothes (Hartline et al. 2016), and Almroth et al. (2018) hypothesized the same case in their work as well. Previously, calculations of fibres lost from fabrics were carried out with water only, as detergent-containing effluent water created problems to analyse as filters appear to clog (Browne et al. 2011). Browne et al. (2011) enhanced rinsing and filtering techniques make for more practical estimation of laundry routines' consequences and a clearer contrast of their findings with earlier works. Very recent evidence has also found that more fibres appear to be emitted from bio-detergent and conditioner-washed garments (Napper and Thompson 2016). These changes in fibre depletion through detergent are possibly based on detergents' basic characteristics; they minimise surface tension. Knitwear, which consists of woven coils, creates a cloth surface that functions as a cage containing units like moveable fibres. Nevertheless, the low surface tension arising from detergent use is practically equal to better wetting, resulting in increased cloth rinsing, and it could release any particles retained in the 3D knitwear system. In contrast to this wetting feature, the important part of detergents is that they serve as dispersants; dirt is dissolved in the solvent water and stored there, thus preventing the cloth from falling back. For fibre pieces, this phenomenon relates; most fibres emitted from the textile are often carried away and have an improved probability of being found in the corresponding analysis.

Studies about microplastics in various environments highlighted the ubiquity of anthropogenic fibres. Dris et al. (2017) assessed that microfiber levels were greater indoors than outdoors. The fibres' deposition rate in indoor environments is between 1586 and 11,130 fibres/day/m^2, leading to an accumulation of fibres in settled dust 190–670 fibres/mg. Concerning fibre category, 67% of the analyzed fibres in indoor environments are made of natural material, primarily cellulosic, while the remaining 33% fibres contain petrochemicals with polypropylene being predominant (Dris et al. 2017). The observed fibres are supposedly too large to be inhaled, but the exposure may occur through dust ingestion, particularly for young children. They did not differentiate natural vs human-made fibres instead they used the word cellulosic. Polypropylene, which is widely used in carpets and other fittings, is the main synthetic fibre. Zhang et al. (2020) studied the microplastic fallout in a dormitory, an office, and a corridor. Among the three sampling sites, the highest average microplastic abundance appeared in the dormitory 9.9×10^3 MPs/m^2/d, followed by the office 1.8×10^3 MPs/m^2/d and the corridor 1.5×10^3 MPs/m^2/d. Microplastic fallout existed mostly in fibres and showed similar polymer compositions to the textile products used in indoor environments.

Study on the adverse consequences of microplastic toxicity on species has generally been restricted to localised interactions, primarily in marine environments, over short periods. Human symptoms from dispersed terrestrial sources are much extra probably to be a part of accumulated exposure (Henry et al. 2019). There have been records of microplastics in an extensive variety of human food and drinks, like seafood (Rochman et al. 2015), drinking water (Kosuth et al. 2018; Mason et al. 2018; Zuccarello et al. 2019), beer (Kosuth et al. 2018; Liebezeit and Liebezeit 2014), salt and sugar (Liebezeit and Liebezeit 2013; Rist et al. 2018) and in the air (Abbasi et al. 2019; Cai et al. 2017; Dris et al. 2017; Levermore et al. 2020; Liu et al. 2019; Kaya et al. 2018; Wright et al. 2019), expanding fears regarding the danger to human health by ingestion and inhalation (Waring et al. 2018). Validation of the possibility of pollution of products of food and beverage (Rist et al. 2018) and any previous evaluations of microplastic digestion of food have put suspicion on the ambiguity of analytical methods (Catarino et al. 2018). Based on foodstuff consumption, the estimated intake of microplastics is 39,000–52,000 particles person^{-1} year^{-1} (Cox et al. 2019). Yet a level of prolonged exposure is an essential portion of everyday human life (Wright and Kelly 2017). However, the latest review of the presented confirmation by (Waring et al. 2018) found that microplastic food chain pollution is liable to pose toxic effects at existing intake amounts and concentrations of microfibers and nanofibers absorption and translocation into the intestinal system and/or the lungs. Moreover, Waring et al. (2018) indicated that more study is needed on the consequences of elevated cumulative amounts of exposure in body cells and health problems like a leaky stomach, permeable blood-brain barrier or long-term intake of polluted foods. Human experimentation is complicated since; first of all, scientific trials are ethically problematic. In comparison, identifying a test sample which has not been subjected to microfibers is challenging, if not impractical, and separating the influences of individual components in large population-based trials is complicated and expensive. Nevertheless, empirical trials are starting to analyse the proof for future human health sensitivity to microplastics. As among most species, impacts are likely to represent a mixture of processes that are physical, chemical, and biological, e.g. pathogen transfer. Serious scientific gaps comprise the long-term possibility of toxicity and inclusive approach of nanoplastic absorption and inhalation (Gasperi et al. 2018; Waring et al. 2018). A degree of bioaccumulation can occur from persistent human exposure to textile micro (nano)fibres as in other species, these fibres can subject humans to compounds correlated with toxicity, carcinogenicity and mutagenicity such as non-reacted monomers, and additives, like PBDEs (Gasperi et al. 2018).

According to Dris et al. (2017) textile ambient fibres are normally too big to be inhaled; however, smaller fibres can be easily inhaled. Inhalation of 26–130 airborne microplastics has been measured on a day^{-1} basis (Prata 2018). It is assessed that a male individual with gentle exercise inhales 272 microplastics per day based on air sampling that used a breathing thermal manikin (Vianello et al. 2019). However, in human respiratory tissues, both cellulose-based and plastic microfibers are being found Pauly et al. (1998). This suggests that certain small filaments may penetrate the lungs and respiratory passages (Gasperi et al. 2018). Revel et al. (2018) suggested that a less significant path of exposure is known to be dermal interaction with microplastics, even though it has been hypothesised that nanoplastics (<100 nm) could cross the skin. An in vitro research by Gasperi et al. (2018) provided support of the resilience and biopersistence of biological fluid plastic fibres, with polypropylene, polyethylene and polycarbonate fibres showing virtually no degradation or surface properties modifications and features after six months in artificial extracellular lung fluid. There is an unclear existence and pathway of potential toxicity results from prolonged inhalation, and precaution is required to understand the evidence on the influence of textile micro and nanofibers which are extrapolated from inorganic fibre inhalation experience, like asbestos.

In brief, minimal studies plus empirical data indicate that current human exposure to synthetic textile microfibers can pose significant toxicity. The possible influences of persistent long-term ingestion and inhalation of nanofibers, furthermore, remain unclear. Knowledge gaps and community concerns pose these important-urgent investigation challenges and provide an argument for restricting persistent plastic microfibers and nanofibers' exposure to the maximum degree possible, especially for vulnerable populations. To track exposure, accurate and consistent methodological criteria are a priority.

27.12 Conclusion and Recommendations

Textiles offer softness to interior spaces, absorb noise, and enhance aesthetics; nevertheless, they can adversely impact indoor air quality because of chemicals used in processing, treatments, and more. Multiple chemical sensitivity is a condition related to low indoor air quality. Exposure to different chemicals found in the atmosphere can cause symptoms of multiple chemical sensitivities. The primary source of the disease is inadequate indoor air quality. Growing consumer consciousness and a sense of social responsibility related to environmental issues have driven the textile industry to develop goods with better environmental benefits. Depending on the latest data available, it is assumed that pollutants from indoor textile materials can disrupt the quality of indoor air or present a potential health hazard under ordinary environmental conditions. The total environmental effects of any textile product can be subdivided into those involved with its processing, servicing, and subsequent disposal. Among other factors, manufacturing concerns include the renewability of the raw resources and the toxicity of the chemicals emitted during production and processing (crop handling, chemical by-products, solvents). The consistency and reliability of the substances used for washing or dry cleaning over the items' lifespan are mainly consumer concerns. Finally, concerns with recycling concern the recyclability and/or biodegradability of the materials. In coping with potential threats, the skin offers the vital first defence system for the body. To preserve human health and normal functions, garment material and the dermally establish a shielding mechanism that maintains a warm and sensory comfort condition. A lack of regulatory relationships between cloth and skin can result in different health issues. All fibres, natural and synthetic, contain various non-fibrous materials in their native state. Still, the vast majority of these are either removed before conversion into consumer articles or present in such small quantities that they are of little significance. Once the fibres are converted into the fabric, however, the situation becomes much different. Fairly large amounts of chemicals are frequently added to fabrics to produce various special effects for the consumer.

The prevalence of plastics in the atmosphere presents a major danger to both human health and natural environments. In fact, plastics of the fibre type have a reasonably high share of these plastics. One of the most substantial suppliers of fibre is the garment industry. The washing of synthetic textiles and garments is a cause of microplastic pollution into the environment. However, mechanical drying as a route for the emission of microplastics is little known. The enhanced quantities of microplastics indoors depend on a larger number of occupants/people inside. The amount of microplastics found indoors refers to plastic items such as carpets, toys, chairs, kitchen equipment, electrical cables, electronics, textiles, indoor decor, cleaning materials, and more. The regular indoor practises and use of plastic goods will eventually contribute to the discharge of microplastics that accumulate in the indoor dust. Future studies on particular microplastic origins are therefore also required to establish specific microplastic forms from each source. Apart from this future work should focus on the intake of fibres via ingestion and inhalation. The penetration through the skin should also be given importance. Skin contact to Ag nanoparticles can also raise a wide variety of detrimental health issues, and further study is worthwhile.

Of our indoor settings, carpets are an important component. They are dynamic, multi-component structures that have major consequences for human exposure and indoor chemistry. In the interior room, the textiles used differed for the purpose of implementing unique treatments by chemical compounds. Such compounds may give fabrics many characteristics that may differ based on the nature of the area. In high temperatures and humidity, certain textile-related compounds are active and reactive and can vary in characteristics when interacting with another compound. Other causes related to indoor person behavior, like cooking, washing and friction, can also enhance the discharge of harmful compounds into the indoor environment, influencing its consistency plus, in turn, affecting the lung system. Chemicals emitted from clothing products that have already been processed are affected by indoor air. Due to prolonged use, these chemicals have different impacts on human health, allowing certain illnesses to change in severity due to human physical and psychological characteristics. The indoor air quality can be significantly influenced by textiles chosen for interiors. Thus, fabrics must be studied prior to selection in order to reduce indoor air emissions. Once suitable textiles have been chosen, steps need to be taken throughout installation and retained to guarantee that the condition of indoor air is not affected during servicing. Context information should be given on the characteristics of synthetic and natural fibres relevant to the quality of indoor air, the application of finishes and the installation and maintenance of textiles: all relative to the quality of indoor air. Worsening of indoor air and human health is caused by clothing processing, fabrics and chemicals used for manufacturing. It is crucial to study chemicals and other substances in textiles, and further research on this topic is required. Garments can contain dangerous chemicals. We need to create awareness among individuals about potentially harmful textiles.

References

Abb M, Heinrich T, Sorkau E, Lorenz W (2009) Phthalates in house dust. Environ Int 35(6):965–970. https://doi.org/10.1016/j.envint.2009.04.007

Abbasi S, Keshavarzi B, Moore F, Turner A, Kelly FJ, Dominguez AO, Jaafarzadeh N (2019) Distribution and potential health impacts of microplastics and micro rubbers in air and street dusts from Asaluyeh County, Iran. Environ Pollut 244(1):153–164. https://doi.org/10.1016/j.envpol.2018.10.039

Aizenshtein EM (2012) Production and use of chemical fibers in 2010. Fibre Chem 43(6):395–405. https://doi.org/10.1007/s10692-012-9372-1

Alcock RE, MacGillivray BH, Busby JS (2011) Understanding the mismatch between the demands of risk assessment and practice of scientists—the case of Deca-BDE. Environ Int 37(1):216–225. https://doi.org/10.1016/j.envint.2010.06.002

Aldag N, Gunschera J, Salthammer T (2017) Release and absorption of formaldehyde by textiles. Cellulose 24(10):4509–4518. https://doi.org/10.1007/s10570-017-1393-8b

Almeida L, Ramos D (2017) Health and safety concerns of textiles with nanomaterials. IOP Conf Ser Mater Sci Eng 254(10): https://doi.org/10.1088/1757-899X/254/10/102002

Almroth BM, Åström L, Roslund S, Petersson H, Johansson M, Persson NK (2018) Quantifying shedding of synthetic fibers from textiles; a source of microplastics released into the environment. Environ Sci Pollut Res 25(2):1191–1199. https://doi.org/10.1007/s11356-017-0528-7

Alwarda R, Zhou S, Abbatt JPD (2018) Heterogeneous oxidation of indoor surfaces by gas-phase hydroxyl radicals. Indoor Air 28(5):655–664. https://doi.org/10.1111/ina.12476

Anderson A, Andrady A, Arthur C, Baker J, Bouwman H, Gall S, Hidalgo-Ruz Valeria, Köhler A, Law KL, Leslie H, Kershaw P, Pahl S, Potemra J, Ryan P, Shim WJ, Thompson R, Takada H, Turra A, Vethaak D, Wyles K (2015) Sources, fate and effects of microplastics in the marine environment: a global assessment. In: GESAMP working group 40, international maritime organization, London. https://doi.org/10.13140/RG.2.1.3803.7925

Arora S, Jain J, Rajwade JM, Paknikar KM (2008) Cellular responses induced by silver nanoparticles: in vitro studies. Toxicol Lett 179(2):93–100. https://doi.org/10.1016/j.toxlet.2008.04.009

Avagyan R, Luongo G, Thorsén G, Östman C (2015) Benzothiazole, benzotriazole, and their derivatives in clothing textiles—a potential source of environmental pollutants and human exposure. Environ Sci Pollut Res 22(8):5842–5849. https://doi.org/10.1007/s11356-014-3691-0

Bakker MI, De Winter-Sorkina R, De Mul A, Boon PE, Van Donkersgoed G, Van Klaveren JD, Baumann BA, Hijman WC, Van Leeuwen SPJ, De Boer J, Zeilmaker MJ (2008) Dietary intake and risk evaluation of polybrominated diphenyl ethers in the Netherlands. Mol Nutr Food Res 52(2):204–216. https://doi.org/10.1002/mnfr.200700112

Barker RH (1975) Additives in fibers and fabrics. Environ Health Perspect 11(6):41–45. https://doi.org/10.1289/ehp.751141

Bashari A, Shakeri M, Shirvan AR, Najafabadi SAN (2018) Functional finishing of textiles via nanomaterials. Nanomater Wet Process Textiles 1–70. https://doi.org/10.1002/9781119459804.ch1

Beesoon S, Genius SJ, Benskin JP, Martin JW (2012) Exceptionally high serum concentrations of perfluorohexane sulfonate in a canadian family are linked to home carpet treatment applications. Environ Sci Technol 46(23):12960–12967. https://doi.org/10.1021/es3034654

Benn T, Cavanagh B, Hristovski K, Posner JD, Westerhoff P (2010) The release of nanosilver from consumer products used in the home. J Environ Qual 39(6):1875–1882. https://doi.org/10.2134/jeq2009.0363

Bhatia SC (2017) Pollution control in textile industry. WPI Publishing (10):1–330 https://doi.org/10.1201/9781315148588

Bornehag C, Lundgren B, Weschler CJ, Sigsgaard T, Hagerhed-engman L, Sundell J (2005) Phthalates in indoor dust and their association with building characteristics. Environ Health Perspect 113(10):1399–1404. https://doi.org/10.1289/ehp.7809

Brigden K, Hetherington S, Wang M, Santillo D, Johnston P (2012) Hazardous chemicals in branded textile products on sale in 27 countries/regions during 201. Greenpeace Research Laboratories Technical Report 06/2012. http://archivo-es.greenpeace.org/espana/Global/espana/report/contaminacion/151012_Technical%20Note_Product%20Testing%20Report_%20With%20Exec%20summary.pdf. Accessed 10 Jan 2021

Brillas E, Martínez-Huitle CA (2015) Decontamination of wastewaters containing synthetic organic dyes by electrochemical methods. An updated review. Appl Catal B 166–167(5):603–643. https://doi.org/10.1016/j.apcatb.2014.11.016

Browne MA, Crump P, Niven SJ, Teuten E, Tonkin A, Galloway T, Thompson R (2011) Accumulation of microplastic on shorelines worldwide: sources and sinks. Environ Sci Technol 45(21):9175–9179. https://doi.org/10.1021/es201811s

Brüschweiler BJ, Küng S, Bürgi D, Muralt L, Nyfeler E (2014) Identification of non-regulated aromatic amines of toxicological concern which can be cleaved from azo dyes used in clothing textiles. Regul Toxicol Pharmacol 69(2):263–272. https://doi.org/10.1016/j.yrtph.2014.04.011

Brüschweiler BJ, Merlot C (2017) Azo dyes in clothing textiles can be cleaved into a series of mutagenic aromatic amines which are not regulated yet. Regul Toxicol Pharmacol 88(8):214–226. https://doi.org/10.1016/j.yrtph.2017.06.012

Building Green (1994) Carpeting, indoor air quality, and the environment. Building Green. https://www.buildinggreen.com/feature/carpeting-indoor-air-quality-and-environment. Accessed 14 Jan 2021

Burkinshaw SM (1995) Chemical principles of synthetic fibre dyeing, 1st edn. Blackie Academic & Professional, New York

Busk L, Ahlborg UG (1982) Retinoids as inhibitors of ortho-aminoazotoluene-induced mutagenesis in the Salmonella/liver microsome test. Mutation Res Lett 104(4–5):225–231. https://doi.org/10.1016/0165-7992(82)90148-8

Cai L, Wang J, Peng J, Tan Z, Zhan Z, Tan X, Chen Q (2017) Characteristics of microplastics in the atmospheric fallout from Dongguan city, China: preliminary research and first evidence. Environ Sci Pollut Res 24(32):24928–24935. https://doi.org/10.1007/s11356-017-0116-x

Catarino AI, Macchia V, Sanderson WG, Thompson RC, Henry TB (2018) Low levels of microplastics (MP) in wild mussels indicate that MP ingestion by humans is minimal compared to exposure via household fibres fallout during a meal. Environ Pollut 237(6):675–684. https://doi.org/10.1016/j.envpol.2018.02.069

Chakraborty JN (2011) An overview of dye fastness testing. In: Handbook of textile and industrial dyeing: principles, processes and types of dyes, vol. 1. Woodhead Publishing Limited, 207–224. https://doi.org/10.1533/9780857093974.1.207
Chen D, Hale RC (2010) A global review of polybrominated diphenyl ether flame retardant contamination in birds. Environ Int 36(7):800–811. https://doi.org/10.1016/j.envint.2010.05.013
Chen SJ, Ma YJ, Wang J, Chen D, Luo XJ, Mai BX (2009) Brominated flame retardants in children's toys: concentration, composition, and children's exposure and risk assessment. Environ Sci Technol 43(11):4200–4206. https://doi.org/10.1021/es9004834
Chino S, Kato S, Seo J, Kim J (2013) Measurement of 2-ethyl-1-hexanol emitted from flooring materials and adhesives. J Adhes Sci Technol 27(5–6):659–670. https://doi.org/10.1080/01694243.2012.690656
Chung K (1983) The significance of azo-reduction in the mutagenesis and carcinogenesis of azo dyes. Mutat Res/Rev Genet Toxicol 114(3):269–281. https://doi.org/10.1016/0165-1110(83)90035-0
Collier S, Storm J, Bronaugh R (1993) Reduction of azo dyes during in vitro percutaneous absorption. Toxicol Appl Pharmacol 118(1):73–79. https://doi.org/10.1006/taap.1993.1011
Coman G, Blattner CM, Blickenstaff NR, Andersen R, Maibach HI (2014) Textile allergic contact dermatitis: current status. Rev Environ Health 29(3):163–168. https://doi.org/10.1515/reveh-2014-0061
Covaci A, Harrad S, Abdallah MAE, Ali N, Law RJ, Herzke D, de Wit CA (2011) Novel brominated flame retardants: a review of their analysis, environmental fate and behaviour. Environ Int 37(2):532–556. https://doi.org/10.1016/j.envint.2010.11.007
Cox KD, Covernton GA, Davies HL, Dower JF, Juanes F, Dudas SE (2019) Human consumption of microplastics. Environ Sci Technol 53(12):7068–7074. https://doi.org/10.1021/acs.est.9b01517
Cox SS, Little JC, Hodgson AT (2002) Predicting the emission rate of volatile organic compounds from vinyl flooring. Environ Sci Technol 36(4):709–714. https://doi.org/10.1021/es010802+
Curtis K, Wilding BC, Hulick A, LaBo K, Schuler K (2015) Flame retardants in furniture, foam, floors by clean and healthy New York, clean water action, conservation Minnesota. https://www.conservationminnesota.org/redesign/wp-content/uploads/SafeMattressReport-final.pdf. Accessed 14 Jan 2021
De Boer J, Wester PG, Van Der Horst A, Leonards PEG (2003) Polybrominated diphenyl ethers in influences, suspended particulate matter, sediments, sewage treatment plants and effluents and biota from the Netherlands. Environ Pollut 122(1):63–74. https://doi.org/10.1016/S0269-7491(02)00280-4
De Groot A, Geier J, Flyvholm MA, Lensen G, Coenraads PJ (2010a) Formaldehyde-releasers: relationship to formaldehyde contact allergy, Part 2: metalworking fluids and remainder. Contact Dermatitis 63(8):129–139. https://doi.org/10.1111/j.1600-0536.2010.01714.x
De Groot A, Geier J, Flyvholm MA, Lensen G, Coenraads PJ (2010b) Formaldehyde-releasers: relationship to formaldehyde contact allergy. Metalworking fluids and remainder. Part 1. Contact Dermatitis 63(3):117–128. https://doi.org/10.1111/j.1600-0536.2010.01714.x
Derden A, Huybrechts D (2013) Brominated flame retardants in textile wastewater: reducing Deca-BDE using best available techniques. J Clean Prod 53(8):167–175. https://doi.org/10.1016/j.jclepro.2013.04.013
Desire A, Narula SA (2018) An overview on corporate response towards sustainability issues in the textile industry. Environ Dev Sustain 20(4):1439–1459. https://doi.org/10.1007/s10668-017-9949-1
Dietert RR, Hedge A (1996) Toxicological considerations in evaluating indoor air quality and human health: impact of new carpet emissions. Crit Rev Toxicol 26(6):633–707. https://doi.org/10.3109/10408449609037480
Domingo JL (1993) Metal—induced developmental toxicity in mammals: a review. J Toxicol Environ Health 42(2):123–141. https://doi.org/10.1080/15287399409531868

Domingo JL (2012) Polybrominated diphenyl ethers in food and human dietary exposure: a review of the recent scientific literature. Food Chem Toxicol J 50(2):238–249. https://doi.org/10.1016/j.fct.2011.11.004

Domingo JL, Martí-Cid R, Castell V, Llobet JM (2008) Human exposure to PBDEs through the diet in Catalonia, Spain: temporal trend. A review of recent literature on dietary PBDE intake. Toxicol 248(1):25–32. https://doi.org/10.1016/j.tox.2008.03.006

Dris R, Gasperi J, Mirande C, Mandin C, Guerrouache M, Langlois V, Tassin B (2017) A first overview of textile fibers, including microplastics, in indoor and outdoor environments. Environ Pollut 221(2):453–458. https://doi.org/10.1016/j.envpol.2016.12.013

Eljarrat E, Labandeira A, Marsh G, Raldúa D, Barceló D (2007) Decabrominated diphenyl ether in river fish and sediment samples collected downstream an industrial park. Chemosphere 69(8):1278–1286. https://doi.org/10.1016/j.chemosphere.2007.05.052

Eljarrat E, Marsh G, Labandeira A, Barceló D (2008) Effect of sewage sludges contaminated with polybrominated diphenyl ethers on agricultural soils. Chemosphere 71(6):1079–1086. https://doi.org/10.1016/j.chemosphere.2007.10.047

El-Naggar ME, Shaarawy S, Hebeish AA (2018) Multifunctional properties of cotton fabrics coated with in situ synthesis of zinc oxide nanoparticles capped with date seed extract. Carbohyd Polym 181(2):307–316. https://doi.org/10.1016/j.carbpol.2017.10.074

Elsayed EM, Attia NF, Alshehri LA (2020) Innovative flame retardant and antibacterial fabrics coating based on inorganic nanotubes. Chem Select 5(10):2961–2965. https://doi.org/10.1002/slct.201904182

EPA (1997) Profile of the textile industry. Environmental Protection Agency, Washington, USA. https://nepis.epa.gov/Exe/ZyPDF.cgi/50000HE9.PDF?Dockey=50000HE9.PDF. Accessed 10 Jan 2021

Esancy JF, Freeman HS, Claxton LD (1990) The effect of alkoxy substituents on the mutagenicity of some aminoazobenzene dyes and their reductive-cleavage products. Mutation Res/Rev Genetic Toxicol 238(1):1–22. https://doi.org/10.1016/0165-1110(90)90036-B

Essel R, Engel L, Ahrens RH, Carus M (2015) Sources of microplastics relevant to marine protection in Germany. Texte 64:1219–1226

Eunomia (2016) Plastics in the marine environment. www.eunomia.co.uk. Accessed 26 Dec 2020

Faccini M, Vaquero C, Amantia D (2012) Development of protective clothing against nanoparticles based on electrospun nanofibers. J Nanomater 2012(6):892–894. https://doi.org/10.1155/2012/892894

Freeman HS (2013) Aromatic amines: use in azo dye chemistry. Front Biosci Landmark 18(1):145–164. https://doi.org/10.2741/4093

Gadkari RR, Ali SW, Joshi M, Rajendran S, Das A, Alagirusamy R (2020) Leveraging antibacterial efficacy of silver loaded chitosan nanoparticles on layer-by-layer self-assembled coated cotton fabric. Int J Biol Macromol 162(11):548–560. https://doi.org/10.1016/j.ijbiomac.2020.06.137

García-Esquinas E, Navas-Acien A, Pérez-Gómez B, Artalejo FR (2015) Association of lead and cadmium exposure with frailty in US older adults. Environ Res 137(2):424–431. https://doi.org/10.1016/j.envres.2015.01.013

Gasperi J, Wright SL, Dris R, Collard F, Mandin C, Guerrouache M, Langlois V, Kelly FJ, Tassin B (2018) Microplastics in air: are we breathing it in? Curr Opin Environ Sci Health 1(2):1–5. https://doi.org/10.1016/j.coesh.2017.10.002

Gewurtz SB, Bhavsar SP, Crozier PW, Diamond ML, Helm PA (2009) Perfluoroalkyl contaminants in window film: indoor/outdoor, urban/rural, and winter/summer contamination and assessment of carpet as a possible source. Environ Sci Technol 43(19):7317–7323. https://doi.org/10.1021/es9002718

Gingell R, Bridges JW, Williams RT (1971) The role of the gut flora in the metabolism of prontosil and neo prontosil in the rat. Xenobiotica 1(4):143–156. https://doi.org/10.1042/bj1290881

Ginsberg G, Toal B, Kurland T (2011) Benzothiazole toxicity assessment in support of synthetic turf field human health risk assessment. J Toxicol Environ Health Part A Curr Issues 74(17):1175–1183. https://doi.org/10.1080/15287394.2011.586943

Guigueno MF, Fernie KJ (2017) Birds and flame retardants: a review of the toxic effects on birds of historical and novel flame retardants. Environ Res 154(4):398–424. https://doi.org/10.1016/j.envres.2016.12.033

Guo H, Murray F, Lee SC, Wilkinson S (2004) Evaluation of emissions of total volatile organic compounds from carpets in an environmental chamber. Build Environ 39(2):179–187. https://doi.org/10.1016/j.buildenv.2003.08.015

Guzzella L, Roscioli C, Binelli A (2008) Contamination by polybrominated diphenyl ethers of sediments from the Lake Maggiore basin (Italy and Switzerland). Chemosphere 73(10):1684–1691. https://doi.org/10.1016/j.chemosphere.2008.06.073

Hale RC, La Guardia MJ, Harvey E, Gaylor MO, Mainor TM (2006) Brominated flame retardant concentrations and trends in abiotic media. Chemosphere 64(2):181–186. https://doi.org/10.1016/j.chemosphere.2005.12.006

Halimoon N, Yin RGS (2010) Removal of heavy metals from textile wastewater using zeolite. Environment Asia 31:124–130. https://doi.org/10.14456/ea.2010.51

Ham JE, Raymond Wells J (2009) Surface chemistry of dihydromyrcenol (2,6-dimethyl-7-octen-2-ol) with ozone on silanized glass, glass, and vinyl flooring tiles. Atmos Environ 43(26):4023–4032. https://doi.org/10.1016/j.atmosenv.2009.05.007

Hartline NL, Bruce NJ, Karba SN, Ruff EO, Sonar SU, Holden PA (2016) Microfiber masses recovered from conventional machine washing of new or aged garments. Environ Sci Technol 50(9):11532–11538. https://doi.org/10.1021/acs.est.6b03045

Hatch KL, Maibach H (1995) Textile dye dermatitis. J Am Acad Dermatol 32(4):631–639. https://doi.org/10.1016/0190-9622(95)90350-X

Hatch KL, Maibach HI (1985) Textile dye dermatitis: a review. J Am Acad Dermatol 12(6):1079–1092. https://doi.org/10.1016/S0190-9622(85)70137-5

Hatch KL, Maibach HI (2000) Textile dye allergic contact dermatitis prevalence. Contact Dermatitis 42(4):187–195. https://doi.org/10.1034/j.1600-0536.2000.042004187.x

Hatch KL, Motschi H, Maibach HI (2003a) Disperse dyes in fabrics of patients patch-test-positive to disperse dyes. Am J Contact Dermatitis 14(4):205–212. https://doi.org/10.1097/01634989-200312000-00006

Hatch KL, Motschi H, Maibach HI (2003b) Textile-dye and colored-textile allergic contact dermatitis. Exogenous Dermatol 2(4):206–209. https://doi.org/10.1159/000076803

He Y, Peng L, Zhang W, Liu C, Yang Q, Zheng S, Bao M, Huang Y, Wu K (2018) Adipose tissue levels of polybrominated diphenyl ethers and breast cancer risk in Chinese women: a case–control study. Environ Res 167(22):160–168. https://doi.org/10.1016/j.envres.2018.07.009

Henry B, Laitala K, Klepp IG (2019) Microfibres from apparel and home textiles: Prospects for including microplastics in environmental sustainability assessment. Sci Total Environ 652:483–494. https://doi.org/10.1016/j.scitotenv.2018.10.166

Hodgson AT (2000) Volatile organic compound concentrations and emission rates in new manufactured and site-built houses. Indoor Air 10(3):178–192. https://doi.org/10.1034/j.1600-0668.2000.010003178.x

Hodgson AT, Wooley JD, Daisy JM (1993) Emissions of volatile organic compounds from new carpets measured in a large-scale environmental chamber. Air Waste 43(3):316–324. https://doi.org/10.1080/1073161X.1993.10467136

Horrocks AR (1986) Flame-retardant finishing of textiles. Rev Prog Color Relat Top 16(1):62–101. https://doi.org/10.1111/j.1478-4408.1986.tb03745.x

Hossain MS, Das SC, Islam JMM, Al Mamun MA, Khan MA (2018) Reuse of textile mill ETP sludge in environmental friendly bricks—effect of gamma radiation. Radiat Phys Chem 151(5):77–83. https://doi.org/10.1016/j.radphyschem.2018.05.020

IARC (2012) Chemical agents and related occupations. In IARC monographs on the evaluation of carcinogenic risks to humans/world health organization, international agency for research on cancer. https://www.ncbi.nlm.nih.gov/books/NBK304416/pdf/Bookshelf_NBK304416.pdf. Accessed 5 Jan 2021

Ibrahim NA (2015) Nanomaterials for antibacterial textiles. In: Nanotechnology in diagnosis, treatment and prophylaxis of infectious diseases. Elsevier Inc, 191–216. https://doi.org/10.1016/B978-0-12-801317-5.00012-8

Jaishankar M, Tseten T, Anbalagan N, Mathew BB, Beeregowda KN (2014) Toxicity, mechanism and health effects of some heavy metals. Interdiscip Toxicol 7(2):60–72. https://doi.org/10.2478/intox-2014-0009

Jura J, Szmyd R, Goralczyk AG, Skalniak L, Cierniak A, Lipert B, Filon FL, Crosera M, Borowczyk J, Laczna E, Drukala J, Klein A (2013) Effect of silver nanoparticles on human primary keratinocytes. Biol Chem 394(1):113–123. https://doi.org/10.1515/hsz-2012-0202

Kaya AT, Yurtsever M, Bayraktar SÇ (2018) Ubiquitous exposure to microfiber pollution in the air. Eur Phys J Plus 133(11):488. https://doi.org/10.1140/epjp/i2018-12372-7

Kelly TJ, Smith DL, Satola J (1999) Emission rates of formaldehyde from materials and consumer products found in California homes. Environ Sci Technol 33(1):81–88. https://doi.org/10.1021/es980592+

KEMI (2014) Chemicals in textiles. Risks to human health and the environment. In: Swedish chemicals agency report. http://www3.kemi.se/Documents/Publikationer/Trycksaker/Rapporter/Report6-14-Chemicals-in-textiles.pdf%5Cn, http://www.kemi.se/en/directly-to/publications/reports. Accessed 14 Jan 2021

Kemmlein S, Herzke D, Law RJ (2003) BFR—Governmental testing programme. Environ Int 29(6):781–792. https://doi.org/10.1016/S0160-4120(03)00112-0

Kemmlein S, Herzke D, Law RJ (2009) Brominated flame retardants in the European chemicals policy of REACH-Regulation and determination in materials. J Chromatogr A 1216(3):320–333. https://doi.org/10.1016/j.chroma.2008.05.085

Khan S, Malik A (2014) Environmental and health effects of textile industry wastewater. In: Malik A, Grohmann E, Akhtar R (eds) Environmental deterioration and human health: natural and anthropogenic determinants. Springer Science & Business Media, Berlin, 55–71. https://doi.org/10.1007/978-94-007-7890-0_4

Kissa E (2001) Fluorinated surfactants and repellents, 2nd edn. CRC Press, Florida

Koelmans AA, Besseling E, Shim WJ (2015) Nanoplastics in the aquatic environment. Critical review. In: Bergmann M, Gutow L, Klages M (eds) Marine anthropogenic litter, 325–340. https://doi.org/10.1007/978-3-319-16510-3_12

Korani M, Rezayat SM, Gilani K, Arbabi Bidgoli S, Adeli S (2011) Acute and subchronic dermal toxicity of nanosilver in guinea pig. Int J Nanomed 6(4):855–862. https://doi.org/10.2147/ijn.s17065

Kosuth M, Mason SA, Wattenberg EV (2018) Anthropogenic contamination of tap water, beer, and sea salt. PLoS ONE 13(4):1–18. https://doi.org/10.1371/journal.pone.0194970

Koutsoyiannis A, Leva P, Kotzias D (2008) VOC and carbonyl emissions from carpets: a comparative study using four types of environmental chambers. J Hazard Mater 152(2):669–676. https://doi.org/10.1016/j.jhazmat.2007.07.058

Kubwabo C, Stewart B, Zhu J, Marro L (2005) Occurrence of perfluorosulfonate and other perfluorochemicals in dust from selected homes in the city of Ottawa, Canada. J Environ Monitor 7(11):1074–1078. https://doi.org/10.1039/b507731c

Langer S, Weschler CJ, Fischer A, Bekö G, Toftum J, Clausen G (2010) Phthalate and PAH concentrations in dust collected from Danish homes and daycare centers. Atmos Environ 44(19):2294–2301. https://doi.org/10.1016/j.atmosenv.2010.04.001

Law RJ, Allchin CR, de Boer J, Covaci A, Herzke D, Lepom P, Morris S, Tronczynski J, de Wit CA (2006) Levels and trends of brominated flame retardants in the European environment. Chemosphere 64(2):187–208. https://doi.org/10.1016/j.chemosphere.2005.12.007

Lee HJ, Yeo SY, Jeong SH (2003) Antibacterial effect of nanosized silver colloidal solution on textile fabrics. J Mater Sci 38(10):2199–2204. https://doi.org/10.1023/A:1023736416361

Levermore JM, Smith TEL, Smith TEL, Kelly FJ, Wright SL (2020) Detection of microplastics in ambient particulate matter using raman spectral imaging and chemometric analysis. Anal Chem 92(13):8732–8740. https://doi.org/10.1021/acs.analchem.9b05445

Liebezeit G, Liebezeit E (2013) Food additives & contaminants: part anon-pollen particulates in honey and sugar. Food Additives Contam Part A 30(12):2136–2140. https://doi.org/10.1080/19440049.2013.843025

Liebezeit G, Liebezeit E (2014) Food additives & contaminants: part synthetic particles as contaminants in German beers. Food Additives & Contam Part A 31(9):1574–1578. https://doi.org/10.1080/19440049.2014.945099

Linares V, Bellés M, Domingo JL (2015) Human exposure to PBDE and critical evaluation of health hazards. Arch Toxicol 89(3):335–356. https://doi.org/10.1007/s00204-015-1457-1

Liu J, Wang Z, Liu FD, Kane AB, Hurt RH (2012) Chemical transformations of nanosilver in biological environments. ACS Nano 6(11):9887–9899. https://doi.org/10.1021/nn303449n.Chemical

Liu K, Wang X, Fang T, Xu P, Zhu L, Li D (2019) Source and potential risk assessment of suspended atmospheric microplastics in Shanghai. Sci Total Environ 675(7):462–471. https://doi.org/10.1016/j.scitotenv.2019.04.110

Liu X, Allen MR, Roache NF (2016) Characterization of organophosphorus flame retardants' sorption on building materials and consumer products. Atmos Environ 140(9):333–341. https://doi.org/10.1016/j.atmosenv.2016.06.019

Lombi E, Donner E, Scheckel KG, Sekine R, Lorenz C, Von Goetz N, Nowack B (2014) Silver speciation and release in commercial antimicrobial textiles as influenced by washing. Chemosphere 111(9):352–358. https://doi.org/10.1016/j.chemosphere.2014.03.116

Lorber M (2007) Exposure of Americans to polybrominated diphenyl ethers. J Exposure Sci Environ Epidemiol 18(4):2–19. https://doi.org/10.1038/sj.jes.7500572

Lorimer JP, Mason TJ, Plattes M, Phull SS, Walton DJ (2001) Degradation of dye effluent. Pure Appl Chem 73(12):1957–1968. https://doi.org/10.1351/pac200173121957

Lucas EL, Bertrand P, Guazzetto S, Donna F, Peli M, Jursa TP, Lucchini R, Smith DR (2015) Impact of ferromanganese alloy plants on household dust manganese levels: implications for childhood exposure. Environ Res 138(4):279–290. https://doi.org/10.1016/j.envres.2015.01.019

Lund A, Van der Velden NM, Persson NK, Hamedi MM, Müller C (2018) Electrically conducting fibres for e-textiles: an open playground for conjugated polymers and carbon nanomaterials. Mater Sci Eng R: Rep 126(05):1–29. https://doi.org/10.1016/j.mser.2018.03.001

Luongo G, Avagyan R, Hongyu R, Östman C (2016a) The washout effect during laundry on benzothiazole, benzotriazole, quinoline, and their derivatives in clothing textiles. Environ Sci Pollut Res 23(3):2537–2548. https://doi.org/10.1007/s11356-015-5405-7

Luongo G, Iadaresta F, Moccia E, Östman C, Crescenzi C (2016b) Determination of aniline and quinoline compounds in textiles. J Chromatogr A 1471(11):11–18. https://doi.org/10.1016/j.chroma.2016.09.068

Luongo G, Thorsén G, Östman C (2014) Quinolines in clothing textiles-a source of human exposure and wastewater pollution? Anal Bioanal Chem 406(12):2747–2756. https://doi.org/10.1007/s00216-014-7688-9

Ma Y, Li P, Jin J, Wang Y, Wang Q (2017) Current halogenated flame retardant concentrations in serum from residents of Shandong Province, China, and temporal changes in the concentrations. Environ Res 155(02):116–122. https://doi.org/10.1016/j.envres.2017.02.010

Malarvannan G, Belpaire C, Geeraerts C, Eulaers I, Neels H, Covaci A (2015) Organophosphorus flame retardants in the European eel in Flanders, Belgium: occurrence, fate and human health risk. Environ Res 140(7):604–610. https://doi.org/10.1016/j.envres.2015.05.021

Malinauskiene L, Bruze M, Ryberg K, Zimerson E, Isaksson M (2013) Contact allergy from disperse dyes in textiles-a review. Contact Dermatitis 68(2):65–75. https://doi.org/10.1111/cod.12001

Martínez MA, Rovira J, Prasad Sharma R, Nadal M, Schuhmacher M, Kumar V (2018) Comparing dietary and non-dietary source contribution of BPA and DEHP to prenatal exposure: a Catalonia (Spain) case study. Environ Res 166(10):25–34. https://doi.org/10.1016/j.envres.2018.05.008

Mason RP, Peterson FJ, Holtzman JL (1977) The formation of an azo anion free radical metabolite during the microsomal azo reduction of sulfonate III. Biochem Biophys Res Commun 75(3):532–540. https://doi.org/10.1016/0006-291x(77)91505-4

Mason SA, Welch VG, Neratko J (2018) Synthetic polymer contamination in bottled water. Front Chem 6(8):407. https://doi.org/10.3389/fchem.2018.00407

Mathalon A, Hill P (2014) Microplastic fibers in the intertidal ecosystem surrounding Halifax Harbor, Nova Scotia. Marine Pollut Bull 81(1):69–79. https://doi.org/10.1016/j.marpolbul.2014.02.018

Mathur N, Bhatnagar P, Bakre P (2006) Assessing mutagenicity of textile dyes from pali (Rajasthan) using ames bioassay. Appl Ecol Environ Res 4(1):111–118. https://doi.org/10.15666/aeer/0401_111118

Matoso E, Cadore S (2012) Determination of inorganic contaminants in polyamide textiles used for manufacturing sport T-shirts. Talanta 88(1):496–501. https://doi.org/10.1016/j.talanta.2011.11.022

Mohamed A, El-Sayed R, Osman TA, Toprak MS, Muhammed M, Uheida A (2016) Composite nanofibers for highly efficient photocatalytic degradation of organic dyes from contaminated water. Environ Res 145(2):18–25. https://doi.org/10.1016/j.envres.2015.09.024

Montazer M, Shamei A, Alimohammadi F (2014) Synthesis of nanosilver on polyamide fabric using silver/ammonia complex. Mater Sci Eng, C 38(1):170–176. https://doi.org/10.1016/j.msec.2014.01.044

Morrison G, Li H, Mishra S, Buechlein M (2015) Airborne phthalate partitioning to cotton clothing. Atmos Environ 115(8):149–152. https://doi.org/10.1016/j.atmosenv.2015.05.051

Muenhor D, Harrad S, Ali N, Covaci A (2010) Brominated flame retardants (BFRs) in air and dust from electronic waste storage facilities in Thailand. Environ Int 36(7):690–698. https://doi.org/10.1016/j.envint.2010.05.002

Nakayama T, Kimura T, Kodama M, Nagata C (1983) Generation of hydrogen peroxide and superoxide anion from active metabolites of naphthyl amines and amino azo dyes: its possible role in carcinogenesis. Carcinogenesis 4(6):765–769. https://doi.org/10.1093/carcin/4.6.765

Napper IE, Thompson RC (2016) Release of synthetic microplastic plastic fibres from domestic washing machines: effects of fabric type and washing conditions. Mar Pollut Bull 112(1–2):39–45. https://doi.org/10.1016/j.marpolbul.2016.09.025

Negev M, Berman T, Reicher S, Sadeh M, Ardi R, Shammai Y (2018) Concentrations of trace metals, phthalates, bisphenol A and flame-retardants in toys and other children's products in Israel. Chemosphere 192(2):217–224. https://doi.org/10.1016/j.chemosphere.2017.10.132

Nguyen T, Saleh MA (2017) Exposure of women to trace elements through the skin by direct contact with underwear clothing. J Environ Sci Health 52(1):1–6. https://doi.org/10.1080/10934529.2016.1221212

Niculescu M, Ghituleasa C, Mocioiu AM, Nicula G, Surdu L, Tănăsescu G (2012) Evaluations of formaldehyde emissions in clothing textiles. Indus Textila 64(2):111–114

Novick RM, Nelson ML, McKinley MA, Anderson GL, Keenan JJ (2013) The effect of clothing care activities on textile formaldehyde content. J Toxicol Environ Health Part A Curr Issues 76(9):883–893. https://doi.org/10.1080/15287394.2013.821439

Oda Y, Watanabe T, Terao Y, Nukaya H, Wakabayashi K (2008) Genotoxic activation of 2-phenyl benzotriazole-type compounds by human cytochrome P4501A1 and *N*-acetyltransferase expressed in *Salmonella typhimurium* umu strains. Mutation Res—Genetic Toxicol Environ Mutagenesis 654(1):52–57. https://doi.org/10.1016/j.mrgentox.2008.04.013

OEKO-TEX (2018) International association for research and testing in the field of textile and leather ecology, 2nd edn. Standards 100 by OEKO-Tex. Zurich, Switzerland. https://doi.org/10.2115/fiber.65.p_225. Accessed 2 Jan 2021

Ongwandee M, Morrison GC (2008) Influence of ammonia and carbon dioxide on the sorption of a basic organic pollutant to carpet and latex-painted gypsum board. Environ Sci Technol 42(15):5415–5420. https://doi.org/10.1021/es071935j

Pattnaik P, Dangayach GS, Bhardwaj AK (2018) A review on the sustainability of textile industries wastewater with and without treatment methodologies. Rev Environ Health 33(2):163–203. https://doi.org/10.1515/reveh-2018-0013

Pauly JL, Stegmeier SJ, Allaart HA, Cheney RT, Zhang PJ, Mayer AG, Streck RJ (1998) Inhaled cellulosic and plastic fibers found in human lung tissue. Cancer Epidemiol Biomark Prev 7(5):419–428

Pedersen H, Hartmann J (2004) Toxic childrens wear by Disney. Greenpeace investigations, Brussels. https://www.yumpu.com/en/document/read/22079589/toxic-childrenswear-by-disney-greenpeace-uk. Accessed 11 Jan 2021

Perelló G, Martí-Cid R, Castell V, Llobet JM, Domingo JL (2009) Concentrations of polybrominated diphenyl ethers, hexachlorobenzene and polycyclic aromatic hydrocarbons in various foodstuffs before and after cooking. Food Chem Toxicol 47(4):709–715. https://doi.org/10.1016/j.fct.2008.12.030

Perelshtein I, Applerot G, Perkas N, Guibert G, Mikhailov S, Gedanken A (2008) Sonochemical coating of silver nanoparticles on textile fabrics (nylon, polyester and cotton) and their antibacterial activity. Nanotechnology 19(24):1–6. https://doi.org/10.1088/0957-4484/19/24/245705

Petersen CR (2016) Triclosan antimicrobial polymers. AIMS Mol Sci 3(1):88–103. https://doi.org/10.3934/molsci.2016.1.88

Piccinini P, Senaldi C, Summa C (2007) European survey on the release of formaldehyde from textiles. EUR 22710 EN. https://publications.jrc.ec.europa.eu/repository/bitstream/JRC36150/6150%20-%20HCHO_survey_final_report.pdf. Accessed 3 Jan 2021

Platzek T, Lang C, Grohmann G, Gi U, Baltes W (1999) Formation of a carcinogenic aromatic amine from an azo dye by human skin bacteria in vitro. Hum Exp Toxicol 88:552–559. https://doi.org/10.1191/096032799678845061

Platzek T (2010) Risk from exposure to arylamines from consumer products and hair dyes. Front Biosci Elite 2E(3):1169–1183. https://doi.org/10.2741/e177

Prata JC (2018) Airborne microplastics: consequences to human health? Environ Pollut 234(3):115–126. https://doi.org/10.1016/j.envpol.2017.11.043

Prevedouros K, Cousins IT, Buck RC, Korzeniowski SH (2006) Critical review sources, fate and transport of perfluorocarboxylates. Environ Sci Technol 40(1):32–44. https://doi.org/10.1021/es0512475

Quadros ME, Pierson R, Tulve NS, Willis R, Rogers K, Thomas TA, Marr LC (2013) Release of silver from nanotechnology-based consumer products for children. Environ Sci Technol 47(15):8894–8901. https://doi.org/10.1021/es4015844

Radetić M (2013) Functionalization of textile materials with silver nanoparticles. J Mater Sci 48(1):95–107. https://doi.org/10.1007/s10853-012-6677-7

Remy F, Collard F, Gilbert B, Compère P, Eppe G, Lepoint G (2015) When microplastic is not plastic: the ingestion of artificial cellulose fibers by macrofauna living in seagrass macro phytodetritus. Environ Sci Technol 49(18):11158–11166. https://doi.org/10.1021/acs.est.5b02005

Revel M, Châtel A, Mouneyrac C (2018) Micro(nano)plastics: a threat to human health? Curr Opin Environ Sci Health 1(2):17–23. https://doi.org/10.1016/j.coesh.2017.10.003

Rezić I, Steffan I (2007) ICP-OES determination of metals present in textile materials. Microchem J 85(1):46–51. https://doi.org/10.1016/j.microc.2006.06.010

Rezić I, Zeiner M, Steffan I (2011) Determination of 28 selected elements in textiles by axially viewed inductively coupled plasma optical emission spectrometry. Talanta 83(3):865–871. https://doi.org/10.1016/j.talanta.2010.10.031

Riaz S, Ashraf M, Hussain T, Hussain MT, Rehman A, Javid A, Iqbal K, Basit A, Aziz H (2018) Functional finishing and coloration of textiles with nanomaterials. Color Technol 134(5):327–346. https://doi.org/10.1111/cote.12344

Rist S, Carney Almroth B, Hartmann NB, Karlsson TM (2018) A critical perspective on early communications concerning human health aspects of microplastics. Sci Total Environ 626(6):720–726. https://doi.org/10.1016/j.scitotenv.2018.01.092

Rochman CM, Tahir A, Williams SL, Baxa DV, Lam R, Miller JT, Teh FC, Werorilangi S, Teh SJ (2015) Anthropogenic debris in seafood: Plastic debris and fibers from textiles in fish and bivalves sold for human consumption. Sci Rep 5(9):1–10. https://doi.org/10.1038/srep14340

Rodríguez-Barranco M, Lacasaña M, Gil F, Lorca A, Alguacil J, Rohlman DS, González-Alzaga B, Molina-Villalba I, Mendoza R, Aguilar-Garduño C (2014) Cadmium exposure and neuropsychological development in school children in southwestern Spain. Environ Res 134(10):66–73. https://doi.org/10.1016/j.envres.2014.06.026

Ross PS, Couillard CM, Ikonomou MG, Johannessen SC, Lebeuf M, Macdonald RW, Tomy GT (2009) Large and growing environmental reservoirs of Deca-BDE present an emerging health risk for fish and marine mammals. Mar Pollut Bull 58(1):7–10. https://doi.org/10.1016/j.marpolbul.2008.09.002

Rovira J, Nadal M, Schuhmacher M, Domingo JL (2015) Human exposure to trace elements through the skin by direct contact with clothing: risk assessment. Environ Res 140(7):308–316. https://doi.org/10.1016/j.envres.2015.03.032

Rovira J, Nadal M, Schuhmacher M, Domingo JL (2017a) Home textile as a potential pathway for dermal exposure to trace elements: assessment of health risks. J Text Inst 108(11):1966–1974. https://doi.org/10.1080/00405000.2017.1302635

Rovira J, Nadal M, Schuhmacher M, Domingo JL (2017b) Trace elements in skin-contact clothes and migration to artificial sweat: risk assessment of human dermal exposure. Text Res J 87(6):726–738. https://doi.org/10.1177/0040517516639816

Roy A, Kordas K, Lopez P, Rosado JL, Cebrian ME, Vargas GG, Ronquillo D, Soltzfus RJ (2011) Association between arsenic exposure and behavior among first-graders from Torreón, Mexico. Environ Res 111(5):670–676. https://doi.org/10.1016/j.envres.2011.03.003

Ryberg K, Goossens A, Isaksson M, Gruvberger B, Zimerson E, Nilsson F, Bjork J, Hindsen M, Bruze M (2009) Is contact allergy to disperse dyes and related substances associated with textile dermatitis? Br J Dermatol 160(12):107–115. https://doi.org/10.1111/j.1365-2133.2008.08953.x

Ryberg K, Isaksson M, Gruvberger B, Hindsén M, Zimerson E, Bruze M (2006) Contact allergy to textile dyes in southern Sweden. Contact Dermatitis 54(6):313–321. https://doi.org/10.1111/j.0105-1873.2006.00733.x

Saini A, Okeme JO, Parnis JM, Mcqueen RH, Diamond ML, Properties C (2016a) From air to clothing: characterizing the accumulation of semi-volatile organic compounds to fabrics in indoor environments. Indoor Air 27(8):631–641. https://doi.org/10.1111/ina.12328

Saini A, Thaysen C, Jantunen L, McQueen RH, Diamond ML (2016b) From clothing to laundry water: investigating the fate of phthalates, brominated flame retardants, and organophosphate esters. Environ Sci Technol 50(17):9289–9297. https://doi.org/10.1021/acs.est.6b02038

Sakr W, Weschler CJ, Fanger PO (2006) The impact of sorption on perceived indoor air quality. Indoor Air 16(2):98–110. https://doi.org/10.1111/j.1600-0668.2005.00406.x

Samberg ME, Oldenburg SJ, Monteiro-Riviere NA (2010) Evaluation of silver nanoparticle toxicity in skin in vivo and keratinocytes in vitro. Environ Health Perspect 118(3):407–413. https://doi.org/10.1289/ehp.0901398

Schaeffer VH, Bhooshan B, Chen SB, Sonenthal JS, Hodgson AJ (1996) Characterization of volatile organic chemical emissions from carpet cushions. J Air Waste Manag Assoc 46(9):813–820. https://doi.org/10.1080/10473289.1996.10467516

Schecter A, Shah N, Colacino JA, Brummitt SI, Ramakrishnan V, Robert Harris T, Päpke O (2009) PBDEs in US and German clothes dryer lint: a potential source of indoor contamination and exposure. Chemosphere 75(5):623–628. https://doi.org/10.1016/j.chemosphere.2009.01.017

Schmidt A, Bach E, Schoomeyer E (2003) The dyeing of natural fibres with reactive disperse dyes in supercritical carbon dioxide. Dyes Pigm 56(1):27–35. https://doi.org/10.1016/S0143-7208(02)00108-0

Schuhmacher M, Kiviranta H, Ruokojärvi P, Nadal M, Domingo JL (2009) Concentrations of PCDD/Fs, PCBs and PBDEs in breast milk of women from Catalonia, Spain: a follow-up study. Environ Int 35(3):607–613. https://doi.org/10.1016/j.envint.2008.12.003

Schuhmacher M, Kiviranta H, Ruokojärvi P, Nadal M, Domingo JL (2013) Levels of PCDD/Fs, PCBs and PBDEs in breast milk of women living in the vicinity of a hazardous waste incinerator: assessment of the temporal trend. Chemosphere 93(8):1533–1540. https://doi.org/10.1016/j.chemosphere.2013.07.071

Shaw SD, Berger ML, Weijs L, Covaci A (2012) Tissue-specific accumulation of polybrominated diphenyl ethers (PBDEs) including Deca-BDE and hexabromocyclododecane (HBCDs) in harbor seals from the northwest Atlantic. Environ Int 44(1):1–6. https://doi.org/10.1016/j.envint.2012.01.001

Shu S, Morrison GC (2011) Surface reaction rate and probability of ozone and alpha-terpineol on glass, polyvinyl chloride, and latex paint surfaces. Environ Sci Technol 45(10):4285–4292. https://doi.org/10.1021/es200194e

Silva IO, Ladchumananandasivam R, Nascimento JHO, Silva KKOS, Oliveira FR, Souto AP, Felgueiras HP, Zille A (2019) Multifunctional chitosan/gold nanoparticles coatings for biomedical textiles. Nanomaterials 9(8):1–21. https://doi.org/10.3390/nano9081064

Simoncic B, Tomsic B (2010) Structures of novel antimicrobial agents for textiles: a review. Text Res J 80(16):1721–1737. https://doi.org/10.1177/0040517510363193

Singer BC, Revzan KL, Hotchi T, Hodgson AT, Brown NJ (2004) Sorption of organic gases in a furnished room. Atmos Environ 38(16):2483–2494. https://doi.org/10.1016/j.atmosenv.2004.02.003

Sollinger S, Levsen K, Wünsch G (1993) Indoor air pollution by organic emissions from textile floor coverings. Climate chamber studies under dynamic conditions. Atmospheric Environ. Part B, Urban Atmosphere 27(2):183–192. https://doi.org/10.1016/0957-1272(93)90004-P

Sollinger S, Levsen K, Wünsch G (1994) Indoor pollution by organic emissions from textile floor coverings: climate test chamber studies under static conditions. Atmos Environ 28(14):2369–2378. https://doi.org/10.1016/1352-2310(94)90491-X

Som C, Wick P, Krug H, Nowack B (2011) Environmental and health effects of nanomaterials in nanotextiles and façade coatings. Environ Int 37(6):1131–1142. https://doi.org/10.1016/j.envint.2011.02.013

Stapleton HM, Dodder NG, Offenberg JH, Schantz MM, Wise SA (2005) Polybrominated diphenyl ethers in house dust and clothes dryer lint. Environ Sci Technol 39(4):925–931. https://doi.org/10.1021/es0486824

Stapleton HM, Kelly SM, Allen JG, Mcclean MD, Webster TF (2008) Measurement of polybrominated diphenyl ethers on hand wipes: estimating exposure from hand-to-mouth contact. Environ Sci Technol 42(9):3329–3334. https://doi.org/10.1021/es7029625

Statista (2020) The statistics portal, value of the leading 10 textile exporters worldwide in 2016, by country (in billion U.S. dollars). https://www.statista.com/statistics/236397/value-of-the-leading-global-textile-exporters-by-country. Accessed 10 Jan 2021

Stefaniak AB, Duling MG, Lawrence RB, Thomas TA, LeBouf RF, Wade EE, Abbas Virji M (2014) Dermal exposure potential from textiles that contain silver nanoparticles. Int J Occup Environ Health 20(3):220–234. https://doi.org/10.1179/2049396714Y.0000000070

Stingley RL, Zou W, Heinze TM, Chen H, Cerniglia CE (2010) Metabolism of azo dyes by human skin microbiota. Physiol Behav 59(1):108–114. https://doi.org/10.1117/12.2549369.Hyperspectral

Sundarrajan S, Chandrasekaran AR, Ramakrishna S (2010) An update on nanomaterials-based textiles for protection and decontamination. J Am Ceram Soc 93(12):3955–3975. https://doi.org/10.1111/j.1551-2916.2010.04117.x

Trop M, Novak M, Rodl S, Hellbom B, Kroell W, Goessler W (2006) Silver-coated dressing acticoat caused raised liver enzymes and argyria-like symptoms in burn patients. J Trauma Injury, Infect Crit Care 60(3):648–652. https://doi.org/10.1097/01.ta.0000208126.22089.b6

Tuzen M, Onal A, Soylak M (2008) Determination of trace heavy metals in some textile products produced in Turkey. Bull Chem Soc Ethiop 22(3):379–384. https://doi.org/10.4314/bcse.v22i3.61213

Uhde E, Varol D, Mull B, Salthammer T (2019) Distribution of five SVOCs in a model room: effect of vacuuming and air cleaning measures. Environ Sci Process Impacts 21(8):1353–1363. https://doi.org/10.1039/c9em00121b

USGAO (2010) Formaldehyde in textiles. In the United States Government Accountability Office, Report to Congressional Committees, Washington, DC 110(4):140–142. https://doi.org/10.1111/j.1478-4408.1994.tb01628.x
Van Loy MD, Riley WJ, Daisey JM, Nazaroff WW (2001) Dynamic behavior of semivolatile organic compounds in indoor air. 2. Nicotine and phenanthrene with carpet and wallboard. Environ Sci Technol 35(3):560–567. https://doi.org/10.1021/es001372a
Ventura-Camargo BC, Marin-Morales MA (2013) Azo dyes: characterization and toxicity–a review. Textiles Light Indus Sci Technol 2(2):85–103
Vianello A, Jensen RL, Liu L, Vollertsen J (2019) Simulating human exposure to indoor airborne microplastics using a breathing thermal manikin. Sci Rep 9(1):1–11. https://doi.org/10.1038/s41598-019-45054-w
Vlachou E, Chipp E, Shale E, Wilson YT, Papini R, Moiemen NS (2007) The safety of nanocrystalline silver dressings on burns: a study of systemic silver absorption. Burns 33(8):979–985. https://doi.org/10.1016/j.burns.2007.07.014
Von Goetz N, Lorenz C, Windler L, Nowack B, Heuberger M, Hungerbühler K (2013) Migration of Ag- and TiO2-(nano)particles from textiles into artificial sweat under physical stress: experiments and exposure modeling. Environ Sci Technol 47(17):9979–9987. https://doi.org/10.1021/es304329w
Wakayama T, Ito Y, Sakai K, Miyake M, Shibata E, Ohno H, Kamijima M (2019) Comprehensive review of 2-ethyl-1-hexanol as an indoor air pollutant. J Occup Health 61(1):19–35. https://doi.org/10.1002/1348-9585.12017
Ward J, Mohapatra SP, Mitchell A (2008) An overview of policies for managing polybrominated diphenyl ethers (PBDEs) in the Great Lakes basin. Environ Int 34(8):1148–1156. https://doi.org/10.1016/j.envint.2008.05.003
Waring MS, Siegel JA (2013) Indoor secondary organic aerosol formation initiated from reactions between ozone and surface-sorbed d -limonene. Environ Sci Technol 47(12):6341–6348. https://doi.org/10.1021/es400846d
Waring RH, Harris RM, Mitchell SC (2018) Plastic contamination of the food chain: a threat to human health? Maturitas 115(6):64–68. https://doi.org/10.1016/j.maturitas.2018.06.010
Wijnhoven SWP, Peijnenburg WJGM, Herberts CA, Hagens WI, Oomen AG, Heugens EHW, Roszek B, Bisschops J, Gosens I, Van De Meent D, Dekkers S, De Jong WH, Van Zijverden M, Sips AJAM, Geertsma RE (2009) Nano-silver: a review of available data and knowledge gaps in human and environmental risk assessment. Nanotoxicology 3(2):109–138. https://doi.org/10.1080/17435390902725914
Wilke O, Jann O, Brödner D (2004) VOC- and SVOC-emissions from adhesives, floor coverings and complete floor structures. Indoor Air Suppl 14(8):98–107. https://doi.org/10.1111/j.1600-0668.2004.00314.x
Windler L, Height M, Nowack B (2013) Comparative evaluation of antimicrobials for textile applications. Environ Int 53(3):62–73. https://doi.org/10.1016/j.envint.2012.12.010
Wöhrle D, Schnurpfeil G, Makarov SG, Kazarin A, Suvorova ON (2012) Practical applications of phthalocyanines: from dyes and pigments to materials for optical, electronic and photo-electronic devices. Macroheterocycles 5(3):191–202. https://doi.org/10.6060/mhc2012.120990w
Won D, Corsi RL, Rynes M (2000) New indoor carpet as an adsorptive reservoir for volatile organic compounds. Environ Sci Technol 34(19):4193–4198. https://doi.org/10.1021/es9910412
Won D, Corsi RL, Rynes M (2001) Sorptive interactions between VOCs and indoor materials. Indoor Air 11(4):246–256. https://doi.org/10.1034/j.1600-0668.2001.110406.x
Wright SL, Kelly FJ (2017) Plastic and human health: a micro issue? Environ Sci Technol 51(12):6634–6647. https://doi.org/10.1021/acs.est.7b00423
Wright SL, Ulke J, Font A, Chan KLA, Kelly FJ (2019) Atmospheric microplastic deposition in an urban environment and an evaluation of transport. Environ Int 136(3):105411. https://doi.org/10.1016/j.envint.2019.105411

Xue CH, Wu Y, Guo XJ, Liu BY, Wang HD, Jia ST (2020) Superhydrophobic, flame-retardant and conductive cotton fabrics via layer-by-layer assembly of carbon nanotubes for flexible sensing electronics. Cellulose 27(6):3455–3468. https://doi.org/10.1007/s10570-020-03013-z

Xue J, Liu W, Kannan K (2017) Bisphenols, benzophenones, and bisphenol a diglycidyl ethers in textiles and infant clothing. Environ Sci Technol 51(9):5279–5286. https://doi.org/10.1021/acs.est.7b00701

Yang L, Yiwei W, Caiying L, Yan Z (2013) Determination of aromatic amines from textiles using dispersive liquid-liquid microextraction. J Sep Sci 36(5):947–952. https://doi.org/10.1002/jssc.201200645

Yetisen AK, Qu H, Manbachi A, Butt H, Dokmeci MR, Hinestroza JP, Skorobogatiy M, Khademhosseini A, Yun SH (2016) Nanotechnology in textiles. ACS Nano 10(3):3042–3068. https://doi.org/10.1021/acsnano.5b08176

Yogui GT, Sericano JL (2009) Polybrominated diphenyl ether flame retardants in the U.S. marine environment: a review. Environ Int 35(3):655–666. https://doi.org/10.1016/j.envint.2008.11.001

Yu C, Crump D (1998) A review of the emission of VOCs from polymeric materials used in buildings. Build Environ 33(6):357–374. https://doi.org/10.1016/S0360-1323(97)00055-3

Zarea EN, Motaharib A, Sillanpääc M (2018) Nanoadsorbents based on conducting polymer nanocomposites with main focus on polyaniline and its derivatives for removal of heavy metal ions/dyes: a review. Environ Res 162(01):173–195. https://doi.org/10.1016/j.envres.2017.12.025

Zdzislaw M, Bolt H, Brockmann W (1991) Tumors of the urinary bladder in painters: a case-control study. Am J Ind Med 19:705–713. https://doi.org/10.1002/ajim.4700190604

Zeilmaker M, van Kranen H, van Veen M, Janus J (2000) Cancer risk assessment of azo dyes and aromatic amines from tattoo bands, folders of paper, toys, bed clothes, watch straps and ink. RIVM Rapport 601503019. https://www.rivm.nl/bibliotheek/rapporten/601503019.pdf. Accessed 10 Jan 2021

Zhang Q, Zhao Y, Du F, Cai H, Wang G, Shi H (2020) Microplastic fallout in different indoor environments. Environ Sci Technol 54(11):6530–6539. https://doi.org/10.1021/acs.est.0c00087

Zuccarello P, Ferrante M, Cristaldi A, Copat C, Grasso A, Sangregorio D, Fiore M, Oliveri Conti G (2019) Exposure to microplastics (<10 Mm) associated to plastic bottles mineral water consumption: the first quantitative study. Water Res 157(6):365–371. https://doi.org/10.1016/j.watres.2019.03.091

Chapter 28
Health Impacts of Building Materials on Construction Workers

Joseph Onyango

Abstract The construction area has a long history of health effects that include injuries, fatal accidents, and other ecological health impacts on workers' wellbeing. The health impact of building and construction has raised considerable interest from the exposure of construction workers in various industries. Safety practices and regulatory compliance are expected to reduce the number of catastrophic health effects on a construction worker. This chapter considers emerging trends on health impacts through a systematic review of emerging themes on hazards that include, nanomaterials, climate change and stress, particles and emissions, and their associated occupational diseases and risk factors. In addition, the chapter provides a legal framework for occupational health and safety. It further demonstrates critical lessons on managing occupational construction risks on long term impact based on applied standards, use of personal protective equipment, and training of construction workers to enhance safety and wellbeing for sustainable development initiatives going forward.

Keywords Construction materials · Construction workers · Ecology · Health impact · Sustainability · Workers safety

28.1 Introduction

The construction sector is a basic contributor to the cycle of advancement across the world. The industry is broad in the sense that it covers not only buildings but also, streets/roads, water system works, schools, residential buildings, health centers, office spaces, and construction activities as the actual set-up on which evolution endeavors and better standards for everyday living are attained (World Bank 1984). More than 350 million individuals of the global population are straightforwardly engaged with this area, and the number is developing at a quick rate in the urban communities. Because of a few occupational wellbeing hazards or risks, the construction area is set the first spot on the list as per working environment mishaps and

J. Onyango (✉)
Strathmore University Business School, Institute of Healthcare Management, Nairobi 00100, Kenya
e-mail: jonyango@strathmore.edu

J. A. Malik and S. Marathe (eds.), *Ecological and Health Effects of Building Materials*,
https://doi.org/10.1007/978-3-030-76073-1_28

injuries (Biswas et al. 2017). Laborers in this industry are presented to various physical, synthetic, natural/biological, mechanical, and psychological perils during their everyday working plans. This possesses a great concern on the ecological health that is also associated with the building and construction materials. This chapter focuses its attention on emerging health hazards on building and construction technologies such as nanomaterials, effects of climate change and stress, particles, and emissions on contractions workers. It also takes into consideration occupational diseases and risk factors. To avert these challenges the legal framework and policy areas that are implied for the regulation and compliance in this industry are highlighted. In conclusion, a review of the management of occupational risks on construction laborers' long term wellbeing impacts among various exchanges that incorporates inferred standard, personal protective equipment (PPE) or individual defensive gear use, training and development for construction laborers on health and safety measures are put into consideration.

28.2 Occupational Health Hazards in the Building and Construction Industry

There is a construction boom across the world and more specifically among emerging economies. This growth over the years has become a concern for development organizations like the World Bank (1984). The industry has brought forward a growing interest in many countries as they develop their building and construction industry that has coincided with increasing concern for the wellbeing and security of the workers. The growing trend of occupational wellbeing hazards in the building and construction industry is similar in developed economies and emerging economies. For instance, as of 2016, there were about 9% of all the occupational fatal injuries and 20% non-fatal injuries associated with construction activities in the United States (Choe and Leite 2017). Across the developed and developing nations, West (2016) posit that the risky attributes of the building sector make it challenging to keep conducive work plans all through the construction period of any project. Subsequently, the dynamic and complex nature of construction tasks makes it not feasible to spot potential construction hazards (Gambatese et al. 2016; Karakhan and Gambatese 2018). A hazard is the potential for hurt and can be characterized by multiplying. This includes health effects that emerge for the materials used in building and construction.

For illustration, Manuele (2005) defined a hazard as "any aspect of technology, task, or activity with a potential for harm." Baradan and Usmen (2006) stated that a hazard is "the potential for an activity or condition to produce harmful effects likes illness and injury." Additionally, Fleming (2009) defined a hazard as "something or anything with an inherent source of energy (such as radiation, electricity, noise, fire, or electricity among others) that can cause harm or damage to individuals, physical assets, or the environment if used incorrectly or applied inefficiently." Thus, the definition provided by Fleming seems relevant in this article as it links working environment dangers dependent on the collaboration of energy with individuals, equipment, materials, temperature, chemical, or mechanical among other sources of energy which are relevant in the construction operations (Albert et al. 2014).

An article titled "Ten common construction site hazards" published by Collier (2017) on High-Speed Training.com outlines the top ten dangers and perils from working in the building or other construction sites and incorporate working at stature (height), moving articles (objects), clamor/noise, hand-arm vibration condition, material and manual handling, electricity, airborne filaments and materials, collapsing trenches, asbestos, and slips, trips, or falls. Asbestos involves fibrous materials which when damaged and released in the air, workers can inhale leading to fatal and serious diseases like lung cancer. Airborne fibers/filaments and materials involve dust on construction locales which are regularly an undetected and poisonous combination of risky materials that can harm the lungs of laborers (Collier 2017).

28.2.1 Emerging Technologies—Nanomaterials

Technology in the construction business has acquired significant interest among researchers. Technology in the construction industry refers to "the collection of innovative tools, machinery, modifications, and software among other materials used in the construction phase of a project to enable advancement in field construction techniques." Such incorporate robotized and semi-mechanized construction gear/equipment (Jones 2018). Today, new advances, for example, the Internet of Things (IoT), computer-generated reality and enlarged reality, and mobile SMS in the context of "Safety Management Software" are taking off in the construction industry. The objective is to integrate key technological advances to improve the efficiency and security of construction laborers (Construction Litigation 2019).

Of interest to scholars and policymakers are nanomaterials which are important for basic innovation for different fields like health, climate, and construction (Sharma et al. 2015). According to available information published by ISO (2017), "nanomaterials are materials that have internal structure or surface structure in the nanoscale." Nanoscale materials have actual components of 1–100 mm (1 mm $= 10^{-9}$ m) (Al-Bayati and Al-Zubaidi 2018). Nanomaterials comprise carbon, metal oxides, and different metals that have been utilized in the construction materials (Lee et al. 2010). Additionally, nanomaterials are added substances and can support construction material properties in certain manners like qualities, functionality, and softness (West et al. 2016).

Nanomaterials are exceptionally responsive materials and involve different synthetic substances and natural changes in the construction sites such as bio-reactions, redox reactions, aggregation, and dissolution (Beaudrie et al. 2015). They are also extremely small in size with greater reactive surface areas. However, the exact influence of nanomaterials is still ambiguous (Lowry et al. 2012). Previously, the use of novel materials, for example, asbestos and lead in construction locales have brought about negative upsetting outcomes (West et al. 2016). As the industry grows, the high chances of openness to dust for construction laborers stay higher than in the remaining sectors. For construction laborers, openness to nanomaterials can occur during the assembling or bundling of construction materials, construction phase, or the activity phase.

Nanomaterials innovation assumes a critical job in the construction business. Segments, for example, covers/insulators, various kinds of utilized cylinders in the structures, glass, and concrete/solid assume a huge function. The application of these nanoparticles in the fundamental structures for the most part builds the mechanical properties and Carbon Nanotubes (CNT) (Daryoush and Darvish 2013). In the structure envelope and inside completions, the utilization of building Nano-coatings is so fundamental in the inside and outside façade of structures. Moreover, nanomaterials are likewise applied in the construction materials as the strengthened solid, self-cleaning, and heatproof glass which controls and saves energy utilization (Daryoush and Darvish 2013).

The well-being outcome of nanotechnology is connected to the impacts of nanomaterials on human health. The major health worries associated with the increase of non-degradable or gradually degradable nanoparticles in body organs. Research demonstrates that nanoparticles that are inhaled accidentally or deliberately cannot be cleaned or purified or removed from the body by white blood cells as compared to larger particles. Thus, there are high chances of these nanoparticles to travel throughout the body of construction individuals via various modes which may impact the highly complex nerve cells and fibers (Nandekar and Rautdesai 2019). That is, nanoparticles can overload phagocytes which are cells that ingest and demolish unfamiliar issues, along these lines setting off pressure response that led to aggravation (inflammation) and debilitate the body's protection against different pathogens (Herzog et al. 2007).

In other studies, findings reveal that laborers' openness to titanium dioxide (20–30 nanometre nm) in the packaging/wrapping process surpassed the suggested openness/exposure limit (DuPont 2007). Contact with nanomaterials exposes individuals to great danger such as lung damage as well as brain damage (Shaw and Handy 2011). In an investigation of 177 nanomaterials much of the time utilized in the US building activities, results show that 29.9% of the nanomaterials were unknown while 7.9% were not revealed. The investigation further uncovered that because of the possible damage of nanomaterials to construction laborers and final clients, construction experts must know all the nanomaterials to which they are exposed to (Al-Bayati and Al-Zubaidi 2018).

At the construction site, nanoparticles can enter construction workers' bodies through the air. A careful examination of previous research further reveals that diseases linked to nanoparticles include asthma, bronchitis, lung cancer, and heart disease among others (Buzea et al. 2007; Thomas et al. 2013). Other toxicities of nanomaterials include carbon nanotubes (Kang et al. 2007) which can result in oxidative stress leading to cell wall damage, quantum dots (Lin et al. 2008) which contains weighty metals like lead and zinc which may be harmful causing cytotoxicity or nucleic acid damage as well as copper oxide nanoparticles which may exert strong oxidative stress and DNA damage in human and other mammals (Blaise et al. 2008; Lee et al. 2009).

28.2.2 Ecological Impact Attributed to Climate Change

The environmental impact of growing construction using modern technologies contributes to climate change associated with a worldwide temperature alteration. Construction activities discharge a lot of carbon dioxide and methane. Considerable infrastructure advancements cause contamination and produce waste that has direct effect on construction workers and other beings in the world. Individuals working in construction sites are especially defenseless against wellbeing dangers and hazards since they have not much legitimate legal protection, a poor safety net, expanded openness to environmental change conditions, and much of the time, monetarily impeded (Bharara et al. 2012). Moreover, working at statures/heights, with hefty overhead loads, functioning large equipment and power instruments, or working under extreme temperature add to the danger of mishaps and injuries/wounds. Notwithstanding these challenges are occupational health hazards and socio-economic stresses because of hot environments. Heat impact and changing atmosphere/climate conditions add to expanding heat openness/exposure for construction workers, especially in an environment with little safety and health frameworks (Dutta et al. 2015).

A review of developing literature from 1988 to 2008 on climate change impacts, their connection with occupational risks/hazards, and their sign with the working people uncovered seven classifications of climate/atmosphere related hazards (Schulte and Chun 2009). They include (1) expanded ambient temperature, (2) ultraviolet exposure, (3) air contamination, (4) changes in the built environment, (5) extreme weather, (6) sector advances and arising ventures, and (7) vector-borne illnesses and extended natural surroundings. As such, the authors explained that climate change has a high capability of expanding the pervasiveness, circulation, and seriousness of known occupational hazards in areas such as construction sites.

Most occupational health hazards related to climate elements can likewise be connected to climate/environmental change. For instance, air temperature, moisture, circulation of air, and heat energy (Parsons 2014). Based on the latest survey of countless epidemiological and laboratory examinations (Kjellstrom et al. 2017), various climate change hazards, affected groups, and the health effects are illustrated in Table 28.1. The author explains that individuals working in positions of the low or substantial workforce in a hot environment are at higher risk, inferable from openness to high natural heat and internal energy production. This may result in clinical health effects (Kjellstrom et al. 2017).

A work environment with extremely or too hot conditions is becoming a big concern for an occupational health problem. Despite considerable protective techniques in multiple fields, many workers, especially those in the construction sites are still presented to unfortunately high temperatures and humidity which makes heat

Table 28.1 Climate change-linked occupational health hazards, susceptible groups, and health impacts

Climate hazard	Vulnerable groups	Health effects
High heat openness (temperature and moisture)	Workers performing physically demanding activities; construction laborers exposed to direct sun	Heat tiredness, heat stroke
	Laborers in heat stress conditions who do not hydrate satisfactory Laborers presented to exceptionally evaporative synthetic compounds like natural solvents, high warmth prompts higher work environment substance openness	Chronic kidney disease Chemical poisoning
Extreme weather, windy conditions	External laborers in affected areas; emergency laborers, and others	Injuries, drowning and dehydration
Ecological conditions indirectly linked to climate	Outdoor workers: construction/firm workers needing to work at dawn and dusk	Vector-borne illnesses, illnesses linked to ecological change
Other indirect climate-related hazards	Low-income individuals with limited health protection; workers with existing non-climate health problems affected by heat	Infectious disease, non-communicable diseases, mental health issues, etc.

Adapted from Kjellstrom et al. (2013, 2017)

strain and heat stroke a significant challenge for wellbeing as well as for work effectiveness (Kjellstrom et al. 2009a, b). There is a possibility that worldwide climate change will expand normal temperatures and move the dispersion of day-by-day top temperature and moderately average humidity. Climate change entails a move in the dispersion of everyday temperatures to incorporate more hot days, and more days with wet bulb globe temperature (WBGT) surpassing the edge for heat resistance in individual (Kjellstrom et al. 2007, 2009b).

Climate change is expanding heat openness/exposure in various regions across the world, with some regions experiencing hot/humid/sticky conditions. Individuals are at specific danger of stress, for example, heat due to the intra-body heat production brought about by actual work (Schulte and Chun 2009). A review of existing literature describing the dangers of occupational heat openness on wellbeing and output shows that heat stress at work is a genuine danger, especially for those working under construction sites (Kjellstrom and Crowe 2011). The rapid urbanization in most emerging economies adds to the area heat openings via a great number of construction activities. Thus, higher workstation heat openness is linked to different clinical impacts and to increase the rate of occupational injuries (Tawatsupa et al. 2013).

As indicated by Smith et al. (2014), heat stress is the warmth gotten over that which the body can endure, without physiological hindrance and stems from three variables; intra-body heat creation from strong active work like construction activities, outside or external warmth/heat, and apparel/clothing that influences heat convection and sweat evaporation. The unfriendly impacts of heat stress incorporate warmth/heat strain bringing about clinical illnesses, wellbeing debilitations, and decreased individual performance and workability. Studies posit that the immediate wellbeing effects of heat openness are generally evaluated utilizing mortality (Gasparrini et al. 2015) and hospital admissions (Hansen et al. 2008a, b). Subsequently, Parsons (2014) revealed that indications of severe heat strain are regularly disregarded as individuals continue to undertake their outdoor activities beyond safe heat exposure.

As such, incidences of cardiovascular mortalities among construction laborers have been reported in countries such as Qatar (Gibson and Pattisson 2014) and over thousands of deaths in France during extreme heat seasons. Additionally, recent epidemic analysis of fatal chronic kidney disease have been uncovered among sugarcane harvesters in Central America (Wesselling et al. 2013) and is associated with daily dehydration because of excessive sweating in the hot work environment. Whereas clinical effects remain the biggest concern, physiological effects also reduce individual performance (Kjellstrom 2009). For instance, professions such as construction labor, mining, transportation, and community works are all affected by particular heat-related challenges. According to Kjellstrom et al. (2009a, b), most buildings under construction lack air ventilation or other friendly cooling systems.

Available information reports that high body temperature beyond 39 °C affects an individual's function of the central nervous system, thus, causing confusion and unconsciousness (De Blois et al. 2015). This implies that such individuals may not make preventive moves as expected. Further, the studies project that physiological strain of reallocated/redistribution blood flow, enhanced heart rate, and noticeable sweating as a result of excessive heat affect the cardiovascular system (De Blois et al. 2015; Kjellstrom et al. 2016). According to Bennett and Michael (2010), vector-borne diseases such as malaria and dengue fever may be rampant in individuals working in fields such as construction without effective protection. This may lead to more incidences of these diseases when climate change conditions worsen.

A cross-sectional examination profiling the heat openings of 1900 laborers from eight industrial sectors assessed the wellbeing impacts of heat stress dependent on the WBGT (Venugopal et al. 2020). The authors revealed that heat-related wellbeing concerns are a significant issue and that, laborers exposed to heathad 2.3 times higher chances of serious health outcomes contrasted with people not exposed to heat. For example, the discoveries detailed that openness to high WBGT brings about negative renal health among salt laborers whereas steel-laborers had a 9% predominance of kidney stones. Different investigations have additionally demonstrated that ongoing kidney infection of unknown causes ascribed to heat stress has arrived at a pandemic level in Central America and India (Johnson et al. 2019), while laborers with physical applying occupations in a work environment like construction have chronic persistent direct heat injury that can cause kidney tissue harm prompting monotonous acute kidney injury (AKI), kidney stones, and adrenal harm/damage (Gonzalez-Quiroz et al. 2018).

28.2.3 Particles and Emissions in the Construction Sites and Its Effect on Workers

Contamination sources resulting from construction works involve unsafe gases, dust, noise, solid, and fluid wastes (Tah and Abanda 2011). Construction activities remain a significant source of environmental contamination in light of particulate matter (PM) emissions which cause antagonistic consequences for human wellbeing and are answerable for different mortalities internationally (Araujo et al. 2014). Concrete structures when approaching the finish of their valuable life can require broad structure works like repair, reinforcing, or even destruction and replacement through new construction. Such construction works are known to deliver huge amounts of coarse particles into the nearby workplace (Kumar et al. 2012).

For instance, previous study results demonstrated that ultrafine particles (UFPs) concentrations exist in the region of the old story strengthened solid structure going through demolitions and is probably going to enhance 1.6 times. Moreover, the UFPs created during the destruction of such houses can convey airborne pathogens or infections into the climate/surroundings, learning centers, residential places, and thickly populated regions, for example, business places (Hansen et al. 2008a, b). Thus, the production of UFPs in the construction sites is an area of concern due to its occupational health hazards among workers. Whereas most dust particles may be too large to be inhaled, a recent study illustrated that dust particles cause eye, nose, and even throat irritation resulting in cardiovascular diseases (Araujo et al. 2014).

According to Heal et al. (2012), openness to PM like PM_{10} (≤ 10 μm), $PM_{2.5}$ (≤ 2.5 μm), and PM_1 (≤ 1 μm) is known to have severe health outcomes on human health. Janssen et al. (2013) demonstrated that excess mortality is a result of individual exposure to PM from sources like industries. Turner et al. (2011) added that excessive inward breath of PM_{10} and $PM_{2.5}$ is connected to various respiratory diseases like lung cancer and asthmatic. Other diseases associated with inhalation of such particles include renal (Spencer-Hwang et al. 2011; Weng et al. 2015), cardiovascular diseases (Brook et al. 2010), and depression challenges among the construction workforce (Haynes and Savage 2007). A similar study reported increased death-related to ischemic coronary illness among construction laborers like plasterers, masons, and welders (Cavallari et al. 2007).

In Europe alone, studies have shown that construction and demolition of buildings contribute more than 450 million tonnes of waste annually, with around 53 million tonnes in the United Kingdom (UK) alone (Rao et al. 2007). Therefore, the increased rate of building demolitions and construction of new buildings could be refined through one or the other collapse or mechanical methods, which all produce a significant amount of PM that may be short-lived but severe. Individuals working in construction sites like cement products are exposed to harmful emissions compared to those working in other areas (Azarmi et al. 2014). Azarmi et al. (2015) and Font et al. (2014) presented evidence to link demolition and building activities such as innovation with PM and degradation of the surrounding air quality which can adversely affect the health of construction workers and those living close to the construction sites.

Kumar et al. (2012) further modeled the release of ultrafine particles (UFPs) and reported that during building works, the crushing of solid 3D shapes, and the destruction of old chunks, the bulk of particles mass concentrations (PMCs) emissions constitutes particles of over 100 nm (nanomaterials) which are harmful to human health. Tong et al. (2018) also developed a probabilistic model to investigate the wellbeing impacts of construction dust on the construction experts in the U.S. Findings reveals that individuals working in the construction zone were presented with the biggest wellbeing hazard especially those working in steel zones, concrete, and floor zones.

Exposure to construction materials with alkaline debasement of paste as the arrangement of 2-ethyl-1-hexanol increases ocular, nasal, and respiratory symptoms (Wieslander et al. 2010). Particles with small diameter arrive at the lower respiratory tract resulting in little airways of the lungs after inward breadth (Tang et al. 2012). In their study on fine and ultrafine particles radiated from laser printers in Germany, Tang et al. (2012) citing previous research (Glytsos et al. 2010) confirmed that openness to emissions from laser printers are associated with several health risks and inflammation diseases among others.

28.3 Emerging Occupational Diseases and Risk Factors for Construction Workers

According to Ray et al. (2015), any construction-related activity is perceived to be an incredibly unsafe process as it very well might be related to continuous happening activity-related wounds/injuries to the construction laborers. Just as in older workers, physically demanding occupations present the threat of more extreme injuries and longer recuperation times for all workers; men and women. Moreover, physical activities related to singular trades may likewise result in laborer carelessness or easy routes (Parida and Ray 2012). Therefore, establishing various occupational diseases in every field is of great concern for developing successful occupational health systems.

The construction industry presents a unique case for establishing occupational diseases and risk factors involved (Ray et al. 2015). Previous studies highlight that the construction zone is perceived as the top hazardous sector and occupational-related illnesses are the fundamental driver of injuries or fatalities among the laborers. It tends to be a consequence of explicit working conditions just as dynamic and complex construction works (Choi 2015). Several studies have been undertaken to explain occupation diseases and risk factors in various construction activities using different methods. Established results vary regarding types and common occupational diseases and risk factors in the construction industry.

A case in point is Lee et al. (2016) who identified hearing loss as a major occupational illness in the construction area in Iran. According to the authors, the affected group of construction workers include (a) machinery operator, (b) concrete laborers, (c) those who work with shotcrete machines and (d) channel builders. Choi (2015) found out damage to the wrist, fingers, and hands, vision issues, soreness, cuts, and being sprained, harm to the ankle, hand, and knee as some of the occupational diseases in the United States. However, this study focused on older construction workers. Nevertheless, Ray et al. (2015) and Blanc et al. (2015) also supported Choi's findings. The two studies conducted in India and Sweden respectively, identified pain in the neck, wrist, ankle, and knee, shoulder, and pain in the back as immediate occupational diseases and risks associated with working in the construction industry.

Additionally, Stocks et al. (2011) identified skin cancer, skin diseases, asthma, musculoskeletal disease, and respiratory disease. In their study conducted in Great Britain, the authors classified the diseases based on the construction professions. For instance, skin cancer is associated with individuals working as painters, laborers, carpenters; skin diseases are related to metal laborers; asthma in welders; and musculoskeletal disease in welders, laborers, special equipment operators (road and construction operators). Haj Aghazadeh et al. (2012) further indicated that musculoskeletal remains a major occupational disease in construction workers in Iran. Considering a synthesis review of the literature, Abbasianjahromi and Talebian (2021) presumed that non-articular illnesses are the most well-known occupational illnesses in construction jobs, more so in countries like the UK, Germany, and China.

In Germany like in Malaysia, hearing misfortune in the construction industry is more well-known than other occupational illnesses. Moreover, musculoskeletal illnesses are the most well-known in Iran and Venezuela with their types being ankle, back, and wrist pain problems. Respiratory diseases and lung diseases are common in the UK and China respectively (Abbasianjahromi and Talebian 2021). In another analysis of chronic occupational analysis and construction lifetime dangers for a period of 45 years working life as an employee/employer, results set up that 15.9% of construction laborers contacted the chronic obstructive pulmonary disease (COPD), 11% acquired parenchymal radiological abnormality, and 73.8% acquired material hearing loss (Ringen et al. 2014).

Van der Molen et al. (2016) further posits that hearing loss and musculoskeletal problems are the most well-known occupational illnesses for laborers in Dutch construction areas. Commotion-induced hearing misfortune and contact dermatitis were additional construction occupational illnesses affecting laborers. The examination further showed that issues like pain in the back, arthrosis, tedious strain wounds, burnout, and frequent obstructive pulmonary illnesses, and asthma remain daily occupational illness challenges for construction laborers. The investigation applied a dynamic prospective cohort over a 5-year duration (2010–2014) assessing occurrence and patterns in the rate of occupational illnesses in the Netherland's construction zone.

The occupational risk factors may differ depending on the occupation in the construction industry. However, the most common risk factors across the construction sites include age, experience, workload, climate (heat stress), physical tiredness, long working hours, repetitive task, static posture, delayed sitting and standing, awkward posture, demanding task, and lack of rest (Parida and Ray 2012). Mersha et al. (2017) found cut and abrasions as the fundamental kinds of occupational wounds/diseases related to construction activities. Most of these injuries occur in fingers and arms because they are the most dynamic body parts straight forwardly presented to machines and sharp hand instruments. The two major causes of injury or risk factors identified were a fall accident and hurt by a sharp instrument.

28.3.1 Health Impacts of Construction Materials and Products on Workers

Building and construction laborers in various industries are occasionally exposed to different types of harmful chemicals in their daily activities, for example, natural and artificial mineral fibers, concrete, quartz, dust, diesel fumes, paints, and thinners/particles that affect their health and wellbeing (Thienen and Spee 2008). Some of the common materials used and their effects are:

- The cement that is attributed to skin disease and skin cancer (Prodan and Bachofen 1998)
- Wood Dust and wood preservatives that contain arsenic chemicals associated with nasal cancer and have various non-malignant respiratory and dermatological effects (Demers et al. 1997);
- Paints and varnishes associated with dermatitis (Estlander et al. 2000);
- Fiber and mineral wools used as thermal and acoustical insulation materials have associated effects on eyes, skin, and upper respiratory tract (Maxim et al. 2003);
- Asbestos associated with pneumoconiosis and other related lung tissues infection and cancers (Cugell and Kamp 2004);
- Epoxy resins associated with carcinogenic effects to cancerous humans (IARC 1990);
- Concrete release agents associated with respiratory complications and other allergies (Terwoert 1999)
- Polyurethane products associated with irritating to the skin and asthma (WHO 2000)
- Organic solvents used in paints, glues, adhesives, cleaning agents, and other substances (Dick 2000)
- Diesel Exhaust from the diesel engines associated with effects on the lungs and respiratory tract leading to lung cancer (EPA 2002).

28.4 Legal Framework: Occupational Safety and Health Administration (OSHA)

According to Strategic Human Resources Management (SHRM) (2021), Occupational Safety and Health Administration (OSHA) guidelines/standards are legitimate structures that portray the strategies that businesses should apply to shield their laborers from dangers and other hazards. There are OSHA principles/standards in different areas like construction activities and general zones. OSHA standards limit the amount of hazardous substances representatives can be presented to, require the usage of certain protected practices and equipment, and induces managers or associations to screen dangers and track worksite injuries and illnesses. For example, OSHA principles/standards involve requirements to avail fall protection and prevent openness to unsafe/harmful substances like asbestos.

Legally, all laborers are qualified for a safe and protected working environment. Accordingly, understanding OSHA guidelines is critical for both construction experts and laborers. That is, it addresses the utilization of the executive standards, correspondence, and human relations factors, safety/wellbeing rules, industry legal frameworks, accident investigation, and technical issues particularly within the job planning stages in the construction area. The rates of occupational injuries and illnesses among construction workers continue to be generally more than in other sectors. In this way, the goal of OSHA is to portray the legal aspects including liability and regulatory necessities of construction safety (Yuan 2012).

Globally, the 10-OSHA standards most regularly remembered for the firm's references incorporate; Scaffolding, fall protection, general and other excavations, ladders, head protection, hazard correspondence, training requirements, construction health and safety, and electrical protection (OSHA Pocket Guide 2005). Other General Safety and Health Provisions include (OSHA):

- General safety and health provisions
- Safety preparing and education
- Recording and reporting of incidents
- First aid and medical attention
- Fire protection and prevention
- Housekeeping (alteration, repairs, form and scrap lumber with protruding nails, and all other debris, shall be kept cleared from work areas)
- Illumination (areas such as aisles, stairs, ramps, and runways shall be lighted with either natural or artificial illumination)
- Sanitation (requirements for drinking water)
- Personal protective equipment
- Acceptable certifications
- Access to worker exposure and medical records
- Laborer emergency action plans.

A review of the historical OSHA standards presented in the 2020 OSHA Field Safety and Health Manual provide other standards such as; office security and well-being, crisis alternate course of action or plan, PPE, hazard correspondence, control of risky energy sources, permit-required confined zones, emergency treatment, respiratory protection, blood-borne pathogens, ergonomics, radiation, electrical safety, insurance/protection during an incident investigation, worker help, clinical administration, decontamination, and policy and procedures for field staff exposure among others (OSHA 2020).

Multiple studies have been conducted to assess OSHA execution in the construction areas concerning enforcement, compliance conduct, and the sufficiency of the principles/standards in addressing safety outcomes. Previous study findings established that enforcement actions among the construction players in the US were significant, with players confronting a high yearly likelihood inspection. Moreover, inspection impacts on changes in firm compliance with OSHA principles were quite modest (Weil 2001). Yassin and Martonik (2004) reported that following the modified scaffold protection principles may stop close to 4.6 death toll, 404 non-fatal wounds, and 2896 lost work-days annually with a total compliance cost estimate of $5.8 million. Additionally, compliance with the new safety guidelines would ensure a secure working environment.

A recent study further examined a quantifiable impact of the updated OSHA 29 CFR 1926 utilizing the Bureau of Statistics data in the US with four key wellbeing measure rates; casualty, days away injury, standardized casualty, and standardized days-away injury (Choe et al. 2016). The authors revealed that for both laborers in the structural iron and steel zones, the casualty rate reduced by 22.2% while the days-away injury rate decreased by 53.7% respectively. This occurred after the execution of the updated OSHA guidelines. Concerning adequacy of OSHA outreach training, Schoenfich et al. (2017) demonstrated that it brought about a 13% non-significant decrease in injury claim rates in general among carpenter laborers in the construction industry.

Information regarding the guideline is the major key for the effective sanctioning of best safety rehearses for the decreased hazardous situations at work. Workers should get proper wellbeing/safety training from their bosses before being assigned to any risky construction activities (Desruisseaux 2019). OSHA guidelines give accentuation on the preparation or training of laborers. Because of the sensitivity to hierarchical, response, substance, cycle, and laborer matters, experts utilize visual guides and avail interpreters and safety rules drafted in laborers' dialect when faced with the language barrier (Demirkesen and Arditi 2015). Sokas et al. (2009) found out that hazard awareness training and wellbeing atmosphere is linked to progress in fall safety attitudes. Thus, OSHA training on hazard awareness results in knowledge and attitude improvement in the construction building trades.

28.5 Managing Occupational Risk of Construction Worker's Health

The building and construction sector has two main classes that possess occupational health hazards to workers. The World Bank (1984) classifies this industry in two main domains as buildings which are linked to accommodations, workplaces, medical clinics, industrial facilities, and so forth and others as government works which involve the irrigation, transport, irrigation, power generation, and so on. All these domains expose construction workers to different types of occupational health hazards that are managed differently. Various sorts of dangerous working conditions, ecological pressure conditions, absence of PPE use, prolonged harmful working postures, low degree of mindfulness on wellbeing risks, and hygiene practices and so on, are together influencing the ailments of the laborers (Biswas et al. 2017), and calls for successful administration of wellbeing and safety of the laborers. Conventional occupational safety management techniques are related to enactment, guideline, safety norms and guidelines, application of safety best practices, analysis reports on accident statistics, investigations and inspections of construction sites, analysis of SMS, and individual conduct. Several studies have been conducted to examine the causes of occupational wounds and death toll (Aneziris et al. 2012). Key to managing occupational wellbeing and safety hazards is the utilization of health and safety standards, PPE, and training of construction works on health and safety are some of the key mechanisms discussed in the section that follows. However, the recent development in managing health hazards by Aneziris et al. (2012) recommends the use of a probabilistic quantified occupational risk model to help with managing hazardous effects on building and construction works health.

28.5.1 The Utility of Implied Health and Safety Standard

Because of the many business-related mishaps/accidents and wounds/injuries in the construction activities, wellbeing program execution has been given numerous contemplations as one of the viable inferred principles/standards to control such challenges. In any case, this relies upon various factors like labourers inclusion, wellbeing prevention and control frameworks, safety management, and assurance from both (Aksorn and Hadikusumo 2008). Experts and owners' qualities on the health safety standards, the level of consistency with work safety principles and guidelines, and health safety skills are other basic elements for the effective execution of occupational-related wellbeing and safety principles in the worksite (Cheng et al. 2010). Therefore, to minimize the overall happening rate of occupational accidents per 1000 persons in the construction zone, the execution of health and safety standards and training is necessary.

One important framework in the construction activities is an SMS (safety management system) which was presented around the 1980s and has since been embraced in numerous different places. SMS utilized in construction work follow different global guidelines: "*BS8800–2004 Guide to Occupational Health and SMS, HS(G)65 – Successful Health and Safety Management, BS OHSAS 18001 – Occupational Health and Safety Management, and AS/NZS 4804:2001 Occupational Health and Safety Management*" (Lam 2003). These guidelines take a gander at (1) general work-related wellbeing and safety policies, objectives, targets, and organization responsibilities, (2) activity design comprising scheduling and advancement, safety manual and methods, and cooperation, (3) project learning including preparing and risk control as well as educative action frameworks, and (4) performance review like performance measures, evaluation, assessment, and executive audit (Robson et al. 2007).

In a systematic review of practices in the administration of climatic heat stress risk in the construction zone, Rowlinson et al. (2014) revealed the requirement for execution of management practices related to heat stress that reflect special climatic change conditions, working practices, and acclimatization inclination by laborers as a component of obligatory regional guidelines which are site-based and not global-based. Advantages of executing management systems in the construction work to address implied standards entails a secure working environment and minimal danger to laborers regarding safety management as a part of best construction management practices/standards (You et al. 2019).

A previous study in 2017 further indicated that good management systems such as SMS prevent injuries and illness, diminished accident-related costs, occasionally attending work, lower turnover, higher performance, and improved confidence among laborers (OshAcademy 2017). Principles and actions, safety preparedness, reporting and investigation of incidents, and safety culture should be considered for the usage of occupational health and safety standards in the worksite to aid in managing accidents and illness related to construction activities (Grote 2012).

28.5.2 Use of Personal Protective Equipment

Personal protection equipment (PPE) refers to "all stuff (comprising clothing bearing protection against the weather and other dust particles) which is anticipated to be put on by any individual at work environment to protect the individual against at least one danger to their wellbeing. PPE incorporates head protectors/helmets, gloves, eye protection, high visibility clothing, and safety footwear among others. As such, PPE is won to minimize openness to various hazards and is also won to shield a person from hot and cold temperatures to safeguard the wellbeing of laborers, especially in the construction industry.

Even though regarded as useful tools for occupational health safety, studies indicate that the use of PPE is still low in other regions. For instance, Izudi et al. (2017) while assessing the utilization of PPE among building construction laborers in Uganda, reported that PPE use was low, thus, the need to implement PPE measures. In the UK, Stocks et al. (2011) reported that some construction laborers still have a rather negative perception towards protective clothing, and what is worse is that, there are few measures to conduct PPE training to rectify the situation among construction workers. However, in other regions such as Nigeria, Tanko and Anigbogu (Tank and Anigbogu 2012) indicate that the vast majority of construction workers understand the need for PPE and want to be protected against accidents, injury, and illness associated with construction activities.

The viability of utilizing the PPE at the work zone would rely upon the experience of the laborers to utilize PPE (Spillane and Oyedele 2013). Available studies report that the use of PPE to make a boundary between the expert and the openness helps in minimizing the monetary effect that occupational health impacts has on people, subsequently, expanding the wellbeing and lifespan of construction laborers (Wright et al. 2019). Further, the study reported that because of the continuing use of PPE, most construction firms find it effective to make it mandatory for their employees to use PPE as part of PPE standard compliance. In the healthcare setting, the utilization of PPE is a significant aspect to shield the workforce from contamination and forestall the spread of pathogens (Reddy et al. 2019).

Since laborer information, perception, and belief regarding the use of PPE in the construction would also affect its effectiveness and use at a personal level (Adeyemi et al. 2015), it is important for construction professionals to create PPE awareness via PPE training of the workers, to make it essential process or compulsory requirement for all construction workers (Cooper 2012). Mohd Amir Shazwan and Ee (2018) concluded that effective use of PPE as a result of awareness through training ultimately results in a reduction in the risk of serious accidents that occurs in the areas of construction. Therefore, organizations must work on the experience, proficiency, skills, attitudes, and beliefs as some of the personal factors that determine the effectiveness of PPE at the construction site.

28.5.3 Training and Development

Safety in the worksite, especially in the construction activities still calls for more attention. The results of the synthesis review in this topic as discussed above revealed the level of fatal injuries and other occupational diseases in the construction industry are quite common than in other industries (Endroyo et al. 2015). Attempts to manage the various occupational risks and diseases are through training and development. In line with this assumption, scholars argue that safety training is important in improving the skills, knowledge, attitudes, and habits of the workforce involved. As a result, both construction professionals and the workforce should focus on preventing accidents and risks because individuals with safety training skills are aware that some risks or occupational diseases are preventable and predictable.

In a proposed model of learning occupational safety and health (OSH) in the construction industry to minimize accidents (Endroyo et al. 2015), findings indicate that to reduce the occupational number of injuries is through learning, training, and development of both professionals and workforce. The model of training method in the construction industry should entail four components; (1) material, containing OSH fundamental theory and its use, (2) training techniques that apply Competency-Based learning, cooperative training, and contextual training, (3) tool/equipment to get the skills, satisfactory media as well as optimum usage of the surroundings, and (4) training assessment in the form of a portfolio. The examination inferred that the learning model can effectively improve the OSH knowledge, skills, and attitude of the workforce (Endroyo et al. 2015).

Ability or skills to identify or prevent risk is resolved in part by safety or risk training. Whereas training with low engagement has been common for years and considered ineffective in the current emerging market, new forms of more engaging training and development are emerging. One such concept is training utilizing a virtual construction site presented with a 3D vivid computer-generated experience power-wall. Virtual reality (VR) is an innovation that applies PCs, programming, and fringe equipment to create a mimicked environment for its operators (Bailenson et al. 2008).

The environment projected in the VR may recreate a genuine or fanciful environment. In this way, an *immersive/vivid virtual environment* (IVE) is a computer-generated environment that gives an individual a feeling of being inside it by drawing in the individual's sense and decreasing or wiping out their view of this present reality (Bailenson, et al. 2008; Webster 2016). In assessing construction safety preparedness utilizing IVR, Sacks et al. (2013) established a significant advantage of VR preparedness for stone cladding jobs and cast-in-situ concrete jobs. The study concluded that VR is more viable regarding keeping up learners' consideration and focus.

Another important aspect of safety training in construction is electronic learning or e-learning. E-learning is conveyed through the web to accomplish connection among students, course, and instructor (Lee and Hwang 2007). Ho and Dzeng (2010) directed an investigation to test the viability of safety training/preparing through e-learning in construction to decide client fulfilment and wellbeing conduct. The examination added that introducing e-learning mode in the construction wellbeing teaching readiness to assist construction experts to use it more autonomously and other media systems like animated training resources, case instruction, and repeated course preparedness, to minimize the common blunder linked to laborers in the construction site and other minor diseases accidents that are avoidable.

Providing training assists construction workers to learn about major parameters like techniques, guidelines, obstacles, and company/worker attitudes in the construction safety-related intervention strategies to manage risks (Goldenhar et al. 2001). Contrasted to little or no training, frequent engaging laborers as a training practice is linked to improved standards of recognition of hazardous elements and positive safety risk understanding. Subsequently, training assists individuals to identify a larger proportion of hazards and how to develop preventive mechanisms (Namian et al. 2016).

28.6 Conclusion

Construction laborers are exposed to multiple physical, chemical, biological, mechanical, and psychological risks during their everyday construction activity schedule. This review paid critical attention to the emerging health hazards such as nanomaterials, effects of climate change and stress, particles, and emissions on construction workers. Across the developed and developing nations, the hazardous attributes of construction activities make it challenging to keep smooth work activities all through the entire process. Besides, nanoscale materials like carbon, metal oxides, and other metals with small surfaces used in the construction industry present bigger risks to the wellbeing and safety of laborers.

Subsequently, occupational safety health hazards and socio-economic stresses are a result of hot environments brought about by climatic change like increased ambient temperature, ultraviolet exposure, air pollution, and changes in the work environment among other issues like extreme weather conditions. Together with particles and emissions, climate change is associated with adverse impacts, for example, hot weather tiredness, chronic kidney diseases, heat stroke, infectious diseases, cardiovascular disease, and depression challenges. Other established emerging occupational risks include skin cancer, skin diseases, asthma, musculoskeletal, and respiratory diseases among others. In providing a legal framework based on OSHA global standards, the chapter concludes by advocating for effective implementation of the utility of implied health and safety standard, use of PPE among employees and stakeholders, as well as safety training and development to improve knowledge and attitude of construction workers regarding safety compliance to reduce occupational health risks and hazards in the construction industry.

References

Abbasianjahromi H, Talebian R (2021) Identifying the most important occupational diseases in the construction industry: a case study of building industry in Iran. Int J Constr Manag 21(3):235–245

Adeyemi HO, Adejuyigbe SB, Ismaila SO, Adekoya AF (2015) Low back pain assessment application for construction workers. J Eng Des Technol 13(3):419–434

Aksorn T, Hadikusumo BH (2008) Critical success factors influencing safety program performance in Thai construction projects. Saf Sci 46(4):709–727

Al-Bayati AJ, Al-Zubaidi HA (2018) Inventory of nanomaterials in construction products for safety and health. J Constr Eng Manag 144(9):06018004

Albert A, Hallowell MR, Kleiner BM (2014) Enhancing construction hazard recognition and communication with energy-based cognitive mnemonics and safety meeting maturity model: multiple baseline study. J Constr Eng Manag 140(2):04013042

Aneziris ON, Topali E, Papazoglou IA (2012) Occupational risk of building construction. Reliab Eng Syst Saf 105:36–46

Araujo IPS, Costa BD, de Moraes RJB (2014) Identification and characterization of particulate matter concentrations at construction jobsites. J Sustain 6:7666–7688

Azarmi F, Kumar P, Mulheron M (2014) The exposure to coarse, fine, and ultrafine particles emissions from concrete mixing, drilling, and cutting activities. J Hazard Mater 279:268–279

Azarmi F, Kumar P, Marsh D, Fuller GW (2015) Assessment of long-term impacts of PM10 and PM2.5 particles from construction works on surrounding areas. Environ Sci Process Impacts 18:208–221

Bailenson JN, Yee N, Blascovich J, Beall AC, Lundblad N, Jin M (2008) The use of immersive virtual reality in the learning sciences: digital transformation of teachers, students, and social context. J Learn Sci 17(1):102–141

Baradan S, Usmen M (2006) Comparative injury and fatality risk analysis of building trades. J Constr Eng Manag 132(5):533–539

Beaudrie CEH, Kandlikar M, Gregory R, Long G, Wilson T (2015) Nanomaterials risk screening: a structured approach to aid decision-making under uncertainty. Environ Syst Decis 35(1):88–109

Bennett CM, Michael AJ (2010) Non-heat related impacts of climate change on working populations. Global Health Action 3:54–60

Bharara K, Sandhu P, Sidhu M (2012) Issues of occupational health and injuries among unskilled female laborers in construction industries: a scenario of Punjab State. Stud Home Commun Sci 6(1):1–6

Biswas G, Bhattacharya A, Bhattacharya R (2017) Occupational health status of construction workers: a review. Int J Med Sci Public Health 6(3):2–7

Blaise C, Gagné F, Ferard JF, Eullaffroy P (2008) Ecotoxicity of selected nano-materials to aquatic organisms. Environ Toxicol Int J 23(5):591–598

Blanc PD, Jarvholm B, Toren K (2015) Prospective risk of rheumatologic disease associated with occupational exposure in a cohort of male construction workers. Am J Med 128(10):1094–1101

Brook RD, Rajagopalan S, Pope CA III, Brook JR, Bhatnagar A, Diez-Roux AV, Holguin F, Hong Y, Luepker RV, Mittleman MA, Peters A, Kaufman JD (2010) Particulate matter air pollution and cardiovascular disease: an update to the scientific statement from the American Heart Association. Circulation 121(21):2331–2378

Buzea C, Pacheco II, Robbie K (2007) Nanomaterials and nanoparticles: sources and toxicity. Biointerphases 2(4):MR17-MR71

Cavallari JM, Eisen EA, Chen JC, Fang SC, Dobson CB, Schwartz J, Christiani DC (2007) Night heart rate variability and particulate exposures among boilermaker construction workers. Environ Health Perspect 115(7):1046–1051

Cheng CW, Leu SS, Lin CC, Fan C (2010) Characteristics analysis of occupational accidents at small construction enterprises. Saf Sci 48(6):698–707

Choe S, Leite F (2017) Assessing safety risk among different construction trades: quantitative approach. J Constr Eng Manag 143(5):04016133

Choe S, Yun S, Leite F (2016) Analysis of the effectiveness of the OSHA steel erection standard in the construction industry. Saf Sci 89:190–200

Choi SD (2015) Aging workers and trade-related injuries in the US construction industry. Saf Health Work 6(2):151–155

Collier E (2017) Ten common construction site hazards. High Speed Training. https://www.highspeedtraining.co.uk/hub/common-construction-hazards/

Construction Litigation (2019) How technology is affecting the construction industry. CMBG3 LAW

Cooper M (2012) Current issues in health and safety training in the UK. J Eur Ind Train 22(9):354–361

Cugell DW, Kamp DW (2004) Asbestos and the pleura: a review. Chest 125(3):1103–1117

Daryoush B, Darvish A (2013) A case study and review of nanotechnology and nanomaterials in green architecture. Res J Environ Earth Sci 5(2):78–84

De Blois J, Kjellstrom T, Agewall S, Ezekowitz JA, Armstrong PW, Atar D (2015) The effects of climate change on cardiac health. Cardiology 131:209–217

Demers PA, Teschke K, Kennedy SM (1997) What to do about softwood? A review of respiratory effects and recommendations regarding exposure limits. Am J Ind Med 31(4):385–398

Demirkesen S, Arditi D (2015) Construction safety personnel's perceptions of safety training practices. Int J Project Manag 33(5):1160–1169

Desruisseaux B (2019) Training effectiveness analysis of OSHA silica and excavation safety standards for construction. Wayne State University Theses 701. https://digitalcommons.wayne.edu/oa_theses/701
Dick F (2000) Solvent neurotoxicity. Occup Environ Med 63(3):221–226
DuPont (2007) Nanomaterial risk assessment worksheet. Light stabilizer
Dutta P, Rajiva A, Andhare D, Azhar GS, Tiwari A, Sheffield P, Climate Study Group (2015) Perceived heat stress and health effects on construction workers. Indian J Occup Environ Med 19(3):151
Endroyo B, Yuwono BE, Mardapi D (2015) Model of learning/training of occupational safety and health (OSH) based on industry in the construction industry. Procedia Eng 25:83–88
EPA (2002) Health assessment document for diesel engine exhaust. US Environmental Protection Agency, Washington DC
Estlander T, Jolanki R, Kanerva L (2000) Handbook of occupational dermatology. Springer, Berlin, Heidelberg, New York
Fleming M (2009) Hazard recognition. By Design ASSE 11–15
Font A, Baker T, Mudway IS, Purdie E, Dunster C, Fuller GW (2014) Degradation in urban air quality from construction activity and increased traffic arising from a road widening scheme. Sci Total Environ 497(8):123–132
Gambatese J, Pestana C, Lee H (2016) Alignment between lean principles and practices and worker safety behavior. J Constr Eng Manag 143(1):04016083
Gasparrini A, Guo Y, Hashizume M, Lavigne E, Zanobetti A, Schwartz J, Tobias A, Tong S, Rocklöv J, Forsberg B, Leone M (2015) Mortality risk attributable to high and low ambient temperature: a multicountry observational study. The Lancet 386(9991):369–375
Gibson O, Pattisson P (2014) Qatar World Cup: 185 Nepalese died in 2013—official records. Guardian 24
Glytsos T, Ondráček J, Džumbová L, Kopanakis I, Lazaridis M (2010) Characterization of particulate matter concentrations during controlled indoor activities. Atmos Environ 44(12):1539–1549
Goldenhar LM, Moran SK, Colligan M (2001) Health and safety training in a sample of open-shop construction companies. J Saf Res 32(2):237–252
Gonzalez-Quiroz M, Smpokou ET, Silverwood RJ, Camacho A, Faber D, Garcia BLR, Oomatia A, Hill M, Glaser J, Le Blond J, Wesseling C (2018) Decline in kidney function among apparently healthy young adults at risk of Mesoamerican nephropathy. J Am Soc Nephrol 29(8):2200–2212
Grote G (2012) Safety management in different high-risk domains—all the same? Saf Sci 50(10):1983–1992
Haj Aghazadeh M, Mohammadian Y, Nourmohammadi M, Zare M (2012) Evaluation of ergonomic risk factors for musculoskeletal disorders in construction workers by PATH. Int J Environ Eng 11(1):43
Hansen AL, Bi P, Ryan P, Nitschke M, Pisaniello D, Tucker G (2008a) The effect of heat waves on hospital admissions for renal disease in a temperate city of Australia. Int J Epidemiol 37:1359–1365
Hansen D, Blahout B, Benner D, Popp W (2008b) Environmental sampling of particulate matter and fungal spores during demolition of a building on a hospital area. J Hosp Infect 70(3):259–264
Haynes R, Savage A (2007) Assessment of the health impacts of particulates from the redevelopment of Kings Cross. Environ Monit Assess 130:47–56
Heal MR, Kumar P, Harrison RM (2012) Particles, air quality, policy, and health. Chem Soc Rev 41(19):6606–6630
Herzog E, Casey A, Lyng FM, Chambers G, Byrne HJ, Davoren M (2007) A new approach to the toxicity testing of carbon-based nanomaterials—the clonogenic assay. Toxicol Lett 174(1–3):49–60
Ho CL, Dzeng RJ (2010) Construction safety training via e-learning: learning effectiveness and user satisfaction. Comput Educ 55(2):858–867
IARC (1990) Monographs on the evaluation of carcinogenic risks to humans, vol 49. IARC Press, Lyon

ISO (2017) Nanotechnologies—plain language explanation of selected terms from the ISO/IEC 8004 series, 1st edn

Izudi J, Ninsiima V, Alege JB (2017) Use of personal protective equipment among building construction workers in Kampala, Uganda. J Environ Public Health

Janssen NAH, Fischer P, Marra M, Ameling C, Cassee FR (2013) Short-term effects of PM2. 5, PM10 and PM2. 5–10 on daily mortality in the Netherlands. Sci Total Environ 463:20–26

Johnson RJ, Wesseling C, Newman LS (2019) Chronic kidney disease of unknown cause in agricultural communities. N Engl J Med 380(19):1843–1852

Jones K (2018) Construction technology is reshaping the industry: construction technology. Construct Connect

Kang S, Pinault M, Pfefferle LD, Elimelech M (2007) Single-walled carbon nanotubes exhibit strong antimicrobial activity. Langmuir 23(17):8670–8673

Karakhan A, Gambatese JA (2018) Hazards and risks in construction and the impact of incentives and rewards on safety outcomes. Pract Period Struct Des Constr 23(2):04018005

Kjellstrom E, Barring D, Jacob D, Jones R, Lenderink G, Schar C (2007) Modelling daily temperature extremes: recent climate and future changes over Europe. Clim Change 81(1):249–265

Kjellstrom T (2009) Editorial: climate change, direct heat exposure, health, and well-being in low and middle-income countries. Global Health Action 2

Kjellstrom T, Crowe J (2011) Climate change, workplace heat exposure, and occupational health and productivity in Central America. Int J Occup Environ Health 17(3):270–281

Kjellstrom T, Lemke B, Otto M (2017) Climate conditions, workplace heat, and occupational health in South East Asia in the context of climate change. WHO South-East Asia J Public Health 6(2):15–21

Kjellstrom T, Lemke B, Venugopal V (2013) Occupational health and safety impacts of climate conditions. In: Pielke R (ed) Elsevier, New York

Kjellstrom T, Briggs D, Freyberg C, Lemke B, Otto M, Hyatt O (2016) Heat, human performance, and occupational health: a key issue for the assessment of global climate change impacts. Annu Rev Public Health 37:97–112

Kjellstrom T, Holmer I, Lemke B (2009a) Workplace heat stress, health, and productivity—an increasing challenge for low and middle-income countries during climate change. Global Health Action 2(1):46–51

Kjellstrom T, Kovats RS, Lloyd SJ, Holt T, Tol RS (2009b) The direct impact of climate change on regional labor productivity. Arch Environ Occup Health 64(4):217–227

Kumar P, Mulheron M, Som C (2012) Release of ultrafine particles from three simulated building processes. J Nanoparticles Res 14(1):771

Lam HC (2003) An investigation into the implementation of safety management systems by Hong Kong construction contractors. PhD thesis, The University of Hong Kong

Lee J, Mahendra S, Alvarez PJ (2010) Nanomaterials in the construction industry: a review of their applications and environmental health and safety considerations. ACS Nano 4(7):3580–3590

Lee J, Mahendra S, Alvarez PJJ (2009) Potential environmental and human health impacts of nanomaterials used in the construction industry. In: Nanotechnology in construction, vol 3. Springer, Berlin, Heidelberg, pp 1–14

Lee JK, Hwang CY (2007) The effects of computer self-efficacy and learning management system quality on e-learner's satisfaction. In: Proceedings of the 2007 European LAMS conference: designing the future of learning. LAMS Foundation Greenwich, pp 73–79

Lee LJH, Lin CK, Hung MC, Wang JD (2016) Impact of work related cancers in Taiwan: estimation with QALY (quality-adjusted life year) and healthcare costs. Prev Med Rep 4:87–93

Lin P, Chen JW, Chang LW, Wu JP, Redding L, Chang H, Yeh TK, Yang CS, Tsai MH, Wang HJ, Kuo YC (2008) Computational and ultrastructural toxicology of a nanoparticle, Quantum Dot 705, in mice. Environ Sci Technol 42(16):6264–6270

Lowry GV, Gregory KB, Apte SC, Lead JR (2012) Transformations of nanomaterials in the environment. Environ Sci Technol 46(13):6893–6899

Manuele F (2005) Risk assessment and hierarchies of control. Prof Saf 50(5):33–39

Maxim LD, Eastes W, Hadley JG, Carter CM, Reynolds JW, Niebo R (2003) Fiberglass and rock/slag wool exposure of professional and do-it-yourself installer. Regul Toxicol Pharmacol 37:28–44
Mersha H, Mereta ST, Dube L (2017) Prevalence of occupational injuries and associated factors among construction workers in Addis Ababa, Ethiopia. J Public Health Epidemiol 9(1):1–8
Mohd Amir Shazwan H, Ee JM (2018) Effectiveness of personal protective equipment (PPE) at construction site. INTI J 1(12):1–12
Namian M, Albert A, Zuluaga CM, Behm M (2016) Role of safety training: impact on hazard recognition and safety risk perception. J Constr Eng Manag 142(12):04016073
Nandekar UP, Raut Desai R (2019) Nanotechnology innovations in construction industry and environmental sustainability. Int J Innov Technol Explor Eng 8(11S):286–292
OshAcademy (2017) Developing a construction safety management system. Beaverton, Oregon, UK, 97006
OSHA (2020) OSHA field safety and health manual. https://www.osha.gov/OshDoc/Directive_pdf/ADM_04-00-001.pdf. Accessed 20 Jan 2021
OSHA Pocket Guide (2005) Worker safety series construction. 1–2
Parida R, Ray PK (2012) Study and analysis of occupational risk factors for ergonomic design of construction work systems. Work J Prevent Assess Rehabil 41(Supplement 1):3788–3794
Parsons K (2014) Human thermal environment. The effects of hot, moderate and cold temperatures on human health, comfort and performance, vol 3. CRC Press, New York
Prodan L, Bachofen G (1998) Cement and concrete in chapter 93: construction, encyclopaedia of occupational health and safety. Geneva: International Labour Organization (ILO)
Rao A, Jha KN, Misra S (2007) Use of aggregates from recycled construction and demolition waste in concrete. Resour Conserv Recycl 50(1):71–81
Ray PK, Parida R, Saha E (2015) Status survey of occupational risk factors of manual material handling tasks at a construction site in India. Procedia Manuf 3:6579–6586
Reddy SC, Valderrama AL, Kuhar DT (2019) Improving the use of personal protective equipment. Applying lessons learned. Clin Infect Dis 69(Supplement 3):S165–S170
Ringen K, Dement J, Welch L, Dong XS, Bingham E, Quinn PS (2014) Risks of a lifetime in construction: part II: chronic occupational diseases. Am J Ind Med 57(11):1235–1245
Robson LS, Clarke JA, Cullen K, Bielecky A, Severin C, Bigelow PL, Irvin E, Culyer A, Mahood Q (2007) The effectiveness of occupational health and safety management system interventions: a systematic review. Saf Sci 45(3):329–353
Rowlinson S, YunyanJia A, Li B, ChuanjingJu C (2014) Management of climatic heat stress risk in construction: a review of practices, methodologies, and future research. Accid Anal Prev 66:187–198
Sacks R, Perlman A, Barak R (2013) Construction safety training using immersive virtual reality. Constr Manag Econ 31(9):1005–1017
Schoenfisch AL, Lipscomb H, Sinyai C, Adams D (2017) Effectiveness of OSHA outreach training on carpenters' work-related injury rates, Washington State 2000–2008. Am J Ind Med 60(1):45–57
Schulte PA, Chun H (2009) Climate change and occupational safety and health: establishing a preliminary framework. J Occup Environ Hyg 6(9):542–554
Sharma N, Ojha H, Bharadwaj A, Pathak DP, Sharma RK (2015) Preparation and catalytic applications of nanomaterials: a review. RSC Adv 5(66):53381–53403
Shaw BJ, Handy RD (2011) Physiological effects of nanoparticles on fish: a comparison of nanomaterials versus ions. Environ Int 37(6):1083–1097
Smith K, Woodward A, Campbell-Lendrum D, Chadee D, Honda Y, Liu Q, Olwoch J, Revich B, Sauerborn R, Aranda C, Berry H (2014) Human health: impacts, adaptation, and co-benefits. In: Climate change 2014: impacts, adaptation, and vulnerability. Part A: global and sectoral aspects. Contribution of Working Group II to the fifth assessment report of the Intergovernmental Panel on Climate Change. Cambridge University Press, pp 709–754

Sokas RK, Jorgensen E, Nickels L, Gao W, Gittleman JL (2009) An intervention effectiveness study of hazard awareness training in the construction building trades. Public Health Rep 124(4):161–168

Spencer-Hwang R, Knutsen SF, Soret S, Ghamsary M, Beeson WL, Oda K, Shavlik D, Jaipaul N (2011) Ambient air pollutants and risk of fatal coronary heart disease among kidney transplant recipients. Am J Kidney Dis 58(4):608–616

Spillane J, Oyedele LJ (2013) Strategies for effective management of health and safety in confined site construction. Austalasian J Constr Econ Build 13(4):50

Stocks S, Turner S, McNamee R, Carder M, Hussey I, Agius R (2011) Occupation and work related ill-health in UK construction workers. Occup Med 61(6):407–415

Strategic Human Resources Management (SHRM) (2021) What are OSHA standards? https://www.shrm.org/resourcesandtools/tools-and-samples/hr-qa/pages/industrystandards.aspx. Accessed 20 Jan 2021

Tah JH, Abanda HF (2011) Sustainable building technology knowledge representation: using semantic web techniques. Adv Eng Inform 25(3):547–558

Tang T, Hurraß J, Gminski R, Mersch-Sundermann V (2012) Fine and ultrafine particles emitted from laser printers as indoor air contaminants in German offices. Environ Sci Pollut Res 19(9):3840–3849

Tank BL, Anigbogu NA (2012) The use of personal protective equipment (PPE) on construction sites in Nigeria. In: Laryea S, Agyepong SA, Leininger R, Hughes W (eds) Proceedings 4th West Africa built environment research (WABER) conference, Abu, Nigeria, pp 1341–1348

Tawatsupa B, Yiengprugsawan V, Kjellstrom T, Berecki-Gisolf J, Seubsman SA, Sleigh A (2013) The association between heat stress and occupational injury among Thai workers: finding of the Thai Cohort Study. Ind Health 51(1):34–46

Terwoert J (1999) Application of vegetable-oil based concrete moulds release agents (VERAs) at construction sites and in precast concrete factories. Chemiewinkel Universiteit van Amsterdam, Amsterdam

Thienen G, Spee T (2008) Health effects of construction materials and construction products. J App Occup Sci 14(13):2–23

Thomas SP, Al-Mutairi EM, De SK (2013) Impact of nanomaterials on health and environment. Arab J Sci Eng 38(3):457–477

Tong R, Cheng M, Zhang L, Liu M, Yang X, Yin W (2018) The construction dust-induced occupational health risk using Monte-Carlo simulation. J Clean Prod 184:598–608

Turner MC, Krewski D, Pope CA, Chen Y, Gapstur SM, Thun MJ (2011) Long-term ambient fine particulate matter air pollution and lung cancer in a large cohort of never-smokers. Am J Respir Crit Care Med 184:1374–1381

Van der Molen HF, de Vries SC, Stocks SJ, Warning J, Frings-Dresen MH (2016) Incidence rates of occupational diseases in the Dutch construction sector. Occup Environ Med 73(5):350–352

Venugopal V, Latha PK, Shanmugam R, Krishnamoorthy M, Johnson P (2020) Occupational heat stress induced health impacts: a cross-sectional study from South Indian working population. Adv Clim Change Res 11(1):31–39

Webster R (2016) Declarative knowledge acquisition in immersive virtual learning environments. Interact Learn Environ 24(6):1319–1333

Weil D (2001) Assessing OSHA performance: new evidence from the construction industry. J Policy Anal Manag J Assoc Public Policy Anal Manag 20(4):651–674

Weng CH, Hu CC, Yen TH, Huang WH (2015) Association between environmental particulate matter and arterial stiffness in patients undergoing hemodialysis. BMC Cardiovasc Disord 15:115

Wesselling C, Crowe J, Hogstedt C, Jakobsson K, Lucas R, Wegman D (2013) Mesoamerican nephropathy. Report from the first international research workshop on men, IRET, Unive Nac. Heredia, Costa Rica

West B (2016) Sorting it out: what is the difference between hazard and risk? The safety, wellbeing & rehab experts. https://www.rrp.com.au/what-is-the-difference-between-the-terms-hazard-and-risk. Accessed 20 Jan 2021

West GH, Lippy BE, Cooper MR, Marsick D, Burrelli LG, Griffin KN, Segrave AM (2016) Toward responsible development and effective risk management of nano-enabled products in the US construction industry. J Nanopart Res 18(2):49

WHO (2000) Diphenylmethane diisocyanate (MDI): concise international chemical assessment document 27. World Health Organisation, Geneva

Wieslander G, Kumlin A, Norback D (2010) Dampness and 2-ethyl-1-hexanol in floor construction of rehabilitation center: health effects in staff. Arch Environ Occup Health 65(1):3–11

World Bank (1984) The construction industry: issues and strategies in developing countries. The World Bank, Washington, DC

Wright T, Adhikari A, Yin J, Vogel R, Smallwood S, Shah G (2019) Issue of compliance with use of personal protective equipment among wastewater workers across the South East region of the United States. Int J Environ Res Public Health 16(11):2009

Yassin AS, Martonik JF (2004) The effectiveness of the revised scaffold safety standard in the construction industry. Saf Sci 42(10):921–931

You NS, Chan DW, Shan M, Sze NN (2019) Implementation of safety management system in managing construction projects: benefits and obstacles. Saf Sci 117:23–32

Yuan L (2012) Learning OSHA construction standards through a participatory approach. In: ASSE professional development conference and exposition. American Society of Safety Engineers

Chapter 29
Bioconcrete: The Promising Prospect for Green Construction

Anita V. Handore, Sharad R. Khandelwal, Rajib Karmakar, Abhijeet S. Jagtap, and Dilip V. Handore

Abstract Since the construction industry is raising at 0.8–1.2%/yr the concrete utilization growth is assumed to 3.7–4.4 billion tonnes by 2050. Cement is the major composite of concrete. However, its manufacturing contributes approximately 10% of global CO_2 emissions and other greenhouse gases. Worldwide, production of cement for concrete makes it 3rd ranking producer of anthropogenic CO_2 followed by transport and energy generation. Despite the importance of cement-concrete for construction, it causes several adverse impacts on environment as well as human health at all stages of its manufacturing. In view of these adverse impacts a new term "Bioconcrete" has been originated. It is the self-healing form of concrete in which an ecofriendly key ingredient i.e. Biocement is added. This healing agent can stay dormant for many years and can become active only in the presence of water. Therefore, in modern civilization, MICP derived Biocement-bioconcrete can be the suitable alternative for eco-friendly construction. This chapter presents the overview of Bio Concrete. It describes the need for Biocement and Bioconcrete and the significant mechanism of Biomineralization/Bioprecipitation. The chapter represents the detailed method of MICP, i.e., Microbially induced calcium carbonate precipitation including all the possible pathways. It emphasizes triggering the metabolic pathways by various microbes for synthesis of Biocement-Bioconcrete. The chapter also discusses various factors influencing performance of MICP. Besides, it illustrates

A. V. Handore (✉)
Research and Development Department, Sigma Wineries Pvt. Ltd., Maharashtra 422112, Maharashtra, India

S. R. Khandelwal
H.A.L. College of Science & Commerce, Nashik 422207, Maharashtra, India

R. Karmakar
Department of Agricultural Chemicals, Bidhan Chandra Krishi Viswavidyalaya, Nadia 741252, West Bengal, India

A. S. Jagtap
Pune Institute of Computer Technology, Pune 411043, Maharashtra, India

D. V. Handore
Research and Development Department, Sigma Wineries Pvt. Ltd., Nashik 422112, Maharashtra, India

J. A. Malik and S. Marathe (eds.), *Ecological and Health Effects of Building Materials*,
https://doi.org/10.1007/978-3-030-76073-1_29

the current limitation for commercialization of such self-healing construction material along with numerous potential applications. Furthermore, it highlights the future perspectives of these materials in the upcoming era as a promising and self-healing ecofriendly material of choice for green construction.

Keywords Biocement · Bioconcrete · Biomineralization · Bioprecipitation · Green construction · MICP · Ureolysis

29.1 Cement-Concrete

Concrete is a highly demanded versatile material of the construction industry. It has been used extensively as qualities like availability, durability, high compressive strength, compatibility with reinforcement bars, and simple preparation are associated with it. It is chosen easily for various applications due to its low cost and can be casted in different shapes and sizes of desire. By the year 2050, worldwide consumption of concrete is expected to rise to 3.7–4.4 billion tones as the growth of the construction industry is rising yearly at 0.8–1.2% (Benhelal et al. 2013).

In spite of its huge importance in construction, concrete affects the environment and human health at all stages of its manufacturing process (Ivanov et al. 2015) which contributes to emission of carbon-di-oxide (approximately 10% of global emission) (Fig. 29.1). Worldwide, production of cement for concrete makes it 3rd ranking producer of anthropogenic CO_2 followed by transport and energy generation. This reflects the exclusive and universal significance of concrete in worldwide construction industry. Therefore, considering the adverse environmental impact of concrete production, various features like extraction of materials causing habitat damage and huge energy consumption, production and transportation may be emphasized.

Cement is the major composite of concrete, which is normally used for binding the aggregates and filling the pore spaces in between fine and coarse particles. The principal hydraulic cement–Portland, has very great demand but being very energy-intensive materials of concrete, this cement is producing greenhouse gases. It is reported that, four GJ (Gigajoule) energy is required to producing 1 tonne of portland cement and manufacturing of portland cement clinker produces approximately one ton of CO_2 into the atmosphere (Malhotra 1999; Mehta 1999).

The top-soil is destroyed and lost due to extensive deforestation as a result of mining raw materials like limestone, clay and fuel including coal. Ninety two percent of the energy is used to produce concrete containing only 9–13% cement causing adverse impact on the health of workers as well on the environment (Kubba 2012).

Besides impact on the environment, concrete forms cracks to allow penetration of aggressive chemicals in the structure. The cracks are an indication of concrete deterioration reducing its durability. At an early age, cracks do not threaten its strength. But for a long time, the concrete lifespan may face serious risk (Jonkers 2011). Worldwide, considerable budget is wasted annually to repair the existing cementitious structures (Cailleux and Pollet 2009). It is stated that \$147 per m^3 of concrete

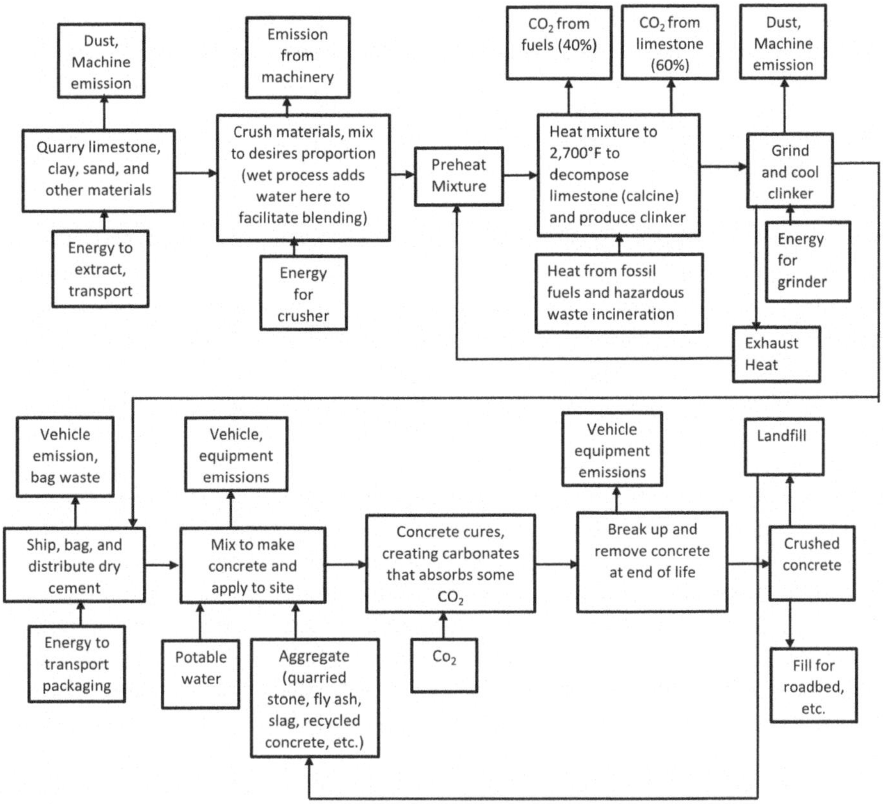

Fig. 29.1 Concrete manufacturing and its impact on environment (adapted and redrawn after: https://www.greenspec.co.uk)

involves the direct cost of maintenance of cracks repair, although production of concrete costs \$65–\$80 per m^3 (Silva 2015). Therefore, preventive measures are very crucial to hinder and finish formation of crack at an early stage.

Research on adverse impacts of building materials on users' health was started in 1970. In this context, gypsum binders, paints and wood based materials along with blocks of silica etc. were introduced for promotion of human health and protection of the environment. Therefore a need to produce and design various green construction materials supporting to decrease pollution (Stanaszek-Tomal 2020; Häkkinen and Belloni 2011; Chen et al. 2019).

In view of environmental and health impact on society by the process of concrete manufacturing, new term "Bioconcrete" has been originated. It is the self-healing form of concrete in which an eco friendly key ingredient i.e. Biocement is added which shows self–healing potential. This healing agent can stay dormant for many years and can become active only in the presence of water. Therefore to overcome the adverse environmental and health impact, Biocement based bio concrete can be the suitable alternative for green construction (Mortensen et al. 2011).

29.2 Biocement

Biocement is the mixture of different component which can alter the pH as well as start binder precipitation, enzymes, live cells of microbes, dead but enzymatically active biomass of microbes catalyzing the changes in pH and or other biochemical reactions as well inorganic constituents makes the binding material.

Presently, it is being explored as a bio-based green material in diverse fields for wide applications. Bio concrete is prepared by adding or bacteria exhibiting the ability of $CaCO_3$ precipitation with the mechanism of Microbial Induced Carbonate Precipitation (MICP). Biocement is an integral part of bioconcrete, produced via MICP mechanism and Calcium carbonate is the appropriate grout for Bio Concrete due to significant compatibility with the cementitious alignments. Due to the property of self-healing along with the up gradation of durability and mechanical features of concrete constructions, it has gained extensive attention from the last few decades. Moreover, coming up as eco friendly–economic technology due to its the repair–maintenance cost and less emissions of carbon dioxide (Castro-Alonso et al. 2019). Although, there are around fifteen different kinds of Biocement, microbial induced calcite precipitation induced Biocement is very popular (Volodymyr and Viktor 2019).

29.3 Biomineralization/Bioprecipitation

Biomineralization/bioprecipitation takes place due to chemical variation in the environment by various microbial activities, resulting in precipitation of minerals (Periasamy et al. 2016). Biomineralization process is well-known and mediated by various microorganisms like algae, fungi, bacteria, protists, and plants etc. Basically, these Bio minerals can be found everywhere, in the form of shells, bone, and teeth, coral reefs, limestone caves offering the amazing solutions for various environmental and engineering issues (Mostafa and Aydin 2019).

Schematic representation of bacterial structure and production of $CaCO_3$ is shown in Fig. 29.2.

Generally, Biomineralization/Bio precipitation of minerals in bacteria occurs via following three mechanisms.

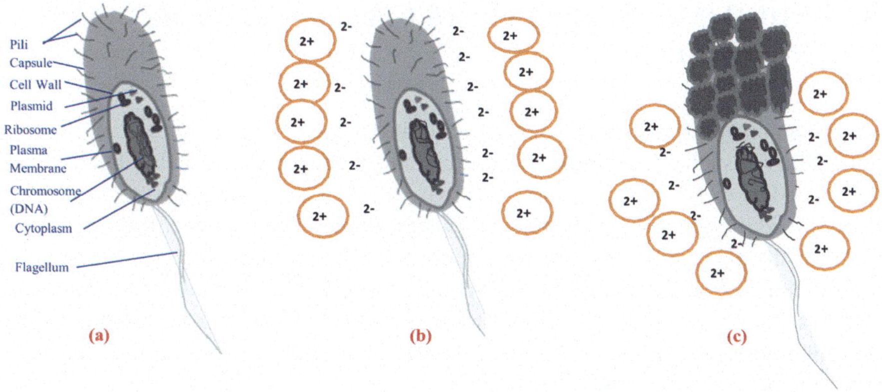

Fig. 29.2 Diagrammatic representation of structure of bacteria and synthesis of $CaCO_3$ **a** structure of bacteria, **b** cell wall of bacteria with negative and positive charged ions, **c** bio minerals synthesis by ions binding to bacterial cell walls (adapted and redrawn after: Mostafa et al. 2016)

29.3.1 BCM (Biologically Controlled Mineralization)

BCM could be intra-cellular i.e. BCMin, extracellular i.e. BCMe and intercellular i.e. BCMint with participation of exopolysaccharides or vesicles (Weiner and Dove 2003; Castro-Alonso et al. 2019).

29.3.2 BIM (Biologically Induced Mineralization)

In BIM, the minerals are indirectly mineralized because of interactions of the metabolic byproduct and the ions associated with their surroundings. These minerals are identified by variations in the size of particulate matter, morphology and reduces the crystallinity (Weiner and Dove 2003; Castro-Alonso et al. 2019).

29.3.3 BMM (Biologically Mediated Mineralization)

In BMM, the mineralization process occurs due to interaction of organic and or inorganic matrices without any intracellular or extracellular bioactivity (Weiner and Dove 2003; Castro-Alonso et al. 2019). However, among the large mineral production via bio-mineralization, precipitation of $CaCO_3$ is of interest to the scientific community because of its excellent bonding capability as well as compatibility with concrete compositions.

29.4 MICP (Microbially Induced Calcium Carbonate Precipitation)

This mechanism attracted the scientific community due to its potential for diverse applicability. Usually, Calcium carbonate crystals are formed by metabolites reaction synthesized by CO_3^{2-}, enriched in Ca^{2+}environment. Four essential factors viz. DIC i.e. dissolved inorganic carbon, presence of sites of nucleation, concentration of Ca^{2+} as well as the pH of medium have been reported as the essential factors affecting the precipitation of $CaCO_3$ (Hammes and Verstraete 2002). Carbonate is also reported in tidal flats, coral reefs, fluviatile tufas, lacustrine whitings, cavecrusts, and hot-spring travertines. Besides, in the sub aerial calcrete, few other coatings, grains, matrices etc. inside the sediments (Riding 2000).

Microbes like heterotrophs and autotrophs show ability for precipitation of $CaCO_3$ by various pathways like urea hydrolysis, reduction of iron, sulphate, denitrification, Photosynthesis, ammonification, Methane oxidation, and anaerobic sulfide oxidation, etc. The urea hydrolysis is reported to be a dominating process during the photosynthesis of terrestrial systems. It is also reported that sulphate reduction is dominant in aquatic habitats (Rusznyak et al. 2012; Zhu and Dittrich 2016). The MICP reactions include the following.

29.4.1 Ureolysis

Hydrolysis of urea in presence of ureases results in production of CO_2 and ammonia, resulting in an increase in pH. The urease enzyme shows Nickel ions at its active site (Krajewska 2009). Ureases are reported to promote. Formation and precipitation of $CaCO_3$ in natural waters, geological sediments and soil etc. (Mobley and Hausinger 1989). The urease enzyme is produced independently without the substrate i.e. induced or constitutive (Mobley et al. 1991). In the MICP, bacterial cells can affect the kind of minerals formed (Douglas and Beveridge 1998; Rodriguez et al. 2012). The *Bacillus* group is reported for higher synthesis and activity of urease (Achal et al. 2015).

29.4.1.1 Mechanism of Urease Synthesis (Mobley and Hausinger 1989)

In the microbial systems, Urease synthesis can be carried out through different modes as in the constitutive mode, the enzyme activity is expressed per cell and it does not depend on any external factor. Whereas, in the inducible mode, the level of enzyme activity is generally expressed per cell and triggered by the existence of inducer molecules such as urea. In case of the Repressible mode, synthesis is depressed under nitrogen limiting conditions when there is occurrence of ammonia or its precursors.

In the Growth Pattern—swarming versus non-swarming growth pattern expresses variation in urease synthesis (Falkinham and Hoffman 1984).

29.4.1.2 MICP by Production of Urease Enzymes

Bacteria synthesize urease enzymes and catalyze urea into NH_4 which results in production of carbonate as well as raises the pH at the cell attachment surface. The carbonate precipitation occurs at seven stages. Following reaction shows spontaneous hydrolysis resulting in carbonic acid formation at the surface of the cell (Hammes et al. 2003; Li et al. 2000; Stocks-Fischer et al. 1999; Ferris et al. 1996; Qian et al. 2010).

$$\begin{aligned}&\text{Urea } [CO\,(NH_2)_2] + \text{Water } [H_2O]\\&\quad\rightarrow \text{Carbamic acid } [NH_2COOH] + \text{Ammonia } [NH_3]\end{aligned} \tag{29.1}$$

$$\begin{aligned}&\text{Carbamic acid } [NH_2COOH] + \text{Water } [H_2O]\\&\quad\rightarrow \text{Ammonia } [NH_3] + \text{Carbonic acid } [H_2CO_3]\end{aligned} \tag{29.2}$$

$$\begin{aligned}&\text{Ammonia } [2NH_3] + \text{Water } [2H_2O]\\&\quad\rightleftharpoons \text{Ammonia ion } [2NH_4^+] + \text{Hydroxide } [2OH^-]\end{aligned} \tag{29.3}$$

$$\begin{aligned}&\text{Carbonic acid } [H_2CO_3]\\&\quad\rightleftharpoons \text{Bicarbonate } [HCO_3^-] + \text{Hydrogen ion } H^+\end{aligned} \tag{29.4}$$

$$\begin{aligned}&\text{Bicarbonate } [HCO_3^-] + \text{Hydrogen ion } H^+ + \text{Hydroxide ion } [2OH^-]\\&\quad\rightleftharpoons \text{Carbonate ion } [CO_3^{2-}] + \text{Water}[\ 2H_2O]\end{aligned} \tag{29.5}$$

$$\begin{aligned}&\text{Calcium ion } [Ca^2+] + \text{cell of Bacteria}\\&\quad\rightarrow \text{cell of Bacteria } - \text{ Calcium ion } [Ca^{2+}]\end{aligned} \tag{29.6}$$

$$\begin{aligned}&\text{Cell of Bacteria } - \text{ Calcium ion } [Ca^{2+}] + \text{Carbonate ion } [CO_3^{2-}]\\&\quad\rightarrow \text{Cell of Bacteria } - \text{ Calcium Carbonate}\end{aligned} \tag{29.7}$$

Figure 29.3 shows bacterial urease activity precipitation of $CaCO_3$ and generation of ATP as: (1) On basis of concentration gradient, urea diffuses into the bacterial cell, (2) urea hydrolysis results into increased because of ammonia production, (3) this ammonia get diffused out of cell as per the concentration gradient and increases cell membrane potential both outside-inside, (4) due to this the proton driving force allows generation of adenosine triphosphate i.e. ATP (Whiffin 2004).

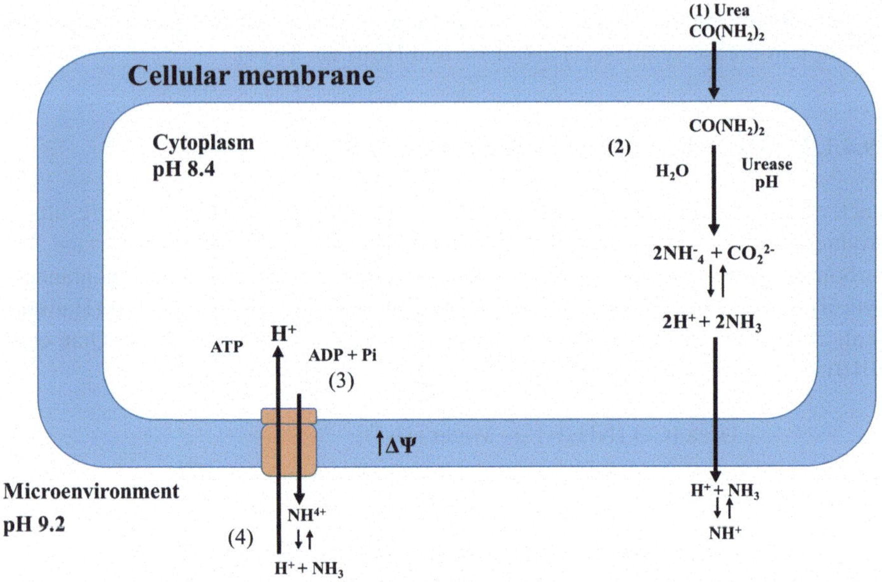

Fig. 29.3 Urease activity with calcium carbonate (adapted and redrawn after: Sandra et al. 2018)

29.4.2 Metabolic Transformation of Organic Compound-Heterotrophic Bacteria

In heterotrophic bacteria aerobic oxidation under alkaline conditions results in increase in pH and carbonate production;

(I) Substrate Lactate

$$\text{Calcium lactate } [CaC_6H_{10}O_6] + \text{Oxygen } [6O_2] \rightarrow \text{Calcium carbonate } [CaCO_3] + \text{Carbon dioxide } [5CO_2] + \text{ Water } [5H_2O] \quad (29.8)$$

$$\text{Carbon dioxide } [5CO_2] + \text{Calcium Hydroxide } [5Ca(OH)_2] \rightarrow \text{ Calcium carbonate } [5CaCO_3] + \text{Water } [5H_2O] \quad (29.9)$$

(II) Substrate Acetate

$$\text{Acetate ion } [CH_3COO^{1-}] + \text{ Oxygen } [2O_2] \rightarrow \text{ Carbon dioxide } [2CO_2] + \text{ Hydroxide } [OH^{1-}] + \text{Water } [H_2O] \quad (29.10)$$

$$\text{Carbon dioxide } [CO_2] + \text{ Hydroxide } [OH^{1-}]$$

$$\rightarrow \text{Bicarbonate } [HCO_3^-] \quad (29.11)$$

$$\text{Bicarbonate } [HCO_3^-] + \text{Hydroxide } [OH^{1-}] \rightarrow \text{Carbonate ion } [CO_3^{2-}] + \text{Water } [H_2O] \quad (29.12)$$

29.4.3 Dissimilatory Nitrate Reduction

Bacterial denitrification leads to increase in pH initiating the MICP without buffering action (Hamdan et al. 2011).

Synthesis of carbonate ion occurs at increased pH and carbon (IV) oxide gas. The reactions are as follow:

$$\text{Organic compound } + \text{Nitrate } [a\ NO_3] + \text{Hydrogen ion } [b\ H^+] \rightarrow \text{Carbon dioxide } [c\ CO_2] + \text{Water } [d\ H_2O] + \text{Nitrogen } [e\ N_2] \quad (29.13)$$

$$\text{Carbon dioxide } [CO_2] + \text{Hydroxide } [2OH^{1-}] \rightarrow \text{Carbonate ion } [CO_3^{2-}] + \text{Water } [H_2O] \quad (29.14)$$

$$\text{Calcium ion } [Ca^{2+}] + \text{Carbonate ion } [CO_3^{2-}] \rightarrow \text{Calcium carbonate } [CaCO_3] \quad (29.15)$$

29.4.4 Dissimilatory Sulphate Reduction

Under the aerobic conditions, unstable sulphides are biologically oxidized to sulphate which are harmful to concrete (Alexendar and Karen 2012). The reactions are as follow:

$$\text{Calcium sulphate } [CaSO_4] + \text{Formaldehyde } [2(CH_2O)] \rightarrow \text{Calcium sulphide } [CaS] + \text{Carbon dioxide } [2CO_2] + \text{Water } [2H_2O] \quad (29.16)$$

$$\text{Calcium sulphide } [CaS] + \text{Water } [2H_2O] \rightarrow \text{Calcium hydroxide } [Ca(OH)_2] + \text{Hydrogen Sulfide } [H_2S] \quad (29.17)$$

$$\text{Carbon dioxide } [2CO_2] + \text{Water } [2H_2O]$$

$$\rightarrow \text{Carbonic acid } [H_2CO_3] \quad (29.18)$$

$$\text{Calcium hydroxide } [Ca(OH)_2] + \text{Carbonic acid } [H_2CO_3] \rightarrow \text{Calcium carbonate } [CaCO_3] + \text{Water } [2H_2O] \quad (29.19)$$

29.4.5 Photosynthesis (Castro-Alanso et al. 2019)

The photosynthetic microalgae and cyanobacteria in aquatic environment lead to $CaCO_3$ precipitation (Table 29.1). The reactions are as follow:

$$\text{Calcium ion } [Ca^2+] + \text{Bicarbonate } [2HCO_3^-] \rightarrow \text{Calcium carbonate } [CaCO_3] + \text{Carbon dioxide } [CO_2] + \text{Water } [H_2O] \quad (29.20)$$

$$\text{Calcium ion } [Ca^2+] + \text{Formate } [HCO_2^-] + \text{Hydroxide } [OH^1-] \rightarrow \text{Calcium carbonate } [CaCO_3] + \text{Water } [H_2O] \quad (29.21)$$

$$\text{Calcium ion } [Ca^{2+}] + \text{Bicarbonate } [2HCO_3^-] \rightarrow \text{Calcium carbonate } [CaCO_3] + \text{Carbon dioxide } [CO_2] + \text{Water } [H_2O] \quad (29.22)$$

The ureolysis pathway for MICP indicating, urea hydrolyzation by the microorganisms in nature (Mobley and Hausinger 1989). This process of hydrolysis is well known for biocement-bio concrete production by bacteria. It is time and cost effective (Dhami et al. 2012).

29.5 Factors Influencing Performance of MICP

29.5.1 Energy Substrates

Nutrients like carbon dioxide, nitrogen, phosphorus, potassium, magnesium, calcium and iron etc. should be provided to microbes during stage of culture and treatment stage so that cell can sustain long enough to support $CaCO_3$ precipitation (Wei-Soon et al. 2012).

Table 29.1 Microorganisms and metabolic pathway for inducing precipitation of $CaCO_3$ (adapted after: Mostafa and Aydin 2019)

Sr. No	Metabolic pathway	Microorganism
1	Sulfate reduction	*Desulfovibrio* sp.
2	Ammonification	*M. xanthus*
3	Denitrification	*C. denitrificans*
		D. nitroreducens
		P. aeruginosa
		D. nitroreducens
		H. halodenitrificans
4	Ureolysis	*B. lentus*
		Bacillus sp.
		B. licheniformis
		B. thuringiensi
		B. aerius U2
		B. cereus
		Halomonas sp.
		K. flava
		L. sphaericus
		P. vulgaris
		S. ginsengisoli
		B.megaterium
		S. pasteurii
		B. sphaericus
5	Organic acid to $CaCO_3$ conversion	*Acinetobacter*
		B. pseudofirmus
		B. cohnii
		B. pumilus
		B. alkalinitrilicus
		B. subtilis
		Micrococcus spp.
		Pseudomonas spp.
6	Photosynthesis	*C. vulgaris*
		S. platensis
		Synechococcus spp.

29.5.2 Urease Positive Bacteria

Aerobic bacteria are most preferable for urea hydrolysis because it releases CO_2 from respiration of the cell, which results in increased pH due to ammonium production.

The Urease positive bacteria of genera *Sporosarcina* spp., *Bacillus* spp., *Sporolactobacillus* spp., *Clostridium* spp., and *Desulfotomaculum* spp. etc. are used. Usually, the common types of *Bacillus* spp. are used as, *B. sphaericus, B. megaterium* and *B. pasteurii.* Among all types of microorganisms used, *Bacillus* species is a widely accepted bacterial species utilized for $CaCO_3$ precipitation. However, the variations in calcite synthesis during MICP are observed in *Bacillus* spp. (Wei-Soon et al. 2012).

29.5.3 Geometric Compatibility of Bacteria

Basically, an appropriate type of bacteria having geometric compatibility is needed for MICP treatment (Wei-Soon et al. 2012). Small pore size is observed to limit free passage of bacteria. This depends on the size of cell and material used (Maier et al. 2009).

29.5.4 Bacterial Cell Concentration

It is reported that higher bacterial concentration increases the synthesis of urease per unit volume (Wei-Soon et al. 2012).

29.5.5 Fixation and Distribution of Bacteria

It is reported that increased salinity in solution encourages flocculation which accelerates the bacterial adsorption as well as withholding capacity in the sand column (Ritvo et al. 2003; Torkzaban et al. 2008). Decreased strength of ions and less strength of adsorption of bacteria in less salinity solution allows transportation over great distances (Harkes et al. 2010). During the fluid fixation, greater flow rate can flush the bacterial cells to comparatively longer distance.

29.5.6 Temperature

Urea hydrolysis is found to occur at higher temperature 30 °C. Increases in temperature do not accelerate decomposition (Nemati et al. 2005). Therefore, it is recommended to exploit bacteria surviving at optimum temperature i.e. approximately 60 °C (Sahrawat 1984).

29.5.7 Reactant Concentration

It is reported that the activity of microbes may be retarded at high salinity thereby, decreases urease synthesis by ureolytic bacteria. Likewise, higher urea and $CaCl_2$ concentration increase the rate of urea decomposition (Rivadeneyra et al. 2004; Nemati et al. 2005; Rivadeneyra et al. 1998).

29.5.8 pH

The optimum pH of urease enzyme is 7.5–8.0 (Stocks-Fischer et al. 1999). It is reported that the activity has decreased above pH 8.0. In the MICP process, pH increases gradually due to release of ammonia, whereas, carbon dioxide released acts as a buffer to control the pH (Wei-Soon et al. 2012).

29.6 Limitations for MICP Derived Biocement-Bioconcrete

Commercial Applications of Biocement-Bio Concrete has few limitations due to following reasons as:

- Product and performance of MICP has found to vary relating to geographic and environmental condition and needs adaptation.
- Application of MICP derived Biocement-Bioconcrete at commercial scale is influenced by different environmental factors.
- Exploitation of potent microbial strain under controlled condition may serve industries to overcome the problem of various hazards and pollution.
- Till date, the designs of any bioconcrete are not included in is or any other codes.
- The field conditions and lab situation do not go hand in hand to monitor the production process.
- Since the construction engineers are less familiar with the microbiological processes, its acceptance is found to affect its commercialization.
- Standard protocols are not properly developed regarding the testing and acceptance criteria.

29.7 Potential Applications of Biocement/Bio Concrete

- The MICP can be utilized to increase the extraction efficiency of the crude oil and used to decrease its leakage as well the top soil contamination which is enriched with various microorganisms. Besides, it can be used as a promising solution for strengthening the soil and consolidation of sand (Mostafa and Aydin 2019).

- The leakages from reservoirs or ponds cause loss of fluid resulting in leakage into the underneath soil or sand. However, this problem can be solved by decreasing the rate of the seepage rate and the permeation of the reservoir via MICP technology (Mostafa and Aydin 2019).
- It can be helpful in removal of calcium, sequestration of radionuclides, heavy metals and Carbon dioxide. It can also be used for removal of inorganic contaminants (Sandra et al. 2018; Zaghloul et al. 2020; Periasamy et al. 2016).
- In the material engineering, ecofriendly technologies are needed to produce materials–composite showing less resource exhaustion and energy consumption. Thereby MICP can be used as an alternative (Periasamy et al. 2016).
- It can be efficiently used for prevention of concrete corrosion and cracks (Zaghloul et al. 2020).
- MICP based biocement shows noteworthy improvement w.r.t. the robustness of constructions and can decrease the amount of cement used in it without compromising its quality in a cost-effective manner (Achal et al. 2014; Sandra et al. 2018).
- MICP process could help to understand numerous natural biological phenomena and applications in geo-microbiological processes viz.: treatment of Ca-rich wastewater, leaching minimization from channels and reservoirs, purification of groundwater, repairing of constructions and in situ improvement of the mechanical properties of soil (Jian et al. 2012).
- MICP derived Biocement-Bioconcrete shows remarkable eco-friendly ability. Also, the Biocementation can achieve better mechanical strengths in comparison to conventional cement (Veerappan and Chandru 2016).
- Moreover, it can be used for manufacturing mould and other biocemented products. It can be applicable in the earth stabilisation for mining, tunneling, and earthquake repair, repairing or deteriorating "masonry and built structures", "instant pavement", "filters, ornamental and structural blocks" etc. (Veerappan and Chandru 2016).
- It increases the compressive strength of concrete and mortar up around 38%. It could make the process economical, by using various industrial by products. Utilization of microbial additives can also increase the long-life of the bio concrete structures. In this way, it can be more efficiently used as compared to an ordinary cement-concrete (Laxmana et al. 2015)
- Various cultural heritage objects are impaireddue to micro cracks causing expenses. In this case, the $CaCO_3$ precipitation derived from bacteria can be proved as the best substitute for micro–crack repairing technique in an environmentally friendly manner (Annamalai et al. 2013).

29.8 Future Perspectives

- The use of urease and carbonic anhydrase enzymes in MICP processes needs to be metabolically controlled.

- The research and development study to formulate hardening of concrete is the need of the hour.
- Extensive research is needed for bond coherence to avoid crack formation.
- It is necessary to genetically engineered bacterial strains which would have long sustainability during ureolysis.
- Successful commercialization of MICP biconcrete process could be designed with alteration in the energy substrates and physiological conditions.
- If the reliable and reproducible research related to MICP derived biocement-bioconcrete production process will successfully achieved and documented, this ecofriendly technology can be efficiently used for large scale commercialization.

Therefore, in the upcoming era, the bioconcrete can come up as a promising and self-healing eco-friendly material of choice for green construction.

29.9 Conclusion

Worldwide, production of cement for concrete makes it 3rd ranking producer of anthropogenic CO_2 followed by transport and energy generation. Bioconcrete is the self-healing form of concrete in which an ecofriendly key ingredient i.e. Biocement is added. This healing agent can stay dormant for many years and can become active only in the presence of water. It is expected that for large production of ecofriendly bioconcrete, development of new competitive bacterial strains is needed. This is because the synthetic cement has several unwanted impacts on Environment as well as human health. In view of this, microbiological and molecular research may overcome the hazardous effect by production of cost-effective Bioconcrete. Therefore, the extensive research related to MICP derived products is the need of the upcoming era. The Bioconcrete could be a promising and self-healing ecofriendly material of choice for green construction.

Acknowledgements The authors are grateful to the Research and Development Department, Sigma Wineries Pvt. Ltd. Nashik. Special thanks to Mr. V.C. Handore, Mrs. Hira V. Handore, and Ms. V. Jagtap for valuable technical assistance.

References

Achal V, Li M, Zhang Q (2014) Biocement, recent research in construction engineering: status of China against the rest of the world. Adv Cem Res 26(5):281–291. https://doi.org/10.1680/adcr.13.00044

Achal V, Mukherjee A, Kumari D, Zhang Q (2015) Biomineralization for sustainable construction—a review of processes and applications. J Mater Civil Eng Earth Sci Rev 23(6):730–734. https://doi.org/10.1016/j.earscirev.2015.05.008

Alexander Q, Karen LS (2012) Hydration of C3A-gypsum systems. Laboratory Construction Materials Station 12, CH-1015, Lausanne, pp 52–56
Annamalai SK, Arunachalam KD, Sathyanarayanan S, Prasanth Kumar VR (2013) Asian J Chem 25:143–146
Benhelal E, Zahedi G, Shamsaei E, Bahadori A (2013) Global strategies and potentials to curb CO_2 emissions in the cement industry. J Clean Prod 51:142–161
Cailleux E, Pollet V (2009) Investigations on the development of self-healing properties in protective coatings for concrete and repair mortars. Paper presented at the 2nd international conference on Self-Healing Materials, Chicago, IL, USA
Castro-Alonso MJ, Montañez-Hernandez LE, Sanchez-Muñoz MA, Macias Franco MR, Narayanasamy R, Balagurusamy N (2019) Microbially induced calcium carbonate precipitation (MICP) and its potential in bioconcrete: microbiological and molecular concepts. Front Mater 6:126. https://doi.org/10.3389/fmats.2019.00126
Chen ZS, Martinez L, Chang JP, Wang XJ, Xiong SHI, Chin KS (2019) Sustainable building material selection-A QFD- and ELECTRE III-embedded hybrid MCDM approach with consensus building. Eng Appl Artif Intell 85:783–807
Dhami NK, Redy MS, Mukherjee A (2012) Improvement in strength properties of ash bricks by bacterial calcite. Ecol Eng 39:31–35
Douglas S, Beveridge TJ (1998) Mineral formation by bacteria in natural microbial communities. FEMS Microbiol Ecol 26(2):79–88. https://doi.org/10.1016/S0168-6496(98)00027-0
Falkinham JO, Hoffman PS (1984) Unique developmental characteristics of the swarm and short cells of *Proteus vulgaris* and *Proteus mirabilis*. J Bacteriol 158:1037–1040
Ferris FG, Stehmeier LG, Kantzas A, Mourits FM (1996) Bacteriogenic mineral plugging. J Can Petrol Technol 35:56–61
HäkkinenT BK (2011) Barriers and drivers for sustainable building. Build Res Inform 39:239–255
Hamdan N, Kavazanjian E, Rittman BE (2011) Carbonate mineral precipitation for soil improvement through microbial denitrification. Adv Geochem Eng Dallas American Society of Civil Engineers 21–23
Hammes F, Verstraete W (2002) Key roles of pH and calcium metabolism in microbial carbonate precipitation. Rev Environ Sci Biotechnol 1(1):3–7
Hammes F, Boon N, Clement G, Villiers JD, Siciliano SD, Verstraete W (2003) Molecular, biochemical and ecological characterisation of a bio-catalytic calcification reactor. Appl Microbiol Biotechnol 62:191–201
Harkes MP, Van Paassen LA, Booster JL (2010) Fixation and distribution of bacterial activity in sand to induce carbonate precipitation for ground reinforcement. Ecol Eng 36:112–117
Ivanov, Chu J, Stadnikov V (2015) Basics of construction microbial biotechnology in biotechnologies and biomimetics for civil engineering. Springer, Berlin, Germany, pp 21–56
Jian C, Viktor S, Volodymyr I (2012) Microbially induced calcium carbonate precipitation on surface or in the bulk of soil. Geomicrobiol J 29(6):544–549
Jonkers HM (2011) Bacteria-based self-healing concrete. Heron 56(1–2):5–16
Krajewska, (2009) Ureases functional catalytic and kinetic properties. J Mol Catal B-Enzym. https://doi.org/10.1016/j.molcatb.2009.01.003
Kubba S (2012) Handbook of green building design and construction: LEED, BREEAM, and Green Globes. Butterworth-Heinemann Ed, Oxford
Laxmana RR, Manjusha A, Arunkumar M (2015) Bio cement—an ecofriendly construction material. Int J Curr Eng Technol 5(4):2880–2883
Li YH, Chen YYM, Burne RA (2000) Regulation of urease gene expression by *Streptococcus salivarius* growth in biofilms. Environ Microbiol 2:169–177
Maier RM, Pepper IL, Gerba CP (2009) Environmental Microbiology, 2nd edn. Elsevier Science, China, p 366
Malhotra VM (1999) Making concrete greener with fly ash. Concr Int 21(5):61–66
Mehta PK (1999) Concrete technology for sustainable development. Concr Int 21(11):47–53

Mobley H, Hausinger (1989) Microbial ureases significance regulation and molecular characterisation. Microbiol Rev 53(1):85–108
Mobley HL, Chippendale GR, Swihart KG, Welch RA (1991) Cytotoxicity of the Hpm Ahemolysin and disease of *Proteus mirabilis* and *Proteus vulgaris* against cultured human renal proximal tubular epithelial cells. Infect Immun 59(6):2036–2042
Mortensen BM, Haber MJ, De Jong JT, Caslake LF, Nelson DC (2011) Effects of environmental factors on microbial induced calcium carbonate precipitation. J Appl Microbiol 111(2):338–349
Mostafa S, Aydin B (2019) Microbially induced calcium carbonate precipitation a widespread phenomenon in the biological world. Appl Microbiol Biotechnol 103:7
Mostafa S, Ali KS, Aydin B (2016) Bioconcrete next generation of self-healing concrete. Appl Microbiol Biotechnol 100:2591–2602
Nemati M, Greene EA, Voordouw G (2005) Permeability profile modification using bacterially formed calcium carbonate comparison with enzymic option. Process Biochem 40:925–933
Periasamy CHK, Yu JS, Jane S (2016) Formations of calcium carbonate minerals by bacteria and its multiple applications. Springerplus 5:250
Qian G, Wang R, Cheng L, Wang J (2010) Theory of microbial carbonate precipitation and its application in restoration of cement-based materials defects. Chinese J Chem 28(5):847–857. https://doi.org/10.1002/cjoc.201090156
Riding R (2000) Microbial carbonates the geological record of calcified bacterial algal mats and biofilms. Sedimentology 47:179–214
Ritvo G, Dassa O, Kochba M (2003) Salinity and pH effect on the colloidal properties of suspended particles in super intensive aquaculture systems. Aquaculture 218:379–386
Rivadeneyra MA, Delgado G, Ramos CA, Delgado R (1998) Biomineralization of carbonates by halomonas euryhaline in solid and liquid media with different salinities crystal formation sequence. Res Microbiol 149:277–287
Rivadeneyra MA, Párraga J, Delgado R, Ramos CA, Delgado G (2004) Biomineralization of carbonates by *Halobacillus trueperi* in solid and liquid media with different salinities. FEMS Microbiol Ecol 48:39–46
Rodriguez NC, Ground SM, Ruiz AE, Gonzalez MT (2012) Influence of substrate mineralogy on bacterial mineralization of calcium carbonate implications for stone conservation. Appl Environ Microb 78(11):4017–4029. https://doi.org/10.1128/AEM.07044-11
Rusznyak A, Akob DM, Nietzsche S, Eusterhues K, Totsche KU, Neu TR (2012) Calcite biomineralization by bacterial isolates from the recently discovered pristine karstic herrenberg cave. Appl Environ Microbiol 78:1157–1167
Sahrawat K (1984) Effects of temperature and moisture on urease activity in semi arid tropical soils. Plant Soil 78:401–408
Sandra PCA, Becerra Jiménez ML, Martínez-Zambrano JJ, Rojas Sarmiento HA (2018) Soil bacteria that precipitate calcium carbonate mechanism and applications of the process. Acta Agron 67(2):277–288
Silva FB (2015) Industrial applications of biological self-healing concrete challenges and economical feasibility. J Commer Biotechnol 21(1):31–38
Stanaszek-Tomal E (2020) Bacterial concrete as a sustainable building material. Sustainability 12(2):696
Stocks-Fischer FS, Galinat JK, Bang SS (1999) Microbiological precipitation of $CaCO_3$. Soil Biol Biochem 31:1563–1571
Torkzaban S, Tazehkand SS, Walker SL, Bradford SA (2008) Transport and fate of bacteria in porous media coupled effects of chemical conditions and pore space geometry. Water Resource Res 44:W04403
Veerappan K, Chandru PA (2016) Brief review on biocement based concrete. Int J Eng Manag Res 6(5):142–144
Volodymyr VS (2019) Sustainable and safe construction biomaterials biocements and biogrouts. Front Biomater 6:177–193

Weiner S, Dove PM (2003) An overview of biomineralization processes and the problem of the vital effect. Rev Mineral Geochem 54:1–29. https://doi.org/10.2113/0540001
Wei-Soon NG, Min-Lee L, Siew-Ling H (2012) An overview of the factors affecting microbial-induced calcite precipitation and its potential application in soil improvement. Int Sch Sci Res Innov 6(2):188–194
Whiffin VS (2004) Microbial $CaCO_3$ precipitation for the production of biocement. Doctoral dissertation, Murdoch University
Zaghloul EH, Ibrahim IYAH, ElBadan DES (2020) Production of biocement with marine bacteria *Staphylococcus epidermidis* EDH to enhance clay water retention capacity. Egypt J Aquat Res. https://doi.org/10.1016/j.ejar.2020.08.005
Zhu T, Dittrich M (2016) Carbonate precipitation through microbial activities in natural environment, and their potential in biotechnology. Front Bioeng Biotechnol 4:4

Chapter 30
Environmental Life Cycle Analysis of Residential Building Materials: A Case Study

Md. Al Sadikul Islam, Md. Ashiquzzaman, Amiu Shadik Utshab, and Nehreen Majed

Abstract Many challenges exist in the construction sector including lack of sufficient regulation, development and deposition of debris, consequent degradation of the surrounding environment in the post-construction phase etc. The residential buildings play a leading role in urban development and subsequent environmental impact. Thus, life cycle assessment (LCA) of residential buildings becomes demanding for paving the way towards sustainable growth through impact characterization. This motivates this study to apply the LCA approach based on the ISO 14,040 Standard for residential building and assess the impact with respect to human carcinogenic, non-carcinogenic, global warming and energy demand etc. categories. For this purpose, the IMPACT 2002+ method utilizing SimaPro software was applied to evaluate the impact from the processes and materials used for the construction of a seven-storied concrete structure building in Dhaka, Bangladesh. The results indicated that among the construction materials and operations, categories such as concrete, steel, brick and electricity consumption exert significant impacts on the environment. Besides comparative impact characterization, this study also illustrates the need for steps to be adopted in the supply chains of the steel, cement and ceramic materials. The use of ceramic materials and the impacts incurred also need to be investigated further due to its frequency in usage.

Keywords Building materials · Environmental impact · Global warming · Life cycle · Residential building

The original version of this chapter was revised: First author name citation has been updated in Springer link. The correction to this chapter is available at https://doi.org/10.1007/978-3-030-76073-1_31

Md. A. S. Islam · Md. Ashiquzzaman · A. S. Utshab · N. Majed (✉)
Department of Civil Engineering, University of Asia Pacific, 74/A, Green Road, Dhaka 1205, Bangladesh
e-mail: nehreen-ce@uap-bd.edu

J. A. Malik and S. Marathe (eds.), *Ecological and Health Effects of Building Materials*,
https://doi.org/10.1007/978-3-030-76073-1_30

30.1 Introduction

At present-day, our planet is facing numerous environmental crises such as global warming, loss of ozone layer, pollution build-up, etc. In recent decades, analysis has demonstrated that the global atmosphere is changing fast (Change 2001) and also suggests that the transition is going on in time (Hulme et al. 2002). Those undesirable problems resulting from our current lifestyle desperately need to be mitigated in order to protect our climate and our planet.

The construction sector plays a significant role in the world's overall utilization of energy. The building industry has a big impact on the overall consumption of natural resources and emitted emissions. A building uses energy throughout its life cycle, from the construction phase until demolition. Energy consumption in buildings is both direct and indirect in their life cycle. Energy is directly utilized for construction work, retrofitting and demolition; the renovation, restoration and technological installations are rendered using indirect energy (Sartori and Hestnes 2007). Major development efforts are taking place world-wide in order to accommodate the influx by 2030 of the world's population into urban areas (Syal et al. 2006). Such a blast in development is an impressive factor to save our assets from exhaustion. Buildings can be classified into residential and non-residential buildings, based on their purpose of construction. Residential houses can also be classified into single-family houses and multifamily houses and those that are used for commercial uses, such as schools, hospitals, offices etc. are non-residential buildings.

Residences reflect a vital development market, substantial resource demand, large expenditure by individuals and families as well as constitute a major cause of pollution. Moreover, residences have the environmental impacts from construction which a lot of times outweigh the economic benefits. The rapid extent in populace and economic growth in Bangladesh have prompted the development of elevated structures in metropolitan zones particularly in Dhaka Metropolitan region. The application of construction methodologies in Bangladesh still follows the traditional system, nonetheless, the conception of the government of Bangladesh about contractors have imposed a manual material Scoring System framework in an industrialized construction system. The Life Cycle Analysis (LCA) brings a performance evaluation approach which essentially allows for an overview of possible impacts of a building over the whole lifecycle, quantifying material contributions and construction processes and their environmental impacts (de Larriva et al. 2014). LCA is a complicated method for application due to building subsystems, the particular features of each project and the impact of user behavior (Bribian et al. 2009; Buyle et al. 2013). Despite the significance of the usage and operating process, the relevance of the effect of the building materials cannot be ignored.

As buildings utilize a high volume of energy during the time of use and occupancy, those phases are considered to be the most impactful phases in its overall life cycle (Asdrubali et al. 2013; Guggemos and Horvath, 2005; Ortiz et al. 2009; Wu et al. 2012; Xing et al. 2008; Zhang et al. 2013). Few types of research have evaluated the contributions of the construction and material processes that define the critical structures and critical materials while practicable solutions can be identified to reduce the effect of these systems and materials (Asdrubali et al. 2013; Heravi et al. 2016; Mosteiro-Romero et al. 2014). Many studies detailing LCA on building construction prioritize assessing CO_2 releases, responsible for global warming opportunities and energy demands (Asdrubali et al. 2013; de Larriva et al. 2014; Motuzienė et al. 2016; Oyarzo and Peuportier 2014). The carbon pollution build-up was predicted to rise from 34 million tons of CO_2 to 58 million tons of CO_2 by 2020 (Wen et al. 2015).

Despite significant advancements with LCA in the civil engineering field, studies conducted in building construction often display variable outcomes, which require greater clarity in methodology and data collection in order to increase the reliability and repetitiveness of such studies (Evangelista et al. 2018). Lack of information on various regional typologies of systematic structure, ordinary basis, construction boundary, transport, water utilization and energy consumption and practical application of LCA pose barriers to its region-specific application (Soust-Verdaguer et al. 2016) such as in Bangladesh.

In general, previous LCAs focused on areas like materials for construction work, energy, water and materials used for operation and maintenance; waste management in construction and demolition etc. (Buchanan and Levine 1999; Matsuno and Betz 2000; Nebel et al. 2006; Reddy and Jagadish 2003; Rivela et al. 2006); LCA analyses with very limited application and scope were carried out for complete building structures (Hondo 2006; Mithraratne and Vale 2004; Thormark 2002; Treloar et al. 2000). A very few environmental life cycle assessments of buildings have been undertaken in Bangladesh with no residential building in specific. This motivated this study to accomplish Environmental Life Cycle Analysis of Residential Building Materials, being a new inquest to draw a representation of environmental impacts of buildings by realizing the most significant contribution of building processes and materials during the project phase of typical residences in Bangladesh.

30.2 Life Cycle Assessment (LCA) and Application

LCA is a method that quantifies and measures the material and energy distribution of a particular application or a product. The first inventory is generally built on the use of the upstream (extraction, processing, transportation and construction) and downstream (deconstruction and disposal) goods or services (Ramesh et al. 2010). The investigation of LCA is based on four phases: goal and scope definition, inventory analysis; impact assessment and interpretation. During a comparative analysis, before the findings are analyzed, equivalence on the structures being compared is measured. When carrying out a comparative analysis, before analyzing the findings based on methodology ISO 14040 (Evangelista et al. 2018) of the structures, the equivalence is calculated.

A building's normal life cycle can be split up into three stages, with every stage comprising multiple life cycle phases—assembly, operation and dismantling/demolition (Optis 2008). The assembly process concerns the acquisition of raw materials by processing or reusing inventory, the manufacturing of these products, the construction of products in-house, the substitution of assemblies and goods in buildings and intermediary transport. Asdrubali et al. (2013) mentioned that as a result of new legislation that has been implemented to minimize building use in operating phases, the value of incarnated energy has increased. The operation process applies to the specifications for heating and fuel, water facilities and other services including all other facilities except substitution of materials. The process of dismantling concerns the processes such the dismantling, decommissioning, re-use, disposal/recycling, transportation measures in construction materials and assemblies. There are multiple unit processes for each of the phases of a life cycle.

In all the phases of the buildings, energy is absorbed directly or indirectly, from the cradle to the grave, and every stage of energy intake is interrelated (incarnated and active energy). A review of literature on such cases found that the energy usage of the buildings in life cycles depend on energy consumption (80–90%) and on construction energy (10–20%) (Ramesh et al. 2010). Construction materials (Carbon energy) in general do not constitute more than 10% of the operating energy (Heater, refrigeration and light included) utilized over 50 years of life-span (Wen et al. 2015).

Buildings must also be analyzed from the point of view of the life cycle. In the present study, the authors primarily focused on the assembly process/materials used for a Residential Building in Dhaka, Bangladesh with special focus on environmental impacts. Chang (2012) claimed that energy used by building machineries such as excavators, bulldozers, loaders, and cranes requires diesel, oil, and electricity. The same study also found out that the energy found in this equipment is mostly consumed by buildings (90 per cent) while only 4 and 6 per cent are consumed by transport and other means respectively. Furthermore, Nässén et al. (2007) has performed case studies leading to an energy input of 91% in material processing, 3% in transport and 6% in the building. The findings, however, may vary attributed to the environment, location and construction process.

30.3 Previous Comparison Studies for LCA of Residential Buildings

Adalberth et al. (2001) carried out LCA on four multi-family buildings constructed in Sweden in 1996. The working unit was considered as the floor area (m^2) for use. The primary objective of this study was to review multiple life cycle processes of all four buildings and to figure out which process has the highest environmental effect and there were variations in the environmental impact associated with the types of buildings and framing choices. An LCA framework developed at the Danish Building Science Institute measured the environmental impact (Peterson 1997). The effects of GWP (Global Warming Potential), AP (Acidification Potential), EP (Eutrophication Potential), and human toxicity were explored in this report. The various phases related to the construction of the buildings under consideration were: development, transportation, erections, construction, reconstruction and demolition while those installation options during construction were identified that have a high environmental impact.

The environmental impacts of steel and concrete structured constructions with LCA were contrasted by Guggemos and Horvath (Guggemos and Horvath 2005). There were two five-story buildings in the mid-West of the U.S. with floor space of 4400 m^2. In this analysis, two approaches were used to determine the environmental period of life of each building across multiple phases: development, design, usage, repair and demolition. The LCA and *EIO*-LCA process was focused on processes. The findings revealed that, due to longer construction processes, the concrete built structure was correlated with more energy consumption and pollution.

Norman et al. (2006) compared high and low-populated dwellings for electricity usage and GHG pollution. It demonstrates that the choice of the functional unit is highly essential for a complete comprehension of the urban density results. The analysis indicated through *EIO*-LCA study that brick, glass, drywall and structural concrete in the buildings were the most important construction materials for generating energy and *GEGs* for both density cases. These four materials together accounted for 60–70 per cent of the global effect of low- and high-density case studies on energy and development.

30.4 Materials and Methodology

The LCA application for a residential building in Bangladesh was performed using the flow of activities as shown in Fig. 30.1. The configuration is based on the methodology ISO 14040 (Evangelista et al. 2018) and the Inverted Cone method (Souza 2014) as shown in Fig. 30.2.

In order to analyze the most important data to recognize the environmental effects of the object of research, the authors adopted the Inverted Cone methodology, which is a technique that guides data surveying and quality management in LCA studies, especially in the inventory. This method advises, in the LCA, to analyze the influence of inventory data on the selection, distribution, and quality management activities to minimize the uncertainties and maximize the efforts against their effect on the results of the report (Fig. 30.2).

30.4.1 Description of the Building Features

The LCA was carried out for seven-story typical residential buildings in Dhaka, Bangladesh. The basic parameters of present residential buildings are displayed in Table 30.1.

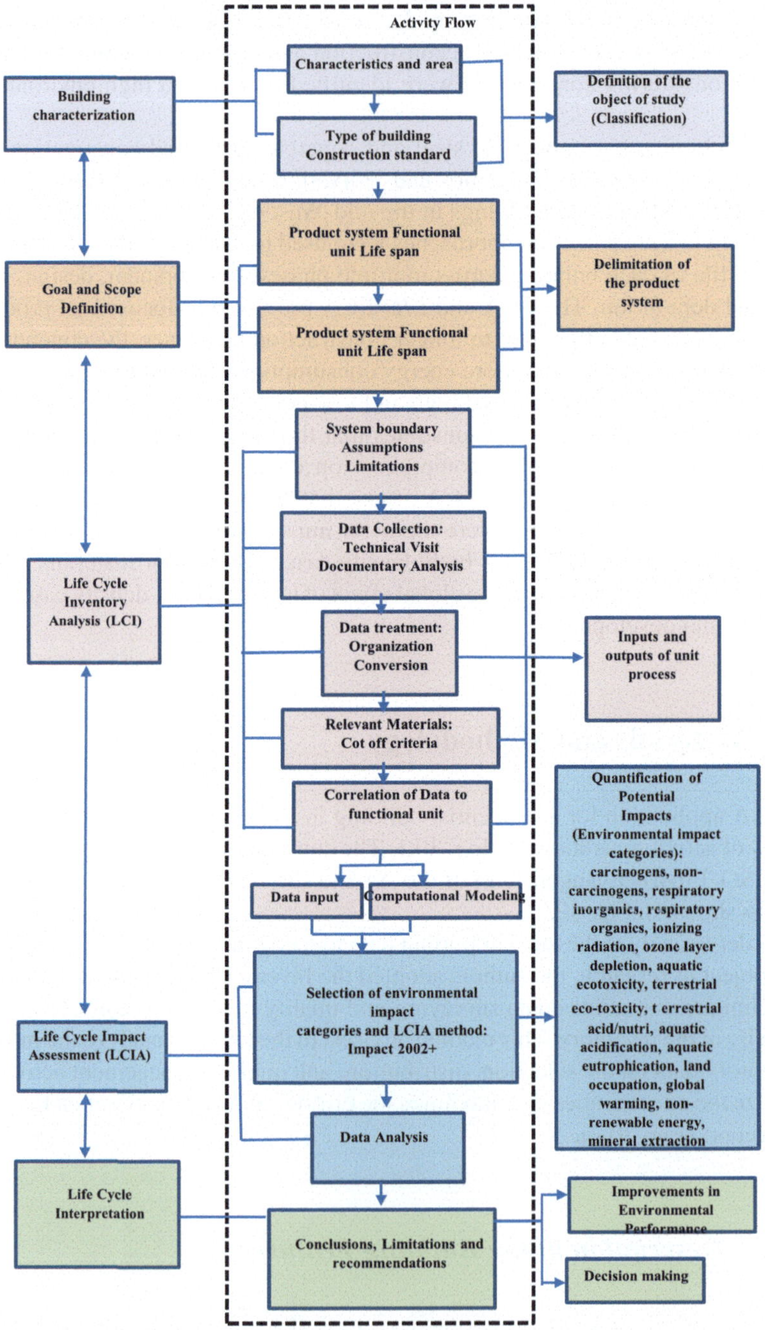

Fig. 30.1 Flowchart of environmental life cycle analysis of residential building of Bangladesh

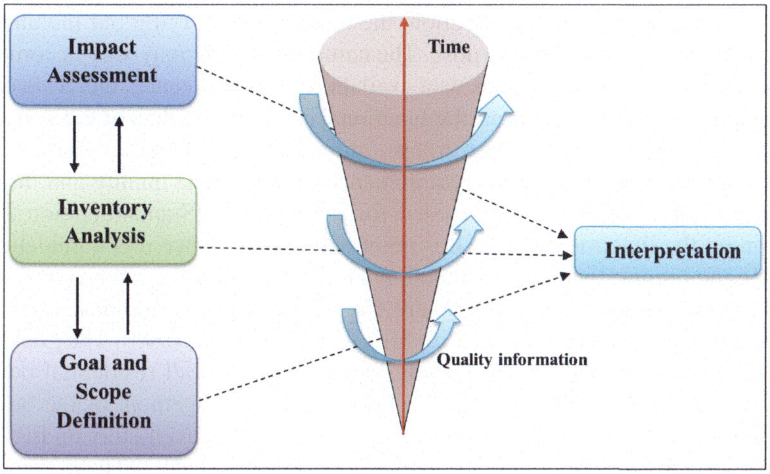

Fig. 30.2 Inverted cone methodology (Adopted after: Souza 2014)

Table 30.1 Basic parameters of the residential building

Building parameters	Specifications
No of floors	Seven
No of Apartment's	13
Ceiling height	2.9 m
Structure	Concrete
Foundation/ basement	Concrete foundation and walls
Walls (interior)	Brick and curtain wall combination (partially glass curtain wall, brick facing)
Flooring finish	Ceramic tiles
Roof	Flat roof, Concrete

30.5 Application of LCA in Case Studies: Goal and Scope

This is the first step to evaluating the life cycle and is a crucial step to establish the purpose of the research and the ways it can impact the LCA results (Selmes 2005). The research shapes the goals and limitations of the study in this process and also makes several essential assumptions: defining device boundaries primarily, with practical units e.g., m^2 floor area; quality of data; and other bounds. The objective description and in-depth exercise eventually determine the course and benchmarks for the analysis which will be subsequently assessed in the interpretation process. There may be certain instances of limited concern within the life cycle of some products, but those may not be included in this stage even by definition of elements of the life cycle as a whole (Selmes 2005). According to ISO 14,040, each LCA aims to define the proposed application, the reasons for carrying out the analysis, the intended audience and the purpose. The nature of a study may vary according to many aspects of the study such as data availability, insignificance, etc. This covers the evaluation system, its features, the functional category, borders of the structures, assigned methods, selected impact groups, impact assessment methodologies, data criteria, expectations, constraints, requirements on initial data quality and the form of critical evaluation and report needed for the analysis (Standardization 2006). Determining the functional unit at this point is important since it is a guideline for all inputs and outputs, and several functional units for buildings can be considered.

The general purpose of sustaining an LCA on the level of buildings is to reduce environmental pressures over the entire life cycle (Fava et al. 1993). The goal of the study was to evaluate the environmental impact of the overall residential building construction process, with a special emphasis on the management of the building materials used. The building under consideration consists of seven-story multi-family building characteristics in Dhaka city, (capital of Bangladesh) erected from traditional materials and building methods in Bangladesh. Table 30.1 displays the major characteristics along with the functional unit that correlates to the sum of net building accessible areas.

Figure 30.1 shows the structure borders, in terms of foremost part and context in a schematic as proposed by Clift et al. (2000), including all the steps that were followed by authors in the present study. The environmental burdens identified in the life cycle inventory (direct, indirect, preventable) were collected by processing the data based on scientific papers along with technical visits and discussions with smaller and intermediate-sized ventures within Dhaka. The remaining pressure inputs which are primarily in indirect forms resulted from the *Ecoinvent 3.6 database*. In the implementation of the methodology and enhancing the system, materials that were replaced by the co-products obtained in the markets were specified and replaced by models that were responsible for relevant allocations in LCA models (Finnveden et al. 2009). The effects of recycled materials which in most cases are ignored were calculated by the approach suggested by Ren et al. (2015) and meanwhile obtained in other LCA studies on wastage regulation methods (Arena et al. 2015; Sévigné-Itoiz et al. 2014). In particular, the overall combination of virgin and recycled products used in the market (60% main steel and 40% secondary steel), as stated in an official survey of the World Steel Association were utilized to replace recycled steel (Vitale et al. 2017).

The LCA was performed with *SimaPro 9.1.1* in accordance with *Impact 2002+* methodology, v2.15 (Jolliet et al. 2003; Vitale et al. 2017) to calculate the effects of the life cycles as have been documented in other studies already applied to the similar sector (Blengini and Garbarino 2010). The LCI findings were related with 15 groups of midpoint categories for impacts such as carcinogens, land occupation, respiratory organics, aquatic eutrophication, non-carcinogens, ionizing radiation, global warming, aquatic ecotoxicity, terrestrial ecotoxicity, respiratory inorganics, ozone layer depletion, aquatic acidification, mineral extraction, non-renewable energy and terrestrial acidification/nitrification. These impact factors were standardized from Jolliet et al. (2003) by dividing the influence per pollution unit by the average impact per year per person of all substances within the particular category for which the characterization factor existed.

30.6 Life Cycle Inventory (LCI)

This phase of the study accounts for the full cradle to the grave life cycle of residential buildings including construction materials, water and electricity use during the construction process. Architectural materials and assemblies included primary materials and all additional materials required for the manufacturing of the product. Transportation of products from the manufacturers to the facilities and the delivery of waste to the final destination were also accounted for. Data from the quantity and the bill of material (Iyer-Raniga and Wong 2012; Monahan and Powell 2011) or approximate quantities from the sketches of the building and field calculated data (Blengini and Di Carlo 2010) were consulted to make estimations on the construction materials. The flow of content and the unit processes selected by the *Ecoinvent v3.6* database were specified during the pre-operational inventory. Water and electricity related data were derived from the bills provided by manufacturers and supplied by the construction firms during the building phase. The normal statistical loss was added to solid waste in order to measure the total mass extracted from building sites by means of material loss in comparison with the overall price of the construction materials that were inventoried. The fact that this parameter has to be measured is attributed to the uncertainty in material failure levels and the lack of monitoring and availability at building sites in Bangladesh. Table 30.2 shows the overall inventory of materials for the residential building.

The distances for the transportation-related cause were consolidated. If the values were not readily available from the companies, average distances were considered for shipping building components to the location and transport of huge solid waste to the ultimate goal. The objective was to consider as twice the distance when the vehicle arrived empty at the building site and returned to the building site. At this point, the consumption of water and electricity for machinery and light and the material for building construction were measured. In case of the building, the transport of such material and the produced wastage were considered in the loss in solid waste created by part of replacement during the upkeep procedure. The estimated energy, water use and the number of inhabitants were also recorded. Different researchers used different approaches to assess transportation-related consumptions. In previous studies, authors have used average distances between factories and building sites based on information from the manufacturers and the contractors (Blengini and Di Carlo 2010) while national averages were consulted in certain instances. Alternatively, Ortiz-Rodríguez et al. (2010) used assumptions to define the distance from the fabricator to the location.

The operating process had a comparatively low share of overall CO_2 emissions owing to the highest environmental impacts (Monahan and Powell 2011; Rossi et al. 2012). In certain installations, approximate amounts were measured based upon

Table 30.2 Inventory of materials for the residential building

Construction material	Unit	Quantity
Reinforcing steel	kg	89,830
Red brick	kg	229,586
Concrete		
Cement	m^3	95
Stone chips	m^3	100
Red brick	m^3	55
Sand	m^3	300
Ceramic tail	kg	18,883
Wood	kg	7105
Glass	kg	2990
Stainless steel	N (DN 370)	502
Water	liter	3,749,267
Transportation (Truck)		
3.5–7.5 ton	t.km	570
16–32 ton	t.km	27,669
Fuel Consumption		
Excavator	liter	670
Backhoe loaders	liter	110
Stone crushers	liter	1470
Pile boring machine	liter	7300
Brick crushing machine	liter	480
Shallow pump	liter	270
Portable welding machine	liter	150
Concrete mixer	liter	128
Jack hammer	liter	60
Needle vibrator	liter	100
Electricity	kWh	75,000

drawings or quantity bills in an inventory. Some studies suggest that about five per cent of the materials are lost on-site during construction due to product vulnerability, maltreatment and incorrect installation of products (Buchan et al. 2012; Rossi et al. 2012). Blengini and Di Carlo (2010) gathered building data through correspondence from the builder, contractors and literature on field measurements.

30.7 Results and Discussion [Life Cycle Impact Assessment (LCIA)]

All the Characterization factors and Damage units of *IMPACT 2002+* are shown in Table 30.3. This was accomplished by utilizing the combination of inventory data with the chosen category. The consequent estimation of indicators for these categories measured the contribution of construction materials on the environmental impacts from the building. This research demonstrated the use of the inventory materials and also defined the key materials corresponding to the critical phase of the construction.

The first commodity (Reinforcing Steel) exerted the greatest contribution in the categories of Global warming, ozone layer depletion, terrestrial ecotoxicity and aquatic acidification. In the category of concrete, ceramic compounds had a similar impact in the field of non-carcinogens and terrestrial acidity was attributed to the heavy resource depletion from clay as raw material extraction in output. There have

Table 30.3 Sources for characterization factors and damage units of IMPACT 2002+ (Adopted after: Ali et al. 2015)

Midpoint impact category	Substance	Damage		
		Category	Unit	Normalized unit
Carcinogens[a]	kg C_2H_3Cl eq	Human health	DALY	Point
Non-carcinogens[a]	kg C_2H_3Cl eq			
Respiratory inorganics[b]	kg PM2.5 eq			
Ionizing radiation[b]	Bq C-14 eq			
Ozone layer depletion[b]	kg CFC-11 eq			
Respiratory organics[b]	kg C_2H_4 eq			
Terrestrial acid/nutri[b]	kg SO_2 eq	Ecosystem quality	PDF.m^2.y	
Land occupation[b]	m^2org.arable			
Aquatic acidification[c]	kg SO_2 eq			
Aquatic eutrophication[c]	kg PO_4 P-lim			
Global warming IPCC	kg CO_2 eq	Climate change	kg CO_2 eq	
Non-renewable energy[d]	MJ or kg Crude oil Eq. (860 kg/m^3)	Resources	MJ primary	
Mineral extraction[b]	MJ or kg iron eq			

DALY- disability adjusted life years; eq- equivalents; y- year; PDF- potentially disappeared fraction of species; a- IMPACT 2002, b- Eco-indicator 99, c- CML 2002, d- Ecoinvent, IPCC- IPCC (2014)

also been important impacts in the present analysis of brick and ceramic products (tile, toilet and ceramic hygiene).

Among all impact categories, particularly reinforcement steel and the ceramic tiles have the greatest contribution towards the impact of terrestrial ecotoxicity from the construction of residential standard buildings. In comparison to the findings of the researchers Chang et al. (2012) and Nässén et al. (2007) the impact results were comparatively lower, with impact from the materials being roughly 90% and the impact from the construction process being 10%. Thus, the present study obtained its major contribution of environmental impact from the materials used. This is also attributed to the heavy energy usage for building machinery, in particular for high diesel-fired mobile cranes and piling machines.

The contribution of each of the construction materials on the environmental effects from the buildings has also been evaluated. Figure 30.3 summarizes the contributed impacts from construction materials for certain midpoint categories. There were impacts during construction due to water and energy consumption and they were minimal for both the levels that were assessed. Concrete exerted enormous contribution to global warming, aquatic acidification, terrestrial ecotoxicity and aquatic eutrophication as shown in Fig. 30.4 which displays the individual impacts from the building materials for certain crucial impact categories. Tiles were most prominent in toxicity categories such as respiratory inorganics, and aquatic eutrophication - due to the heavy use of ceramic materials. Because of the heavy utilization of concrete in the building structure and steel building characteristics, it had the highest impact with contributions of about 50% in almost all the segments. During resource degradation with 65% contribution, coated materials had the most impact due to their composition of ceramic materials.

The future impacts of transport are all significant, as expected. However, compared to the major and overwhelmingly harmful impacts of waste disposal, those are fairly minimal. The same observation could be made for the results of the glass used, but only with respect to the exceptional contribution in '*Respiratory Inorganics*'. For most of the mid-point categories, energy consumed during the construction period had an effect, primarily due to electricity being consumed over vast quantities during the construction phase. It is important to remember that the mass impact from reinforced steel is more than 70 per cent; it also makes the highest contribution to health-related impacts (over 40% for "*Carcinogens*" and 50% for "*Respiratory organics*").

The recovery of steel also produces positive environmental effects, primarily from the category of "*Carcinogens*", that are significantly linked to the recycling process emissions of aromatic hydrocarbons as defined in the relevant statement from *Ecoinvent 3.6* (Classen et al. 2009). The cumulative findings indicate that the improvement in the amount and characteristic of the remaining part emerging from recycling greatly increases the overall environmental efficiency of the residential building being studied. Figure 30.5 demonstrates a comparative impact scenario of Damage assessment for the materials from their respective life cycles. Interestingly every material that was considered in the present study for the residential building had a high contribution to Resources (Fig. 30.5).

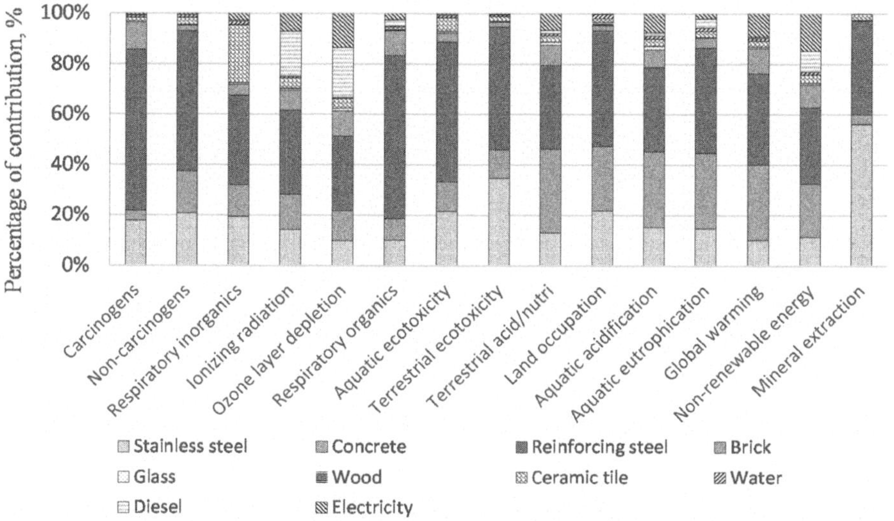

Fig. 30.3 Comparative impact assessment of residential building's material from their respective life cycles

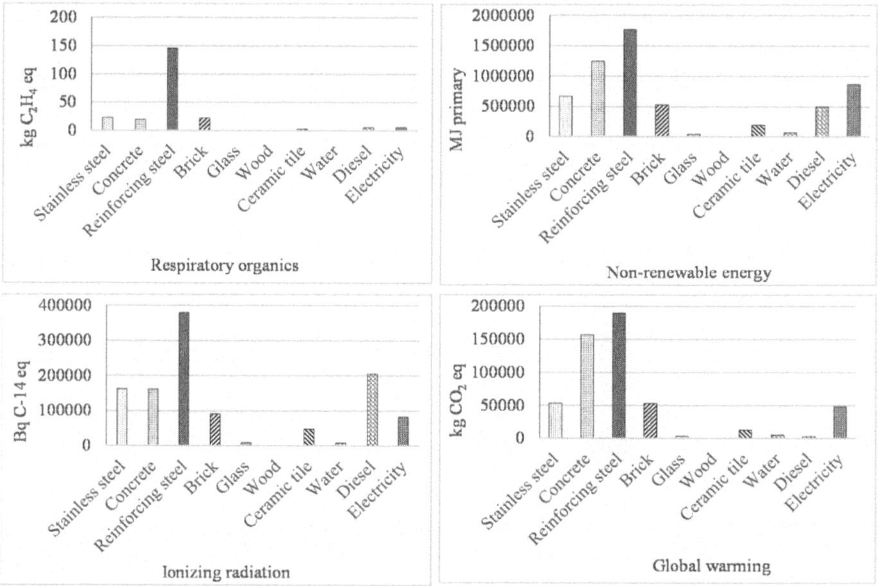

Fig. 30.4 Results of LCIA in the base case scenario with respect to crucial impact categories

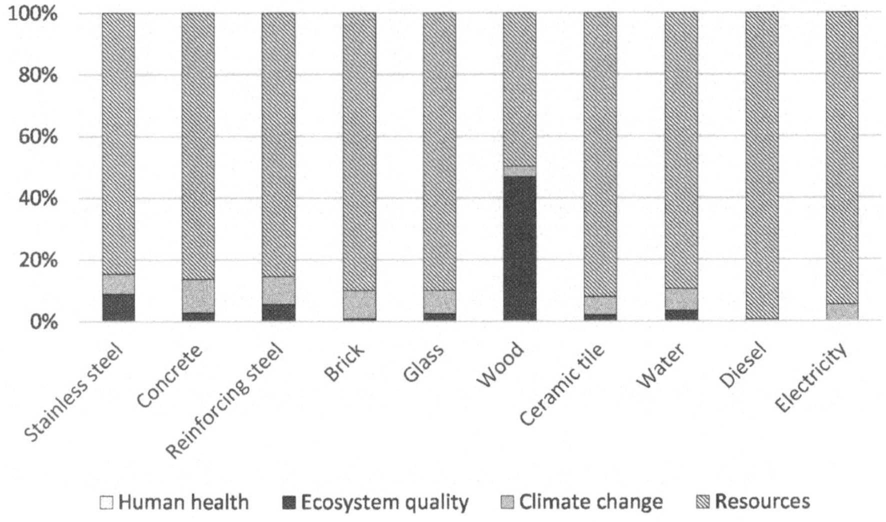

Fig. 30.5 Comparative scenario of damage assessment among building materials

The same observation could be made for Climate change. Every material that contributed to Climate change contributed to Resources as well. Wood, reinforcing steel, concrete, water, glass, ceramic tile and stainless steel seem to exert a significant impact on ecosystem quality. The comparative scenario of the impacts after normalization for all the materials are presented in Fig. 30.6 The normalized findings only involve the most important influence groups, so that the histogram is more readable (Vitale et al. 2017).

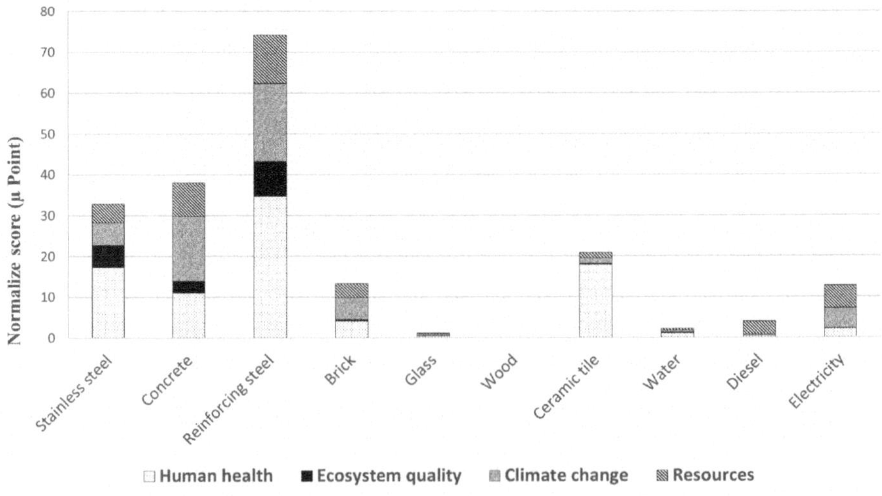

Fig. 30.6 Comparative scenario of normalization among building materials

Supporting the previous analysis obtained from the demonstration of impacts exerted by materials in different impact categories, normalized impacts also show that reinforcing steel is the highest contributor (>70 μpt) while stainless steel also exert significant (>30 μpt) impact on the environment. Concrete has a substantial impact (>35 μpt) as well on normalization with an average score. Considerable contributions from both brick (>10 μpt) and ceramic tile (15 μpt) are also observed on the normalization plot. Figure 30.6 also demonstrates that stainless steel, concrete, reinforcing steel, brick, ceramic tile and electricity have an impact on Human health with stainless steel contributing the highest. On the other hand, environmental impact for resources received a major contribution from reinforcing steel, stainless steel, concrete, electricity, ceramic tile and brick.

In summary, an LCA study of a residential building in Bangladesh revealed that the major impacts were associated with construction materials. Single score plots demonstrated that the reduction of total impacts would necessitate the reduction of sources of respiratory inorganics to protect human health, restrict global warming emissions to prevent climate change and reduce consumption of non-renewable resources. Thus, the sustainability of the application of residential building has to be concomitant with the reduced usage of chemicals and energy-intensive processes.

30.8 Interpretation

Since steel recycling could play a potentially important role in sustainable resource utilization, Fig. 30.6 concentrates on the normalized impacts of alternate applications which can be used for the evaluation of the efficiency of this method from the environmental perspective (see Sect. 36.7.1). This case is separable, according to the World Steel Associations scenario, that recycled steel will replace the primary and secondary strain mixture used by industry (Vitale et al. 2017) in which "steel bis" requires that only primary steel substitutes for recycled material. The above shows low efficiency, with only positive environmental effects, primarily in relation to the strain produced by the process of electric arc furnaces used to recycle steel waste. The "steel bis' scenario has also been used previously by other studies (Rigamonti et al. 2009). At most, main steel that is manufactured solely by the built-in steel procedure called *'Blast Furnace + Simple Oxygen Furnace'* is replaced by the tool, but it does not represent the current market conditions. The data analyzed mentioned in Fig. 30.5 may indicate a minor contribution to the total efficiency of recycled inert fractions. The European Community (Directive 2008), on the other hand, firmly promotes the use of recycled aggregate. This indicated that the conclusions drawn on the environmental efficiency of this method should be closely investigated. Based on the above analysis, three contrasting alternatives could be visualized: the scenario of the present study assumes 30 percent of calcareous and 70 percent of sand; the second case supposes that reclaimed inert can be considered as replacement of 30 percent of the gravel (rather than crushed calcareous, gravel is used as natural aggregates in southern Italy) and 70 per cent of the sand; the third scenario introduces the related presumption as the base case, but with a significant difference, i.e. 50 per cent for both sand and crushed limestone. The results show that the conclusions taken in this case in hand are protective enough and that a more marked influence on prevented impacts can be accomplished only by making a very small change in the relative fractions of the raw and fine commodity raw materials.

Vitale et al. (2017) compares the effects of the volume of waste submitted to recycling care on total environmental efficiency. The key reason is a targeted demolition that can recover about 80 percent of the demolition material. The alternative scenarios of the standard than the selective demolition phase of the building can recover quantities of material below the 70% target (Directive 2008). Both estimates are thought to be 25–50% of the overall materials that are created respectively by demolition. The first value (recover quantities of material below the 70%) is the average value between the Italian building and demolition screening design and the official Italian Environmental Protection Agency survey (Vitale et al. 2017), that are explicitly for south Italy (Vitale et al. 2017). The second value (25% to 50% of the overall materials) is reservedly lower from the previous value, in view of the poor reliability of this sort of data, as indicated recently in a special workshop arranged by the European Commission. The consistency and control parameters of residues were also considered to be applicable in all the situations.

The findings confirmed that an improvement in the amount of recycling will considerably boost the overall efficiency of the building's end-of-life level of impact. The situation concerns a renunciation of inert fraction recycling, as it is occurring in many EU countries. The most effective recycling-based solution for the present study as expected facilitates betterment in general. However, the primary position of the metal recycling seems obvious: its contribution represents approximately 75% of the overall possible impacts, 25% from "*Aquatic Ecotoxicity*" while 20% of the impacts from "*Non-renewable Resources*". These results are broadly consistent with those presented in other articles on Italian regions (Blengini 2009; Blengini and Garbarino 2010). These effects, by contrast, cannot be considered to be an indication of the negligible input of inert material recovery which is generally agreed as important (Manfredi et al. 2011) while it was emphasized with regards to land occupation (see Sect. 30.7).

30.9 Conclusion

Life cycle measurement is less advanced in the field of building construction rather than in the other fields, but researchers are focusing on increasing opportunities to consider LCA as a method for decision making in the design process. It is clear that LCA is well explained, with well-established methodologies open to consumers, however, it has several challenges in the mechanisms, which would impact the research agenda for the future. In the construction industry, there is no standardization and huge lacking in storing records. Researchers work tirelessly to address the issue, but data collection becomes challenging when international representation in the dataset becomes difficult which is crucial for comparative life cycle assessments. International agreement and benchmarking on structure, protocol and tools taking multicultural factors into account should be developed to make it possible to compare among LCA outcomes.

The research comprised both on-site and off-site operations, and the description and quantification in various process units of the construction materials. The environmental implications related to a residential building construction process, with a special focus on comparative comparison among environmental impacts, were examined and quantified in an evaluation of the attributive life cycle. The evaluated contributions of each of the materials to the overall impact on the environment demonstrated the vital significance of recycling various wastes sources, particularly those of steel reinforcement. The positive contribution of recycling in construction also appears to be interesting, especially with regards to the possible effect of non-renewable resources. The building has a larger influence than that with the same standard with multi-family housing. The same standard refers to high-quality housing compared with low-quality housing, regardless of the size of the regions, the number of inhabitants and the demand profile. The identified material hotspots also need to account for the contributions from energy and carbon secretions, mostly for prefabricated concrete, strengthened steel and horticulture.

Among all the materials that were under consideration, reinforcement still, concrete and stainless steel contributed 37.2%, 19.1% and 16.4% respectively, towards overall environmental impact. To reduce the effects of concrete, use of environmentally sustainable materials could be suggested for construction such as fly ash, which will, according to the American Concrete Institute (ACI), minimize 13 to 15% of total CO_2 emissions from fly ash in a standard concrete mix by combining fly ash with 25% of Portland cement (Flower and Sanjayan 2007). Considering CO_2 emissions from the cement manufacturing non-energy phase, the greenhouse gas emissions in the processing stages can be potentially increased by around 8 per cent. Nevertheless, a thorough analysis of fly ash output in the Bangladeshi climate and the atmosphere should be carried out.

This research has shown that the most commonly used LCA tools for residential building are approximate and incompatible with anything other than Climate Change categories examined. It has been suggested that the *ReCiPe* method of LCA represents the LCA community's current view as to when and how to report (Dahlbo et al. 2013). However, the analysis indicates significant difficulties in comprehending the results in the construction industry. With regards to this issue, *ReCiPe* was also promoted in relation to the consistency of the endpoint, which is said to conform with the midpoint and single score evaluations (Dong and Ng, 2014). It seems like both LCA tools will be able to achieve very different outcomes by estimating endpoints and single point results.

The assessment of contribution from the building construction revealed that the architectural project and building the framework had the greatest effect on subsystems. The subsystems with the highest environmental effects in residential building constitute the foundation, construction, brick (masonry), and coating. The environmental effects are thus measured according to the value, with the highest proportion

(13.9%) of CO_2 emissions, water extraction at 5.7%, while the lowest (0.02%) being the proportion for acidification.

In order to produce more specific outcomes from Climate and energy related consequences for example the effects exerted from multistoried residential buildings, further research needs to be carried out in Bangladesh. The importance of sustainable development and green buildings must be realized by the construction industries in Bangladesh. The key effort is to educate and inspire the stakeholders through the local authorities and the government. In addition to minimizing environmental effects, residential buildings exert influence upon construction efficiency during the activity and demolition process. In addition, the cost analysis of the life cycle can also be carried out on the aspect of constructing residential buildings using recycled material, as construction costs are the main concerns of the developers of the real estate business.

References

Adalberth K, Almgren A, Petersen EH (2001) Life cycle assessment of four multi-family buildings. Int J Low Energy Sustain Build 2:1–21

Ali AAMM, Negm AM, Baby MF (2015) Environmental life cycle assessment of a residential building in Egypt: a case study. Procedia Technol 19:349–356

Arena U, Ardolino F, Di Gregorio F (2015) A life cycle assessment of environmental performances of two combustion-and gasification-based waste-to-energy technologies. Waste Manag 41:60–74

Asdrubali F, Baldassarri C, Fthenakis V (2013) Life cycle analysis in the construction sector: guiding the optimization of conventional Italian buildings. Energy Build 64:73–89

Blengini GA (2009) Life cycle of buildings demolition and recycling potential: a case study in Turin Italy. Build Environ 44(2):319–330

Blengini GA, Di Carlo T (2010) The changing role of life cycle phases subsystems and materials in the LCA of low energy buildings. Energy Build 42(6):869–880

Blengini GA, Garbarino E (2010) Resources and waste management in Turin (Italy): the role of recycled aggregates in the sustainable supply mix. J Clean Prod 18(10–11):1021–1030

Bribian IZ, Uson AA, Scarpellini S (2009) Life cycle assessment in buildings: state-of-the-art and simplified LCA methodology as a complement for building certification. Build Environ 44(12):2510–2520

Buchan RD, Fleming E, Grant F (2012) Estimating for builders and surveyors. Routledge

Buchanan AH, Levine SB (1999) Wood-based building materials and atmospheric carbon emissions. Environ Sci Policy 2(6):427–437

Buyle M, Braet J, Audenaert A (2013) Life cycle assessment in the construction sector: a review. Renew Sustain Energy Rev 26:379–388

Chang Y, Ries RJ, Lei S (2012) The embodied energy and emissions of a high-rise education building: a quantification using process-based hybrid life cycle inventory model. Energy Build 55:790–798

Change IC (2001) The scientific basis. Cambridge University Press, Cambridge, UK

Classen M, Althaus HJ, Blaser S, Tuchschmid M, Jungbluth N, Doka G, Faist Emmenegger M, Scharnhorst W (2009) Life cycle inventories of metals. Final report ecoinvent data, vol 2, no 10. EMPA Dübendorf, Swiss Centre for Life Cycle Inventories Dübendorf, CH

Clift R, Doig A, Finnveden G (2000) The application of life cycle assessment to integrated solid waste management: part 1—Methodology. Process Saf Environ Prot 78(4):279–287

Dahlbo H, Koskela S, Pihkola H, Nors M, Federley M, Seppälä J (2013) Comparison of different normalised LCIA results and their feasibility in communication. Int J Life Cycle Assess 18(4):850–860

de Larriva RA, Rodríguez GC, López JMC, Raugei M, Palmer PF (2014) A decision-making LCA for energy refurbishment of buildings: conditions of comfort. Energy Build 70:333–342

Directive EU (2008) 98/EC of the European parliament and of the council of 19 November 2008 on waste and repealing certain Directives (Text with EEA relevance). Off J Eur Union 312:3–30. https://eur-lex.europa.eu/LexUriServ/LexUriServ.do?uri=OJ:L:2008:312:0003:0030:en:PDF

Dong YH, Ng ST (2014) Comparing the midpoint and endpoint approaches based on ReCiPe—a study of commercial buildings in Hong Kong. Int J Life Cycle Assess 19(7):1409–1423

Evangelista PPA, Kiperstok A, Torres EA, Gonçalves JP (2018) Environmental performance analysis of residential buildings in Brazil using life cycle assessment (LCA). Constr Build Mater 169:748–761

Fava J, Consoli F, Denison R, Dickson K, Mohin T, Vigon B (1993) A conceptual framework for life cycle impact assessment workshop report society of environmental toxicology and chemistry (SETAC). Foundation for Environmental Education, Pensacola

Finnveden G, Hauschild MZ, Ekvall T, Guinée J, Heijungs R, Hellweg S, Koehler A, Pennington D, Suh S (2009) Recent developments in life cycle assessment. J Environ Manage 91(1):1–21

Flower DJM, Sanjayan JG (2007) Greenhouse gas emissions due to concrete manufacture. Int J Life Cycle Assess 12(5):282–288

Guggemos AA, Horvath A (2005) Comparison of environmental effects of steel-and concrete-framed buildings. J Infrastruct Syst 11(2):93–101

Heravi G, Nafisi T, Mousavi R (2016) Evaluation of energy consumption during production and construction of concrete and steel frames of residential buildings. Energy Build 130:244–252

Hondo H (2006) A method for technology selection considering environmental and socio-economic impacts (11 p). Int J Life Cycle Assess 11(6):383–393

Hulme M, Turnpenny J, Jenkins GJ (2002) Climate change scenarios for the United Kingdom: the UKCIP02 briefing report. School of Environmental Sciences, University of East Anglia, Tyndall Centre for Climate Change Research

IPCC (2014) AR5 synthesis report: climate change 2014, fifth assessment report of intergovernmental panel for climate change, summary for policy makers. IPCC Switzerland

Iyer-Raniga U, Wong JPC (2012) Evaluation of whole life cycle assessment for heritage buildings in Australia. Build Environ 47:138–149

Jolliet O, Margni M, Charles R, Humbert S, Payet J, Rebitzer G, Rosenbaum R (2003) IMPACT 2002: a new life cycle impact assessment methodology. Int J Life Cycle Assess 8(6):324–330

Manfredi S, Pant R, Pennington DW, Versmann A (2011) Supporting environmentally sound decisions for waste management with LCT and LCA. Int J Life Cycle Assess 16(9):937–939

Matsuno Y, Betz M (2000) Development of life cycle inventories for electricity grid mixes in Japan. The International Journal of Life Cycle Assessment 5(5):295–305

Mithraratne N, Vale B (2004) Life cycle analysis model for New Zealand houses. Build Environ 39(4):483–492

Monahan J, Powell JC (2011) An embodied carbon and energy analysis of modern methods of construction in housing: a case study using a life cycle assessment framework. Energy Build 43(1):179–188

Mosteiro-Romero M, Krogmann U, Wallbaum H, Ostermeyer Y, Senick JS, Andrews CJ (2014) Relative importance of electricity sources and construction practices in residential buildings: a Swiss-US comparison of energy related life-cycle impacts. Energy Build 68:620–631

Motuzienė V, Rogoža A, Lapinskienė V (2016) Construction solutions for energy efficient single-family houses based on its life cycle multi-criteria analysis: a case study. J Clean Prod 112:532–541

Nässén J, Holmberg J, Wadeskog A, Nyman M (2007) Direct and indirect energy use and carbon emissions in the production phase of buildings: an input–output analysis. Energy 32(9):1593–1602

Nebel B, Zimmer B, Wegener G (2006) Life cycle assessment of wood floor coverings-a representative study for the german flooring industry (11 p). Int J Life Cycle Assess 11(3):172–182
Norman J, MacLean HL, Kennedy CA (2006) Comparing high and low residential density: life-cycle analysis of energy use and greenhouse gas emissions. J Urban Plan Dev 132(1):10–21
Optis M (2008) Incorporating life cycle assessment into the LEED green building rating system, Doctoral dissertation
Ortiz-Rodríguez O, Castells F, Sonnemann G (2010) Life cycle assessment of two dwellings: one in Spain, a developed country, and one in Colombia, a country under development. Sci Total Environ 408(12):2435–2443
Ortiz O, Bonnet C, Bruno JC, Castells F (2009) Sustainability based on LCM of residential dwellings: a case study in Catalonia Spain. Build Environ 44(3):584–594
Oyarzo J, Peuportier B (2014) Life cycle assessment model applied to housing in Chile. J Clean Prod 69:109–116
Peterson EH (1997) Database and inventory tool for building components and buildings environmental parameters. SBI Report 275
Ramesh T, Prakash R, Shukla KK (2010) Life cycle energy analysis of buildings: an overview. Energy Build 42(10):1592–1600
Reddy BVV, Jagadish KS (2003) Embodied energy of common and alternative building materials and technologies. Energy Build 35(2):129–137
Ren J, Manzardo A, Mazzi A, Zuliani F, Scipioni A (2015) Prioritization of bioethanol production pathways in China based on life cycle sustainability assessment and multicriteria decision-making. Int J Life Cycle Assess 20(6):842–853
Rigamonti L, Grosso M, Sunseri MC (2009) Influence of assumptions about selection and recycling efficiencies on the LCA of integrated waste management systems. Int J Life Cycle Assess 14(5):411–419
Rivela B, Hospido A, Moreira T, Feijoo G (2006) Life cycle inventory of particleboard: a case study in the wood sector (8 p). Int J Life Cycle Assess 11(2):106–113
Rossi B, Marique AF, Glaumann M, Reiter S (2012) Life-cycle assessment of residential buildings in three different European locations, basic tool. Build Environ 51:395–401
Sartori I, Hestnes AG (2007) Energy use in the life cycle of conventional and low-energy buildings: a review article. Energy Build 39(3):249–257
Selmes DG (2005) Towards sustainability: direction for life cycle assessment. Heriot-Watt University
Sevigné-Itoiz E, Gasol CM, Rieradevall J, Gabarrell X (2014) Environmental consequences of recycling aluminum old scrap in a global market. Resour Conserv Recycl 89:94–103
Soust-Verdaguer B, Llatas C, García-Martínez A (2016) Simplification in life cycle assessment of single-family houses: a review of recent developments. Build Environ 103:215–227
Souza HHDS (2014) Assessment of the life cycle and lifetime influence of the wooden and concrete posts of the electricity distribution system
Standardization IO (2006) Environmental management: life cycle assessment; principles and framework, vol 14044. ISO
Syal M, Hastak M, Mullens M, Sweaney A (2006) United States-India collaborative research directions in urban housing and supporting infrastructure. J Archit Eng 12(4):163–167
Thormark C (2002) A low energy building in a life cycle—its embodied energy, energy need for operation and recycling potential. Build Environ 37(4):429–435
Treloar GJ, Love PED, Faniran OO, Iyer-Raniga U (2000) A hybrid life cycle assessment method for construction. Constr Manag Econ 18(1):5–9
Vitale P, Arena N, Di Gregorio F, Arena U (2017) Life cycle assessment of the end-of-life phase of a residential building. Waste Manage 60:311–321
Wen TJ, Siong HC, Noor ZZ (2015) Assessment of embodied energy and global warming potential of building construction using life cycle analysis approach: case studies of residential buildings in Iskandar Malaysia. Energy Build 93:295–302

Wu HJ, Yuan ZW, Zhang L, Bi J (2012) Life cycle energy consumption and CO_2 emission of an office building in China. Int J Life Cycle Assess 17(2):105–118
Xing S, Xu Z, Jun G (2008) Inventory analysis of LCA on steel-and concrete-construction office buildings. Energy Build 40(7):1188–1193
Zhang X, Shen L, Zhang L (2013) Life cycle assessment of the air emissions during building construction process: a case study in Hong Kong. Renew Sustain Energy Rev 17:160–169

Correction to: Environmental Life Cycle Analysis of Residential Building Materials: A Case Study

Md. Al Sadikul Islam, Md. Ashiquzzaman, Amiu Shadik Utshab, and Nehreen Majed

Correction to:
Chapter 30 in: J. A. Malik and S. Marathe (eds.), *Ecological and Health Effects of Building Materials*, https://doi.org/10.1007/978-3-030-76073-1_30

The original version of the chapter was inadvertently published with incorrect author name citation in Springer Link. The citation name of author "Md. Al Sadikul Islam" has been updated.

The correction chapter and the book have been updated with the change.

The updated version of this chapter can be found at
https://doi.org/10.1007/978-3-030-76073-1_30

J. A. Malik and S. Marathe (eds.), *Ecological and Health Effects of Building Materials*,
https://doi.org/10.1007/978-3-030-76073-1_31

Index

J. A. Malik and S. Marathe (eds.), *Ecological and Health Effects of Building Materials*, https://doi.org/10.1007/978-3-030-76073-1